The Chloroplast

Advances in Photosynthesis and Respiration

VOLUME 31

The scope of our series, beginning with volume 11, reflects the concept that photosynthesis and respiration are intertwined with respect to both the protein complexes involved and to the entire bioenergetic machinery of all life. *Advances in Photosynthesis and Respiration* is a book series that provides a comprehensive and state-of-the-art account of research in photosynthesis and respiration. Photosynthesis is the process by which higher plants, algae, and certain species of bacteria transform and store solar energy in the form of energy-rich organic molecules. These compounds are in turn used as the energy source for all growth and reproduction in these and almost all other organisms. As such, virtually all life on the planet ultimately depends on photosynthetic energy conversion. Respiration, which occurs in mitochondrial and bacterial membranes, utilizes energy present in organic molecules to fuel a wide range of metabolic reactions critical for cell growth and development. In addition, many photosynthetic organisms engage in energetically wasteful photorespiration that begins in the chloroplast with an oxygenation reaction catalyzed by the same enzyme responsible for capturing carbon dioxide in photosynthesis. This series of books spans topics from physics to agronomy and medicine, from femtosecond processes to season long production, from the photophysics of reaction centers, through the electrochemistry of intermediate electron transfer, to the physiology of whole organisms, and from X-ray crystallography of proteins to the morphology or organelles and intact organisms. The goal of the series is to offer beginning researchers, advanced undergraduate students, graduate students, and even research specialists, a comprehensive, up-to-date picture of the remarkable advances across the full scope of research on photosynthesis, respiration and related processes.

For other titles published in this series, go to
www.springer.com/series/5599

The Chloroplast

Basics and Applications

Edited by

Constantin A. Rebeiz
Rebeiz Foundation for Basic Research, Champaign, USA

Christoph Benning
Michigan State University, East Lansing, USA

Hans J. Bohnert
University of Illinois at Urbana-Champaign, Urbana, USA

Henry Daniell
Central Florida University, Orlando, USA

J. Kenneth Hoober
Arizona State University, Tempe, USA

Hartmut K. Lichtenthaler
University of Karlsruhe, Germany

Archie R. Portis
University of Illinois at Urbana-Champaign, Urbana, USA

and

Baishnab C. Tripathy
Jawaharlal Nehru University, New Delhi, India

Springer

Library of Congress Control Number: 2010926249

ISBN 978-90-481-8530-6 (HB)
ISBN 978-90-481-8531-3 (e-book)

Published by Springer,
P.O. Box 17, 3300 AA Dordrecht, The Netherlands.

www.springer.com

Cover Image:

The cover shows an idealized section of a chloroplast encircled by a schematic representation of the cotton chloroplast genome. Superimposed are pictures of an Arabidopsis plant (top left), the structure of a cytochrome complex (top right), a chlorophyll molecule (bottom left) and a Rubisco structure (bottom right). Cover design and graphic assistance by Eric T. Portis

Printed on acid-free paper

[redacted]

This book is dedicated to the memory of

Sam Granick*

For his ground-breaking discoveries that paved the way for our understanding of the chlorophyll biosynthetic pathway

* Sam Granick (February 16, 1909–April 29, 1977) was born in New York City; he obtained his B.S., M.S. and Ph.D. degrees, all from the University of Michigan. With Keith R. Porter, he made the first electron micrograph of a chloroplast and discovered several Mg-intermediates of the chlorophyll biosynthetic pathway. He was a member of both the United States National Academy of Sciences and the American Academy of Arts and Science.

From the Series Editor

Advances in Photosynthesis and Respiration

Volume 31: The Chloroplast

Basics and Applications

Welcoming Thomas D. Sharkey (as my co-series editor), and the new consulting editors

First of all, I sincerely thank the following nine past consulting editors for their suggestions and cooperation during the planning and production of volumes 20–30 of the Advances in Photosynthesis and Respiration (AIPH) series: Julian Eaton-Rye (Dunedin, New Zealand); Christine H. Foyer (Harpenden, U.K); David B. Knaff (Lubbock, Texas, USA); Anthony L. Moore (Brighton, UK); Sabeeha Merchant (Los Angeles, California, USA); Krishna Niyogi (Berkeley, California, USA); William Parson (Seattle, Washington, USA); Agepati Raghavendra, (Hyderabad, India); and Gernot Renger (Berlin, Germany). In addition, I wish to express my heartfelt thanks to Julian Eaton-Rye, Christine H. Foyer, David B. Knaff, Sabeeha Merchant, William Parson and Agepati Raghavendra for co-editing books in the AIPH series.

Beginning with volume 31, we not only have a new board of consulting editors, but a new Co-series Editor, Thomas D. Sharkey (Michigan State University, East Lansing, MI, USA. His biography and photograph are depicted at the end of this Editorial). I welcome Tom Sharkey, as well as all the ten consulting editors: Elizabeth (Lisa)Ainsworth (United States Department of Agriculture, Urbana, IL, USA); Basanti Biswal (Sambalpur University, Jyoti Vihar, Orissa, India); Robert (Bob) E. Blankenship (Washington University, St Louis, MO, USA); Ralph Bock (Max–Planck Institute, Postdam-Golm, Germany); Julian J. Eaton-Rye (University of Otago, Dunedin, New Zealand); Wayne Frasch (Arizona State University, Tempe, AZ, USA); Johannes Messinger (Umeå University, Umeå, Sweden); Masahiro Sugiura (Nagoya City University, Nagoya, Japan); Davide Zannoni (University of Bologna, Bologna, Italy); and Lixin Zhang (Institute of Botany, Beijing, China). Blankenship and Eaton-Rye are not really new as they have not only served on the AIPH Board in the past, but have co-edited one book each in the AIPH series. Further, Tom Sharkey was a co-editor of volume 9 in the AIPH series (see list of published volumes).

The First 30 Volumes of the AIPH Series

- *Volume 30* (2009): ***Lipids in Photosynthesis: Essential and Regulatory Functions***, edited by Hajime Wada and Norio Murata, both from Japan. Twenty chapters, 506 pp, Hardcover, ISBN: 978-90-481-2862-4; e-book, ISBN: 978-90-481-2863-1
- *Volume 29* (2009): ***Photosynthesis in silico: Understanding Complexity from Molecules***, edited by Agu Laisk, Ladislav Nedbal, and Govindjee, from Estonia, The Czech Republic, and USA. Twenty chapters, 508 pp, Hardcover, ISBN: 978-1-4020-9236-7
- *Volume 28* (2009): ***The Purple Phototrophic Bacteria***, edited by C. Neil Hunter, Fevzi Daldal, Marion C. Thurnauer and J. Thomas Beatty, from UK, USA and Canada. Forty-eight chapters, 1014 pp, Hardcover, ISBN: 978-1-4020-8814-8
- *Volume 27* (2008): ***Sulfur Metabolism in Phototrophic Organisms***, edited by Christiane Dahl, Rüdiger Hell, David Knaff and Thomas Leustek, from Germany and USA. Twenty-four chapters, 551 pp, Hardcover, ISBN: 978-4020-6862-1
- *Volume 26* (2008): ***Biophysical Techniques in Photosynthesis***, Volume II, edited by Thijs Aartsma and Jörg Matysik, both from The Netherlands.

Twenty-four chapters, 548 pp, Hardcover, ISBN:978-1-4020-8249-8

- *Volume 25* (2006): ***Chlorophylls and Bacteriochlorophylls: Biochemistry, Biophysics, Functions and Applications***, edited by Bernhard Grimm, Robert J. Porra, Wolfhart Rüdiger, and Hugo Scheer, from Germany and Australia. Thirty-seven chapters, 603 pp, Hardcover, ISBN: 978-1-40204515-8
- *Volume 24* (2006): ***Photosystem I: The Light-Driven Plastocyanin: Ferredoxin Oxidoreductase***, edited by John H. Golbeck, from USA. Forty chapters, 716 pp, Hardcover, ISBN: 978-1-40204255-3
- *Volume 23* (2006): ***The Structure and Function of Plastids***, edited by Robert R. Wise and J. Kenneth Hoober, from USA. Twenty-seven chapters, 575 pp, Softcover, ISBN: 978-1-4020-6570-5; Hardcover, ISBN: 978-1-4020-4060-3
- *Volume 22* (2005): ***Photosystem II: The Light-Driven Water:Plastoquinone Oxidoreductase***, edited by Thomas J. Wydrzynski and Kimiyuki Satoh, from Australia and Japan. Thirty-four chapters, 786 pp, Hardcover, ISBN: 978-1-4020-4249-2
- *Volume 21* (2005): ***Photoprotection, Photoinhibition, Gene Regulation, and Environment***, edited by Barbara Demmig-Adams, William W. Adams III and Autar K. Mattoo, from USA. Twenty-one chapters, 380 pp, Softcover, ISBN: 978-1-4020-9281-7; Hardcover, ISBN: 978-1-4020-3564-7
- *Volume 20* (2006): ***Discoveries in Photosynthesis***, edited by Govindjee, J. Thomas Beatty, Howard Gest and John F. Allen, from USA, Canada and UK. One hundred and eleven chapters, 1304 pp, Hardcover, ISBN: 978-1-4020-3323-0
- *Volume 19* (2004): ***Chlorophyll a Fluorescence: A Signature of Photosynthesis***, edited by George C. Papageorgiou and Govindjee, from Greece and USA. Thirty-one chapters, 820 pp, Softcover, ISBN: 978-90-481-3882-1; Hardcover, ISBN: 978-1-4020-3217-2
- *Volume 18* (2005): ***Plant Respiration: From Cell to Ecosystem***, edited by Hans Lambers and Miquel Ribas-Carbo, from Australia and Spain. Thirteen chapters, 250 pp, Hardcover, ISBN: 978-14020-3588-3
- *Volume 17* (2004): ***Plant Mitochondria: From Genome to Function***, edited by David Day, A. Harvey Millar and James Whelan, from Australia. Fourteen chapters, 325 pp, Hardcover, ISBN: 978-1-4020-2399-6
- *Volume 16* (2004): ***Respiration in Archaea and Bacteria: Diversity of Prokaryotic Respiratory Systems***, edited by Davide Zannoni, from Italy. Thirteen chapters, 310 pp, Hardcover, ISBN: 978-14020-2002-5
- *Volume 15* (2004): ***Respiration in Archaea and Bacteria: Diversity of Prokaryotic Electron Transport Carriers***, edited by Davide Zannoni, from Italy. Thirteen chapters, 350 pp, Hardcover, ISBN: 978-1-4020-2001-8
- *Volume 14* (2004): ***Photosynthesis in Algae***, edited by Anthony W. Larkum, Susan Douglas and John A. Raven, from Australia, Canada and UK. Nineteen chapters, 500 pp, Hardcover, ISBN: 978-0-7923-6333-0
- *Volume 13* (2003): ***Light-Harvesting Antennas in Photosynthesis***, edited by Beverley R. Green and William W. Parson, from Canada and USA. Seventeen chapters, 544 pp, Hardcover, ISBN: 978- 07923-6335-4
- *Volume 12* (2003): ***Photosynthetic Nitrogen Assimilation and Associated Carbon and Respiratory Metabolism***, edited by Christine H. Foyer and Graham Noctor, from UK and France. Sixteen chapters, 304 pp, Hardcover, ISBN: 978-07923-6336-1
- *Volume 11* (2001): ***Regulation of Photosynthesis***, edited by Eva-Mari Aro and Bertil Andersson, from Finland and Sweden. Thirty-two chapters, 640 pp, Hardcover, ISBN: 978-0-7923-6332-3
- *Volume 10* (2001): ***Photosynthesis: Photobiochemistry and Photobiophysics***, authored by Bacon Ke, from USA. Thirty-six chapters, 792 pp, Softcover, ISBN: 978-0-7923-6791-8; Hardcover: ISBN: 978-0-7923-6334-7
- *Volume 9* (2000): ***Photosynthesis: Physiology and Metabolism***, edited by Richard C. Leegood, Thomas D. Sharkey and Susanne von Caemmerer, from UK, USA and Australia. Twenty- four chapters, 644 pp, Hardcover, ISBN: 978-07923-6143-5
- *Volume 8* (1999): ***The Photochemistry of Carotenoids***, edited by Harry A. Frank, Andrew J. Young, George Britton and Richard J. Cogdell,

from UK and USA. Twenty chapters, 420 pp, Hardcover, ISBN: 978-0-7923-5942-5

- *Volume 7* (1998): ***The Molecular Biology of Chloroplasts and Mitochondria in Chlamydomonas***, edited by Jean David Rochaix, Michel Goldschmidt-Clermont and Sabeeha Merchant, from Switzerland and USA. Thirty-six chapters, 760 pp, Hardcover, ISBN: 978-0-7923-5174-0
- *Volume 6* (1998): ***Lipids in Photosynthesis: Structure, Function and Genetics***, edited by Paul-André Siegenthaler and Norio Murata, from Switzerland and Japan. Fifteen chapters, 332 pp, Hardcover, ISBN: 978-0-7923-5173-3
- *Volume 5* (1997): ***Photosynthesis and the Environment***, edited by Neil R. Baker, from UK. Twenty chapters, 508 pp, Hardcover, ISBN: 978- 07923-4316-5
- *Volume 4* (1996): ***Oxygenic Photosynthesis: The Light Reactions***, edited by Donald R. Ort, and Charles F. Yocum, from USA. Thirty-four chapters, 696 pp, Softcover: ISBN: 978-0-7923- 3684-6; Hardcover, ISBN: 978-0-7923-3683-9
- *Volume 3* (1996): ***Biophysical Techniques in Photosynthesis***, edited by Jan Amesz and Arnold J. Hoff, from The Netherlands. Twenty-four chapters, 426 pp, Hardcover, ISBN: 978-0-7923-3642-6
- *Volume 2* (1995): ***Anoxygenic Photosynthetic Bacteria***, edited by Robert E. Blankenship, Michael T. Madigan and Carl E. Bauer, from USA. Sixty-two chapters, 1331 pp, Hardcover, ISBN: 978-0-7923-3682-8
- *Volume 1* (1994): ***The Molecular Biology of Cyanobacteria***, edited by Donald R. Bryant, from USA. Twenty-eight chapters, 916 pp, Hardcover, ISBN: 978-0-7923-3222-0

Further information on these books and ordering instructions can be found at http://www. springer. com/series/5599. Contents of volumes 1–29 can be found at <http://www.life.uiuc. edu/govindjee/ photosynSeries/ttocs.html>.

Special discounts are available to members of the International Society of Photosynthesis Research, ISPR http://www.photosynthesis-research.org/>: See http://www.springer.com/ ispr

Read-on-line website information is now available for several books in the Photosynthesis and Respiration Series

For ease in finding the books on a specific topic, I provide below the volume number and its topic followed by the web site where all information on the Table of Content, titles of chapters and authors are available. In addition, free downloads of the front matter and the Index of the books are available at the sites. It is my understanding that Springer plans to provide such information on many more books in the Series.

Volume 30 (Lipids II)
http://www.springerlink.com/content/978-90-481-2862-4

Volume 29 (Photosynthesis in silico)
http://www.springerlink.com/content/978-1-4020-9236-7

Volume 28 (Photosynthetic Bacteria II)
http://www.springerlink.com/content/978-1-4020-8814-8

Volume 27 (Sulfur Metabolism)
http://www.springerlink.com/content/978-1-4020-6862-1

Volume 26 (Biophysical Techniques II)
http://www.springerlink.com/content/978-1-4020-8249-8

Volume 23 (The Plastids)
http://www.springerlink.com/content/978-1-4020-4060-3

Volume 21 (Photoprotection and Photoinhibition)
http://www.springerlink.com/content/978-1-4020-3564-7

Volume 19 (Chlorophyll a Fluorescence)
http://www.springerlink.com/content/978-1-4020-3217-2

Volume 16 (Bacterial Respiration II)
http://www.springerlink.com/content/978-1-4020-2002-5

Volume 31: The Chloroplast: Basics and Applications

Now, I turn my attention to the current book. I am indeed delighted to announce the publication, in the *Advances in Photosynthesis and Respiration*

(AIPH) Series, of *The Chloroplast: Basics and Applications.* Eight distinguished authorities have edited this volume; the editor-in-chief of this volume is Constantin A. Rebeiz (USA). His co-editors (arranged alphabetically) are Christoph Benning (USA), Hans J. Bohnert (USA), Henry Daniell (USA), J. Kenneth Hoober (USA), Hartmut Lichtenthaler (Germany), Archie R. Portis (USA) and Baishnab C. Tripathy (India).

A Background for This Book

Improvement in photosynthetic efficiency would require (a) a thorough knowledge of the biosynthesis of photosynthetic membrane components such as heme, chlorophylls, carotenoids, quinones, and lipids; (b) a thorough knowledge of photosynthetic membrane apoprotein biosynthesis; (c) a deeper understanding of the biosynthesis and regulation of the assembly of pigment-apoprotein complexes; and (d) a deeper knowledge of various facets of carbon fixation. Ways to increase agricultural productivity may include bioengineering of chloroplasts with higher photosynthetic efficiency and ones that are more adapted to deal with stress and/or alteration of the kinetic properties of the primary CO_2-assimilating enzyme, Rubisco. This enzyme is a multisubunit protein located in the chloroplasts that is encoded by genes located in both the nucleus and the chloroplast. Chloroplast genetic engineering is an exciting field of research since it is an environmentally friendly approach and has been successfully used to engineer several valuable plant traits, including herbicide resistance, disease resistance, insect resistance, drought/salt tolerance, and phytoremedition. In the past, chloroplast transformation had been successful in only a few crops, including tobacco, potato and tomato. More recently, highly efficient plastid transformation has been achieved in soybean, carrot, and cotton, via somatic embryogenesis using species-specific vectors. In addition to biotechnological applications, plastid transformation has also been extensively used to study chloroplast biochemistry and molecular biology. Recent advances and future challenges have been addressed in this book; it is meant to foster closer cooperation between biochemists, and molecular biologists, working on porphyrin, pigment, lipid and chloroplast apoprotein, as well as leading scientists involved in photosynthesis research, and biotechnologists involved in chloroplast transformation. We hope that future research will focus attention on "Chloroplast Bioengineering" as an integrated novel field of research.

Authors of This Book

The current book contains 25 chapters (see the content, and the preface of this book) written by authors from seven countries (Australia; France; Germany; India; Japan; Spain; and USA). I thank each and everyone (arranged alphabetically): Kinya Akashi (Japan); Thomas J. Bach (France); Christoph Benning (USA); Hans J. Bohnert (USA); Albert Boronat (Spain); Jeffrey W. Cary (USA); Min Chen (Australia); William A. Cramer (USA); Dring N. Crowell (USA); Henry Daniell (USA); Christoph Docktor (Germany); Peter Dormann (Germany); Natalia Dudareva (USA); Laura L. Eggink (USA); Esther Gerber (France); Ruth Grene (USA); Bernhard Grimm (Germany) Andrea Hemmerlin (France); Kenneth Hoober (USA); Hisashi Ito (Japan); Gunnar Jeschke (Germany); Gamini Kanangara (Germany); Kocchi Kobayashi (Japan); Kaori Kohzuma (Japan); Vladimir Kolossov (USA); Itzhak Kureck (USA); Anthony W. D. Larkum (Australia); Pighua Li (USA); Hartmut Lichtenthaler (Germany); Lu Liu (USA); Chikahiro Miyake (Japan); Yuri Nakajima Munekage (Japan); Dinesh A. Nagegowda (USA); Yuki Nakamura (Japan); Yoshihiko Nanasato (Japan); Krishna Niyogi (USA); Hiroyuki Ohta (Japan); Gopal K. Pattanayak (USA) Harald Paulsen (Germany); Qungang Qi (USA); Kanniah Rajasekaran (USA); Constantin A. Rebeiz (USA); David Rhodes (USA); Michel Rohmer (France); Tracey A. Ruhlman (USA); Sergei Savikhin (USA); Robert E. Sharwood (Australia); Toshiharu Shikanai (Japan); Mie Shimojima (Japan); N. D. Singh (USA); Anchalee Sirikhachornkit (USA); Kantaro Takahara (Japan); Ayumi Tanaka (Japan); Riyouchi Tanaka (Japan); Baishnab C. Tripathy (India); Henry E. Valentin (USA); Aleksei Volkov (Germany); Dieter von Wettstein (USA); Spencer Whitney (Australia); Liusheng Yan (USA); E. Yamashita (Japan); Akiho Yokota (Japan); and Genhai Zhu (USA).

Future Advances in Photosynthesis and Respiration and Other* Related Books

The readers of the current series are encouraged to watch for the publication of the forthcoming books (not necessarily arranged in the order of future appearance):

- **Abiotic Stress Adaptation in Plants: Physiological, Molecular and Genomic Foundation* (Editors: Ashwani Pareek, Sudhir K. Sopory, Hans J. Bohnert and Govindjee). This book has now been released; see http://www.springer.com/life+sciences/book/978-90-481-3112-2; its table of content is at: http://www.springerlink.com/content/978-90-481-3111-2
- *C-4 Photosynthesis and Related CO_2 Concentrating Mechansims* (Editors: Agepati S. Raghavendra and Rowan Sage)
- *Photosynthesis: Perspectives on Plastid Biology, Energy Conversion and Carbon Metabolism* (Editors: Julian Eaton-Rye and Baishnab Tripathy)
- *Functional Genomics and Evolution of Photosynthetic Systems* (Editors: Robert Burnap and Willem Vermaas)
- **The Bioenergetic Processes of Cyanobacteria: From Evolutionary Singularity to Ecological Diversity* (Editors: Guenter A. Peschek, Christian Obinger, and Gernot Renger)
- *Chloroplast Biogenesis: During Leaf Development and Senescence* (Editors: Basanti Biswal, Karin Krupinska and Udaya Chand Biswal
- *The Structural Basis of Biological Energy Generation* (Editor: Martin Hohmann-Marriott)

In addition to the above contracted books, the following topics are under consideration:

- Artificial Photosynthesis
- ATP Synthase and Proton Translocation
- Biohydrogen Production
- Carotenoids II
- Cyanobacteria
- The Cytochromes
- Ecophysiology
- Evolution of Photosynthesis
- Genomics of Chloroplasts and Mitochondria
- Global Aspects of Photosynthesis
- Green Bacteria and Heliobacteria
- Interactions between Photosynthesis and other Metabolic Processes
- Limits of Photosynthesis
- Photosynthesis, Biomass and Bioenergy

If you have any interest in editing/co-editing any of the above listed books, or being an author, please send me an E-mail at gov@illinois.edu, and/or to Tom Sharkey (tsharkey@msu.edu). Suggestions for additional topics are also welcome.

In view of the interdisciplinary character of research in photosynthesis and respiration, it is my earnest hope that this series of books will be used in educating students and researchers not only in Plant Sciences, Molecular and Cell Biology, Integrative Biology, Biotechnology, Agricultural Sciences, Microbiology, Biochemistry, Chemical Biology, Biological Physics, and Biophysics, but also in Bioengineering, Chemistry, and Physics.

I take this opportunity to thank and congratulate Tino Rebeiz and his co-editors for their outstanding editorial work; they have done a fantastic job not only in editing, but also in organizing this book for Springer, and for their highly professional dealing with the typesetting process and their help in preparing this editorial. I have already thanked (see above) all the 66 authors of this book: without their authoritative chapters, there would be no such volume. I give special thanks to Juno Martina George for directing the typesetting of this book: her expertise has been crucial in bringing this book to completion. We owe Jacco Flipsen, Noeline Gibson and André Tournois (of Springer) thanks for their friendly working relation with us that led to the production of this book. Thanks are also due to Jeff Haas (Director of Information Technology, Life Sciences, University of Illinois at Urbana-Champaign, UIUC), Feng Sheng Hu (Head, Department of Plant Biology, UIUC) and my dear wife, Rajni Govindjee for constant support.

October 2, 2009

Govindjee
Founding Series Editor, *Advances in Photosynthesis and Respiration*
University of Illinois at Urbana-Champaign
Department of Plant Biology
Urbana, IL 61801-3707, USA

E-mail: gov@illinois.edu
URL: http://www.life.uiuc.edu/govindjee

Govindjee, Series Editor

Govindjee, born in 1932, obtained his B.Sc. (Chemistry, Biology) and M.Sc. (Botany, Plant Physiology) in 1952 and 1954, from the University of Allahabad, India. He learned his Plant Physiology from Prof. Shri Ranjan, who was a former student of Felix Frost Blackman. For his Ph.D. (in Biophysics, 1960), he studied under the pioneers of photosynthesis: Professors Robert Emerson and Eugene Rabinowitch, both at the University of Illinois at Urbana-Champaign (UIUC), IL, U.S.A. He considers Eugene Rabinowitch as his main role model and his mentor. Govindjee is best known for his research on the mechanisms of excitation energy transfer, light emission, the primary photochemistry and the electron transfer in Photosystem II (PS II). His research, with many collaborators, has included the discovery of a short-wavelength form of chlorophyll (Chl) *a* functioning in the Chl *b*-containing system, now called PS II; of the two-light effects in Chl *a* fluorescence and in NADP (nicotinamide dinucleotide phosphate) reduction in chloroplasts (Emerson Enhancement). Further, he has worked on the existence of different spectral fluorescing forms of Chl *a* and the temperature dependence of excitation energy transfer down to 4 K; basic relationships between Chl *a* fluorescence and photosynthetic reactions; unique role of bicarbonate on the acceptor side of PS II, particularly in the protonation events involving the Q_B binding region; the theory of thermoluminescence in plants; picosecond measurement on the primary photochemistry of PS II; and the use of Fluorescence Lifetime Imaging Microscopy (FLIM) of Chl *a* fluorescence in understanding photoprotection against excess light. Govindjee's current focus is on the 'History of Photosynthesis Research,' in 'Photosynthesis Education,' and in the 'Possible Existence of Extraterrestrial Life.' He is the founding Historical Corner Editor of 'Photosynthesis Research', and the founding Series Editor of 'Advances in Photosynthesis and Respiration'. He has served on the faculty of the UIUC for ~40 years. Since 1999, he has been Professor Emeritus of Biochemistry, Biophysics and Plant Biology at the same institution. His honors include: Fellow of the American Association of Advancement of Science (AAAS); Distinguished Lecturer of the School of Life Sciences, UIUC; Fellow and Life member of the National Academy of Sciences (India); President of the American Society for Photobiology (1980–1981); Fulbright Scholar and Fulbright Senior Lecturer; Honorary President of the 2004 International Photosynthesis Congress (Montréal, Canada); the 2006 recipient of the Lifetime Achievement Award from the Rebeiz Foundation for Basic Research, and the 2007 recipient of the 'Communications Award' of the International Society of Photosynthesis Research (ISPR) and the 2008 Lifetime Achievement Award of the college of liberal Arts and Sciences, University of Illinois at Urbana-Champaign. He is a member of the Board of Directors of the Rebeiz Foundation for Basic Research. Further information on Govindjee and his research can be found at his web site: http://www.life.illinois.edu/govindjee/. He can always be reached by e-mail at gov@illinois.edu.

Thomas D. Sharkey, Co-Series Editor

Tom Sharkey obtained his Bachelor's degree in Biology in 1974 from Lyman Briggs College, a residential science college at Michigan State University, East Lansing, Michigan. After 2 years as a research technician, Tom entered a Ph.D. program in the federally funded Plant Research Laboratory at Michigan State University under the mentorship of Klaus Raschke and graduated in 1980 after 3 years and 3 months. Post-doctoral research was carried out with Graham Farquhar at the Australian National University, in Canberra, where he coauthored a landmark review on photosynthesis and stomatal conductance that continues to receive much attention even today. For 5 years, he worked at the Desert Research Institute, followed by 20 years as a professor of botany at the University of Wisconsin in Madison. In 2008, Tom became professor and chair of the Department of Biochemistry and Molecular Biology at Michigan State University. Tom's research interests center on the exchange of gases between plants and the atmosphere. The biochemistry and biophysics underlying carbon dioxide uptake and isoprene emission from plants form the two major research topics in his laboratory. Among his contributions are measurement of the carbon dioxide concentration inside leaves, an exhaustive study of short-term feedback effects on carbon metabolism, as well as a significant contribution to the elucidation of the pathway by which leaf starch breaks down at night. In the isoprene research field, Tom is recognized as the leading advocate for *thermotolerance of photosynthesis* as the explanation for why plants emit isoprene. In addition, his laboratory has cloned many of the genes that underlie isoprene synthesis, and he has published many papers on the biochemical regulation of isoprene synthesis. Tom has co-edited two books: T.D. Sharkey, E.A. Holland and H.A. Mooney (Eds.) *Trace Gas Emissions from Plants*, Academic, San Diego, CA, 1991; and R.C. Leegood, T.D. Sharkey, and S, von Caemmerer (Eds.) *Physiology and Metabolism*, Advances in Photosynthesis (and Respiration), Volume 9 of this Series, Kluwer (now Springer), Dordrecht, 2000. Tom is a "highly cited researcher" according to the Thomson Reuters Institute for Scientific Information. We welcome Tom as he adds new dimensions to our AIPH Series which is now in its 16th year and has completed its 30th volume in 2009.

Contents

Preface

As the industrial revolution that has been based on utilization of fossil fuels nears its end [R.A. Ker (2007) Even oil optimists expect energy demand to outstrip supply. Science 317: 437], the next industrial revolution will most likely need development of alternate sources of clean energy. In addition to the development of hydroelectric power, these efforts will probably include the conversion of wind, sea wave motion and solar energy [Solar Day in the Sun (2007) Business week, October 15, pp 69–76] into electrical energy. The most promising of those will probably be based on the full usage of solar energy. The latter is likely to be plentiful for the next 2–3 billion years. Most probably, that usage will take advantage of (a) the physical conversion of solar to electric energy [E. Pooley (2007) The Last temptation of Al Gore. Time, May 28], (b) solar energy converting systems [M. Hambourger, F. Gary, G.F. Moore, D.M. Kramer, D. Gust, A.L. Moore and T.A. Moore (2009) Biology and Technology for Photochemical Fuel Production. Chem. Soc. Rev. 38: 25–35], and (c) the biological engineering of higher photosynthetic efficiencies for the generation of fixed carbon.

The world population of about six billion is expected to increase to nine billion by the year 2030. It may increase even further by the end of this century. Worldwide, there has been a progressive decline in cereal yield and at present the annual rate of yield increase is below the rate of population increase [C. Somerville and J. Briscoe (2001) Genetic Engineering and Water. Science 292: 2217]. Since it will be difficult to increase the land area under cultivation without serious environmental consequences, the increased demand for food and fiber will have to be met by higher agricultural plant productivity.

With the imminent depletion of finite oil reserves [L. Eugene (2007) From peak oil to dark ages? Business Week, June 25], there has been considerable interest in the conversion of biomass into biofuels. For such a technology to become financially competitive, biomass transportation should not exceed a radius of about 25 miles. In other words, a bioconversion plant should be built in the center of every 2,000 miles of farmland. Such a caveat would be greatly alleviated by higher photosynthetic efficiencies and more biomass production per unit area.

According to Times Magazine (April 30, 2007 issue), one fifth of the US corn crop is presently converted into ethanol, which is considered to burn cleaner than gasoline and to produce less greenhouse gases. In order to meet a target of 35 billion gallons of ethanol produced by the year 2017, the entire US corn crop would need to be turned into fuel. But crops such as corn and sugarcane cannot yield enough to produce all the needed fuel. Furthermore, even if all available starch is converted into fuel, it would only produce about 10% of our gasoline needs [R.F. Service (2007) Biofuels researchers prepare to reap a new harvest. Science 315: 1488–1491]. Also, corn is used for a multitude of food products and diverting too much of the corn crop to ethanol would cause dramatic rises in the cost of food. All these calculations are based on a photosynthetic conversion rate of about 0.2–0.4% of incident radiation. If photosynthetic efficiency were to be increased to approach the 12% upper theoretical limit of photosynthetic energy conversion to fixed carbon [S. Lein and A. San Pietro (1975) An inquiry into biophotolysis of water to produce hydrogen, RANN, pp 50; X-G. Zhu, S.P. Long and D.R. Ort (2008) What is the maximum efficiency with which photosynthesis can convert solar energy into biomass? Current Opinion in Biotechnology 19: 153–159], there would be more than enough land to meet all of the world corn needs [C.A. Rebeiz, V.L. Kolossov and K.K. Kopetz (2004). Chloroplast Bioengineering 1. Photosynthetic Efficiency, Modulation of the Photosynthetic Unit Size, and the Agriculture of the Future. In: Agricultural Applications of Green Chemistry (W. M. Nelson, ed) ACS Symposium Series 887, pp 81–105].

According to the US Department of Energy [R.F. Service (2007), cited above], the US could convert 1.3 billion dry tons of biomass a year to 60 billion gallons of ethanol with little impact on food or timber harvests and in the process meet 30% of the nation's transportation fuel. But converting cellulose and hemicellulose to fuel is far more difficult than starting with simple sugar as is done in Brazil with sugarcane or in the USA with corn starch. This is due to the fact that agricultural and forest wastes are composed of cellulose, a polymer of six-carbon glucose, and of hemicellulose, a branched polymer composed of xylose and other five-carbon sugars and lignin. To convert biomass to ethanol, the complex carbohydrates must first be made available. However there is no naturally occurring organism that can convert xylose and other 5-carbon sugars to ethanol [R.F. Service (2007), cited above]. If this problem were solved, the conversion of biomass to ethanol would be greatly alleviated by higher photosynthetic conversion efficiencies.

In this respect it is interesting to mention the large scale effort of British Petroleum in conjunction with the University of California at Berkeley and the University of Illinois at Urbana-Champaign in pursuing the development of biofuels. Also, efforts focused on breeding more efficient fiber plants, Miscanthus and switchgrass [E. Kintish (2008) Sowing the seeds for high energy plants. Science 320: 478–483], Jatropha seeds for an oil-based biofuel [T. Padget and F. Myers (2009) Time Magazine, February 9, p 50], and Rutabagas (C. Benning, Personal communication, 2009) are under way. Additional efforts are also dealing with the possibility of converting algal biomass into biofuels [G. Edmonson (2007) Here comes pond scum power. Business week, December, pp 65–66].

In December 2006, the National Aeronautic and Space Organization (NASA) announced plans to establish a permanent international base camp on the moon by the year 2024. Sir Richard Branson stated [C.B. Thomas and N. Island (2007) The space cowboys, Time March 5: 52–58] that prices on Virgin Galactic, his space company, will eventually drop so that millions can go into space. If this heralds the beginning of extra-terrestrial space colonization, it is time to start thinking of alternatives to conventional agriculture by developing systems capable of converting solar energy, water as a source of electrons, and carbon dioxide into food fiber and energy at rates far above the photosynthetic rates presently experienced. Due to the eventual dwindling of water resources around the globe [C. Somerville and J. Briscoe (2001), cited above], Genetic engineering and the dearth of water in deserts and in space stations, the use of water as a source of electrons instead of an evapotranspirant would be a much needed feature of this putative agriculture.

The conversion of solar energy, water, as a source of electrons, and carbon dioxide into carbohydrates at rates that approach the upper photosynthetic theoretical limits was raised in a 1974 essay [C.A. Rebeiz (1974) Cell-free agriculture: fiction or realty? Illinois Research 16: 3–4] following the reporting of the biosynthesis of chlorophyll in vitro [C.A. Rebeiz and P. Castelfranco (1971) Protochlorophyll biosynthesis in a cell-free system from higher plants. Plant Physiol. 47: 24–32; C.A. Rebeiz and P. Castelfranco (1971) Chlorophyll biosynthesis in a cell-free system from higher plants. Plant Physiol. 33–37]. In a 1974 exchange of letters with Melvin Calvin, this possibility was debated by Calvin and Rebeiz, and Calvin suggested that the ultimate system that will achieve that goal will be a synthetic one.

Now, 38 years later, we believe that enough biochemical and molecular biological knowledge has accumulated to render this dream amenable to experimentation. As Confucius stated, "A journey of a thousand miles has to start with the First step." Indeed conventional agriculture is one of the few human activities that have not undergone a revolution to join other activities such as flying, defying gravity by landing on the moon, and crossing underwater the polar cap. Indeed, Craig Venter recently declared that he is about to succeed in creating an artificial microbe [J. Carey (2007) On the brink of artificial life. Business Week, June 27 p 40]. Other research teams around the country are following suit.

In this book, an effort is made to take Confucius's first steps. An attempt will be made to explore that first step in the light of present knowledge of the chemistry, biochemistry, and molecular biology of the greening process. We believe, that in the near future, the bioengineering of improved photosynthetic efficiency would require thorough knowledge of the biosynthesis of photosynthetic membrane components such as heme, chlorophylls, carotenoids, quinones, and lipids, thorough knowledge of photosynthetic membrane apoprotein biosynthesis, deeper understanding of the biosynthesis and regulation of the assembly of pigment–apoprotein complexes, and deeper knowledge of various facets of photon capture and carbon fixation. Another acknowledged way to increase agricultural productivity is to bioengineer chloroplasts more adapted to deal with stress and to alter the kinetic properties of the primary CO_2–assimilating enzyme, Rubisco, a multi-subunit protein located in the chloroplasts that is encoded by genes located in both the nucleus and chloroplast. The results of initial attempts at the bioengineering of Rubisco indicate that (a) increased knowledge of the factors controlling and coordinating gene expression in these two organelles, (b) better understanding of the significance of post-translational modifications of this enzyme, (c) increased knowledge of the assembly process, and (d) facile methods for gene replacement in both the chloroplast and the nucleus may be necessary for success. Also, chloroplast genetic engineering is an exciting field of research. It has several advantages over other genetic engineering techniques, among which are high levels of foreign gene expression, transgene containment, reduced cellular toxicity of the foreign proteins, lack of gene silencing or position effects that are current concerns of nuclear genetic engineering. It is an environmentally friendly approach and has been successfully used to engineer several valuable plant traits such as herbicide, disease and insect resistance, drought and salt tolerance, and phytoremedition. In the past, chloroplast transformation had been successful in only a few crops, including tobacco, potato and tomato. More recently, highly efficient soybean, carrot, and cotton plastid transformation has been achieved via somatic embryogenesis using species-specific vectors. In addition to biotechnological applications, plastid transformation has also been extensively used to study chloroplast biochemistry and molecular biology. Recent advances and future challenges will be addressed.

In Chapter 1, C.A. Rebeiz discusses the possible relationships between the multibranched chlorophyll (Chl) biosynthetic pathway, the assembly of Chl–protein complexes and a blueprint for the bioengineering of higher photosynthetic efficiencies.

In Chapter 2, V.L. Kolossov and Rebeiz describe the evidence for the occurrence of multiple 4-vinyl Chl reductase enzymes in higher plants. These enzymes are at the origin of the Chl biosynthetic heterogeneity in green plants.

In Chapter 3, B. Grimm describes the control of the metabolic flow in tetrapyrrole biosynthesis. He emphasizes the regulation of expression of the enzymes in the Mg-branch of tetrapyrrole biosynthesis.

In Chapter 4, R. Tanaka, H. Itoh and A. Tanaka discuss the regulation of the Chlorophyll (Chl) cycle that is at the basis of the interconversion of Chl *a* and *b* in higher plants.

In Chapter 5, C.G. Kannangara and D. von Wettstein review their pioneering work about the molecular biology of Mg-chelatase. This important enzyme signals the start of the Mg-branches of tetrapyrrole biosynthesis.

In Chapter 6, W.A. Cramer, S. Savikhin, J. Yan and E. Yamashita discuss the enigmatic Chl *a* molecule in the cytochrome *b6f* complex. They propose that this Chl may be involved in the trans-membrane conformational changes dependent upon the plastoquinone redox state.

In Chapter 7, H.K. Lichtenthaler reviews the plastidic deoxy-xylulose 5-phosphate/methylderithrol 4-phosphate (DOX/MEP) pathway of isoprenoid and pigment biosynthesis in plants.

In Chapter 8, A. Boronat discusses the regulatory role of the deoxy-xylulose 5-phosphate/methylderithrol 4-phosphate (GOX/MEP) pathway in the plastid isoprenoid biosynthesis.

In Chapter 9, E. Gerber, A. Hemmerlin, D.N. Crowell, M. Rohmer and T.J. Bach discuss the role of plastids in the geranylgeranylation of proteins. This process is of fundamental importance in the metabolism of eukaryotic cells.

In Chapter 10, D.A. Nagegowda, D. Rhodes and N. Dudareva describe the role of the deoxy-xylulose 5-phosphate/methylderithrol 4-phosphate (DOXP-MEP) pathway in the rhythmic emission of volatiles in plants. This process plays a major role in pollinator attraction, defense, communication and interaction of plants with the surrounding environment.

In Chapter 11, H. Valentin and Q. Qi describe the biological function and recent advances in the bioengineering of the enhancement of oil seed vitamin E in plastids.

In Chapter 12, C. Benning describes the biosynthesis and molecular biology of phosphatidylglycerol and sulfoquinovosyldiacylglycerol. These polar lipids provide the basic building blocks for the biosynthesis of the chloroplast membranes in all plants and cyanobacteria.

In Chapter 13, Y. Nakamura, K. Kobayashi, M. Shimojima and H. Ohta discuss the biosynthesis and function of monogalactosyldiacylglycerol in chloroplasts. This galactolipid amounts to about 50 % of the photosynthetic membrane lipids.

In Chapter 14, P. Dörmann discusses the synthesis and function of the galactolipid digalactosyldiacylglycerol, which is related to photosynthetic efficiency.

In Chapter 15, J.K. Hoober, L.L. Eggink, M. Chen and A.W.D. Larkum discuss the chemistry and biology of the light-harvesting Chl–protein complex II (LHCII). This Chl-protein complex is an important component of the bulk of light collection during photosynthesis.

In Chapter 16, H. Paulsen, C. Dockter, A. Volkov and G. Jeschke discuss the folding and pigment binding of the light-harvesting Chl–protein complex II (LHCII).

In Chapter 17, T.A. Ruhlman, J.W. Cary and K. Rajasekharan discuss the use of the plastid genome as a platform for the expression of microbial resistance genes. This technology promises to cut losses incurred in the production-conception continuum in plants.

In Chapter 18, N.D. Singh and H. Daniell describe the field of chloroplast genetic engineering. They emphasize the advantages of this technique over nuclear transformations.

In Chapter 19, R.E. Sharwood and S.M. Whitney discuss the bioengineering of the sunflower Rubisco subunits into tobacco chloroplasts.

In Chapter 20, G. Zhu, I. Kurek and L. Liu describe the bioengineering of photosynthetic enzymes involved in CO_2 assimilation by gene shuffling in order to enhance photosynthetic efficiency.

In Chapter 21, R. Grene, P. Li and H.J. Bohnert discuss the effects of elevated CO_2 and ozone on photosynthetic efficiency.

In Chapter 22, T. Shikanai discusses the regulation of photosynthetic electron transport with the aim of allowing green plants to avoid photodamage and to compete favorably with other plants in the field.

In Chapter 23, Y. Nanasato, K. Takahara, K. Kohzuma, Y.N. Munekage, A. Yokota, K. Akashi and C. Miyake discuss the mechanisms of drought and high-light stress tolerance in the wild watermelon xerophyte as a model system for the bioengineering of drought resistance in plants.

In Chapter 24, A. Sirikhachornkit and K.K. Niyogi discuss antioxidants and photo-oxidative stress responses in plants and algae.

In Chapter 25, B.C. Tripathy and G.K. Pattanayak discuss singlet oxygen-induced oxidative stress in plants.

This volume on the chloroplast is meant to provide the basis for chloroplast bioengineering and therefore foster closer cooperation between scientists who study porphyrins, isoprenoids, lipids, chloroplast proteins, biochemists and molecular biologists as well as graduate students and leading scientists involved in photosynthesis research, and biotechnologists involved in chloroplast bioengineering. We hope that the book will focus attention on chloroplast biochemistry, molecular biology and bioengineering as an integrated field of research, act as a magnet for attracting new talents and funding, and further the implementation of more efficient photosynthesis.

C.A. Rebeiz was supported by the Rebeiz Foundation for Basic Research; H.J. Bohnert was supported by the WCU program, Government of Korea, US Department of Agriculture, and by University of Illinois institutional grants; H. Daniell was supported by NIH grant Ro1 GM 63879 and USDA grant 3611-21000-021-02S; and A.R. Portis was supported by the United States Department of Agriculture and grants from the United States Department of Energy.

Many thanks to Eric Portis and Carole Rebeiz for helping with the book cover.

Constantin A. Rebeiz
Rebeiz Foundation for Basic Research,
Champaign USA

Christoph Benning
Department of Biochemistry and Molecular
Biology, Michigan State University,
Biochemistry 215, East Lansing, USA

Hans J. Bohnert
Department of Plant Biology and Department of
Crop Sciences, University of Illinois,
at Urbana-cChampaign, USA

Henry Daniell
Department of Molecular Biology and
Microbiology, Central Florida University,
Orlando, USA

J. Kenneth Hoober
School of Life Sciences,
Arizona State University,
Tempe, USA

Hartmut K. Lichtenthaler
Molecular Biology and Biochemistry of Plants,
University of Karlsruhe, Germany

Archie R. Portis
University of Illinois, Urbana, USA

Baishnab C. Tripathy
School of Life Sciences,
Jawaharlal Nehru University,
New Delhi, India

The Editors

Constantin A. Rebeiz

Constantin A. Rebeiz (Tino) was born on July 11, 1936, in Beirut Lebanon, where his family has been in the area since the fourth century AD. After 14 years of French schooling in a private French school, he joined the American University of Beirut, where he obtained a BS degree in General Agricultural Sciences in 1959. In 1960, he obtained an MS degree with Julian Crane in Horticulture from the University of California at Davis (UC Davis) for producing parthenocarpic cherries and peaches. In 1964, he obtained a Ph.D. in Plant Physiology from UC Davis with Paul Castelfranco for the discovery of the extra-mitochondrial β-oxidation of long chain fatty acids. From 1965 to 1969 he headed the Department of Biological Sciences at the Lebanese National Agricultural Research Institute. There he co-founded the Lebanese Association for the Advancement of Sciences. In 1972, He joined the University of Illinois at Urbana-Champaign (UIUC) as an associate professor of Plant Biochemical Physiology. In 1984 he received the John P Trebellas Endowment for Biotechnological Research and moved his Laboratory of Plant Biochemistry and Photobiology into a newly remodeled Laboratory space in the Agricultural Biotechnology Building; then in 1992 his laboratory was moved into the new Edward R. Madigan Building. Tino retired as a Professor Emeritus in May, 2005. He is currently the President of the Rebeiz Foundation for Basic Research which is involved in the promotion of chloroplast research and bioengeneering nationally and internationally, and is on the editorial board of Analytical Biochemistry. Tino's research in tetrapyrrole biochemistry and chemistry (see <http://www.vlpbp.org>) spans the fields of botany, plant physiology, preparative methodologies, analytical biochemistry, biochemistry, chemistry, the development of pesticides, biomedical research as well as chloroplast bioengeneering. His tetrapyrrole biochemical work led to the discovery of novel Mg-porphyrins, prochlorophylls, chlorophylls, the discovery of the multibranched chlorophyll biosynthetic pathway, and the development of a blueprint for bioengineering chloroplasts with higher photosynthetic efficiencies. His pioneering work on the development of photodynamic pesticides has led to the development of δ-aminolevulinic-dependent photodynamic cancer treatment and skin keratoses that are used worlwide in the medical community. Many laboratories around the world are now developing δ-aminolevulinic-dependent non-invasive photodynamic cancer treatment strategies for several types of cancer. Among the awards received by Tino are the Siemens award (1957), The University of Illinois Funk award (1985), selection by science digest as one of America's outstanding innovators responsible for the 100 top technological achievements (1985), the University of Illinois College of Agriculture Senior Faculty award for excellence in research (1985), the US presidential Green Chemistry Challenge Award (1999), and the American University of Beirut Scientific Achievement award during commemoration of the 50th Anniversary of the College of Agriculture and Food Sciences (2002).

Christoph Benning

Christoph Benning was born in Soest, Germany, in 1960. Christoph received his MS from the University of Freiburg, Germany, and his Ph.D., with Chris Sommerville, from the Michigan State University at East Lansing, Michigan, USA. He was an Independent Group Leader at the Institute for Gene Biological Research in Berlin and an Assistant Professor at the Freie University of Berlin from 1993 until 1998, He is now a professor of biochemistry and molecular biology at Michigan State University. Beginning January 2010, he is editor-in-chief of The Plant Journal and he serves on the Board of Directors of the Rebeiz Foundation for Basic Research. His research has focused on the biochemistry of lipids and the control of lipid metabolism in plants, algae, and photosynthetic bacteria. Current areas of his research are the regulation of photosynthetic membrane lipid assembly, lipid trafficking between the endoplasmic reticulum (ER) and the plastid, and the regulation of storage lipid biosynthesis. Studying lipid metabolism is inherently challenging due to the complexity of lipid pathways in photosynthetic organisms. To meet this challenge Christoph applies genetic and genomic approaches in suitable model organisms. The nature of the problems studied also demands the application of state-of-the-art techniques of enzyme biochemistry, analytical chemistry, and cell biology. His discoveries encompass proteins and their mode of function in sulfolipid, phospholipid, betaine lipid and galactolipid biosyntheses, transport proteins involved in ER-to-plastid lipid trafficking, and transcription factors and enzymes critical for storage lipid biosynthesis. These findings have advanced our general understanding of lipid metabolism in photosynthetic organisms. In addition, work by Christoph has led to the development of patentable tools and methods that will be useful for the engineering of improved oil yield in crop plants and algae for the purpose of biofuel production.

Hans J. Bohnert

Hans J Bohnert was born in Heilbronn, Germany, in 1944. Hans is currently professor of plant biology and crop sciences at the University of Illinois at Urbana-Champaign (UIUC), Illinois. In addition, he is a faculty member in the Institute for Genomic Biology (IGB) at the UIUC; and Director of the Keck Center for Comparative and Functional Genomics, also at the UIUC. Since 2009, he is also a visiting professor at Gyeongsang National University, Jinju, Korea, Adjunct Research fellow at the Korea Research Institute of Bioscience and Biotechnology, a visiting scientist at Shandong Normal University, Jinan, China; and at King Abdullah University of Science and Technology (KAUST), Jeddah, Saudi Arabia. Hans obtained his Staatsexamen Diploma and Ph.D. (Biology, Chemistry, and Biochemistry) from the University of Heidelberg, Germany. He was a fellow at the European Molecular Biology Laboratory (EMBL), Heidelberg, and at the Max–Planck Institute for Plant Breeding, Cologne, Germany. After habilitation at the University of Düsseldorf, he was a Dozent (associate professor) at Düsseldorf, and after 1983, served in the Department of Biochemistry, University of Arizona, Tucson, Arizona, as an associate professor and then professor. In addition, he served, from 1995 to 1997, as a Program Director of the Integrative Plant Biology Program of the US National Science Foundation. His research interests focus on plant abiotic stress genomics, genome structure and gene expression in plants evolutionarily adapted to extreme habitats, and in developing and applying tools for integrating genomic datasets. His honors include: Heisenberg-Fellowship awarded by the Deutsche Forschungsgemeinschaft; Research Scientist Award of the College of Agriculture, University of Arizona; and a Senior Research Fellowship by the Smithsonian/Carnegie-Mellon Foundation. He is a Fellow of the American Society of Plant Biology and a member of the Board of Directors of the Rebeiz Foundation for Basic Research. Information about his recent work can be found at http://www.life.uiuc.edu/bohnert.

Henry Daniell

Henry Daniell was born in Salem, India. He obtained an MS from the University of Madras in 1971 and a Ph.D. in 1980 from Madurai Kamaraj University in the combined fields of Biophysics, Biochemistry and Microbiology with Professor A. Gnanam. From 1980 to 1982 he was a Postdoctoral trainee in C. A. Rebeiz's Laboratory. Henry is presently a Pegasus Professor in the Department of Molecular Biology and Microbiology, at the College of Medicine, University of Central Florida (UCF). He is a Fellow of the American Association for the Advancement of Science and holds the first University Board of Trustee Chair in Life Sciences at UCF. He is a member of the Italian Academy of Sciences and a member of the Board of Directors of the Rebeiz Foundation for Basic Research. Henry also won an American Diabetes Association Award for his outstanding contributions. He is recognized for pioneering chloroplast genetic engineering as a new platform to produce low cost vaccines and biopharmaceuticals to eliminate the expensive fermentation based technologies currently in use. He developed and advanced the concept to create vaccines against biothreat agents such as anthrax and plague and several other infectious diseases identified by the World Health Organization and Centers for Disease Control. Oral delivery of chloroplast-derived proinsulin proved to be a powerful tool to prevent the onset of type 1 diabetes. His laboratory has determined the complete chloroplast genome sequences of more than 20 crop plants. His research has been featured on the cover of several high impact scientific journals, and ranked by Nature Biotechnology among the top ten inventions of this decade and among Biomed Central's Hot 100 authors in the world. His research led to the formation of UCF's first biotechnology company for commercial development of his patented technology. He has served as a consultant to the United Nations, to several NIH or National Academy of Science panels and is a senior editor of several scientific journals. For more details visit <http://daniell.ucf.edu>.

J. Kenneth Hoober

J. Kenneth Hoober (Ken) was born in Lancaster, Pennsylvania, in 1938. Ken received a PhD degree in biochemistry from the University of Michigan in 1965 and obtained additional postdoctoral training at Vanderbilt University with Stanley Cohen and at the Rockefeller University with Philip Siekevitz and George Palade. While at the Rockefeller University, he began working on chloroplast development in the green alga *Chlamydomonas reinhardtii*, with a focus on the mechanisms of thylakoid membrane biogenesis. He held faculty positions in biochemistry at Rutgers University Medical School for 3 years and then at Temple University School of Medicine for 20 years. In 1987, he had an exceptional experience working at the Carlsberg Laboratory in Copenhagen with Gamini Kannangara, Simon Gough and Diter von Wettstein. He was chair of the Department of Biochemistry during his last 3 years at Temple University. In 1991, he accepted the position of chair of the Department of Plant Biology at Arizona State University, which he held until 2003. He, along with his students in his group, showed convincingly that chlorophyll *a/b*-binding proteins were initially integrated into membranes of the chloroplast envelope. An emphasis in this project was the role of chlorophyll *b* in the assembly of light-harvesting complexes. He retired in 2004 but continued working as Professor Emeritus. Since 2006, he has been co-founder and chief scientific officer of Susavion Biosciences, Inc., a start-up biotechnology company that is focused on peptides as pharmaceutical compounds. Ken is a member of the Board of Director and Vice-President of the Rebeiz Foundation for Basic Research

Hartmut K. Lichtenthaler

Hartmut K. Lichtenthaler was born in 1934 in Weinheim, Germany. Hartmut studied pharmacy, chemistry and botany at the Universities of Karlsruhe and Heidelberg, where he received his Ph.D. in Plant Biology in 1961 with Professor August Seybold. From 1962 to 1964, he was a postdoctoral fellow with Nobel laureate Melvin Calvin at the University of California Berkeley, where he revealed the complete pigment and lipid composition of the photosynthetic thylakoid membrane. After his return to Germany he was an associate professor of plant physiology at the University of Münster. In 1970, he was appointed full professor of plant physiology and plant biochemistry at the University of Karlsruhe, Germany (now, Karlsruhe Institute of Technology, KIT), where he is an emeritus professor since 2001. In 1978, he was a founding member of the Federation of European Societies of Plant Biology (FESPB) and was FESPB President from 1984 to1986. Since the 1979s he has strongly promoted European scientific cooperation and exchange of scientists in photosynthesis, plant physiology and plant biochemistry with particular emphasis on integrating Eastern European scientists from the Czech Republic, Hungary, Poland, and Russia into the European Federation of Plant Biology. He has also maintained close collaborations with researches in the USA, France and several other countries. His main research fields are: function of prenylquinones in chloroplasts; plastoglobuli and the fine structure of chloroplasts; adaptation of photosynthetic function and chloroplast structure to sun and shade conditions; localization of carotenoids in thylakoid pigment proteins; mode of action of herbicides in photosynthesis and plant fatty acid biosynthesis; and chlorophyll fluorescence imaging of plants. During 1993 to 1999, he detected and established, in close collaboration with M. Rohmer, Strasbourg, the plastidic DOXP/MEP pathway of isoprenoid, carotenoid and isoprene biosynthesis. For his pioneering research, Hartmut received three honorary doctoral degrees (Mendel University, Brno, Czech Republic; ELTE University, Budapest and St. Istvan University, Gödöllö, Hungary) and multiple scientific honors and in 2001, the Cross of Merits of the Federal Republic of Germany.

Archie R. Portis

Archie R. Portis, Jr was born in 1949 in Winston-Salem, NC, USA. Archie obtained a BS degree in organic chemistry from Duke University in 1971 and a Ph.D. degree in biochemistry from Cornell University in 1976 under the direction of Richard E. McCarty. He conducted postdoctoral studies with Hans W. Heldt at the Universität München, Munich, Germany and with. D. Papahadjopoulos at the Roswell Park Memorial Institute in Buffalo, NY before joining the US Department of Agriculture (USDA) as a plant physiologist on the University of Illinois at Urbana-Champaign (UIUC) campus. Archie also held faculty appointments in the Departments of Crop Science and Plant Biology at the UIUC before he retired from active research in 2008. He has had many peer-reviewed publications with students and colleagues and his research career was centered on the regulation of photosynthetic carbon metabolism. Following his discovery of the regulatory and chaperone-like protein, Rubisco activase, in 1985 with Michael E. Salvucci and William (Bill) L. Ogren, his research became mainly focused on physiological, biochemical, and molecular biological studies of this protein with the aim of increasing the efficiency of photosynthesis. His pioneering research contributions were recognized by his election as a Fellow of the American Association for the Advancement of Science in 1997. Archie is a member of the Board of Directors of the Rebeiz Foundation for Basic Research.

Baishnab C. Tripathy

Baishnab Charan Tripathy was born on January 5, 1952 in Kutilo, Cuttack, located on the east coast of India. After 11 years of primary, middle and high school education in a rural setting, he joined Ravenshaw College, Cuttack in 1967, and subsequently Buxi Jagabandhu Bidyadhar (BJB) College, Bhubaneswar in 1969, where he received a BSc (Major: Botany) in 1971. Then, he obtained an MSc. (Botany) in 1973 from Utkal University, Orissa, India. In 1981, he was awarded a Ph.D. degree, from Jawaharlal Nehru University (JNU), New Delhi, India, for his work on the *primary production and role of metal ions on the primary processes of photosynthesis* under the supervision of Prasanna Mohanty. During 1981–1983, he was a post-doctoral research associate, with Elizabeth Gross and John S. Rieske, at the Ohio State University, Columbus, Ohio, USA, where, he worked on *the role of cations on chlorophyll a fluorescence and on electron transport from plastocyanin to cytochrome c through complex III of mitochondria*. From December, 1983 till March, 1987, he worked, with Constantin A. Rebeiz, at the University of Illinois at Urbana-Champaign, Illinois, USA on *chloroplast biogenesis and biochemistry of chlorophylls*. In 1987, he joined as an assistant professor of photobiology in the School of Life Sciences, JNU, where, he was promoted to associate professor and finally to professor in 2002. At JNU, Baishnab teaches and does research in plant physiology and photobiology. His research work has significantly contributed to our knowledge of plant tetrapyrroles and chloroplast biogenesis. His work on plant senescence in relation to chlorophyll metabolism has demonstrated the presence of different sets of enzymes i.e., kinetin-sensitive and kinetin-insensitive that are responsible for the degradation of chlorophyll and protochlorophyllide, respectively. His demonstration of the role of the meristematic layer in the intra- and inter-cellular signaling system in relation to chloroplast development of wheat and rice plants, grown under red or red plus blue light, are novel and interesting. At the National Aeronautics and Space Administration (NASA, USA), he has published research papers on photosynthetic responses of plants grown in space. In addition to research articles, he has contributed several chapters to books, and has edited books on photosynthesis. He has mentored a large number of Ph.D., M.Phil and MSc students. Baishnab is on the editorial board of "*Horticulture, Environment and Biotechnology*". He is a member of the Board of Directors of the Rebeiz Foundation for Basic Research, is a fellow of the National Academy of Sciences, India, of the Indian National Science Academy and of the National Academy of Agricultural Sciences, India.

Contributors

Kinya Akashi, Graduate School of Biological Sciences, Nara Institute of Science and Technology (NAIST), Ikoma, Nara 630-0192, Japan

Thomas J. Bach, Institut de Biologie Moléculaire des Plantes, Centre National de la Recherche Scientifique (Unité Propre de Recherche 2357 conventionée à l'Université de Strasbourg), Département Reseaux Metaboliques, Institut de Botanique, 28 rue Goethe F-67083 Strasbourg, France

Christoph Benning, Department of Biochemistry and Molecular Biology, Michigan State University, East Lansing, MI 48824, USA

Hans J Bohnert, Center for Comparative and Functional Genomics, University of Illinois at Urbana-Champaign, Urbana, IL 61801, USA

Albert Boronat, Department of Biochemistry and Molecular Biology, Faculty of Biology, University of Barcelona, 08028-Barcelona, Spain

Min Chen, School of Biological Sciences, University of Sydney, NSW 2006, Australia

Jeffrey W. Cary, Agricultural Research Service, Southern Regional Research Center, USDA, 1100 Robert E. Lee Blvd., New Orleans, LA 70124, USA

William A. Cramer, Department of Biological Sciences, Purdue University, West Lafayette, IN 47907-2054, USA

Dring N. Crowell, Department of Biological Sciences, Idaho State University, 650 Memorial Drive, Pocatello, ID 83209-8007

H. Daniell, Department of Molecular Biology and Microbiology, College of Medicine, University of Central Florida, Biomolecular Science Building, Room 336, Orlando, FL 32816-2364, USA

Christoph Dockter, Institut für Allgemeine Botanik, Universität Mainz, Müllerweg 6, D-55099 Mainz, Germany

Peter Dörmann, Institute of Molecular Physiology and Biotechnology of Plants, Karlrobert-Kreiten-Strasse 13, 53115 Bonn, Germany

Natalia Dudareva, Department of Horticulture and Landscape Architecture, Purdue University, West Lafayette, IN 47907, USA

Laura L. Eggink, School of Life Sciences and Center for the Study of Early Events in Photosynthesis, Arizona State University, Tempe, AZ 85287-4501, USA

Esther Gerber, Institut de Biologie Moléculaire des Plantes, Centre National de la Recherche Scientifique (Unité Propre de Recherche 2357 conventionée à l'Université de Strasbourg), Département Reseaux Metaboliques, Institut de Botanique, 28 rue Goethe F-67083 Strasbourg, France

Ruth Grene, Department of Plant Pathology, Virginia Tech, Blacksburg, VA 24061, USA

Bernhard Grimm, Institute of Biology/ Plant Physiology, Humboldt University, Philippstr.13, Building 12, 10115 Berlin, Germany

Andréa Hemmerlin, Institut de Biologie Moléculaire des Plantes, Centre National de la Recherche Scientifique (Unité Propre de Recherche 2357 conventionée à l'Université de Strasbourg), Département Reseaux Metaboliques, Institut de Botanique, 28 rue Goethe F-67083 Strasbourg, France

J. Kenneth Hoober, School of Life Sciences and Center for the Study of Early Events in Photosynthesis, Arizona State University, Tempe, AZ 85287-4501, USA

Hisashi Ito, Institute of Low Temperature Science, Hokkaido University, N19W8, Kita-ku, Sapporo 060-0819, Japan

Gunnar Jeschke, Institut für Physikalische Chemie, Universität Konstanz, 78475 Konstanz, Germany

C. Gamini Kannangara, Department of Crop and Soil Sciences, Washington State University, Pullman, WA99164, USA

Koichi Kobayashi, Graduate School of Bioscience and Biotechnology, Tokyo Institute of Technology, 4259-B-65 Nagatsuta-cho, Midori-ku, Yokohama 226-8501, Japan

Kaori Kohzuma, Graduate School of Biological Sciences, Nara Institute of Science and Technology (NAIST), Ikoma, Nara 630-0192, Japan

Vladimir L. Kolossov, Institute for Genomic Biology, University of Illinois, Urbana, IL 61801, USA

Itzhak Kurek, Pioneer Hi-Bred International Inc., DuPont Agriculture & Nutrition, 700A Bay Road, Redwood City, CA 94063, USA

Anthony W. D. Larkum, School of Biological Sciences, University of Sydney, NSW 2006, Australia

Pinghua Li, Departments of Plant Biology and of Crop Sciences, University of Illinois at Urbana-Champaign, Urbana, IL 61801, USA

Hartmut K. Lichtenthaler, Botanisches Institut (Molecular Biology and Biochemistry of Plants), University of Karlsruhe, Kaiserstr. 12, D-76133 Karlsruhe, Germany

Lu Liu, Pioneer Hi-Bred International Inc., DuPont Agriculture & Nutrition, 700A Bay Road, Redwood City, CA 94063, USA

Chikahiro Miyake, Department of Biological and Environmental Science, Faculty of Agriculture, Kobe University, 1-1 Rokkodai-cho, Nada-ku, Kobe, 657-8501, Japan

Qungang Qi, Monsanto Co., 700 Chesterfield Parkway West, Chesterfield, MO 63017, USA

Yuri Nakajima Munekage, Graduate School of Biological Sciences, Nara Institute of Science and Technology (NAIST), Ikoma, Nara 630-0192, Japan

Dinesh A. Nagegowda, Department of Horticulture and Landscape Architecture, Purdue University, West Lafayette, IN 47907, USA

Yuki Nakamura, Yong Loo Lin School of Medicine, Centre for Life Sciences, Department of Biochemistry, National University of Singapore, 28 Medical Drive, Singapore 117456, Republic of Singapore

Yong Loo, Lin School of Medicine, Centre for Life Sciences, Department of Biochemistry, National University of Singapore, 28 Medical Drive, Singapore 117456, Republic of Singapore

Yoshihiko Nanasato, Graduate School of Biological Sciences, Nara Institute of Science and Technology (NAIST), Ikoma, Nara 630-0192, Japan

Krishna K. Niyogi, Department of Plant and Microbial Biology, University of California, Berkeley, CA 94720-3102, USA

Hiroyuki Ohta, Research Center for the Evolving Earth and Planets, Tokyo Institute of Technology, 4259-B-65 Nagatsuta-cho, Midori-ku, Yokohama 226-8501, Japan

Gopal K. Pattanayak, School of Life Sciences, Jawaharlal Nehru University, New Delhi-110067, India

Harald Paulsen, Institut für Allgemeine Botanik, Universität Mainz, Müllerweg 6, D-55099 Mainz, Germany

Kanniah Rajasekaran, Agricultural Research Service, Southern Regional Research Center, USDA, 1100 Robert E. Lee Blvd., New Orleans, LA 70124, USA

Constantin A. Rebeiz, Rebeiz Foundation for Basic Research, 2209 Edgewater Place, Champaign, IL 61822, USA

David Rhodes, Department of Horticulture and Landscape Architecture, Purdue University, West Lafayette, IN 47907, USA

Michel Rohmer, Université de Strasbourg CNRS, Institut de Chimie (Unité Mixte de Recherche UMR 7177), 4 rue Blaise Pascal, F-67070 Strasbourg, France (MR)

Tracey A. Ruhlman, Agricultural Research Service, Southern Regional Research Center, USDA, 1100 Robert E. Lee Blvd., New Orleans, LA 70124, USA

Sergei Savikhin, Departments of Physics, Purdue University, West Lafayette, IN 47907-2054, USA

Toshiharu Shikanai, Department of Botany, Graduate School of Science, Kyoto University, Kitashirakawa, Sakyo-ku, Kyoto, 606-8502, Japan

Robert E. Sharwood, Molecular Plant Physiology, The Australian National University, GPO box 475, Canberra, ACT 2601, Australia

Mie Shimojima, Center for Biological Resources and Informatics, Tokyo Institute of Technology, 4259-B-65 Nagatsuta-cho, Midori-ku, Yokohama 226-8501, Japan

N.D. Singh, Department of Molecular Biology and Microbiology, College of Medicine, University of Central Florida, Biomolecular Science Building, Room 336, Orlando, FL 32816-2364, USA

Anchalee Sirikhachornkit, Department of Plant and Microbial Biology, University of California, Berkeley, CA 94720-3102, USA

Kentaro Takahara, Graduate School of Biological Sciences, Nara Institute of Science and Technology (NAIST), Ikoma, Nara 630-0192, Japan

Ayumi Tanaka, Institute of Low Temperature Science, Hokkaido University, N19W8, Kita-ku, Sapporo 060-0819, Japan

Ryouichi Tanaka, Institute of Low Temperature Science, Hokkaido University, N19W8, Kita-ku, Sapporo 060-0819, Japan

Baishnab C. Tripathy, School of Life Sciences, Jawaharlal Nehru University, New Delhi-110067, India

Henry E. Valentin, Monsanto Co., Calgene Campus, 1920 Fifth Street, Davis, CA 95616, USA

Aleksei Volkov, Max-Planck-Institut für Polymerforschung, Ackermannweg 10, D-55128 Mainz, Germany

Diter von Wettstein, Department of Crop and Soil Sciences and School of Molecular Biosciences, Washington State University, Pullman, WA 99164 and Institute of Phytopathology and Applied Zoology, IFZ, Heinrich-Buff-Ring 26-32, D-35392 Giessen, Germany

Spencer M. Whitney, Molecular Plant Physiology, The Australian National University, GPO box 475, Canberra, ACT 2601, Australia

E. Yamashita, Institute of Protein Research, Osaka University, Toyonaka, Osaka 560-0043, Japan

Jiusheng Yan, Section of Neurobiology, School of Biological Sciences, University of Texas, Austin, TX 78713, USA

Akiho Yokota, Graduate School of Biological Sciences, Nara Institute of Science and Technology (NAIST), Ikoma, Nara 630-0192, Japan

Genhai Zhu, Pioneer Hi-Bred International Inc., 4010 Point Eden Way, Hayward, CA 94545-3721

Author Index

Chapter 1

Investigation of Possible Relationships Between the Chlorophyll Biosynthetic Pathway, the Assembly of Chlorophyll–Protein Complexes and Photosynthetic Efficiency

Constantin A. Rebeiz
Rebeiz Foundation for Basic Research, 2209 Edgewater Place, Champaign, IL 61822, USA

Author for Correspondence, e-mail: crebeiz@uiuc.edu

Constantin A. Rebeiz et al. (eds.), The Chloroplast: Basics and Applications, pp. 1–24.

Summary

The world population of about six billion is expected to increase to nine billion by the year 2030. It may reach 18 billion by the end of the century. Worldwide, there has been a progressive decline in cereal yield and at present the annual rate of yield increase is below the rate of population increase. Since it will be difficult to increase the land area under cultivation without serious environmental consequences, the increased demand for food and fiber will have to be met by higher agricultural plant productivity. Plant productivity depends in turn upon photosynthetic efficiency. We have reason to believe that agricultural productivity can be significantly increased by alteration of the photosynthetic unit size. On the basis of recent advances in the understanding of the chemistry and biochemistry of the greening process and significant advances in molecular biology, we believe that alteration of the PSU size has become a realistic possibility. The thorough understanding of photosynthetic membrane assembly requires a deeper knowledge of the coordination of chlorophyll (Chl) and thylakoid apoprotein biosynthesis. As a working model for future investigations we have proposed three Chl–thylakoid apoprotein biosynthesis models namely a single-branched Chl biosynthetic pathway (SBP)-single location model, a SBP-multilocation model, and a multi-branched Chl biosynthetic pathway (MBP)-sublocation model. Rejection or validation of these models was probed by determination of resonance excitation energy transfer between various tetrapyrrole intermediates of the Chl biosynthetic pathway and various thylakoid Chl–protein complexes. The occurrence of resonance excitation energy transfer between several Chl precursors namely protoporphyrin IX (Proto) Mg-Proto + Mg-Proto monomethyl ester [Mp(e)] and divinyl (DV) and monovinyl (MV) Pchlide *a* and several Chl–protein complexes was demonstrated in situ. It was possible to calculate the distances separating Proto Mp(e) and Pchl(ide) *a* donors from Chl *a* acceptors. The calculated distances were incompatible with the operation of the SBP-single location Chl–protein biosynthesis model but were compatible with the operation of the MBP-sublocation model. The compatibility of the MBP-sublocation model of Chl–thylakoid protein complexes assembly opened the way for testing the hypothesis of whether certain Chl biosynthetic routes are indeed involved in the formation of specific Chl–protein complexes. The experimental strategy to tackle this issue was described. Based on the results so far acquired it was suggested that the greening process may now be manipulated to bioengineer genetically modified plants with a smaller PSU i.e. with more PSU units having fewer antenna Chl per unit thylakoid area, and ensuing higher photosynthetic efficiencies.

Abbreviations: ALA – δ-aminolevulinic acid; Chl – chlorophyll; Chlide – *a* chlorophyllide *a*; CP29 – an inner light harvesting Chl antenna of photosystem II, with molecular mass of 29 kDa; CP47 – a core antenna of photosystem II, with molecular mass of 47 kDa; Dpy – 2,2′-dipyridyl; DV – divinyl; LHC – light harvesting chlorophyll antenna; LHCI-680 – an LHC antenna of Photosystem I; LHCI-730 – another LHC antenna of photosystem I; LW – long wavelength; MP – Mg-Protoporphyrin IX; Mpe – Mg-Proto monomethyl ester; Mp(e) – Mg-Proto and/or Mpe; MV – monovinyl; Pchlide – *a* protochlorophyllide *a*; Pchl(ide) – Pchlide and/or Pchlide ester; PETS – photosynthetic electron transport system; Proto - protoporphyrin IX; PS - photosystem; PSI – photosystem I; PSII – photosystem II; PSU – photosynthetic unit; RC – reaction center; s – second; SBP – single-branched pathway; MBP – multibranched-pathway; SW – short wavelength

Note: Unless preceded by MV or DV, tetrapyrroles are used generically to designate metabolic pools that may consist of MV and/or DV components.

I Introduction

The world population of about six billion is expected to increase to nine billion by the year 2030. It may reach 18 billion by the end of the century. Worldwide there has been a progressive decline in cereal yield and at present the annual rate of yield increase is below the rate of population increase. Since it will be difficult to increase the land area under cultivation without serious environmental consequences, the increased demand for food and fiber will have to be met by higher agricultural plant productivity.

With the imminent depletion of finite oil reserves there has been considerable interest in the conversion of biomass into biofuels. For such an undertaking to become financially competitive, biomass transportation should not exceed a radius of about 25 miles. In other words a bioconversion plant should be built in the center of every 2,000 $miles^2$ of farmland. Such a caveat would be greatly alleviated by higher photosynthetic efficiencies.

In December 2006 the National Aeronautic and Space Organization (NASA) announced plans to establish a permanent international base camp on the moon by the year 2024. Sir Richard Branson recently stated that prices on Virgin Galactic, his space company, will eventually drop so that millions can go into space (Booth-Thomas, 2007). If this heralds the beginning of extra-terrestrial space colonization, it is time to start thinking of alternatives to conventional agriculture by developing systems capable of conversion of solar energy, water as a source of electrons, and carbon dioxide, into food fiber and energy at rates far above the photosynthetic rates presently experienced.

In this chapter, I will attempt to explore the above possibilities in the light of present knowledge of the chemistry, biochemistry, and molecular biology of the greening process.

II Agricultural Productivity and Photosynthetic Efficiency

Life in the biosphere is carbon based. All molecules needed for life are made up of a carbon skeleton which is complemented by organic elements such as O, H, N, and inorganic elements such as K, P, Ca, Fe, etc. The carbon O and H of organic compounds originate in CO_2 and H_2O. The carbon skeleton is assembled via the process of photosynthesis which essentially converts solar energy into chemical energy (Blankenship, 2002). Nitrogen originates in NH_3, and inorganic elements originate in the rocks of the biosphere and are incorporated into the carbon skeleton via enzymatic reactions. Chemical energy consists of the covalent bond energy embedded into the carbon – carbon skeleton as well as the high energy bonds of ATP and NADPH which are formed during the process of photosynthesis.

The carbon cycle essentially describes how photosynthesis supports organic life in the biosphere. The carbon skeleton formed via the process of photosynthesis is converted into the simple and complex food consumed by organic life. The needed energy for enzymatic interconversions and biosynthetic processes is provided by ATP and NADPH. The organic matter of dead biota is converted in turn into CO_2, H_2O, and inorganic elements by bacterial activity then the cycle starts all over again.

At issue then, is whether agricultural productivity at today's levels of photosynthetic efficiency is high enough to feed a growing world population and meet the needs of space colonization. Since plants form food by conversion of solar energy, CO_2 and H_2O into chemical energy via the process of photosynthesis, it ensues that agricultural productivity depends in turn upon photosynthetic efficiency. Let us therefore briefly dissect the efficiency components of the photosynthetic process.

Photosynthetic efficiency is controlled by intrinsic and extrinsic factors (Lien and San Pietro 1975). Extrinsic factors include the availability of water, CO_2, inorganic nutrients, ambient temperature, and the metabolic and developmental state of the plant. The most important intrinsic factor is the efficiency of the photosynthetic electron transport system (PETS).

A The Primary Photochemical Act of Photosystem I (PS I) I and II

Conversion of solar energy into chemical energy is the results of two photochemical acts that take place in photosystem I (PSI) (Golbeck, 2006) and PSII (Wydrzynski and Satoh, 2006). The primary photochemical act of PSII is initiated by the absorption of light by antenna Chl *a* and *b*. The absorbed photons are conveyed to special Chls in the PSII reaction center. There, light energy is used to generate a strong oxidant which liberates oxygen from water. It also generates a weak reductant which together with the plastoquinone electron acceptor pools serve for temporary storage of the electrons extracted from water.

The primary photoact of PSI is also initiated by the absorption of light mostly by antenna Chl *a* and here too, the absorbed photons are conveyed to special Chls in the PSI reaction center. There light energy generates a weak oxidant $P700^+$ which receives electrons from the plastoquinone pools via cytochrome *f* and plastocyanin. It also generates a strong reductant A_0^- which donates electrons to $NADP^+$ via a series of electron carriers and converts it to NADPH. The photochemical acts of PSII and PSI, and the flow of electrons between PSII and PSI are depicted in Fig. 1.

B Conversion of Carbon Dioxide into Carbohydrates

During electron and proton flow, energy rich ATP and NADPH are formed. The energy of NAPDH and ATP is used for the enzymic conversion of CO_2 into carbohydrates which are in turn converted into a variety of organic molecules (Leegood et al. 2000). In summary the efficiency of food formation by green plants depends to a great extent on the efficiency of NADPH and ATP formation which depends in turn on the efficiency of the PETS.

The rest of this chapter will therefore be devoted to a discussion of the efficiency of the PETS and possible alterations in the circuitry of the chloroplast that may lead to a higher efficiency of the PETS and higher plant productivity under field conditions.

C Theoretical Maximal Energy Conversion Efficiency of the Photosynthetic Electron Transport System of Green Plants

This discussion is essentially extracted from a 1975 "Research Applied for National Needs" report (Lien and San Pietro 1975). At the maximal quantum efficiency of one, two photons are required to move one electron across the potential difference of about 1.25 V between the Mn cluster hole and A_0 (Fig. 1) The maximal

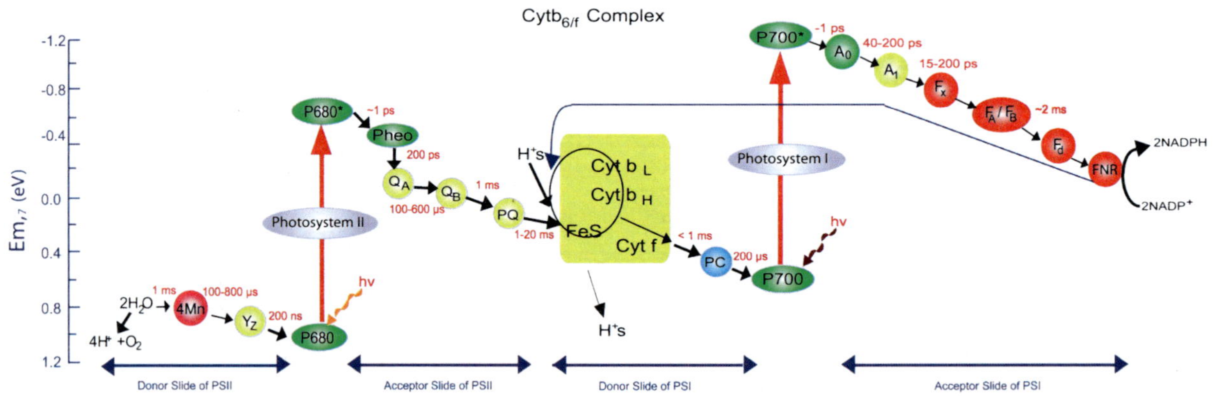

Fig. 1. The Z-scheme of oxygenic photosynthesis for electron transfer from water to oxidized nicotinamide adenine dinucleotide phosphate (NADP). The symbols are: Mn = Mn cluster; Y = Tyrosine –161 on D1 protein; P680 = a pair of Chls the reaction center (RC) Chls of PS II having one of its absorption bands at 680 nm, P680* = excited P680; Pheo = the primary electron acceptor of PS II; Q_A = the primary plastoquinone electron acceptor of PS II; Q_B = secondary plastoquinone electron acceptor of PS II; PQ = plastoquinone pool; FeS = Rieske iron sulfur protein; Cyt f = cytochrome *f*; CytbII = high potential cytochrome *b6*; PC = plastocyanin; P700 = a pair of Chl *a* and Chl *a′* the RC Chls of PS I; P700* = excited P700; A_0 = primary electron acceptor of PS I a Chl monomer; A_1 = secondary electron acceptor of PS I vitamin K; F_x F_A and F_B = three different iron sulfur centers; Fd = ferredoxin; and FNR = ferredoxinNADP reductase. Approximate estimated times = various steps are also noted on the figure. A *circular path* in the Cyt b6*f* complex symbolizes the existence of a Q-cycle; and a *dotted line* from the electron acceptor side of PS I to the PQ/ Cyt b6*f* region symbolizes the existence of a cyclic flow around PS I under certain conditions. Reproduced from Govindjee (2004); colored version is from Satoh et al. (2005).

efficiency of the photochemical reactions leading to the formation of the Mn cluster hole and A_0 is given by:

$$E = 1.25 \text{ eV}/2 \text{ h}\nu \quad (1)$$

where
E = efficiency of PETS
eV = energy units in electron volts
hν = energy of the absorbed photon in eV

Since the red 680 nm photons absorbed by PSI and PSII have an energy of 1.83 eV, it ensues from Eq. 1 that

$$E = 1.25 \text{ eV}/2 * 1.83 \text{ eV} = 0.34 \text{ eV} \quad (2)$$

Therefore the absolute maximal efficiency of the PETS under red light is

$$(0.34 \text{ eV}/1.25 \text{ eV}) * 100 = 27\% \quad (3)$$

Under natural white light although the Chl concentration in photosynthetic membranes is high enough to result in near total absorption of all incident photosynthetically active photons between 400 and 700 nm, under normal weather conditions, these photons represent only about 44.5% of the total incident solar radiation. Therefore under these circumstances, the possible overall maximal energy conversion efficiency amounts to:

$$(27\% * 44.5\%)/100 = 12\% \quad (4)$$

D Actual Energy Conversion Efficiency of the PETS of Green Plants Under Field Conditions

Under field conditions, the average net photosynthetic efficiency results in a net agricultural productivity in the range of 2–8 t of dry organic matter per acre per year (Lien and San Pietro, 1975). This corresponds to a solar conversion efficiency of 0.1–0.4% of the total average incident radiation. Therefore the discrepancy between the 12% maximal theoretical efficiency of the PETS and the agricultural photosynthetic efficiency observed under field conditions ranges from

$$(12\%/0.4\%) * 100 = 3000\% \text{ to} \quad (5)$$

$$(12\%/0.1\%) * 100 = 12000\% \quad (6)$$

III Molecular Basis of the Discrepancy Between the Theoretical Maximal Efficiency of the Photosynthetic Electron Transport Chain and the Actual Solar Conversion Efficiency of Photosynthesis Under Field Conditions

The discrepancy between the 12% theoretical maximal efficiency of the PETS and the actual 0.1–0.4% solar conversion efficiency of photosynthesis observed under field conditions can be attributed to (a) factors extrinsic to the PETS, and (b) to intrinsic rate limitations of the PETS (Lien and San Pietro, 1975).

A Contribution of Extrinsic Photosynthetic Electron Transport System Parameters to the Discrepancy between the Theoretical Photosynthetic Efficiency of 12% and the Actual Photosynthetic Field Efficiency of 0.1–0.4%

Photosynthetic efficiency under field conditions is directly or indirectly affected by extrinsic factors such as ambient weather conditions availability of water CO_2 and inorganic nutrients as well as the metabolic and developmental state of the plant. Some of those factors are under human control while others are not. They do contribute nevertheless to variations in photosynthetic efficiency under field conditions. The rest of this discussion will focus upon the impact of intrinsic factors that affect the PETS and photosynthetic efficiency.

B Contribution of Intrinsic Photosynthetic Electron Transport Chain Parameters to the Discrepancy Between the Theoretical Photosynthetic Efficiency of 12% and the Actual Photosynthetic Field Efficiency of 0.1–0.4%

The 12% theoretical efficiency of the PETS assumes that under natural conditions, PSI and PSII operate at a maximal quantum efficiency close to one. In other words, it is assumed that every absorbed photon is completely converted into energy without losses (Lien and San Pietro, 1975).

Using the conventional figure of 200 light harvesting Chl molecules per RC per Photosystem

(PS) i.e. for a photosynthetic unit (PSU) size of 200 per PS under the moderate light intensities of a shady sky (about 1/10 of full sunlight) each RC would receive about 200 photons s^{-1} (Lien and San Pietro, 1975). In other words, each RC would receive about 200 hits or excitons s^{-1}. Under these conditions, in order to maintain a quantum efficiency of one, the slowest dark reaction of the entire PS must have a turnover rate of 200 s^{-1} (Lien and San Pietro, 1975).

Under full sunlight, which is about tenfold higher than in the shade the turnover rate of the limiting dark-reaction should be 200*10 = 2,000 s^{-1}. This turnover rate corresponds to a rate of O_2 evolution of about 9,000 μmol of O_2 per mg Chl h^{-1}. Yet the maximal rate of O_2 evolution observed during a Hill reaction, which results in the oxidation of H_2O and the release of O_2 under saturating light intensities and other optimal conditions, rarely exceeds 5–10% of the above value. In other words, it is equal to the optimal rate of O_2 evolution observed in the shade (Lien and San Pietro, 1975).

Furthermore extensive kinetic studies have demonstrated that the rate limiting steps of the PETS do not reside in the initial photochemical reactions that take place in the RC, but reside within the redox-carriers i.e. the electron transport chains connecting PSII to PSI. The discrepancy between the capacity of the photon gathering apparatus, i.e. the antenna Chl–protein complexes and the capacity of the rate-limiting dark reactions has been named the antenna/PS Chl mismatch (Lien and San Pietro, 1975).

IV Correction of the Antenna/ Photosystem Chlorophyll Mismatch

The first and most important effect of the antenna/ PS Chl mismatch is one of reduced quantum conversion efficiencies at light intensities above shade levels. The second effect relates to the photodestructive effects of the excess photons collected by antenna Chl but not used in the initial photochemical acts. The energy of these unused photons leads to photodestruction of the PETS which must be repaired at a cost (Lien and San Pietro, 1975; Demmis-Adams et al., 2006).

Early on, the possible correction of the antenna/ PS mismatch attracted the interest and curiosity of the photosynthesis community. It was suggested that one way of correcting the mismatch was by reducing the size of the PSU which may be achieved by growing plants with chloroplasts having less antenna and more RC Chl per unit thylakoid area (Lien and San Pietro 1975). Research performed in the early 1970s failed however in its effort to alter significantly the PSU size in algal cell cultures (Lien and San Pietro, 1975).

Now, on the basis of a deeper understanding of the chemistry and biochemistry of the greening process, which was achieved during the past 45 years, we have reason to believe that alteration of the PSU has become a realistic possibility (Rebeiz et al., 2004).

V What Kind of Scientific Knowledge Is Needed to Bioengineer a Reduction in the Photosynthetic Unit Size

Thorough metabolic and molecular biological knowledge in the following fields is needed for successful research aimed at the bioengineering of a reduced PSU size namely: (a) Chl, (b) lipid, (c) carotenoid, (d) plastoquinone, (e) chloroplast apoprotein, and (f) assembly of pigment-protein complexes. Because of space limitations, the remainder of this discussion will focus only on the Chl and apoprotein components of chloroplasts as well as on the assembly of Chl–protein complexes. The reader is referred to Frank et al. (1999) and Siegenthaler and Murata (1998) for background reading on carotenoids and Lipids.

A State of the Art in Our Understanding of Chlorophyll Biosynthesis

During the past 45 years, it has become apparent that contrary to previous beliefs, the Chl biosynthetic pathway is not a simple single-branched pathway but a complex multibranched pathway that consist of at least 15 carboxylic and two fully esterified biosynthetic routes (Rebeiz et al., 2003). The single and multibranched carboxylic pathways are briefly discussed below.

1 The Single-Branched Chl Biosynthetic Pathway Does Not Account for the Formation of All the Chlorophyll in Green Plants

The single-branched Chl biosynthetic pathway is depicted in Fig. 2. It consists of a linear sequence

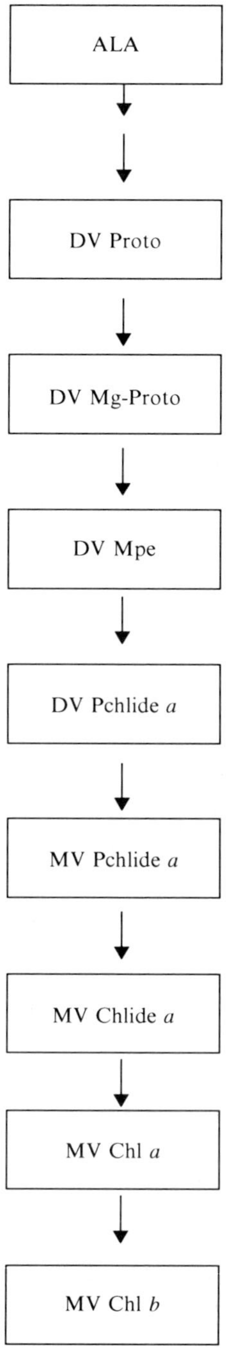

Fig. 2. The single branched Chl biosynthetic pathway reproduced from Rebeiz et al. (1994). DV = Divinyl, Proto = protoporphyrin IX, MV = monovinyl, Mpe = Mg-Proto monomethyl ester, Pchlide *a* = protochlorophyllide *a*, Chlide *a* = chlorophyllide *a*.

of biochemical reactions that convert divinyl (DV) protoporphyrin IX (Proto) to monovinyl (MV) Chl *a* via DV Mg-Proto, DV Mg-Proto monomethyl ester (Mpe), DV protochlorophyllide a (Pchlide a), MV Pchlide a, and MV Chlorophyllide a (Chlide a). The salient features of this pathway are (a) the assumption that DV Pchlide *a* does not accumulate in higher plants, but is a transient metabolite which is rapidly converted to MV Chl *a* via MV Pchlide *a* and (b) that the formation and accumulation of MV tetrapyrroles between Proto and Mpe and DV tetrapyrroles between Pchlide *a* and Chl *a* does not take place (Rebeiz et al., 1994). All in all, experimental evidence gathered over the past 43 years indicates that only a small fraction of the total Chl of green plants is formed via this pathway (Rebeiz et al., 2003).

2 The Chlorophyll of Green Plants Is Formed Via a Multibranched Biosynthetic Pathway

Since the seminal review of Smith and French (1963), our understanding of the Chl biosynthetic pathway has changed dramatically. Several factors have contributed to this phenomenon among which: (a) development of systems capable of Chl and thylakoid membrane biosynthesis in organello and in vitro (Rebeiz and Castelfranco, 1971a, b; Daniell and Rebeiz, 1982a, b, Rebeiz et al., 1984; Kolossov et al., 1999), (b) powerful analytical techniques that allowed the qualitative and quantitative determination of various intermediates of the pathway (Rebeiz, 2002), (c) recognition that the greening process proceeds differently in etiolated and green tissues in darkness and in the light and in plants belonging to different greening groups (Carey and Rebeiz, 1985; Ioannides et al., 1994; Abd-El-Mageed et al., 1997) and (d) recognition of the probability that the structural and functional complexity of thylakoid membranes is rooted in a multibranched heterogeneous Chl biosynthetic pathway (Rebeiz et al., 1999). Chlorophyll biosynthetic heterogeneity refers either (a) to spatial biosynthetic heterogeneity (b) to chemical biosynthetic heterogeneity or (c) to a combination of spatial and chemical biosynthetic heterogeneities (Rebeiz et al., 2003). Spatial biosynthetic heterogeneity refers to the biosynthesis of an anabolic tetrapyrrole or end product by identical sets of enzymes at several different locations of the thylakoid membrane. On the other hand chemical biosynthetic heterogeneity refers to the biosynthesis of an anabolic tetrapyrrole or end product at several different locations of the thylakoid membrane

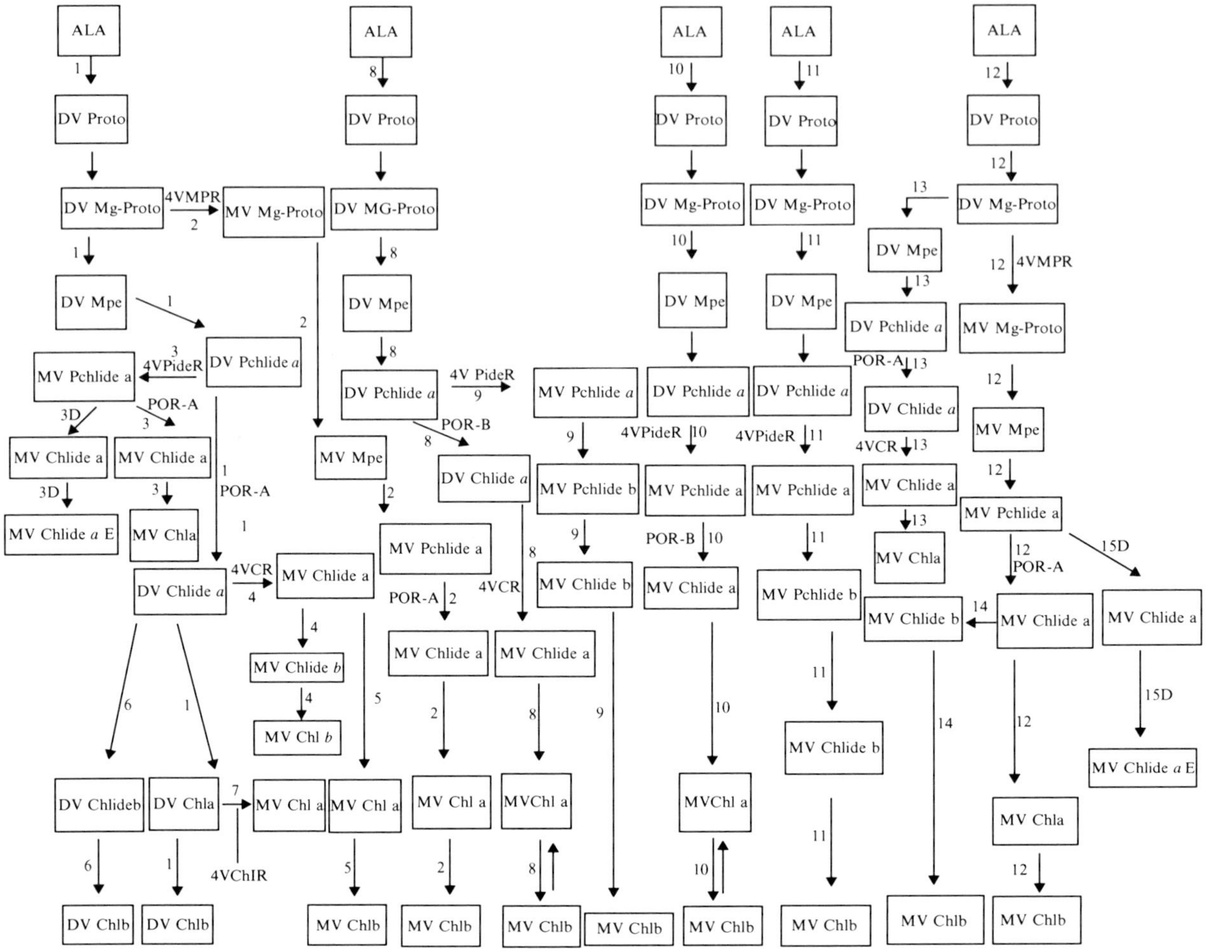

Fig. 3. Integrated Chl a/b biosynthetic pathway depicting 15 carboxylic routes. To facilitate understanding of the text various biosynthetic routes are designated by numbers 1-15. ALA = delta-aminolevulinic acid, 4VMPR: [4-vinyl] Mg-Proto reductase, 4VPideR = [4-vinyl] protochlorophyllide *a* reductase, 4VCR = [4-vinyl] chlorophyllide *a* reductase, 4VChlR = [4-vinyl] Chl reductase, POR = chlide *a* oxidoreductase, D = reaction occurring in darkness. Other abbreviations are as in Fig. 2. *Arrows* joining DV and MV routes refer to reactions catalyzed by [4-vinyl] reductases. Various biosynthetic routes are designated by *Arabic numerals* (Reproduced from Rebeiz et al., 2003).

via different biosynthetic routes each involving at least one different enzyme. Figure 3 organizes all known carboxylic Chl biosynthetic reactions into a logical scheme made up of 15 different biosynthetic routes.

B Thylakoid Apoprotein Biosynthesis

The biosynthesis of thylakoid apoproteins is a very complex phenomenon. Some apoproteins are coded for by nuclear DNA are translated on cytoplasmic ribosomes and are transported to developing chloroplasts. Other apoproteins are coded for by plastid DNA and are translated on chloroplast ribosomes. A detailed discussion of chloroplast apoprotein biosynthesis is beyond the scope of this discussion. The reader is referred to Sundqvist and Ryberg (1993) for a comprehensive discussion of this topic. For the purpose of this discussion it suffices to say that a PSU is an extremely complex structure that consists of many highly folded thylakoid and soluble proteins as well as membrane-bound pigment protein complexes having different functions in the light and dark steps of photosynthesis (Golbeck, 2006; Wydrzynski and Satoh 2006). An early visualization of a linear model of a PSU in the unfolded state is depicted in Fig. 4.

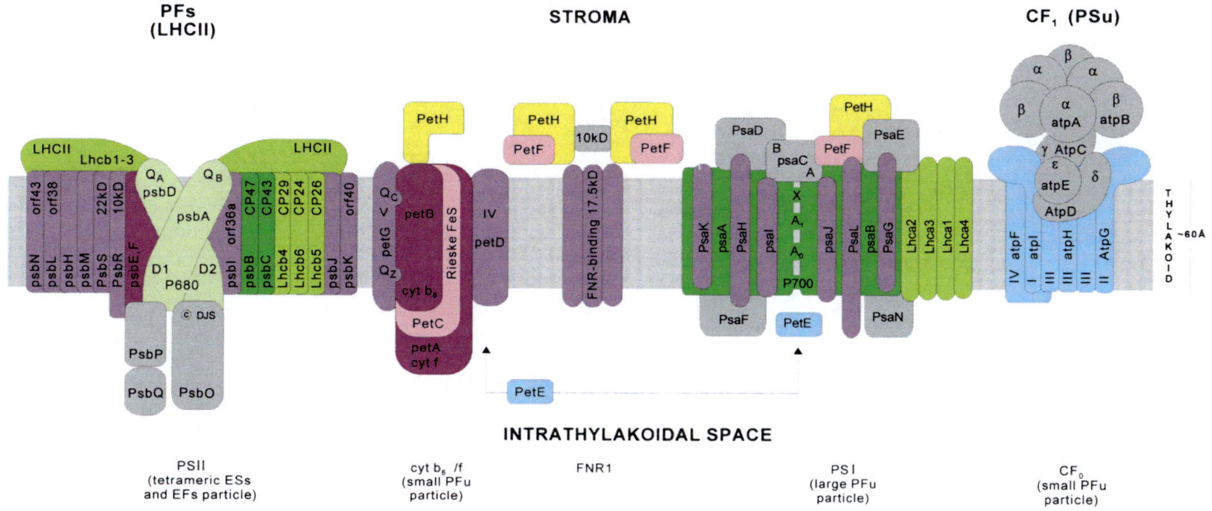

Fig. 4. Schematics of a linear model of a PSU in the unfolded state (Reproduced from von Wettstein et al., 1995).

C Assembly of Chlorophyll–Protein Complexes

Success in the bioengineering of smaller PSUs resides in a thorough understanding of how the Chl and thylakoid apoprotein biosynthetic pathways are coordinated to generate a specific functional Chl–protein complex. It is known for example that an apoprotein formed in the cytoplasm or in the chloroplast has to pick up Chl molecules, has to fold properly, and has to wind up in the right place on the thylakoid in order to become a functional Chl–protein complex having a specific role in photosynthesis. What is unknown is how an apoprotein formed in the cytoplasm or in the chloroplast becomes associated with Chl to become a specific Chl–protein complex of PSI, PSII, or light harvesting Chl–protein complex II (LHCII) having a specific function in photosynthesis. We have recently addressed this issue by examining three possible models for that scenario which have been referred to as: (a) the single-branched Chl biosynthetic pathway (SBP)-single location model (b) the SBP-multilocation model, and (c) the multi-branched Chl biosynthetic pathway (MBP)-sublocation model. The models take into account the structure and dimension of the PSU (Bassi et al., 1990; Allen and Forsberg, 2001; Anderson, 2002; Staehelin, 2003), the biochemical heterogeneity of the Chl biosynthetic pathway (Kolossov and Rebeiz, 2001; Rebeiz, 2002), and the biosynthetic and structural complexity of thylakoid membranes (Sundqvist and Ryberg, 1993). The three models are described below.

1 Assembly of Chlorophyll–Protein Complexes: The Single-Branched Chlorophyll Biosynthetic Pathway (SBP)-Single Location Model

The SBP-single location model is depicted schematically in Fig. 5. Within the PSU, this model accommodates only one Chl–apoprotein biosynthesis center and no Chl–apoprotein biosynthesis subcenters. Within the Chl–apoprotein biosynthesis center, Chl *a* and *b* are formed via a single-branched Chl biosynthetic pathway (Fig. 2) at a location accessible to all Chl-binding apoproteins. The latter will have to access that location in the unfolded state pick up a complement of MV Chl a and/or MV Chl b, and undergo appropriate folding. Then the folded Chl–apoprotein complex has to move from the central location to a specific PSI, PSII, or Chl a/b LHC-protein location within the Chl–apoprotein biosynthesis center over distances of up to 225 Å in the linear continuous array model (Bassi et al., 1990), or over larger distances in

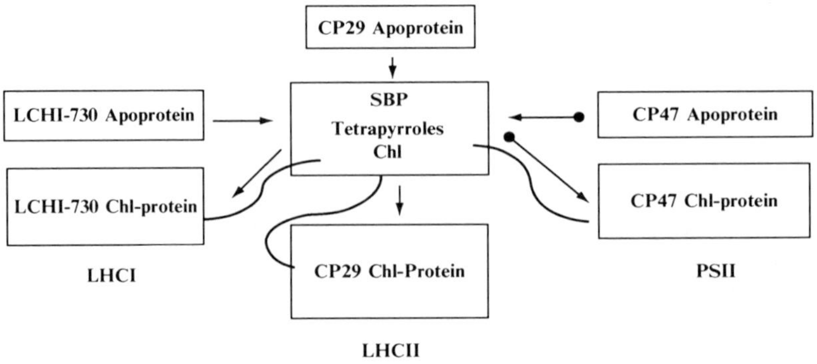

Fig. 5. Schematics of the SBP-single location model. As an example, the functionality of the model was illustrated with the use of three apoproteins namely CP29, LCHI-730 and CP47. SBP = single-branched Chl biosynthetic pathway; PS II = photosystem II; LHCII, the major light-harvesting Chl–protein complex of PS II; LHCI, one of the LHC antennae of PS I; CP47 and CP29, two PSII antennae, LHCI-730, the LHC antenna of PS I. Curved lines indicate putative energy transfer between tetrapyrroles and a Chl–protein complex (Adapted from Kopetz et al., 2004).

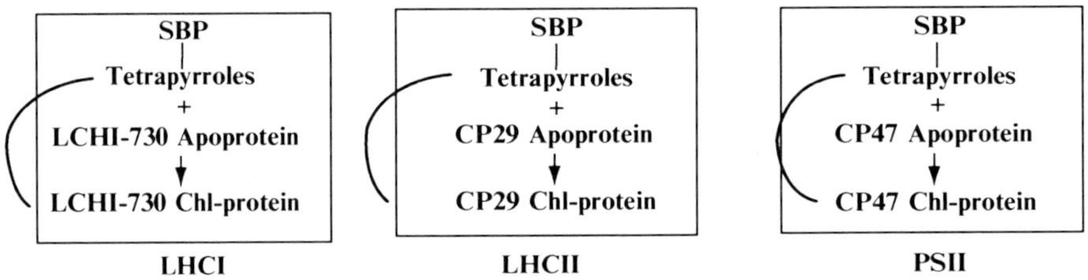

Fig. 6. Schematics of the SBP-Multilocation model. All abbreviations and conventions are as in Fig. 5 (Adapted from Kolossov and Rebeiz, 2003.

the laterally heterogeneous models (Allen and Forsberg, 2001; Anderson, 2002; Staehelin, 2003), to become part of PSI, PSII, or LHCII. In this scenario no resonance excitation energy transfer is observed between anabolic tetrapyrroles and various Chl–protein complexes since resonance excitation energy transfer takes place only over distances shorter than 100 Å (Calvert and Pitts, 1967).

2 Assembly of Chlorophyll–Protein Complexes: The Single- Branched Chlorophyll Biosynthetic Pathway-Multilocation Model

The SBP-Multilocation model is depicted schematically in Fig. 6. In this model, every biosynthetic location within the PSU is considered to be a Chl, apoprotein thylakoid biosynthesis center. In every Chl–apoprotein biosynthesis location, a complete single-branched Chl *a/b* biosynthetic pathway (Fig. 2) is active. Association of Chl *a* and/or Chl *b* with specific PS I, PS II or LHC apoproteins at any location is random. In every Chl–apoprotein biosynthesis center, distances separating metabolic tetrapyrroles from the Chl–protein complexes are shorter than in the SBP-single-location model. Because of the shorter distances separating the accumulated tetrapyrroles from Chl–protein complexes resonance excitation energy transfer between various tetrapyrroles and Chl–apoprotein complexes within each center may be observed. However formation of MV Mg-Proto (Mp) and its MV methyl ester (Mpe) is not observed in any pigment–protein complex. This is because the single-branched Chl biosynthetic pathway does not account for MV Mpe biosynthesis.

3 Assembly of Chlorophyll–Protein Complexes: The Multi-Branched Chlorophyll Biosynthetic Pathway (MBP)-Sublocation Model

The MBP-sublocation model is depicted schematically in Fig. 7. In this model the unified

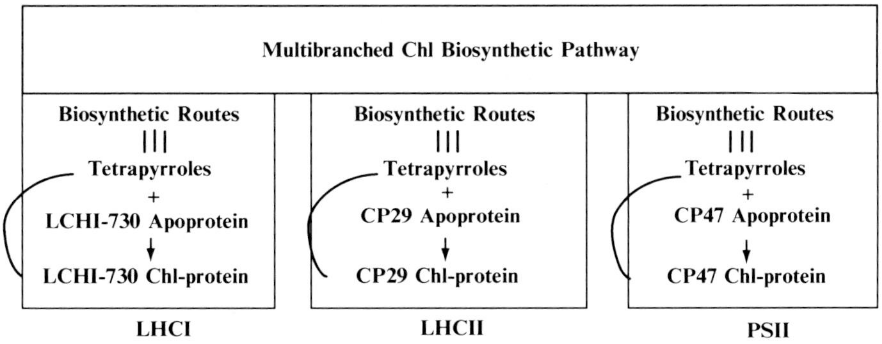

Fig. 7. Schematics of the MBP-sublocation model. All abbreviations and conventions are as in Fig. 5 (Adapted from Kolossov and Rebeiz 2003). Because of the shorter distances separating the accumulated tetrapyrroles from Chl–protein complexes within each subcenter, resonance excitation energy transfer between various metabolic tetrapyrroles and Chl is readily observed. In this model, both MV and DV Mp(e) may be present in some pigment-protein complexes in particular if more than one Chl biosynthetic route are involved in the Chl formation of a particular Chl–protein complex.

multibranched Chl *a/b* biosynthetic pathway (Rebeiz et al., 2003) is visualized as the template of a Chl–protein biosynthesis center where the assembly of PS I, PS II and LHC takes place. The multiple Chl biosynthetic routes are visualized individually or in groups of one or several adjacent routes as Chl–apoprotein biosynthesis subcenters earmarked for the coordinated assembly of individual Chl–apoprotein complexes. Apoproteins destined to some of the subcenters may possess specific polypeptide signals for specific Chl biosynthetic enzymes peculiar to that subcenter such as 4-vinyl reductases, formyl synthetases or Chl *a* and *b* synthetases. Once an apoprotein formed in the cytoplasm or in the plastid reaches its subcenter destination and its signal is split off, it binds nascent Chl formed via one or more biosynthetic routes as well as carotenoids and other isoprenoids. During pigment binding, the apoprotein folds properly and acts at that location while folding, or after folding, as a template for the assembly of other pigment-proteins. Because of the shorter distances separating the accumulated tetrapyrroles from Chl–protein complexes within each subcenter, resonance excitation energy transfer between various metabolic tetrapyrroles and Chl is readily observed. In this model, both MV and DV MgProto, and Mpe may be present in some pigment–protein complexes, in particular if more than one Chl biosynthetic route are involved in the Chl formation of a particular Chl–protein complex.

D Which Chl–Thylakoid Apoprotein Assembly Model Is Validated by Experimental Evidence

In order to determine which Chl–thylakoid apoprotein assembly model is likely to be functional during thylakoid membrane formation, we tested the compatibility of the three aforementioned models with resonance excitation energy transfer calculations between anabolic tetrapyrrole of the Chl biosynthetic pathway and various thylakoid Chl–protein complexes. Since the goal was to determine whether various Chl biosynthetic routes contribute differentially to the formation of various Chl–protein complexes, it was conjectured that resonance excitation energy transfer determinations between various intermediates of the Chl biosynthetic routes and pigment-protein complexes may shed some light upon the proximity and thus possible involvement, of a Chl biosynthetic route with the formation of a particular pigment-protein.

Indeed fluorescence resonance excitation energy transfer involves the transfer of excitation energy from an excited donor "D*" to an unexcited acceptor "A" (Turro, 1965; Calvert and Pitts, 1967; Lakowicz, 1999, Clegg, 2004). The transfer is the result of *dipole–dipole interaction* between donor and acceptor and does not involve the exchange of photons. The rate of energy transfer depends upon (a) the extent of *overlap* of the emission spectrum of the donor and the absorption spectrum of the acceptor, (b) the relative orientation of the donor and acceptor *transition dipoles*, and (c) distances between donor and acceptor molecules. The probability of resonance excitation energy transfer between donors and acceptors increases with time but is inversely proportional to the sixth power of the fixed distance separating the centers of the donor and acceptor molecules. It has been estimated that *dipole–dipole energy transfer* between donor and acceptor molecules may occur up to distances of 50–100 Å (Calvert and Pitts, 1967).

Therefore in order to test the functionality of the various Chl–protein assembly models discussed above (Figs. 5–7) using Fluorescence resonance excitation energy transfer tools, it was necessary to develop appropriate analytical tools. First it was mandatory to determine whether resonance excitation energy transfer between various anabolic tetrapyrroles and Chl–protein complexes did take place in green tissues. If this proved to be the case it was then necessary to develop tools to calculate the distances separating various anabolic tetrapyrroles from specific Chl–protein complexes in the thylakoid membranes. It was conjectured that the close proximity of an anabolic tetrapyrrole to a specific Chl–protein complex may indicate a possible biosynthetic relationship between a certain Chl biosynthetic route and a Chl–protein complex. Further experimentation discussed below would then be aimed at determining specific relationships between specific Chl–biosynthetic routes and certain Chl–protein complexes. Such a relationship would then be the basis for attempting to bioengineer chloroplasts with smaller PSU size (vide infra).

1 Can Resonance Excitation Energy Transfer Between Anabolic Tetrapyrroles and Chlorophyll–Protein Complexes be Demonstrated?

Demonstration of fluorescence resonance excitation energy transfer from anabolic tetrapyrroles to Chl–protein complexes faced several hurdles. Chloroplasts contain only trace amounts of anabolic tetrapyrroles, the fluorescence of which is masked by high Chl content. It was therefore necessary to (a) boost the chloroplast content of anabolic tetrapyrroles, (b) select appropriate and convenient in situ Chl *a* acceptors, (c) enhance the detection of putative resonance energy transfer between donors and acceptors by correction for the possible occurrence of endogenous resonance excitation energy transfers, and (d) determine the location of anabolic tetrapyrrole excitation peaks.

Thus, attempts were made to determine the possibility of resonance excitation energy transfer from three tetrapyrrole donors to the Chl *a* of various Chl–protein complexes, namely: from Proto, Mpe and MV and DV Pchlide *a* (Kolossov and Rebeiz, 2003; Rebeiz et al., 2005).

Divinyl Proto is a common precursor of heme and Chl. It is the immediate precursor of DV Mg-Proto. As such, it is an early intermediate along the Chl biosynthetic chain. Biosynthetically, it is several steps removed from the Chl end product (Rebeiz et al., 2003). Mg-Proto is a mixed DV-MV dicarboxylic tetrapyrrole pool consisting of DV and MV Mg-Proto (Rebeiz et al., 2003). It is the precursor of DV and MV Pchlide a. The protochlorophyll(ide) [(Pchl(ide)] of higher plants consists of about 95% Pchlide a and about 5% Pchlide *a* ester. The latter is esterified with long chain fatty alcohols at position 7 of the macrocycle. While Pchlide *a* ester consists mainly of MV Pchlide *a* ester, Pchlide *a* consists of DV and MV Pchlide *a* components. The latter are the immediate precursors of DV and MV Chlide *a*, the precursors of DV and MV Chl *a* (Rebeiz et al., 2003). Demonstration of resonance excitation energy transfer from three tetrapyrrole donors to the Chl *a* of various Chl–protein complexes is described below.

(a) Induction of Tetrapyrrole Accumulation

Various levels of Proto, Mp(e) and MV and DV Pchlide *a*. accumulation were achieved by incubation of excised tissues with various concentrations of δ-aminolevulinic acid (ALA) in the absence and presence of various concentrations of 2, 2′-dipyridyl (Dpy) for various lengths of time (Tripathy and Rebeiz, 1986). Cucumber cotyledons were used for the induction of Proto, Mp(e) and divinyl (DV) Pchlide *a*, while barley leaves were used for the induction of Proto, Mp(e) and MV Pchlide *a* accumulation (Tripathy and Rebeiz, 1986).

Both green and etiolated tissues were used. Per milliliter of undiluted chloroplast suspension from 3 g of tissue, tetrapyrrole accumulation was linear up to 6,600, 1,500 and 1,200 pmol for Pchlide *a*, Proto and Mpe in cucumber and up to 3,000, 1,100 and 550 in barley. In other words the derived data spanned non-saturating and saturating tetrapyrrole accumulation conditions (Kolossov et al., 2003).

(b) Selection of Appropriate Chlorophyll a Acceptors

The task of selecting appropriate Chl a-protein acceptors was facilitated by the fluorescence properties of green plastids. At 77 K emission spectra of isolated chloroplasts exhibit fluorescence maxima at 683–686 nm (~F685), 693–696 nm (~F695), and 735–740 nm (~F735). It is believed that at 77 K, the fluorescence emitted at ~F685 is a mixture, arising from the Chl *a* of light harvesting chlorophyll–protein II, (LHCIIb, F680), and in one of the pigment protein complexes (CP 43) in the core of PSII (Gilmore, 2004). That emitted at ~F695 nm is believed to originate mainly from the Chl *a* of the core of PS II (*CP47, CP29*, two PSII antennae (Bassi et al., 1990; Gilmore, 2004). That emitted at ~F735 nm is believed to originate primarily from the Chl *a* of *LHCI-730*, a PSI antenna (Bassi et al., 1990). Since these emission maxima are readily observed in the fluorescence emission spectra of green tissues at 77 K, and are associated with definite thylakoid Chl *a*–protein complexes, it was conjectured that they would constitute a meaningful resource for monitoring excitation resonance energy transfer between anabolic tetrapyrroles and representative Chl *a*–protein complexes.

(c) Acquisition of In Situ Emission and Excitation Spectra at 77 K

In situ emission and excitation spectra were recorded on tissue homogenates or isolated plastids as described elsewhere (Kolossov et al., 2003).

(d) Generation of Reference In Situ tetrapyrrole Excitation Spectra

To better locate the wavelength regions where resonance excitation energy transfer bands may be observed in situ, excitation spectra of in situ accumulated Proto, Mp(e) and Pchlide *a* were generated. These spectra were recorded at the in situ emission maxima of Proto, Mp(e) and Pchlide *a* in dark-prepared homogenates of etiolated cucumber cotyledons or barley leaves preincubated with ALA and Dpy in darkness (Kolossov et al., 2003). The etiolated tissues lacked Chl and Chl-dependent endogenous excitation resonance energy transfer bands, but exhibited pronounced excitation bands corresponding to accumulated Proto, Mp(e), and Pchlide *a*. Since the in situ excitation spectrum of a given tetrapyrrole was recorded at the emission maximum of that tetrapyrrole, the most pronounced excitation maximum in the excitation profile corresponded to that particular tetrapyrrole. Other apparent excitation maxima and shoulders of lesser magnitude originated in the other accumulated tetrapyrroles (Kolossov et al., 2003).

(e) Processing of Acquired Excitation Spectra

To eliminate slight differences in the Chl concentration of diluted, control and treated samples, excitation spectra of every control and treated pair were normalized to a value of one fluorescence unit at a wavelength of 499 nm. This is a wavelength region outside the Soret excitation region of Chl and tetrapyrroles. Thus by correcting for minute dilution differences by normalization to the same value at 499 nm, the difference spectra became more representative of the real differences between control and treated samples. The normalized spectra were smoothed five times. For detection of resonance excitation energy transfer bands, the control spectrum (incubation without added ALA) was subtracted from the spectrum of treated tissues (i.e. tissues incubated with ALA (Kolossov et al., 2003).

(f) Demonstration of Resonance Excitation Energy Transfer Between Anabolic Tetrapyrroles and Chlorophyll–Protein Complexes

To monitor the possible occurrence of resonance excitation energy transfer between the accumulated anabolic tetrapyrroles and the Chl *a*–protein complexes, excitation spectra were recorded at 77° at the respective emission maxima of the selected Chl *a* acceptors, namely at ~685 ~695 and ~735 nm. It was conjectured that if resonance excitation energy transfers were to be observed between the tetrapyrrole donors and the selected Chl *a* acceptors, definite excitation maxima would be observed after processing the acquired spectra as described above. These excitation maxima would correspond to absorbance maxima of the various tetrapyrrole donors and would correspond to the peaks of the resonance excitation energy transfer bands (Kolossov et al., 2003).

By reference to in situ emission and excitation spectra, assignment of in situ excitation maxima of various anabolic tetrapyrroles was unambiguous except for a few cases at the short wavelength (SW) and long wavelength (LW) extremes of excitation bands. It was most surprising to observe a considerable diversity in the various intra-membrane environments of Proto, Mp(e) and Pchl(ide) *a*. In other words instead of observing one resonance excitation energy transfer maximum per tetrapyrrole donor, several resonance excitation energy transfer maxima were observed (Table 1). This was highly compatible with the notion of Chl biosynthetic heterogeneity (Kolossov et al., 2003).

Finally, the detection of pronounced excitation resonance energy transfer from Proto, Mp(e) and Pchl(ide) *a* to Chl *a* F685, F695, and F735 indicated that these anabolic tetrapyrrole donors are within distances of 100 Å or less of the immediate Chl *a* acceptors. Indeed resonance excitation energy transfer is insignificant at distances larger than 100 Å since dipole–dipole energy transfer may only occur up to a separation distance of 50–100 Å only (Calvert and Pitts, 1967). The impact of short distances separating anabolic tetrapyrroles from Chl *a*-protein complexes upon the topography of the Chl biosynthetic pathway and the assembly of Chl–protein complexes will be addressed below.

2 *Development of* Analytical Tools *for Measuring Distances Separating Various Chlorophyll–Protein Complexes from Anabolic Tetrapyrroles*

With the demonstration of resonance excitation energy transfer from anabolic tetrapyrroles to Chl–protein complexes in chloroplast thylakoids, it became possible to develop analytical tools that made it possible to determine the distances separating Proto, Mp(e) and Pchl(ide) *a* donors from Chl *a* acceptors. This in turn was a prerequisite for testing the functionality of the thylakoid-protein assembly models discussed above.

(a) Determination of the Molar Extinction Coefficients of Total Chl a In Situ at 77 K

In order to calculate resonance excitation energy transfer distances separating Proto, Mp(e) and Pchl(ide) a donors from Chl *a* acceptors in situ, molar extinction coefficients of total Chl a and various Chl a acceptors needed to be determined. The molar extinction coefficient of total Chl *a* at 77 K was determined on green tissue filtrates as described in Kopetz et al. (2004).

(b) Estimation of the Molar Extinction Coefficients of Chl a ~F685, ~F695 and ~F735 at 77 K

The Chl a species used in the calculation of resonance excitation energy transfer from Proto, Mp(e) and Pchl(ide) a donors to Chl a acceptors in situ were Chl a (E670F685) (i. e. Chl a ~F685) which amounts to about 26% of the total Chl a absorbance under the Chl a envelope, Chl a (E677F695) (i.e. Chl a ~ F695) which amounts to about 32% of the total Chl a absorbance area, and Chl a (E704F735) (i.e. Chl a ~ F735) which amounts to about 2% of the total Chl a absorbance area (French et al., 1972). In this context, E refers to the absorbance and F to the emission maxima of the Chl a species in situ at 77 K. The assignment of emission F values to the absorbance E values was based on the mirror image symmetry of the red absorbance and fluorescence emission maxima of Chl a. The molar extinction coefficient of the various Chl a acceptors were estimated from the molar extinction coefficients of total Chl a at 77 K in situ and the relative areas and half band widths of the various Chl a species under the total Chl a envelope (Kopetz et al., 2004).

(c). Calculation of Distances R Separating Anabolic Tetrapyrroles from Various Chl a–protein Complexes

The efficiency of resonance excitation energy transfer "E" from donors D to acceptors A are directly related to the distance R separating donors from acceptors (Lakowicz, 1999, 2006) by the following equation:

$$E = R_0^6 / \left(R_0^6 + R6\right) \tag{7}$$

Equation 7 can be rewritten as:

$$R^6 = \left(R_0^6 - ER_0^6\right) / E$$

Table 1. Mapping of excitation resonance energy transfer maxima to Chl *a* F686, Chl *a* F694 and Chl *a*F738-742 in situ. A *dash* represents missing data. Undil = donor concentration before dilution Dil = donor concentration after dilution s = shoulder; p = peak. Only the barley spectra depicted in Fig. 8 were recorded at the observed peak of Chl *a* emission at F642 nm (Adapted from Kolossov et al., 2003).

Plant species	Major donor	Undil donor Conc (pmol/ml suspension)	Dil donor Conc (pmol/ml suspension)	Excitation resonance energy maxima to Chl *a* F686 (nm)	Chl *a* F694 (nm)	Chl *a* F738	Conc ALA (nM)	Conc Dpy	Incub (h)
Cucumber	Proto	1,620	54	397p 402p 410p 415p	390s 400p 409p	390s 395s 408p 417p	4.5	3.7	6
Cucumber	Proto	1,242	83	387p 402p 412p	392p 406p	388p 399p 403p 410p 415p	20	4	6
Cucumber	Proto	1,374	92	390p 399p 405p 412p	399p 409p 412s	399p 400p 416p	20	0	6
Cucumber	Proto	5,640	376	395p 404s 411p 416p	395p 406p 414p	393p 400s 407p	20	16	6
Cucumber	Proto	3,138	1,046	402s 411p	404p 410s 416p	399s 405s 411p	20	0	12
Barley	Proto	390	13	391 398s 404s 411p	389s 395p 406p 414p	390s 393p 400s 406p 412p 416s	4.5	3.7	6
Barley	Proto	1,492	61	389p 396s 404s 410p 412p	396p 406p 412p	389s 395p 406s 410p	20	16	6
Barley	Proto	966	64	395s 400p 405s 413p	389p 397s 403p 412p	388s 393p 400s 406p 412p	20	0	6
Barley	Proto	1,015	68	389p 396p 412p 413s	389p 398p 409p	396s 400p 412p 414s	20	4	6
Cucumber	Mp(e)	390	26	419p 431p	422p 429p 434p	–	20	0	6
Cucumber	Mp(e)	2,490	83	422p 432p	420p 425p	417p 424s 427s 429p	4.5	3.7	6
Cucumber	Mp(e)	1,374	9[illegible]	418s 424p 433p	419p 426p	414p 423p	20	4	6
Cucumber	Mp(e)	1,854	185	421p 427s 430s	421p 428s	421p 430p	20	0	12
Cucumber	Mp(e)	1,854	618	421p 427s 430s	421p 427s 430s	421p 430p	20	0	12
Barley	Mp(e)	300	10	420p 428s	424p 430s	426s 432s	4.5	3.7	6
Barley	Mp(e)	162	11	423p	418p 422s 427p	422s 426p 431s	20	0	6
Barley	Mp(e)	378	25	423p 428s	418p 430p	426s 432p	20	4	6
Cucumber	DV Pchlide *a*	1,998	133	438p 446p 453s 460s 467p	440s 448p 454s 460p	448p 453p 461p	20	4	6
Cucumber	DV Pchlide *a*	4,590	153	443p 449p 457p	436s 442p 453p 463p	439p 453p 457p 460p	4.5	3.7	6
Cucumber	DV Pchlide *a*	6,180	4[illegible]2	437p 444s 452p 458p	435p 441p 451p 462p	437p 447s 454s 457p 463s	20	0	6
Cucumber	DV	6,180	1030	438s 447p 452p 456s 462s	441s 447p 452p 459p	436p 448s 454s 458p p	20	0	6
Cucumber	DV Pchlide *a*	6,180	1435	435p 447s 453p 460s	438s 445s 452p 456s 460s 462s	436s 444s 452s 458p 462s	20	0	12
Cucumber	DV Pchlide *a*	14,352	4784	440s 449p 455s 460s	434p 440s 447s 452p 459s	434s 440p 447s 462p	20	0	12
Barley	MV Pchlide *a*	780	26	434s 441p 452p 460p	438s 445p 449p 463p	440p 449p 458s 468p	4.5	3.7	6
Barley	MV Pchlide *a*	1554	1[illegible]4	439p 445s 450p 458p 463s	436s 447p 455p 463s	440p 450p 458p	20	4	6
Barley	MV Pchlide *a*	2,900	193	439s 444p 451p 462p 467p	435p 440s 446p 453p 460p	438s 453p 457p 464s	20	0	6

or as

$$R6 = \left(R_0^6 / E\right) - R_0^6 \quad (8)$$

where

R = distance separating donor D from acceptor A

R_0 = critical separation of donor and acceptor for which energy transfer from excited donor *D** to unexcited acceptor *A* and emission from *D** to the ground state *D* amount to 50% i.e. are equally probable

E = efficiency of resonance excitation energy transfer from donor to acceptor

Therefore in order to calculate R, the distance separating donor *D* from acceptor *A* it was necessary to calculate first, R_0 the critical separation of donor and acceptor for which energy transfer from excited donor *D** to unexcited acceptor *A* and emission from *D** to the ground state are equally probable, as well as E, the efficiency of resonance excitation energy transfer from donor to acceptor.

(d) Calculation of R_0

As described by Eq. 8 calculation of R_0 is needed for the calculation of R the distance separating donors from acceptors. According to Forster (1960), for practical applications R_0 can be calculated from an approximate equation where the emission spectrum of the donor is expressed in terms of its absorption spectrum by using the approximate mirror-image symmetry of these spectra namely, after various manipulations (Kopetz et al., 2004):

$$R_0^6 \cong \left[\left(1.2055e^{-33}\right)\kappa^2\right]\left[(\tau_D J_\upsilon)/(\nu_0^2)\right] \quad (9)$$

where

κ = orientation dipole

τ_D = actual mean lifetime of the excited donor i.e. of the excited sensitizer

$$J_\upsilon = \text{Overlap integral} = \int_0^\infty \varepsilon_A(\nu)\,\varepsilon_D\,(2\nu_0 - \nu)\,d\nu. \quad (10)$$

ν_0 = the wavenumber of the 0–0′ transition of the donor, which is approximated by the arithmetic mean in wavenumbers of the donor absorption and fluorescence maxima (Förster, 1960).

(e) Calculation of κ, the Orientation Dipole

Determination of the orientation dipole κ is needed for the calculation of R_0 (see Eq. 9).The rate of resonance excitation energy transfer from a donor D to an acceptor A depends upon the orientation of the donor and acceptor dipoles and is independent of the polarity of the medium. For systems with random dipoles, κ^2 assumes a value of about 0.67 (Förster, 1960; Calvert and Pitts, 1967). In this work, as in other reported work (Wu and Stryer, 1972) a random dipole orientation of 0.67 will be assumed. See Kopetz et al. (2004) for validation.

(f) Calculation of the Overlap Integral J_υ at 77K

Calculation of the overlap integral J_υ, is needed for the calculation of R_0 (see Eq. 9). The efficiency of resonance excitation energy transfer between accumulated tetrapyrroles donors and Chl *a* acceptors depends in a large measure on the overlap between the red fluorescence vibrational bands of tetrapyrrole donors, and the red absorbance bands of Chl acceptors. The overlap between the vibrational bands of the Proto, Mp(e) and Pchlide *a* donors and the absorbance bands of the Chl *a* acceptors was complete (Kopetz et al., 2004).

The overlap integral J_υ, normalized by the area of the corrected emission spectrum, can be calculated from the following formula (Lakowicz, 1999):

$$\begin{aligned} J(\lambda) &= \int_0^\infty F_D(\lambda)\varepsilon_A(\lambda)\lambda^4(d\lambda) \\ &= \int_0^\infty F_D(\lambda)\varepsilon_A(\lambda)\lambda^4(d\lambda) / \int_0^\infty F_D(\lambda)d\lambda \end{aligned} \quad (11)$$

where

$F_D(\lambda)$ = Corrected fluorescence emission intensity at every wavelength

ε_A (λ) = molar extinction coefficient of the acceptor as a function of wavelength λ

λ^4 = wavelength in nm in the emision-absorbance overlap region raised to the fourth power

$\int_0^\infty F_D(\lambda)d\lambda$ = area of the corrected emission spectrum.

Calculated J_ν values for Proto-Chl *a*, Mp(e)-Chl *a*, and Pchlide *a*-Chl *a* donor-acceptor pairs for barley and cucumber are reported in Kopetz et al. (2004)

(g) Calculation of ν_0, the Mean Wavenumber of Absorption and Fluorescence Peaks of the Donor at 77 K

Calculation of ν_0, the mean wavenumber of absorption and fluorescence maxima of donors, is needed for the calculation of R_0 (Eq. 9). It was determined as follows. The donor tetrapyrroles were adsorbed to chloroplast lipoproteins prepared from green barley leaves or cucumber cotyledons as described elsewhere (Kopetz et al., 2004). Their absorbance and fluorescence emission spectra were recorded at 77 K. The absorbance and fluorescence emission maxima were then converted to wavenumbers, and ν_0, the arithmetic means of the two wavenumbers were calculated. Calculated ν_0 values for Proto, Mp(e) and Pchlide *a* are reported in Kopetz et al., (2004).

(h) Calculation of τ_0, the Inherent Fluorescence Lifetime of Donors at 77 K

Determination of the inherent fluorescence lifetime of the donors, τ_0, is needed for the calculation of the actual mean fluorescence lifetimes of the excited donors, τ_D (see below). The latter is needed for the calculation of R_0 (Eq. 9).

The inherent radiative lifetime of a donor, τ_0, is the inherent radiative lifetime of its excited state. It is the mean time it would take to deactivate the excited state in the absence of radiationless processes such as internal conversion (i.e. heat dissipation) and intersystem crossing (i.e. conversion from a singlet to a triplet excited state) (Turro, 1965). The measured fluorescence lifetime of an excited donor, τ_D, is determined by the sum of the rates of all processes depopulating the donor excited state. Therefore in cases where other unimolecular processes (such as intersystem crossing) or bimolecular processes (such as resonance excitation energy transfer), compete with fluorescence, the observed radiative lifetime τ_D, will be proportionally less than the inherent radiative lifetime τ_0 (Calvert and Pitts, 1967). The latter can be calculated as follows (Turro, 1965):

$$\tau_0 = 3.5E^8 / (\upsilon_m^2)(\varepsilon_m)(\Delta\upsilon_{1/2}). \quad (12)$$

where

υ_m = Soret absorbance maximum of the donor in wavenumbers

ε_m = molar extinction coefficient at the Soret absorbance maximum of the donor

$\Delta\upsilon_{1/2}$ = half bandwidth of the Soret absorbance band of the donor in wavenumbers.

Calculated τ_0 values for Proto, Mp(e) and Pchlide *a* are reported in Kopetz et al., (2004).

(i) Calculation of Fy_{Da} the Relative Fluorescence Yield of Tetrapyrrole Donors in the Presence of Chl Acceptors In Situ at 77 K

Determination of the relative fluorescence yield of donors in the presence of acceptors, Fy_{da}, is also needed for the calculation of τ_D, the actual mean fluorescence lifetimes of excited donors (see below). The latter is needed for the calculation of R_0 (Eq. 9). It is also needed for the calculation of the efficiency of energy transfer, E, which is needed for the calculation of R, the distance separating donor from acceptor molecules.

The absolute fluorescence quantum yields of many compounds have been determined with considerable precision. For example rhodamine B in ethanol at low concentrations, exhibits an absolute fluorescence quantum yield of 0.69 (Calvert and Pitts, 1967). Compounds like rhodamine B are used in turn as *actinometers*, for determination of the relative fluorescence quantum yield of other compounds as described by Kopetz et al. (2004).

The relative fluorescence quantum yield Fy_D of a fluorescent donor *D* is related to the corrected intensity of an absolute actinometer CFI_{act} by the following equation (Kopetz et al., 2004)

$$Fy_D = (CFI_D)(0.69)/(CFI_{act}) \quad (13)$$

where

Fy_D = relative fluorescence quantum yield of donor *D* in the absence of an acceptor in a particular solvent, at a particular temperature

CFI_{act} = corrected intensity of the actinometer at its fluorescence emission maximum at a particular temperature, such as room temperature, recorded in a cell having the same path length as the cell used for recording the fluorescence spectrum of donor *D*. In this case the actinometer is rhodamine B dissolved in ethanol

Equation 13 and the rhodamine B dissolved in ethanol at room temperature can be used for the determination of the relative fluorescence quantum yield of any fluorescent compound or donor in the presence of an acceptor in any solvent at any temperature.

Calculated Fy_{Da} values for Proto, Mp(e) and Pchlide *a* are reported by Kopetz et al., (2004).

(j) Calculation of τ_D, the Actual Mean Fluorescence Lifetime of the Excited Donor in the Presence of Acceptor at 77 K

The actual mean fluorescence lifetimes of excited donors in the presence of Chl acceptors τ_D, is needed for the calculation of R_0 (Eq. 9). The actual mean fluorescence lifetime of excited donors τ_D, is related to the relative fluorescence yield of donors in the presence of acceptors, Fy_{Da}, by the following equation (Turro, 1965):

$$\tau_D = (Fy_{Da})(\tau_0) \quad (14)$$

where

Fy_{Da} = relative fluorescence yield of the donor in the presence of acceptor

τ_0 = inherent radiative lifetime of the donor in the absence of acceptor

Calculated τ_D values for Proto, Mp(e) and Pchlide *a* are reported by Kopetz et al. (2004).

(k) Calculation of R_0 for Proto, Mp(e) and Pchlide a donors-Chl a Acceptors Pairs at 77 K

The critical separations of donors from acceptors, R_0, for which energy transfer from excited donors D* to acceptors A and emission from excited acceptors A* to the ground state amounts to 50% i.e. are equally probable, are needed for the calculation of R, the distances separating donors D from acceptors A. As described by Eq. 9 R_0^6 is given by

$$R_0^6 \cong \left[\left(1.2055E^{--33}\right)k^2\right]\left[(\tau_D J_\upsilon)/(\nu_0^2)\right]$$

For example, for the Proto-Chl *a* ~F685 pair in barley chloroplast membranes, the following values for the various expressions in the above equation are:

$\kappa^2 = 0.67$
$\tau_D = 3.0649^{-10}$ s
$J_\upsilon = 3.32E^{12}$ cm^3 mol^{-1}
$\nu_0 = 15{,}800$ cm^{-1}

By substituting the above values in the above equation the latter reduces to:

$$R_0^6 \cong \left(1.2055E^{-33}\right)(0.67)/\left[\left(3.0649E^{-10}s\right)\left(3.32E^{12}cm^3.mole^{-1}\right)/\left(15800cm^{-1}\right)^2\right]$$

or

$$R_0^6 \cong 3.28854E^{-39}$$

and

$$R_0 \cong 38.5624E^{-8}cm \;\; i.e.\, 38.56\mathring{A}.$$

Calculated R_0^6 and R_0 values for the Proto, Mp(e) and Pchlide *a* donors-Chl *a* acceptor pairs are reported by Kopetz et al., (2004).

(l) Calculation of E, the Efficiency of Energy Transfer In Situ at 77 K

The efficiency of energy transfer, E, is needed for the calculation of R, the distances separating donor from acceptor molecules. It is calculated from the following equation (Lakowicz, 1999):

$$E = 1 - Fy_{D_a}/Fy_D \quad (15)$$

where

Fy_{D_a} = relative fluorescence yield of donors *D* in the presence of acceptors *A*

Fy_D = relative fluorescence yield of donors *D* in the absence of acceptors *A*

According to Calvert and Pitts (1967), Fy_{Da} is given by

$$Fy_{Da} = [(CFI_{Da})(\Sigma_{act})(C_{act})/(CFI_{act})(e_{Da})(C_{Da})]\,Qy_{act} \quad (16)$$

where

CFI_{Da} = corrected fluorescence intensity of the fluorescence emission bands of donors *D* in the presence of acceptors A in a particular solvent, at a particular temperature

CFI_{act} = corrected fluorescence intensities of the fluorescence emission band of the actinometer

ε_{act} = molar extinction coefficient of the actinometer

ε_{Da} = molar extinction coefficient of donors D in the particular solvent at the particular temperature

C_{act} = concentration of the actinometer

C_{Da} = concentration of donors D in the particular solvent at the particular temperature

QY_{ACT} = absolute fluorescence quantum yield of the actinometer

Likewise for donors D in the absence of acceptors (Kopetz et al., 2004), Fy_D is given by

$$Fy_D = [(CFI_D)(\varepsilon_{act})(C_{act})/(CFI_{act})(e_D)(C_D)]\,Qy_{act} \quad (17)$$

If the concentration of the donors are adjusted so that $(\varepsilon_{Da})(C_{Da}) = (\varepsilon_D)(C_D)$, then Fy_{Da}/Fy_D reduces to CFI_{Da}/CFI_D, and

$E = 1 - Fy_{Da}/Fy_{D}$ transforms into

$$E = 1 - CFI_{Da}/CFI_{D} \quad (18)$$

Calculated efficiencies of energy transfer values E for Proto Mp(e) and MV and DV Pchlide *a* as calculated from Eq. 18 and reported by Kopetz et al. (2004.).

(m) Calculation of the Distances That Separate Proto, Mp(e), DV Pchlide a, and MV Pchlide a from Various Chl a Acceptors in the Photosynthetic Unit

The distances separating Proto, Mp(e), and DV and MV Pchlide *a* from Chl *a* acceptors were determined from Eq. 8 and were compared to current concepts of the photosynthetic unit size and structure (Allen and Forsberg, 2001; Anderson, 2002; Staehelin, 2003) and the proposed Chl–thylakoid biogenesis models (Figs. 5–7). The calculated distances separating Proto, Mp(e) and DV and MV Pchlide *a* from various Chl *a* acceptors in situ are reported in Table 2.

Distances separating anabolic tetrapyrroles from various Chl–protein complexes ranged from a low of 16.26 Å for Proto-Chl *a* ~F735 separation in cucumber, to a high of 40.91 Å for Proto-Chl *a*~F695 separation in barley (Table 2). The magnitude of these distances is compatible with the observation of the intense resonance excitation energy transfer reported in Table 1.

3 Testing the Functionalities of the Various Chl–Thylakoid Biogenesis Models

The early concept of a PSU consisting of about 500 antenna Chl per reaction center has evolved into two pigment systems each with its own reaction center and antenna Chl (Anderson, 2002). The early visualization of the two photosystems consisted of various pigment–protein complexes arrayed into a linear PSU (the continuous array model), about 450 Å in length and 130 Å in width (Bassi et al., 1990). In the PSU, the LHCII was depicted as being shared between the two photosystems. More recent models favor the concept of a laterally heterogeneous PSU (Allen and Forsberg, 2001; Anderson, 2002; Staehelin, 2003). In these models LHCII shuttles between PS I and PS II upon phosphorylation and dephosphorylation (Allen and Forsberg, 2001). Furthermore while PSII is mainly (but not exclusively) located in appressed thylakoid domains, PSI is located in non-appressed stroma thylakoids grana margins, and end membranes (Anderson, 2002; Staehelin, 2003).

In cucumber a *DDV-LDDV* plant species (Abd-El-Mageed et al., 1997), the distances that separated Proto from the Chl *a* acceptors were shorter than those that separated Mp(e) and DV Pchlide *a* from the acceptors. Since Proto is an earlier intermediate of Chl *a* biosynthesis than Mp(e) and Pchlide *a*, it indicates that in cucumber, the Chl *a*-protein biosynthesis region is a highly folded entity where linear distances separating intermediates from end products bear little meaning. On the other hands, in barley, a DMV-LDMV plant species (Abd El Mageed et al., 1997), distances separating Proto from various Chl *a* acceptors were generally longer than those separating Mp(e) and MV Pchlide *a* from the Chl *a* acceptors. This in turn suggests that tetrapyrrole-protein folding in cucumber, a DDV-LDDV plant species is different than in barley a *DMV-LDMV* plant species.

In all cases, it was observed that while distances separating metabolic tetrapyrroles from Chl *a* F670F685 and Chl *a* F677F695 were in the same range, those separating Chl *a* E704F735 from anabolic tetrapyrroles were much shorter (Table 2). As may be recalled, it is believed that the fluorescence emitted at F685 nm arises from the Chl *a* of the

Table 2. Calculated distances R, that separate proto, Mp(e) and Pchlide *a* donors from Chl *a*-protein acceptors in barley and cucumber chloroplasts at 77 K in situ (Adapted from Kopetz et al., 2004).

Chl *a*	Proto		Mp(e) R (Å)		MV Pchlide *a*	DV Pchlide *a*
Species	Barley	Cucumber	Barley	Cucumber	Barley	Cucumber
Chl *a* F685 (LHCII-680 + CP 43)	38.83	30.07	35.60	38.74	37.73	35.22
Chl *a* F695 (CP47) + CP29)	41.23	29.94	35.15	41.53	39.41	30.60
Chl *a* F735 (LHCI-730)	22.72	16.52	19.36	23.76	23.32	22.11

light-harvesting Chl–protein complexes (LHCII and Cp43), that emitted at F695 nm originates mainly from the PSII antenna Chl *a* (CP47 and/or CP29), while that emitted at F735 nm originates primarily from the PSI antenna Chl *a* (LHC-730) (Gilmore, 2004, Bassi et al., 1990). This in turn suggests that in the Chl *a*-protein biosynthesis region(s), protein folding is such that the PS I antenna Chl *a* (LHCI-730) is much closer to the terminal steps of anabolic tetrapyrrole biosynthesis than the LHCII and CP 43 Chl–protein complexes or the CP47 and/or CP29 PSII antenna Chl *a* complexes.

(a) The Single-Branched Pathway-Single Location Model Is Not Compatible with Resonance Excitation Energy Transfer Between Anabolic Tetrapyrrole Donors and Chl a-Proteins Acceptors in Chloroplasts

The SBP-single location model is incompatible with the continuous array and the laterally heterogeneous PSU models. Indeed the SBP-single location model calls for Chl *a* and *b* to be formed via a single-branched Chl biosynthetic pathway at a location accessible to all Chl-binding apoproteins. The latter will have to access that location in the unfolded state, pick up a complement of MV Chl *a* and/or MV Chl *b*, and undergo appropriate folding. Then the folded Chl–apoprotein complex has to move from the central location to a specific PS I, PS II, or Chl *a*/*b* LHC-protein location within the Chl–apoprotein biosynthesis center over distances of up to 225 Å in the continuous array model, or over larger distances in the laterally heterogeneous model, in order to become part of PS I, PS II or LHCII. If this were the case, then no resonance excitation energy transfer would be observed between anabolic tetrapyrroles and the various Chl–protein complexes. Furthermore, the distances separating anabolic tetrapyrroles from various Chl–protein complexes would be much larger than the values reported in Table 2.

(b) The SBP-Multilocation Model Is Not Compatible with the Realities of Chl Biosynthesis in Green Plants

Although the shorter distances separating anabolic tetrapyrroles from Chl–protein complexes reported in Table 2 are compatible with the SBP-multilocation model, overwhelming experimental evidence argues against the operation of a single-branched Chl biosynthetic pathway in plants (Rebeiz et al., 2003). For example, the SBP does not account for the formation of *MV Mpe* in green plants. Furthermore, according to the single-branched pathway, DV Pchlide *a* is supposed to be a transient metabolite that does not accumulate in higher plants (Jones, 1963), yet definite resonance excitation energy transfers from DV Pchlide *a* to Chl *a*-,-protein acceptors were observed (Table 1). All in all, that leaves the MBP-sublocation model alternative as a working hypothesis.

(c) The MBP-Sublocation Model Is Compatible with the Realities of Chl Biosynthesis in Green Plants, and with Resonance Excitation Energy Transfer Considerations Between Anabolic Tetrapyrrole Donors and Chl a-protein Acceptors in Chloroplasts

As was mentioned earlier an overwhelming body of experimental evidence discussed in Rebeiz et al. (2003) supports the operation of a multibranched Chl biosynthetic pathway in green plants. In the MBP-Sublocation model, the unified multibranched Chl *a*/*b* biosynthetic pathway is visualized as the template of a Chl–protein biosynthesis center where the assembly of PS I, PS II and LHC takes place (Rebeiz et al., 1999, 2003). The multiple Chl biosynthetic routes are visualized individually, or in groups of one or several adjacent routes, as Chl–apoprotein biosynthesis subcenters earmarked for the coordinated assembly of individual Chl–apoprotein complexes. Apoproteins destined to some of the subcenters may possess specific polypeptide signals for specific Chl biosynthetic enzymes peculiar to that subcenter, such as 4-vinyl reductases, formyl synthetases or Chl *a* and *b* synthetases. Once an apoprotein formed in the cytoplasm or in the plastid reaches its subcenter destination and, its signal is split off, and binds carotenoids and nascent Chl formed via one or more biosynthetic routes. During pigment binding, the apoprotein folds and acts at that location, while folding or after folding, as a template for the assembly of other pigment–proteins. This model is compatible with the lateral heterogeneity of the PSU and can account for the observed resonance excitation energy transfer (Table 1), and the short distances separating anabolic tetrapyrroles from Chl–protein complexes in the distinct PSI, PSII and shuttling LHCII entities that compose the PSU (Table 2).

E Guidelines and Suggestions to Bioengineer Plants with Smaller Photosynthetic Unit Size

The compatibility of the MBP-sublocation model of Chl–thylakoid protein assembly opens the way for testing the hypothesis of whether certain Chl biosynthetic routes are indeed involved in the formation of specific Chl–protein complexes. Below are outlined some guidelines and suggestions for tackling this issue.

The experimental strategy involves a two pronged experimental approach. In a first approach, a variety of higher and lower plant mutants that lack specific Chl–protein complexes could be used to determine which specific Chl biosynthetic route(s) is/are missing from the mutant Chl biosynthetic pathway. In this manner it may be possible to link a particular Chl biosynthetic route to a specific Chl–protein complex formation.

Likewise in the second approach functional PS I, and PS II particles as well as LHCII preparations could be isolated from wild types and mutants using mild detergents and the putative Chl biosynthetic routes associated with a particular preparation could be determined. In this manner it may also be possible to link particular Chl biosynthetic routes with the lateral heterogeneity of the PSU.

1 Selection of Mutants

A literature search of higher and lower plant mutants deficient in specific Chl–protein complexes reveals a rather large number of such mutants. Final selection of specific mutants for specific studies will therefore depend on the nature of the missing Chl–protein complexes and availability of plant material. Below are listed some of the candidate mutants:

(a) Mutants of Higher Plants Other Than Arabidopsis

Chlorina--f2 viridis-m^{29} *viridis-n*^{34} *and viridis-zd*^{69} of Barley (Machold et al., 1979; Henry-Landis et al., 1983; White and Green, 1987; Preiss and Thornber, 1995).Chl *b*-less barley mutant (Mullet et al., 1980; Bellemare et al., 1982).

*Viridis-zb*63 and *viridis-h*15 of barley(Hiller et al., 1980).

Qy/+ hcf3/hcf3 of maize (Polacco, 1984).

*hcf**-3 nuclear maize mutant (Leto et al., 1985).

hcf1-2-3-6-19- 38-42-44-50-101-102-103-104-108-111 in maize (Miles, 1994).

U374 mutant of sweet clover (Markwell et al., 1985).

C*ab4Bst*EII, C*ab4.23,*and C*ab4.3* mutants of tomato (Huang et al., 1992).

(b) Arabidopsis Mutants

A large number of mutants with defects in or elimination of chlorophyll/protein complexes in *Arabidopsis thaliana* have already been identified (H. Bohnert, personal communication, 2009). Also some interesting mutants can be obtained from the US or European seed banks.

(c) Lower Plant Mutants

Y-1 mutant of *Chlamydomonas reinhardi* (Gershoni et al., 1982; Ish-Shalom and Ohad, 1983; Hoober, 1990).

Gr1BSL, G1BU, and O4BSL of *Euglena gracilis* (Cunningham and Schiff, 1986).

PS I-less/*apc*E^{-} *Synechocystis* sp. PCC 6803 mutant (Shen and Vermaas, 1994).

2 Preparation of Photosynthetic Particles

A large volume of literature dealing with the preparation of various photosynthetic particles is readily available. The most recent review of various procedures for the preparation of PS I, PS II, LHCII, and a variety of smaller Chl–protein complexes has been reported in Paulsen and Schmid (2002). Procedures described there can be complemented by original standalone procedures available in the photosynthesis literature.

3 Determination of Biosynthetic Routes Functional in a Specific Mutant or Photosynthetic Particle

Partial and full biosynthetic routes that are functional in various mutants and isolated photosynthetic particles and complexes can be determined by various techniques described in Rebeiz 2002). These techniques have been developed over a period of 3.5 decades and are routinely used on daily basis by several scientists. Cold and ^{14}C-substrated can be prepared and used as described by Rebeiz (2002). Wild types and mutants can be light or dark adapted in order to poise them in the DV or MV modes (Carey et al., 1985) prior to subplastidic particle isolation. Single or multistep reaction sequences can be executed by feeding appropriate substrates in well defined cofactor media capable of supporting nearly all the reactions described in Fig. 3.

Epilogue

As suggested above, future research dealing with the bioengineering of smaller PSU sizes will have to focus on the MBP-sublocation Chl *a*-thylakoid protein biosynthesis model. The first order of business will have to deal with determining which Chl biosynthetic routes gives rise to PS I, PS II, and LHCII Chl–protein complexes. Appropriate genes of the greening process may then be manipulated by molecular biological techniques to bioengineer genetically modified plants with a smaller PSU, i.e., PSU units having more RC complexes and fewer antenna Chl per unit thylakoid area. Nevertheless this type of agriculture using genetically modified plants with smaller PSU sizes and higher photosynthetic conversion efficiencies will still be at the mercy of extrinsic factors and weather uncertainties.

In our opinion the ultimate agriculture of the future should consist of bioreactors populated with bioengineered, highly efficient photosynthetic membranes, with a small PSU size, and operating at efficiencies that approach the 12% maximal theoretical efficiency of the PETS that may be observed under white light, or the 27% maximal theoretical efficiency that may be achieved under red light. Such conditions may be set up during space travel or in large space stations (Rebeiz et al., 1982). The photosynthetic product may well be a short chain carbohydrate such as glycerol that can be converted into food fiber and energy. In the meanwhile let us not forget that a journey of 10,000 miles starts with the first step.

Acknowledgment

The writing of this chapter was supported by the Rebeiz Foundation for Basic Research.

References

Abd El Mageed HA, El Sahhar KF, Robertson KR, Parham R and Rebeiz CA (1997) Chloroplast Biogenesis 77. Two novel monovinyl and divinyl light-dark greening groups of plants and their relationship to the chlorophyll *a* biosynthetic heterogeneity of green plants. Photochem Photobiol 66: 89–96

Allen JF and Forsberg J (2001) Molecular recognition in thylakoid structure and function. Trends Plant Sci 6: 317–326

Amindari SM, Splittstoesser WE and Rebeiz CA (1995) Photodynamic effects of several metabolic tetrapyrroles on isolated chloroplasts. In: Heitz JR and Downum KR (eds) Light Activated Pest control, Vol 616. American Chemical Society, Washington, DC, pp. 217–246

Anderson JM (2002) Changing concepts about the distribution of photosystem I and II between grana-appressed and stroma-exposed thylakoid membranes. Photsynth Res 73: 157–164

Bassi R, Rigoni F and Giacometti GM (1990) Chlorophyll binding proteins with antenna function in higher plants and green algae. Photochem Photobiol 52: 1187–1206

Bellemare G, Bartlett SG and Chua NH (1982) Biosynthesis of chlorophyll *a*/*b*-binding polypeptides in wild type and the Chlorina f2 mutant of barley. J Biol Chem 257: 7762–7767

Blankenship RE (2002) Molecular Mechanism of Photosynthesis. Blackwell, Malden, MA

Booth-Thomas C (2007) The space cowboys. Time, March 5: 51–58

Calvert JG and Pitts JN (1967) Photochemistry. Wiley, New York

Carey EE and Rebeiz CA (1985) Chloroplast Biogenesis 49. Difference among angiosperms in the biosynthesis and accumulation of monovinyl and divinyl protochlorophyllide during photoperiodic greening. Plant Physiol 79: 1–6

Carey EE, Tripathy BC and Rebeiz CA (1985) Chloroplast biogenesis 51. Modulation of monovinyl and divinyl protochlorophyllide biosynthesis by light and darkness in vitro. Plant Physiol 79: 1059–1063

Clegg RM (2004) Nuts and bolts of excitation energy migration and energy transfer. In: Papageorgiou GC and Govindjee (eds) Chlorophyll *a* Fluorescence: A Signature of Photosynthesis, Vol 19. Advances in Photosynthesis and Respiration Series. Springer, Dordrecht, pp. 83–105

Cunningham FXJ and Schiff J (1986) Chlorophyll-protein complexes from *Euglena gracilis* and mutants deficient in chlorophyll *b*. Plant Physiol 80: 23–238

Daniell H and Rebeiz CA (1982a) Chloroplast Culture IX. Chlorophyll(ide) *a* biosynthesis in vitro at rates higher than in vivo. Biochem Biophys Res Commun 106: 466–470

Daniell H and Rebeiz CA (1982b) Chloroplast Culture VIII. A new effect of kinetin in enhancing the synthesis and accumulation of protochlorophyllide in vitro. Biochem Biophys Res Commun 104: 837–843

Demmis-Adams B, Adams III WW and Mattoo KD (eds) (2006) Photoinhibition, Gene Regulation and Environment. Advances in Photosynthesis and Respiration Series, Vol 9. Springer, Dordrecht

Förster TH (1960) Transfer mechanisms of electronic excitation energy. Radiat Res Suppl 2: 326–339

Frank HA, Young AJ, Britton G and Cogdell RJ (eds) (1999) The Photochemistry of Carotenoids. Advances in Photosynthesis and Respiration Series, Vol 8. Springer, Dordrecht

French CS, Brown JS and Lawrence MC (1972) Four universal forms of chlorophyll *a*. Plant Physiol 49: 421–429

Gershoni JM, Shochat S Malkin S and Ohad I (1982) Functional organization of the chlorophyll-containing complexes of *Chlamydomonas reinhardi*. Plant Physiol 70: 637–644

Golbeck JH (ed) (2006) Photosystem I: The Light-Driven Plastocyanin:Ferredoxin Oxidoreductase. Advances in Photosynthesis and Respiration Series, Vol 24. Springer, Dordrecht

Govindjee (2004) Chlorophyll *a* fluorescence: a bit of basics and history. In: Papageorgiou PC and Govindjee (eds) Chlorophyll *a* Fluorescence. A Signature of Photosynthesis, Vol 19. Advances in Photosynthesis and Respiration Series, Springer, Dordrecht, pp. 1–41

Gilmore, AM (2004) Excess light stress: probing excitation dissipation mechanisms through global analysis of time- and wavelength-resolved chlorophyll a fluorescence In: Papageorgiou PC and Govindjee (eds) Chlorophyll *a* Fluorescence. A Signature of Photosynthesis, Vol 19. Advances in Photosynthesis and Respiration Series. Springer, Dordrecht Springer, Dordrecht, pp. 555–581

Henry-Landis E, Dalgaard-Mikkelsen J and lindberg-MØller B (1983) Pigment and acyl lipid composition of photosystem I and II vesicles and of photosynthetic mutants in barley. Carlsberg Res Commun 48: 131–148

Hiller RG, Lindberg-MØller B and Hoyer-Hansen G (1980) Characterization of six putative photosystem I mutants in barley. Carlsberg Res Commun 45: 315–328

Hoober KJ (1990) Accumulation of chlorophyll *a/b*-binding polypeptides in *Chlamydomonas reinhardi y-1* in the light or dark at 38 C. Evidence for proteolytic control. Plant Physiol 92: 419–426

Huang L Adam Z and Hoffman NE (1992) Deletion mutants of chlorophyll *a/b* binding proteins are efficiently imported into chloroplasts but do not integrate into thylakoid membranes. Plant Physiol 99: 247–255

Ioannides IM, Fasoula DM, Robertson KR and Rebeiz CA (1994) An evolutionary study of chlorophyll biosynthetic heterogeneity in green plants. Biochem Syst Ecol 22: 211–220

Ish-Shalom D and Ohad I (1983) Organization of chlorophyll–protein complexes of photosystem I in *Chlamydomonas reinhardi*. Biochem Biophys Acta 722: 498–507

Jones OTG (1963) Magnesium 2, 4-divinyl phaeoporphyrin a_5 monomethyl ester a protochlorophyll-like pigment produced by *Rhodopseudomonas spheroides*. Biochem J 89: 182–189

Kolossov VL and Rebeiz CA (2001) Chloroplast Biogenesis 84. Solubilization and partial purification of membrane-bound [4-vinyl] chlorophyllide *a* reductase from etiolated barley leaves. Anal Biochem. 295: 214–219

Kolossov VL and Rebeiz CA (2003) Chloroplast Biogenesis 88. Protochlorophyllide *b* occurs in green but not in etiolated plants. J Biol Chem 278: 49675–49678

Kolossov V, Ioannides IM, Kulur S and Rebeiz CA (1999) Chloroplast biogenesis 82: development of a cell-free system capable of the net synthesis of chlorophyll(ide) *b*. Photosynthetica 36: 253–258

Kolossov VL Kopetz K and Rebeiz CA (2003) Chloroplast Biogenesis 87: evidence of resonance excitation energy transfer between tetrapyrrole intermediates of the chlorophyll biosynthetic pathway and chlorophyll *a*. Photochem Photobiol 78: 184–196

Kopetz KJ, Kolossov VL and Rebeiz CA (2004) Chloroplast Biogenesis 89: development of analytical tools for probing the biosynthetic topography of photosynthetic membranes by determination of resonance excitation energy transfer distances separating metabolic tetrapyrrole donors from chlorophyll *a* acceptors. Anal Biochem 329: 207–219

Lakowicz JR (1999) Principles of Fluorescence Spectroscopy. Kluwer/Plenum, New York

Lakowicz JR (2006) Principles of Fluorescence Spectroscopy. Springer, New York

Leegood RC, Sharkey TD and von-Camerer S (eds) (2000) Photosynthesis: Physiology and Metabolism. Advances in Photosynthesis and Respiration Series, Vol 9. Springer, Dordrecht

Leto K J Bell E and McIntosh L (1985) Nuclear mutation leads to an accelerated turnover of chloroplast-encoded 48 kD polypeptides in thylakoids lacking photosystem II. The EMBO J 4: 1645–1653

Lien S and San Pietro A (1975) An inquiry into the biophotolysis of water to produce hydrogen. RANN Reports, Indiana University, IN, pp. 1–50

Machold O, Simpson DJ and Lindberg-Moller B (1979) Chlorophyll-proteins of thylakoids from wild-type and mutants of barley (*Hordeum vulgare* L.). Carlsberg Res Commun 44: 235–254

Markwell J, Webber AN and Lake B (1985) Mutants of sweetclover (*Melilotus alba*) lacking chlorophyll *b*. Plant Physiol 77: 948–951

Miles D (1994) The role of high chlorophyll fluorescence photosynthesis mutants in the analysis of chloroplast thylakoid membrane assembly and function. Medica 39: 35–45

Mullet JE, Burke J and Arntzen CJ. (1980) A developmental study of photosystem I peripheral chlorophyll proteins. Plant Physiol 65: 823–827

Paulsen H and Schmid VHR (2002) Analysis and reconstitution of chlorophyll-proteins. In: Smith AG and Witty M (eds) Heme Chlorophyll and Bilins. Methods and Protocols. Humana Press, Totowa, NJ, pp. 235–253

Polacco ML (1984) Chl (A/B) light harvesting complex assembly in maize: genetic evidence that it may compete with PSII for Chl. Curr Topics Plant Biochem Physiol 3: 167

Preiss S and Thornber JP (1995) Stability of the apoproteins of light-harvesting complex I and II during biogenesis of thylakoids in the chlorophyll *b*-less barely mutant *Chlorina f2*. Plant Physiol 107: 709–717

Rebeiz CA (2002) Analysis of intermediates and end products of the chlorophyll biosynthetic pathway. In: Smith AG and Witty M (eds) Heme Chlorophyll and Bilins. Methods and Protocols. Humana Press, Totowa, NJ, pp. 111–155

Rebeiz CA and Castelfranco P (1971a) Protochlorophyll biosynthesis in a cell-free system from higher plants. Plant Physiol 47: 24–32

Rebeiz CA and Castelfranco P (1971b) Chlorophyll biosynthesis in a cell-free system from higher plants. Plant Physiol 47: 33–37

Rebeiz CA, Daniell H and Mattheis JR (1982) Chloroplast bioengineering: the greening of chloroplasts in vitro. In: Scott CD (ed) Biotechnology and Bioengineering Symposium 12.Wiley, New York, pp. 414–439

Rebeiz CA, Montazer-Zouhoor A and Daniell H (1984) Chloroplast culture X: thylakoid assembly in vitro. Isr J Bot 33: 225–235

Rebeiz CA, Parham R, Fasoula DA and Ioannides IM (1994) Chlorophyll biosynthetic heterogeneity. In: Chadwick DJ and Ackrill K (eds) The Biosynthesis of the Terapyrrole Pigments. Wiley, New York, pp. 177–193

Rebeiz CA, Ioannides IM, Kolossov V, and Kopetz KJ (1999) Chloroplast Biogenesis 80. Proposal of a unified multibranched chlorophyll *a/b* biosynthetic pathway. Photosynthetica 36: 117–128

Rebeiz CA, Kolossov VL, Briskin D and Gawienowski M (2003) Chloroplast biogenesis: chlorophyll biosynthetic heterogeneity multiple biosynthetic routes and biological spin-offs. In: Nalwa HS (ed) Handbook of Photochemistry and Photobiology, Vol 4. American Scientific Publishers Los Angeles, CA, pp. 183–248

Rebeiz CA, Kopetz KJ and Kolossov VL (2005) Chloroplast Biogenesis. Probing the relationship between chlorophyll biosynthetic routes and the topography of chloroplast biogenesis by resonance excitation energy transfer determinations. In: Pessarakli M (ed) Handbook of Photosynthesis, 2nd edn. Marcel Dekker, New York, pp. 55–95

Satoh K, Wydrzynski T and Govindjee (2005) Introduction to Photosynthesis. In: Wydrzynski T and Satoh K (eds) Photosystem II: The light-driven water: plastoquinone oxidoreductase, pp. 11–22. Springer, Dordrecht

Shen G and Vermaas WFJ (1994) Mutation of chlorophyll ligands in the chlorophyll-binding CP47 protein as studied in a *Synechocystis* sp. PCC 6803 photosystem-less background. Biochemistry 33: 7379–7388

Smith JHC and French CS (1963) The major accessory pigment in photosynthesis. Ann Rev Plant Physiol 14: 181–224

Siegenthaler PA and Murata N (eds) (1998) Lipids in Photosynthesis: Structure, Function and Genetics. Advances in Photosynthesis and Respiration Series, Vol 6. Springer, Dordrecht

Staehelin LA (2003) Chloroplast structure: from chlorophyll granules to supra-molecular architecture of thylakoid membranes. Photsynth Res 76: 185–196

Sundqvist C and Ryberg M (eds) (1993) Pigment–Protein Complexes in Plastids: Synthesis and Assembly. Academic, New York

Tripathy BC and Rebeiz CA (1986) Chloroplast Biogenesis. Demonstration of the monovinyl and divinyl monocarboxylic routes of chlorophyll biosynthesis in higher plants. J Biol Chem 261: 13556–13564

Turro NJ (1965) Molecular Photochemistry. W. A. Benjamin, London

von Wettstein D, Gough S and Kannangara CG (1995) Chlorophyll biosynthesis. Plant Cell 7: 1039-1057

White MJ and Green BR (1987) Polypeptides belonging to each of the three major chlorophyll *a+b* protein complexes are present in a chlorophyll-*b*-less barley mutant. Eur J Biochem 165: 531–535

Wu CW and Stryer L (1972) Proximity relationship in Rhodopsin. Proc Natl Acad Sci 69: 1104–1108

Wydrzynski TI and Satoh K (eds) (2006) Photosystem II-The Light-Driven Water plastoquinone Oxidoreductase. Advances in photosynthesis and Respiration, Vol 22. Springer, Dordrecht

Chapter 2

Evidence for Various 4-Vinyl Reductase Activities in Higher Plants

Vladimir L. Kolossov
Institute for Genomic Biology, University of Illinois, Urbana, IL 61801, USA

Constantin A. Rebeiz*
Rebeiz Foundation for Basic Research, 2209 Edgewater Place, Champaign, IL 61822, USA

*Author for Correspondence, e-mail: crebeiz@uiuc.edu

Constantin A. Rebeiz et al. (eds.), The Chloroplast: Basics and Applications, pp. 25–38.

Summary

[4-Vinyl] Mg-protoporphyrin reductase, [4-vinyl] Mg-protoporphyrin monoester reductase, [4-vinyl] protochlorophyllide *a* reductase, and [4-vinyl] chlorophyllide *a* reductase activities were solubilized from isolated barley plastid membranes. Solubilization enhanced the 4-vinyl reductase activities. Preliminary comparative investigations were performed on [4-vinyl] protochlorophyllide *a* reductase and [4-vinyl] chlorophyllide *a* reductase. Both activities were localized in the inner plastid membranes and were missing from the plastid envelopes. Like [4-vinyl] chlorophyllide *a* reductase, [4-vinyl] protochlorophyllide *a* reductase exhibited an absolute requirement for NADPH, and both activities were insensitive to NADP and nicotinamide. However, [4-vinyl] protochlorophyllide *a* reductase activity appeared to be less sensitive to light treatments than [4-vinyl] chlorophyllide *a* reductase, since after 4 h of illumination the ratio of [4-vinyl] chlorophyllide *a* reductase/[4-vinyl] protochlorophyllide *a* reductase activities dropped from 67 to 43. This in turn suggested that [4-vinyl] chlorophyllide *a* reductase and [4-vinyl] protochlorophyllide *a* reductase activities may be catalyzed by two different enzymes. Also, the following information suggests that the various 4-vinyl reductase activities were catalyzed by different enzymes; (a) the different 4-vinyl reductase activities exhibited different activations following solubilization, and (b) the ratio of [4-vinyl] chlorophyllide reductase/[4-vinyl] protochlorophyllide *a* reductase activities was different in variously purified fractions. Contrary to previous beliefs, it became evident that [4-vinyl] chlorophyllide *a* reductase was as active at the end of the dark phase as in the middle of the light phase of the photoperiod, in dark-monovinyl-light divinyl-light-dark monovinyl plants such as barley, as well as in dark-divinyl-light divinyl-light-dark divinyl plants such as cucumber. The carboxylic Chl biosynthetic routes were revised to accommodate the above observations. As a consequence two different pathways comprising appropriate biosynthetic routes are presented for dark-monovinyl-light divinyl-light-dark monovinyl plants such as barley, and dark-divinyl-light divinyl-light-dark divinyl plants such as cucumber.

Abbreviations: ALA – δ-aminolevulinic acid; Chaps – 3-[(3-cholamidopropyl)dimethylammonio]-1-propanesulfonate; Chl – chlorophyll; Chlide – *a* chlorophyllide *a*; DDV-LDV-LDDV – dark divinyl-light divinyl-light-dark divinyl; DMV-LDV-LDMV – dark monovinyl-light divinyl-light-dark monovinyl; DV – divinyl; 2,2′Dpy – 2,2′-dipyridyl; HEAR – hexane-extracted acetone residue; KEGG – Kyota encyclopedia of genes and genomes; Mpe - Mg-Proto monomethyl ester; Mp(e) – Mg-Proto and/or Mpe; MV – monovinyl; Pchlide - *a* protochlorophyllide *a*; Proto - protoporphyrin IX; 4VCR – [4-vinyl] chlorophyllide *a* reductase; 4VChlR – [4-vinyl] chlorophyll *a* reductase; 4VMPR – [4-Vinyl] Mg-protoporphyrin reductase; 4VMpeR – [4-vinyl] Mg-protoporphyrin monoester reductase; 4VPideR-[4-vinyl] protochlorophyllide *a* reductase; 4VR – [4-vinyl] reductase; TAIR – The Arabidopsis Information Resource

Note: Unless proceeded by MV or DV, tetrapyrroles are used generically to designate metabolic pools that may consist of MV and/or DV components. It has come to our attention that the various 4VCR enzymes may be coded for by a limited number of genes whose expression is modified by various regulatory factors that bind to a particular gene and modify its expression. Alternatively, a single mutation as in the Nec 2 corn mutant could disrupt multiple enzymes due to a mutation in a single regulatory factor (Wang et al., 1999; Rebeiz et al., 2010).

I Introduction

Chlorophylls are ones of the most abundant pigments in nature. Its biosynthesis in higher plants takes place exclusively in the chloroplasts where chlorophyll–protein complexes serve as major players in the photosynthetic process (Sundqvist and Ryberg 1993).

Most of the enzymes responsible for chlorophyll (Chl) biosynthesis have been identified and fairly well characterized. However, [4-vinyl] reductase (4VR), enzyme(s) involved in the vinyl side chain reaction of Chls, and serving as branching points to the biosynthetic heterogeneity of the Chl biosynthetic pathway are still poorly understood (Rebeiz et al. 2003). Under various experimental conditions, all known intermediates of the chlorophyll biosynthetic pathway between protoporphyrin IX (Proto) and Chl have been found in the monovinyl (MV) and divinyl (DV) forms (Rebeiz et al. 1983, 1994, 2003; Shioi and Sasa 1983; Whyte and Griffiths, 1993). The DV intermediates have vinyl groups, at position 2 and 4 of the macrocycle (Fig. 1). In contrast, the MV intermediates have one vinyl and one ethyl group at positions 2 and 4 of the macrocycle respectively (Fig. 1). In higher plants, the end products of the Chl biosynthetic heterogeneity are invariably MV Chl *a*

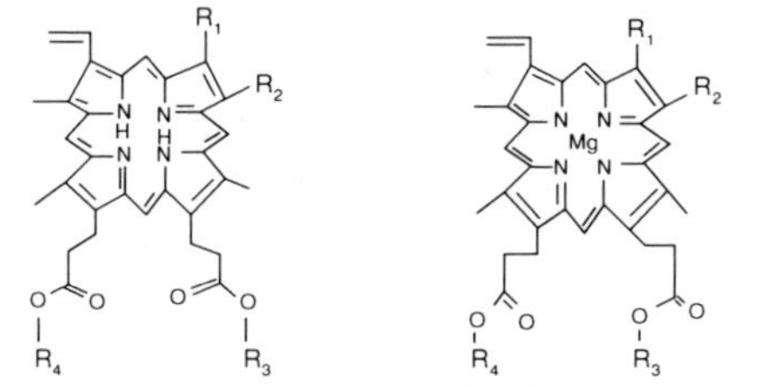

1. Protoporphyrin IX **2. Mg-Protoporphyrin LX**

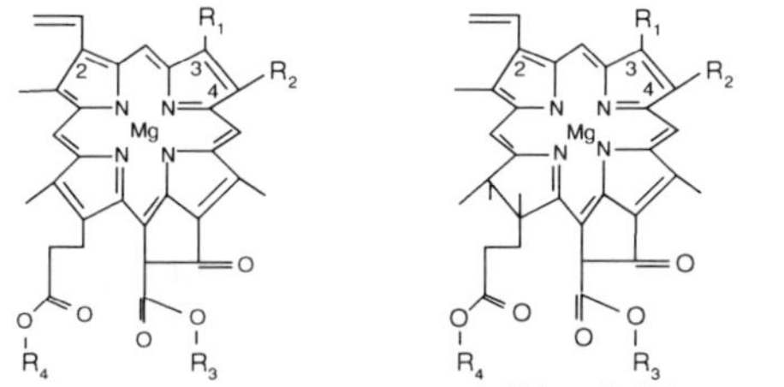

3. Protochlorophyllide a **4. Chlorophyllide a**

	R$_1$	R$_2$	R$_3$	R$_4$	Compound
1	Methyl	Vinyl	H	H	DV Proto
2a	Methyl	Vinyl	H	H	DV Mg-Proto
2b	Methyl	Ethyl	H	H	MV Mg-Proto
2c	Methyl	Vinyl	Methyl	H	DV Mpe
2d	Methyl	Ethyl	Methyl	H	MV Mpe
3a	Methyl	Vinyl	Methyl	H	DV Pchlide a
3b	Methyl	Ethyl	Methyl	H	MV Pchlide a
4a	Methyl	Vinyl	Methyl	H	DV chlide a
4b	Methyl	Ethyl	Methyl	H	MV Chlide a
4c	Methyl	Vinyl	Methyl	Phytol	DV Chla
4d	Methyl	Ethyl	Methyl	Phytol	MV Chla

Fig. 1. Chemical structures of various DV and MV tetrapyrroles.

and *b*. The lethal Nec 2 maize mutant however (Bazzaz 1981) forms only DV Chl *a* and *b*, and a pale green Arabidopsis mutant which accumulates mostly DV Chl *a* and *b* with smaller amounts of MV Chl has been reported (Nagata et al. 2005). However, in the prochlorophyte picophytoplankton of tropical oceans, DV Chl *a* and *b* are the primary photosynthetic pigments (Chisholm et al. 1992; Goerike and Repeta 1992).

The origin and physiological significance of the MV and DV chemical heterogeneity of Chl metabolic precursors in higher plants is poorly understood. The debate revolves around the question of whether there are multiple 4VR enzymes unique to a specific DV tetrapyrrole substrate (Suzuki and Bauer 1995; Rebeiz et al. 2003), or one enzyme exhibiting broad substrate specificity (Whyte and Griffiths, 1993). Furthermore, while the conversion of DV protochlorophyllide (Pchlide) *a* and DV chlorophyllide (Chlide) *a* to their MV counterparts is gaining broad recognition, the occurrence of 4VR activity towards DV Proto, DV Mg-Proto and DV Mg-Proto monoester (Mpe) has been questioned as the MV intermediates are found in smaller amounts under natural conditions (Rudiger 2003). This fact has been emphasized in the Kyota encyclopedia of genes and genomes (KEGG) database (http://www.genome.jp/kegg/pathway/map/map00860.html), while the Arabidopsis Information Resource (TAIR) database recognizes only the vinyl side chain reduction at the Pchlide *a* level (http://www.arabidopsis.org:http://www.plantcyc.org:1555/ARA/NEW-IMAGE?type=PATHWAY&object=CHLOROPHYLL-SYN).

The classical single-branched chlorophyll biosynthetic pathway proposed by Granick (1950) and modified by Jones (1963) assumed the rapid reduction of the vinyl side chain to ethyl at position 4 of the Pchlide *a* macrocycle by a putative 4-vinyl Pchlide *a* reductase. This in turn made DV Pchlide *a*, a transient species that does not accumulate in higher plants. However, the detection of accumulated DV Pchlide *a* in DV plant species has refuted this statement (Carey and Rebeiz 1985; Carey et al. 1985). Also, in 1973, Ellsworth and Hsing (1973), reported the reduction of DV Mpe to MV Mpe by a soluble NADH-dependent enzyme in etiolated wheat homogenates. However, to our knowledge, no one, including ourselves has been able to confirm these results.

Evidence is currently accumulating that 4VR activities in higher plants occur at the levels of Mg-Proto, Pchlide *a*, Chlide *a* and Chl *a* (Rebeiz et al. 2003). The most potent among the 4VRs is a membrane-bound NADPH-dependent 4-vinyl chlorophyllide *a* (4VCR) activity. The first evidence of conversion of DV Chlide *a* to MV Chlide *a* was reported in vivo in etiolated cucumber exposed to light (Duggan and Rebeiz 1982a, b). [4-Vinyl] chlorophyllide *a* activity was later detected in isolated cucumber etioplasts (Duggan and Rebeiz 1982b). It was also detected in plastid membranes from etiolated cucumber, maize and barley (Parham and Rebeiz 1992, 1995). It was also demonstrated that 4VCR activity decreased upon illumination in etiolated barley and cucumber (Abd-El-Mageed et al., 1997; Kolossov and Rebeiz 2001). 4VCR was recently solubilized and partially purified from etiolated barley leaves (Kolossov and Rebeiz 2001).

The conversion of DV Pchlide *a* to MV Pchlide *a* was observed in isolated etioplasts from cucumber and barley (Tripathy and Rebeiz 1986). DV Pchlide *a* reduction by a putative [4-vinyl] Pchlide *a* reductase (4VPideR) was demonstrated in plastids prepared from green(ing) barley, but in cucumber the conversion did not take place on the barley time-scale of incubation (Tripathy and Rebeiz 1986,1988). Whyte and Griffiths (1993) speculated that the differences in rates of 4-vinyl reduction between mono- and dicotyledonous plants may be accounted for by their NADPH redox state. Extensive work on the selective inhibition of Chl biosynthesis between DV-Pchlide and MV-Pchlide in *Rhodobacter Sphaeroides* grown in the presence of nicotinamide was summarized by Richards (1993).

In addition to 4VCR and 4VPideR activities, a week putative membrane-bound [4-vinyl] Mg-Proto reductase (4VMPR) activity was demonstrated by conversion of DV Mg-Proto to MV Mg-Proto in etiochloroplasts and etiochloroplast membranes prepared from etiolated barley (Kim and Rebeiz 1996). At that time, the authors were not able to demonstrate 4VMPR activity in etiolated cucumber (Kim and Rebeiz 1996).

More recently, the possible role of a putative 4-vinyl Chl *a* reductase (4VChlR) was proposed to account for the transient fall of DV Chl *a* and corresponding rise in the MV Chl *a* content following illumination of etiolated cucumber cotyledons (Adra and Rebeiz 1998).

Finally, involvement of the plastid stroma in the regulation of vinyl side chain reduction has been emphasized (Kim et al. 1997).

Despite obvious progress in the study of 4VR activities, little is known about gene(s) coding for these enzyme(s). The first identification of a 4VR gene(s), *bchJ*, was reported in *Rhodobacter capsulatus*, a photosynthetic bacterium. Disruption of the *bchJ* gene of *Rhodobacter capsulatus* altered the ratio of MV and DV Pchlide *a* accumulation in a strain deficient in the reduction of Pchlide *a* to Chlide *a* (Suzuki and Bauer 1995). However, the presence or absence of the *bchJ* gene which was assumed to be responsible for conversion of DV Pchlide *a* to MV Pchlide *a* did not altered the massive accumulation MV and DV Mpe (Suzuki and Bauer 1995). Furthermore, the protein product of the *bchJ* gene expressed in *E. coli* was unable to convert DV Pchlide *a* to MV Pchlide *a* (Kolossov and Rebeiz 2005; Chew and Bryant 2007). More recently, Chew and Bryant (2007) reported a gene (*bci*A) in *Chlorobium tripidum* that coded for vinyl reduction to ethyl at the eighth position of DV Pchlide *a*, which was converted to MV Pchlide *a*. However the authors did not test the specificity of the coded enzyme for other reductase activities.

Also, a 4VR gene (At5g18660) from an *Arabidopsis* mutant has been reported (Nagata et al. 2005). The expression product of the gene in *E. coli* catalyzed the conversion of DV Chlide *a* to MV Chlide *a* suggesting that At5g18660 codes for 4VCR (Nagata et al. 2005).

In this case too, the authors did not test the specificity of the coded enzyme for other reductase activities. The low similarity of At5g18660 to higher plants and its absence from the complete genomic sequence of the unicellular red alga *Cyanidioschyzon merolae,* as well as the results of Chew and Bryant (2007) lead to the conclusion that other putative 4VR genes may be present.

At this point, the number of 4VR genes is uncertain, and the presence of MV Chls in the *Arabidopsis* mutant presumes the occurrence other 4VR isoform(s) and/or gene(s), although incomplete inactivation of the gene product by a single mutation of At5g18660 cannot be ruled out (Nagata et al. 2005). Furthermore, although the pigment composition of the Nec 2 maize mutant suggests that a single gene product is responsible for 4-vinyl reduction (Beale 1999), the possibility that the maize gene encodes a regulator for the divinyl reduction cannot be excluded (Nagata et al. 2005).

In this study, we demonstrate in vitro, the occurrence of NADPH-dependent membrane-bound 4VR activities in etiolated barley, namely: (a) [4-Vinyl] Mg-protoporphyrin reductase (4VMPR) that converts DV Mg-Proto to MV Mg-Proto; (b) [4Vinyl] Mg-protoporphyrin monoester reductase (4VMpeR) that converts DV Mpe to MV Mpe; (c) [4-Vinyl] protochlorophyllide *a* reductase (4VPideR) that converts DV Pchlide *a* to MV Pchlide *a*; and (d) [4-Vinyl] chlorophyllide *a* reductase (4VCR) that converts DV Chlide *a* to MV Chlide *a*. In contrast to previous reports (Ellsworth and Hsing 1973), and for the first the time, 4VMpeR activity is unambiguously demonstrated as a membrane-bound enzyme, In addition, the conversion of both DV Pchlide *a* and DV Chlide *a* to their MV counterparts is achieved in greening and green barley and cucumber. [4-Vinyl] chlorophyllide reductase remains the most potent 4VR activity in etioplast and chloroplast membranes.

II Materials and Methods

A Plant Material

Barley (*Hordeum vulgare*, Hi Barley Brand) seeds were purchased from Illini FS Inc. (Urbana, IL). Cucumber (*Cucumis sativus* var. Beit alpha) seeds were purchased from Hollar Seeds (Rocky Ford, CO). Germination was carried out in plastic trays containing wet vermiculate either in darkness or in a growth chamber illuminated with 1,000 W metal halide lamps (211 W m^{-2}) under a 14 h light/10 h dark photoperiod. Etiolated or green tissue was harvested after 5–7 days of growth at 28°C.

B Light Pretreatment

In some cases, etiolated barley seedlings were exposed to 4 h of illumination in the growth chamber prior to etiochloroplasts isolation.

C Chemicals

DV Mg-Proto was purchased from Porphyrin Products (Logan, UT) and the BCA protein assay kit from Pierce (Rockford, IL). NADH, NADPH, NADP, Nicotinamide, Chaps and 2,2′-dipyridyl (Dpy) were purchased from Sigma Chemical Co. (St. Louis, MO) and δ-aminolevulinic acid (ALA) was purchased from Biosynth International (Naperville, IL).

D Preparation of Divinyl Protochlorophyllide a

DV Pchlide *a* was prepared from etiolated cucumber cotyledons that were induced to accumulate exclusively DV Pchlide *a*, as described elsewhere (Duggan and Rebeiz 1982a). DV Pchlide *a* was extracted as described previously (Kolossov and Rebeiz 2001).

E Preparation of Divinyl Chlorophyllide a

DV Pchlide *a*-enriched cotyledons were subjected to a 2.5 ms actinic white light flash that converted DV Pchlide *a* to DV Chlide *a* (Duggan and Rebeiz 1982a). This was followed by immediate extraction of DV Chlide *a* as described above for DV Pchlide *a*.

F Preparation of Divinyl Mg-Protoporphyrin Mono Methyl Ester

DV Mpe was prepared from 5-day-old etiolated cucumber cotyledons, incubated with ALA and Dpy in darkness (Tripathy and Rebeiz 1986). Etiolated cotyledons were excised with hypocotyl hooks and were incubated in darkness at 28°C for 20 h, in deep Petri dishes (9 × 9 cm). Each Petri dish contained 5 g of cotyledons, 10 ml of an aqueous solution containing 4.5 mM ALA, 3.7 mM Dpy, 228 mM methanol and was adjusted to pH 6.2. DV Mpe was extracted as described for DV Pchlide *a*. Separation of DV Mpe from DV Mg-Proto and MV from DV Pchlide *a* was achieved by chromatography on thin layers of Silica gel H as described elsewhere (Rebeiz 2002).

G Isolation of Crude and Purified Plastids

Crude and Percoll-purified plastids were isolated from green cucumber, etiolated, greening and green barley tissue. Fifty gram batches of barley seedlings or cucumber cotyledons were homogenized in a pre-chilled Waring Blender in 250–300 ml of homogenization buffer. The later consisted of 500 mM sucrose, 15 mM Hepes, 10 mM Tes, 1 mM EDTA, and 1 mM $MgCl_2$ adjusted with KOH to pH 7.7 at room temperature.

Some isolation of crude barley chloroplasts were achieved by hand-ground homogenization of 6 g batches in pre-cooled mortars. Plastids were prepared essentially as described elsewhere (Kolossov and Rebeiz 2001). Both crude and purified plastids were used for 4VR activity assays.

H Preparation of Plastid Membranes and Stroma

Plastid membranes were prepared from crude and Percoll-purified organelles as described elsewhere (Kolossov and Rebeiz 2001). Supernatants containing stroma were used immediately as indicated in specific experiments.

I Preparation of Envelope Membranes

Envelope membranes from Percoll-purified etiochloroplasts were prepared according to Douce et al. (1973). Etiochloroplast pellets corresponding to 130 g of barley tissue were suspended in 1 ml of homogenization buffer. Plastids were osmotically lysed by adding 20 ml of swelling buffer consisting of 10 mM Tricine-NaOH, 4 mM $MgCl_2$, pH 7.6. Aliquots (11 ml) of the suspension were layered on top of a discontinuous sucrose gradients and centrifuged according to Douce et al. (1973). Various layers were collected in 6 ml volumes.

J Solubilization of [4-Vinyl] Reductase(s) by 3-[(3-Cholamidopropyl) dimethylammonio]-1-Propanesulfonate

Membranes from Percoll-purified and crude plastids were resuspended in solubilization buffer at a rate of 2 mg protein per ml. The solubilization buffer consisted of 20 mM Tris-HCl, 1 mM EDTA, 10% of Glycerol and 4 mM 3-[(3-Cholamidopropyl) dimethylammonio]-1-Propanesulfonate (Chaps) adjusted to a pH of 7.7 at room temperature. The remaining steps were performed as described elsewhere (Kolossov and Rebeiz 2001).

K Assay of [4-Vinyl] Reductase Activities

All reactions excluding the 4VCR assay were carried out under a green safelight that did not photoconvert Pchlide *a* to Chlide *a*. 4VCR activity was determined as described by Kolossov and Rebeiz (2001) with the exclusion of 20 mM citrate monohydrate. 4VPideR and 4VMpeR activities were determined in the same buffer with the following adjustments: (a) the sample amount was increased four- to 15-fold i.e. up to 200 μl per assay; (b) the assay was allowed to proceed for 60 min, (c) DV Pchlide and DV Mpe were added to a final concentration of 0.3 μM. The buffer for the 4VMPR assay was decreased twofold by dilution with distilled water to facilitate better extraction of the Mg-Proto pool from the hexane-extracted acetone residue (HEAR) preparation. The amount of loaded DV Mg-Proto was the same as for DV Mpe. Incubations were terminated by precipitation with 7 ml of acetone:0.1N NH_4OH.

L Protein Determination

Protein content was determined by BCA (Smith et al. 1985).

M Extraction and Determination of the Amounts of Divinyl and Monovinyl Tetrapyrroles

Tetrapyrrole determination by room temperature and 77 K spectrofluorometry, were as described elsewhere (Kolossov and Rebeiz, 2001; Rebeiz 2002). The only exception was the analysis of Mg-Proto. The pH of the HEAR was adjusted to 4.5 just before Mg-Proto extraction by ether. Otherwise only minor amounts of the total Mg-Proto pool were extracted by ether.

III Results

A Experimental Strategy

Kim et al. (1997) reported that an interaction of plastid membranes, stroma, and NADPH was involved in the regulation of DV and MV Pchlide *a* biosynthesis. Although the combination of plastid stroma and plastid membranes resulted in higher total Pchlide *a* formation from exogenous DV Mg-Proto, the presence of stroma in the reaction mixture resulted in a five- to 16-fold decrease in the MV/DV Pchlide *a* ratio. This in turn suggested an inhibition of 4VR reactions between DV Mg-Proto and Pchlide *a* by the plastid stroma. It was therefore conjectured that the study of 4VR activities in the absence of the plastid stroma, i.e. in isolated plastid membranes, may give a deeper insight into 4VR activities and may unmask some additional undetected 4VR activities. As a consequence efforts were made to solubilize 4VR activities from the plastid membranes in order to perform preliminary comparative studies.

B Detection of [4-Vinyl]Protochlorophyllide a Reductase, [4-Vinyl]Mg-Protoporphyrin Monoester Reductase and [4-Vinyl]Mg-Protoporphyrin Reductase Activities in addition to [4-vinyl]Chlorophyllide a Reductase in Barley Etiochloroplast Membranes

Initial attempts at detecting 4VMg-ProtoR and 4VPideR activities in isolated etiochloroplast membranes under 4VCR incubation conditions (Kolossov and Rebeiz 2001) were unsuccessful. After further experimentation, and adjustment of incubation conditions, it became possible to detect 4VMg-ProtoR and 4VPideR activities in the same membrane preparation (Table 1). In addition it became possible to detect 4VMpeR activity. The latter was first reported by Ellsworth and Hsing in a supernatant of etiolated wheat leaves homogenates (Ellsworth and Hsing 1973), but was never confirmed by others (Rebeiz et al. 2003).

In order to detect 4VMg-ProtoR, 4VMpeR, and 4VPideR activities in the same preparation that exhibited 4VCR activity the following adjustments in incubation conditions were made to the 4VMg-ProtoR, 4VMpeR, and 4VPideR assays, (a) the incubation time for 4VMg-ProtoR, 4VMpeR, and 4VPideR was raised from 5 to 60 min, (b) the sample load was raised about tenfold; and, (c) the exogenous substrate concentration was lowered four to eight times. The incubation conditions for 4VCR were kept unchanged (Kolossov and Rebeiz 2001). All 4VR assays were carried out in the same buffer system. For 4VMPR however, the buffer was diluted 1:1 (v/v) with distilled water in order to facilitate the extraction of Mg-Proto into diethyl ether. The 4VR activities of isolated barley etioplast membranes exhibited the following order of activities: 4VCR >>> 4VPideR >> 4VMpeR ≥ 4VMg-ProtoR (Table 1).

C Solubilization of [4-Vinyl]Protochlorophyllide a Reductase, [4-Vinyl]Mg-Protoporphyrin Monoester Reductase and [4-Vinyl]Mg-Protoporphyrin Reductase Activities from Barley Etiochloroplast Membranes by Chaps

Since it was conjectured that a comparative study of various 4VR activities would be facilitated by solubilization of 4VR activities, efforts were made to solubilize all 4VR activities from barley etiochloroplast membranes. To this effect 4 mM Chaps was used. Such a concentration was successfully used in the past to solubilize 4VCR (Kolossov and Rebeiz 2001). Solubilization of the four 4VR activities was successfully achieved. As previously reported for 4VCR (Kolossov and Rebeiz 2001), solubilization dramatically increased the activities of all four 4VRs (Table 1).

D 4-Vinyl Side Chain Reduction Occurs Before Isocycle Ring Formation in Photoperiodically-Grown Barley

The detection of 4VMg-ProtoR and 4VMpeR in addition to 4VPideR activity suggested that vinyl reduction occurs at the level of DV Pchlide *a* as well as at the levels of DV Mg-Proto and DV Mpe. This notion has been questioned in a recent review that stated that the heterogeneity of intermediates prior to Pchlide *a* is doubtful because these intermediates are found only in trace amounts under natural conditions (Rudiger 2003).

Table 1. Detection of various 4VR activities in isolated plastid membranes and solubilized fractions prepared from etiolated barley leaves. Percoll-purified etiochloroplasts were isolated from barley seedlings under laboratory light. The membranes and Chaps-solubilized fractions were prepared as described in Materials and Methods. Values are means of two to three replicates ± standard deviation.

	Net change in			
	MV Mg-Proto in 60 min	MV Mpe in 60 min	MV Pchlide *a* in 60 min	MV Chlide *a* in 5 min
Membrane fraction	(pmol/mg protein)			
Membranes	29.2 ± 14.1	48.2 ± 19.7	243.8 ± 16.8	47,451 ± 345.1
Solubilized fraction	155.9 ± 30.6	235.9 ± 30.6	1,375.1± 132.6	92,211± 345.1
Solubilized fraction as % of Membrane activity	533.9	489.4	564.0	194.33
4VCR/4VR activities	591	391	67	1

To reconfirm the formation of MV Mg-Proto and MV Mpe in significant amounts in green tissues, the biosynthesis of these intermediates in photoperiodically grown barley leaves was induced as follows. Excised green barley leaves were incubated with ALA and Dpy in darkness for various periods of time. ALA acted as a source of substrate while Dpy partially inhibited the conversion of the Mg-tetrapyrroles to Pchlide *a* (Duggan and Gassman 1974), thus allowing the accumulation of measurable amounts of the Mg-intermediates. The formation of MV Mg-Proto + Mpe [Mp(e)] was unambiguous (Table 2). The Mp(e) pool consisted mainly of Mpe and smaller amounts of Mg-Proto (data not shown).

E [4-Vinyl] Chlorophyllide a Reductase and [4-Vinyl]Protochlorophyllide a Reductase Activities do not Occur in Barley Etiochloroplast Envelope Membranes

Next, distribution of 4VR activities between plastid envelopes and inner plastid membranes was investigated. To this effect, the two most potent 4VR activities, namely 4VPideR and 4VCR were monitored in various subplastidic membrane fractions. The envelope membranes were completely devoid of 4VR activity. All activity was detected in the inner membrane fractions (Table 3). 4VCR and 4VPideR activities were about equally distributed between fractions 3 and 4 which represented presumably prolamellar bodies and prothylakoid membranes respectively.

F [4-Vinyl] Protochlorophyllide a Reductase Activity Is Detectable in Greening Barley

It was previously reported that the 4VCR of barley etiochloroplasts lost about 40% of its activity after 4 h of illumination of etiolated barley seedlings (Kolossov and Rebeiz 2001). To determine whether 4VPideR activity responded in a similar manner to illumination as 4VCR, etiolated barley seedlings were illuminated for 4 h, etiochloroplasts were isolated, lysed and the plastid membranes were solubilized with 4 mM Chaps. 4VCR and 4VPideR activities were next determined. As shown in Table 4, 4VPideR activity appeared to be less sensitive to illumination than 4VCR, since after 4 h of illumination the ratio of 4VCR/4PideR activities dropped from 67 to 43 (Tables 1 and 4).

G NADPH, but Not NADH is a Cofactor for [4-Vinyl]Chlorophyllide Reductase and [4-Vinyl]Protochlorophyllide Reductase Solubilized from Etiochloroplast Membranes of Etiolated Barley Exposed to Light for 4 h

NADPH requirement for 4VCR activity has been already demonstrated for etioplast membranes of

Table 2. MV and DV tetrapyrrole accumulation in green barley leaves incubated for various periods of time in darkness with 4.5 mM ALA and 3.7 mM Dpy. Vales are means of two replicates ± standard deviation.

Time in darkness (h)	Proto	Mpe				Pchlide *a*			
	Total	Total	MV	DV	DV/MV	Total	MV	DV	DV/MV
	(nmol/g fresh weight or DV/MV ratio)								
0	54 ± 41	119± 37	20	99	5	277 ± 99	222	55	0.25
2	7,235 ± 2086	3235 ± 78	125	3,110	25	13,260 ± 806	10450	2,810	0.27
4	4,585 ± 1534	4,780 ± 608	370	4,410	12	16,520 ± 78	14,160	2,360	0.17
6	11,390 ± 1,800	7,270± 161	820	6,450	9	19,580 ± 240	17,470	2,110	0.11
8	11,260 ± 410	11,150 ± 1,620	1,850	9,300	5	22,770 ± 2,880	20,900	1,870	0.09

Table 3. Intraplastidic localization of 4VPideR and 4VCR activities. Percoll-purified etiochloroplasts were prepared from etiolated barley seedlings under laboratory light. Envelope and inner membranes were prepared as described in Materials and Methods. Nd = not determined.

Fraction	Membrane type	Net change in MV Pchlide *a* in 60 min	Net change in MV Chlide *a* in 5 min	4VCR/4VpideR activities
		(pmol/mg protein		
2	Envelopes	0	0	–
3	Inner membranes	512.8	25,445	49.6
4	Inner membranes	405.3	31,182	76.9
	Pellet	nd	3,850	–

cucumber, barley and corn [Parham and Rebeiz 1992]. Here it is shown that solubilized 4PideR activity also exhibited an absolute requirement for NADPH in etiolated barley seedlings exposed to 4 h of illumination (Table 4).

H The Presence of NADP or Vitamin B_3 in the Incubation Buffer Has No Effect on the Activities of [4-Vinyl]Chlorophyllide a Reductase and [4-Vinyl]Protochlorophyllide a Reductase of Greening Barley

It was reported by Shioi et al. (1988), that vitamin B3 inhibited the conversion of DV Pchlide *a* to MV Pchlide *a*, while Whyte and Griffiths (1993) reported the inhibition of DV Chlide *a* conversion to MV Chlide *a* by NADP. The effects of these cofactors on the activities of 4VPideR and 4VCR solubilized from greening barely etiochloroplasts are described in Table 5. In both cases, no inhibition of 4VR activities was observed.

I Demonstration of [4-Vinyl] Protochlorophyllide a Reductase and [4-Vinyl] Chlorophyllide a Reductase Activities in Barley Chloroplast Membranes

It was previously assumed that 4VCR activity disappears or decreases to undetectable levels in photoperiodically grown plants (Abd-El-Mageed et al. 1997). That conclusion was based on experimentation involving isolated chloroplasts having a full complement of Chl. It has now become apparent that in addition to 4VR inhibition by the plastid stroma, the high concentration of Chl interfered with the 4VR spectrofluorometric assays. Upon solubilization of the 4VR activities from chloroplast membranes, the stromal inhibition was relieved, most of the Chl was left behind in the membranes, and the 4PideR and 4VCR activities became unmasked as reported in Table 6.

Table 4. Absolute dependence of 4VPideR and 4VCR activities upon NADPH, and response to illumination. Etiolated barley seedlings were illuminated for 4 h with metal halide light (211 W m^{-2}) prior to etiochloroplast isolation. 4VPideR and 4VCR activities were assayed as described under Materials and Methods by incubation of the Chaps-solubilized fraction with appropriate substrates in the presence and absence of NADPH and NADH. Vales are means of two replicates ± standard deviation.

	Net change in MV Pchlide *a* in 60 min	Net change in MV Chlide *a* in 5 min	4VCR/4VpideR activities
Cofactor	(pmol/mg protein) or 4VCR/4VPideR ratio		
None	0	0	0
0.5 mM NADPH	270 ± 21	11,607 ± 198	43
1 mM NADH	0	0	0

Table 5. 4VPideR and 4VCR are not inhibited by NADP and nicotinamide. Solubilized fractions were prepared from plastid membranes of crude etiochloroplasts prepared from etiolated barley leaves exposed to light for 4 h

		Net change in MV Pchlide *a* In 60 min	Net change in MV Chlide *a* in 5 min
Experiment	Additional cofactor	(pmol/mg protein)	
A	None		11,607
	5 mM NADP		11,102
	5 mM Nicotinamide		11,321
B	None	250.5	
	5 mM NADP	222.3	
	5 mM Nicotinamide	244.6	

Table 6. Detection of 4VPideR and 4VCR activities in green barley. Membrane fractions were prepared from Percoll-purified chloroplasts. Values are means of two replicates ± standard deviation. nd, not determined

	Net change in MV Pchlide *a* in 60 min	Net change in MV Chlide *a* in 5 min
Plastid fraction	(pmol/mg protein)	
Inner membranes	nd	1,682 ± 154
Solubilized fraction	191 ± 80	7,650 ± 884
Inner membranes/solubilized fractions	0	0.22

Table 7. Effects of various light treatments on 4VCR activity of barley membranes prepared from crude plastids. Values are means of two replicates ± standard deviation.

	Length of illumination (h)			
	0	2	4	Photoperiodic illumination
	(nmol MV Chlide *a* formed in 5 min/mg protein)			
Development stage	Etioplasts	Etiochloroplasts	Etiochloroplasts	Chloroplasts
4VCR activity	14.9 ± 0.4	10.0 ± 0.03	9.2 ± 0.3	1.68 ± 0.15
Percentage of 0 h illumination	100	67	62	11

Table 8. Effect of the phase of the photoperiod on 4VCR activity in isolated chloroplast membranes. Plastid membranes were prepared from percoll-purified barley and cucumber chloroplasts. Values are means of 2 replicates ± standard deviation.

Phase of the photoperiod	Barley	Cucumber
	Net change in MV Chlide *a* in 5 min	
	(pmoles/mg protein)	
End of the dark phase	3604 ± 148	477 ± 61
Middle of the light phase	4166 ± 348	515 ± 127
Dark phase/Light phase	0.87	0.93

J Effects of Various Light Treatments on [4-Vinyl] Clorophyllide a Reductase Activity

It was reported elsewhere that Illumination reduced the activity of 4VCR (Kolossov and Rebeiz 2001). However, the effects of various light treatments on solubilized 4VCR activity were not reported.

In a first set of experiments the effects of various lengths of illumination of etiolated barley seedlings on solubilized 4VCR were compared to the effects of photoperiodic growth. It became obvious that 4VCR activity was highest in solubilized preparations from etiolated tissues and steadily decreased as greening progressed (Table 7). However although 4VCR activity dropped about tenfold in photoperiodically grown tissues, it never disappeared completely from green tissues and remained there at a steady-state level (Table 7).

In a second set of experiments, the effects of the phase of the photoperiod on solubilized 4VCR activity of photoperiodically-grown barley and cucumber were compared. Barley and cucumber belong to two different greening plant groups, specifically dark-monovinyl-light-divinyl-light-dark-monovinyl (DMV-LDV-LDMV) and dark-divinyl-light-divinyl-light-dark-divinyl (DDV-LDV-LDDV) respectively (Abd-El-Mageed et al. 1997). In daytime and at night, barley and cucumber form Pchlide *a* mainly via MV and DV biosynthetic routes respectively. At that time, and on the basis of available experimental evidence, it was proposed that in photoperiodically grown green tissues, 4VCR was active at the end of the dark phase, and at the beginning of the light phase of the photoperiod only in DDV-LDV-LDDV but not in DMV-LDV-LDMV plants species (Rebeiz et al. 2003). It is shown here that this hypothesis has to be reconsidered since 4VCR was as active at the end of the dark phase as in the middle of the light phase of the photoperiod, in DMV-LDV-LDMV plants such as barley, as well as in DDV-LDV-LDDV plants such as cucumber (Table 8). The difference in 4VCR activities between barley and cucumber reflects mostly the higher amount of proteins solubilized from cucumber plastid membranes. Indeed, total 4VCR activity per gram of fresh tissue was similar for both plant species.

IV Discussion

In the absence of plastid stroma, 4VPideR, 4VMpeR and 4VMg-ProtoR activities, in addition to 4VCR activity, became readily measurable. This became possible because the regulatory-inhibitory effects of the plastid stroma (Kim et al. 1997) were bypassed (Table 1). Treatment with Chaps solubilized all four 4VRs activities and enhanced considerably their in vitro activities (Table 1). All efforts to detect 4V-Proto reductase activity were unsuccessful (data not shown).

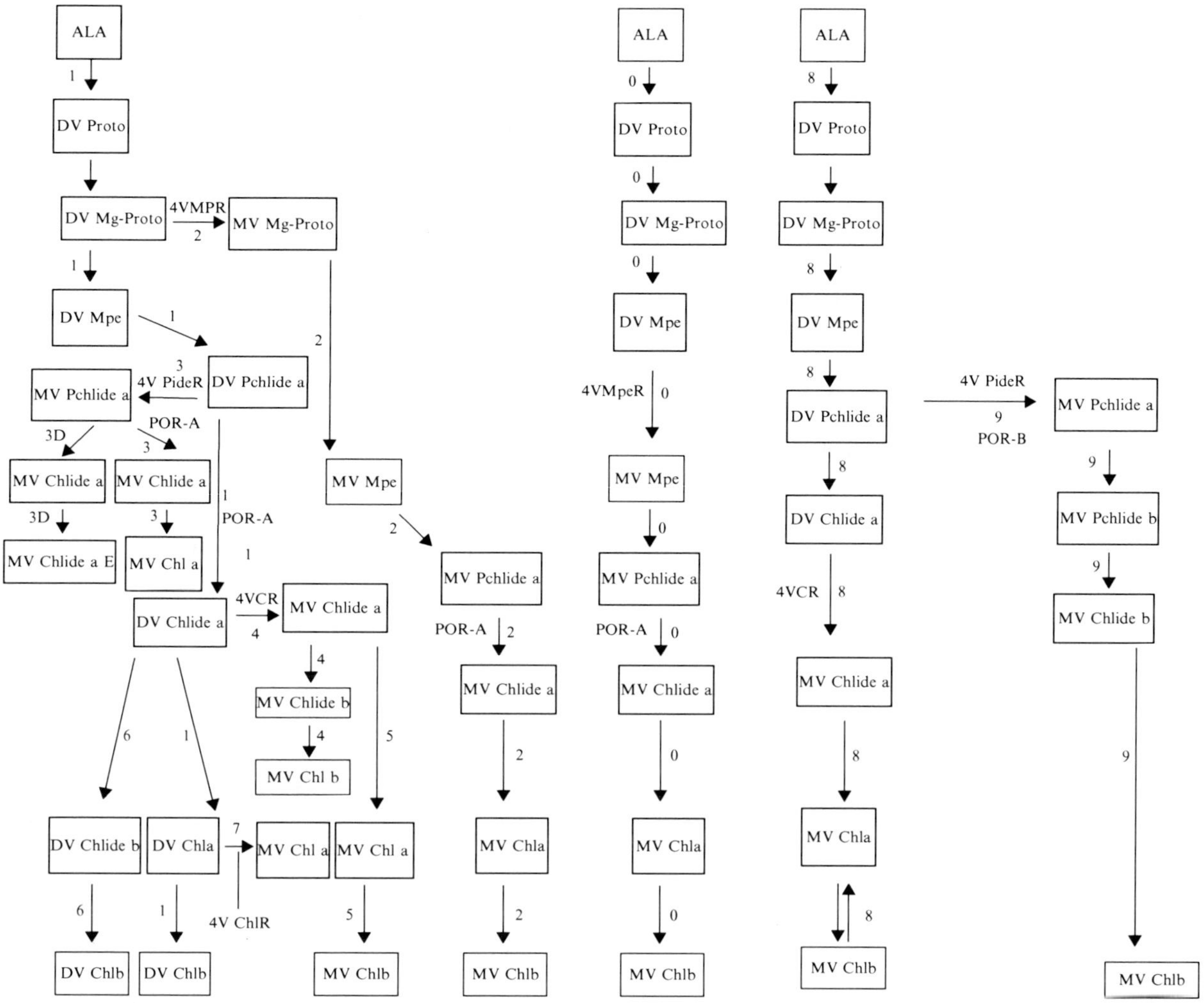

Fig. 2. Monocarboxylic divinyl Chl *a* and *b* biosynthetic route of DDV-LDV-DLDV plants adapted from Rebeiz et al. (2003). DV = divinyl; MV = monovinyl; ALA = δ-aminolevulinic acid; Proto = protoporphyrin IX; Mpe = Mg-Proto monomethyl ester; Pchlide = protochlorophyllide; Chlide = chlorophyllide; Chl = chlorophyll; 4VMPR: [4-vinyl] Mg-Proto reductase; 4VMpeR = [4-vinyl] Mg-Proto monoester reductase; 4VPideR = [4-vinyl] protochlorophyllide *a* reductase; 4VCR = [4-vinyl] chlorophyllide *a* reductase; 4VChlR = [4-vinyl] Chl reductase; POR = Pchlide *a* oxidoreductase. Arrows joining DV and MV routes refer to reactions catalyzed by [4-vinyl] reductases. The new biosynthetic route labelled 0 was called for by the discovery of 4VMpeR. All other routes are designated by Arabic numerals as described in Rebeiz et al. (2003).

Detection of 4VR activities at every step between Mg-Proto and Chlide *a* (Table 1) is compatible with the reported detection of MV Mg-Proto and MV Mpe in etiolated tissues (Belanger and Rebeiz 1982). It justifies extending the DV-MV Chl biosynthetic heterogeneity to the level of DV Mg-Proto as described in (Kolossov et al. 2003; Rebeiz et al. 2003).

The following information suggests that the various 4VR activities are catalyzed by different enzymes; (a) the different 4VR activities exhibited different activations following solubilization (Table 1), (b) the ratio of 4VCR/4VR activity was different in variously purified fractions (Table 3), and (c) 4VPideR activity appeared to be less sensitive to illumination than 4VCR (Tables 1 and 4). Further investigations of solubilized 4VR activities focused on a comparison of 4VPideR and 4VCR properties. It was readily demonstrable that both activities were localized in the inner plastid membranes and were absent from the plastid envelope of etiolated barley (Table 3). The

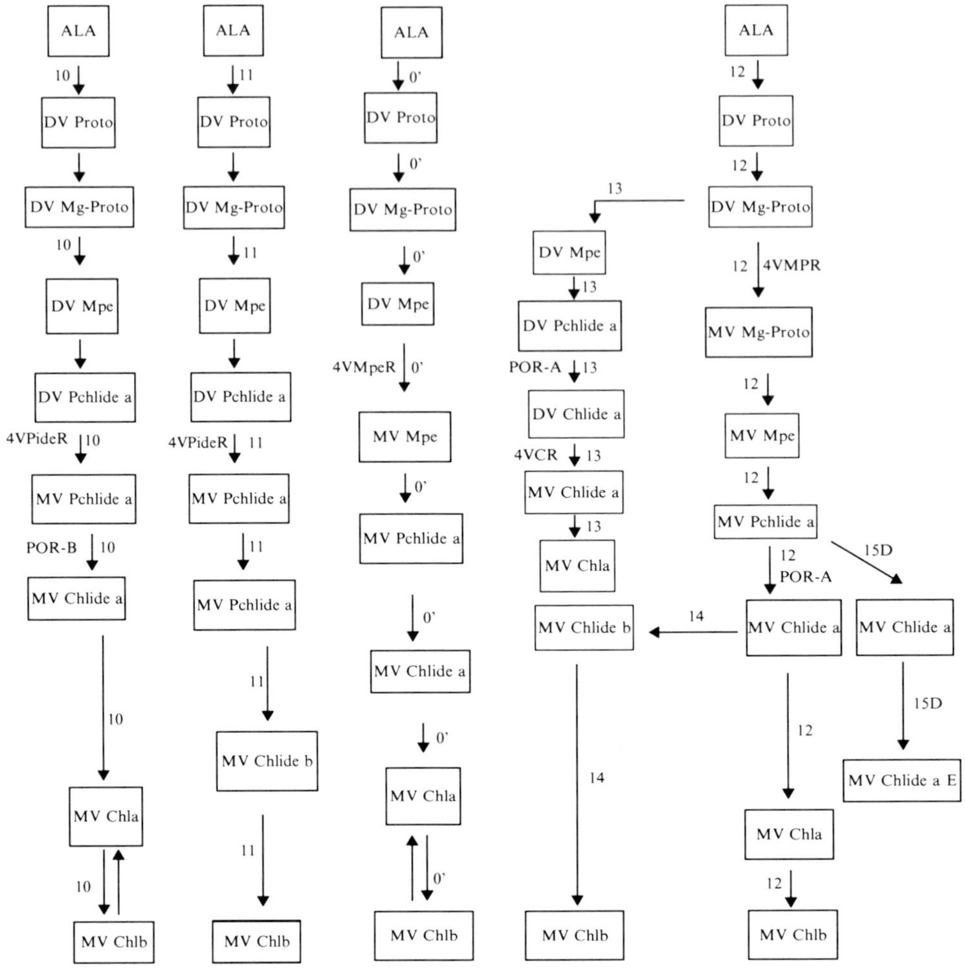

Fig. 3. Monocarboxylic monovinyl Chl *a* and *b* biosynthetic route of DMV-LDV-DLMV plants adapted from Rebeiz et al. (2003). The new biosynthetic routes labelled 0' and 14 were called for by the discovery of 4VMpeR and 4VCR activity in barley chloroplasts. All other routes are numbered as described in Rebeiz et al. (2003). D = reaction occurring in darkness. All abbreviations are as in Fig. 2.

significance of this observation to the overall biosynthetic heterogeneity of the Chl biosynthetic pathway is presently unclear.

It has been reported elsewhere that 4VCR had an absolute requirement for NADPH (Parham and Rebeiz 1992). Likewise 4VPideR exhibited a similar requirement for NADPH (Table 4). However contrary to previous assertions by Shioi et al. (1988), and Whyte and Griffiths (1993) 4VPideR and 4VCR were not inhibited either by NADP or by Nicotinamide (Table 5).

The differential response of 4VPideR and 4VCR to illumination resulted in interesting observations. Solubilization of 4VCR made it possible to detect for the first time its activity in green barley leaves (Table 6). It was previously assumed that 4VCR was not active in green DMV-LDV-LDMV plant species such as barley, but was active in green DDV-LDV-LDDV plant species such as cucumber (Rebeiz et al. 2003). This led to the omission of biosynthetic routes involving 4VCR in green DMV-LDV-LDMV plant species in the light (Rebeiz et al. 2003). In this work it was clearly demonstrated that 4VCR was as active in green barley, a DMV-LDV-LDMV plant species, as in green cucumber in darkness, at the end of the dark phase of the photoperiod, in the light, and in the middle of the light phase of the photoperiod (Table 8). This observation called for a revision of the carboxylic Chl biosynthetic routes in DMV-LDV-LDMV plant species reported by Rebeiz et al. (2003). In order to accomodate all new observations including the detection of 4VMpeR activity, the Chl carboxylic biosynthetic pathway was split into two parts (Figs. 2 and 3). Figures 2 and 3 depict the reactions that occur in DDV-LDV-LDDV and DMV-LDV-LDMV plants (Abd-El-Mageed et al., 1997). The various biosynthetic routes depicted in the two figures were discussed in Rebeiz et al. (2003) as well as in this chapter. The new biosynthetic routes are given the numerical 0, and 0′ and 14.

Acknowledgment

The writing of this chapter was supported by the Rebeiz Foundation for Basic Research

References

Abd-El-Mageed HA, El Sahhar KF, Robertson KR, Parham R and Rebeiz CA (1997) Chloroplast Biogenesis 77. Two novel monovinyl and divinyl light-dark greening groups of plants and their relationship to the chlorophyll *a* biosynthetic heterogeneity of green plants. Photochem Photobiol 66: 89–96

Adra AN and Rebeiz CA (1998) Chloroplast Biogenesis 81. Transient formation of divinyl chlorophyll *a* following a 2.5 ms light flash treatment of etiolated cucumber cotyledons. Photochem Photobiol 68: 852–856

Bazzaz MB (1981) New chlorophyll chromophores isolated from a chlorophyll deficient mutant of maize. Photobiochem Photobiophys 2: 199–207

Beale SI (1999) Enzymes of chlorophyll biosynthesis. Photosynth Res 60: 43–73

Belanger F C and Rebeiz C A (1982) Chloroplast Biogenesis: Detection of monovinyl magnesium protoporphyrin monoester and other monovinyl magnesium porphyrins in higher plants. J Biol Chem 257: 1360–1371

Carey EE and Rebeiz CA (1985) Chloroplast Biogenesis 49. Difference among angiosperms in the biosynthesis and accumulation of monovinyl and divinyl protochlorophyllide during photoperiodic greening. Plant Physiol 79: 1–6

Carey EE, Tripathy BC and Rebeiz CA (1985) Chloroplast biogenesis 51. Modulation of monovinyl and divinyl protochlorophyllide biosynthesis by light and darkness *in vitro*. Plant Physiol 79: 1059–1063

Chew AGM and Bryant DA (2007) Characterization of a plant-like protochlorophyllide *a* divinyl reductase in green sulfur bacteria. J Biol Chem 282 : 2967–2975

Chisholm SW, Frankel S, Goerike R, Olson R, Palenic R, Urbach B, Waterbury JB and Zettler ER (1992) *Prochlorococcus marinus* nov. gen. sp.: an oxyphototrophic marine prokaryote containing divinyl chlorophyll *a* and *b*. Arch Mikrobiol 157: 297–300

Douce R, Holtz B and Benson AA (1973) Isolation and properties of the envelope of spinach chloroplast. J Biol Chem 258: 7215–7222

Duggan JX and Gassman M (1974) Induction of porphyrin biosynthesis in etiolated bean leaves by chelators of iron. Plant Physiol 53: 206–215

Duggan JX and Rebeiz CA (1982a) Chloroplast Biogenesis 37: induction of chlorophyllide *a* (E459F675) accumulation in higher plants. Plant Sci Lett 24: 27–37

Duggan JX and Rebeiz CA (1982b) Chloroplast Biogenesis 42. Conversion of DV chlorophyllide *a* to monovinyl chlorophyllide *a in vivo* and *in vitro*. Plant Sci. Lett 27: 137–145

Ellsworth RK and Hsing AS (1973) The reduction of vinyl-side chains of Mg-protoporphyrin IX monomethyl ester in vitro. Photosynthetica 313: 119–129

Goerike R and Repeta D (1992) The pigments of *Prochlorococcus marinus*. The presence of divinyl-chlorophyll *a* and *b* in a marine prochlorophyte. Limnol Oceanogr 37: 425–433

Granick S (1950) Magnesium vinyl pheoporphyrin a_5 another intermediate in the biological synthesis of chlorophyll. J Biol Chem 183: 713–730

Jones OTG (1963) Magnesium 2,4-divinyl phaeoporphyrin a_5 monomethyl ester, a protochlorophyll-like pigment produced by *Rhodopseudomonas spheroides*. Biochem J 89: 182–189

Kim JS and Rebeiz CA (1996) Origin of chlorophyll *a* biosynthetic heterogeneity in higher plants. J Biochem Mol Biol 29: 327–334

Kim JS, Kolossov V and Rebeiz CA (1997) Chloroplast Biogenesis 76: Regulation of 4-vinyl reduction during conversion of divinyl Mg-protoporphyrin IX to monovinyl protochlorophyllide *a* is controlled by plastid membrane and stromal factors. Photosynthetica 34: 569–581

Kolossov VL and Rebeiz CA (2001) Chloroplast Biogenesis 84. Solubilization and partial purification of membrane-bound [4-vinyl] chlorophyllide *a* reductase from etiolated barley leaves. Anal Biochem 295: 214–219

Kolossov VL, Kopetz KJ and Rebeiz CA (2003) Chloroplast Biogenesis 87: evidence of resonance excitation energy transfer between tetrapyrrole intermediates of the chlorophyll biosynthetic pathway and chlorophyll *a*. Photochem Photobiol 78: 184–196

Nagata N, Tanaka R, Satoh S and Tanaka A (2005) Identification of a vinyl reductase gene for chlorophyll synthesis in *Arabidopsis thaliana* and implications for the evolution of prochlorococcus species. Plant Cell 17: 233–240

Parham R and Rebeiz CA (1992) Chloroplast Biogenesis: [4-vinyl] chlorophyllide *a* reductase is a divinyl chlorophyllide *a*-specific NADPH-dependent enzyme. Biochemistry 31: 8460–8464

Parham R and Rebeiz CA (1995) Chloroplast Biogenesis 72: a [4-vinyl] chlorophyllide *a* reductase assay using divinyl chlorophyllide *a* as an exogenous substrate. Anal Biochem 231: 164–169

Rebeiz CA (2002) Analysis of intermediates and end products of the chlorophyll biosynthetic pathway. In: Smith A and Witty M (eds) Heme Chlorophyll and Bilins, Methods and Protocols. Humana Press, Totowa, NJ, pp 111–155

Rebeiz CA, Parham R, Fasoula DA and Ioannides IM (1994) Chlorophyll biosynthetic heterogeneity. In: Chadwick DJ and Ackrill K (eds) The Biosynthesis of the Terapyrrole Pigments. Wiley, New York, pp. 177–193

Rebeiz CA, Kolossov VL, Briskin D and Gawienowski M (2003) Chloroplast Biogenesis: chlorophyll biosynthetic heterogeneity , multiple biosynthetic routes and biological spin-offs. In: Nalwa HS (ed) Handbook of Photochemistry and Photobiology. American Scientific Publishers, Los Angeles, CA, pp. 183–248

Rebeiz CA, Wu SM, Kuhadje M, Daniell H and Perkins EJ (1983) Chlorophyll *a* biosynthetic routes and chlorophyll *a* chemical heterogeneity. Mol Cell Biochem 58: 97–125

Rebeiz M, Pool JE, Kassner VA, Aquadro CF and Carroll SB (2010) Stepwise modification of a molecular enhancer underlies adaptation in a *Drosophila* population. Science 326: 1663–1667

Richards WR (1993) Biosynthesis of the chlorophyll chromophore of pigmented thylakoid proteins. In: Sundqvist C Ryberg M (eds) Pigment–Protein Complexes in Plastids: Synthesis and Assembly, Academic, New York, pp. 91–178

Rudiger W (2003) The last steps of chlorophyll synthesis. In: Kadish KM, Smith KM (eds) Chlorophylls and Bilins: Biosynthesis, Synthesis, and Degradation. Elsevier, New York, pp. 71–108

Shioi Y and Sasa T (1983) Compositional heterogeneity of protochlorophyllide ester in etiolated leaves of higher plants. Arch Biochem Biophys 220: 286–292

Shioi Y, Doi M and Bodi B (1988) Selective inhibition of Chlorophyll biosynthesis by nicotinbamide. Arc Biochem Biophys 267: 69–74

Smith PK, Krohn RI, Hremanson GT, Mallia AK, Gardner FH, Provenzano EK, Fujimoto EK, Goeke NM and Klenk DC (1985) Measurement of protein using bicinchonic acid. Anal Biochem 150: 76–85

Sundqvist C and Ryberg M (eds) (1993) Pigment–Protein Complexes in Plastids: Synthesis and Assembly. Academic, New York

Suzuki JY and Bauer CE (1995) Altered monovinyl and divinyl protochlorophyllide pools in bchJ mutants of Rhodobacter capsulatus. Possible monovinyl substrate discrimination of light-independent protochlorophyllide reductase. J Biol Chem 270: 3732–3740

Tripathy BC and Rebeiz CA (1986) Chloroplast Biogenesis. Demonstration of the monovinyl and divinyl monocarboxylic routes of chlorophyll biosynthesis in higher plants. J Biol Chem 261: 13556–13564

Tripathy BC and Rebeiz CA (1988) Chloroplast Biogenesis 60. Conversion of divinyl protochlorophyllide to monovinyl protochlorophyllide in green(ing) barley, a dark monovinyl/light divinyl plant species. Plant Physiol 87: 89–94

Wang RL, Stec A, Hey J, Lukens L and Doebley J (1999) The limits of selection during maize domestication. Nature 398: 236–239

Whyte BJ and Griffiths TV (1993) 8-Vinyl reduction and chlorophyll a biosynthesis in higher plants. Biochem J 291: 939–944

Chapter 3

Control of the Metabolic Flow in Tetrapyrrole Biosynthesis: Regulation of Expression and Activity of Enzymes in the Mg Branch of Tetrapyrrole Biosynthesis

Bernhard Grimm
Institute of Biology/Plant Physiology, Humboldt University, Philippstr. 13, Building 12, 10115 Berlin, Germany

Summary

The complex regulation, metabolism and physiology of the Mg branch of tetrapyrrole biosynthesis have developed into an attractive research area. Any change in plant development as well as in growth and environmental conditions provokes changes in metabolic activities of chlorophyll synthesis including de novo synthesis of proteins and cofactors, as well as protein modification and degradation. Transcriptional and different posttranslational control mechanisms have been reported for chlorophyll synthesis, which underscore the need for a very dynamic and flexible regulatory system. In the following paragraphs the enzymes of the Mg branch and their unique catalytic reactions are introduced. The control of expression and posttranslational modification of enzymes, the association of enzymes with cofactors and other compounds and the assembly into protein complexes as well as the activation mechanisms of these enzymes will be surveyed. Furthermore, the review discusses particular regulatory incentives originating from the Mg branch to be advantageous for the entire tetrapyrrole biosynthetic pathway and, hence, for chloroplast development. It also comments on the significant gaps in our understanding of the regulatory mechanisms of the Mg branch of tetrapyrrole biosynthesis.

Author for Correspondence, e-mail: bernhard.grimm@rz.hu-berlin.de

Constantin A. Rebeiz et al. (eds.), The Chloroplast: Basics and Applications, pp. 39–53.

I Introduction

Among the four tetrapyrrole end products protoheme, siroheme, phytochromobiline and chlorophyll, the latter molecule accumulates most in plants. Around 10^9 t of chlorophyll per year are produced and degraded from marine and terrestrial cyanobacteria, algae and plants (Hendry et al., 1987). Chlorophyll is the essential light absorbing pigment for photosynthesis. The indispensable functions of chlorophyll in photosynthesis are (i) light harvesting in the photosynthetic antenna complexes, (ii) energy transfer from the antenna to the reaction centres of photosystem I and II, in which (iii) charge separation take place by oxidizing the final chlorophyll and transferring electrons to acceptors of the electron transport chain.

These chlorophyll-related functions of light driven energy conversion in the photosynthetic complexes are facilitated by the simultaneous availability of chlorophyll-binding proteins, the two types of pigments, chlorophyll and carotenoids and specific lipid species in the lipid bilayer of the thylakoid membranes (Eichacker et al., 1996; Plumley and Schmidt, 1987). All components most likely have to be present in stoichiometric amounts to be assembled into functional pigment–protein complexes. These partly assisted and partly self-organized processes enable stabilisation of these protein complexes in the thylakoid membranes (Horn et al., 2007, Melkozernov et al., 2006; Saenger et al., 2002). When any of the constituents of a pigment–protein complex is not available or pigments or lipids are lacking, the protein component is degraded. Consequently, biosynthesis of chlorophyll is tightly synchronized with the formation of the proteineous components of the pigment–protein-complexes, which are encoded either in the plastid or the nuclear genome (further details are given in Chapters 15 and 16).

In the biosynthetic pathway of tetrapyrroles 5-aminolevulinic acid (ALA) serves as the first committed precursor for all tetrapyrrole end-products. Eight molecules of ALA are finally assembled into an instable linear tetrapyrrole intermediate, hydroxymethylbilane, before the first cyclic tetrapyrrole, uroporphyrinogen III is enzymatically made. From ALA up to the synthesis of uroporphyrinogen III the linear pathway is identical in all organisms. At this step the tetrapyrrole synthesising pathway bifurcates in plants into the siroheme synthesising branch and the porphyrin branch. At the protoporphyrinogen biosynthesis step, the molecule is oxidized to fluorescent protoporphyrin IX, in which magnesium or iron cations are inserted leading to Mg protoporphyrin and protoheme, respectively. The Mg branch leads to the formation of chlorophyll via a complex biosynthetic pathway, as described in Chapters 1 and 2 of this book.

In the following paragraphs the enzymes of the Mg branch are described and the unique enzymatic reactions are briefly introduced. The subsequent paragraphs encompass recent results on the control of expression and posttranslational modifications of these enzymes. Their regulatory roles in tetrapyrrole biosynthesis and chloroplast development and functionality are also surveyed. I would finally like to point out that tetrapyrrole biosynthesis has been reviewed on several occasions. I intend in this chapter to consider more recent publications in the field of Mg porphyrin metabolism and refer the reader to other reviews dealing with the same issue, in which the entire pathway have been surveyed (Eckhardt et al., 2004; Moulin and Smith, 2005; Papenbrock and Grimm, 2001; Rebeiz et al., 2003, Tanaka and Tanaka, 2007; and several chapters of Vol. 25 of the series "Chlorophylls and Bacteriochlorophylls: Biochemistry, Biophysics, Functions and Applications, edited by Grimm et al., 2006). The formation of chlorophyll *a* and *b* in the "chlorophyll cycle" is described by Tanaka and Tanaka in Chapter 4 of this book.

II Mg Protoporphyrin IX Chelatase

A Structure and Catalytic Activity

The first committed step in the Mg branch is the insertion of the magnesium ion into protoporphyrin IX by magnesium protoporphyrin IX chelatase (Mg chelatase). This enzymatic step channels protoporphyrin into the chlorophyll-synthesizing pathway. The ferrochelatase formally competes for the same substrate and chelates ferrous ion into protoporphyrin IX to form protoheme. But control

Abbreviations: ALA – 5-aminolevulinic acid; MTF –S-Adenosyl-L-methionine:Mg protoporphyrin IX methyltransferase

of expression and activity as well as location of the two different chelatases in different subcompartments in the plastids contributes to well-adjusted distribution of protoporphyrin IX for chlorophyll and heme biosynthesis (see below).

The structure, enzymatic and regulatory characteristics of Mg chelatase, makes it a very special enzyme. It is present in bacteriochlorophyll *a*-producing prokaryotes and in chlorophyll-producing cyanobacteria and higher plants and consists of three different subunits: BchI/ChlI (38–46 kDa), BchD/ChlD (60–87 kDa), and BchH/ChlH (140–150 kDa) (Gibson et al., 1995, 1996; Jensen et al., 1996a, b; Papenbrock et al., 1997). With the exception of *Euglena gracilis* and certain red algae, which contain the *ChlI* homologous gene in the chloroplast genome, all other photosynthetically active eukaryotes possess nuclear-encoded genes for the three Mg chelatase subunits.

Magnesium chelatase activity has been demonstrated from dicots and monocots isolated chloroplasts and in vitro by combining the recombinant individual subunits from *Rhodobacter, Synechocystis* PCC 6803 and *Chlorobium vibrioforme.* For a detailed survey on Mg chelatase the following reviews are recommended: Walker and Willows, 1997; Willows, 2003. A stoichiometry of 4 BchI:1 BchD and 2 CHLI:1 CHLD:4 CHLH was found to be optimum in in vitro reconstitution assays with the three bacterial recombinant subunits (Willows et al., 1996; Hansson et al., 1999; Jensen et al., 1999). Generally, reconstitution of activity required all three subunits, although experiments with truncated CHLD and a fusion of N-terminal CHLI half and C-terminal CHLD half restored partial activity (Gräfe et al., 1999).

The current ideas about the reaction mechanism of Mg chelation suggest various steps, in which the subunits interact successively for an intact Mg chelatase complex. Initially, an ATP-dependent double-ring structure is formed consisting of six CHLI and CHLD subunits (Walker and Willows, 1997; Fodje et al., 2001). Then this stacked double hexameric ring forms a ternary structure with the CHLH subunit, which inserts ATP-dependently Mg^{2+} into the ring structure of bound protoporphyrin IX. Therefore, it is hypothesized that ATP hydrolysis by the CHLH is involved in the binding process of the porphyrin to the subunit. Finally, release of the ATP-binding site of CHLI enables disassembly of Mg chelatase involving ATP hydrolysis (Fodje et al., 2001).

Exploring the substrate specificity and inhibitory effects of substrate analogues on Mg chelatase activity using in organello assays revealed that protoporphyrin IX is the preferred substrate and to lesser extent deuteroporphyrin IX, mesoporphyrin IX, 3-ethyl, 8-vinyl protoporphyrin IX and 3-vinyl, 8-ethyl protoporphyrin IX (Walker and Weinstein, 1991). Accumulation of Mg protoporphyrin or its monomethylester strongly inhibits pea Mg chelatase activity suggesting a possible regulatory role of product inhibition as soon as Mg porphyrins accumulate and are not further metabolized (Averina et al., 1996; see below). Additionally, a physiological role is also considered for pheophorbide as a strong inhibitor of Mg chelatase as this catabolite is formed during leaf senescence, when chlorophyll synthesis slows down and its degradation is induced (Pöpperl et al., 1997). Other tetrapyrrole products, such as protochlorophyllide, chlorophyllide, chlorophyll, zinc pheophorbide, phaeophytin, heme or magnesium–protoporphyrin had apparently very little effect on the Mg chelatase activity (Gibson et al., 1995; Willows, 2003; Willows and Beale, 1998).

CHLI possesses ATPase activity (Willows and Beale, 1998) and belongs to a functionally diverse family of AAA^+ proteins (ATPase associated with various cellular activities) and is proposed to form a hexameric ring structure with a diameter of the ring of about 110 Å (Fodje et al., 2001). The N-terminal domain of BchI contains Walker A and Walker B motifs, which are generally present in nucleotide triphosphate-hydrolysing enzymes (Hansson et al., 2002). An integrin I-domain-binding motif was proposed in the C-helical domain of I and H subunits, which could allow interaction of all subunits through the D subunit and, hence permits a linkage between porphyrin metalation by CHLH/BchH and ATP hydrolysis by CHLI/BchI (Fodje et al., 2001).

The CHLD subunit forms a stable complex with the CHLI subunit in the presence of Mg^{2+} and ATP (Petersen et al., 1998; Willows and Beale, 1998). Structural similarity exists between the AAA domain of CHLI and the N-terminus of CHLD. This could be derived from gene duplication and fusion during evolution. The C-terminus is connected through a proline-rich motif and contains an integrin I domain. Both are suggested to be required for the two-stage assembly including the hexameric ring formation and interaction between the two subunits CHLD and CHLI and, the subsequent addition of the CHLH (Fodje et al., 2001, Axelsson et al., 2006). In fact protein–protein interaction has been demonstrated between CHLD and the other two subunits. Thereby, truncated versions of the CHLD confirm the binding capacity to both other subunits and a core CHLD peptide consisting of only 50 amino acids out of the entire protein sequence was sufficient to maintain a residual in vitro wild-type Mg chelatase activity of 20% (Gräfe et al., 1999). A dominant role for the formation of the whole multimeric Mg chelatase complex was proposed from studies with wild-type and mutant CHLD.

The CHLH subunit is with approximately 1.200 amino acid residues an exceptionally large catalytic subunit that binds protoporphyrin IX and releases Mg protoporphyrin IX. An NMR and electron microscopy structure analysis was recently reported and revealed a 3D reconstruction to a resolution of 25 Å. From these studies it was concluded that the protoporphyrin binding involves the N- and C-terminal parts of the H-subunit (Sirijovski et al., 2008).

B Control of Expression, Activity and Localisation

Light, the circadian clock, phytohormones like cytokinins, and leaf development induce the expression of the *CHLH* gene and to certain extent the *CHLI* gene (Jensen et al., 1996b; Kruse et al., 1997; Lake and Willows, 2003; Nakayama et al., 1998; Papenbrock et al., 1999). The content of *CHLH* mRNA increases during diurnal growth conditions of tobacco plants before the onset of light and reaches its maximum within the first 2 h of the light period, while the *CHLD m*RNA content inversely oscillates over a 24-h light-dark cycle (Papenbrock et al., 1999). A rapid light-induced increase of Mg chelatase activity was observed after transition from dark to light. But posttranslational modifications of Mg chelatase subunits for this rapid light-dependent activation of the enzyme are likely, in spite of the apparent enhanced capacities of Mg chelatase activity and the increasing expression of *CHLH* after transition from light to dark.

Four and 11 cysteine residues are present in the sequences of the *Synechocystis* PCC6803 CHLI and CHLH subunits, respectively. It was shown that cysteine residues modifying reagents, such as *N*-ethylmaleimide, were potent inhibitors of both the ATPase activity and the overall Mg chelatase activity (Walker and Weinstein, 1991, Jensen et al., 2000) indicating the importance of one or more cysteine residues for catalysis. Interestingly, the ATPase activity of CHLI was completely protected from *N*-ethylmaleimide inactivation, if either Mg-ATP, ATP or ADP were present. On the other hands, CHLH could be partially protected from *N*-ethylmaleimide inactivation only if ATP, Mg^{2+} and protoporphyrin IX were present. *Arabidopsis* CHLI undergoes a thioredoxin-dependent reduction resulting in two intramolecular disulphide bridges. The ATPase activity of oxidized recombinant CHLI was entirely abolished. As is known for high Mg chelatase activity these findings are compatible with the required reducing conditions, which prevail during photosynthesis (Ikegami et al., 2007).

The huge ATP requirement of the catalytic Mg insertion was exemplified in *Arabidopsis thaliana* null mutants of the two genes encoding the plastidic ATP/ADP transporter. When the nocturnal plastid import of ATP produced from mitochondrial respiration is inhibited, short day and low light growth conditions lead to deficient Mg chelatase activity resulting in protoporphyrin accumulation and formation of necrotic leaf lesions (Reinhold et al., 2007).

The complex molecular organization of the stepwise functional assembly of the Mg chelatase complex as well as its location at the branch point of chlorophyll or heme synthesis make interaction plausible with various protein partners. The interdependency of CHLH to the other two Mg chelatase subunits has been demonstrated by various methods including the yeast-two-hybrid-system and affinity chromatography. Previous attempts to fractionate protein complexes with Mg chelatase activity from both barley and *Rhodobacter sphaeroides* revealed association, or at least co-purification of the BchD/CHLD protein with ribosomes (Kannangara et al., 1997). In this context, the interaction between cyanobacterial CHLH and the sigma factor SigE is worth mentioning (Osanai et al., 2007) and leads to speculation about the regulatory balances of chlorophyll synthesis and light-induced activation of genes encoding chlorophyll-binding proteins.

CHLH interacts also with Mg protoporphyrin IX methyltransferase (Alawady et al., 2005) suggesting that both enzyme activities are coupled for optimized metabolic channelling and prevention of phototoxicity of free Mg porphyrins. The GUN4 protein binds to CHLH (Larkin et al., 2003) and acts in activation of the recombinant cyanobacterial subunits in in vitro Mg chelatase assays most likely by lowering the threshold Mg^{2+} concentration (Davison et al., 2005; Verdecia et al., 2005). In spite of the significant effect on Mg chelatase activity, GUN4 does not appear to be essential for in vitro Mg chelatase activity. The *GUN4* gene was first identified through a screen for mutants with nuclear transcriptional activities, which are uncoupled from plastid development (Larkin et al., 2003). Although the physiological functions of GUN4 remain unclear, its biochemical properties make it a distinguished partner in the control of tetrapyrrole metabolism and, presumably, an essential factor for chloroplast biogenesis and photoautotrophic growth (see also below).

More recently, CHLH has been identified as one the three known receptors for abscisic acid (Shen et al., 2006). As CHLH mediates abscisic acid signalling as a positive regulator in seed germination, post-germination growth and stomatal movement, it raises the question of physiological relevance. Abscisic acid-mediated abiotic stress response could speculatively correlate with modified photosynthetic activities or a change in chlorophyll contents, which could always be mediated through Mg chelatase as a highly regulated and regulating protein complex (see below).

Mg chelatase subunits can exhibit Mg^{2+}-dependent changes in their distribution between stroma and membranes of chloroplasts (Gibson et al., 1996; Nakayama et al., 1998; Luo et al., 1999). CHLH is largely soluble at low Mg^{2+} concentrations and assembles with the envelope membrane at Mg^{2+} concentration above 5 mM. The association of CHLD with envelope membranes shows a similar Mg^{2+} dependency. However, while CHLI was detected in the soluble fraction regardless of the Mg^{2+} concentration, the importance of the subcompartmental distribution of Mg chelatase for modulation of its activity remains open. Light-induced increase of Mg^{2+} concentrations stimulates activities of enzymes of the Calvin–Benson-cycle and correlates most likely with an increased Mg chelation. The bivalent cation could also function in complex formation of Mg chelatase with other enzymes of tetrapyrrole biosynthesis, which are bound or anchored to membranes.

C Analysis of Mutants and Transgenic Plants

Mg chelatase-deficient mutants of photosynthetic prokaryotes and green algae show a highly photosensitive phenotype due to the accumulation of protoporphyrin IX. Chlorophyll-free *Chlamydomonas reinhardtii* mutants with a deficiency of CHLH or CHLD could grow only heterotrophically in the dark (Chekounova et al., 2001; von Gromoff et al., 2008). In contrast, plant mutants with reduced content or a point mutation of one of the Mg chelatase subunits are characterized by a pale green phenotype. Consequently, a knock-out mutation would entirely abolish Mg chelatase activity indicating that these proteins are essential for photoautotrophic growth. The pale green phenotype is explained by a feedback-controlled decrease of upstream activities in the ALA biosynthesis part of the pathway (Papenbrock et al., 2000a).

Twenty barley mutants at three genetic loci termed *xantha-f, xantha-g* and *xantha-h* were identified as orthologs of the Arabidopsis *CHLH, CHLD* and *CHLI* genes, respectively. The chlorophyll-deficient yellow phenotype and the deficient Mg chelatase activity are consistent with their mutant genotype (Jensen et al., 1996b). The barley *chlorina-125, -157,* and *-161* mutants contain point mutations in the *CHLI* gene and display a semidominant phenotype due to simultaneous expression of mutant and wild-type *CHLI*. A small proportion of wild-type hexamers rescues the heterozygous plants but can not provide sufficient Mg chelatase activity for wild-type chlorophyll levels (Hansson et al., 1999, 2002). Barley *Xantha-h* mutants lack not only XAN-H (CHLI) but additionally the XAN-G (CHLD) subunit (Hansson et al., 1999; Petersen et al., 1999a).

Several allelic mutants have been reported for each Mg chelatase subunit of *Arabidopsis thaliana*. Apart from the initially characterized *Arabidopsis* mutants *cs* and *ch-42* having genetic lesions in the *CHLI-1* gene (Koncz et al., 1990), a second *CHLI-2* gene was identified, which is proposed to be the reason for the small, but significant content of chlorophyll in a *ChlI-1* mutant background (Rissler et al., 2002). However, the contribution of CHLI-2 to the formation of an intact hexameric CHLI ring which is mandatory for a functional Mg chelatase complex might depend on the abundance of the inhibitory mutant CHLI-1 isoform (Apchelimov et al., 2007; Soldatova et al., 2005). Among several allelic *A. thaliana ChlH* mutants, two mutants designated *cch* (conditional *chlorina*) and *gun5* (genomes uncoupled) have a single missense mutation and contain a substitution of a single amino acid residue, respectively. Both mutants show slightly reduced chlorophyll content (Mochizuki et al., 2000). Among pale green *chlorina* rice (*Oryza sativa*) mutants a single missence mutation was identified in the *CHLH* and *CHLD* genes of two *chlorina* mutants, respectively (Zhang et al., 2006). An interruption of the rice *CHLH* gene by T-DNA insertion resembled phenotypically chlorophyll deficient mutants of other plant species with Mg chelatase deficiency (Jung et al., 2003). It will be attractive to perform global transcript analysis of rice mutants with decreased expression of genes encoding the three subunits and activity of Mg chelatase. The comparison of these transcript profiles with transcriptome data from Mg chelatase deficient *Arabidopsis* mutants facilitates assessment as to what extent modified metabolic activities of the Mg branch triggers changes of the nuclear gene expression.

Transgenic plants with inactivated expression of *CHLH* and *CHLI* by antisense RNA synthesis display reduced green pigmentation and plant growth. The loss of chlorophyll correlates with diminished Mg chelatase activities, which did not result in protoporphyrin IX accumulation. Lower heme contents in these transgenic lines make possible redirection of non-metabolized Proto into the heme-synthesizing pathway very unlikely. The changes in porphyrin steady state levels are explained by feedback-mediated reduction of ALA synthesis. The diminished flow of metabolites in the tetrapyrrole biosynthetic pathway is accompanied by reduced transcript levels for genes encoding glutamyl-tRNA reductase, glutamate 1-semialdehyde aminotransferase, the other Mg chelatase subunits and Mg protoporphyrin methyltransferase (Papenbrock et al., 2000a, b; own unpublished data). Silencing of the tobacco *CHLH* gene by infection with tobacco mosaic virus vectors containing *CHLH* inserts resulted not only in strongly decreased levels of *CHLH*-mRNA, but also of *CHLD*-mRNA (Hiriart et al., 2002).

III S-Adenosyl-L-Methionine:Mg Protoporphyrin IX Methyltransferase

This monomeric 25–27 kDa enzyme catalyzes the transfer of a methyl group to the carboxyl group of the 13-propionate side-chain of Mg protoporphyrin to yield Mg protoporphyrin monomethylester. Consistent with previous suggestions that the Mg chelatase: Mg protoporphyrin methyltransferase complex channels Mg protoporphyrin, the tobacco methyltransferase has been shown to physically interact with the CHLH subunit of Mg chelatase (Alawady et al., 2005). Moreover, the recombinant *R. capsulatus, Synechocystis* and tobacco Mg protoporphyrin methyltransferase has been reported to be stimulated upon addition of the BchH/CHLH subunit of Mg chelatase, respectively (Alawady et al., 2005; Hinchigeri et al., 1997, Shepherd et al., 2005). In contrast, Sawicki and Willows could not confirm the stimulatory effect of BchH on the Mg protoporphyrin methyltransferase (2007) and assume that depending on the assay conditions with bacterial proteins, stimulatory effects can be masked during long-time scale assays. Soluble extracts from *E. coli* co-expressing BchM and the three Mg chelatase subunits converted protoporphyrin to Mg protoporphyrin monomethylester (Jensen et al., 1999b). Mg chelatase assays with protein extracts from yeast strains expressing the tobacco CHLI, H and D subunits lead unexpectedly to Mg protoporphyrin IX monomethylester, which was explained by unspecific methyltransferase activity in the yeast extract (Gräfe et al., 1999). The first plant *CHLM* cDNA sequence encoding Mg protoporphyrin methyltransferase was described for *Arabidopsis thaliana.* The protein contains an N-terminal plastid transit sequence. A hydrophobic region is suggested in the N-terminal half of the mature protein to be responsible for anchoring the protein to both the envelope membrane and the thylakoids (Block et al., 2002). *CHLM* antisense RNA expression in transgenic tobacco plants results additionally in reduced transcript levels for several highly regulated enzymes of tetrapyrrole biosynthesis, which are associated with reduced ALA synthesis rate and Mg chelatase activity. In contrast, the ferrochelatase activity was elevated.

Overexpression of the methyltransferase causes the reverse effect: increased transcriptional activity of genes involved in ALA synthesis and Mg chelatase activity and reduced ferrochelatase activity (Alawady and Grimm, 2005). An *Arabidopsis chlM* null mutant shows an *albino* phenotype with absence of all chlorophyll-containing proteins. In contrast to the *CHLM* antisense tobacco lines more *CHLH* transcript and protein accumulated in the *Arabidopsis chlm* mutant (Pontier et al., 2007). It is possible that CHLH is stabilized when its enzymatic product Mg protoporphyrin IX is still bound and not directed to the next enzyme.

IV Mg Protoporphyrin IX Monomethylester Cyclase

The complex catalytic formation of the isocyclic ring E of Mg porphyrins results in the conversion of Mg protoporphyrin IX monomethylester to divinyl and monovinyl protochlorophyllide (Rebeiz et al., 2003). In this catalytic reaction oxygen can be incorporated by two different mechanisms, which are used by either an oxygen-independent or an oxygen-dependent cyclase. During anaerobiosis in the light the introduced oxygen atom at the $C13^1$ position is derived from water by a hydratase (Porra et al., 1995), while under aerobic conditions the same 13^1-oxogroup was derived from molecular oxygen introduced by oxygenase (Walker et al., 1988). Oxygen is introduced by a stereospecific hydroxylation at the β–position of the propionate side chain of ring C followed by oxidation of the hydroxyl group to a carbonyl group and oxidative ring formation at the C-α position leading to the cyclopentanon ring E (Bollivar and Beale, 1996). The oxidation is apparently facilitated by protection of the carboxylate group through methylation in the proceeding enzymatic step.

Mutant gene analysis of the photosynthetic operon of Rhodobacter species revealed the *bchE* gene to be involved in the oxygen-independent cyclase reaction (Bollivar et al., 1994). The purple bacterium *Rubrivivax gelatinosus* synthesizes bacteriochlorophyll *a* under both aerobic and anaerobic conditions (Pinta et al., 2002; Ouchane et al., 2004). Gene disruption of *orf358* (*acsF = aerobic cyclisation system Fe-containing protein)* in this purple bacterium causes under high oxygenation condition accumulation of Mg protoporphyrin monomethylester indicating that the encoded protein is involved in the subsequent oxidative cyclization step (Pinta et al., 2002). The deletion of the *bchE* gene impairs bacteriochlorophyll synthesis of *Rubrivivax* under low oxygen concentration and in anaerobic conditions (Ouchane et al., 2004).

Sequence comparison between *acsF* and other homologous genes revealed similarity to the *PNZIP* gene in *Pharbitis nil*, which shows characteristic induced expression by phytochrome and the endogenous clock (Zheng et al., 1998), the genes *Crdl* and the *Cth1* from *C. reinhardtii, Chl27* from *Arabidopsis,* and the two cyanobacterium *Synechocystis PCC 6803 chlA* genes. *Crd1* was initially identified by a mutant screen of copper deficiency showing a chlorotic phenotype with reduced photosystem I function (Moseley et al., 2000, 2002), *Crd1* is expressed under low aeration and/or low copper conditions and *Cth1* is expressed under oxygenated and copper-sufficient conditions. The *Synechocystis* genome contains two *acsF* homologous genes designated $chlA_I$ and $chlA_{II}$ and three *bchE*-like genes (Minamizaki et al., 2008). Interestingly, the second *chlA* gene is apparently required for growth under low oxygen conditions, while the $chlA_I$ protein is essential for normal aerobic growth conditions. The *Synechocystis BchE* like gene products could not be related to the cyclase reaction (Minamizaki et al., 2008).

The aerobic and anaerobic cyclase proteins have no structural similarity. While the oxygen-dependent AcsF protein is characterized by a metal-binding site consisting of two domains D/EX_nEXXH and belongs to the family of diiron carboxylate proteins (Berthold and Stenmark, 2003), the BchE protein possess a CXXXCXXC domain, which is conserved in different proteins involved in anaerobic processes. On the basis of the iron-requirement for the functional cyclase Gough et al. (2000) suggested a vitamin B12-dependent mechanism for this enzyme.

Moreover, studies of the cyclase reaction with the two barley mutants *xantha l-35* and *viridis K-23* revealed the need of at least two plastidal protein fractions for the cyclization reaction (Walker and Willows, 1997). Structural comparison of AcsF with other monooxygenase related proteins suggests that this protein functions as a Mg protoporphyrin monomethylester hydroxylase (Pinta et al., 2002; Berthold and Stenmark, 2003).

V Divinyl Reductase

In some cases, the chlorophyll and bacteriochlorophyll precursors protochlorophyllide and chlorophyllide are initially formed as 3,8-divinyl derivates, before the 8-vinyl group is reduced to an ethyl group. For a detailed review consult Rebeiz et al. (2003). The marine *Prochlorococcus* species which accumulate 8-vinyl chlorophyll *a* and *b* are the exception. Until recently, it was not clear whether parallel routes for the reduction of the divinyl to the monovinyl derivative of Mg porphyrins requires different enzymes or is catalyzed by one enzyme only (see Chapter 2 of this book). Furthermore, it has been suggested that the reduction of the 8-vinyl group can probably occur at any stage from protoporphyrin IX to chlorophyllide *a* (Tripathy and Rebeiz, 1988, Chapter 2 of this book). By map-based cloning of a divinyl chlorophyll accumulating *Arabidopsis* mutant the *DVR* gene encoding 8-vinyl reductase was identified. The recombinant protein was successfully tested for the conversion of the C8-vinyl group of chlorophyllide to an ethyl group on ring B (Nagata et al., 2005). Previous disruption of the *R. capsulatus bchJ* gene caused accumulation of divinyl protochlorophyllide as well as monovinyl Mg protoporphyrin monomethylester or chlorophyllide (Suzuki and Bauer, 1995). Separate inactivation of the homologous *bchJ* and *DVR* genes from *Chlorobium tepidum,* which are designated *CT2014* and *CT1063*, revealed that the latter gene encodes the C8 divinyl reductase, while the former gene encodes a protein with a likely role in substrate channelling or control of chlorophyll synthesis (Chew and Bryant, 2007). Interestingly, among all photosynthetic species no homologous *DVR* gene could be identified in several cyanobacteria species including Synechocystis PCC6803, which accumulate monovinyl protochlorophyllide *a*. Selecting possible open reading frames for proteins with reductase function in genomes of unicellular organisms without the homologous *DVR* gene resulted in the identification of the *slr1923* gene. Inactivation of this gene resulted in accumulation of 3,8-divinyl chlorophyll (Ito et al., 2008).

VI Regulatory Aspects of Mg Porphyrin Synthesis

This section attempts to outline regulatory mechanisms, which have been described for the gene expression and enzyme activities operating at the beginning of the Mg porphyrin biosynthetic pathway. Protoporphyrin is directed into the Mg and the Fe branches and is distributed in different amounts for chlorophyll and heme synthesis depending on plant development, growth conditions and diurnal changes. These dynamic variations in the demand for both tetrapyrrole end products indicate a tight control of the metabolic flow into the branched pathway.

It is generally suggested that acclimation to environmental conditions as well as regulatory feed-back and feed-forward loops in the metabolic pathway control the synthesis of tetrapyrroles. The initial steps of the Mg branch and its metabolic intermediates are imbedded in these regulatory networks and are associated with plastid-derived signalling for communicating the plastidic physiological and developmental states to the nucleus for the adjustment of transcriptional activities of nuclear genes. Considering all different regulatory changes of expression and activity at each enzymatic step of tetrapyrrole biosynthesis, the question arises as to how are all control processes combined into a balanced metabolic pathway using a hierarchy of different regulatory levels up to a main control switch? Also, how can the whole set of exogenous and endogenous stimuli be sensed and integrated into a functioning regulatory network?

Plants have involved many different strategies to respond to environmental changes. These changes range from long term developmental processes to fine-tuned adjustments of the functionality of metabolic activities within seconds. The short term regulatory mechanisms make an impact on single proteins or multienzymatic protein complexes. Changes of metabolic activities are the result of transcriptional activation and de-novo synthesis of proteins. Superimposed on the transcriptional control, which have been intensively studied in the last decades, a distinct pattern of posttranslational modifications alters the functioning of proteins up to the state of protein degradation.

While transcript and protein contents for almost each enzymatic step of tetrapyrrole biosynthesis have been determined, the elucidation of posttranslational control of the pathway has just begun and new findings are eagerly awaited for. Most progress was achieved by the analysis of Mg chelatase assembly and activation (Axelsson et al., 2006; Jensen et al., 1996b; Willows et al., 1996; see also the Section II.A). Apart from studies of the assembly of single enzymes, such as the Mg chelatase complex, the proposed concerted action of enzymes of ALA synthesis, the rate-limiting step of the tetrapyrrole-synthesizing pathway, the formation of a multi-enzymatic complex at the beginning of the Mg branch and the regulation of monovinyl and divinyl branches of chlorophyll biosynthesis await still further elucidation (Moser et al., 2001; Rebeiz et al., 2003, Shepherd et al., 2005).

The clearest evidence of exogenous stimuli of tetrapyrrole biosynthesis and of the dynamic and flexible response of plants has been shown for light and the light-induced transcriptional activation of genes encoding enzymes of chlorophyll biosynthesis. The control is modulated in response to light intensity, quality and the oscillating 24 h day-night cycle (He et al., 1994; McCormac and Terry, 2002a, b; Mohanty et al., 2006; Papenbrock et al., 1999). These plant responses to light seem to be always very compatible with the experimental conditions, the different developmental states of plants, their genotype, as well as other environmental factors. Therefore, the responses to a certain environmental factor, such as light, have sometimes a lesser or greater impact, because many other factors are involved in metabolic control, such as temperature, mechanical stress (wind), drought, soil, nutrients or pathogens. All of these factors, even if they have only an indirect impact, affect the need of photosynthetic pigments, and activities along the tetrapyrrole biosynthetic pathway. It should be acknowledged that former studies of regulatory mechanisms of tetrapyrrole biosynthesis still follow correlative principles and that a big gap still exists in the understanding of regulatory interdependencies. Thus it remains open, how control is governed, how the expression of each enzymatic step is integrated into the concerted action of the entire pathway, and how the combined action of transcriptional up to post-translational control mechanisms regulates the adjusted supply of active enzymes, which play a role in combination with other enzymes to ensure the availability of appropriate amounts of metabolites. Moreover, it is not clear, whether different kind of regulatory signals have separate signalling pathways all the way down to the transcriptional activation or inactivation or whether positive and adverse changes in the environment are converged earlier to a master switch, as suggested by Koussevitzky et al. (2007) and affect tetrapyrrole synthesis and catabolic pathways.

In view of so many different questions, the answers are still fairly preliminary. For example the effects on tetrapyrrole metabolism may have additional regulatory impact on other plant functions, like photosynthesis, stress defence, etc. Moreover, such a regulatory impact leading to integration of various stimuli for a single combined response would imply that plants can detect metabolic perturbations independently from the primary causes and can respond accordingly. Then, it is assumed that the compensatory regulatory reactions result always from a vectorial sum of adverse, inconvenient or stimulatory exogenous conditions and include a combination of different regulatory processes. Having these ideas in mind, the observed cross tolerance against different environmental stimuli in response to a distinct factor appears in a different light.

Apart from the integrated processing of environmental factors the synergistic effect of informations derived from different endogenous stimuli or from organelles poses another challenge to the cell. Phytohormone-stimulated expression of genes for chlorophyll biosynthesis is rarely investigated. A few reports on cytokinine-directed expression in tetrapyrrole biosynthesis are known (Fusada et al., 2005; Jung, 2004; Kusnetsov et al., 1998; Shen et al., 2006; Sood et al., 2005; Yaronskaya et al., 2006). Upon exogenous cytokinin supply, transcript levels up to enzyme activities of regulatory enzymatic steps were found to be elevated. It is interesting that those genes and their encoded proteins are cytokinin sensitive, which are generally accepted to be highly regulated by other factors and to have regulatory functions in the pathway, such as glutamyl-tRNA reductase, Mg chelatase subunits and protochlorophyllide oxidoreductase. Effects of phytohormones, such as abscisic acid, ethylene, brassinosteroids or jasmonate, on greening processes, plastid development, photoreceptor signalling and senescence have been reported, although the direct correlation to chlorophyll and heme metabolism has mostly not been drawn. It is expected that future studies will reveal the contribution of phytohormones to the control of this metabolic pathway as it has been assumed in the past and will underline the indispensable role of tetrapyrroles.

It has to be acknowledged that all plastids are not evenly exposed to the exogenous conditions. As a consequence, photosynthetic and metabolic activities of all plastids make a contribution, although these dues are not equally in each plastid.

A cellular network has to combine and integrate all plastid activities including the supply of metabolites and signalling components from plastids. As result transcriptional activities are modulated to provide a sufficient amount of cellular components for different subcompartments of the cell. This interplay of retrograde and anterograde signals ensures the adjustment of nuclear gene activities in response to environmental factors and organellar development and activities (Baier and Dietz, 2005; Beck, 2005; Nott et al., 2006, Pesaresi et al., 2007, Rodermel and Park, 2003).

Regarding other endogenous signalling mechanisms, chloroplast biogenesis has been always connected to chlorophyll biosynthesis. Modulation of the initial steps of the Mg branch has been demonstrated to elicit also plastid-to-nucleus signals for coordination of nuclear gene expression in response to the physiological and developmental state of the plastid tetrapyrrole metabolism, i.e. tetrapyrrole-mediated signaling (Surpin et al., 2002; Beck, 2005). At least three concepts of tetrapyrrole-mediated signaling are proposed and they are generally based on the respective experimental design, which reflects differences in plant development, genotype, growth conditions and organisms.

1. The *gun* (genome uncoupled) phenotype is based on elevated *Lhcb* transcript levels of 5 days old *Arabidopsis* seedlings, which were treated with the inhibitor of carotenoid synthesis, Norflurazon. Four out of five published *gun* mutants were found to be involved in tetrapyrrole biosynthesis (Susek et al., 1993). The contribution of the GUN4 protein as an activator of tetrapyrrole biosynthesis has been described in a previous paragraph (see Section II.B). Its signalling function is conclusive as GUN4 binds protoporphyrin and Mg protoporphyrin and is not essential for enzymatic reactions in the Mg branch. Both tetrapyrrole intermediates are photoreactive and their accumulation is intentionally prevented in light and oxygen exposed photosynthetically active organisms. Consequently, a perturbed signalling pathway from plastids to nucleus was related to the action of *gun4* point mutations (Larkin et al., 2003).

The other three *gun* mutant genes code for enzymes of tetrapyrrole biosynthesis: heme oxygenase (GUN2), phytochromobiline synthase (GUN3) and the CHLH subunit of Mg chelatase (GUN5) (Mochizuki et al., 2000; Susek et al., 1993). Thus, CHLH appears to have at least three different functions: Mg insertion into protoporphyrin for chlorophyll biosynthesis, chloroplast-to-nucleus signal transduction and perception of abscisic acid signals (Mochizuki et al., 2000; Shen et al., 2006; see also Section II).

Using the *gun* mutant screen, the identification of these genes, which are involved in tetrapyrrole biosynthesis, has naturally implicated involvement of tetrapyrrole intermediates in the regulatory interorganellar communication. Following this concept it was suggested that these repressive plastid signals are emitted, when chloroplast activities are impaired, and participate in modulation of transcriptional activity at a very early state of chloroplast biogenesis (Koussevitzky et al., 2007; Ruckle et al., 2007). However it remains debatable whether these plastid signals, which are required for the coordination of nuclear gene expression and plastid function during an early development of photodynamically-affected seedlings, can be extrapolated for any predictions of continuous interactions between nucleus and plastids in adult plants.

2. The gradual modifications of enzymes in the Mg porphyrin pathway perturb tetrapyrrole biosynthesis in transgenic plants with gene inactivation by antisense RNA synthesis or ectopic overexpression of genes involved in Mg porphyrin synthesis. Reduced activity of Mg chelatase or Mg protoporphyrin methyltransferase affects also the activity of neighbouring enzymes, but also causes the attenuation of ALA synthesis. These decreased enzyme activities are accompanied with reduced transcript levels of genes encoding proteins of these enzymatic steps. Overexpression of methyltransferase is correlated with increased transcript levels and activities for ALA synthesis and Mg chelatase (Alawady and Grimm, 2005). It is suggested that metabolic activities of the Mg branch are continuously sensed and transmitted to the cytoplasm for fine-tuned regulation of nuclear gene expression. Thus, it might be sensible to suggest that tetrapyrrole-mediated signals communicate always the current status of the tetrapyrrole biosynthetic pathway. The final proof of concept could be provided when inactivation or stimulation of gene expression and enzyme activity of these metabolic steps, which are attributed to plastid-derived signalling, e.g. the enzyme of the Mg branch, can intentionally be induced at any

developmental state of plants. Then, transcript profiles should be instantaneously determined after the induced modulation of expression and activity of the Mg branch.

3. A third potential tetrapyrrole-derived signalling pathway was described in *Chlamydomonas reinhardtii*. Light-induced expression of the nuclear gene encoding the chaperone HSP70 is mediated through Mg protoporphyrin and its monomethylester. The experimental evidences were provided by feeding Mg porphyrins to cultures of *Chlamydomonas* and analysis of mutants affected in expression of different Mg chelatase subunits (Chekounova et al., 2001; Kropat et al., 1997, von Gromoff et al., 2008). While supply of Mg porphyrins and, as recently been shown, hemin can compensate for light-induced expression of the *HSP70* gene and to certain extent also of the *HEMA, CHLH and CHLI* genes encoding glutamyl-tRNA reductase and the two Mg chelatase subunits, respectively, the induction of these genes was abolished in *chlH and chlD* mutants. It is suggested that light-induced signalling for the activation of gene expression includes Mg porphyrins (Beck, 2005).

In conclusion, control of tetrapyrrole biosynthesis plays an important role for the development of functioning chloroplasts. Thus it is seems appropriate that an essential metabolic pathway as tetrapyrrole biosynthesis responds in combination to many different aspects of environmental changes and to endogenous stimuli. The integration of numerous potential signals to a meaningful response remains still unknown. The coordination of nuclear and plastidic gene expression and the adjustment to metabolic and developmental activities in different compartments of the cells still belongs to a highly attractive research field. Future studies will contribute to minimize the current gap in understanding of these regulatory processes.

References

Alawady AE and Grimm B (2005) Tobacco Mg protoporphyrin IX methyltransferase is involved in inverse activation of Mg porphyrin and protoheme synthesis. Plant J 41: 282–290

Alawady A, Reski R, Yaronskaya E and Grimm B (2005) Cloning and expression of the tobacco *CHLM* sequence encoding Mg protoporphyrin IX methyltransferase and its interaction with Mg chelatase. Plant Mol Biol 57: 679–691

Apchelimov AA, Soldatova OP, Ezhova TA, Grimm B and Shestakov SV (2007) The analysis of the ChlI 1 and ChlI 2 genes using acifluorfen-resistant mutant of *Arabidopsis thaliana*. Planta 225: 935–943

Averina NG, Yaronskaya EB, Rassadina VV and Walter G (1996) J Photochem Photobiol 36: 17

Axelsson E, Lundqvist J, Sawicki A, Nilsson S, Schröder I, Al-Karadaghi S, Willows RD and Hansson M (2006) Recessiveness and dominance in barley mutants deficient in Mg-chelatase subunit D, an AAA protein involved in chlorophyll biosynthesis. Plant Cell 18: 3606–3616

Baier M and Dietz KJ (2005) Chloroplasts as source and target of cellular redox regulation: a discussion on chloroplast redox signals in the context of plant physiology. J Exp Bot 56: 1449–1462

Beck CF (2005) Signaling pathways from the chloroplast to the nucleus. Planta 222: 743–756

Berthold DA and Stenmark P (2003) Membrane-bound diiron carboxylate proteins. Ann Rev Plant Biol 54: 497–517

Block MA, Tewari AK, Albrieux C, Maréchal E and Joyard J (2002) The plant S-adenosyl-L-methionine:Mg-protoporphyrin IX methyltransferase is located in both envelope and thylakoid chloroplast membranes. Eur J Biochem 269: 240–248

Bollivar DW and Beale SI (1996) The chlorophyll biosynthetic enzyme Mg-protoporphyrin IX monomethyl ester (oxidative) cyclase. Plant Physiol 112: 105–114

Bollivar DW, Suzuki JY, Beatty JT, Dobrowolski JM and Bauer CE (1994) Directed mutational analysis of bacteriochlorophyll *a* biosynthesis in *Rhodobacter capsulatus*. J Mol Biol 237: 622–640

Chekounova E, Voronetskaya V, Papenbrock J, Grimm B and Beck CF (2001) Characterization of Chlamydomonas mutants defective in the H subunit of Mg-chelatase. Mol Genet Genom 266: 363–373

Chew AG and Bryant DA (2007)Characterization of a plant-like protochlorophyllide a divinyl reductase in green sulfur bacteria. J Biol Chem 282: 2967–2975

Davison PA, Schubert HL, Reid JD, Iorg CD, Heroux A, Hill CP and Hunter CN (2005) Structural and biochemical characterization of Gun4 suggests a mechanism for its role in chlorophyll biosynthesis. Biochemistry 44: 7603–7012

Eckhardt U, Grimm B and Hörtensteiner S (2004) Recent advances in chlorophyll biosynthesis and breakdown in higher plants. Plant Mol Biol 56: 1–14

Eichacker LA, Helfrich M, Rüdiger W and Müller B (1996) Stabilization of chlorophyll a-binding apoproteins P700, CP47, CP43, D2, and D1 by chlorophyll a or Zn-pheophytin a. J Biol Chem 271: 32174–32179

Fodje MN, Hansson A, Hansson M, Olsen JG, Gough S, Willows RD and Al-Karadaghi S (2001) Interplay between an AAA module and an integrin I domain may regulate the function of magnesium chelatase. J Mol Biol 311: 111–122

Fusada N, Masuda T, Kuroda H, Shimada H, Ohta H and Takamiya K (2005) Identification of a novel cis-element exhibiting cytokinin-dependent protein binding in vitro in the 5'-region of NADPH-protochlorophyllide oxidoreductase gene in cucumber. Plant Mol Biol 59: 631–645

Gibson LCD, Willows RD, Kannangara CG, von Wettstein D and Hunter CN (1995) Magnesium-protoporphyrin chelatase of *Rhodobacter sphaeroides*: Reconstitution of activity by combining the products of the *bchH*, *-I*, and *-D* genes expressed in *Escherichia coli*. Biochemistry 92: 1941–1944

Gibson LCD, Marrison JL, Leech RM, Jensen PE, Bassham DC, Gibson M and Hunter CN (1996) A putative Mg chelatase subunit from *Arabidopsis thaliana cv* C24. Plant Physiol 111: 61–71

Gough SP, Petersen BO, Duus JO (2000) Anaerobic chlorophyll isocyclic ring formation in Rhodobacter capsulatus requires a cobalamin cofactor. Proc Natl Acad Sci USA 97: 6908–6913

Gräfe S, Saluz HP, Grimm B and Hänel F (1999) Mg-chelatase of tobacco: the role of the subunit CHL D in the chelation step of protoporphyrin IX. Proc Natl Acad Sci USA 96: 1941–1946

Grimm, B., Porra R, Rüdiger, W and Scheer H. (2006) Chlorophylls and Bacteriochlorophylls: Biochemistry, Biophysics, Functions and Applications Series: Advances in Photosynthesis and Respiration, Vol.25. Springer, Dordrecht

Hansson A, Kannangara CG, von Wettstein D and Hansson M (1999) Molecular basis for semidominance of missense mutations in the XANTHA-H (42-kDa) subunit of magnesium chelatase. Proc Natl Acad Sci USA 96: 1744–1749

Hansson A, Willows RD, Roberts TH and Hansson M (2002) Three semidominant barley mutants with single amino acid substitutions in the smallest magnesium chelatase subunit form defective AAA+ hexamers. Proc Natl Acad Sci USA 99: 13944–13949

He ZH, Li J, Sundqvist C and Timko MP (1994) Leaf developmental age controls expression of genes encoding enzymes of chlorophyll and heme biosynthesis in pea (*Pisum sativum* L.). Plant Physiol 106: 537–546

Hendry GAF, Houghton JD and Brown SB (1987) Chlorophyll degradation. A biological enigma. New Phytol 107: 255–302

Hinchigeri SB, Hundle B and Richards WR (1997) Demonstration that the BchH protein of *Rhodobacter capsulatus* activates S-adenosyl-L-methionine:magnesium protoporphyrin IX methyltransferase. FEBS Lett 407: 337–342

Hiriart J-B, Lehto K, Tyystjärvi, Junttila T and Aro E-M (2002) Suppression of a key gene involved in chlorophyll biosynthesis by means of virus-inducing gene silencing. Plant Mol Biol 50: 213–224

Horn R, Grundmann G and Paulsen H (2007) Consecutive binding of chlorophylls a and b during the assembly in vitro of light-harvesting chlorophyll-a/b protein (LHCIIb)J Mol Biol 366: 1045–1054

Ikegami A, Yoshimura N, Motohashi K, Takahashi S, Romano PG, Hisabori T, Takamiya K and Masuda T (2007) The CHLI1 subunit of Arabidopsis thaliana magnesium chelatase is a target protein of the chloroplast thioredoxin. J Biol Chem 282: 19282–19291

Ito H, Yokono M, Tanaka R and Tanaka A (2008) Identification of a novel vinyl reductase gene essential for the biosynthesis of monovinyl chlorophyll in synechocystis sp. PCC6803. J Biol Chem 283: 9002–9011

Jensen PE, Gibson LCD, Henningsen KW and Hunter CN (1996a) Expression of the *chlI, chlD,* and *chlH* genes from the cyanobacterium *Synechocystis* PCC6803 in *Escherichia coli* and demonstration that the three cognate proteins are required for magnesium-protoporphyrin chelatase activity. J Biol Chem 271: 16662–16667

Jensen PE, Willows RD, Petersen BL, Vothknecht UC, Stummann BM, Kannangara CG, von Wettstein D and Henningsen KW (1996b) Structural genes for Mg-chelatase subunits in barley: Xantha-*f*, *-g* and *-h*. Mol Gen Genet 250: 383–394

Jensen PE, Gibson LCD and Hunter CN (1999a) ATPase activity associated with the magnesium-protoporphyrin IX chelatase enzyme of *Synechocystis* PCC6803: evidence for ATP hydrolysis during Mg2+ insertion, and the MgATP-dependent interaction of the ChlI and ChlD subunits. Biochem J 339: 127–134

Jensen PE, Gibson LC, Shephard F, Smith V and Hunter CN (1999b) Introduction of a new branchpoint in tetrapyrrole biosynthesis in Escherichia coli by co-expression of genes encoding the chlorophyll-specific enzymes magnesium chelatase and magnesium protoporphyrin methyltransferase. FEBS Lett 455: 349–354

Jensen PE, Reid JD and Hunter CN (2000) Modification of cysteine residues in the ChlI and ChlH subunits of magnesium chelatase results in enzyme inactivation. Biochem J 352: 435–441

Jung S (2004) Effect of chlorophyll reduction in *Arabidopsis thaliana* by methyl jasmonate or norflurazon on antioxidant systems. Plant Physiol Biochem 42: 225–231

Jung KH, Hur J, Ryu CH, Choi Y, Chung YY, Miyao A, Hirochika H and An G (2003) Characterization of a rice chlorophyll-deficient mutant using the T-DNA gene-trap system. Plant Cell Physiol 44: 463–472.

Kannangara CG, Vothknecht UC, Hansson M and von Wettstein D (1997) Magnesium chelatase: association with ribosomes and mutant complementation studies identify barley subunit Xantha-G as a functional counterpart of Rhodobacter subunit BchD. Mol Gen Genet 254: 85–92

Koussevitzky S, Nott A, Mockler TC, Hong F, Sachetto-Martins G, Surpin M, Lim J, Mittler R and, Chory J (2007) Signals from chloroplasts converge to regulate nuclear gene expression. Science 316: 715–719

Koncz C, Mayerhofer R, Koncz-Kalman Z, Nawrath C, Reiss B, Redei GP and Schell J (1990) Isolation of a gene encoding a novel chloroplast protein by T-DNA tagging in Arabidopsis thaliana. EMBO J 9: 1337–1346

Kropat J, Oster U, Rudiger W and Beck CF (1997) Chlorophyll precursors are signals of chloroplast origin involved in light induction of nuclear heat-shock genes. Proc Natl Acad Sci USA 94: 14168–14172

Kruse E, Mock HP and Grimm B (1997) Isolation and characterisation of tobacco (*Nicotiana tabacum*) cDNA clones encoding proteins involved in magnesium chelation into protoporphyrin IX. Plant Mol Biol 35: 1053–056

Kusnetsov V, Herrmann RG, Kulaeva ON and Oelmüller R (1998) Cytokinin stimulates and abscisic acid inhibits greening of etiolated Lupinus luteus cotyledons by affecting the expression of the light-sensitive protochlorophyllide oxidoreductase. Mol Gen Genet 259: 21–28

Lake V and Willows RD (2003) Rapid extraction of RNA and analysis of transcript levels in Chlamydomonas reinhardtii using real-time RT-PCR: magnesium chelatase chlH, chlD and chlI gene expression. Photosynth Res 77: 69–76

Larkin RM, Alonso JM, Ecker JR, Chory J (2003) GUN4, a regulator of chlorophyll synthesis and intracellular signaling. Science 299: 902–906

Luo M, Weinstein JD and Walker CJ (1999) Magnesium chelatase subunit D from pea: Characterization of the cDNA, heterologous expression of an enzymatically active protein and immunoassay of the native protein. Plant Mol Biol 41: 721–731

McCormac AC and Terry MJ (2002) Light-signalling pathways leading to the co-ordinated expression of HEMA1 and Lhcb during chloroplast development in *Arabidopsis thaliana*. Plant J 32: 549–559

McCormac AC and Terry MJ (2002) Loss of nuclear gene expression during the phytochrome A-mediated far-red block of greening response. Plant Physiol 130: 402–414

Melkozernov AN, Barber J and Blankenship RE (2006) Light harvesting in photosystem I supercomplexes. Biochemistry 45: 331–345

Minamizaki K, Mizoguchi T, Goto T, Tamiaki H and Fujita Y (2008) Identification of two homologous genes, chlAI and chlAII, that are differentially involved in isocyclic ring formation of chlorophyll a in the cyanobacterium Synechocystis sp. PCC 6803. J Biol Chem 283: 2684–2692

Mochizuki N, Brusslan JA, Larkin R, Nagatani A and Chory J (2000) Arabidopsis genomes uncoupled 5 (GUN5) mutant reveals the involvement of Mg-chelatase H subunit in plastid-to-nucleus signal transduction. Proc Natl Acad Sci USA 98: 2053–2058

Mohanty S, Grimm B and Tripathy BC (2006) Light and dark modulation of chlorophyll biosynthetic genes in response to temperature. Planta 224: 692–699

Moseley J, Quinn J, Eriksson M and Merchant S (2000) The Crd1 gene encodes a putative di-iron enzyme required for photosystem I accumulation in copper deficiency and hypoxia in Chlamydomonas reinhardtii. EMBO J 19: 2139–2151

Moseley JL, Page MD, Alder NP, Eriksson M, Quinn J, Soto F, Theg SM, Hippler M and Merchant S. (2002) Reciprocal expression of two candidate di-iron enzymes affecting photosystem I and light-harvesting complex accumulation. Plant Cell 14: 673–688

Moser J, Schubert WD, Beier V, Bringemeier I, Jahn D and Heinz DW (2001) V-shaped structure of glutamyl-tRNA reductase, the first enzyme of tRNA-dependent tetrapyrrole biosynthesis. EMBO J 20: 6583–6890

Moulin M and Smith AG (2005) Regulation of tetrapyrrole biosynthesis in higher plants. Biochem Soc Trans 33: 737–742

Nagata, N, Tanaka, R, Satoh, S, and Tanaka A (2005) Identification of a vinyl reductase gene for chlorophyll synthesis in *Arabidosis thaliana* and implications for the evolution of prochlorococcus species. The Plant Cell 17: 233–240

Nakayama M, Masuda T, Bando T, Yamagata H, Ohta H and Takamiya K-i (1998) Cloning and expression of the soybean *chlH* gene encoding a subunit of Mg-chelatase and localization of the Mg 2+ concentration-dependent ChlH protein within the chloroplast. Plant Cell Physiol 39: 275–284

Nott A, Jung HS, Koussevitzky S and Chory J (2006) Plastid-to-nucleus retrograde signaling. Annu Rev Plant Biol 57: 739–759

Osanai T, Imashimizu M, Seki A, Sato S, Tabata S, Ikeuchi M and Tanaka K (2007) An H subunit of Mg-chelatase ChlH represses transcriptional activity of a sigma factor SigE by protein–protein interaction. Poster on 7th International Conference on Tetrapyrrole Photoreceptors in Photosynthetic Organisms, Kyoto, Dec 9–14, 2007

Ouchane S, Steunou AS, Picaud M and Astier C (2004) Aerobic and anaerobic Mgprotoporphyrin monomethyl ester cyclases in purple bacteria: a strategy adopted to bypass the repressive oxygen control system. J Biol Chem 279: 6385–6394

Papenbrock J and Grimm B (2001) Regulatory network of tetrapyrrole biosynthesis – studies for intracellular signaling involved in metabolic and developmental control of plastids. Planta 213: 667–681

Papenbrock J, Gräfe S, Kruse E, Hänel F and Grimm B (1997) Mgchelatase of tobacco: Identification of a *Chl D* cDNA sequence encoding a third subunit, analysis of the interaction of the three subunits with the yeast two-hybrid system, and reconstitution of the enzyme activity by co-expression of recombinant CHL D, CHL H and CHL I. Plant J 12: 981–990

Papenbrock J, Mock HP, Kruse E and Grimm B (1999) Expression studies in tetrapyrrole biosynthesis – inverse maxima of magnesium chelatase and ferrochelatase. Planta 208: 264–273

Papenbrock J, Mock H-P, Tanaka R, Kruse E and Grimm B (2000a) Role of magnesium chelatase activity in the early steps of the tetrapyrrole biosynthetic pathway. Plant Physiol 122: 1161–1169

Papenbrock J, Pfündel E, Mock H-P and Grimm B (2000b) Decreased and increased expression of the subunit CHL I diminishes Mg chelatase activity and reduces chlorophyll synthesis in transgenic tobacco plants. Plant J 22: 155–164

Pesaresi P, Schneider A, Kleine T and Leister D (2007) Interorganellar communication. Curr Opin Plant Biol 10: 600–606

Petersen BL, Jensen PE, Gibson LC, Stummann BM, Hunter CN and Henningsen KW (1998) Reconstitution of an active magnesium chelatase enzyme complex from the bchI, -D, and -H gene products of the green sulfur bacterium Chlorobium vibrioforme expressed in *Escherichia coli*. J Bacteriol 180: 699–704

Pinta V, Picaud M, Reiss-Husson F and Astier C (2002) Rubrivivax gelatinosus acsF (previously orf358) codes for a conserved, putative binuclear-iron-cluster-containing protein involved in aerobic oxidative cyclization of Mg-protoporphyrin IX monomethylester. J Bacteriol 184: 746–753

Plumley FG and Schmidt GW (1987) Reconstitution of chlorophyll a/b light-harvesting complexes: Xanthophyll-dependent assembly and energy transfer. Proc Natl Acad Sci USA 84: 146–150

Pontier D, Albrieux C, Joyard J, Lagrange T and Block MA (2007) Knock-out of the magnesium protoporphyrin IX methyltransferase gene in Arabidopsis. Effects on chloroplast development and on chloroplast-to-nucleus signaling. J Biol Chem 282: 2297–2304

Pöpperl G, Oster U, Blos I and Rüdiger W (1997) Magnesium chelatase of Hordeum vulgare L is not activated by light but inhibited by pheophorbide <http://apps.isiknowledge.com/full_record.do?product=UA&search_mode=GeneralSearch&qid=8&SID=R2CGeb8opLMLLAenNmG&page=5&doc=42&colname=WOS&cacheurlFromRightClick=no>. Z Naturf 52: 144–152

Porra RJ, Schäfer W, Katheder I and Scheer H (1995) The derivation of the oxygen atoms of the 13´-oxo and 3-acetyl groups of bacteriochlorophyll a from water in *Rhodobacter sphaeroides* cells adapting from respiratory to photosynthetic conditions: evidence for an anaerobic pathway for the formation of isocyclic ring E. FEBS Lett 371: 21–24

Rebeiz C, Kollosov, V.L, Briskin D, and Gawienowski M (2003) Chloroplast biogenesis: chlorophyll biosynthetic heterogeneity, multiple biosynthetic, routes and biological spin-offs. In Nalwa HS (ed) Handbook of Phtochemistry and Photobiology, Vol. 4. American Scientific Publishers, Los Angeles, pp. 183–247

Reinhold T, Alawady A, Grimm B, Beran KC, Jahns P, Conrath U, Bauer J, Reiser J, Melzer M, Jeblick W and Neuhaus HE (2007) Limitation of nocturnal import of ATP into Arabidopsis chloroplasts leads to photooxidative damage. Plant J 50: 293–304

Rissler HM, Collakova E, DellaPenna D, Whelan J and Pogson BJ (2002) Chlorophyll biosynthesis. Expression of a second chl I gene of magnesium chelatase in Arabidopsis supports only limited chlorophyll synthesis. Plant Physiol 128: 770–779

Rodermel S and Park S (2003) Pathways of intracellular communication: tetrapyrroles and plastid-to-nucleus signaling. Bioessays 25: 631–636

Ruckle ME, DeMarco SM and Larkin RM (2007) Plastid signals remodel light signaling networks and are essential for efficient chloroplast biogenesis in Arabidopsis. Plant Cell 19: 3944–3960

Saenger W, Jordan P and Krauss N (2002) The assembly of protein subunits and cofactors in photosystem I. Curr Opin Struct Biol 12: 244–254

Shen YY, Wang XF, Wu FQ, Du SY, Cao Z, Shang Y, Wang XL, Peng CC, Yu XC, Zhu SY, Fan RC, Xu YH and Zhang DP (2006) The Mg-chelatase H subunit is an abscisic acid receptor. Nature 443: 823–826

Shepherd M, McLean S and Hunter CN (2005) Kinetic basis for linking the first two enzymes of chlorophyll biosynthesis. FEBS J 272: 4532–4539

Sirijovski N, Lundqvist J, Rosenbäck M, Elmlund H, Al-Karadaghi S, Willows RD and Hansson M (2008) Substrate-binding model of the chlorophyll biosynthetic magnesium chelatase BchH subunit. J Biol Chem 283: 11652–11660

Soldatova O, Apchelimov A, Radukina N, Ezhova T, Shestakov S, Ziemann V, Hedtke B and Grimm B (2005) An Arabidopsis mutant that is resistant to the protoporphyrinogen oxidase inhibitor acifluorfen shows regulatory changes in tetrapyrrole biosynthesis. Mol Genet Genom 273: 311–318

Sood S, Gupta V and Tripathy BC (2005) Photoregulation of the greening process of wheat seedlings grown in red light. Plant Mol Biol 59: 269–287

Surpin M, Larkin RM and Chory J (2002) Signal transduction between the chloroplast and the nucleus. Plant Cell 14: S327–338

Susek RE, Ausubel FM and Chory J (1993) Signal transduction mutants of Arabidopsis uncouple nuclear CAB and RBCS gene expression from chloroplast development. Cell 74: 787–799

Suzuki JY and Bauer CE (1995) Altered monovinyl and divinyl protochlorophyllide pools in bchJ mutants of Rhodobacter capsulatus. Possible monovinyl substrate discrimination of light-independent protochlorophyllide reductase. J Biol Chem 270: 3732–3740

Tanaka R, Tanaka A (2007) Tetrapyrrole biosynthesis in higher plants. Annu Rev Plant Biol 58: 321–346

Tripathy BC and Rebeiz CA (1988) Chloroplast biogenesis 60: conversion of divinyl protochlorophyllide to monovinyl protochlorophyllide in green(ing) barley, a dark monovinyl/light divinyl plant species. Plant Physiol 87: 89–94

Verdecia MA, Larkin RM, Ferrer JL, Riek R, Chory J and Noel JP (2005) Structure of the Mg-chelatase cofactor GUN4 reveals a novel hand-shaped fold for porphyrin binding. PLoS Biol 3: e151.

von Gromoff E, Alawady A, Meinecke L, Grimm B and Beck CF (2008) Heme, a plastid-derived regulator of nuclear gene expression in Chlamydomonas. Plant Cell 20: 552–567

Walker CJ and Weinstein JD (1991) Further characterization of the magnesium chelatase in isolated developing cucumber chloroplasts : substrate specificity, regulation, intactness, and ATP requirements. Plant Physiol 95: 1189–1196

Walker CJ and Willows RD (1997) Mechanism and regulation of Mg-chelatase. Biochem J 327: 321–333

Walker CJ, Mansfield KE, Rezzano IN, Hanamoto CM, Smith KM and Castelfranco PA (1988) The magnesium-protoporphyrin IX (oxidative) cyclase system. Studies on the mechanism and specificity of the reaction sequence. Biochem J 255: 685–692

Wilde A, Mikolajczyk S, Alawady A, Lokstein H and Grimm B (2004) The gun4 gene is essential for cyanobacterial porphyrin metabolism. FEBS Lett 571: 119–123

Willows RD (2003) Biosynthesis of chlorophylls from protoporphyrin IX. Nat Prod Rep 20: 327–341

Willows RD and Beale SI (1998) Heterologous expression of the Rhodobacter capsulatus BchI, -D, and -H genes that encode magnesium chelatase subunits and characterization of the reconstituted enzyme. J Biol Chem 273: 34206–34213

Willows RD, Gibson LCD, Kanangara CG, Hunter CN and von Wettstein D (1996) Three separate proteins constitute the magnesium chelatase of *Rhodobacter sphaeroides*. Eur J Biochem 235: 438–445

Yaronskaya E, Vershilovskaya I, Poers Y, Alawady AE, Averina N and Grimm B (2006) Cytokinin effects on tetrapyrrole biosynthesis and photosynthetic activity in barley seedlings. Planta 224: 700–709

Zheng CC, Porat R, Lu P and O'Neill SD (1998) PNZIP is a novel mesophyll-specific cDNA that is regulated by phytochrome and the circadian rhythm and encodes a protein with a leucine zipper motif. Plant Physiol 116: 27–35

Zhang H, Li J, Yoo JH, Yoo SC, Cho SH, Koh HJ, Seo HS and Paek NC (2006) Rice Chlorina-1 and Chlorina-9 encode ChlD and ChlI subunits of Mg-chelatase, a key enzyme for chlorophyll synthesis and chloroplast development. Plant Mol Biol 62: 325–337

Chapter 4

Regulation and Functions of the Chlorophyll Cycle

Ryouichi Tanaka, Hisashi Ito, and Ayumi Tanaka*
Institute of Low Temperature Science, Hokkaido University, N19W8, Kita-ku, Sapporo 060-0819, Japan

Summary

The chlorophyll cycle refers to the interconversion of chlorophyll *a* and chlorophyll *b* that occurs within the chloroplasts of higher plants. The forward reaction that converts chlorophyll *a* to *b* is catalyzed by chlorophyllide *a* oxygenase. The backward reaction from chlorophyll *b* to *a* is catalyzed by a recently identified enzyme, chlorophyll *b* reductase, and an unidentified enzyme, which is hypothetically named

*Author for Correspondence, e-mail: ayumi@pop.lowtem.hokudai.ac.jp

Constantin A. Rebeiz et al. (eds.), The Chloroplast: Basics and Applications, pp. 55–77.

7-hydroxymethyl chlorophyll *a* reductase. The chlorophyll cycle plays two important physiological roles: (a) the chlorophyll cycle adjusts the ratio of the peripheral antenna to the core antenna, and (b) it facilitates the degradation of chlorophyll *b*, light-harvesting complexes, and thylakoid membranes during leaf senescence. In this article, we summarize the research history, the functions, and the regulation of the chlorophyll cycle. Furthermore, we discuss the evolution of the chlorophyll cycle. Recent progresses in evolutionary aspect are emphasized: those include (a) the discovery of the *Prochlorococcus* gene for chlorophyllide *a* oxygenase, (b) the presumable gene separation which occurred in the Prasinophyceae, and (c) the duplication and diversification of the genes encoding chlorophyll *b* reductase. In addition, our current understanding of the regulatory mechanisms of chlorophyll *a* to *b* conversion is described and covers topics such as: (a) transcriptional regulation of chlorophyllide *a* oxygenase in response to light intensities, (b) the mechanism of chlorophyll *b*-dependent feedback regulation of chlorophyllide *a* oxygenase which mechanism involves Clp protease, and (c) the close coordination of the chlorophyll cycle activity and the construction of the photosynthetic machinery. Finally, recent progress in the study of chlorophyll *b* to *a* conversion is summarized, the major topic of which is the identification of the genes encoding two isoforms of chlorophyll *b* reductase.

I Introduction

A Distribution of Chlorophyll b

Chlorophyll (Chl) and bacteriochlorophyll (BChl) are unique molecules that harvest light energy and drive electron transfer. The biosynthesis of these molecules is a prerequisite for virtually all photosynthetic organisms. Various types of BChl molecules such as BChl *a*, *b*, *c*, *d*, *e* and *g* are produced in photosynthetic bacteria. On the other hand, several types of chlorophyll including chlorophyll *a*, *b*, *c* and *d* are synthesized in oxygenic photosynthetic organisms (Green and Durnford, 1996). Typically, photosynthetic organisms belonging to different classes have different sets of BChl or Chl molecules. For example, green algae, land plants and special groups of cyanobacteria have Chl *a* and Chl *b* (Fig. 1a and Table 1). Conversely, many other oxygenic photosynthetic organisms only possess Chl *a*. Most BChl and Chl molecules have different spectra from each other (Fig. 1b). This capability to absorb light at different wavelength enables these organisms to adapt to different ecological niches. Among all of the BChl and Chl species, the metabolic pathways of Chl *a* and Chl *b* are the best understood. These pigments have garnered the greatest attention, simply because they are the pigments that enable land plants to carry out photosynthesis. In this review, we primarily focus on the metabolism of Chl *a* and Chl *b*, and will pay particular attention to the interconversion system of these pigments.

Chl needs to bind to photosynthetic proteins to carry out their functions. The function of each pigment is not only dependent on their optical properties, but also on the proteins they bind to. In most photosynthetic organisms, Chl binds to specific proteins to form light-harvesting antenna and reaction center systems. The antenna systems are divided into two groups, that are named core antenna and peripheral antenna (Table 1) (Barber et al., 2000). In oxygenic photosynthetic organisms, a core antenna contains only Chl *a*. *Acaryochloris marina* is the single exception to this, since it contains Chl *d* in its core antennae (Chen et al., 2002). The structure of a core antenna is common in all oxygenic photosynthetic organisms. On the other hand, a peripheral antenna contains various chlorophylls depending on the species (Ting et al., 2002). Unlike core antenna, the structures of peripheral antenna are also highly variable among photosynthetic organisms. Chlorophyll *b* occurs in both eukaryotic and prokaryotic oxygenic photosynthetic organisms. In prokaryotes, three cyanobacterial groups, *Prochlorothrix*, *Prochloron* and *Prochlorococcus*, have Chl *b* (Table 1) (Ting et al., 2002). Based upon their pigment compositions, these organisms are called Prochlorophytes. Among them, *Prochlorococcus* species are unique since they have unusual 3,8-divinyl Chl *b* molecules in which the ethyl

Abbreviations: BChl – bacteriochlorophyll; CAO – chlorophyllide *a* oxygenase; Chl – chlorophyll; CP1 – the core complex of the photosystem I; GFP – green fluorescent protein; HM – 7-hydroxymethyl; LHC – light-harvesting complexSAM S-adenosyl-L-methionine; PS – photosystem PCB prochlorophytes chlorophyll *b* binding proteins

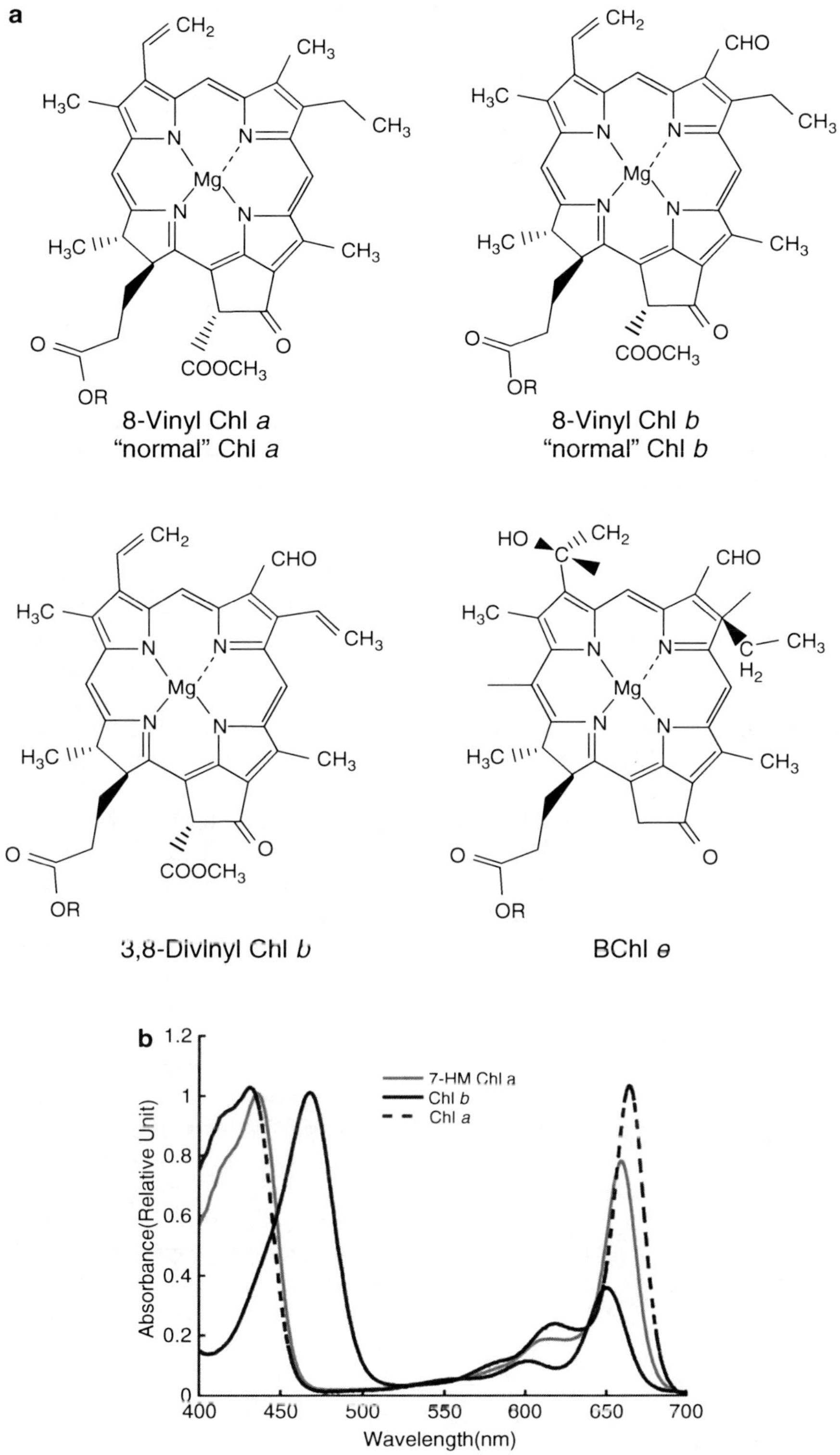

Fig. 1. Structures and spectra of Chl molecules. (**a**) Structures of Chl *a*, Chl *b*, 3,8-divinyl Chl *b* and BChl *c*. R denotes a phytyl group. (**b**) Spectra of Chl *a*, 7-hydroxymetyl Chl *a* and Chl *b* in methanol.

Table 1. Representative organisms that produce 3,8-divinyl chlorophyll *b* or 3-vinyl chlorophyll *b*. MV: 3,vinyl-8,ethyl chlorophyll *b*, DV: 3,8-divinyl chlorophyll *b*, PCB: Prochlorophytes chlorophyll binding proteins, LHC: Light harvesting (chlorophyll *a/b*-binding) complex proteins.

Common category	Class	Genus	Type of chlorophyll	Type of peripheral antenna
Prochlorophytes		*Prochloron*	MV	PCB
		Prochlorothrix	MV	PCB
		Prochlorococcus	DV	PCB
Prasinophyte	Prasinophyceae	*Ostreococcus*	MV	LHC
Chlorophyte	Chlorophyceae	*Chlamydomonas*	MV	LHC
		Dunaliella	MV	LHC
Angiosperms	Brassicaceae	*Arabidopsis*	MV	LHC
	Poaceae	*Oryza*	MV	LHC

group at the C8 position of the macrocycle is replaced by a vinyl group (Fig. 1a) (Rippka et al., 2000). In Prochlorophytes, Chl *b* is primarily bound to specific proteins named Prochlorophytes Chl *b* binding (PCB) proteins that are phylogenetically related to PsbC (also termed CP47) of the photosystem (PS) II core antenna (van Der Staay et al., 1998). It is important to note that it is unclear whether Chl *b* is exclusively localized in PCB in Prochlorophytes. The existence of Chl *b* in the core antenna complexes cannot be excluded, since the amount of Chl *b* in the peripheral antenna does not appear to be sufficient to explain the very low value of the observed Chl *a/b* ratio in the PSI complexes (Garczarek et al., 1998). Furthermore, the core antenna complexes of other organisms are capable of binding a substantial amount of Chl *b* under certain conditions (Hirashima et al., 2006). Therefore, the potential distribution of Chl *b* in the core antenna complexes in Prochlorophytes should be further examined in future investigations. Among eukaryotes, green algae and land plants synthesize Chl *b*. In these organisms, the majority of Chl *b* molecules bind to proteins which belong to the light-harvesting complex (LHC) superfamily (Table 1) (Green and Durnford, 1996). It is also important to note that a small amount of Chl *b* occurs in the PSI core antenna complexes as reported elsewhere (Satoh et al., 2001).

B Establishment of the Chl Cycle

1 Chl b Synthesis

Among the various steps in the Chl biosynthetic pathway, the route for Chl *b* synthesis was the last step to be identified. Since the discovery of Chl *b*-less mutants of *Arabidopsis* in which mutations in a single genetic locus brought about Chl *b* deficiency without significant effect in Chl *a* biosynthesis, it was hypothesized that Chl *b* was synthesized from Chl *a* (Hirono and Redei, 1963). Subsequent physiological experiments supported the concepts that were gained from the genetic analysis. Chl *a* appeared immediately after the onset of illumination of etiolated seedlings, however, Chl *b* was synthesized only after the accumulation of Chl *a* (Virgin, 1960). More direct evidence for the synthesis of Chl *b* from Chl *a* was obtained by experiments in greening cucumber cotyledons (Tanaka and Tsuji, 1981). When intermittently illuminated cotyledons, in which Chl *a* predominantly accumulated, were treated with a solution containing a high concentration of calcium ions, Chl *b* begun to accumulate, which was accompanied by a concomitant decrease in the Chl *a* level in the dark, thereby indicating that Chl *a* is converted to Chl *b*. Two groups independently suggested a biochemical reaction leading to the synthesis of Chl *b* by the experiments using stable isotope of oxygen (Schnegurt and Beale, 1992; Porra et al., 1994). They found that molecular oxygen is incorporated into the formyl group of Chl *b*. These results led the authors to propose that 7-hydroxymethyl (HM) Chl *a* (Fig. 2a) is an intermediate molecule of Chl *b* synthesis. However, the enzyme responsible for Chl *b* synthesis had not yet been isolated. In addition, the exact pathway had not been revealed for an extended period of time, because in vitro attempts to measure the Chl *b* synthesizing activity had not been successful. Finally, a gene responsible for Chl *b* synthesis was identified by molecular genetics using Chl *b*-less mutants of *Chlamydomonas reinhardtii*

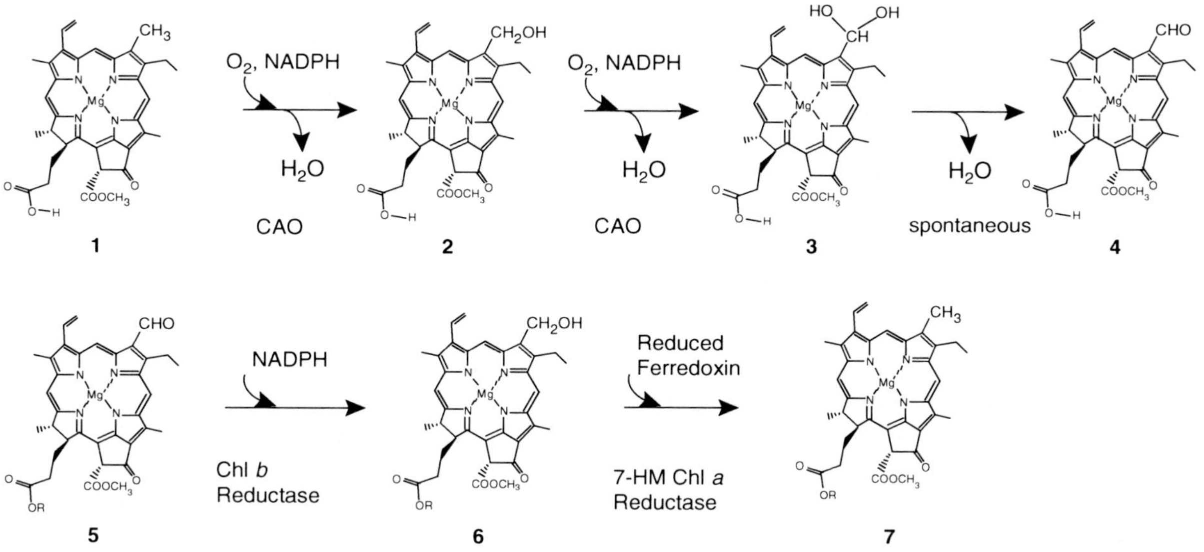

Fig. 2. Proposed biochemical reactions of the Chl cycle. (**a**) Proposed pathway for the conversion of Chl *a* to *b* by Oster et al. (2000). (**b**) Proposed pathway for the conversion of Chl *b* to *a* by Scheumann et al. (1998). Numbers in the figure indicate the following compounds: 1 – chlorophyllide *a*; 2 – HM-chlorophyllide *a*; 3 – 7,7′-dihydroxymethyl chlorophyllide *a*; 4 – chlorophyllide *b*; 5 – chlorophyllide *b*; 6 – HM-chlorophyllide *a*; 7 – Chl *a*. R denotes a phytyl group.

(Tanaka et al., 1998). The identified gene encoding a Rieske-type oxygenase was named Chl *a* oxygenase (CAO). This name was later modified to chlorophyllide *a* oxygenase based upon the results of biochemical studies (Oster et al., 2000). These findings were consistent with the stable isotope experiments that were described above (Schnegurt and Beale, 1992; Porra et al., 1994). Before identification of the *CAO* gene, alcohol dehydrogenase was also proposed to convert HM-Chl *a* to Chl *b* (Porra et al., 1994). However, enzymatic studies with recombinant proteins expressed in *E. coli* finally demonstrated that CAO carries out the two successive oxygenation reactions and converts Chl *a* to Chl *b* without the involvement of any additional enzymes (Fig. 2a) (Oster et al., 2000).

2 Chl b to Chl a Conversion

It was originally thought that the conversion of Chl *b* to Chl *a* could not occur because the reduction of a formyl group to a methyl group is generally perceived as a very difficult chemical reaction (Shlyk, 1971). However, some reports suggested that it would be possible that the conversion of Chl *b* to Chl *a* could occur in vivo. Kupke and Huntington (Kupke and Huntington, 1963) discovered that an increase in Chl *b* was coupled with a concomitant loss of Chl *a* in bush bean. They hypothesized that Chl *a* arose from Chl *b* (Kupke and Huntington, 1963). In addition to our research group, a few other laboratories have reported similar observations with greening cucumber cotyledons (Tanaka and Tsuji, 1983) and tobacco (Ikegami et al., 1984). In order to draw the conclusion that Chl *b* can be converted to Chl *a*, the following assumptions were made: (a) That Chl is not synthesized in the dark due to the lack of light-independent protochlorophyllide in angiosperms and (b) that the Chl content must be exactly measured. At this time, the presence of light-independent protochlorophyllide reductase was only proposed to exist in angiosperms (Masuda and Takamiya, 2004). However, its existence in angiosperms was contradicted by the analyses of the genome sequences of *Arabidopsis* (Arabidopsis Genome Initiative, 2000) and rice (International Rice Genome Sequencing Project, 2005). Furthermore, in these studies, Chl contents were determined only by spectroscopic techniques whose accuracy was not sufficient to draw the conclusion that Chl *a* can also be synthesized from Chl *b*. Convincing evidence for the conversion of Chl *b* to Chl *a* was obtained by biochemical studies using isolated etioplasts (Ito et al., 1993;

Ito et al., 1994). When intact etioplasts were incubated with chlorophyllide *b*, a phytol-less moiety of Chl *b,* an accumulation of Chl *a* and HM-Chl *a* was observed. In addition, when HM-chlorophyllide *a* was used as a substrate, Chl *a* accumulated as well (Ito et al., 1996). These experiments unequivocally demonstrated that Chl *a* is produced from Chl *b* via HM-Chl *a* in etioplasts. This could be concluded since the background Chl level in etioplasts was negligible and only the exogenously added Chl derivatives could be tracked. Subsequent enzymatic studies clarified that the Chl *b* to *a* conversion occurred via a two-step reaction (Ito et al., 1996; Scheumann et al., 1996). In the first step Chl *b* is converted to HM-Chl *a* (Fig. 2b). This step is catalyzed by Chl *b* reductase that requires NADPH. In the second step HM-Chl *a* is converted to Chl *a*. This reaction is catalyzed by 7-hydoroxymethyl-Chl *a* reductase that uses reduced ferredoxin as its electron donor (Scheumann et al., 1998). Recently, the gene responsible for the first reaction was identified by a genetic dissection of a stay-green mutant of rice, *nyc1* (Kusaba et al., 2007). The identification of the gene finally settled a longstanding argument concerning the existence of a biochemical conversion of Chl *b* to *a*. It is important to note that the gene for the HM-Chl *a* reductase has not been identified yet, despite the researchers' efforts to discover it.

3 Why Is the Interconversion of Chl a and Chl b Called the Chl Cycle?

The interconversion of Chl *a* and Chl *b* is not merely a reversible reaction in the Chl biosynthetic pathway. This interconversion has important and specific physiological roles like the famous xanthophyll cycle (Niyogi, 1999). In this section, we summarized several observations or hypotheses that distinguish this interconversion system from the rest of the Chl metabolic pathway.

(a) The Chl cycle comprises the last steps of the Chl biosynthetic pathway as well as the first steps of Chl degradation.
(b) The forward and the backward reactions of the Chl cycle are catalyzed by different enzymatic systems as described above.
(c) The Chl cycle is controlled by its specific regulatory mechanisms.
(d) The Chl cycle plays key roles in many physiological events such as acclimation to light intensities and leaf senescence.
(e) The Chl cycle is found only in green plants but not in cyanobacteria.

In this article, we review the research on the Chl cycle, highlighting recent progress at the molecular level. In addition, we describe the functions of the Chl cycle in relation to the construction of the photosynthetic apparatus. Furthermore, we discuss the evolutionary aspects of the cycle.

II Pathway and Enzymes of the Chlorophyll (Chl) Cycle

A Pathway of the Chl Cycle

Despite our increasing understanding of the Chl cycle, there are still several major questions that remain unanswered. The biggest unanswered question regarding the pathway itself is whether the phytylated (Chl) or unphytylated (chlorophyllide) moieties are the true substrates of the Chl cycle. Studies with the recombinant CAO enzyme produced in *E. coli* partially answered this question (Oster et al., 2000). When chlorophyllide *a* was reacted with the recombinant enzyme, chlorophyllide *b* and a small amount of 7-hydroxymehtyl chlorophyllide *a* were formed. Furthermore, the enzyme more efficiently converted 7-hydroxymehtyl chlorophyllide *a* to chlorophyllide *b* (Oster et al., 2000). In contrast, when Chl *a* was added to the reaction mixture of CAO as a substrate, neither Chl *b* nor 7-hydoroxymethyl *a* was formed (Oster et al., 2000). These results indicate that chlorophyllide *a* is the preferred substrate of CAO. However, the possibility that CAO uses Chl *a* in vivo cannot be excluded. Since Chl *a* is poorly dissolved in an aqueous solution without detergent, it is possible that Chl *a* may associate with some proteins without detergent within the in vitro assay system, which may in turn have limited the accessibility of Chl *a* to CAO. Therefore, an in vivo experiment might be necessary to draw a conclusion which substrate is preferred by CAO. We note that a series of in vivo experiments carried out by Rebeiz and coworkers supported the presence of multiple routes for chlorophyll *b* biosynthesis (Rebeiz et al., 1999, 2003).

If we assume that CAO only uses unphytylated substrates, two additional reactions are required for the conversion of Chl *a* to *b* as illustrated in Fig. 3b. Chl *a* should be first de-phytylated by

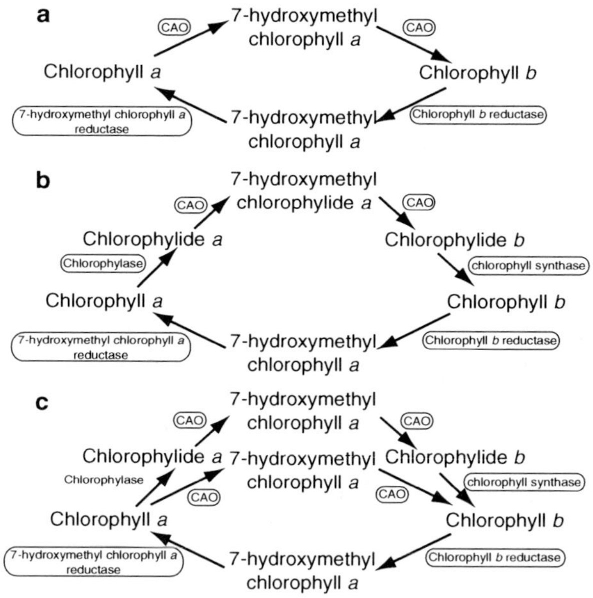

Fig. 3. Tentative models for the pathway of the Chl cycle. (**a**) A simple model for the Chl cycle in which 7-HM Chl *a* is the intermediate of the conversion of Chl *a* to *b*. (**b**) A modified model for the Chl cycle in which 7-HM chlorophyllide *a* is the intermediate of the conversion of Chl *a* to *b*. (**c**) A double-track model for the Chl cycle in which both 7-HM Chl *a* and 7-HM chlorophyllide *a* are the intermediates of the conversion of Chl *a* to *b*.

chlorophyllase. CAO then converts chlorophyllide *a* to HM-chlorophyllide *a* and subsequently to chlorophyllide *b*. Finally, Chl synthase converts chlorophyllide *b* to Chl *b*. One could argue that in the biosynthetic pathway to Chl *a*, chlorophyllide *a* should be always formed as an intermediate. Thus, it is reasonable to consider that chlorophyllide *a* is used as a substrate. However, since the conversion of Chl *b* to *a* was observed without de novo synthesis of total Chl in several different species as described above (Kupke and Huntington, 1963; Tanaka and Tsuji, 1983; Ikegami et al., 1984), some requirements should exist for the conversion of Chl *b* into Chl *a* in plants. Therefore it is tempting to speculate that CAO uses Chl *a* as its substrate in vivo (Fig. 3a). It would also be possible to assume that the conversion of Chl *a* to *b* is a double-track pathway (Fig. 3c). According to this model, chlorophyllide *a* is the substrate of CAO during the de novo synthesis of Chl *b* during leaf development (Oster et al., 2000). On the other hand, Chl *a* is the substrate of CAO during re-arrangement of the photosynthetic machinery upon light illumination as discussed in Section V.

Another confusing question regarding CAO is whether protochlorophyllide *a* functions as a substrate of CAO or not. Several lines of evidence exist which indicate that protochlorophyllide *a* is converted into protochlorophyllide *b*. Firstly, several reports have described the presence of protochlorophyllide *b* in greening seedlings (Shedbalkar et al., 1991; Kolossov and Rebeiz, 2003; Reinbothe et al., 2003). Etioplasts also have been reported to exhibit activity for the conversion of protochlorophyllide *a* to protochlorophyllide *b* (Reinbothe et al., 2003). Furthermore, a significant accumulation of protochlorophyllide *b* was observed in a cyanobacterial strain that was transformed with the *Arabidopsis CAO* gene (Xu et al., 2001). However, the appearance of protochlorophyllide *b* was delayed several days by that of chlorophyllide *b* (Xu et al., 2001). Thus, it is possible that protochlorophyllide *b* accumulated as a consequence of a disturbed Chl metabolism in the cyanobacterial transformant. In vitro experiments by Oster et al. (Oster et al., 2000) demonstrated that CAO was not able to convert protochlorophyllide *a* into *b* in an observable rate within the assay system. Moreover, we found that protochlorophyllide *a* accumulated instead of protochlorophyllide *b* in CAO-overexpressing *Arabidopsis* plants after a dark period of up to several days. Upon illumination of these plants Chl *b* was detected within a minute (Y. Yamasato, R. Tanaka, A. Tanaka, unpublished results). Taken together, it would be reasonable to conclude that protochlorophyllide *a* is not a preferred substrate of CAO. In other words, it is not likely that protochlorophyllide *b* is a major intermediate of Chl *b* synthesis in higher plants. This conclusion is supported by in vivo analyses of chlorophyll biosynthetic routes by Rebeiz and coworkers. Their studies suggest that the biosynthetic routes involving protochlorophyllide *b* may account for only two out of more than a dozen of possible routes (Rebeiz et al., 2003). In contrast to the case of CAO, a preferred substrate for the Chl *b* reductase seems clear. A recombinant Chl *b* reductase that was expressed in *E. coli* reduced the formyl group of both Chl *b* and chlorophyllide *b* to a HM group with a similar efficiency (Horie et al., 2009). In addition, Kusaba et al. (Kusaba et al., 2007) found that during leaf senescence, Chl *b* accumulated in the rice *nyc1* mutant which lacks an isoform of Chl *b* reductase. These data indicated that the reduction of Chl *b* precedes the dephytylation step in plants. If this is not the case, chlorophyllide *b* should have accumulated in the *nyc1* mutant.

As for HM reductase, only circumstantial evidence is available for its substrate specificity. When isolated plastids were incubated with HM-chlorophyllide *a*, this substance was immediately esterified with a phytol molecule within plastids to form HM-Chl *a*. Subsequently, Chl *a* started to accumulate with a concomitant decrease in HM-Chl *a* (Ito et al., 1996). Similarly, when etiolated rye leaves were infiltrated with chlorophyllide *b*, Chl *a* increased with the concomitant decrease in Chl *b* after the completion of the esterification of exogenous chlorophyllide *b* (Rudoi et al., 1982). Collectively, the conversion of the formyl group to the methyl group is likely to occur at the level of phytylated Chl in vivo.

B Enzymes of the Chl Cycle

1 Chlorophyllide a Oxygenase

The *Chlamydomonas CAO* gene encodes a methyl-monooxygenase-like sequence with a putative [2Fe–2S] Rieske center and a mononuclear-iron binding motif. There are several identified methyl-monooxygenase proteins that catalyze biological reactions similar to that of Chl *a* to *b* (Harayama et al., 1992; Junker et al., 1997). For example, a bacterial methyl-monooxygenase, TsaM, is involved in the conversion of the methyl group of 4-toluenesulfonate to a hydroxylmethyl group and then to a formyl group to form 4-sulfo-benzaldehyde (Junker et al., 1997). Most of the known monooxygenase proteins only catalyze the methyl to the HM reaction. The subsequent reaction forming a formyl group is catalyzed by a separate alcohol dehydrogenase. In contrast, the results of genetic analysis of Chl *b*-less mutants *Chlamydomonas reinhardtii* (Tanaka et al., 1998) as well as the study of *Arabidopsis thaliana* mutants (Hirono and Redei, 1963) indicate that a single genetic locus (the *CAO* gene) is responsible for the Chl *b* biosynthesis. After the *CAO* genes were identified from *Chlamydomonas* (Tanaka et al., 1998) as well as from *Arabidopsis* (Espineda et al., 1999), enzymatic studies using the recombinant CAO enzymes quickly eliminated any concern regarding the involvement of a yet-unidentified protein in the synthesis of Chl *b* (Oster et al., 2000). When chlorophyllide *a* was reacted with the recombinant CAO protein in the presence of reduced ferredoxin, chlorophyllide *b* as well as a small amount of HM-chlorophyllide *a* were formed (Fig. 2a) (Oster et al., 2000). This result clearly showed that the presumed dehydrogenase-like protein is unnecessary, and that CAO alone is able to convert chlorophyllide *a* to chlorophyllide *b*. In addition, the authors demonstrated that CAO converts an analog of HM-chlorophyllide *a*, Zn-HM-pheophorbide *a*, into Zn-pheophorbide *b*. These data provide further evidence for the ability of CAO to perform both the first and the second step of the reaction. This conclusion was also supported by a study with a cyanobacterium, *Synechocystis* PCC6803, which lacks a *CAO* gene in nature (Xu et al., 2001). Transformation of this organism with the *Arabidopsis CAO* gene resulted in the successful production of Chl *b* (Xu et al., 2001).

A detailed reaction mechanism of CAO was proposed by Oster et al (2000) (Fig. 2a) (Oster et al., 2000). According to their model, CAO catalyzes two successive hydroxylations at the 7-methyl group of chlorophyllide *a* to form a dihydroxyl group in the same position, which is spontaneously converted to an aldehyde group. In the course of the reaction, CAO consumes two molecules of NADPH, as well as two molecules of molecular oxygen, in a ferredoxin dependent manner. The conserved Rieske center and non-heme-iron binding motifs of CAO are likely to be involved in the electron transport from ferredoxin to molecular oxygen. Eggink et al. observed the formation of a radical during the CAO reaction (Eggink et al., 2004). This radical was quenched by the addition of chlorophyllide *a* to the reaction mixture. This result suggested the involvement of a radical in the reaction. The authors suggested that the radical might be derived from a conserved tyrosine residue in the C-terminal region of CAO (Eggink et al., 2004). Determination of the three-dimensional structure of CAO may help elucidate the exact reaction chemistry of CAO that is necessary to accomplish the two successive reactions. It should be noted that such an activity for the two successive oxygenation reactions seems to be unique to CAO. We speculate that it would be important for plants to decrease the potential risk of releasing these molecules into cellular environments by combining these two reactions in a single enzyme. This may be the case since the substrate, the intermediate, and the product of the CAO reaction are all photosensitizers (Nyman and Hynninen, 2004).

2 Chl b Reductase

Biochemical studies have shown that the conversion of Chl *b* to Chl *a* is carried out by two different enzymes, namely, Chl *b* reductase and HM-Chl *a* reductase (Fig. 2b). The first enzyme catalyzes the reduction of the formyl group of Chl *b* to a hydroxymethyl group. This enzyme is localized in the thylakoid membrane and uses NADPH as a reductant (Ito et al., 1996). Identification of the gene encoding Chl *b* reductase from *Oryza sativa* (rice) revealed that the enzyme belongs to a family of short-chain dehydrogenase/reductases (Kusaba et al., 2007). Interestingly, two genes for Chl *b* reductase were found in the genomes of *Arabidopsis* and rice (Kusaba et al., 2007). The paralogue of *NYC1* was named *NOL* (for NYC1-like). The amino acid sequences encoded by the two genes are similar to each other. However, only the amino acid sequence of NYC1 contains three putative transmembrane domains.

3 HM-Chl a Reductase

HM-Chl *a* reductase catalyzes the reduction of the hydroxymethyl group of HM-Chl *a* to a methyl group to form Chl *a*. In nature, only a few similar reactions exist. It is possible that the dearth of similar reactions is because the strong carbon-oxygen bond of a hydroxymethyl group is not easy to break. A reaction mechanism involving a unique electronic rearrangement of the tetrapyrrole molecule has been proposed (Folly and Engel, 1999). However, at the present time, this mechanism has not yet been confirmed. Ito and coworkers were the first to show the existence of this activity in isolated chloroplasts (Ito et al., 1993, 1994, 1996; Scheumann et al., 1996). Later, Scheumann et al. (1998) have demonstrated that this reaction is ferredoxin-dependent and associated with thylakoid membranes. The gene for this enzyme has not yet been identified.

III Diversity and Evolutionary Aspects of Chlorophyllide *a* Oxygenase

A Diversity of CAO Sequences

Among various molecular species of BChl and Chl, only three species contain the 7-formyl group, namely, 3,8-divinyl Chl *b*, 3-vinyl Chl *b* and BChl *e* (Fig. 1a). As described in the Introduction, 3-vinyl Chl *b* ("normal" Chl *b*) is produced by green algae, land plants and a few groups of the Prochlorophytes. All of these organisms have *CAO* genes in their genomes. Tomitani et al. (1999) found that these genes form a single clade, even though the putative chloroplast ancestor and the groups of Prochlorophytes are dispersed in the phylogeny of cyanobacteria. Within this CAO clade, the tree topology is consistent with that of the generally accepted one (Tomitani et al., 1999). Thus, it is likely that the common ancestor of cyanobacteria had a *CAO* gene and, was able to synthesize Chl *b*, even though the majority of cyanobacteria of today lost the ability to synthesize Chl *b*.

3,8-Divinyl Chl *b* is only produced by a group of marine cyanobacteria, *Prochlorococcus* species. The genomes of these organisms contain a gene that is distantly related to *CAO* (Dufresne et al., 2003; Rocap et al., 2003). Satoh and Tanaka (2006) transformed *Synechocystis* PCC6803, an organism that is unable to produce Chl *b,* with the *CAO* homologue of the *Prochlorococcus*. The authors successfully detected Chl *b* production in the transformant. This result clearly demonstrated that the *Prochlorococcus CAO* homologue encodes a functional Chl *b* synthesizing enzyme. One possible explanation for the weak similarity between *Prochlorococcus CAO* genes and other *CAO* genes is that *Prochlorococcus CAO* evolved from the common cyanobacterial *CAO,* but, due to the absence of the DNA-repair systems in *Prochlorococcus* (Dufresne et al., 2003), its nucleotide sequence was rapidly changed and no longer exhibited a higher similarity to other *CAO* genes. Another possible explanation is that, because these organisms use divinyl-type chlorophylls instead of the normal monovinyl type, the *CAO* gene sequence was rapidly modified to optimize the protein structure for these unusual pigments.

BChl *e* is produced by several organisms belonging to a group of green sulfur bacteria, *Chlorobium.* However, the genome sequences of these organisms do not contain CAO homologues (data not shown). The absence of CAO homologues in these organisms makes biological sense. Since they are anoxygenic organisms, the utilization of an oxygen molecule would be difficult. It is likely that the biosynthesis of BChl *e* is accomplished by an oxygen-independent reaction. Since oxygenic and anoxygenic photosynthetic organisms often use different enzymes for the same oxidation reactions, one may speculate that the synthesis of BChl *e* falls into this category. An example of a reaction catalyzed by different enzymes is the oxidation of Mg-protoporphyrin IX monomethyl ester to form 3,8-divinyl-protochlorophyllide *a*. In photosynthetic eukaryotic organisms, this reaction is catalyzed by a di-iron-motif containing cyclase that utilizes molecular oxygen. In anoxygenic photosynthetic organisms, however, the same reaction is catalyzed by an enzyme called BchE that does not use oxygen as the oxygen acceptor (Gough et al., 2000; Pinta et al., 2002; Tottey et al., 2003; Rzeznicka et al., 2005). However, most cyanobacteria and some photosynthetic bacteria (Ouchane et al., 2004) appear to contain both types of cyclase. The existence of aerobic and anaerobic types of enzymes for the reactions of coproporphyrinogen III oxidase and protoporphyrin IX oxidase are also found in nature (for a review see Tanaka and Tanaka, 2007). Given the exclusive distribution of the BChl *e* synthesizing enzyme in a specific group of *Chlorobium* species, the gene for this enzyme may be identified through a bioinformatics analysis. Using a recently developed genome-comparison tool (Satoh and Tanaka, 2006), we compared the whole-genome sequences between *Chlorobium* species that produce BChl *e* and those that are unable to produce this pigment. We found that all of the BChl *e*-synthesizing *Chlorobium* species contain a gene encoding a radical S-adenosyl-L-methionine (SAM) enzyme. On the other hand, the remainder of *Chlorobium* species lacked homologues of this gene (S. Satoh, M. Yokono and A. Tanaka, unpublished results). Bryant and Frigaard (2006) recently reached a similar conclusion. By using a different bioinformatics method, they concluded that a radical SAM enzyme is most likely involved in BChl *e* biosynthesis. Furthermore, they claimed that they identified a dehydrogenase gene adjacent to the gene encoding the radical SAM enzyme (Bryant and Frigaard, 2006). It is likely that the pair of the radical SAM enzyme and dehydrogenase catalyze the first and second step of a methyl to formyl transformation, respectively. Adjacent location of these genes may indicate they are simultaneously transcribed in the same transcriptional unit. The whole genome searches in the cyanobacterial or eukaryotic genomes failed to identify the homologues of the radical SAM-encoding gene and of the dehydrogenase gene (Bryant and Frigaard, 2006). This result implies that these enzymes are not present in current oxygenic photosynthetic organisms. Nevertheless, the possibility cannot be excluded that these two enzymes participated in Chl *b* synthesis instead of CAO when the organisms initially acquired Chl *b* in their evolutionary history.Despite the strong similarity between the CAO sequences from cyanobacteria and those of the C domain (see section IIIB) from higher plants, analysis of the Prasinophyceae CAO sequences led to a striking discovery (T. Chikuni and I. Inouye, personal communication, 2007). Since the phylogenetic analyses revealed that the Prasinophyceae is one of the ancient groups occurring at the base of the Chlorophyta, it was expected that its CAO sequence would be moderately similar to both cyanobacterial and higher plants' sequences (Courties et al., 1998). Surprisingly, it was found that the CAO sequence was apparently split into two sequences in *Ostreococcus* species belonging to the Prasinophyceae. In the genomes of *Ostreococcus tauri* (Derelle et al., 2006) and *Ostreococcus lucimarinus* (Palenik et al., 2007), two different genes are present which exhibit sequence similarity to known CAO genes. One of them, named *OtCAO1* and *OlCAO1* in *O. tauri* and *O. lucimarinus,* respectively, by T. Chikuni and I. Inouye (personal communication), encodes a protein with similarity to the entire C domain, but its non-heme iron binding motif is incomplete (Fig. 4a and b). In contrast, the other gene, named *OtCAO2* and *OlCAO2* in *O. tauri* and *O. lucimarinus*, respectively, encodes a protein with similarity to the second half of the C domain. Its predicted sequence contains a putative non-heme iron biding site, but it lacks a motif for a Rieske center (Fig. 4a and b). Note that the categorization of *CAO1* and *CAO2* genes in the *Ostreococcus* genomes does not correspond to that of the rice *CAO1* and *CAO2* genes described by Lee et al. (2005). Although no experimental evidence is presently available, it is tempting to speculate that a heteromeric complex of CAO1 and CAO2 proteins carry out Chl *b* synthesis in the Prasinophyceae. In this hypothetical scenario, the Rieske center in CAO2 accepts

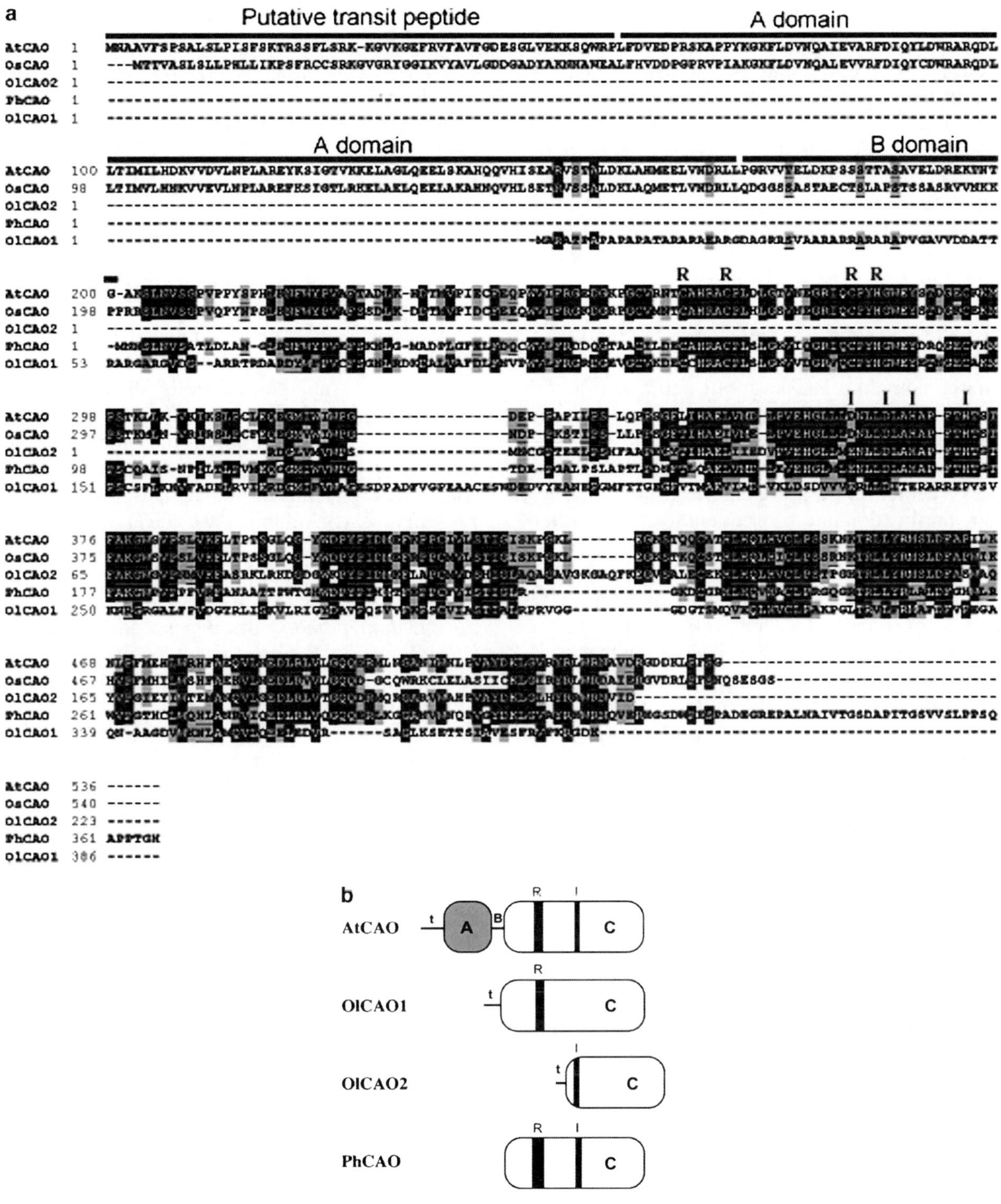

Fig. 4. Comparison of amino acid sequences of CAO from various organisms. (**a**) Alignment of amino acid sequences of CAO from *Arabidopsis thaliana* (AtCAO), *Oryza sativa* (OsCAO), *Ostreococcus lucimarinus* CCE9901 (OlCAO1 and OlCAO2), and *Prochlothrix hollandica* (PhCAO). The regions corresponding to the putative transit peptide, the A and B domains are indicated by *labels* and *black bars* above the regions. The region *without a black bar* corresponds to the C domain. Essential amino acid residues in the Rieske binding motif and the non-heme iron binding motif are indicated by R and I, respectively, on each residue. Amino acid residues conserved among four organisms are shaded in *black*. Amino acid residues substituted for those with similar biochemical properties are shaded in *gray*. The alignment was performed with the ClustalW software (Thompson et al., 1994). The accession numbers for the sequences are as follows: AtCAO, BAA82484; OsCAO, AF284781; OlCAO1, ABO96992; OlCAO2, ABO99103; PhCAO, BAD02269. (**b**) A schematic overview of CAO domains from representative organisms including *Prochlorothrix hollandica, Arabidopsis thaliana,* and *Ostreococcus lucimarinus*.

electrons from ferredoxin, and transfers them to the non-heme iron in the CAO1 where the oxygenation of chlorophyllide *a* is carried out. Another possible scenario is that these two proteins function independently and some unknown factors participate in the electron transfer between CAO1 and CAO2. The CAO proteins in *Ostreococcus* species would serve as an interesting material for the future biochemical study of the enzyme.

It is not clear why the *Ostreococcus* species has two *CAO* genes and how these genes have evolved. Since other known CAO and other related oxygenases contain a Rieske motif and a non-heme iron binding site within a single polypeptide, it is likely that an ancestor of *Ostreococcus* species had a CAO protein which contained both a Rieske center and a non-heme iron binding site. Gene duplication and subsequent losses of either an N-terminus encoding part, or a C terminus-encoding part of the ancestral sequence may have resulted in the evolution of two distinct *CAO* genes, *CAO1* and *CAO2,* respectively. This interpretation is supported by the observation that both CAO1 and CAO2 sequences share a common part that corresponds to the middle part of the C domain of higher plants CAO proteins (Fig. 4a and b). Taken together, we emphasize that the variability of CAO sequences among different organisms illustrates a flexible alteration of gene structure that may enable the variation of pigment synthesis and its regulatory mechanisms.

B Domain Structure of CAO

Figure 4a shows an alignment of amino acid sequences of CAO from various species. It is clear from this alignment, that the CAO sequence from *Prochlorothrix hollandica* (367 residues) is much shorter than those of *Arabidopsis* and rice (501 and 508 residues, respectively and without a predicted transit peptide) (Fig. 4a). Within higher plants' sequences, the first 114 residues are highly conserved. In contrast, the subsequent 30 residues are less conserved, and with an obvious trend which shows an enrichment in hydrophilic amino acids (Fig. 4a). Based on the comparison of amino acid sequences, we assumed that higher plant CAO sequences are composed of three domains, that we named A, B and C domains, respectively (Nagata et al., 2004). By transforming a *Synechocystis* strain with the *CAO* gene from *P. hollandica* encoding only the C domain, we demonstrated that the C domain alone is sufficient to catalyze Chl *b* synthesis (Nagata et al., 2004). Subsequently, Yamasato et al. (2005) showed that introduction of the sequence encoding the C domain of *Arabidopsis* CAO into a *CAO* knockout mutant of *Arabidopsis, chlorina1 (ch1)*, restored the Chl *b*-synthesizing ability in this mutant. These data clearly indicated that the C domain of *Arabidopsis* CAO is also sufficient for the reaction in higher plants. In the same report, Yamasato et al. provided evidence which suggested that the A domain plays a key role in controlling the protein level of CAO, and consequently, the level of Chl *b* biosynthesis. This topic is described in greater detail in the next section. On the other hand, we have found that the B domain is neither necessary for the catalysis, nor for the regulation of the CAO protein level (Sakuraba et al., 2007). It is possible that it serves as a linker for the stabilization of the overall protein structure of CAO.

The CAO sequences of two Chlorophyceae species, *Chlamydomonas reinhardtii* and *Dunaliella salina* have an N-terminal extension whose size is equivalent to those of the A and B domains of higher plants (Nagata et al., 2004). However, this region does not exhibit any significant sequence similarity to either the A or B domain of higher plants (Yamasato et al., 2005). Nevertheless, since the N-terminal extension of Chlorophyceae CAO is longer than the average size of transit peptides (Nagata et al., 2004), it would be reasonable to assume that it may have certain functions like the A and B domains from higher plants. The region within the *Chlamydomonas* CAO corresponding to the B domain of higher plant CAO is predicted to be highly hydrophilic, it is therefore tempting to speculate that the region functions as a linker.

C Distribution of Chl b Reductase

Although searches in the public sequence databases identified *NYC1/NOL* homologues in green plants such as *A. thaliana,* and *C. reinhardtii,* and in *Ostreococcus* species, none have been identified in the sequenced genomes of the Prochlorophytes (Kusaba et al., 2007). Considering that the reduction of Chl *b* is the first step of Chl *b* breakdown, it is possible that the Prochlorophytes do not degrade Chl *b*. Instead, it is possible that they excrete Chl *b* to their surrounding environments. On the other hand, these organisms may contain a Chl *b* reductase that is different from NYC1/NOL.

In contrast to higher plants, *C. reinhardtii* and the Prasinophyceae apparently possess only one copy of a *NYC1/NOL* homologue. Using phylogenetic analyses alone makes it difficult to discriminate whether these *NYC1/NOL* homologues are *NYC1* or *NOL,* because the algal sequences are similar to the *NYC1* and *NOL* genes from higher plants. One remarkable feature of the algal *NYC1/NOL* homologues is that they do not encode any putative membrane-spanning regions. Thus, we assumed that the ancestral Chl *b* reductase was a soluble protein or a membrane-bound protein. In this regard, the ancestral protein becomes more similar to NOL. During evolution, the algal gene for Chl *b* reductase may have been duplicated and one of the duplicates may have acquired a sequence that encodes three membrane-spanning regions. If we consider that Chl *b* reductase should obtain its substrate (Chl *b*) directly or indirectly from LHC, which binds the majority of Chl *b* in higher plants, it would be feasible to speculate that an ancestral *NOL* (or *NYC1*) obtained a DNA sequence that encodes membrane-spanning regions to facilitate better interaction of the gene product with LHC. Future biochemical studies should be able to clarify whether any interaction does occur between LHC and Chl *b*

IV Regulation of the Chl Cycle

A Regulation of the Chl a to b Conversion

1 Transcriptional Control

To the best of our knowledge, the activity of the Chl *a* to *b* conversion is regulated at the transcriptional level of CAO as well as through the protein stability of CAO. The steady state levels of the *CAO* mRNA were examined in *A. thaliana* (Espineda et al., 1999; Harper et al., 2004; Tanaka and Tanaka, 2005), tobacco (Pattanayak et al., 2005), and in a green alga, *Dunaliella salina* (Masuda et al., 2002; Masuda et al., 2003). In all of these organisms, *CAO* mRNA levels were lower under stronger light conditions than under weaker light conditions. These observations are compatible with the higher Chl *a* to *b* ratios that occur under strong light conditions. Since it is generally believed that higher Chl *a* to *b* ratios reflect smaller photosynthetic antenna sizes, the authors concluded that the changes in the steady state *CAO* mRNA level was well correlated with the antenna size of these photosynthetic organisms. It is important to note that the expression patterns of *CAO* are parallel to those of the *Lhcb* genes encoding the major light-harvesting proteins of photosystem (PS) II (Masuda et al., 2002, 2003; Harper et al., 2004; Tanaka and Tanaka, 2005).

2 The Signal Transduction Pathway

Masuda et al. further examined the signal transduction pathway leading to the regulation of the steady-state level of *CAO* mRNA in *D. salina* (Masuda et al., 2003). They found that inhibition of the photosynthetic electron transport by DCMU (3-(3,4-dichlorophenyl)-1,1-dimethylurea) or DBMIB (dibromothymoquinone) did not effect the rapid changes in the *CAO* mRNA levels. Conversely, a series of inhibitors which block the Ca^{2+}-dependent signal transduction pathway in the cytosol exhibited a clear impact on the expression of *CAO.* These observations suggested that a short-term (within a few hours) transcriptional control of *CAO* in response to light intensities were primarily mediated by a signal transduction pathway, which was initiated by the perception of light via a photoreceptor in the cytosol, rather than by the excitation pressure of the photosynthetic electron transport chain in the chloroplasts of this alga.

Similarly, the transcription of *CAO* appeared to be controlled by a photoreceptor-mediating the signal transduction pathway in higher plants. Micro-array data describing the gene-expression profiles of higher plants under various environments indicated a strong correlation between *CAO* and *Lhc* gene expression (Jen et al., 2006). Since *Lhc* genes are known to be regulated by two different groups of photoreceptors, namely phytochrome and cryptochrome, (Quail, 2002), it would be reasonable to assume that *CAO* is under the control of these photoreceptors. By using a selected probe set of a DNA array, Matsumoto et al (Matsumoto et al., 2004) were able to effectively analyze the expression profiles of all genes encoding Chl-biosynthesis enzymes that were available at the time of analysis. The researchers found that expression of *CAO* correlated well with that of *HemA* and *ChlH,* two genes which encode two key enzymes of Chl biosynthesis, namely glutamyl-tRNA reductase and a subunit of the magnesium chelatase, respectively. The *CAO, Lhc, HemA* and *ChlH* genes exhibited light-induced expression patterns with oscillations that were indicative of a circadian rhythm response. These results indicated that all of these genes were under the control of the common photoreceptors and the circadian regulators. Such a coordination of gene expression for *CAO, Lhc,* and other key enzymes of Chl biosynthesis may facilitate an efficient coupling of Chl biosynthesis with the construction of the photosynthetic machinery. It should be noted that gene expression of *CAO* and *Lhc* are not proportionally related to the protein levels of these gene products. This disparity indicates that the transcription of these genes is not the rate-limiting step for the synthesis of their respective protein products. In other words, the coordinated expression of these genes may function to assure that there is a sufficient supply of their gene products. Instead, regulation at a post-transcriptional level may be the determining factor which regulates the real level for protein accumulation. Actually, CAO activity is more strictly regulated at the post-transcriptional level than it is at the transcriptional level. This phenomenon will be described in greater detail in the following section.

3 Post-transcriptional Control

As described in the previous section, we suspected that the A and B domains of CAO enzymes in higher plants have additional functions (Nagata et al., 2004; Yamasato et al., 2005). Therefore, we overexpressed several variants of CAO-coding regions which lacked one or two of the CAO domains in the background of a CAO deficient mutant of *Arabidopsis*. This was carefully performed as a means to dissect the function of each domain. Overexpression of the full-length CAO in *Arabidopsis* only resulted in a small effect on the synthesis of Chl *b*. In comparison to wild type plants, the mutants exhibited a reduction in their Chl *a* to *b* ratio from 3.0 in wild type to 2.8 in the transgenic plants (Tanaka et al., 2001; Tanaka and Tanaka, 2005; Yamasato et al., 2005). In contrast, overexpression of the truncated CAO which lacked the A domain markedly enhanced the synthesis of Chl *b*. In these transgenic plants, the Chl *a* to *b* ratio was reduced down to 2.2 in the transgenic plants (Yamasato et al., 2005). In accordance with this result, a high level accumulation of CAO was detected in the A-domain-less overexpressor. On the other hand, the CAO protein did not accumulate to a detectable level in the full-length CAO overexpressor. These results indicated that the A domain has a role in controlling the Chl *b* level by repressing the accumulation of the CAO protein.

Interestingly, when the A domain was overexpressed in wild-type *Arabidopsis*, the protein failed to accumulate. On the other hand, when it was overexpressed in the Chl *b*-less mutant, *ch1*, the protein accumulated to a significant amount (Yamasato et al., 2005). These observations indicate that Chl *b* accumulation induces the degradation of the A domain. Alternatively, one may speculate that CAO becomes destabilized when CAO catalyzes the reaction of Chl *a*-to-Chl *b* conversion. However, this hypothesis is unlikely, because the A domain was degraded in wild type despite that this domain alone didn't posses the Chl *b* synthesizing activity. In other words, the aforementioned experiments indicate that Chl *b* molecules supplied by the intrinsic CAO protein were likely to induce the degradation of the A domain. Therefore, we speculated that the A domain functions as a kind of a degradation tag that induces the destabilization of CAO when Chl *b* accumulated to a certain level. At present, it is not clear how Chl *b* accumulation triggers the degradation of CAO.

In order to understand the mechanism of CAO destabilization, Nakagawara et al. (2007) screened a mutagenized pool of *Arabidopsis* for a suppressor mutant in which the activity for CAO destabilization was impaired. They successfully isolated a mutant which exhibited a higher level of accumulation of CAO and subsequently named it *de-regulated accumulation of CAO (dca)1*. The mutation was mapped to the *ClpC1* gene locus which encodes a subunit of a plastid Clp protease. These data indicate that Clp protease is responsible for the degradation of CAO. However, since the effects of the *dca1* mutation were limited only to the cotyledonous stage of *Arabidopsis*, it should be noted that another protease might be involved in the degradation of CAO.

B Regulation of the Chl b to a Conversion

In contrast to our extensive understanding regarding the regulatory mechanism of the Chl *a* to *b* conversion, our knowledge pertaining to the other half of the Chl cycle is very limited. Ito et al. (1993, 1994) and Scheumann et al. (1996) detected the activity of the Chl *b* to *a* conversion in plastids isolated from etiolated seedlings where the biosynthesis of Chl had not yet occurred. In addition, Scheumann et al. (1999) found this activity in young oat seedlings as well as in senescent barley leaves. Therefore, it is plausible that the Chl *b* to *a* activity exists throughout multiple developmental stages of leaves. The activity that was detected during the vegetative stage may be related to the adjustment of the Chl *a* to *b* ratio that is optimal for the construction of the photosynthetic machinery. On the other hand, the activity that is present during the senescence of leaves may play a role in the degradation of Chl *b*. This topic will be discussed in a later section.

The mechanisms which control the activity of the Chl *b* to *a* conversion are still not well understood. In particular, nothing is known about the regulation of HM-Chl *a* reductase. This lack of knowledge is primarily due to the fact that this enzyme has not yet been identified. Since the genes encoding Chl *b* reductase have only been recently identified, researchers have now begun to examine its regulatory mechanisms. Kusaba et al. (2007) found that the *NYC1* gene encoding the membrane-spanning isoform of Chl *b* reductase is expressed during leaf senescence in rice. Microarray data which are deposited in the public database at the Nottingham Arabidopsis Stock Centre (NASC) for the expression profiles of Arabidopsis genes (analyzed at the Genevestigator web site https://www.genevestigator.ethz.ch/) also supported Kusaba's results by showing the induction of *NYC1* expression during leaf senescence in *Arabidopsis*. According to the same database, the expression of the other isoform, *NOL,* is weaker but more persistent throughout developmental stages than that of *NYC1* in *Arabidopsis*. It is possible that the membrane-spanning NYC1 functions during leaf senescence, while NOL operates constitutively to adjust Chl *a* to *b* ratios in leaves. Little is known about the signal transduction pathway which leads to the expression of *NYC1* and *NOL*.

It is even unclear whether the expression patterns of these genes are reflected by their protein levels. In other words, little is known about the post-transcriptional control of the Chl *b* reductase activity. It is possible that the activity of this enzyme is regulated by other proteins. One possible candidate for such a regulatory protein is the stay-green related (SGR) protein, which was recently identified through analysis of a stay-green mutant of rice and *Arabidopsis* (Park et al., 2007; Ren et al., 2007). Mutations in the *SGR* gene that led to a phenotype similar to that of *NYC1* in rice, resulted in delayed degradation of thylakoid membrane and LHC proteins (Park et al., 2007). In addition, the expression pattern of the *SGR* gene is correlated with that of *NYC1* (data not shown). Therefore, it is possible to speculate that the SGR protein is an enhancer of NYC1 activity. This idea may explain why both proteins are essential for the degradation of Chl *b,* LHC and thylakoid membrane. On the other hand, NOL does not appear to require any additional protein for its activity. This was concluded because the recombinant NOL protein that was expressed in *E. coli* showed high Chl *b* reductase activity (Horie et al., 2009).

V Roles of the Chl Cycle in the Construction of the Photosynthetic Apparatus

The Chl cycle has two main roles related to the physiology of plants. It adjusts Chl *a* to *b* ratios that are optimal for the construction of the photosynthetic machinery, and it provides the first and prerequisite step for Chl *b* degradation during leaf senescence. In the following section, we first describe the evidence that shows the essential role of the Chl cycle in the construction of the photosynthetic apparatus. We then attempt to construct a model which illustrates how plants coordinate Chl biosynthesis and the construction of the photosynthetic machinery. Finally, we will discuss how it is possible that the Chl cycle can regulate the degradation of thylakoid membranes during leaf senescence.

A Coordination of the Chl cycle and the Construction of the Photosynthetic Apparatus

Many photosynthetic proteins bind Chl molecules. For example, the light-harvesting core complex protein of photosystem II (PS II), CP47, binds 16 Chl *a* molecule (Barber, 2006). On the other hand, the light-harvesting complex II (LHCII) binds 8 Chl *a* and 6 Chl *b* molecules (Liu et al., 2004). The proteins in the core complex of PSII and PSI only bind Chl *a*, and those in the peripheral antenna complexes bind both Chl *a* and Chl *b*. Since plants often alter the ratios of the core complexes to the peripheral antenna complexes as a means of acclimating to varying environments, the demand for Chl *a* and *b* molecules should change accordingly. Therefore, it would be reasonable to assume that the Chl cycle activity is controlled in order to meet the demand for different Chl *a* to *b* ratios in the photosynthetic apparatus.

There are several lines of evidence showing that the rate of Chl *a* to *b* conversion is partly determined by the balance between the synthesis of Chl *a* and that of the core complex proteins (Fig. 5). When etiolated cucumber seedlings were fed 5-aminolevulinic acid, a precursor of Chl biosynthesis, before the initiation of greening, Chl biosynthesis was boosted by more than two times. In addition, the accumulation of Chl *b* as well as LHCII was markedly increased (Tanaka et al., 1992). Similar effects were observed when Chl synthesis was enhanced by treatment with cytokinin (Tanaka et al., 1993). Tzinas and Argyroudi-Akoyunoglou found that inhibition of plastid protein synthesis also increased the accumulation of Chl *b* and LHCII in *Phaseolus vulgaris* (Tzinas and Argyroudi-Akoyunoglou, 1988). They interpreted this effect as the competition for available Chl *a* molecules during the synthesis of both the core complex and Chl *b*. It is not clear whether this is a real competition or whether it is simply a mechanism employed by plants to control the activity of the Chl cycle which is dependent upon the synthesis of the core complex. It is reasonable that a priority would be given for the synthesis of the core complexes. On the other hand, increasing Chl *b* synthesis without a sufficient amount of the core complex would be useless because Chl *b* ultimately transfers its absorbed energy to the core complexes.

Given that CAO retrieves Chl *a* molecules that are not incorporated into the core complexes, a few questions remain regarding "where" and "how" the enzyme does so. There is no doubt that the construction of the core complexes occurs in thylakoid membranes (Zhang et al., 2001). In contrast, it is proposed that the assembly of LHC and Chl molecules occur in the envelope membrane (Hoober and Eggink, 2001; Reinbothe et al., 2006). Therefore, it is crucial to determine the exact location of LHC assembly as a means to understand the mechanism of how CAO retrieves Chl *a* and passes it to LHCII. It is also proposed that most of the enzymes catalyzing the late steps of Chl biosynthesis, including CAO, occur in both thylakoid and envelope membranes (Eckhardt et al., 2004; Yamasato et al., 2005).

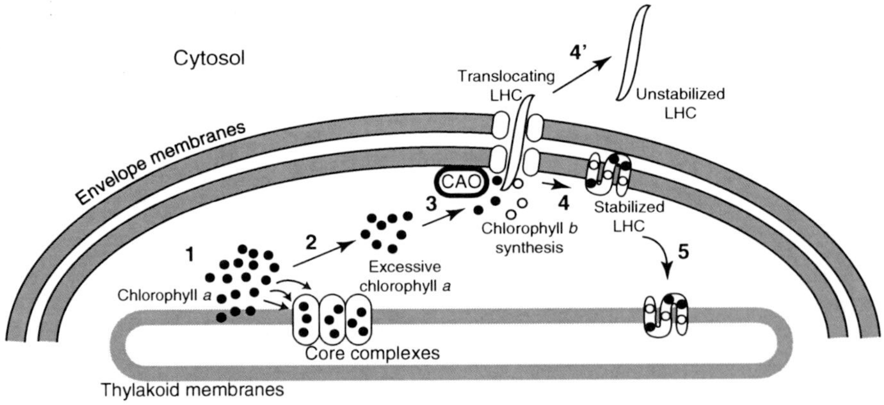

Fig. 5. Hypothetical model showing the coordination of the Chl *a* to *b* conversion and the construction of the photosynthetic apparatus. 1. Chl *a* is synthesized in envelope and/or thylakoid membranes. Part of the Chl *a* binds to the newly synthesized apoproteins of the core complexes on thylakoid membranes. 2. Excessive Chl *a* molecules that are not incorporated in the core complexes may be transferred to envelope membranes by unknown mechanisms. 3. Chl *a* interacts with the apoproteins of LHC that are being translocated across envelope membranes. CAO converts approximately half of the Chl *a* molecules that are bound to translocating LHC. 4. LHC is transformed into the mature form by binding to both Chl *a* and Chl *b*. 4′. If the amount of LHC surpasses that of Chl molecules, the excess amount of LHC would be moved from envelope membranes to the cytosol. 5. Finally, LHC in its mature form is transferred to thylakoid membranes.

CAO doesn't simply pass its product, Chl *b*, to LHC proteins. Instead, it plays an essential role in the construction of LHC. There are several lines of evidence showing that Chl *b* synthesis controls LHC accumulation. First of all, a few research groups independently reported that LHC accumulation is not controlled at a transcriptional level (Flachmann and Kuehlbrandt, 1995; Flachmann, 1997; Andersson et al., 2003), even though a minimal level of LHC transcripts are required to maintain the synthesis of LHC precursor (Ruban et al., 2003). It is likely that the accumulation of LHC is regulated by the assembly with pigments, especially with Chl *b*. This idea is supported by observations with Chl *b*-less mutants of barley or *Arabidopsis*, in which the accumulation of major isoforms of LHC were markedly decreased (Harrison et al., 1993; Krol et al., 1995; Bossmann et al., 1997; Tanaka and Tanaka, 2005). Through biochemical studies, Kuttkat et al. (1997) demonstrated that LHC was only stabilized by the presence of a Chl *b* derivative namely Zn-pheophytin *b*. On the other hand, when an in-vitro synthesized LHC precursor was incubated with isolated plastids, a similar Chl *a* derivative (Zn-pheophytin *a*) did not show any stabilization effect. Furthermore, it was observed that enhanced Chl *b* biosynthesis in CAO-overexpressing transgenic plants led to a slight or moderate increase in the steady-state level of LHC proteins (Tanaka et al., 2001; Tanaka and Tanaka, 2005; Hirashima et al., 2006). Therefore, it is likely that CAO controls the accumulation of LHC, up to a certain level, through the synthesis of Chl *b*. It should also be noted that additional factors other than Chl *b* likely exist and function to influence the level of LHC. This statement is based on the observations that the level of Chl *b* is not always proportional to that of LHC (Pattanayak et al., 2005; Hirashima et al., 2006). In the model of Chl *b*-dependent LHC accumulation, LHC precursors that fail to assemble with Chl *b* should be removed from the thylakoid membrane. In this respect, it has been proposed that premature LHC molecules are degraded by a metalloprotease (Lindahl et al., 1995; Yang et al., 2000; Zelisko et al., 2005). Alternatively, it is possible that they are removed by excretion from the chloroplasts (Hoober and Eggink, 1999; Hoober and Eggink, 2001). Molecular entities that are responsible for either mechanism have not yet been identified.

It would be noteworthy that the Chl *b*-dependent LHC stabilization mechanism has several advantages for plant cells. This mechanism ensures that all excessive Chl molecules for the core complexes are incorporated into LHC proteins. This is quite essential because unbound Chl molecules readily transfer their excited energy to molecular oxygen (Nyman and Hynninen, 2004). This would consequently result in the generation of a highly-toxic singlet oxygen. In this sense, LHC appears to function as a safe shelter for Chl molecules. The degradation of pigment-less LHC may be also important for plant cells. This is suggested because if such a pigment-less LHC is incorporated into the photosynthetic apparatus, it should block an efficient transfer of energy between functional LHC and core complexes.

The functional role of CAO is not only limited to the stabilization of LHC. It is also proposed that the enzyme is involved in the proper distribution of Chl *a* and *b* to the core complexes and LHC. When a cyanobacterial CAO was overexpressed in *Arabidopsis*, Hirashima et al. reported that a substantial amount of Chl *b* was incorporated into the core complexes (Hirashima et al., 2006). The Chl *a* to *b* ratios in the leaves of the overexpressors decreased to 1.1 from the wild-type value of 3.0. The ratios obtained from CP1, the core complex of PSI, decreased from 10.6 to 1.58. Surprisingly, there was no loss of the photosynthetic activity in the overexpressor when compared with wild type (Hirashima et al., 2006). These results showed that CP1 could accommodate a substantial amount of Chl *b* without a significant loss of activity. It is therefore suggested that some unknown mechanism prevents Chl *b* from binding to CP1 in the wild type. Therefore, it is possible that CAO plays a role in preferentially distributing Chl *b* to LHC. Since a population of CAO is localized in the envelope membrane in wild type plants, we speculate that this population preferentially distributes Chl *b* to LHC. This speculation is based upon the observation with the cyanobacterial CAO overexpressing plants in which CAO is mainly localized in thylakoid membranes (Hirashima et al., 2006). It is likely that this localization is primarily due to the lack of the A domain. It is also possible that the incorrect subcellular localization of the cyanobacterial CAO may have resulted in the loss of selectivity in the distribution of Chl *b*.

Another interesting aspect of CAO-LHC interaction was reported by Xu et al. (2001). In their study, they demonstrated that LHC enhances the activity of CAO in a yet-unidentified manner. A strain of cyanobacteria which lacked PSI was transformed with the *Arabidopsis CAO* gene. They found that the transformant produced only a small amount of Chl *b*. Surprisingly, co-transformation with a gene encoding LHC markedly increased Chl *b* synthesis up to 50% of the total Chl in the same strain. Chl *b* was incorporated in the photosynthetic apparatus, but did not lead to any detectable decrease in photosynthetic activity. Since the accumulation of LHC in this strain was below a detectable level, the authors interpreted these results as indicating that LHC did not function as a reservoir of Chl *b*, but that it functioned as an enhancer of CAO activity. Eggink and Hoober proposed a mechanistic model implicating a role of CAO in the stabilization of LHC (Hoober and Eggink, 1999, 2001). According to their model, chlorophyllide *a* binds a specific sequence, called a retention motif, that is EXXNXR or EXXHXR within the LHC sequence. This binding occurs while a premature LHC protein is being translocated from the cytosol into the envelope membrane (Fig. 5). Then, chlorophyllide *a* is converted to chlorophyllide *b* by CAO. In their model, this step is a prerequisite to hold the translocating LHC and to prevent it from being sorted out to the cytosol. Recently, Reinbothe et al. (2006) reported that chloroplasts isolated from a *CAO*-deficient mutant of *Arabidopsis* had reduced import ability for LHC proteins to approximately 0.6% of the wild-type level. Furthermore, they showed a direct interaction of CAO with LHC. Taken together, their experiments strongly support the model of Eggink and Hoober (Hoober and Eggink, 1999, 2001). However, many questions still remain unanswered regarding the mechanism of how Chl *b* regulates LHC stabilization and how CAO and LHC interact with each other. What role do the CAO proteins that are localized in the thylakoid membrane play? Where is Chl *a* synthesized and how is it incorporated into the core complex? Since Chl *a* molecules that are not incorporated into the core complex in thylakoid membranes are finally used by CAO in the envelope membrane, how is Chl *a* translocated from thylakoid to envelope membranes? In order to answer these questions, further investigations on the spatial distribution of CAO and the LHC precursor, and on their interaction are warranted.

B Construction and Deconstruction of the Photosynthetic Apparatus and Its Coordination with the Chl b to a Conversion System

The conversion of Chl *b* to *a* is also coordinated with the construction of the photosynthetic machinery. During the 1980s, two groups independently found that a decrease in the Chl *b* contents concurrently occurred with an increase in the Chl *a* contents in darkness without a net change in the total Chl contents in cucumber (Tanaka and Tsuji, 1983) and tobacco (Ikegami et al., 1984). Interestingly, the changes in the Chl *a* to *b* ratios are accompanied by an increase in the amount of the core complex proteins. These changes are also accompanied by a decrease in the amount of LHC. These observations were interpreted as the consequence of Chl *b* to *a* conversion. It is possible that such changes in Chl *a* to *b* ratios within a relatively short term (from several hours up to 1 day) may be part of the response of plants to changing environments, including a response to higher or lower light intensities.

Recently, Kusaba et al. reported another example that showed the coordination between the conversion of the Chl *b* to *a* and the degradation of LHC (Kusaba et al., 2007). As described in a previous section, Chl *b* is necessary for the stabilization of LHC. In the *nyc1* mutant of rice, deficiency in the Chl *b* to *a* conversion impairs LHC degradation during senescence (Kusaba et al., 2007). This observation suggested that the conversion of Chl *b* to *a* is necessary to initiate the degradation of LHC. In summary, coordinated changes in Chl biosynthesis, synthesis of the core complex proteins, the Chl cycle activity, and the stabilization of LHC may play key roles in controlling the composition of the photosynthetic machinery.

The second role of the Chl cycle is to provide a route for the degradation of Chl *b*. Since at least one enzyme of the Chl degradation pathway (pheophorbide *a* oxygenase) catalyzes the breakdown of the *a* form of the Chl catabolites (Hortensteiner et al., 1995), it is thought that plants must transform the *b* form into the *a* form prior to this step. Since the enzymes that convert Chl *a* into pheophorbide *a* do not appear to posses a substrate specificity toward the *a* form, it is conceivable that inhibition of Chl *b* to *a* conversion may result in the accumulation of pheophorbide *b*. In *nyc1*, the inability to catalyze the Chl *b* to *a* conversion led to an accumulation of Chl *b* instead of pheophorbide *b* (Kusaba et al., 2007). This observation indicates that Chl *b* is not accessible to chlorophyllase, the enzyme which catalyzes the immediate downstream reaction of the conversion of Chl *b* to *a* (Hortensteiner, 2006). Therefore, we can conclude that Chl *b* reductase functions as a prerequisite for making Chl *b* catabolites accessible to downstream enzymes. If we take into an account the remarkably slow degradation of LHC in rice *nyc1* mutants compared with wild type, it is possible that Chl *b* reductase is involved in the removal of Chl *b* molecules from LHC. It is likely that the detachment of the pigment probably transforms LHC into its protease-sensitive form. Actually, we recently published a data which shows that Chl *b* reductase can react with Chl *b* molecules that are still bound to LHC (Horie et al., 2009).

The study of Kusaba et al. (2007) also suggested that the Chl *b* reductase reaction is required not only for LHC degradation, but also for thylakoid membrane disintegration during leaf senescence. In the *nyc1* mutant, the degradation of thylakoid membranes was markedly retarded. In addition, both the population and the degree of appressed regions were significantly increased during senescence (Kusaba et al., 2007). These observations were most likely attributed to delayed LHC degradation. This conclusion is in accordance with the proposal that both the negative and positive segments of the stromal surface of LHC produce a major force to stack thylakoid membranes (Standfuss et al., 2005). The degradation of LHC and thylakoid membranes provides a vital source of nitrogen that has to be remobilized from leaves to the reproductive organs during transition from the vegetative to the reproductive phase of the plant life cycle (Thomas, 1997). Therefore, it is reasonable to conclude that Chl *b* reductase plays an important role during leaf senescence.

Epilogue

Although the study on the interconversion of Chl *a* and *b* has a long history, our understanding on its reaction chemistry, its regulatory mechanisms and its evolution is limited. Identification of HM-Chl *a* reductase might be the first step toward gaining a better understanding of the all aspects of the Chl cycle. Subsequently, biochemical analyses of the interaction of the Chl cycle enzymes and photosynthetic proteins may reveal how the photosynthetic machinery acclimates to different light conditions. Furthermore, studies of the diversity of the enzymes involved in the Chl cycle may reveal a part of the evolutionary history of photosynthetic organisms.

Acknowledgements

We thank Prof. Isamu Inouye and Ms. Tomoko Chikuni for helpful discussion and for sharing unpublished results with us. We are grateful to Ms. Junko Kishimoto for illustrations. We acknowledge the financial support from the Grant-in-Aid for Creative Scientific Research (17GS0314) to AT and the Grant-in-Aid for Scientific Research (19687003) to RT from the Japanese Ministry of Education, Culture, Sports, Science and Technology.

References

Andersson J, Wentworth M, Walters RG, Howard CA, Ruban AV, Horton P and Jansson S (2003) Absence of the Lhcb1 and Lhcb2 proteins of the light-harvesting complex of photosystem II – effects on photosynthesis, grana stacking and fitness. Plant J 35: 350–361

Arabidopsis Genome Initiative (2000) Analysis of the genome sequence of the flowering plant *Arabidopsis thaliana*. Nature 408: 796–815

Barber J (2006) Photosystem II: an enzyme of global significance. Biochem Soc Trans 34: 619–631

Barber J, Morris E and Buchel C (2000) Revealing the structure of the photosystem II chlorophyll binding proteins, CP43 and CP47. Biochim Biophys Acta 1459: 239–247

Bossmann B, Knoetzel J and Jansson S (1997) Screening of chlorina mutants of barley (Hordeum vulgare L.) with antibodies against light-harvesting proteins of PS I and PS II: absence of specific antenna proteins. Photosynth Res 52: 127–136

Bryant DA and Frigaard NU (2006) Prokaryotic photosynthesis and phototrophy illuminated. Trends Microb 14: 488–496

Chen M, Quinnell RG and Larkum AW (2002) The major light-harvesting pigment protein of *Acaryochloris marina*. FEBS Lett 514: 149–152

Courties C, Perasso R, Chretiennot-Dinet MJ, Gouy M, Guillou L and Troussellier M (1998) Phylogenetic analysis and genome size of *Ostreococcus tauri* (Chlorophyta, Prasinophyceae). J Phycol 34: 844–849

Derelle E, Ferraz C, Rombauts S, Rouze P, Worden AZ, Robbens S, Partensky F, Degroeve S, Echeynie S, Cooke R, Saeys Y, Wuyts J, Jabbari K, Bowler C, Panaud O, Piegu B, Ball SG, Ral JP, Bouget FY, Piganeau G, De Baets B, Picard A, Delseny M, Demaille J, Van de Peer Y and Moreau H (2006) Genome analysis of the smallest free-living eukaryote *Ostreococcus tauri* unveils many unique features. Proc Natl Acad Sci USA 103: 11647–11652

Dufresne A, Salanoubat M, Partensky F, Artiguenave F, Axmann IM, Barbe V, Duprat S, Galperin MY, Koonin EV, Le Gall F, Makarova KS, Ostrowski M, Oztas S, Robert C, Rogozin IB, Scanlan DJ, Tandeau de Marsac N, Weissenbach J, Wincker P, Wolf YI and Hess WR (2003) Genome sequence of the cyanobacterium Prochlorococcus marinus SS120, a nearly minimal oxyphototrophic genome. Proc Natl Acad Sci USA 100: 10020–10025

Eckhardt U, Grimm B and Hortensteiner S (2004) Recent advances in chlorophyll biosynthesis and breakdown in higher plants. Plant Mol Biol 56: 1–14

Eggink LL, LoBrutto R, Brune DC, Brusslan J, Yamasato A, Tanaka A and Hoober JK (2004) Synthesis of chlorophyll b: localization of chlorophyllide a oxygenase and discovery of a stable radical in the catalytic subunit. BMC Plant Biol 4: 5

Espineda C, Linford A, Devine D and Brusslan J (1999) The AtCAO gene, encoding chlorophyll a oxygenase, is required for chlorophyll b synthesis in *Arabidopsis thaliana*. Proc Natl Acad Sci USA 96: 10507–10511

Flachmann R (1997) Composition of photosystem II antenna in light-harvesting complex II antisense tobacco plants at varying irradiances. Plant Physiol 113: 787–794

Flachmann R and Kuehlbrandt W (1995) Accumulation of plant antenna complexes is regulated by post-transcriptional mechanisms in tobacco. Plant Cell 7: 149–160

Folly P and Engel N (1999) Chlorophyll b to chlorophyll a conversion precedes chlorophyll degradation in Hordeum vulgare L. J Biol Chem 274: 21811–21816

Garczarek L, Van Der Staay GWM, Thomas JC and Partensky F (1998) Isolation and characterization of Photosystem I from two strains of the marine oxychlorobacterium Prochlorococcus. Photosynth Res 56: 131–141

Gough SP, Petersen BO and Duus JO (2000) Anaerobic chlorophyll isocyclic ring formation in Rhodobacter capsulatus requires a cobalamin cofactor. Proc Natl Acad Sci USA 97: 6908–6913

Green BR and Durnford DG (1996) The chlorophyll-carotenoid proteins of oxygenic photosynthesis. Annu Rev Plant Physiol Plant Mol Biol 47: 685–714

Harayama S, Kok M and Neidle EL (1992) Functional and evolutionary relationships among diverse oxygenases. Ann Rev Microb 46: 565–601

Harper AL, von Gesjen SE, Linford AS, Peterson MP, Faircloth RS, Thissen MM and Brusslan JA (2004) Chlorophyllide a oxygenase mRNA and protein levels correlate with the chlorophyll a/b ratio in *Arabidopsis thaliana*. Photosynth Res 79: 149–159

Harrison MA, Nemson JA and Melis A (1993) Assembly and composition of the chlorophyll a-b light-harvesting complex of barley (Hordeum vulgare L.): Immunochemical analysis of chlorophyll b-less and chlorophyll b-deficient mutants. Photosynth Res 38: 141–151

Hirashima M, Satoh S, Tanaka R and Tanaka A (2006) Pigment shuffling in antenna systems achieved by expressing prokaryotic chlorophyllide a oxygenase in Arabidopsis. J Biol Chem 281: 15385–15393

Hirono Y and Redei GP (1963) Multiple allelic control of chlorophyll b level in *Arabidopsis thaliana*. Nature 197: 1324–1325

Hoober J and Eggink L (2001) A potential role of chlorophylls b and c in assembly of light-harvesting complexes. FEBS Lett 489: 1–3

Hoober JK and Eggink LL (1999) Assembly of light-harvesting complex II and biogenesis of thylakoid membranes in chloroplasts. Photosynth Res 61: 197–215

Horie Y, Ito H, Kusaba M, Tanaka R and Tanaka A (2009) Participation of chlorophyll *b* reductase in the initial step of the degradation of light-harvesting chlorophyll *a/b*-protein complexes in *Arabidopsis*. J Biol Chem 284: 17449–17456

Hortensteiner S (2006) Chlorophyll degradation during senescence. Annu Rev Plant Biol 57: 55–77

Hortensteiner S, Vicentini F and Matile P (1995) Chlorophyll breakdown in senescent cotyledons of rape, brassica-napus L – enzymatic cleavage of phaeophorbide-a in-vitro. New Phytol 129: 237–246

Ikegami I, Kamiya A and Hase E (1984) Dark formation of chlorophyll in cultured tobacco cells. Plant Cell Physiol 25: 343–348

International Rice Genome Sequencing Project (2005) The map-based sequence of the rice genome. Nature 436: 793–800

Ito H, Ohtsuka T and Tanaka A (1996) Conversion of chlorophyll b to chlorophyll a via 7-hydroxymethyl chlorophyll. J Biol Chem 271: 1475–1479

Ito H, Takaichi S, Tsuji H and Tanaka A (1994) Properties of synthesis of chlorophyll a from chlorophyll b in cucumber etioplasts. J Biol Chem 269: 22034–22038

Ito H, Tanaka Y, Tsuji H and Tanaka A (1993) Conversion of chlorophyll b to chlorophyll a by isolated cucumber etioplasts. Arch Biochem Biophys 306: 148–151

Jen CH, Manfield IW, Michalopoulos I, Pinney JW, Willats WG, Gilmartin PM and Westhead DR (2006) The Arabidopsis co-expression tool (ACT): a WWW-based tool and database for microarray-based gene expression analysis. Plant J 46: 336–348

Junker F, Kiewitz R and Cook AM (1997) Characterization of the p-toluenesulfonate operon tsaMBCD and tsaR in Comamonas testosteroni T-2. J Bacteriol 179: 919–927

Kolossov VL and Rebeiz CA (2003) Chloroplast biogenesis 88. Protochlorophyllide b occurs in green but not in etiolated plants. J Biol Chem 278: 49675–49678

Krol M, Spangfort MD, Huner NPA, Oquist G, Gustafsson P and Jansson S (1995) Chlorophyll a/b-binding proteins, pigment conversions, and early light-induced proteins in a chlorophyll b-less barley mutant. Plant Physiol 107: 873–883

Kupke DW and Huntington JL (1963) Chlorophyll a appearance in the dark in higher plants: analytical notes. Science 140: 49–51

Kusaba M, Ito H, Morita R, Iida S, Sato Y, Fujimoto M, Kawasaki S, Tanaka R, Hirochika H, Nishimura M and Tanaka A (2007) Rice NON-YELLOW COLORING1 is involved in light-harvesting complex II and grana degradation during leaf senescence. Plant Cell 19: 1362–1375

Kuttkat A, Edhofer I, Eichacker L and Paulsen H (1997) Light-harvesting chlorophyll a/b-binding protein stably inserts into etioplast membranes supplemented with Zn-pheophytin a/b. J Biol Chem 272: 20451–20455

Lee S, Kim JH, Yoo ES, Lee CH, Hirochika H and An G (2005) Differential regulation of chlorophyll a oxygenase genes in rice. Plant Mol Biol 57: 805–818

Lindahl M, Yang DH and Andersson B (1995) Regulatory proteolysis of the major light-harvesting chlorophyll a/b protein of photosystem II by a light-induced membrane-associated enzymic system. Eur J Biochem 231: 503–509

Liu Z, Yan H, Wang K, Kuang T, Zhang J, Gui L, An X and Chang W (2004) Crystal structure of spinach major light-harvesting complex at 2.72 ANG resolution. Nature (London) 428: 287–292

Masuda T, Polle JE and Melis A (2002) Biosynthesis and ting recovery of the green alga *Dunaliella salina* from irradiance stress. Plant Physiol 128: 603–614

Masuda T and Takamiya K (2004) Novel insights into the enzymology, regulation and physiological functions of light-dependent protochlorophyllide oxidoreductase in angiosperms. Photosynth Res 81: 1–29

Masuda T, Tanaka A and Melis A (2003) Chlorophyll antenna size adjustments by irradiance in *Dunaliella salina* involve coordinate regulation of chlorophyll a oxygenase (CAO) and Lhcb gene expression. Plant Mol Biol 51: 757–771

Matsumoto F, Obayashi T, Sasaki-Sekimoto Y, Ohta H, Takamiya K and Masuda T (2004) Gene expression profiling of the tetrapyrrole metabolic pathway in Arabidopsis with a mini-array system. Plant Physiol 135: 2379–2391

Nagata N, Satoh S, Tanaka R and Tanaka A (2004) Domain structures of chlorophyllide a oxygenase of green plants and Prochlorothrix hollandica in relation to catalytic functions. Planta 218: 1019–1025

Nakagawara E, Sakuraba Y, Yamasato A, Tanaka R and Tanaka A (2007) Clp protease controls chlorophyll b synthesis by regulating the level of chlorophyllide a oxygenase. Plant J 49: 800–809

Niyogi K (1999) Photoprotection revised: genetic and molecular approaches. Annu Rev Plant Physiol Plant Mol Biol 50: 333–359

Nyman ES and Hynninen PH (2004) Research advances in the use of tetrapyrrolic photosensitizers for photodynamic therapy. J Photochem Photobiol B: Biol 73: 1–28

Oster U, Tanaka R, Tanaka A and Rudiger W (2000) Cloning and functional expression of the gene encoding the key enzyme for chlorophyll b biosynthesis (CAO) from *Arabidopsis thaliana*. Plant J 21: 305–310

Ouchane S, Steunou AS, Picaud M and Astier C (2004) Aerobic and anaerobic Mg-protoporphyrin monomethyl ester cyclases in purple bacteria: a strategy adopted to bypass the repressive oxygen control system. J Biol Chem 279: 6385–6394

Palenik B, Grimwood J, Aerts A, Rouze P, Salamov A, Putnam N, Dupont C, Jorgensen R, Derelle E, Rombauts S, Zhou K, Otillar R, Merchant SS, Podell S, Gaasterland T, Napoli C, Gendler K, Manuell A, Tai V, Vallon O, Piganeau G, Jancek S, Heijde M, Jabbari K, Bowler C, Lohr M, Robbens S, Werner G, Dubchak I, Pazour GJ, Ren Q, Paulsen I, Delwiche C, Schmutz J, Rokhsar D, Van de Peer Y, Moreau H and Grigoriev IV (2007) The tiny eukaryote Ostreococcus provides genomic insights into the paradox of plankton speciation. Proc Natl Acad Sci U S A 104: 7705–7710

Park SY, Yu JW, Park JS, Li J, Yoo SC, Lee NY, Lee SK, Jeong SW, Seo HS, Koh HJ, Jeon JS, Park YI and Paek NC (2007) The senescence-induced staygreen protein regulates chlorophyll degradation. Plant Cell 19: 1649–1664

Pattanayak GK, Biswal AK, Reddy VS and Tripathy BC (2005) Light-dependent regulation of chlorophyll b biosynthesis in chlorophyllide a oxygenase overexpressing tobacco plants. Biochem Biophys Res Commun 326: 466–471

Pinta V, Picaud M, Reiss-Husson F and Astier C (2002) Rubrivivax gelatinosus acsF (previously orf358) codes for a conserved, putative binuclear-iron-cluster-containing protein involved in aerobic oxidative cyclization of Mg-protoporphyrin IX monomethylester. J Bacteriol 184: 746–753

Porra RJ, Schafer W, Cmiel E, Katheder I and Scheer H (1994) The derivation of the formyl-group oxygen of chlorophyll-B in higher-plants from molecular-oxygen – achievement of high enrichment of the 7-formyl-group oxygen from O-18(2) greening maize leaves. Eur J Biochem 219: 671–679

Quail PH (2002) Photosensory perception and signalling in plant cells: new paradigms? Curr Opin Cell Biol 14: 180–188

Rebeiz CA, Ioannides IM, Kolossov V and Kopetz KJ (1999) Chloroplast biogenesis 80. Proposal of a unified multibranched chlorophyll a/b biosynthetic pathway. Photosynthetica 36: 117–128

Rebeiz CA, Kolossov VL, Briskin D and Gawienowski M (2003) Chloroplast Biogenesis: Chlorophyll biosynthetic heterogeneity, multiple biosynthetic routes and biological spin-offs. In: Nalwa HS (ed) Handbook of Photochemistry and Photobiology, Vol 4. American Scientific Publishers, Los Angels, CA, pp. 183–248

Reinbothe C, Bartsch S, Eggink LL, Hoober JK, Brusslan J, Andrade-Paz R, Monnet J and Reinbothe S (2006) A role for chlorophyllide a oxygenase in the regulated import and stabilization of light-harvesting chlorophyll a/b proteins. Proc Natl Acad Sci USA 103: 4777–4782

Reinbothe S, Pollmann S and Reinbothe C (2003) In situ conversion of protochlorophyllide b to protochlorophyllide a in barley. Evidence for a novel role of 7-formyl reductase in the prolamellar body of etioplasts. J Biol Chem 278: 800–806

Ren G, An K, Liao Y, Zhou X, Cao Y, Zhao H, Ge X and Kuai B (2007) Identification of a novel chloroplast protein AtNYE1 regulating chlorophyll degradation during leaf senescence in Arabidopsis. Plant Physiol 144: 1429–1441

Rippka R, Coursin T, Hess W, Lichtle C, Scanlan DJ, Palinska KA, Iteman I, Partensky F, Houmard J and Herdman M (2000) Prochlorococcus marinus Chisholm et al 1992 subsp pastoris subsp nov strain PCC 9511, the first axenic chlorophyll a2/b2-containing cyanobacterium (Oxyphotobacteria). Int J System Evol Microb 50: 1833–1847

Rocap G, Larimer FW, Lamerdin J, Malfatti S, Chain P, Ahlgren NA, Arellano A, Coleman M, Hauser L, Hess WR, Johnson ZI, Land M, Lindell D, Post AF, Regala W, Shah M, Shaw SL, Steglich C, Sullivan MB, Ting CS, Tolonen A, Webb EA, Zinser ER and Chisholm SW (2003) Genome divergence in two Prochlorococcus ecotypes reflects oceanic niche differentiation. Nature 424: 1042–1047

Ruban A, Wentworth M, Yakushevska A, Andersson J, Lee P, Keegstra W, Dekker J, Boekema E, Jansson S and Horton P (2003) Plants lacking the main light-harvesting complex retain photosystem II macro-organization. Nature 421: 648–652

Rudoi AB, Vezitskii A and Shlyk AA (1982) Enzymatic system transforming chlorophyllide into chlorophyll in etiolated leaves using exogenous substrates. Biokhimiia (Moscow, Russia) 47: 733–739

Rzeznicka K, Walker CJ, Westergren T, Kannangara CG, von Wettstein D, Merchant S, Gough SP and Hansson M (2005) Xantha-l encodes a membrane subunit of the aerobic Mg-protoporphyrin IX monomethyl ester cyclase involved in chlorophyll biosynthesis. Proc Natl Acad Sci U S A 102: 5886–5891

Sakuraba Y, Yamasato A, Tanaka R and Tanaka A (2007) Functional analysis of N-terminal domains of Arabidopsis chlorophyllide a oxygenase. Plant Physiol Biochem 45: 740–749

Satoh S, Ikeuchi M, Mimuro M and Tanaka A (2001) Chlorophyll b expressed in cyanobacteria functions as a light-harvesting antenna in photosystem I through flexibility of the proteins. J Biol Chem 276: 4293–4297

Satoh S and Tanaka A (2006) Identification of chlorophyllide a oxygenase in the Prochlorococcus genome by a comparative genomic approach. Plant Cell Physiol 47: 1622–1629

Scheumann V, Ito H, Tanaka A, Schoch S and Rudiger W (1996) Substrate specificity of chlorophyll(ide) b reductase in etioplasts of barley (Hordeum vulgare L.). Eur J Biochem 242: 163–170

Scheumann V, Schoch S and Rudiger W (1998) Chlorophyll a formation in the chlorophyll b reductase reaction requires reduced ferredoxin. J Biol Chem 273: 35102–35108

Scheumann V, Schoch S and Rudiger W (1999) Chlorophyll b reduction during senescence of barley seedlings. Planta 209: 364–370

Schnegurt MA and Beale SI (1992) Origin of the chlorophyll b formyl oxygen in Chlorella vulgaris. Biochemistry 31: 11677–11683

Shedbalkar VP, Ioannides IM and Rebeiz CA (1991) Chloroplast biogenesis. Detection of monovinyl protochlorophyll(ide) b in plants. J Biol Chem 266: 17151–17157

Shlyk A (1971) Biosynthesis of chlorophyll b. Ann Rev. Plant Physiol 22: 169–184

Standfuss J, Terwisscha van Scheltinga AC, Lamborghini M and Kuhlbrandt W (2005) Mechanisms of photoprotection and nonphotochemical quenching in pea light-harvesting complex at 2.5 A resolution. EMBO J 24: 919–928

Tanaka A, Ito H, Tanaka R, Tanaka NK, Yoshida K and Okada K (1998) Chlorophyll a oxygenase (CAO) is involved in chlorophyll b formation from chlorophyll a. Proc Natl Acad Sci USA 95: 12719–12723

Tanaka A and Tsuji H (1981) Changes in Chlorophyll-a and Chlorophyll-B Content in Dark-Incubated Cotyledons Excised from illuminated seedlings – the effect of calcium. Plant Physiol 68: 567–570

Tanaka A and Tsuji H (1983) Formation of chlorophyll-protein complexes in greening cucumber cotyledons in light and then in darkness. Plant Cell Physiol 24: 101–108

Tanaka R, Koshino Y, Sawa S, Ishiguro S, Okada K and Tanaka A (2001) Overexpression of chlorophyllide a oxygenase (CAO) enlarges the antenna size of photosystem II in *Arabidopsis thaliana*. Plant J 26: 365–373

Tanaka R and Tanaka A (2005) Effects of chlorophyllide a oxygenase overexpression on light acclimation in *Arabidopsis thaliana*. Photosynth Res 85: 327–340

Tanaka R and Tanaka A (2007) Tetrapyrrole biosynthesis in higher plants. Annu Rev Plant Biol 58: 321–346

Tanaka Y, Tanaka A and Tsuji H (1992) Stabilization of apoproteins of light-harvesting chlorophyll-a/b protein complex by feeding 5-aminolevulinic acid under intermittent illumination. Plant Physiol Bioehm 30: 365–370

Tanaka Y, Tanaka A and Tsuji H (1993) Effects of 5-aminolevulinic acid on the accumulation of chlorophyll b and apoproteins of the light-harvesting chlorophyll a/b-protein complex of photosystem II. Plant Cell Physiol 34: 465–472

Thomas H (1997) Tansley review no 92 – chlorophyll: a symptom and a regulator of plastid development. New Phytol 136: 163–181

Thompson JD, Higgins DG and Gibson TJ (1994) CLUSTAL W: improving the sensitivity of progressive multiple sequence alignment through sequence weighting, position-specific gap penalties and weight matrix choice. Nucl Acids Res 22: 4673–4680

Ting CS, Rocap G, King J and Chisholm SW (2002) Cyanobacterial photosynthesis in the oceans: the origins and significance of divergent light-harvesting strategies. Trends Microb 10: 134–142

Tomitani A, Okada K, Miyashita H, Matthijs H, Ohno T and Tanaka A (1999) Chlorophyll b and phycobilins in the common ancestor of cyanobacteria and chloroplasts. Nature 400: 159–162

Tottey S, Block MA, Allen M, Westergren T, Albrieux C, Scheller HV, Merchant S and Jensen PE (2003) Arabidopsis CHL27, located in both envelope and thylakoid membranes, is required for the synthesis of protochlorophyllide. Proc Natl Acad Sci USA 100: 16119–16124

Tzinas G and Argyroudi-Akoyunoglou JH (1988) Chloramphenicol-induced stabilization of light-harvesting complexes in thylakoids during development. FEBS Lett 229: 135–141

van Der Staay GWM, Yurkova N and Green BR (1998) The 38 kDa chlorophyll a/b protein of the prokaryote Prochlorothrix hollandica is encoded by a divergent pcb gene. Plant Mol Biol 36: 709–716

Virgin HI (1960) Pigment transformations in leaves of wheat after irradiation. Physiol Plant 13: 155–164

Xu H, Vavilin D and Vermaas W (2001) Chlorophyll b can serve as the major pigment in functional photosystem II complexes of cyanobacteria. Proc Natl Acad Sci USA 98: 14168–14173

Yamasato A, Nagata N, Tanaka R and Tanaka A (2005) The N-terminal domain of chlorophyllide a oxygenase confers protein instability in response to chlorophyll b accumulation in Arabidopsis. Plant Cell 17: 1585–1597

Yang DH, Paulsen H and Andersson B (2000) The N-terminal domain of the light-harvesting chlorophyll a/b-binding protein complex (LHCII) is essential for its acclimative proteolysis. FEBS Lett 466: 385–388

Zelisko A, Garcia-Lorenzo M, Jackowski G, Jansson S and Funk C (2005) AtFtsH6 is involved in the degradation of the light-harvesting complex II during high-light acclimation and senescence. Proc Natl Acad Sci USA 102: 13699–13704

Zhang L, Paakkarinen V, Suorsa M and Aro EM (2001) A SecY homologue is involved in chloroplast-encoded D1 protein biogenesis. J Biol Chem 276: 37809–37814

Chapter 5

Magnesium Chelatase

C. Gamini Kannangara
Department of Crop and Soil Sciences
Washington State University, Pullman, WA99164, USA

Diter von Wettstein*
Department of Crop and Soil Sciences and School of Molecular Biosciences,
Washington State University, Pullman, WA99164, USA and
Institute of Phytopathology and Applied Zoology, IFZ,
Heinrich-Buff-Ring 26–32, D-35392 Giessen, Germany

Summary

In chlorophyll biosynthesis, Mg^{2+} is inserted into protoporphyrin IX in an ATP-dependent process catalyzed by three soluble proteins. The three proteins involved in the magnesium insertion have average molecular weights of 40, 70 and 140 kDa. In bacteriochlorophyll synthesizing organisms they are designated as BchI, BchD and BchH and in chlorophyll synthesizing organisms they are referred to as ChI, ChD and ChH. The 40 kDa subunit is compared with the Golgi membrane protein NSF-D2, Heat shock locus protein HslU and the δ′ subunit of the DNA polymerase III (PolIII-δ′). The 70 kDa subunit and its complex formation with the 40 kDa subunit are also described; along with the 140 kDa subunit, which is referred to as the H subunit and in barley, corresponds to the *Xantha – F* gene product. The Gun4 protein, a porphyrin binding stimulator protein of the magnesium chelatase is also discussed.

*Author for Correspondence, e-mail: diter@wsu.edu

Constantin A. Rebeiz et al. (eds.), The Chloroplast: Basics and Applications, pp. 79–88.

I Introduction

In chlorophyll biosynthesis, Mg^{2+} is inserted into protoporphyrin IX in an ATP-dependent process catalyzed by three soluble proteins. The three proteins involved in the magnesium insertion are referred to as subunits of the enzyme magnesium chelatase (Fig. 1) and have average molecular weights of 40, 70 and 140 kDa (Gibson et al., 1995; Willows et al., 1996). In bacteriochlorophyll synthesizing organisms they are designated as BchI, BchD and BchH and in chlorophyll synthesizing organisms they are referred to as ChI, ChD and ChH. Seedling lethal mutants of barley that accumulate protoporphyrin IX upon feeding 5-aminolevulinate, i.e. *xantha-f, -g,* and *–h*, carry mutations in the magnesium chelatase subunits.

II The 40 kDa Subunit

This protein corresponds to the *Xantha–H* gene product of barley. It is an approximately 40 kDa protein consisting of ca. 350 amino acids. The primary amino acid sequence of the 40 kDa subunit is highly conserved in bacteria and in plants and shows an ATP-binding domain. It belongs to the family of AAA+ proteins. The AAA+ proteins are a fascinating group of ATPases containing ATP-binding domains (Martin et al., 2005; Werbeck et al., 2008). Upon binding and subsequent hydrolysis of ATP these proteins undergo conformational changes that are needed for their activity. Such conformational changes lead to

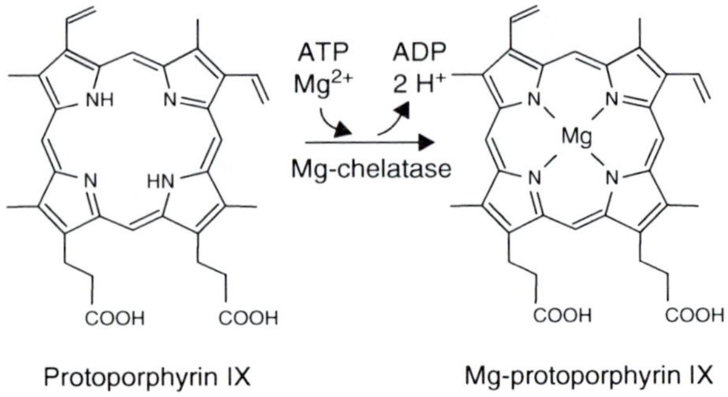

Fig. 1. The incorporation of Mg^{2+} into protoporphyrin IX by Mg-chelatase. This reaction requires three soluble proteins (subunits) of 40, 70 and 140 kDa.

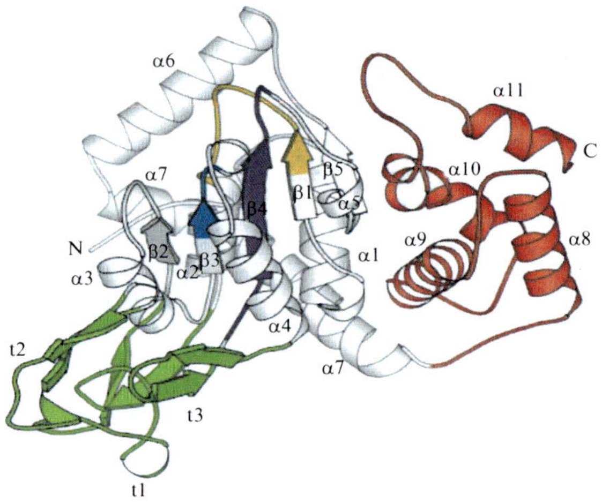

Fig. 2. Ribbon diagram of the 3-D structure of the 40 kDa subunit of *Rhodobacter capsulatus*. *Gray*, N-terminal domain; *red*, C-terminal domain; *yellow*, P-loop; *blue*, Walker B motif; *purple*, sensor 1 region; *green*, insertion specific for Bch I (Fodje et al., 2001).

remodelling of substrates, promote unfolding of proteins for proteolysis, and disassemble protein aggregates and protein complexes. AAA proteins are known to form hexamers in the presence of ATP and Mg^{2+}. The crystal structure of *Rhodobactor capsulatus* BchI has been determined by the multiple wavelength anomalous dispersion (MAD) method and refined to a crystallographic R-factor of 22.2% and to an R-free of 24.5% at 2.1 Å resolution (Fodje et al., 2001). The ribbon diagram of the structure of *Rhodobactor capsulatus* BchI is given in Fig. 2. The N-terminal domain (gray) is constructed around a scaffold of five twisted strands of parallel β-sheets flanked by α-helices to form a α/β/α nucleotide binding fold,

Abbreviations: Chl – chlorophyll; Proto – protochlorophyll; Bchl – bacteriochlorophyll; BchI, BchD, BchH – subunits of Mg-Proto chelatase in Bchl synthesizing organisms; ChI, ChD, ChH – subunits of Mg-Proto chelatase in Chl synthesizing organisms; MAD – multiple wavelength anomalous dispersion; HslU – chaperone part of the bacterial homologue of the mammalian proteasome; PolIII-δ′ – delta subunit of the clamp-loader complex of *E. coli* DNA polymerase III; NSF-D2 – N-ethylmaleimide-sensitive fusion protein D2; M_1-L-M_2 – linker domain that contains 36% praline and 26% aspartate + glutamate residues; ITA2 – human collagen receptor; ITAX – human leucocyte adhesion receptor; ITAE – integrin α-E precursor of mouse; CAMA – cartilage matrix protein of chicken; MIDAS – metal ion dependent adhesion site; N-dom – N-terminal domain; C-dom – C-terminal domain; Gun – 4 protein a porphyrin binding stimulator protein of the magnesium chelatase

the Rossman fold. It includes a P-loop (yellow), a Walker B motif (blue) and a sensor 1 region (purple). A long α-helix with a kink in the middle (α6–α7) makes the connection to the C-terminal domain and is shown in red. The part of the molecule shown in green is specific for BchI and is not found in other AAA+ proteins. It contains a β- hairpin t2 within the α-helix 3 and additional hairpins t1 between α-helix 2 and strand β2 and t3 between α-helix 4 and β4. This region is seen protruding from the N-terminal domain. The surface potential of BchI reveals a positively charged groove opening to the surface of the molecule.

III Comparision of 40 kDa Subunit with the Golgi Membrane Protein NSF-D2, Heat Shock Locus Protein HslU and the δ′ Subunit of the DNA Polymerase III (PolIII-δ′)

The NSF-D2, i.e. N-ethylmaleimide-sensitive fusion protein D2 is a Golgi membrane located ATPase required for protein transport from endoplasmic reticulum to Golgi (Block et al., 1988; Lenzen et al., 1998). HslU is the chaperone part of the a bacterial homologue of the mammalian proteasome, where it activates a protease by hydrolyzing ATP (Song et al., 2003). The PolIII-δ′ is the delta' subunit of the clamp-loader complex of *Escherichia coli* DNA polymerase III. Together with five other subunits, which also contain AAA+ domains, the PolIII-δ′ forms a ring that catalyzes the assembly of the ring shaped protein complexes (clamps) on to DNA. These clamps encircle the duplex DNA and slide freely along the DNA. The DNA polymerase III attached to the clamp allows it to copy the template while sliding along the DNA molecule (Anderson et al., 2007; O' Donnel, 2006).

The 3-D structure of BchI shows significant similarity to NSF-D2, HslU and PolIII-δ′(Fig. 3). Same color coding used in Fig. 2 is employed to indicate, the N-terminal domain (gray), the P-loop (yellow), Walker B motif (blue) and the C-terminal domain(red). The nucleotide binding fold and the P-loop of BchI are conserved when compared with the corresponding folds of NSF-D2, HslU and PolIII-δ′. However, unlike these proteins the C-terminal helical domain of BchI shows extensive contact with the N-terminal domain. The portion of BchI that is seen protruding from the N-terminal domain (shown in green in Fig. 2) corresponds to the so called intermediate domain of HslU. This domain in HslU is involved in the interactions with the heat shock locus V protease of *E. coli* and other bacteria (Song et al., 2003). The amino acid sequences of the regions depicted in green in Bch I and in HslU are highly conserved. The mode in which ATP binds to BchI has been revealed by superimposition of the P-loops of Bch I and HslU. The two arginines R208 and R289 of BchI correspond to the arginine finger that is invoved in ATP-dependent conformational

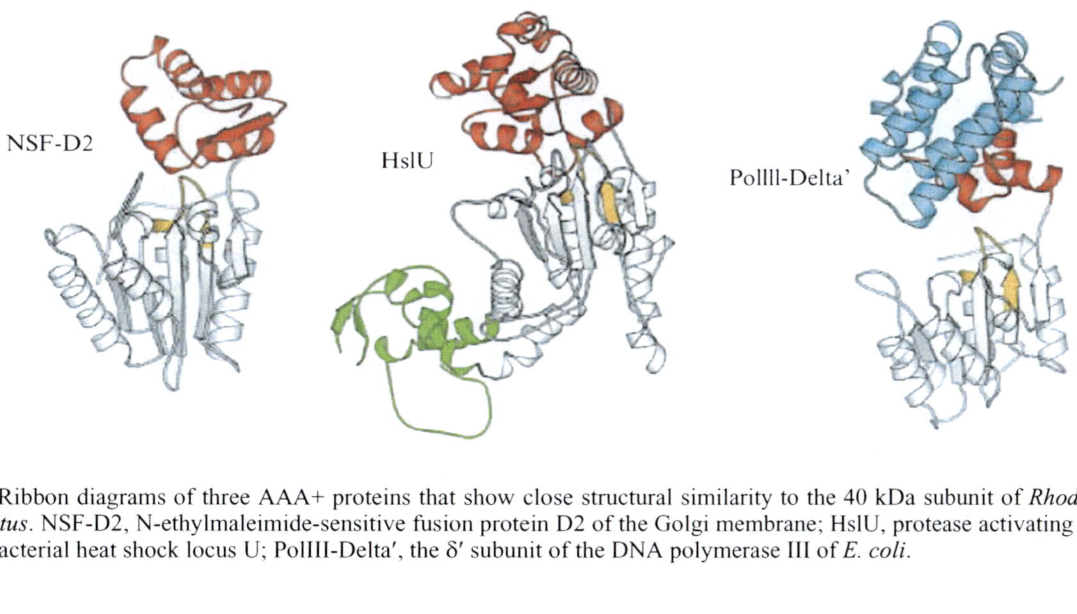

Fig. 3. Ribbon diagrams of three AAA+ proteins that show close structural similarity to the 40 kDa subunit of *Rhodobacter capsulatus*. NSF-D2, N-ethylmaleimide-sensitive fusion protein D2 of the Golgi membrane; HslU, protease activating ATPase of the bacterial heat shock locus U; PolIII-Delta′, the δ′ subunit of the DNA polymerase III of *E. coli*.

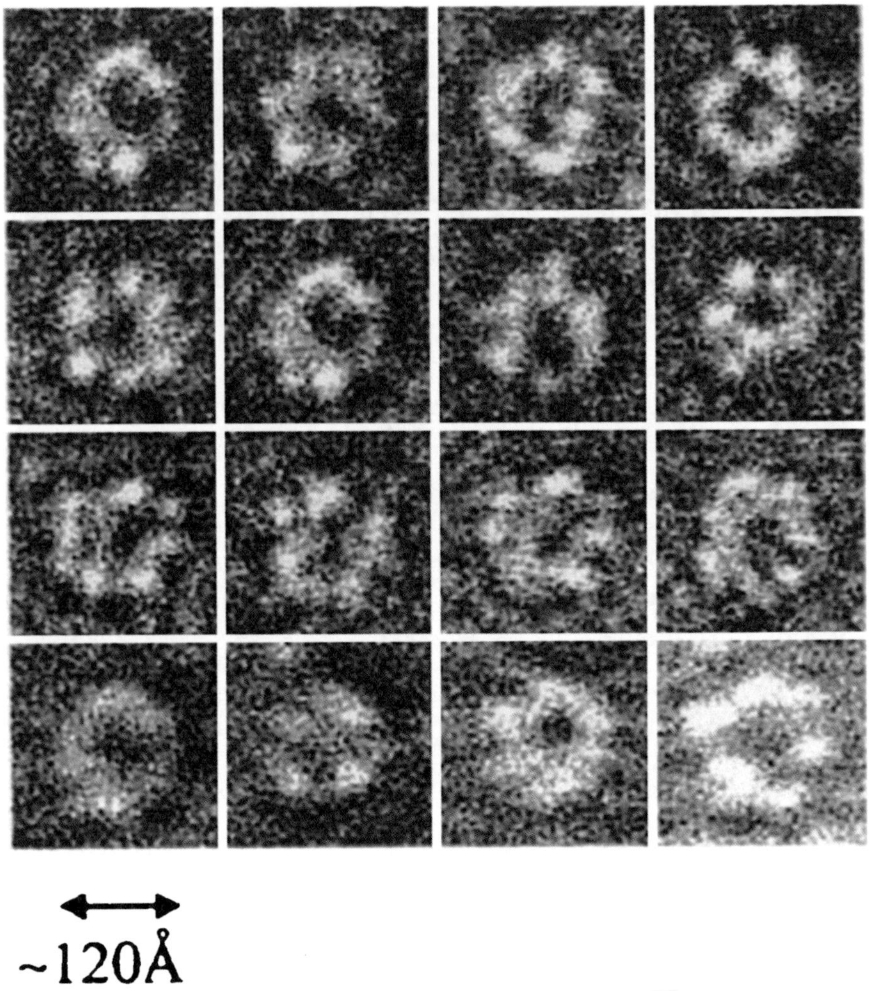

Fig. 4. Negatively stained *Rhodobactor capsulatus* 40 kDa subunit hexamers as seen in electron micrographs (Fodje et al., 2001).

changes in other AAA+ proteins. The hexamer formation has been analysed in several AAA proteins with diverse functions (Erzberger and Berger, 2006, Roll-Mecak and Vale, 2008). Like other AAA proteins BchI forms hexamers in the presence of ATP and $MgCl_2$ (Fig. 4).

Seven mutant alleles of the 40 kDa subunit of barley, *xantha-h*, have been described (Jensen et al., 1996, Hansson et al., 1999). Of these four are recessive and three are semi-dominant. The homozygous mutants of *xantha-h* are unable to insert magnesium into protoporphyrin IX. The recessive mutations prevent the transcription of the gene while the semi-dominant mutations consist of single amino acid changes (D270N, R289K and L111F) (Hansson et al., 2002). The heterozygous semi-dominant mutants can form some chlorophyll and are yellow green with a characteristic aurea phenotype (Fig. 5).

The three missense mutations reside close to the ATP binding site of the 40 kDa subunit. One of them, L111 is located on one side of the wedge shaped structure while D207 and K289 are on the opposite side. The K289 is in the sensor 1 region. The three altered amino acids residues are located at the interface between two adjoining subunits in the AAA hexamer.

The three corresponding amino acid changes have been made in the *R. capsulatus* gene encoding the 40 kDa subunit in order to determine whether the mutant subunits can form hexamers and whether mixed hexamers with wild type

40 kDa subunit can support chelation of Mg^{2+} to protoporphyrin IX (Hansson et al., 2002). The mutated genes were expressed in *E. coli* and products were purified. In the presence of ATP all three mutated 40 kDa subunits could form oligomers, but magnesium chelatase activity was inhibited by the addition any one of the three mutated subunits. The ATPase activity of wild type 40 kDa subunit was not affected by the addition of mutated subunits. A hexamer build up from any mutant subunit will not have a wild type subunit interface (Fig. 6). However, a hexamer built with wild type and mutant subunits will have both wild type and mutant interfaces. ATP hydrolysis probably occurs at the subunit interface. It has been shown that the AAA+ protein chaperone ClpB forms a very dynamic complex, reshuffling subunits on a time scale comparable to steady-state ATP hydrolysis (Werbeck et al., 2008).

Fig. 5. Seedlings developing from a single barley spike of a heterozygous plant carrying the *xantha-h-*^{clorina125} mutation. Of the 21 seedlings developed from the spike 3 were *green* (wild type), 13 were *light green* (heterozygous mutants) and five were *yellow* (homozygous mutants) (Hansson et al., 1999).

IV The 70 kDa Subunit and Its Complex Formation with the 40 kDa Subunit

Even though the crystal structures of the two higher molecular weight 70 and 140 kDa subunits of the magnesium chelatase are yet to be achieved a significant insight pertaining to their structure has been obtained by sequence comparisons coupled with electron microscopy and single particle reconstructions.

In plants and algae the 70 kDa subunit is produced with a transit peptide for import into developing chloroplasts (Fig. 7). The N-terminal domain of the 70 kDa subunit has high amino acid sequence homology to the 40 kDa subunit, containing Walker A and B and sensor 1, but lacks helix $\alpha 2$, strand $\beta 2$ and linker helix $\alpha 7$. The absence of

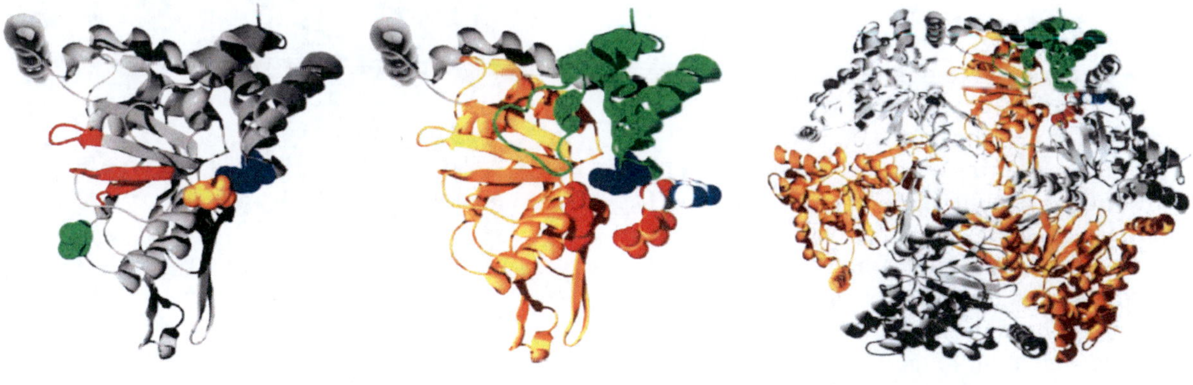

Fig. 6. On the *left* the structure of *R. capsulatus* 40 kDa subunit is shown highlighting the Walker AB motif and sensor 1 region in *red*. Mutated amino acid residues are: *blue*, L111F; *yellow*, D207N and *green* R289K. *Center*: the helical region in *gray* connects the N-terminal domain (*yellow*) with the C-terminal domain (*green*). The arginine finger is colored *red* and the sensor arginine *blue*. An ATP molecule is modeled to show the interface where neighboring subunits interact in the hexamer. On the *right* the structure of the hexamer is shown. The mutated residues lie at the interface between two neighboring subunits (Hansson et al., 2002).

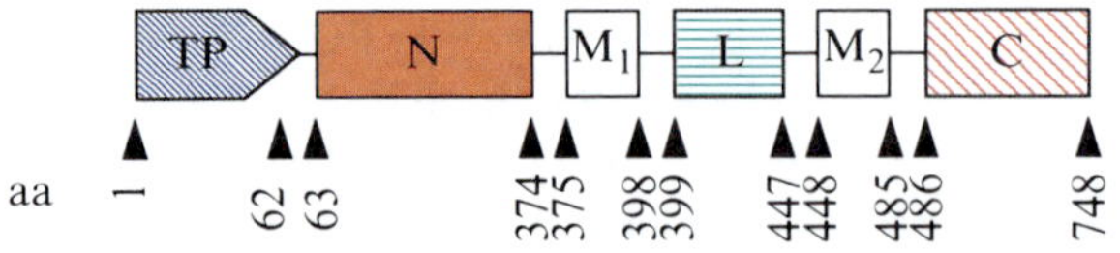

Fig. 7. Domains recognized in the primary sequence of the 70 kDa subunit. TP, transit peptide; N, N-terminal domain that has homology to 40 kDa subunit; M_1-L-M_2, proline rich linker; C, C-terminal domain (Papenbrock et al., 1997).

the β2 strand and the two α helices in the 70 kDa subunit probably prevents its assembly into mixed hexamer rings with 40 kDa subunits. M_1-L-M_2 is the linker domain, and contains 36% proline and 26% aspartate + glutamate residues. The negatively charged linker domain of the 70 kDa subunit could bind to the positively charged grove on the surface of the 40 kDa subunit. The 263 terminal amino acid residues of the C-terminal domain show sequence homology to the integrin I domain of the human collagen receptor (ITA2), human leukocyte adhesion receptor (ITAX), integrin α-E precursor of mouse (ITAE) and cartilage matrix protein of chicken (CAMA) (Fodje et al., 2001).A homology-built model of the integrin I domain of the 70 kDa subunit shows the metal ion dependent adhesion site (MIDAS) with the residues coordinating the Mg^{2+} ion bound to them (Fig. 8). The 70 kDa also revealed hexameric ring formation when analyzed by electron microscopy of negatively stained preparations (Fig. 9) (Elmlund et al., 2008).

The 70 and the 40 kDa subunits associate to form a complex in the presence of Mg:ATP. The three dimensional structure of this complex has been studied using *R. capsulatus* subunits by analyzing 29,400 single-particle images collected from 18 electron micrographs of a specimen preserved in vitrified ice with the aid of EMAN software (Fig. 10) (Elmlund et al., 2008).

Two hexameric rings (trimers of homodimers), one from the I subunit and the other from the D subunit are assembled in a ca. 660 kDa bipartite complex. Each ring is built from the AAA module of the subunits producing a trimer of homodimers. The I subunit hexameric ring is smaller and has a hole in the center while the D subunit hexamer is large and has no central hole. The trigonal crystal form of the homologous AAA protein heat shock locus U from *Escherichia coli* has served as a base for further insight into the structure of the ID complex. Dimerization of I subunit extends nucleotide

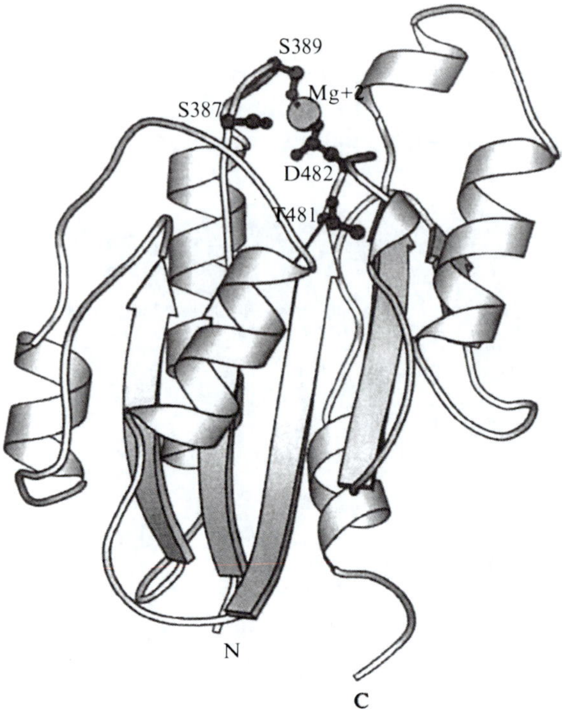

Fig. 8. A homology-built model of the C-terminal domain of *R. capsulatus* 70 kDa subunit. The integrin I domain with a coordinated Mg^{2+} (Fodje et al., 2001).

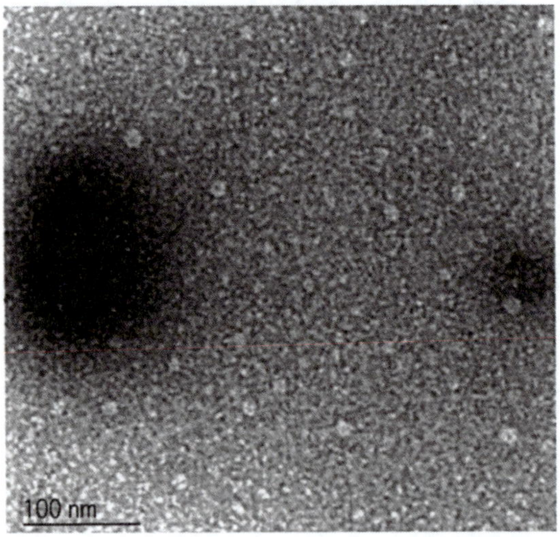

Fig. 9. Hexamers of *Rhodobactor capsulatus* 70 kDa subunit as visualized in negatively stained electron microscopy preparations (Elmlund et al., 2008).

binding interface and the α-helices of the four-helix bundle domain and the groove formed in the nucleotide binding subunit interface substantially

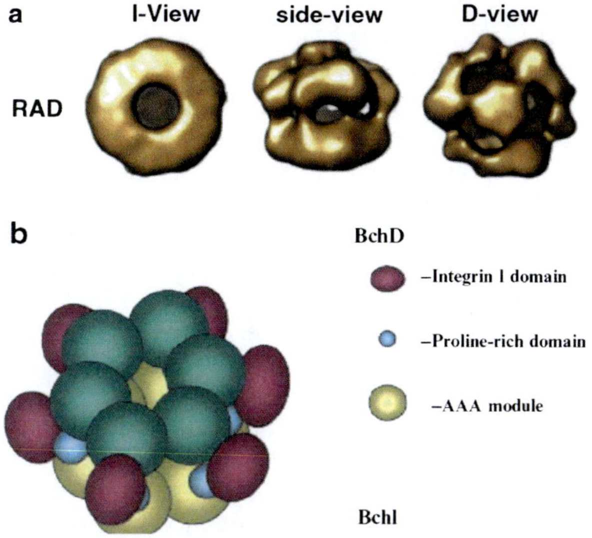

Fig. 10. (**a**) EMAN refinement of single particles and the predicted model for the organization of 40 and 70 kDa complex (Elmlund et al., 2008). (**b**) Hexameric ring of BchD (70 kDa) is associated closely with the hexameric ring of Bch I (Fodje et al., 2001).

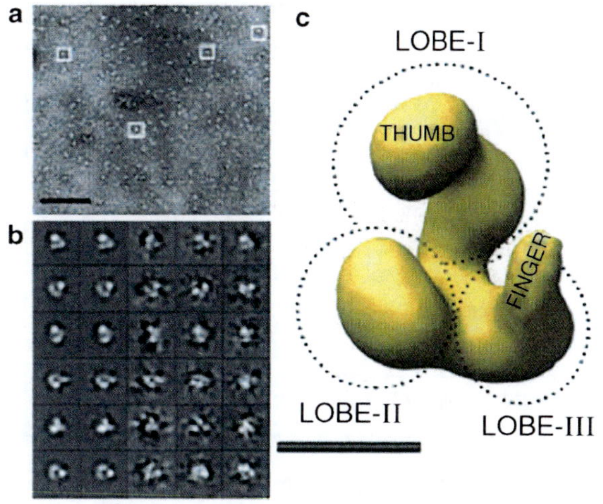

Fig. 11. (**a** and **b**) Negatively stained particles of *R. capsulatus* 140 kDa subunit BchH as seen in electron micrographs. (**c**) 3-D reconstruction (Sirijovski et al., 2008).

assist the docking. In the ID complex the I hexamer is tightly packed and stabilized by interactions with elements of the D subunits, positioned in the intersections between the I monomers. The quaternary structure of the Mg-chelatase ID complex is very similar to the arrangement of other members of the AAA+ family.

V The 140 kDa Subunit

This is referred to as the H subunit and in barley corresponds to the *Xantha-F* gene product. It is approximately 140,000 Da consisting of 1,200–1,400 amino acids depending on the organism. The 140 kDa subunit binds protoporphyrin IX with a molar ratio of 1:1. (Willows et al., 1996; Karger et al., 2001; Viney et al., 2007; Gibson et al., 1995). In negatively stained electron micrographs, the *R. capsulatus* 140 kDa subunit BchH is approximately 100 Å in diameter and shows no oligomerization (Sirijovski et al., 2008) (Fig. 11a and b). The three dimensional reconstruction of protoporphyrin-free BchH revealed three major lobes, lobe I, II and III joined to a flattened base (Fig. 11c). Lobe I has a thumb-shaped structure at its tip and stands perpendicular to lobes II and III. A poorly defined finger-shaped structure protrudes from the tip of lobe III. Three-dimensional reconstructions of protoporphyrin bound BchH revealed the basic three lobed structure but with the thumb domain of lobe I fused with the finger domain of lobe III.

When protoporphyrin is bound to the 140 kDa subunit of *R. capsulatus* the protein was protected from proteolysis by Glu-C protease. The protoporphyrin-free subunit produced a 45 kDa C-terminal peptide after a 16 h of digestion with Glu-C protease.

The N-terminal domain comprising the amino acids 1–733 [N-dom] and the C-terminal domain comprising the amino acids 734–1,194[C-dom] has been expressed separately in *E. coli*, purified and analyzed by electron microscopy. The lobe III and its finger domain were absent in the polypeptide representing the N-terminal half of the 140 kDa subunit. Thus the lobe III represents the C-terminal half of the molecule while the lobe I with its thumb domain and II represented 1–733 amino acid in the N-terminal half. The thumb structure of the N-terminal domain (N-dom) and the finger structure of C-terminal domain (C-dom) of the 140 kDa subunit fuse upon binding of protoporphyrin IX (Sirijovski et al., 2008) (Fig. 12).

The catalytic insertion of Mg^{2+} to protoporphyrin IX is initiated by assembling the complex containing six subunits of 40 kDa and six subunits of 70 kDa arranged in two hexameric rings.

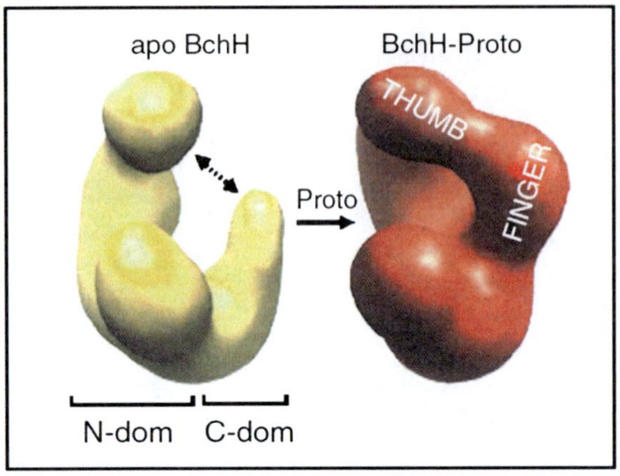

Fig. 12. The thumb structure of N-terminal domain (N-dom) and the finger structure of C-terminal domain (C-dom) of the 140 kDa subunit fuse upon binding of protoporphyrin IX (Sirijovski et al., 2008).

Mg-ATP is required for this assembly but whether ATP is hydrolyzed during the formation of this complex has not been established. The 40–70 kDa complexes interact with the 140 kDa subunit carrying the protoporphyrin IX during the metal insertion step. ATP hydrolysis is needed for Mg^{2+} insertion. The decrease fluorescence emission at 617 nm from protoporphyrin IX bound to 140 kDa subunit and the concomitant increase of fluorescence from Mg – protoporphyrin IX at 580 nm has been studied after excitation at 295 nm (Viney et al., 2007). The observed rate constant for formation of an enzyme-product complex shows a hyperbolic dependence on enzyme concentration indicating fast substrate binding followed by slower conversion of an enzyme-substrate complex to an enzyme–product complex.

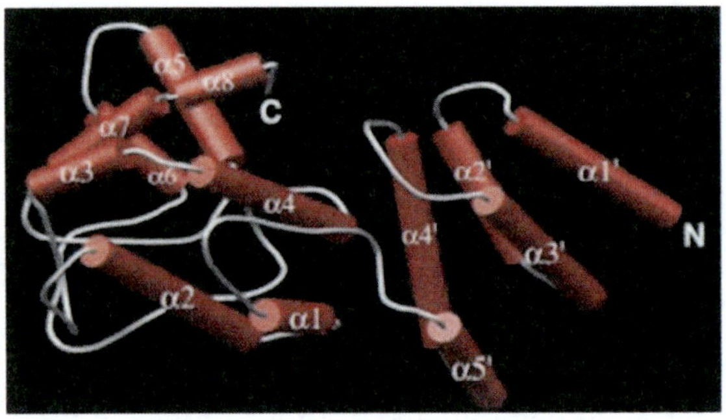

Fig. 13. Structure of Gun4 protein of *Synechocystis*. This 22 kDa protein stimulates magnesium chelatase activity when added to an enzyme assay with containing the 40, 70 and 140 kDa subunits (Verdecia et al., 2005).

VI The Gun4 Protein

This is a porphyrin binding stimulator protein of the magnesium chelatase. It is approximately 18–22 kDa in size. Recombinant Gun4 from *Synechocystis* and *Thermosynechococcus elongatus* has been prepared in *E. coli* and added to in vitro magnesium chelatase enzyme assays with 40, 70 and 140 kDa subunits of *Synechocystis.* Gun4 protein was found to stimulate the insertion of magnesium into deuteroporphyrin (Larkin et al., 2003; Davison et al., 2005; Verdecia et al., 2005). The crystal structure for Gun4 of *Synechocystis* (Fig. 13) as well as *Thermosynechococcus elongates*

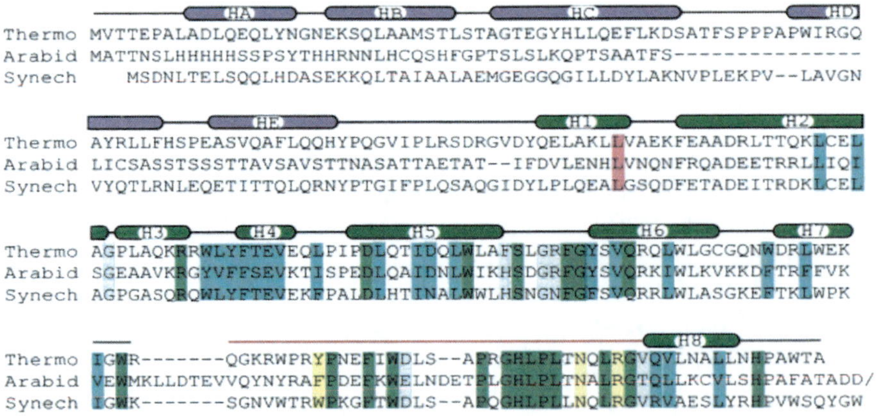

Fig. 14. Comparison of primary amino acid sequences of Gun4 proteins from *Thermosynechococcus elongates*, *Synechocystis* and *Arabidopsis thaliana* (Davison et al., 2005).

has been obtained. Both structures reveal two bundles of helixes linked by a loop. The C-terminal domains of GUN4 protein of different species show significant homology and are composed of eight helices (Fig. 14). The N-terminal domain is made of five helices.

Chloroplast development is under nuclear control (von Wettstein et al., 1971). The observations that suggest the presence of a control mechanism by which chloroplasts regulate expression of nuclear genes have been reviewed (Nott et al., 2006). One of the signaling pathways considered is that magnesium protoporphyrin IX or a protein, that binds to it, functions as a signaling molecule from the plastid to the nucleus. At the present time it is not certain whether such a signaling pathway exists (Mochizuki et al., 2008).

Acknowledgment

This content of this chapter was researched during a stay of C.G. K. at the Institute of Phytopathology and Applied Zoology, IFZ, of the Justus Liebig University, Giessen. D.v.W. acknowledges the support of a Mercator Professorship from the Deutsche Forschungsgemeinschaft. We thank Professor Karl-Heinz Kogel IFZ, Justus Liebig University Giessen for his generous support.

References

Anderson SG, Williams CR, O'Donnel M and Bloom LB (2007) A function for the psi subunit in loading the *Escherichia coli* DNA polymerase sliding clamp. J Biol Chem 282: 7035–7045

Block MR, Glick BS, Wilcox CA, Wieland FT and Rothman JE (1988) Purification of an N-ethylmaleimide-sensitive protein catalyzing vesicular transport. Proc Natl Acad Sci USA 85: 7852–7856

Davison PA, Schubert HL, Reid JD, Iorg CD, Heroux A and Hunter CN (2005) Structural and biochemical characterization of Gun4 suggests a mechanism for its role in chlorophyll biosynthesis. Biochemistry 44: 7603–7612

Elmlund H, Lundqvist J, Al-Karadaghi S, Hansson M, Hebert H and Lindahl M (2008) A new cryo-EM single-particle ab initio reconstruction method visualizes secondary structure elements in an ATP-fueled AAA+ motor. J Mol Biol 375: 934–947

Erzberger JP and Berger JM (2006) Evolutionary relationships and structural mechanisms of AAA+ proteins. Annu Rev Biophys Biomol Struct 35: 93–114

Fodje MN, Hansson A, Hansson M, Olsen JG, Gough S, Willows RD and Al-Karadaghi S (2001) Interplay between an AAA module and an Integrin I domain may regulate the function of magnesium chelatase. J Mol Biol 311: 111–122

Gibson LCD, Willows RD, Kannangara CG, von Wettstein D and Hunter CN (1995) Magnesium-protoporphyrin chelatase: reconstitution of activity by combining the products of the *bchH*, *-I*, and *-D* genes expressed in *Escherichia coli*. Proc Natl Acad Sci USA 92: 1941–1944

Hansson A, Kannangara CG, von Wettstein D and Hansson M (1999) Molecular basis for semidominance of missense mutations in the Xantha-H (42kDa) subunit of magnesium chelatase. Proc Natl Acad Sci USA 96: 1744–1749

Hansson A, Willows RD, Roberts TH and Hansson M (2002) Three semidominant barley mutants with single amino acid substitutions in the smallest magnesium chelatase subunit form defective AAA+ hexamers. Proc Natl Acad Sci USA 99: 13944–13949

Jensen PE, Willows RD, Petersen BL, Vothknecht UC, Stummann BM, Kannangara CG, von Wettstein D and Henningsen KW(1996) Structural genes for Mg-chelatase subunits in barley: *Xantha-f*, *-g* and *-h*. Mol Gen Genet 250: 383–394

Karger GA, Reid JD and Hunter CN (2001) Characterization of the binding of deuteroporphyrin IX to the magnesium chelatase H subunit and spectroscopic properties of the complex. Biochemistry 40: 9291–9299

Larkin RM, Alonso JM, Ecker JR and Chory J (2003) Gun4, Regulator of chlorophyll synthesis and intracellular signaling. Science 299: 902–906

Lenzen CU, Steinmann D, Whiteheart SW and Weis WI (1998) Crystal structure of the hexamerization domain of N-ethylmaleimide-sensitive fusion protein. Cell 94: 525–536

Martin A, Baker TA and Sauer RT (2005) Rebuilt AAA+ motors reveal operating principles for ATP-fuelled machines. Nature 437: 1115–1120

Mochizuki N, Tanaka R, Tanaka A, Masuda T and Nagatani A (2008) The steady state level of Mg-protoporphyrin IX is not a determinant of plastid-to nucleus signaling in *Ara bidopsis*. Proc Natl Acad Sci USA 105: 15184–15189

Nott A,. Jung H-S, Koussevitzky S and Chory J (2006) Plastid-to-nucleus retrograde signaling. Ann Rev Plant Biol 57: 739–759

O' Donnel M (2006). Replisome architecture and dynamics in *Escherichia coli*. J Biol Chem 281: 10653–10656

Papenbrock J, Gräfe S, Kruse E, Hänel F and Grimm B (1997) Mg-Chelatase of tobacco: identification of a Cchl D cDNA sequence encoding a third subunit, analysis of the interaction of three subunits with the yeast two-hybrid system, and reconstitution of the enzyme activity by co-expression of recombinant CHL D, CHL H and CHL I. Plant J 12: 981–990

Papenbrock J, Gräfe S, Kruse E, Hänel and Grimm B (1997) Mg-Chelatase of tobacco: identification of a Cchl D cDNA sequence encoding a third subunit, analysis of the interaction of three subunits with the yeast two-hybrid system, and reconstitution of the enzyme activity by co-expression of recombinant CHL D, CHL H and CHL I. Plant J 12: 981–990

Roll-Mecak A and Vale RD (2008) Structural basis of microtubule severing by the hereditary spastic paraplegia protein spastin. Nature 451: 363–367

Sirijovski N, Lundqvist J, Rosenback M, Elmlund H, Al-Karadaghi S, Willows RD and Hansson M (2008) Substrate-binding model of the chlorophyll biosynthetic magnesium chelatase BchH subunit. J Biol Chem 283: 11652–11660

Song HK, Bochtler M, Azim MK, Hartmann C, Huber R and Ramachandran R (2003) Isolation and characterization of the prokaryotic proteasome homolog HslVU (ClpQY) from Thermotoga maritima and the crystal structure of HslV. Biophys Chem 100: 437–452

Verdecia MA, Larkin RM, Ferrer J-L, Riek R, Chory J and Noel JP (2005) Structure of the Mg-chelatase cofactor GUN4 reveals a novel hand-shaped fold for porphyrin binding. PloSBiol 3: e151

Viney J, Davison PA, Hunter CN and Reid JD (2007) Direct measurement of metal-ion chelation in the active site of the AAA$^+$ ATPase magnesium chelatase. Biochemistry 46: 12788–12794

von Wettstein D, Henningsen KW, Boynton JE, Kannangara CG and Nielsen OF (1971) The genetic control of chloroplast development in Barley. In: Boardman NK, Linnane AW and Smillie RM (eds) Autonomy and Biogenesis of Mitochondria and Chloroplasts, North Holland, Amsterdam, pp. 205–223

Werbeck ND, Schlee S and Reinstein J (2008) Coupling and dynamics of subunits in the hexameric AAA+ chaperone ClpB. J Mol Biol 378: 178–190

Willows RD, Gibson LCD, Kannangara CG, Hunter CN and von Wettstein D (1996) Three separate proteins constitute the magnesium chelatase of Rhodobacter sphaeroides. Eur J Biochem 235: 438–443

Chapter 6

The Enigmatic Chlorophyll *a* Molecule in the Cytochrome b_6f Complex

William A. Cramer*
Department of Biological Sciences, Purdue University, West Lafayette, IN 47907-2054, USA

Sergei Savikhin
Department of Physics, Purdue University, West Lafayette, IN 47907-2054, USA

Jiusheng Yan
Section of Neurobiology, School of Biological Sciences, University of Texas, Austin, TX 78713, USA

E. Yamashita
Institute of Protein Research, Osaka University, Toyonaka, Osaka 560-0043, Japan

Summary

The structure of the cytochrome b_6f complex, which functions in the "dark" reactions of oxygenic photosynthesis contains two pigment molecules, one molecule of chlorophyll a and β-carotene whose functions are not clear. It is proposed that the function of (a) the chlorophyll is to sense the passage of plastoquinone/quinol and, through close contact with a trans-membrane α-helix of the b_6f subunit IV, to induce trans-membrane conformational changes dependent upon the PQ/PQH_2 redox state. (b) It is proposed that the β-carotene has two functions: (i) situated 14 Å from the chlorophyll, it can exert its classical function of quenching the excited triplet state of the chlorophyll through an unprecedented mechanism proposed to involve long distance diffusion of singlet oxygen; (ii) inserted through the peripheral picket fence structure at the periphery of the b_6f complex, β-carotene exerts a toothpick-like function in the assembly of the complex.

* Author for Correspondence, e-mail: waclab@purdue.edu

Constantin A. Rebeiz et al. (eds.), The Chloroplast: Basics and Applications, pp. 89–93.

I Introduction: On the Presence of Two Pigment Molecules in the Cytochrome b_6f Complex

The cytochrome b_6f complex mediates electron transport coupled to proton pumping between the two reaction center complexes of oxygenic photosynthesis (Cramer et al., 2006). Thus, the b_6f complex mediates the "dark reactions" of photosynthesis. However, each monomer of the complex contains one chlorophyll a molecule and one β-carotene. The function of this chlorophyll *a* is enigmatic, because no light reaction is required for enzymatic turnover of the isolated b_6f complex. The rate of light absorption by the Chl *a* in the presence of normal daylight solar intensities is 5–10 photons s^{-1}, a factor of ~100 smaller than the rate of turnover (~ 5–10 × 10^2 s^{-1}) of the b_6f complex. Then, what is the purpose of these two pigment molecules in the cytochrome b_6f complex?

II Crystal Structures of the Cyt b_6f Complex: The Environment of the Bound Chlorophyll

Crystal structures of the b_6f complex from cyanobacteria (Kurisu et al., 2003; Yamashita et al. 2007) and the green alga, *Chlamydomonas reinhardii* (Stroebel et al., 2003) show the b_6f complex to be a hetero-oligomeric integral membrane complex, a symmetric dimer with a molecular weight of approximately 220,000 (Fig. 1a). The monomer consists of eight tightly bound polypeptide subunits. The four relatively large (MW = 18–32 kDa) subunits, cytochrome *f*, cytochrome b_6, the Rieske iron–sulfur cluster, and "subunit IV," which constitute the core of the complex, bind seven prosthetic groups, four hemes, one iron–sulfur [2Fe–2S] cluster, and two non-covalently bound pigment molecules, one chlorophyll *a* and β-carotene. The unit stoichiometry of chlorophyll

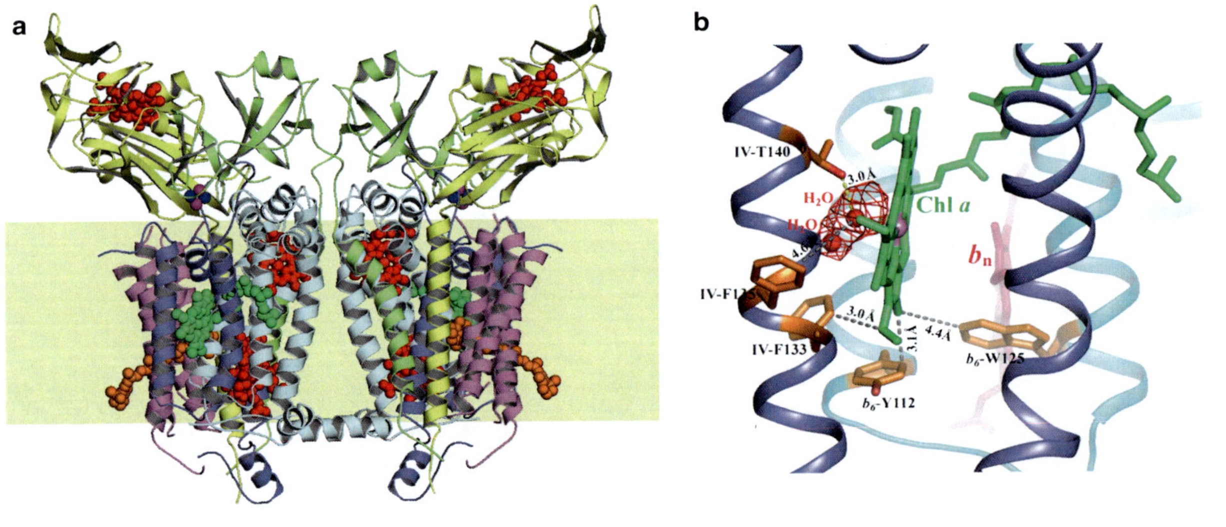

Fig. 1. (**a**)Structure (view parallel to the membrane plane) of the cytochrome b_6f complex from the thermophilic cyanobacterium *M. laminosus* (PDB accession: 2E74). Overall dimensions of the dimer in this profile: 100 × 120 Å. Color code for polypeptide subunits: cyt b_6, *cyan*; subunit IV, *purple*; cyt *f*, *yellow*; ISP, *pale green*; small subunits (petG, -L, -M, and -N), *violet*. Prosthetic groups displayed in space-filling format are chlorophyll (*green*), β-carotene (*orange*), all four hemes (*red*), and the [2Fe-2S] cluster (Fe, *blue*; sulfur, *yellow*). The membrane bilayer is shown in wheat. (**b**) Binding niche of the chlorophyll porphyrin ring in the b_6f complex. Residues that have been mutated are shown in *sticks*; their distance to the porphyrin ring is shown. Labels B, C, F, and G refer to trans-membrane helices in the cyt b_6 (B, C) or subunit IV (F, G) subunit. Unique coordination of the chlorophyll Mg by two water molecules is shown (Modified from those shown in Yan et al., 2008).

Abbreviations: Chl–chlorophyll; Cyt–cytochrome; HPLC–high performance liquid chromatography

a per cytochrome *f* or per monomer had previously been discovered through biochemical analysis of the b_6f complex from higher plants (Huang et al., 1994) and *C. reinhardtii* (Pierre et al., 1997). After the initial studies had established the presence of the Chl *a* in the complex, a search was made for a carotenoid in the complex that would be needed to protect the system from excited state singlet $^1O_2$. The formation of $^1O_2$, whose excited paired spin singlet state results in an increased ability to inflict oxidative damage in regions of the molecular structures to which it can diffuse, occurs through energy transfer from the triplet excited state of the chlorophyll. The generation of $^1O_2$ in light-harvesting chlorophyll proteins is typically prevented by the presence of carotenoid molecules proximal (~4 Å) to the individual chlorophylls, ensuring rapid triplet energy transfer to a carotenoid. Indeed, further reverse phase HPLC analysis of purified b_6f complex showed β-carotene to be present at a stoichiometry of one per chlorophyll *a* in the b_6f complex from spinach thylakoids, the green alga *C. reinhardtii*, and *M. laminosus* (Zhang et al., 1999).

The porphyrin ring of the chlorophyll molecule is inserted between the F and G helices of subunit IV in a region enriched with aromatic residues in both helices (Fig. 1b). The plane of the Chl ring is approximately perpendicular to the membrane plane, and parallel to and separated by 5.5 Å, edge-edge, from heme b_n, the non-covalently bound *b*-heme on the electronegative side of the complex. The 20 carbon phytyl chain of the chlorophyll is wrapped around the G trans-membrane helix of subunit IV and extends through a portal into the inter-monomer quinone exchange cavity (Cramer et al., 2006; Kurisu et al., 2003; Yamashita et al., 2007; Yan and Cramer, 2004). The 3.0 Å crystal structure of the native cyanobacterial b_6f complex shows two water molecules that provide the axial fith ligand of the Chl *a*, with the first water within hydrogen bond distance of the hydroxyl of the Thr140 in subunit IV (Yamashita et al., 2007; Yan et al., 2008). Liganding of a Chl Mg ion by two waters is unusual, if not unique. According to the structure of the cyanobacterial native complex (Yamashita et al., 2007; Yan et al., 2008), four aromatic residues are in close proximity (within 5 Å) of the Chl *a* porphyrin ring. The side chains of Tyr105 and Trp118 in the *M. laminosus* cytochrome b_6 subunit, and Phe133 and Phe135 from subunit IV are, respectively, 3.1, 4.4, 3.0 and 4.6 Å (edge to edge) from the porphyrin ring (Fig. 1b). These aromatic residues are conserved in cyanobacteria, algae, and higher plants. The aromatic residues Phe133 and Phe135 of subunit IV facilitate or participate in electron exchange that shortens the Chl fluorescence lifetime by a factor of 20–25 from approximately 5 ns in solution to 200 ps, thereby reducing the yield of Chl triplet state and singlet oxygen formation (Dashdorj et al. 2005).

III Additional Function(s) of the Bound Chlorophyll

This "wrapping" of the Chl *a* phytyl tail around trans-membrane helix G of subunit IV and the inter-helix position of the Chl ring resemble the proposed structure-stabilizing role of lipids, and its role in the assembly, of the yeast bc_1 complex (Palsdottir and Hunte, 2004). From the mutagenesis studies on the aromatic residues near the chlorophyll, it was inferred that the Chl molecule has a role in stabilization of the structure of the b_6f complex (Yan et al., 2008). Two functions of the Chl have been proposed: (i) one involves the phytyl chain of the Chl that is wrapped around the G helix and thereby inserted through the "portal" that connects the p-side of the complex to the inter-monomer quinone exchange cavity (Cramer et al., 2006; Kurisu et al., 2003; Yamashita et al., 2007). From the crystal structures of a co-complex with the p-side quinone-analogue inhibitor, tridecyl-stigmatellin, it was inferred that this portal provides the pathway for entry/exit of the plasto-quinol (-quinone) associated with electron transfer from the quinol to the Rieske [2Fe–2S] protein concomitant with proton transfer to the p-side aqueous phase. Thus, it has been hypothesized that (a) the Chl phytyl tail can sense the passage of PQ/PQH_2 through the portal and (b) via its wrapped state around trans-membrane helix G, it can exert a torque on this helix and thereby transmit a signal through perturbation of this helix to the n-side of the complex, where it could affect the conformational state of the kinase for the light-harvesting chlorophyll protein (LHC) The LHC kinase is an enzyme on the n-side of the membrane whose activity could be regulated by the state of the quinone on the p-side. Such a role of bound quinone in trans-membrane signaling and activation of the LHC kinase has been proposed (Vener et al., 1997). (ii) A second proposal for the function of the bound Chl, which is not conceptually dissimilar, is that it has a sensor role in the regulation of photosynthetic state transitions (de Lacroix de Lavalette et al. 2008).

IV Additional Function of the β-Carotene

The 9-*cis* beta-carotene in the complex is inserted like a toothpick between the single trans-membrane helices of the small PetL and PetM subunits and is separated by 14 Å from the chlorophyll *a* ring (Kurisu et al., 2003; Stroebel et al., 2003). This distance appears to be much too long for efficient transfer of triplet state energy from the Chl, which requires overlap of wave functions and therefore should not be effective over distances >4 Å (Dexter, 1953). However, energy transfer from the triplet excited state of the chlorophyll *a* to the β-carotene in isolated b_6f complex has been observed (Kim et al., 2005). It was proposed that the long distance transfer mechanism would occur through diffusion of $^1O_2$ through a hydrophobic O_2 channel (Kim et al., 2005). The question arises as to why the spatial separation of β-carotene and chlorophyll *a* is so large if the sole function of the β-carotene is to quench the chlorophyll excited triplet state.

To answer the above question, it is inferred that the β-carotene has a function in addition to that of protection against $^1O_2$ damage, that of a facilitator of assembly of the b_6f complex in which the β-carotene acts as a "toothpick" in the membrane-based assembly (Cramer et al., 2005). Thus, the 15 Å long β-carotene molecule is inserted into the complex through the "picket-fence" at the periphery of the complex formed by the four small subunits, PetG, L, M, and N. These small subunits are essentially hydrophobic sticks," each of which has one to two positively charged amino acid residues near its C-terminus on the electrochemically negative (n; stromal) side of the complex. Thus, the insertion of these asymmetrically charged hydrophobic sticks into the membrane from the n-side can be thermodynamically spontaneous (Krishtalik and Cramer, 1995). Concomitant with the insertion into the membrane of the "picket fence" boundary structure of the complex, the β-carotene inserted into the PetG-L-M-N picket fence would utilize its "toothpick function" to guide the membrane insertion of the polytopic subunit IV, with which it is in contact in the assembled structure of the complex. The insertion of subunit IV into the membrane would initiate the assembly of the core of the complex.

Acknowledgments

The studies described in this article were supported by grants from the NIH GM-038323 (W. A. C.) and NSF MCB-0516939 (S. S.).

Note: Literature survey completed 12/24/08.

References

Cramer WA, Yan J, Zhang H, Kurisu G and Smith JL (2005) Structure of the cytochrome b6f complex: new prosthetic groups, Q-space, and the 'hors d'oeuvres hypothesis' for assembly of the complex. Photosynth Res 85: 133–143

Cramer WA, Zhang H, Yan J, Kurisu G and Smith JL (2006) Trans-membrane traffic in the cytochrome b_6f complex. Annu Rev Biochem 75: 769–790

Dashdorj N, Zhang H, Kim H, Yan J, Cramer WA and Savikhin S (2005) The single chlorophyll *a* molecule in the cytochrome b_6f complex: unusual optical properties protect the complex against singlet oxygen. Biophys J 88: 4178–4187

de Lacroix de Lavalette A, Finazzi G and Zito F (2008) b6f-Associated chlorophyll: structural and dynamic contribution to the different cytochrome functions. Biochemistry 47: 5259–5265

Dexter DL (1953) A theory of sensitized luminescence in solids. J Chem Phys 21: 836–850

Huang D, Everly RM, Cheng RH, Heymann JB, Schägger H, Sled V, Ohnishi T, Baker TS and Cramer WA (1994) Characterization of the chloroplast cytochrome b_6f complex as a structural and functional dimer. Biochemistry 33: 4401–4409

Kim H, Dashdorj N, Zhang H, Yan J, Cramer WA and Savikhin S (2005) An anomalous distance dependence of intra-protein chlorophyll-carotenoid triplet energy transfer. Biophys J 89: 28–30

Krishtalik LI and Cramer WA (1995) On the physical basis for the *cis*-positive rule describing protein orientation in biological membranes. FEBS Lett 369: 140–143

Kurisu G, Zhang H, Smith JL and Cramer WA (2003) Structure of the cytochrome b_6f complex of oxygenic photosynthesis: tuning the cavity. Science 302: 1009–1014

Palsdottir H and Hunte C (2004) Lipids in membrane protein structures. Biochim Biophys Acta 1666: 2–18

Pierre Y, Breyton C, Lemoine Y, Robert B, Vernotte C and Popot JL (1997) On the presence and role of a molecule of chlorophyll a in the cytochrome b_6f complex. J Biol Chem 272: 21901–21908

Stroebel D, Choquet Y, Popot J-L and Picot D (2003) An atypical heam in the cytochrome b_6f complex. Nature 426: 413–418

Vener AV, van Kan PJ, Rich PR, Ohad I and Andersson B (1997) Plastoquinol at the quinol oxidation site of reduced cytochrome bf mediates signal transduction between light and protein phosphorylation: thylakoid protein kinase deactivation by a single-turnover flash. Proc Natl Acad Sci USA 94: 1585–1590

Yamashita E, Zhang H and Cramer WA (2007) Structure of the cytochrome b_6f complex: quinone analogue inhibitors as ligands of heme c_n. J Mol Biol 370: 39–52.

Yan J and Cramer WA (2004) Molecular control of a bimodal distribution of quinone-analogue inhibitor binding sites in the cytochrome b(6)f complex. J Mol Biol 344: 481–493

Yan J, Dashdorj N, Baniulis D, Yamashita E, Savikhin S and Cramer WA (2008) On the structural role of the aromatic residue environment of the chlorophyll a in the cytochrome b6f complex. Biochemistry 47: 3654–3661

Zhang H, Huang D and Cramer WA (1999) Stoichiometrically bound beta-carotene in the cytochrome b_6f complex of oxygenic photosynthesis protects against oxygen damage. J Biol Chem 274: 1581–1587

Chapter 7

The Non-mevalonate DOXP/MEP (Deoxyxylulose 5-Phosphate/Methylerythritol 4-Phosphate) Pathway of Chloroplast Isoprenoid and Pigment Biosynthesis

Hartmut K. Lichtenthaler
Botanisches Institut (Molecular Biology and Biochemistry of Plants), University of Karlsruhe, Kaiserstr. 12, D-76133 Karlsruhe, Germany

Summary

The plastidic pathway for isopentenyl diphosphate (IPP) and isoprenoid biosynthesis, the DOXP/MEP pathway, possesses an essential function in the biosynthesis of thylakoidal prenylllipids chlorophylls (phytyl side-chain), carotenoids, plastoquinone-9, phylloquinone K1 and α-tocopherol, as well as for monoterpenes and diterpenes. Here we review its detection, its enzymes, intermediates, genes, cofactor requirements and inhibitors. This plastidic isoprenoid pathway is also contrasted against the acetate/mevalonate(MVA) pathway of isoprenoid biosynthesis operating in the cytosol for the biosynthesis of sterols, sesquiterpenes and polyterpenes. The DOXP/MEP pathway can specifically be inhibited by fosmidomycin and the acetate/MVA pathway by statins (e.g. mevinolin). By applying specifically marked substrates (^{2}H-deoxyxylulose, ^{14}C-DOXP, ^{3}H-MVA or ^{14}C-MVA) and specific inhibitors a cross-talk between both IPP yielding cell pathways was detected that operates preferentially in the chloroplast-to-cytosol direction. The DOXP/MEP pathway is also involved in the biosynthesis of the volatile hemiterpenes isoprene and methylbutenol. The distribution of the DOXP/MEP pathway in photosynthetic organisms (bacteria, algae, higher plants) and its putative origin in anoxigenic photosynthetic bacteria is reviewed.

Author for Correspondence, e-mail: hartmut.lichtenthaler@bio.uka.de

Constantin A. Rebeiz et al. (eds.), The Chloroplast: Basics and Applications, pp. 95–118.

The DOXP/MEP pathway does not occur in *Archaea*, fungi or animals. The use of etiolated seedlings with their light-induced pigment accumulation as a test system for inhibitors of the DOXP/MEP pathway in the search of active ingredients against pathogenic bacteria and the malaria parasite *Plasmodium* is discussed.

I Introduction

Plants, animals and micro-organisms contain various primary and partially secondary isoprenoid compounds (or natural products) that are made of the C_5-units of 'active isoprene', known as isopentenyl diphosphate (IPP). These isoprenoid compounds have also been termed terpenoids with hemiterpenes, monoterpenes, sesquiterpenes, diterpenes, triterpenes and tetraterpenes as C_5-, C_{10}-, C_{15}-, C_{20}-, C_{30}-, and C_{40}-isoprenoids, respectively. Depending on the plant there exist hemiterpenes (C_5, e.g. isoprene), monoterpenes (C_{10}, e.g. menthol), sesquiterpenes (C_{15}, e.g. farnesol, bisabolol), diterpenes (C_{20}, e.g. camphorene, taxol, ginkgolides), triterpenes (C_{30}, e.g. *Avena* saponins, oleandric acid), tetraterpenes (C_{40}, e.g. ß-carotene, lutein and secondary carotenoids, such as astaxanthin, crocetin) and polyterpenes $(C_5H_8)_n$ as found in rubber latex, gutta-percha or in the chyle latex of various plant families, such as Euphorbiaceae or Asteraceae. The 'biogenetic isoprene rule', i.e. the composition of such natural products from C_5-building blocks, was first detected in 1885 by Wallach (1885), and the head-to-tail addition of 'active C_5 units' was pointed out by Ruzicka (1938) and Ruzicka et al. (1953). In the early 1950s acetate (Little and Bloch, 1950) and acetyl-CoA (Lynen et al., 1951) were detected as precursors, mevalonic acid (MVA) as an intermediate (Wolf et al., 1956), and isopentenyl-diphosphate (IPP) as the active cellular biosynthetic C_5-unit (Chaykin et al., 1958).

For more than 3 decades it had been accepted that all isoprenoids of living cells were synthesized from acetyl-CoA via the classical acetate/MVA pathway (e.g. McGarvey and Croteau, 1995) that provides IPP and via an isomerase its isomer dimethylallyl-diphosphate (DMAPP). The latter is the starter molecule to which more IPP units are condensed in order to yield the C_{10}-, C_{15}-, C_{20}-prenyl diphosphates and higher homologues used for biosynthesis of the different single (isoprene, sterols, carotenoids), or mixed isoprenoid compounds (chlorophylls, prenylquinones) occurring in plant cells (Goodwin, 1977; Lichtenthaler 1977). This is summarized in the general scheme for the biosynthesis of plant isoprenoids (Fig. 1) and also shows the localization of individual plant isoprenoids in different cell compartments. The functional lipophilic isoprenoids of biomembranes, such as sterols in cytosolic biomembranes, ubiquinones in mitochondrial membranes as well as carotenoids, plastoquinone-9, phylloquinone K1, α-tocopherol and chlorophylls (phytyl side-chain) in the photochemically active thylakoids of chloroplasts, have been termed prenyllipids (Goodwin, 1977; Lichtenthaler, 1977) to distinguish them from the bilayer-forming membrane lipids, i.e. the glycerolipids that comprise the phospho-, galacto- and sulfolipids (as reviewed by Lichtenthaler, 2007; see also Lichtenthaler and Park, 1963). Despite the fact that in plants more and more inconsistencies showed up both in labeling of plastidic isoprenoids (e.g. carotenoids, phytol, plastoquinone-9) and in inhibitor studies with mevinolin (a specific inhibitor of the MVA pathway), as reviewed by Lichtenthaler et al. (1997a, b) and Lichtenthaler (1998), the classical acetate/MVA pathway was regarded as the only biosynthetic pathway for IPP biosynthesis in all living

Abbreviations: A – antheraxanthin; a + b – total chlorophylls; a/b – ratio of chlorophyll a to b; CDP-ME – CDP-methyl-D-erythritol; CDP-ME2P – CDP-methyl-D-erythritol-2-phosphate; Chl – chlorophyll; Chls/Cars – weight ratio of chlorophylls to carotenoids, also known as (a + b)/(x + c); c – carotenes; DMAPP – dimethylallyl diphosphate; DOXP – 1-desoxy-D-xylulose-4-phosphate; DOXP/MEP – pathway plastidic 1-desoxy-D-xylulose-4-phosphate/2-C-methylerythritol 5-phosphate – pathway; DXR – DOXP reductoisomerase; DXS – DOXP synthase; GAP – glyceraldehyde-3-phosphate; GGPP – geranylgeranyl diphosphate; HMBPP – 4-hydroxy-3-methyl-2-(E)-butenyl diphosphate; IPP – isopentenyl diphosphate; MBO – 2-methyl-3-buten-2-ol; MEcPP – 2-C-methyl-D-erythritol-2,4-cyclodiphosphate; MEP – 2-C-methylerythritol 5-phosphate; MVA mevalonic acid; PPFD – photosynthetic photon flux density; P – osmiophilic plastoglobuli; x + c – total carotenoids; x – xanthophylls; Z – zeaxanthin

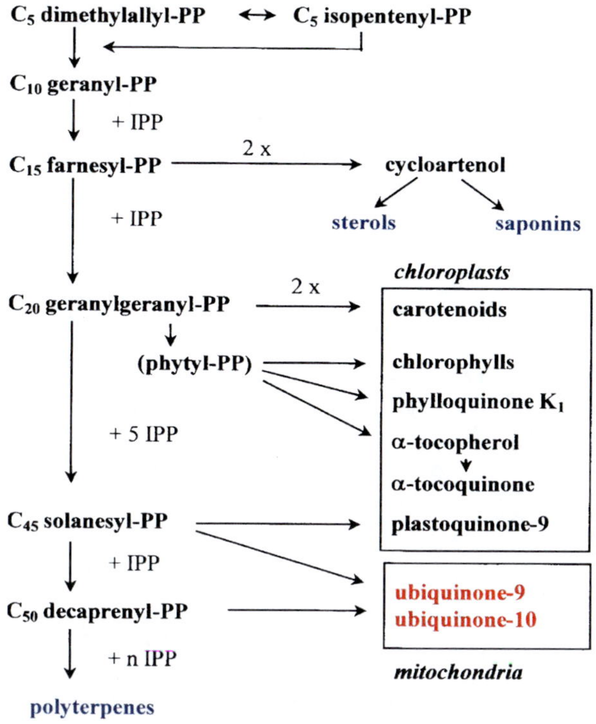

Fig. 1. Biosynthetic relationship and compartmentation of different plant isoprenoids and prenyllipids in plants. The carbon skeleton of pure prenyl compounds (carotenoids, sterols, polyprenols, saponins) is made solely of active isoprenoid C_5 units with DMAPP as a starter molecule to which various IPP molecules are added in a head-to-tail condensation. Mixed prenyl compounds (chlorophylls, phylloquinone K1, a-tocopherol, a-tocoquinone, plastoquinone-9 and ubiquinone-9 and -10) contain a prenyl-side chain that is bound to a non-isoprenoid nucleus (in chlorophylls a porphyrin ring, and in prenylquinones a benzo- or naphthoquinone ring). Sterols and saponins arise by tail-to-tail dimerization of two farnesyl chains, and carotenoids by tail-to-tail dimerization of two geranylgeranyl chains. Saponins are not present in all plants but are found e.g., in *Avena* seedlings (Based on Lichtenthaler, 1998).

organisms, including plants. The individual steps of the cytosolic acetate/MVA pathway of IPP biosynthesis are shown in Fig. 2 (left side) and are commented below in a separate paragraph.

Only as recently as the early 1990s, when labeling with ^{13}C-glucose combined with high resolution NMR spectroscopy and deuterium-labeling complemented by mass spectroscopy, a second, biochemically fully independent, non-mevalonate IPP-biosynthesis pathway, known today as plastidic DOXP/MEP pathway, was detected (Lichtenthaler et al., 1997a, b; Lichtenthaler, 1999, 2000; Rohmer 1999). This plastidic isoprenoid biosynthesis pathway is also referred to as MEP pathway. The first hint came from bacteria where an unusual ^{13}C-labeling pattern was detected (Rohmer et al., 1993). In a close scientific cooperation by Lichtenthaler and Rohmer between 1993 and 1999 this DOXP/MEP pathway of isoprenoid biosynthesis was shown to occur in all photosynthetic oxygen evolving organisms, such as green algae (Lichtenthaler et al., 1995; Schwender et al., 1995, 1996) and higher plants as first reported by Lichtenthaler et al. (1997a, b), then confirmed by Arigoni et al. (1997). These findings were confirmed by various other authors and research groups as reviewed by Lichtenthaler (1998, 1999, 2000) and Rohmer (1999). The detection of the DOXP/MEP pathway demonstrated that green plants (with the exception of some green algae) possess two IPP synthesizing pathways, (i) the acetate/MVA pathway in the cytosol, e.g. for sterol biosynthesis and (ii) the DOXP/MEP pathway in plastids, e.g. for biosynthesis of carotenoids, the phytyl chain of chlorophylls, and the prenyl side-chains of phylloquinone K1, α-tocopherol, α-tocoquinone and plastoquinone-9.

II The Cytosolic Acetate/Mevalonate (MVA) Pathway of Isopentenyl Pyro phosphate (IPP) Biosynthesis and Its Inhibition

The present knowledge of the biochemical enzymatic steps of the classical acetate/MVA pathway proceeding in the cytosol is shown in Fig. 2 (left side). It starts from 3 acetyl-CoA, requires six enzymes, 2 NADPH and 3 ATP to finally yield isopentenyl-diphosphate (IPP). All enzymes have been cloned from plants (see Bach et al., 1999). The regulatory step is the HMG-CoA reductase (enzyme 3) that can specifically be blocked by mevinolin, as first shown for plants by Bach and Lichtenthaler (1982, 1983a, b), and various other statins (cerivastatin, lovostatin), whereby the active part of the inhibitor mevinolin is a structural analogue of the endogenous substrate intermediate mevaldyl-CoA thiohemiacetal as shown in Fig. 3a. The final product of this pathway, IPP, is transferred via IPP isomerase to its isomer DMAPP, the starter molecule for terpenoid biosynthesis to which, depending on the final terpenoid product, one or several IPP molecules are added in a head-to-tail condensation response.

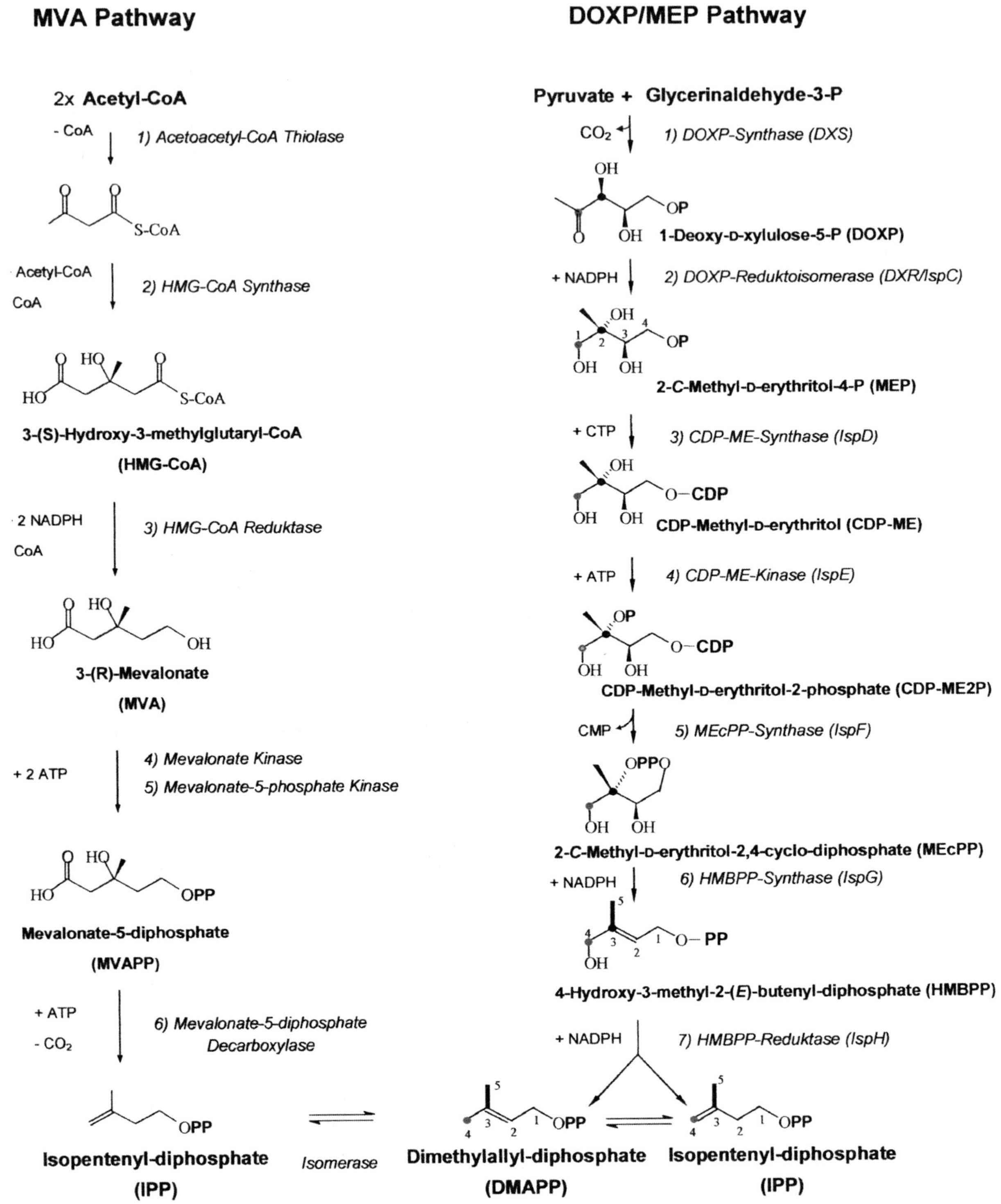

Fig. 2. Scheme of the two biosynthetic pathways of plants for isopentenyl diphosphate (IPP) and isoprenoid biosynthesis: the cytosolic acetate/mevalonate (MVA) pathway and the chloroplastidic 1-deoxy-D-xylulose-5-phosphate/methyl-D-erythrithol (DOXP/MEP pathway). The enzymes of both pathways are numbered. In the plastidic DOXP/MEP pathway the genes of the corresponding enzymes are indicated in *parentheses*.

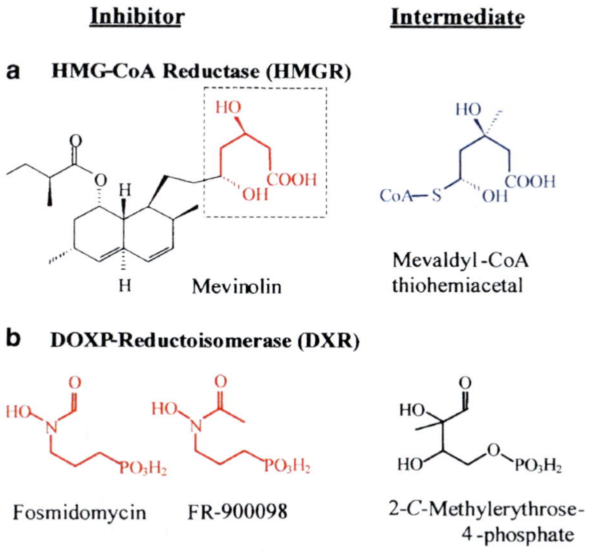

Fig. 3. Inhibitors of the acetate/MVA pathway (mevinolin) and of the DOXP/MEP pathway (fosmidomycin and its methyl derivative FR-900098) for the biosynthesis of isopentenyl diphosphate (IPP). The structural analogy of the active inhibitor (site) with the natural enzyme intermediates is indicated in both cases.

In the case of sterol biosynthesis the farnesyl (C_{15}) chains of two farnesyl diphosphates are combined in a tail-to-tail addition to yield the C_{30} isoprenoid squalene, which is transformed by various enzymatic modifications to form the numerous types of special sterols occurring in different plants.

III The Plastidic DOXP/MEP Pathway of IPP and Its Inhibition

In contrast to the cytosolic acetate/MVA pathway, the non-mevalonate DOXP/MEP pathway of IPP biosynthesis in plastids starts from pyruvate and glyceraldehyde-3-phosphate, comprises seven enzymes and requires three ATP equivalents (ATP or CTP) and three NADPH as shown in Fig. 2 (right half).

The first enzyme of this pathway is the DOXP synthase (DXS), a thiamine-dependent transketolase-type enzyme, that condenses pyruvate with glyceraldehyde-3-phosphate yielding DOXP, 1-desoxy-D-xylulose-4-phosphate, which forms the first C_5 chain of this pathway. The DOXP synthase is specifically inhibited by 5-ketoclomazone, an oxidation product of the herbicide clomazone (Müller et al., 2000; Zeidler et al., 2000). Clomazone is active only in those plants that are able to oxidize it to 5-ketoclomazone, which is the proper active ingredient. Since 5-ketoclomazone, formed in plants from the applied herbicide clomazone, undergoes further metabolic breakdown in intact plant cells, it is only effective when applied in somewhat higher doses. DOXP, the first product of the plastidic isoprenoid pathway, is reduced by *enzyme 2*, the DOXP reductoisomerase (DXR), to 2-C-methylerythritol 5-phosphate (MEP) (Schwender et al., 1999; Müller et al., 2000). This enzymatic step comprises, besides the NADPH catalyzed reduction step, an intramolecular rearrangement of the carbon skeleton. This DOXP reductoisomerase can efficiently be blocked by fosmidomycin and its derivative FR-900098 as was first and independently shown for plants (Schwender et al., 1999; Zeidler et al., 1998) and for bacteria (Kuzuyama et al., 1998). , This enzyme 2 seems to represent an essential regulatory step of the plastid-bound DOXP/MEP pathway. Fosmidomycin is a highly specific inhibitor and a structural analogue of 2-C-methylerythrose 4-phosphate, the intermediate in the enzymatic reaction of DXR, as shown in Fig. 3b.

Enzyme 3, the CDP-ME synthase catalyzes the activation of MEP by CTP to form CDP-methyl-D-erythritol (CDP-ME) (Herz et al., 2000; Rohdich et al., 1999). CDP-ME is phosphorylated by CDP-ME kinase (enzyme 4) to CDP-ME2P, which is further modified by MEcPP synthase (enzyme 5) to 2-C-methyl-D-erythritol 2,4-cyclodiphosphate (MEcPP). The latter is reduced by HMBPP synthase (enzyme 6) to 4-hydroxy-3-methyl-2-(*E*)-butenyl-diphosphate (HMBPP) as was demonstrated by several authors (Altincicek et al., 2001; Campos et al., 2001; Seemann et al., 2002; Querol et al., 2002). Evidence for the seventh enzyme, the HMBPP reductase came from several groups (Adam et al., 2002; Altincicek et al., 2002; Rohdich et al., 2002). The HMBPP reductase (gene: *IspH,* formerly *lyt B*), yields both active isoprenoid C_5 diphosphate substrates required for isoprenoid biosynthesis: IPP and its isomer DMAPP (usually in a ratio of 5:1 or 3:1) as indicated in Fig. 2 (right half). Depending on the cellular metabolic demand IPP and DMAPP can be interconverted by an IPP isomerase. The seven enzymes involved in the DOXP/MEP pathway have been isolated and their genes, given here in the sequence enzyme 1 to enzyme 7, are *DXS, IspC (DXR), IspD (ygbP), IspE (ychB), IspF (ygbB), IspG (cpE)* and IspH (*lytB*). All genes have been cloned in plants and bacteria (as reviewed for the first five enzymes by Lichtenthaler, 2000, 2004; and for all seven enzymes by Rodriguez-Concepción and Boronat (2002). Except for the first enzyme DXS, the genes have been renamed and are now addressed as *IspC* to *IspH*, the former names are indicated above in parentheses. The genes of the DOXP/MEP pathway are bound to the nucleus, yet the proteins operate in the plastids. In contrast to the bacterial DOXP/MEP enzymes, the plant enzymes possess an additional transit peptide sequence that directs them to their proper organelle, the plastid.

IV Labeling Experiments of Chloroplast Prenyllipids

The DOXP/MEP pathway, the isoprenoid biosynthesis pathway of all photosynthetic organisms, was detected by applying labeled [1-^{13}C] glucose under sterile conditions to algae (Schwender et al., 1995, 1996) and higher plants (Lichtenthaler et al., 1997a, b). The ^{13}C labeling of β-carotene, phytol (side-chain of chlorophylls) and the nonaprenyl side-chain of plastoquinone-9 did not exhibit however, the expected ^{13}C-labeling pattern typical for a biosynthesis of these prenyl chains via the cytosolic MVA pathway. Instead, a fully different labeling pattern was detected as shown in Fig. 4. The IPP C_5-units of the plastidic isoprenoids did not possess the expected ^{13}C-enrichment in the C-atoms C-2, C-4 and C-5, but clearly showed specific labeling in the two C-atoms C-1 and C-5.

From the ^{13}C/^{13}C coupling constants in labeling studies with uniformly labeled glucose, [U-$^{13}C_6$] glucose, it appeared that a thiamine bound C_2-unit from pyruvate is transferred to glyceraldehyde-3-phosphate to yield DOXP, which is followed by an intramolecular skeletal rearrangement of carbon atoms to form IPP (see review Lichtenthaler, 1998). Based on the ^{13}C-labeling results we considered desoxyxylulose phosphate (DOXP) as a potential intermediate in the new non-mevalonate isoprenoid biosynthetic pathway of plants. Therefore, further experiments were performed with deuterium-labeled deoxyxylulose (D-DOX or ^{2}H-DOX). D-DOX was applied to plants in the form of its methylxyluloside, since the latter is absorbed better by plant cells than D-DOX and is quickly hydrolyzed in the cytosol to D-DOX. In fact, the deuterium-label of D-DOX was specifically incorporated into the proper methyl groups of the diterpene phytol (in the red alga *Cyanidium* and the green algae *Scenedesmus* and *Chlamydomonas*, as well as in the higher plant *Lemna*) and also into the volatile isoprene emitted by leaves of *Populus*, *Chelidonium* and

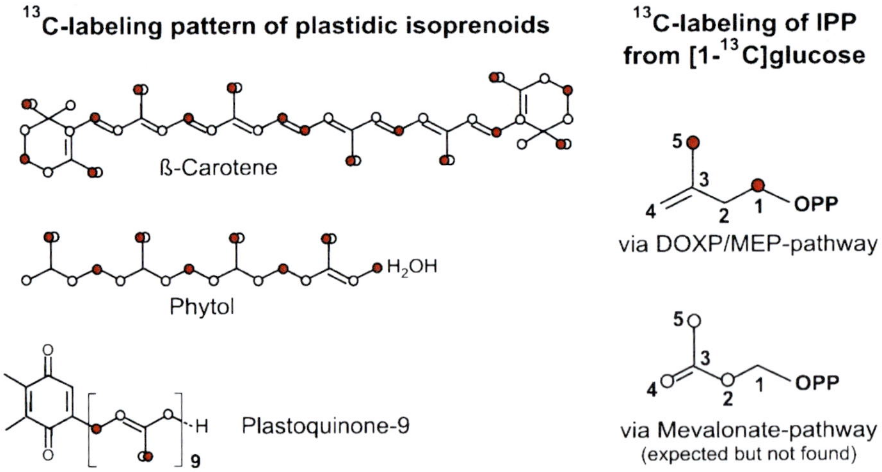

Fig. 4. ^{13}C-labeling pattern (**a**, left side) of plastidic isoprenoids (β-carotene, phytol, plastoquinone-9) indicated by red dots in algae and higher plants when growing on [1-^{13}C]glucose. (**b**, right side) ^{13}C-labeling of isopentenyl diphosphate (IPP) according to the DOXP/MEP pathway (*red dots, upper part*) and anticipated via the acetate/MVA pathway (*open circles, lower part*). The originally expected labeling from the MVA pathway could not be detected.

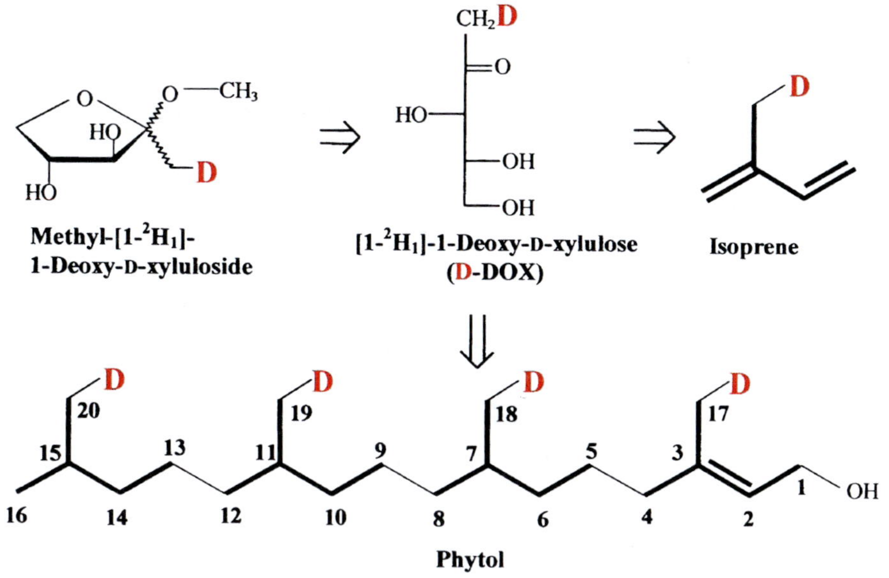

Fig. 5. Specific incorporation of deuterium-labeled [1-^{2}H]-deoxyxylulose (D-DOX) into plastidic isoprenoids: volatile hemiterpene isoprene and diterpene phytol (side-chain of chlorophylls). The deuterium-label '**D**' was found in the methyl group of isoprene emitted and in the four methyl groups of phytol (Based on Schwender et al., 1997).

Salix (Fig. 5) as proven by high resolution NMR and mass spectroscopy (Schwender et al., 1997). The data unambiguously indicated that DOXP is an intermediate in the new IPP forming pathway. Today it is known that a cytosolic xylulose kinase catalyzes the phosphorylation of applied 1-deoxy-D-xylulose into DOXP as precursor of the plastidial isoprenoid pathway (Hemmerlin et al., 2006), which is transported by a recently described plastidic transporter for pentose phosphates, the xylulose 5-phosphate translocator, into chloroplasts (Flügge and Gao, 2005), where it is used as precursor for the biosynthesis of plastidic isoprenoids.

V Compartmentation of Isoprenoid Biosynthesis in Plants

With the labeling techniques mentioned above and various additional investigations the existence of two independent cellular pathways in plants was established (Lichtenthaler et al., 1997a, b) as shown in Fig. 6. In cells of higher plants and algae the two IPP producing biochemical pathways operate in parallel. The MVA pathway in the cytoplasm is responsible for the biosynthesis of sterols, sesquiterpenes, polyterpenes and can efficiently be blocked by mevinolin, whereas the accumulation of plastidic isoprenoids is not affected (Bach and Lichtenthaler, 1983b). The DOXP/MEP pathway of IPP formation operates in the chloroplast and other plastid forms, and provides the C_5-units for the biosynthesis of the tetraterpenoids carotenoids, the diterpene phytol (side-chain of chlorophylls), the nona-prenyl side-chain of plastoquinone-9, the biosynthesis of isoprene (Zeidler et al., 1997) and other plastidic isoprenoids, such as particular monoterpenes, e.g. menthone, pulegone, geraniol and thymol (Eisenreich et al., 1997), and special diterpenes, e.g. ginkgolide A (Schwarz, 1994) or marrubiin (Knöss et al., 1997), see also Lichtenthaler and Zeidler (2002) and Nagegowda et al. (2010) for further literature. This DOXP/MEP pathway and consequently the biosynthesis of carotenoids, chlorophylls (phytol side-chain) can efficiently be blocked by the herbicide fosmidomycin (Zeidler et al., 1998) and by 5-ketoclomazone (Müller et al., 2000), whereas the biosynthesis of the cytosolic sterols is not affected.

Like chloroplasts, plant mitochondria are cell organelles *sui generis* with their own genetic and protein synthesis apparatus. They contain ubiquinones, usually with a nonaprenyl (Q-9) or a decaprenyl side-chain (Q-10) bound to a non-isoprenoid benzoquinone ring. The question was

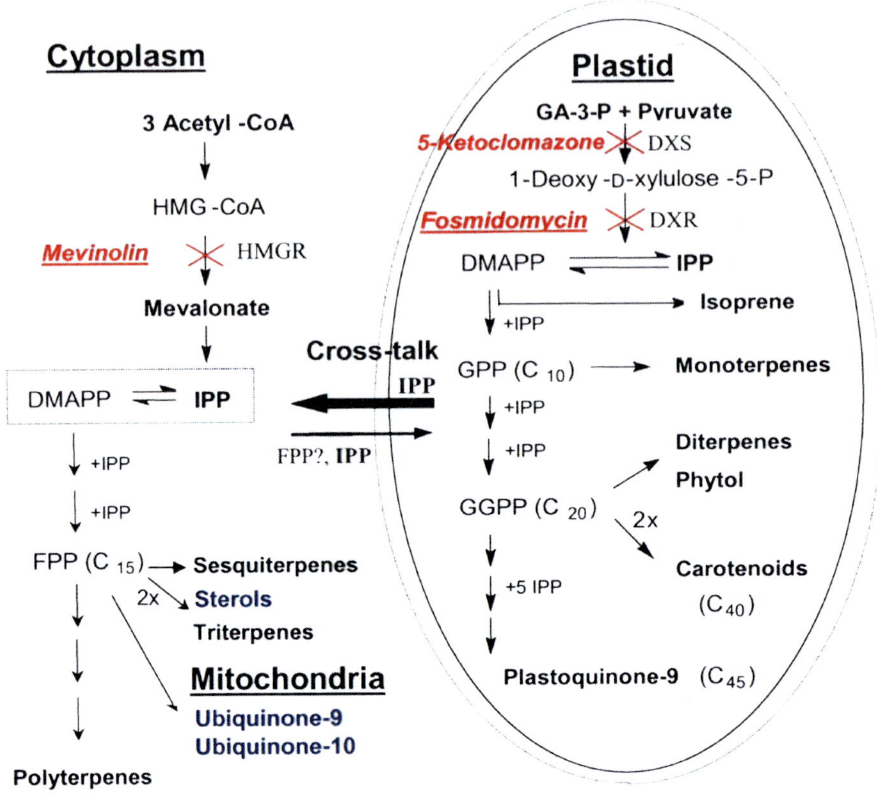

Fig. 6. Scheme showing the compartmentation of the two isoprenoid biosynthesis pathways in the plant cell: (1) the chloroplastidic DOXP/MEP pathway for the biosynthesis of the active C_5-units (IPP) for chlorophylls (phytyl side-chain), carotenoids and prenylquinones (isoprenoid side-chains), and (2) the cytosolic acetate/mevalonate pathway of IPP biosynthesis for the formation of sterols and the prenyl side-chain of the mitochondrial ubiquinones. The specific inhibition of the acetate/mevalonate pathway by *mevinolin* (target: HMG–CoA reductase = HMGR) and of the DOXP/MEP pathway by *fosmidomycin* (target: DOXP-reductase) is indicated. Possible cross-talk between the two cellular biosynthetic isoprenoid pathways, which primarily consists of an export of active C_5 units from chloroplasts to the cytosol (Lichtenthaler 2007), is accentuated. (Based on Lichtenthaler et al. 1997a, b; Lichtenthaler, 1999, 2007). DMAPP = dimethylallyl diphosphate, IPP = isopentenyl diphosphate, FPP = farnesyl diphosphate, GPP = geranyl diphosphate, GGPP = geranylgeranyl diphosphate.

whether mitochondria possess their own IPP biosynthetic system or are dependent, e.g. on the IPP biosynthesis of the cytosol. In non green tobacco cell cultures it was shown that the cytosolic sterols and the prenyl side-chain of Q-10 came from the same IPP pool synthesized via MVA (Disch et al., 1998b) indicating that the acetate/MVA pathway is responsible also for the biosynthesis of the mitochondrial ubiquinones. This also demonstrates that a third IPP biosynthetic pathway does not exist in plant cells. However, in view of the two cellular IPP producing pathways the question arises whether IPP or any other isoprenoid chain produced by the IPP pathway of one cellular compartment can be used by the isoprenoid biosynthesis machinery of the other compartment. Is there a cooperation or cross-talk between both cellular isoprenoid pathways? Are there transport systems for isoprenoids from chloroplasts to the cytosol or vice versa? In fact, there are several early and also more recent observations on this topic as reviewed below.

VI Branching Point of DOXP/MEP Pathway with Other Chloroplast Pathways

Besides the photosynthetic carbon reduction cycle (Calvin cycle) and chlorophyll biosynthesis, the DOXP/MEP pathway for isoprenoid biosynthesis, chloroplasts possess several other biosynthetic activities, such as de novo fatty acid biosynthesis

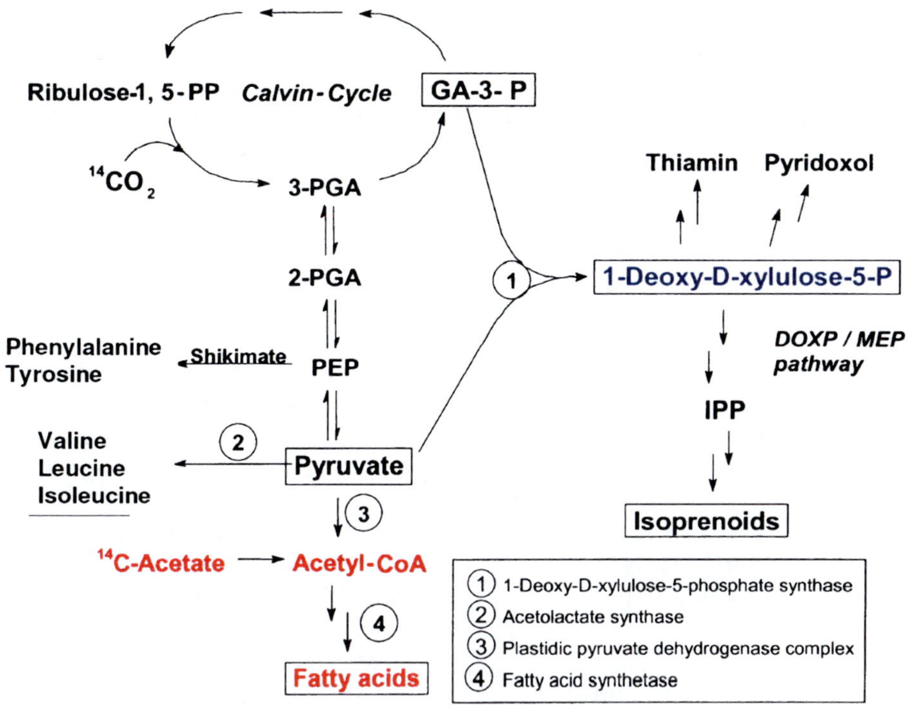

Fig. 7. Biosynthetic pathways and metabolic branching points in chloroplasts. The flow of carbon metabolites from the photosynthetic CO_2 fixing carbon reduction cycle (Calvin cycle) into different end products, such as isopentenyl diphosphate (IPP), plastidic isoprenoids, amino acids, fatty acids as well as thiamine and pyridoxal is indicated. The central role of glyceraldehyde-3-phosphate (GA-3-P) and pyruvate in the DOXP/MEP pathway for biosynthesis of IPP and plastidic isoprenoids is emphasized (Based on Lichtenthaler, 1999). 2-PGA and 3-PGA = 2-phospho- and 3-phosphoglyceric acid, respectively; PEP = phosphoenol-pyruvate.

(Stumpf, 1984) and biosynthesis of particular amino acids, such as aromatic amino acids and branched-chain amino acids. These pathways partially compete for the same substrates, and all are directly or indirectly dependent on the metabolite flow from intermediates or products of the Calvin cycle as is shown in Fig. 7. The early observations that ^{14}C-labeled CO_2, GA-3-P, and pyruvate are better precursors of plastidic isoprenoids than ^{14}C-acetate or ^{14}C-MVA (Braithwaite and Goodwin, 1960; Goodwin, 1958) are now well understood in view of the operation of the DOXP/MEP pathway, in which GA-3-P and pyruvate are direct substrates of the DOXP synthase. In the Calvin cycle $^{14}CO_2$ is rapidly transformed to 3-phosphoglyceric acid (3-PGA) and reduced to glyceraldehyde-3-phosphate (GA-3-P) (Calvin and Bassham, 1962). Pyruvate can be formed in the chloroplast, at least in some plants, such as spinach, directly from photosynthetically fixed carbon via 3-phosphoglyceric acid (3-PGA) and 2-PGA (Schulze-Siebert et al., 1984; Schulze-Siebert and Schulze, 1987). However, pyruvate is also formed in the chloroplast as a byproduct of the ribulose-bisphosphate carboxylase/oxygenase activity (Andrews and Kane, 1991). Thus, the ^{14}C-label of applied $^{14}CO_2$ will quickly show up via GA-3-P and pyruvate in ß-carotene and other plastidic isoprenoids. Also, the fast labeling, formation and emission of isoprene from photosynthetically fixed $^{13}CO_2$ (Delwiche and Sharkey, 1993; Schnitzler et al., 2004) are easily explained by the origin of the starting substrates of the DOXP/MEP pathway, GA-3-P and pyruvate from the Calvin cycle. In contrast, exogenously applied ^{14}C-acetate is rapidly incorporated into fatty acids via the plastidic de novo fatty acid synthetase, but not into β-carotene or other plastidic isoprenoids because acetate is not a substrate of the DOXP/MEP pathway. Phosphoenol pyruvate (PEP) is a substrate of the shikimic acid pathway that occurs in plastids as well. Pyruvate is a substrate of the plastidic biosynthesis of branched-chain amino acids, such as valine, leucine and isoleucine.

When the plastidic pyruvate dehydrogenase complex yielding acetate is blocked by the inhibitor acetylmethylphosphinate (AMPI) (Golz and Lichtenthaler, 1994), biosynthesis and accumulation of the isoprene of leaves is enhanced by 60% at 10^{-4} M of AMPI applied to leaf pieces (Zeidler, 2001) and by 30% at 5×10^{-4} M AMPI. These results indicate that there exist several branching points in the metabolite flow of the Calvin cycle products 3-PGA and GA-3-P as well as pyruvate to different end products that require a fine regulation of chloroplast metabolism. Depending on the daily changing climatic conditions (temperature, irradiance), enhanced or reduced isoprene emission can be one valve for a fine-regulation of carbon flow and chloroplast metabolism (Lichtenthaler, 2007). However, there also exist particular regulations of the activity of the DOXP/MEP pathway and its enzymes as reviewed by Boronat (2010).

VII Cross-Talk Between Both Cellular Isoprenoid Pathways

Can the two IPP pathways cooperate or complement each other if necessary? Is there a cross-talk between the IPP pathways in the cytosol and the plastids? In the original ^{13}C-labeling experiments with algae and higher plants, grown photoheterotrophically on [1-^{13}C]glucose (see Fig. 4), a labeling of photosynthetic isoprenoids was not observed (Schwender et al., 1996; Lichtenthaler et al., 1997a). In fact, the labeling occurred exclusively via the plastidic DOXP/MEP pathway of IPP biosynthesis. However, one has to consider in this respect that, using the high resolution NMR spectroscopy, a low labeling of less than 6% of the plastidic isoprenoids via the cytosolic MVA pathway would not have been detectable. This observation clearly indicated that if a contribution of the cytosolic MVA isoprenoid pathway to the labeling of plastidic isoprenoids should exist it could only occur at very low rates.

However, the situation is different when labeling of cellular isoprenoids is performed with precursors of the plastidic DOXP/MEP pathway. Thus, in the red alga *Cyanidium* containing both isoprenoid pathways a large part of the deuterium-label of [1-^{2}H] DOX, a precursor of plastidic IPP, was not only incorporated into phytol, but also into the cytosolic sterols (Schwender et al. 1997) indicating export of isoprenoid C_5 units. The question of cooperation and crosstalk was further investigated by studying the incorporation of labeled intermediates of each IPP pathway using ^{14}C-DOX, a precursor of the DOXP/MEP pathway, and tritium-labeled mevalonolactone (^{3}H-MVL), a substrate readily used by the cytosolic acetate/MVA pathway of isoprenoid biosynthesis. Two typical isoprenoids were selected: phytol (DOXP/MEP pathway) and sterols (MVA pathway). The results with two algae and a higher plant, all of which possess the two IPP forming pathways, are shown in Table 1. As expected ^{14}C-DOX was incorporated at high rates into plastidic phytol (side-chain of chlorophylls), and ^{3}H-MVL was well incorporated into cytosolic sterols of two algae (*Klebsormidium* and *Mesostigma*) and a higher plant (*Lemna gibba*). Surprisingly, the label of ^{14}C-DOX showed up in the cytosolic sterols as well (to a high extent in both algae and to a lower extent in *Lemna*) indicating that isoprenoid C_5 units must have been exported from the chloroplast to the cytosol, where they were used for biosynthesis of farnesyl diphosphate and incorporated into sterols (Schwender et al., 2001). In contrast to ^{14}C-DOX, the label of ^{3}H-MVL was detected in the plastidic phytol of all three organisms to a very low extent only. In fact, the incorporation of ^{14}C-DOX into phytol of *Klebsormidium, Mesostigma*

Table 1. Cross labeling of the plastidic isoprenoid phytol (side-chain of chlorophylls) and the cytosolic sterols from precursors of the plastidic DOXP/MEP pathway and the cytosolic mevalonate pathway in two algae and a higher plant (*Lemna gibba*) under photosynthetic conditions. ^{14}C-labeled deoxyxylulose (^{14}C-DOX) and tritium-labeled mevalonolactone (^{3}H-MVL) were applied as tracers. The radioactivity of phytol and sterols is given in decays per minute. The labeling of the cytosolic sterols from ^{14}C-DOX indicates export of isoprenoid units from the chloroplast to the cytosol (Based on Schwender et al., 2001).

	Applied precursor		Ratio
Organism/isoprenoid	^{14}C-DOX	^{3}H-MVL	^{14}C/^{3}H
Klebsormidium flaccidum			
Phytol	257.7	2.7	95.4
Sterols	114.2	174.9	0.7
Mesostigma viride			
Phytol	1,600.0	14.9	107.4
Sterols	618.9	535.7	1.2
Lemna gibba			
Phytol	359.6	15.4	23.4
Sterols	28.3	46.5	0.6

and *Lemna* was 95, 107 and 23 times higher, respectively, than the incorporation of ^{3}H-MVL into phytol (Table 1). These data clearly demonstrated the presence of cross-talk between the two isoprenoid pathways of the plant cell, whereby the transport of isoprenoid compounds proceeded almost exclusively into the chloroplast-to-cytosol direction whereas the transport from cytosol to chloroplasts occurred only at extremely low rates. At the developmental stage and time of investigation sterol formation in *Lemna* was considerably lower than in the two algae, but also in this case ^{14}C-DOX was used at 60% of the incorporation rate of the direct cytosolic MVA pathway precursor. These results, obtained with labeled precursors in the light, apparently indicate that under photosynthetic conditions a direct carbon flow takes place from CO_2 via the Calvin cycle, GA-3-P, and pyruvate into plastidic IPP and finally into cytosolic sterols as indicated in Fig. 6.

Further confirmation for cross-talk between the plastidic and cytosolic isoprenoid pathways, as shown in Table 1, came from various other observations and authors. Hemmerlin et al. (2003) applied specific inhibitors of both isoprenoid pathways, mevinolin for the MVA pathway and fosmidomycin for the DOXP/MEP pathway, and for the first time they used the term 'cross-talk' for the cooperation observed between both isoprenoid pathways. In this study 1-deoxy-D-xylulose (DOX) complemented growth inhibition by mevinolin in the low mM concentration range, whereas growth inhibition by fosmidomycin was, however, only partially overcome by MVA. These results confirmed that the export of isoprenoid C_5 units from chloroplasts proceeds better than the import. In another inhibitor experiment Laule et al. (2003) applied fosmidomycin and lovostatin and found changes in pigment and sterol content, e.g. a transient increase or decrease of chlorophyll and carotenoid levels or sterol levels. However, a clear cross-talk, i.e. an isoprenoid metabolite flow from cytosol to chloroplasts or vice versa, was not shown. By contrast, in a more recent investigation Dudareva et al. (2005) unambiguously showed that in the case of volatile terpenoids, being formed in the chlorophyll-free epidermis of snapdragon petals, the DOXP/MEP pathway provided not only the IPP precursors for plastidic monoterpene synthesis, but also for cytosolic sesquiterpene biosynthesis. Also, in this research, the transport of IPP or C_5 isoprenoid compounds occurred unidirectionally from the plastids to the cytosol, an observation that has been further documented in a recent review (Nagegowda et al., 2010). Moreover, Hampel et al. (2005) demonstrated in carrot leaves and carrot roots that monoterpenes are synthesized exclusively via the DOXP/MEP pathway, whereas sesquiterpenes are generated by the classical MVA pathway as well as the DOXP/MEP pathway. The flow of isoprenoid C_5 units from plastids into cytosolic sterols and sesquiterpenes is presumably mediated by the plastidial unidirectional proton symport system allowing the transport of IPP and, at a lower rate, also that of farnesyl diphosphate (Bick and Lange, 2003).

VIII Earlier Observations on Cooperation of Both Isoprenoid Pathways

A certain cross-talk, i.e. cooperation between both cellular isoprenoid pathways, had been described previously (Lichtenthaler, 1998, 1999). One example is the very low labeling rate of plastidic isoprenoids from applied ^{14}C-MVA, whereas sterols are labeled at high rates. This had already been detected in 1958 by Goodwin (1958) and was later noticed by a great number of other authors. This early observation indicated that some isoprenoid C_5-units formed from MVA in the cytosol were transferred to chloroplasts. Another finding showed that, besides the acetate/MVA pathway, another possibility existed to label an isoprenoid C_5-unit. When studying the ^{13}C-labeling of the diterpene ginkgolide from ^{13}C-glucose, three isoprenoid C_5-units were found to be labeled via the MVA pathway and the fourth isoprenoid unit via the DOXP/MEP pathway (Schwarz, 1994). Unfortunately this PhD-thesis was not published and the results became known to the scientific community much later. In addition, in the liverwort *Heterocyphus planus*, the first three isoprene units of phytol and other diterpenes showed some label applied from ^{13}C-MVA, whereas the fourth unit was not labeled (Nabeta et al. 1995, 1997). Both observations pointed to the transfer of cytosolic farnesyl diphosphate (FPP) or a C_{15} isoprenoid compound into the plastid compartment, where it was condensed – as we understand today – with a DOXP/MEP pathway derived IPP to the C_{20} isoprenoid phytol. Also, an observation in chamomile indicated cross-talk between both isoprenoid pathways: in labeling studies of cytosolic sesquiterpenes the first two C_5-units were derived from ^{13}C-glucose via the DOXP/MEP pathway, and the third C_5-unit was labeled by either the plastidic or the MVA pathway (Adam et al., 1998). Yet, one has to consider that, in such experiments mentioned here, the concentration of the applied precursor compounds was usually much higher than the endogenous cellular pool sizes of undisturbed cells. This then can indicate also an artificial import of isoprenoid C_5 units from cytosol to plastids that may not proceed under normal physiological conditions.

In *Arabidopsis* seedlings gibberelins were predominantly synthesized through the DOXP/MEP pathway, whereas the MVA pathway played a major role in the biosynthesis of campesterol (Kasahara et al., 2002). However, consistent with some cross-talk between the two isoprenoid pathways, phenotypic defects caused by the block of the MVA and MEP pathways were partially rescued by exogenous application of the precursors MEP and MVA, respectively. A further observation showed that MVA can partially restore chloroplast and etioplast development in *Arabidopsis* lacking the non-mevalonate pathway (Nagata et al., 2002). A certain type of cross-talk between both isoprenoid pathways was also found in cell cultures of *Catharanthus roseus*, where sitosterol was predominantly labeled via the cytosolic MVA pathway, whereas the plastidic phytol was labeled to a high extent (ca. 60%) via the DOXP/MEP pathway but to ca. 40% also via the MVA pathway (Schuhr et al. 2003). So far, only this observation is in contrast to the above mentioned results, where a preferentional flow of isoprenic C_5-units from plastid to cytosol was detected. However, in the case of non-green or faintly green plant tissue as in *Catharanthus* one has to consider that photoheterotrophic cell cultures usually exhibit a rather limited photosynthetic performance and may not be representative for the metabolite flow in fully green intact plants.

Although cross-talk between both cellular isoprenoid pathways and compartments appears to be possible in both directions, the question arises if a block of one isoprenoid pathway can be compensated for by the C_5 metabolites of the second isoprenoid pathway. In their early work on mevinolin effects on plants by Bach and Lichtenthaler (1982, 1983a, b), the cytosolic MVA and sterol biosynthesis was blocked by the inhibitor mevinolin and the plants finally died, whereas the accumulation of chlorophylls (phytol), carotenoids and plastoquinone-9 in the chloroplasts of these plants was not affected. This finding clearly demonstrated that plastids could not export adequate amounts of isoprene C_5-units to support sufficient cytosolic sterol biosynthesis required for growth. Further, in an *Arabidopsis* mutant, where a nuclear gene (*CLA1*) apparently encoding for the first enzyme DOXP synthase of the plastidic DOXP/MEP pathway is missing (Mandel et al., 1996), the development of chloroplasts and the growth of plants is blocked. From all these data it appears that a cross-talk between both isoprenoid pathways is principally possible. A full or adequate compensation of the missing activity of one isoprenoid pathway by the second isoprenoid pathway of the other cell compartment does apparently not occur under physiological standard conditions of growth.

IX Distribution of the DOXP/MEP and the MVA Pathways in Photosynthetic Algae and Higher Plants

The compartmentalization of isoprenoid biosynthesis with the plastidic DOXP/MEP pathway and the cytosolic MVA pathway of IPP formation (Fig. 6) exists in all higher plants tested so far (Lichtenthaler 1998, 1999, 2000). According to the endosymbiosis theory, chloroplasts derived from previously free-living photosynthetic cyanobacteria that were taken up during evolution by a flagellate in a primary endosymbiotic event (Delwiche, 1999) as is shown in Fig. 8. In fact, present Cyanobacteria possess the DOXP/MEP pathway of IPP formation as has been shown for *Synechocystis* (Disch et al., 1998a; Lichtenthaler et al., 1997b; Proteau, 1998).

The different photosynthetic algae groups that developed during evolution are distinguished according to their photosynthetic pigment apparatus: (i) those with chlorophyll a and b (usually referred to as 'green algae' but representing a polyphyletic group), (ii) chlorophyll a and phycobilisomes (red algae, rhodophytes), and others (iii) with chlorophyll a and c (heterokontophytes).

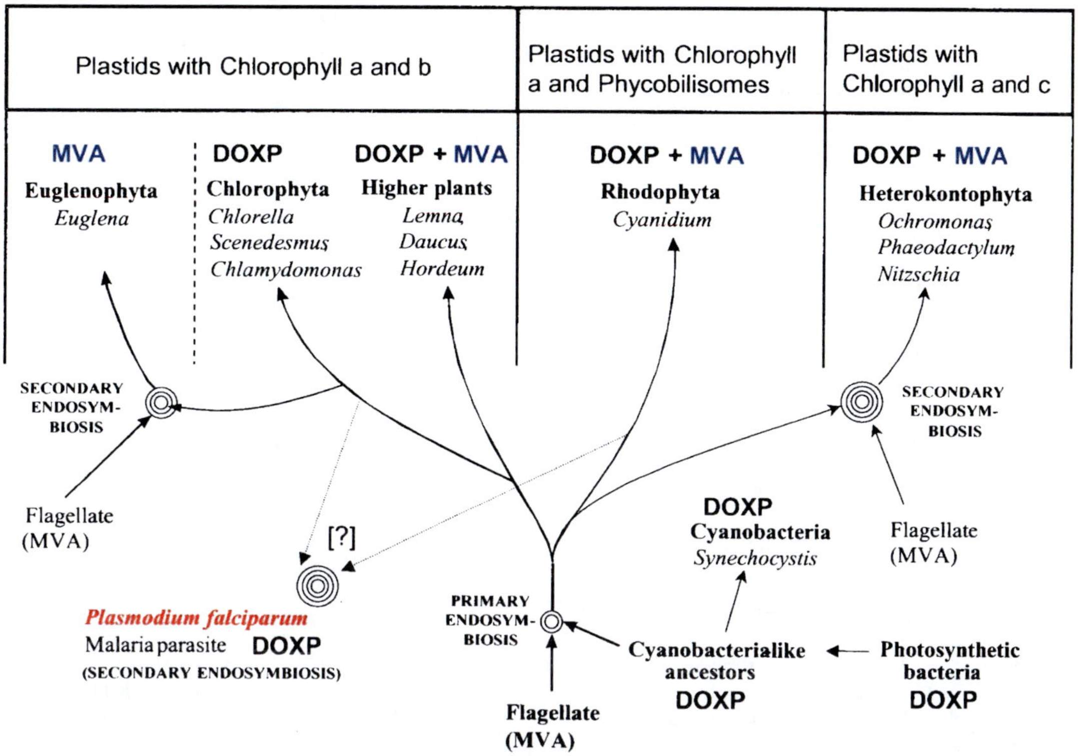

Fig. 8. Putative evolution of photosynthetic algae and higher plants with the indication of one or two cellular isoprenoid pathways of IPP biosynthesis: either the MVA + DOXP pathways or one of the two pathways only, such as the MVA pathway (*Euglena*) or the DOXP pathway (Chlorophyta). *Plasmodium falciparum*, the malaria-causing parasite, has lost its MVA pathway during evolution and is dependent on the DOXP pathway of its apicoplast a former endosymbiont, either a former Chlorophyte or Rhodophyte. Primary and secondary endosymbiotic events in the course of the evolution (e.g. Van den Hoek et al., 1995) leading to chloroplasts (and in the case of *Plasmodium* to an apicoplast) with 2, 3, or 4 biomembranes are accentuated; based on Lichtenthaler (1999) with the inclusion of recent observations.

Most algae groups possess two cellular isoprenoid biosynthesis pathways as higher plants: the DOXP/MEP pathway for the biosynthesis of carotenoids, phytol and isoprene, and the MVA pathway for biosynthesis of the cytosolic sterols (Disch et al., 1998a; Lichtenthaler 1999; Lichtenthaler et al., 1997a; Schwender et al., 1997). This also applies to the marine diatoms *Nitschia* and *Phaeodactylum* belonging to the heterokontophytes (Cvejić and Rohmer, 2000). An exception of this rule are the chlorophytes proper, such as *Chlorella*, *Scenedesmus* and *Chlamydomonas* (Schwender et al., 1995, 1996), where not only plastidic isoprenoids are synthesized via the DOXP/MEP pathway but the cytosolic sterols as well. A detailed investigation of this unexpected phenomenon resulted in the detection that the commonly termed 'green algae', a polyphyletic group, consisting of the *Streptophyta* (including higher plants) with both IPP pathways, and the *Chlorophyta* proper have lost their MVA pathway for IPP and isoprenoid biosynthesis (Schwender et al., 2001). This is indicated in Fig. 9. Chlorophyta have not only lost their acetate/MVA isoprenoid pathway during evolutionary events, but also various cytosolic enzymes of their sugar metabolism (Schnarrenberger et al., 1990). That chlorophyta form a group separate from the streptophytes has also been confirmed by differences in the 18S-rRNA composition and ultra-structural characteristics of their cells and flagellae (Friedl, 1997). Further evidence for the absence of the MVA pathway is the fact that statins (mevinolin, cerivastatin) do not inhibit the growth of chlorophyta, and that genes of the MVA pathway could not be detected (Schwender et al., 2001).

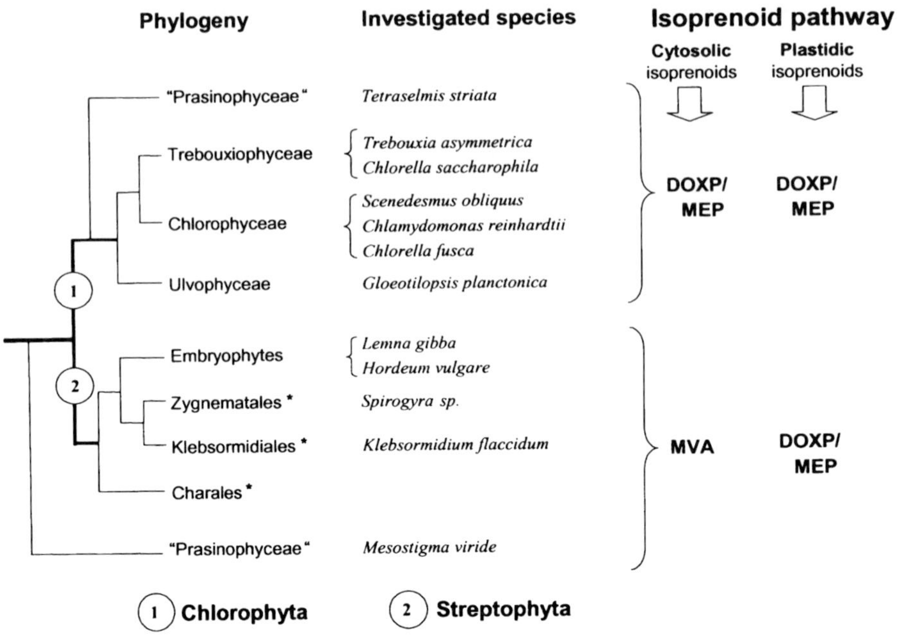

Fig. 9. Spreading of the DOXP/MEP pathway and the acetate/MVA pathways for biosynthesis of plastidic and cytolic isoprenoids in (1) Chlorophyta and (2) Streptophyta including higher plants (based on Schwender et al., 2001). Both isoprenoid pathways are present in Streptophyta, whereas in Chlorophyta, a phylogenic separate deduced group, has only the plastidic DOXP/MEP pathway and uses it for the biosynthesis of both plastidic and cytosolic isoprenoids as is indicated on the right side of the figure. *Mesostigma viride*, in turn, a common 'ancestor' of chlorophytes and streptophytes, shown in the figure as first organism on the evolutionary scale, exhibits both isoprenoid pathways. The distribution of only the DOXP/MEP pathway in the proper chlorophytes (1) and of both isoprenoid pathways in the Streptophyta (2) shown here correlates with the current phylogeny of 'green algae' and higher plants based on 18S-rRNA sequences, ultrastructural characteristics of flagellate cells and of cell mitosis (Friedl, 1997). The former designation "Prasinophyceae", as a green algae clade is no longer valid since it is not a monophyletic group and also included *Mesostigma* which does not belong to the Chlorophyta. These clades of the former group of 'green algae' are often summarized as "Charophyceae", but do not form a monophyletic group. (Based on Disch et al., 1998a; Lichtenthaler et al., 1997a, b).

Moreover, in the pigment-free chlorophyte *Prototheca wickerhamii* ergosterol is synthesized via the DOXP/MEP pathway (Zhou and Nes, 2000). In contrast, the prasinophyte *Mesostigma viride,* which is a common precursor of Chlorophyta and Streptophyta, contains both the DOXP/MEP and the acetate/MVA pathway for IPP biosynthesis (Fig. 9), thus indicating that chlorophytes and streptophytes, including higher plants, are derived from the same evolutionary ancestor with both isoprenoid biosynthesis pathways.

Another exception from the existence of two cellular IPP biosynthesis routes are Euglenophyta. In contrast to all the other photosynthetic organisms, the flagellate *Euglena gracilis* forms sterols and plastidic isprenoids (carotenoids, phytol) exclusively via the classical acetate/MVA pathway (Disch et al., 1998a). *Euglena* originated from a colorless flagellate that incorporated a green alga (possibly a chlorophyte) in a secondary endosymbiosis, during which the plastidic DOXP/MEP pathway of the endosymbiont was lost (see Fig. 8). That a secondary endosymbiosis does not necessarily have to lead to a loss of one of the two IPP producing pathways of the endosymbiont can be seen in the example of the algae group heterokontophytes, in which both IPP producing pathways have been preserved (see Fig. 8).

A special photosynthetic algae group represents the Chlorarachniophytes (e.g. *Bigelowiella natans*), green amoeboflagellate algae that acquired their plastids by a secondary endosymbiosis of a green alga (Archibald et al., 2003). Like the plastids of heterokonts, haptophytes, apicomplexa and cryptomonads, their plastids are surrounded by four envelope membranes, whereas the plastids of euglenids and dinoflagellates are surrounded by only three envelope membranes (Archibald et al., 2003). In contrast to heterokonts, the photosynthetic apparatus of *Bigelowiella* contains chlorophyll a and b as do euglenids, green algae and higher plants; thus, it must be derived from a 'green alga'. It would be of great scientific interest to know if *Bigelowiella* contains both IPP producing pathways, the DOXP/MEP and MVA pathway, or if it has lost one of them during or after the secondary endosymbiosis of a green alga.

A very particular taxonomic group are the *Apicomplexa,* comprising the sporozoa parasites *Plasmodium falciparum* and *Toxoplasma gondii.* These organisms contain a special cell organelle, the apicoplast, a non-green plastid with a genome similar to that of chloroplasts of green algae and an envelope consisting of four membranes. The apicomplexa evolved either from a secondary or even a tertiary endosymbiosis (Delwiche, 1999), possibly by the incorporation of a green alga (Funes et al., 2002) or a red alga (Archibald et al., 2003; Wilson, 2002) (Fig. 8). Whether Apicomplexa, Heterocontophyta, Haptophyta, Cryptophyta and Dinoflagellates have a common ancestor (Archibald et al., 2003; Wilson, 2002) is still a matter of debate. In any case, the parasite *Plasmodium* possesses the DOXP/MEP pathway of IPP and isoprenoid biosynthesis (Jomaa et al., 1999), and its development can be blocked by fosmidomycin, whereas the MVA pathway is missing. This also applies to *Toxoplasma gondii* (Eberl et al., 2003). All these data show that two ancient and distinct isoprenoid biosynthesis pathways evolved in livings organisms (see also Lange et al., 2000).

X Evolutionary Aspects of the DOXP/MEP Pathway

The genes of the DOXP/MEP pathway have not only been found in higher plants, algae and the protozoan parasites *Plasmodium* and *Toxoplasma*, but also in many photosynthetic bacteria, such as *Synechocystis* and *Synechococcus leopoliensis* (both blue-green bacteria with oxygenic photosynthesis) as well as *Chlorobium tepidium* (green sulphur bacteria) and *Rhodobacter capsulatus* (purple bacteria) with anoxigenic photosynthesis. However, also heterotrophic bacteria, such as the pathogenic bacteria *Escherichia coli*, *Haemophilus influenzae*, *Helicobacter pylori*, *Mycobacterium tuberculosis*, *Mycobacterium leprae*, *Chlamydia pneumonia*, *Yersinia pestis* and *Vibrio cholerae* possess the DOXP/MEP pathway (as reviewed by Lichtenthaler et al., 2000b; Lichtenthaler, 2004). The acetate/MVA pathway of isoprenoid biosynthesis is, however, not present in these bacteria. In contrast, *Archaea*, which form an independent evolutionary group, do not have the DOXP/MEP pathway, their isoprenoid biosynthesis is based on the classical or a modified MVA pathway (see Lichtenthaler, 2004 for further literature). This also applies to the chloroplast-free heterotrophic animals and fungi belonging to the *Eucarya*. From the data available it appears that during evolution the DOXP/MEP pathway may have originated in photosynthetic bacteria, possibly those with anaerobic photosynthesis based on one photosystem. The latter seem to have evolved to the oxigenic cyanobacteria with two photosystems that, like algae and higher plants, emit oxygen in their photosynthetic process. Moreover, during evolution various photosynthetic bacteria may have lost their pigment apparatus and with it their competence for photosynthetic quantum conversion. Therefore, one has to consider that the present heterotrophic bacteria with the DOXP/MEP pathway may have derived from former photoautotrophic bacteria. With the loss of their photosynthetic competence such formerly phototropic bacteria may have survived by turning into pathogens that live and multiply in a host organism. These possible evolutionary developments are summarized in Fig. 10, but require further investigations.

XI Biosynthesis of Isoprene and Methylbutenol

Isoprene: This volatile hemiterpene (C_5H_8, 2-methyl-1,3-butadiene) is emitted by many green plants, including mosses, ferns, gymnosperms and angiosperms (Kesselmeier and Staudt, 1999; Sharkey and Yeh, 2001). The light and temperature dependent isoprene emission by leaves preferentially occurs at high irradiance and high rates at temperatures above 28°C, such as full sun light, when the photosynthesis process with the photosynthetic light reactions and associated

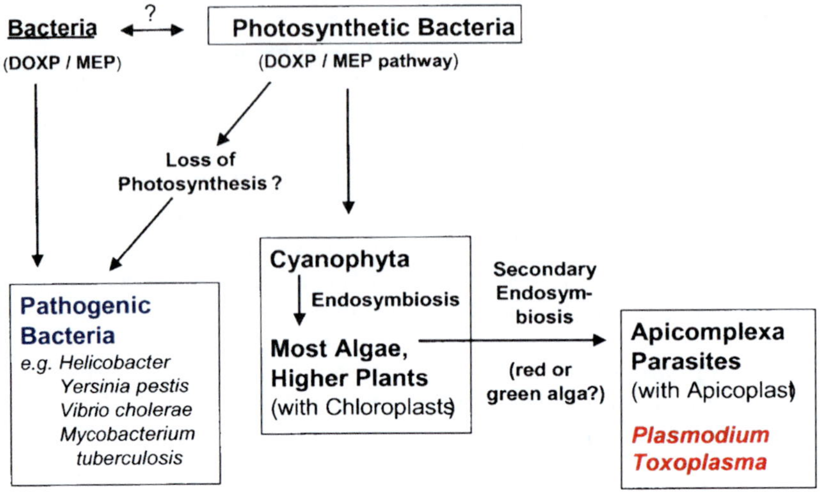

Fig. 10. Putative evolution and transfer of the DOXP/MEP pathway of IPP biosynthesis from early photosynthetic or heterotrophic bacteria to pathogenic bacteria, cyanobacteria, and by endosymbiosis to algae, higher plants, and the parasitic *Apicomplexa*.

electron transport reactions is completely light saturated. Isoprene emission from plants amounts to hundreds of millions of metric tons to the global atmosphere, the estimations range from 180 to 450 × 10^{12} g carbon per year worldwide (e.g. Rasmussen and Khalil, 1998; Sharkey, 1996). The fast appearance of ^{13}C-labeling in isoprene from photosynthetically fixed $^{13}CO_2$ (Delwiche and Sharkey, 1993; Schnitzler et al. 2004) indicates that isoprene biosynthesis must be closely connected to intermediates of the Calvin–Benson cycle. In fact, the biosynthesis of the volatile C_5-hydrocarbon isoprene proceeds via the plastidic DOXP/MEP pathway for isoprenoid biosynthesis from GA-3-P and pyruvate (Fig. 6), as has been shown by the specific incorporation of deuterium-labeled 1-deoxy-D-xylulose (^{2}H-DOX) into isoprene (Fig. 5). This has been verified by gas chromatography combined with mass spectrometry (GC-MS) and byhigh resolution NMR spectroscopy (Zeidler et al., 1997; Schwender et al., 1997). Biosynthesis and emission of isoprene from illuminated plant leaves are efficiently blocked by the herbicide fosmidomycin that inhibits the second enzyme DXR of the DOXP/MEP pathway of plastidic IPP and isoprenoid formation (Zeidler et al., 1998). Isoprene is released from DMAPP in a single enzymatic step via the plastidic isoprene synthase (Silver and Fall, 1995) that exists in a thylakoid-bound form (Wildermuth and Fall, 1996) and stromal isoforms in the chloroplast (Wildermuth and Fall, 1998). The enzyme isoprene synthase is related to monoterpene synthases found in other plants (Sharkey et al., 2005). Its K_m is ten- to 100-fold higher for its dimethylallylic diphosphate substrate DMAPP than related monoterpene synthases for geranyl diphosphate (Wolfertz et al., 2004). The regulation of the light and temperature dependent isoprene emission apparently proceeds via the relative activity of the DOXP/MEP pathway and possibly via the concentration of DMAPP (Rosenstiel et al., 2002; Wolfertz et al., 2004).

Methylbutenol: In western North America the needles of several pines (*Pinus ponderosa*, *P. contorta*, *P. sabiniana*) do not emit isoprene itself but a partially oxidized form, the hemiterpene 2-methylen-3-buten-2-ol (MBO), in a light and temperature-dependent manner (Harley et al., 1998; Schade et al., 2000). The biosynthesis of the C_5 structure of the volatile MBO proceeds also via the plastidic DOXP/MEP pathway as has been determined by a high-rate incorporation of deuterium-labeled deoxy-D-xylulose (^{2}H-DOX) into MBO as verified by mass spectrometry (Zeidler and Lichtenthaler, 2001) and can be inihibited by fos midomycin. Like isoprene, MBO is formed from DMAPP in one step via the enzyme MBO synthase. Isoprene synthase and MBO synthase use the same substrate DMAPP, but the chemical mechanism for cleavage of the C_5 carbon structure from the diphosphate of DMAPP is different, thus yielding different end products (Lichtenthaler, 2007). The enhanced *de novo* biosynthesis of isoprene and MBO consumes, per each molecule formed, 3 ATP and 3 NADPH that are generated in the photosynthetic light reactions. This keeps the two photosynthetic photosystems 'busy' and intact, avoiding overreduction and photo-oxidative damage of the photosynthetic apparatus at excess light conditions. Thus, isoprene and methylbutenol emissions may be a 'safety valve', similar to the process of photorespiration, to protect the photosynthetic pigment apparatus with its photosystems and light-harvesting pigment-proteins (Thornber, 1975; Lichtenthaler et al., 1982a, b) against photo-oxidation.

XII Level of Chlorophylls, Carotenoids and Prenylquinones in Sun and Shade Leaves

Chlorophylls, carotenoids and prenylquinones are essential components of the photochemically active thylakoids of chloroplasts (Lichtenthaler and Calvin, 1964; Lichtenthaler and Park 1963). The photosynthetic apparatus of green leaf tissue, like the leaf structure, is very adaptive to the incident irradiance. This is documented by different relative amounts of photosynthetic pigments and prenylquinones. Thus, sun and shade leaves of trees as well as leaves of high-light (HL) and low-light (LL) plants differ considerably in their quantitative levels of photosynthetic pigments, electron carriers, chloroplast ultrastructure and photosynthetic rates (Anderson et al., 1995; Boardman 1977; Givnish 1988; Lichtenthaler et al. 1981a, 2000a; Meier and Lichtenthaler 1981; Sarijeva et al., 2007; Wild et al. 1986; Zhang et al., 1995). Leaves that develop under high irradiance (sun leaves and HL-leaves) possess sun-type chloroplasts that are adapted to much higher rates of photosynthetic CO_2 assimilation per leaf area and on a chlorophyll basis (Lichtenthaler 1981) in comparison to shade leaves or leaves from LL-plants (reviewed by Lichtenthaler and Babani, 2004; Sarijeva et al., 2007). Sun leaves are characterized by a higher Chl a + b content per leaf area, higher values for the Chl a/b ratios, and a much lower level of light-harvesting Chl a/b proteins (LHCII). As , a consequence, they exhibit a lower degree of thylakoids stacking (lower amounts of appressed membranes) than shade leaves and LL-plants (Lichtenthaler et al., 1981a, 1982b). For this reason, sun-type chloroplasts possess lower and narrower grana thylakoid stacks as well as rather large osmiophilic plastoglobuli (Lichtenthaler, 1968; Lichtenthaler and Sprey 1966) as compared to shade-type chloroplasts which exhibit much higher and broader grana stacks and only few small plastoglobuli (Fig. 11).

The range of the differential levels of chlorophylls, carotenoids and prenylquinones are given for sun and shade leaves of beech (Table 2). Sun leaves possess on a leaf area basis a much higher content of chlorophylls (a + b), carotenoids (x + c), and prenylquinones as compared to shade leaves. However, also the pigment ratios are quite different. The Chl a/b ratio of sun leaves of beech, and also of seven other deciduous tree species, ranges from 2.9 to 3.6 in comparison to values of 2.4–2.9 for shade leaves. The higher carotenoid content of sun leaves and HL-leaves on a Chl a + b basis is documented by significantly lower values for the weight ratio of chlorophyll/carotenoids (a + b)/(x + c) (range: 3.9–4.9) as compared to shade leaves with significantly higher values (range: 5.1–7.0). Particularly remarkable are the high levels of excess plastoquinone-9 (mostly the reduced form plastoquinol-9) and excess α-tocopherol in sun leaves that are deposited in the osmiophilic plastoglobuli (Lichtenthaler, 1968, 1969a, b) of the chloroplast stroma, which steadily increase in size during the course of the vegetation period. In spring, after the completion of leaf development and the greening process (usually at the end of May), the level of

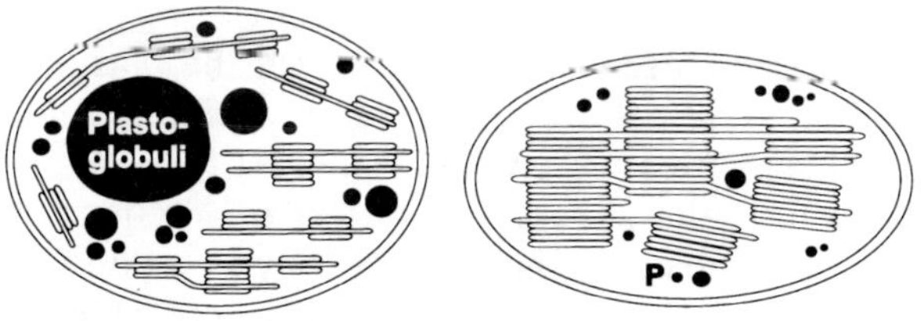

Fig. 11. Differences in ultrastructure, plastoglobuli content, thylakoid frequency, width and stacking degree of thylakoids in sun-type chloroplasts and shade-type chloroplasts of sun and shade leaves of trees, and of leaves of herbaceous plants grown at high-light or low-light conditions. P – Plastoglobuli. (Based on Lichtenthaler 1981; Lichtenthaler et al. 1981b, 1982b; Meier and Lichtenthaler 1981).

Table 2. Standard levels and range of chlorophyll (a + b), total carotenoids (x + c) and plastidic prenylquinones (in mg m^{-2} leaf area), pigment weight ratios, Chl a/b and Chls/carotenoids, Chls/Cars or (a + b)/(x + c) in fully developed green sun and shade leaves of beech (*Fagus sylvatica* L.) in mid-July to the end of August. Sun leaves and sun chloroplasts exhibit particularly high levels of plastoquinone-9 and α–tocopherol, which are deposited in the osmiophilic plastoglobuli of the chloroplast stroma. In the case of plastoquinone-9 and α-tocopherol the relative levels in thylakoids and osmiophilic plastoglobuli are indicated separately. Mean of 16 determinations (pigments) and nine determinations (prenylquinones) from at least eight leaves of four beech trees from different years. The photosynthetic pigments, Chls *a* and *b* as well as total carotenoids x + c, were determined spectrophotometrically (Lichtenthaler, 1987), and the prenylquinones via HPLC.

	Sun leaf	Shade leaf
Pigments		
Chlorophyll a + b	450–650	370–460
Carotenoids x + c	105–140	65–90
Percentage of β-carotene (provitamin A)	32–36	21–27
Pigment ratios		
Chl a/b mean	3.2	2.7
Range	2.9–3.6	2.4–2.9
Chls/Cars mean	4.5	5.7
Range	3.9–4.9	5.1–7.0
Prenylquinones		
Plastoquinone-9		
In thylakoids	34	19
In plastoglobuli	60–110	4–10
α-Tocopherol (vitamin E)		
In thylakoids	4–6	3–5
In plastoglobuli	40–60	4–5
Phylloquinone (vitamin K1)	4–8	3–5
α-Tocoquinone	5–7	4–5

plastoquinone-9 and α-tocopherol (vitamin E) is still low and corresponds to the amounts present in the thylakoids. However, with increasing vegetation time the excess amounts of both prenyl compounds rise. This indicates that the plastidic DOXP/MEP pathway of IPP and isoprenoid biosynthesis after the greening process provide, in particular, the nonaprenyl side-chain for plastoquinone biosynthesis and the phytyl-chain for α–tocopherol biosynthesis. In addition, at high irradiance stress there occurs de novo biosynthesis of β-carotene, xanthophyll cycle carotenoids, such as zeaxanthin, and enormous amounts of isoprene requiring the DOXP/MEP pathway of IPP and isoprenoid biosynthesis to be active throughout the vegetation period (Lichtenthaler, 2007). Plastoquinone-9, α-tocopherol and minor amounts of α-tocoquinone are not only found in thylakoids and plastoglobuli, but in low amounts also in the plastid envelope membranes (Lichtenthaler et al., 1981b) as recently reviewed by Lichtenthaler (2007).

XIII Inhibition of Chlorophyll and Carotenoid Biosynthesis by 5-Ketoclomazone

It had been indicated above that the DOXP/MEP pathway of IPP biosynthesis can be blocked by 5-ketoclomazone (target: DOXP-synthase) (Müller et al., 2000; Zeidler et al., 2000) and also by fosmidomycin (target: DOXP-reductoisomerase) (Zeidler et al., 1998). Blocking one of these two enzymes by an inhibitor results in a dose-dependent, considerable reduction of the biosynthesis and accumulation of chlorophylls and carotenoids. In fact, the greening process of leaves is retarded and one obtains an incomplete photosynthetic apparatus that remains in the biosynthetic transition stage. This is shown for 5-ketoclomazone in Table 3.

When the herbicide clomazone is applied, in the field, the actual inhibitor is 5-ketoclomazone. Since in barley the herbicide clomazone is readily oxidized to its derivative 5-ketoclomazone, clomazone is as effective in inhibiting pigment accumulation as the actual active ingredient 5-ketoclomazone. The accumulation of both photosynthetic pigment classes, chlorophylls (phytyl side-chain) and carotenoids are inhibited. The accumulation of carotenoids is inhibited to a higher extent than that of chlorophylls (Table 3). Moreover, among the two chlorophylls the accumulation of Chl b is inhibited to a higher degree than that of Chl a, resulting in unusually high values of 9.5 and 9.7 at a concentration of 10^{-4} M for the ratio Chl a/b. Similarily high Chl a/b ratios are also obtained in the first few hours of the illumination of etiolated seedlings, since during the endogenous light-induced biosynthetic program of the photosynthetic apparatus, the biosynthesis and accumulation of Chl b lag ca. 1 or 2 h behind that of Chl a (See Sisler and Klein, 1963; Babani and Lichtenthaler, 1996; Lichtenthaler, 1969c). This is also the reason, why in control leaves (controls 1 and 2) the Chl a/b ratio did not decline to the expected range of 2.8–3.3 of fully greened leaf tissue. It has also been shown

Table 3. Inhibition of chlorophyll and carotenoid biosynthesis and accumulation in 5-day old greening barley seedlings during a 20 h light period by the herbicide clomazone and its derivate 5-ketoclomazone. The latter is a specific inhibitor of the DOXP synthase, the first enzyme of the DOXP/MEP pathway of the plastidic isoprenoid biosynthesis. In barley clomazone is readily oxidized to 5-ketoclomazone and also functions as inhibitor. The inhibition causes a considerable increase in the weight ratios of pigments Chl a/b and Chls/carotenoids (a + b)/(x + c). The pigment content and ratios are those of the newly accumulated pigments in the light, without the carotenoid amounts of the seedlings kept in darkness. For better comparison purposes, the pigment levels and ratios of the controls are indicated in bold print (mean of three determinations).

	Chlorophylls		Carotenoids		Pigment ratios	
	μg/g FW (% inhib.)		μg/g FW (% inhib.)		a/b	(a + b)/x + c)
Control 1	**175.4**	(0%)	**32.1**	(0%)	**3.8**	**5.5**
+Clomazone						
10^{-5} M	88.7	(49%)	11.3	(65%)	5.3	7.9
10^{-4} M	17.9	(90%)	1.7	(95%)	9.7	10.5
Control 2	**128.8**	(0%)	**16.6**	(0%)	**4.7**	**7.8**
+5-Ketoclomazone						
10^{-5} M	82.3	(36%)	4.8	(89%)	5.4	17.1
10^{-4} M	9.4	(93%)	0.5	(97%)	9.5	18.8

in an aurea mutant of tobacco (Su/su), which is relatively poor in Chl b due to its retarded accumulation, that the Chl a/b ratio exhibits high values during leaf development, , which slowly and continuously decrease with increasing age of leaves (Schindler et al., 1994).

Furthermore, the weight ratio of chlorophylls to carotenoids, ratio (a + b)/(x + c), is considerably increased after clomazone and 5-ketoclomazone application and shows extremely high values of 10.5 and 18.8 at inhibitor levels of 10^{-4} M, respectively (Table 3). This result is an additional indication that the development and assembling of a fully functional photosynthetic apparatus with normal pigment relations is not possible under inhibitor action. Similar inhibitory results with increased pigment ratios, as shown here for clomazone and 5-ketoclomazone, have also been obtained when the inhibitor fosmidomycin was applied (Zeidler et al., 1998). The high sensitivity of the light-induced chlorophyll and carotenoid biosynthesis in leaves of etiolated seedlings towards 5-ketoclomazone and fosmidomycin demonstrates that the greening of etiolated seedlings is an excellent test system for the detection of potential inhibitors of the DOXP/MEP pathway as well as the detection of new ingredients (Lichtenthaler et al., 2000b; Zeidler et al., 2000) against pathogenic bacteria and the malaria parasite *Plasmodium falciparum*, which is dependent on the DOXP/MEP pathway of its apicoplast.

XIV Conclusion

The non-mevalonate DOXP/MEP pathway of plastidic isopentenyl diphosphate and isoprenoid biosynthesis is well established with its seven enzymes and corresponding genes, and is widely spread in all photosynthetic organisms (with the exception of *Euglena* that has lost its DOXP/MEP pathway during evolution). This isoprenoid pathway channels photosynthetically reduced carbon into the prenyllipids of photochemically active thylakoids, such as chlorophylls, carotenoids and prenylquinones. In addition, it channels photosynthetic carbon into natural plant products, e.g. mono- and diterpenes as well as the volatile hemiterpenes isoprene and methylbutenol. Moreover, it contributes to the biosynthesis of cytosolic sesquiterpenes (e.g. Nagegowda et al., 2010, Chapter X this book) under photosynthetic light conditions by export of active isoprenoid C_5-units for sterol biosynthesis in the cytosol (as shown here and by Lichtenthaler, 2007) as well as in flower petals with non-green plastids. Also we have essential insights into the cellular regulation of the DOXP/MEP pathway (see Boronat 2010) and we know much about the possible cross-talk between the two cellular isoprenoid pathways as shown above. Furthermore, the importance of plastids and the DOXP/MEP pathway in protein prenylation and regulation of cell metabolism is becoming evident (Gerber et al., 2010). Thus, the basic knowledge for the modification of the DOXP/MEP pathway and its role in controlling metabolic pathways in plants and microorganisms is available for a future controlled production of chloroplast isoprenoids, such as β-carotene (provitamin A), α-tocopherol (vitamin E), and particular mono- and diterpenes of medical, and pharmaceutical and agricultural interest. This knowledge opens up new possibilities for chloroplast biotechnology including the search for new active ingredients against pathogenic bacteria, the malaria parasite and other protozoan parasites.

Acknowledgements

I am grateful to several of my former Ph.D. students and group members who carried out parts of the research on plant isoprenoid biosynthesis (T. J. Bach, C. Müller, J. Schwender, J. Zeidler). I also wish to thank Ms Sabine Zeiler for the excellent implementation of pigment determinations, and to Ms Gabrielle Johnson for English language assistance.

References

Adam KP, Thiel R, Zapp J and Becker H (1998) Involvement of the mevalonic acid pathway and the glyceraldehyde-pyruvate pathway in terpenoid biosynthesis of the liverworts *Riccio carpus natans* and *Conocephalum conicum*. Arch Biochem Biophys 354:181–187

Adam P, Hecht S, Eisenreich W, Kaiser J, Gräwert T, Arigoni D, Bacher A and Rohdich F (2002) Biosynthesis of terpenes: studies on 1-hydroxy-2-methyl-2-(*E*)-butenyl 4-diphosphate reductase. Proc Natl Acad Sci USA 99:12108–12113

Altincicek B, Kollas A-K, Sanderbrand S, Wiesner J, Hintz M, Beck, E and Jomaa H (2001) *GcpE* is involved in the 2-*C*-methyl-D-erythritol 4-phosphate pathway of isoprenoid biosynthesis in *Escherichia coli*. J Bacteriol 183:2411–2416

Altincicek B, Duin EC, Reichenberg A, Hedderich R, Kollas A-K, Hintz M, Wagner S, Wiesner J, Beck E and Jomaa H (2002) LytB protein catalyzes the terminal step of the 2-*C*-methyl-D-erythritol-4-phosphate pathway of isoprenoid biosynthesis. FEBS Lett 532:437–440

Anderson JM, Chow WS and Park Y-I (1995) The grand design of photosynthesis: Acclimation of the photosynthetic apparatus to environmental cues. Photosynth Res 46:129–139

Andrews TJ and Kane HJ (1991) Pyruvate is a by-product of catalysis by ribulosebisphosphate carboxylase/oxigenase. J Biol Chem 266:9447–9452

Archibald JM, Rogers MB, Toop M, Ishida K and Keeling PJ (2003) Lateral gene transfer and the evolution of plastid targeted proteins in the secondary plastid-containing alga *Bigelowiella natans*. Proc Natl Acad Sci USA 100:7678–7683

Arigoni D, Sagner S, Latzel C, Eisenreich W, Bacher A and Zenk MH (1997) Terpenoid biosynthesis from 1-deoxy-D-xylulose in higher plants by intramolecular skeletal rearrangement. Proc Natl Acad Sci USA 94:10600–10605

Babani F and Lichtenthaler HK (1996) Light-induced and age-dependent development of chloroplasts in etiolated barley leaves as visualized by determination of photosynthetic pigments, CO_2 assimilation rates and different kinds of chlorophyll fluorescence ratios. J Plant Physiol 148:555–566

Bach TJ and Lichtenthaler HK (1982) Mevinolin a highly specific inhibitor of microsomal 3-hydroxy-methyl-glutaryl-coenzyme A reductase of radish plants. Z Naturforsch Part C 37: 46–50

Bach TJ and Lichtenthaler HK (1983a) Mechanisms of inhibition by mevinolin (MK 803) of microsome-bound radish and of partially purified yeast HMG-CoA reductase (EC. 1.1.1.34). Z Naturforsch 37c:212–219

Bach T J and Lichtenthaler HK (1983b) Inhibition by mevinolin of plant growth, sterol formation and pigment accumulation. Physiol Plant 59:50–60

Bach TJ, Boronat AJ, Campos NJ, Ferrer AJ and Vollack K-UJ (1999) Mevalonate biosynthesis in plants. Crit Rev Biochem Mol Biol 34:107–122

Bick JA and Lange BM (2003) Metabolic cross talk between cytosolic and plastidial pathways of isoprenoid biosynthesis: unidirectional transport of intermediates across the chloroplast envelope membrane. Arch Biochem Biophys 415:146–154

Boardman N (1977) Comparative photosynthesis of sun and shade plants. Annu Rev Plant Physiol 28:355–377

Boronat A (2010) The methylerythritol 4-phosphate pathway: regulatory role in plastid isoprenoid biosynthesis. In: Rebeiz CA, Benning C, Daniel H, Hoober K, Lichtenthaler HK, Portis A, Tripathy B (eds) The Chloroplast: Basics and Application. Springer, Dordrecht, The Netherlands, pp. 119–126

Braithwaite GD and Goodwin TW (1960) Studies on carotenogenesis. 27. Incorporation of [2-^{14}C]acetate, D,L-[2-^{14}C]mevalonate and $^{14}CO_2$ into carrot-root preparations. Biochem J 76:194–197

Calvin M and Bassham JA (1962) The Photosynthesis of Carbon Compounds. WA Benjamin, New York

Campos N, Rodríguez-Concepción M, Seemann M, Rohmer M and Boronat A (2001) Identification of *gcpE* as a novel gene of the 2-*C*-methyl-D-erythritol 4-phosphate pathway for isoprenoid biosynthesis in *Escherichia coli*. FEBS Lett 488:170–173

Chaykin S, Law J, Philipps AH and Bloch K (1958) Phosphorylated intermediates in the synthesis of squalene. Proc Natl Acad Sci USA 44: 998–1004

Cvejić JH and Rohmer M (2000) CO_2 as main carbon source for isoprenoid biosynthesis via the mevalonate-independent methylerythritol 4-phosphate route in the marine diatoms *Phaeodactylum tricornutum* and *Nitzschia ovalis*. Phytochem 53:21–28

Delwiche CF (1999) Tracing the thread of plastid diversity through the tapestry of life. Am Nat 154:S164–S177

Delwiche CF and Sharkey TD (1993) Rapid appearance of ^{13}C in biogenic isoprene when $^{13}CO_2$ is fed to intact leaves. Plant Cell Environ 16:587–591

Disch A, Schwender J, Müller C, Lichtenthaler HK and Rohmer M (1998a) Distribution of the mevalonate and glyceraldehyde phosphate/pyruvate pathways for isoprenoid biosynthesis in unicellular algae and the cyanobacterium *Synechocystis* PCC 6714. Biochem J 333:381–388

Disch A, Hemmerlin A, Bach TJ and Rohmer M (1998b) Mevalonate-derived isopentenyl diphosphate is the biosynthetic precursor of ubiquinone prenyl side chain in tobacco BY-cells. Biochem J 331:615–621

Dudareva N, Andersson S, Orlova I, Gatto N, Reichelt M, Rhodes D, Boland W and Gershenzon J (2005) The non-mevalonate pathway supports both monoterpene and sesquiterpene formation in snapdragon flowers. Proc Natl Acad Sci USA 102:93–938

Eberl M, Hintz M, Reichenberg A, Kollas A-K, Wiesner J and Jomaa H (2003) Microbial isoprenoid biosynthesis and human γδ T cell activation. FEBS Lett 544:4–10

Eisenreich W, Sagner S, Zenk MH and Bacher A (1997) Monoterpene essential oils are not of mevalonate origin. Tetrahedron Lett 39:3889–3892

Flügge UI and Gao W (2005) Transport of isoprenoid intermediates across chloroplast envelope membranes. Plant Biol 7:91–97

Friedl T (1997) The evolution of the green algae. In: D Bhattacharya (ed) Origin of Algae and Their Plastids. Springer, Vienna, pp. 87–101

Funes S, Davidson E, Reyes-Prieto A, Magallon S, Herion P, King MP and Gonzalez-Halphen D (2002) A green algal apicoplast ancestor. Science 298:2155

Gerber E, Hemmerlin A, Crowell DN, Rohmer M and Bach TJ (2010) The role of plastids in protein geranylgeranylation in tobacco BY-2 cells. In: Rebeiz CA, Benning C, Daniel H, Hoober K, Lichtenthaler HK, Portis A, Tripathy B (eds) The Chloroplast Basics and Application. Springer, Dordrecht, The Netherlands, pp. 127–137

Givnish TJ (1988) Adaptation to sun vs. shade: a whole plant perspective. Aust J Plant Physiol 15:63–92

Golz A and Lichtenthaler HK (1994) Inhibition of the plastidic pyruvate dehydrogenase complex in isolated plastids of oat. Z Naturforsch 49c:421–426

Goodwin TW (1958) Studies on carotenogenesis. 25. The incorporation of $^{14}CO_2$, [2-^{14}C]acetate and D,L-[2-^{14}C] mevalonic acid into β-carotene by illuminated etiolated maize seedlings. Biochem J 70:612–617

Goodwin TW (1977) The prenyllipids of the membranes of higher plants. In: Tevini M and Lichtenthaler HK (eds), Lipids and Lipid Polymers in Higher Plants. Springer-Verlag, Berlin, pp. 29–44

Hampel D, Mosandl A and Wüst M (2005) Biosynthesis of mono- and sesquiterpenes in carrot roots and leaves (*Daucus carota* L.): metabolic cross talk of cytosolic mevalonate and plastidial methylerythritol phosphate pathways. Phytochemistry 66:305–311

Harley P, Fridd-Stroud V, Greenberg J, Guenther A and Vasconcellos P (1998) Emission of 2-methyl-3-buten-2-ol by pines: a potential large source of reactive carbon to the atmosphere. J Geophys Res D 103:25479–25486

Hemmerlin A, Hoeffler J-F, Meyer O, Tritsch D, Kagan IA, Grosdemange-Billiard C, Rohmer M and Bach TJ (2003) Plastidial methylerythritol phosphate pathways in tobacco bright yellow-2 cells. J Biol Chem 278:26666–26676

Hemmerlin A, Tritsch D, Hartmann M, Pacaud K, Hoeffler J-F, van Dorsselaer A, Rohmer M and Bach TJ (2006) A cytosolic *Arabidopsis* D-xylulose kinase catalyzes the phosphorylation of 1-deoxy-D-xylulose into a precursor of the plastidial isoprenoid pathway. Plant Physiol 142:441–457

Herz S, Wungsintaweekul J, Schuhr CA, Hecht S, Lüttgen H, Sagner S, Fellermeier MA, Eisenreich W, Zenk MH, Bacher A and Rohdich F (2000) Biosynthesis of terpenoids: YgbB protein converts 4-diphosphocytidyl-2C-methyl-D-erythritol 2-phosphate to 2C-methyl-D-erythritol 2,4-cyclodiphosphate. Proc Natl Acad Sci USA 97:2486–2490

Jomaa H, Wiesner J, Sanderbrand S, Altincicek B, Weidemeyer C, Hintz M, Türbachova I, Eberl M, Zeidler J, Lichtenthaler HK, Soldati D and Beck E (1999) Inhibitors of the nonmevalonate pathway of isoprenoid biosynthesis as antimalarial drugs. Science 285:1573–1576

Kasahara H, Hanada A, Kuzuyama T, Takagi M, Kamiya Y and Yamaguchi S (2002) Contribution of the mevalonate and methylerythritol phosphate pathways to the biosynthesis of gibberellins in *Arabidopsis*. J Biol Chem 277:45188–45194

Kesselmeier J and Staudt M (1999) Biogenic volatile organic compounds (VOC): an overview on emission, physiology and ecology. J Atmos Chem 33:23–88

Knöss W, Reuter B and Zapp J (1997) Biosynthesis of the labdane diterpene marrubiin in Marubium vulgare via a non-mevlonate pathway. Biochem J 326:449–454

Kuzuyama T, Shimizu T and Seto H (1998) Fosmidomycin, a specific inhibitor of 1-deoxy-d-xylulose 5-phosphate reductoisomerase in the nonmevalonate pathway for terpenoid biosynthesis. Tetrahedron Lett 39:7913–7916

Lange BM, Rujan T, Martin W and Croteau RB (2000) Isoprenoid biosynthesis: The evolution of two ancient and distinct pathways across genomes. Proc Natl Acad Sci USA 97:13172–13177

Laule O, Fürholz A., Chang S, Zhu T, Wang X, Heifetz PB, Gruissem W and Lange BM (2003) Crosstalk between cytosolic and plastidial pathways of isoprenoid biosynthesis in *Arabidopsis thaliana*. Proc Natl Acad Sci USA 100: 6866–6871

Lichtenthaler HK (1968) Plastoglobuli and the fine structure of plastids. Endeavour XXVII:144–149

Lichtenthaler HK (1969a) Die Bildung überschüssiger Plastidenchinone in den Blättern von *Ficus elasticus* Roxb. Z Naturforsch 24b:1461–1466

Lichtenthaler HK (1969b) Localization and functional concentrations of lipoquinones in chloroplasts. In: Metzner H (ed), Photosynthesis Research, Vol I. International Union of Biological Sciences, Tübingen, Germany, pp. 304–314

Lichtenthaler HK (1969c) Light-stimulated synthesis of plastid quinones and pigments in etiolated barley seedlings. Biochim Biophys Acta 184:164–172

Lichtenthaler HK (1977) Regulation of prenylquinone biosynthesis in higher plants. In: Tevini M and Lichtenthaler HK (eds) Lipids and Lipid Polymers in Higher Plants. Springer, Berlin, pp. 29–44

Lichtenthaler HK (1981) Adaptation of leaves and chloroplasts to high quanta fluence rates. In: Akoyunoglou G (ed) Photosynthesis VI. Balaban Internat Science Service, Philadelphia, PA, pp. 273–287

Lichtenthaler HK (1987) Chlorophylls and carotenoids, the pigments of photosynthetic biomembranes. In: Douce R, Packer L (eds) Methods Enzymol 148. Academic, New York, pp. 350–382

Lichtenthaler HK (1998) The plants' 1-deoxy-D-xylulose-5-phospate pathway for biosynthesis of isoprenoids. Fett/Lipid 100:128–138

Lichtenthaler HK (1999) The 1-deoxy-D-xylulose-5-phosphate pathway of isoprenoid biosynthesis in plants. Annu Rev Plant Physiol Plant Mol Biol 50:47–65

Lichtenthaler HK (2000) The non-mevalonate isoprenoid biosynthesis: enzymes, genes and inhibitors. Biochem Soc Trans 28:787–792

Lichtenthaler HK (2004) Evolution of carotenoid and sterol biosynthesis in photosynthetic and non photosynthetic organisms. In: Biacs P and Gerely P (eds) Proceedings of the 16th Plant Lipid Symposium. Mete Publisher, Budapest, pp. 11–24. www.mete.mtesz.hu/pls/proceedings/index.htm

Lichtenthaler HK (2007) Biosynthesis, accumulation and emission of carotenoids, α-tocopherol, plastoquinone and isoprene in leaves under high photosynthetic irradiance. Photosynth Res 93. doi: 10.1007/s11120-007–9204-y

Lichtenthaler HK and Babani F (2004) Light adaptation and senescence of the photosynthetic apparatus. Changes in pigment composition, chlorophyll fluorescence parameters and photosynthetic activity. In: Papageorgiou GC and Govindjee (eds) Chlorophyll Fluorescence: A Signature of Photosynthesis. Springer, Dordrecht, pp. 713–736

Lichtenthaler HK and Calvin M (1964) Quinone and pigment composition of chloroplasts and quantasome aggregates from *Spinacia oleracea*. Biochim Biophys Acta 79:30–40

Lichtenthaler HK and Park RB (1963) Chemical composition of chloroplast lamellae from spinach. Nature 198:1070–1072

Lichtenthaler HK and Sprey B (1966) Über die osmiophilen globulären Lipideinschlüsse der Chloroplasten. Z Naturforsch 21b:690–697

Lichtenthaler HK and Zeidler J (2002) Isoprene and terpene biosynthesis. In: Gasche R, Papen H and Rennenberg H (eds) Trace Gas Exchange in Forest Ecosystems. Kluwer, Dordrecht, pp. 79–99

Lichtenthaler HK, Buschmann C, Döll M, Fietz H-J, Bach T, Kozel U, Meier D and Rahmsdorf U (1981a) Photosynthetic activity, chloroplast ultrastructure, and leaf characteristics of high-light and low-light plants and of sun and shade leaves. Photosyn Res 2:115–141

Lichtenthaler HK, Prenzel U, Douce R and Joyard J (1981b) Localization of prenylquinones in the envelope of spinach chloroplasts. Biochim Biophys Acta 641:99–105

Lichtenthaler HK, Prenzel U and Kuhn G (1982a) Carotenoid composition of chlorophyll-carotenoid-proteins from radish chloroplasts. Z Naturforsch 37c:10–12

Lichtenthaler HK, Kuhn G, Prenzel U, Buschmann C and Meier D (1982b) Adaptation of chloroplast-ultrastructure and of chlorophyll-protein levels to high-light and low-light growth conditions. Z Naturforsch 37c:464–475

Lichtenthaler HK, Schwender J, Seemann M and Rohmer M (1995) Carotenoid biosynthesis in green algae proceeds via a novel biosynthetic pathway. In: Mathis P (ed) Photosynthesis: From Light to Biosphere. Kluwer, Amsterdam, pp. 115–118

Lichtenthaler HK, Schwender J, Disch A and Rohmer M (1997a) Biosynthesis of isoprenoids in higher plant chloroplasts proceeds via a mevalonate independent pathway. FEBS Lett 400:271–274

Lichtenthaler HK, Rohmer M and Schwender J (1997b) Two independent biochemical pathways for isopentenyl diphosphate (IPP) and isoprenoid biosynthesis in higher plants. Physiol Plant 101:643–652

Lichtenthaler HK, Babani F, Langsdorf G and Buschmann C (2000a) Measurement of differences in red chlorophyll fluorescence and photosynthetic activity between sun and shade leaves by fluorescence imaging. Photosynthetica 38:521–529

Lichtenthaler HK, Zeidler J, Schwender J and Müller C (2000b) The non-mevalonate isoprenoid biosynthesis of plants as a test-system for new herbicides and drugs against pathogenic bacteria and the malaria parasite. Z Naturforsch 55c:305–313

Little HN and Bloch K (1950) Studies on the utilization of acetic acid for the biological synthesis of cholesterol. J Biol Chem 138:33–46

Lynen F, Reichart E and Rueff L (1951) Zum biologischen Abbau der Essigsäure VI, "aktivierte Essigsäure", ihre Isolierung aus Hefe und ihre chemische Natur. Liebigs Ann Chem 574:1–32

Mandel MA, Feldmann KA, Herrera-Estrella L, Rocha-Sosa M and León P (1996) *CLA1*, a novel gene required for chloroplast development, is highly conserved in evolution. Plant J 9:649–658

McGarvey DJ and Croteau R (1995) Terpenoid metabolism. Plant Cell 7:1015–1026

Meier D and Lichtenthaler HK (1981) Ultrastructural development of chloroplasts in radish seedlings grown at high and low light conditions and in the presence of the herbicide bentazon. Protoplasma 107:195–207

Müller C, Schwender J, Zeidler J and Lichtenthaler HK (2000) Properties and inhibition of the first two enzymes of the non-mevalonate pathway of isoprenoid biosynthesis. Biochem Soc Trans 28:794–795

Nabeta K, Ishikawa T and Okuyama H (1995) Sesqui- and di-terpene biosynthesis from ^{13}C labelled acetate and mevalonate in cultured cells of *Heteroscyphus planus*. J Chem Soc Perkin Trans I: 3111–3115

Nabeta K, Kawae T, Saitoh T and Kikuchi T (1997) Synthesis of chlorophyll a and ß-carotene from ^{2}H and ^{13}C-labelled mevalonates and ^{13}C-labeled glycin in cultured cells of liwerworts *Heterocyphus planus* and *Lophocolea heterophylla*. J Chem Soc Perkin Trans 1:261–267

Nagata N, Suzuki M, Yoshida S and Muranaka T (2002). Mevalonic acid partially restores chloroplast and etioplast development in *Arabidopsis* lacking the non-mevalonate pathway. Planta 216:345–350

Nagegowda DA, Rhodes D and Doudareva N (2010) The role of the methyl-erythritol-phosphate pathway in rhythmic emission of volatiles. In: Rebeiz CA, Benning C, Daniel H, Hoober K, Lichtenthaler HK, Portis A, Tripathy B (eds) The Chloroplast: Basics and Application. Springer, Dordrecht, The Netherlands, pp. 139–153

Proteau, P.J. (1998) Biosynthesis of phytol in the cyanobacterium Synechocystis spec. UTEX 2470: utilization of the non-mevalonate pathway. J. Nat Prod 61:841–843.

Querol J, Campos N, Imperial S, Boronat A and Rodríguez-Concepcíon M (2002) Functional analysis of the *Arabidopsis thaliana* GCPE protein involved in plastid isoprenoid biosynthesis. FEBS Lett 514: 343–346

Rasmussen RH and Khalil MAK (1998) Isoprene over the Amazon Basin. J Geophys Res 93:1417–1421

Rodríguez-Concepción M and Boronat A (2002) Elucidation of the methylerythritol phosphate pathway for isoprenoid biosynthesis in bacteria and plastids. A metabolic milestone achieved through genomics. Plant Physiol 130:1079–1089

Rohdich F, Wungsintaweekul J, Fellermeier M, Sagner S, Herz S, Kis K, Eisenreich W, Bacher A and Zenk MH (1999) Cytidine 5'-triphosphate-dependent biosynthesis of isoprenoids: YgbP protein of *Escherichia coli* catalyzes the formation of 4-diphosphocytidyl-2-*C*-methylerythritol. Proc Natl Acad Sci USA 96: 11758–11763

Rohdich F, Hecht S, Gärtner K, Adam P, Krieger C, Amslinger S, Arigoni D, Bacher A and Eisenreich W (2002) Studies on the nonmevalonate terpene biosynthetic pathway: metabolic role of IspH (LytB) protein. Proc Natl Acad Sci USA 99:1158–1163

Rohmer M (1999) The discovery of a mevalonate-independent pathway for isoprenoid biosynthesis in bacteria, algae and higher plants. Nat Prod Rep 16:565–574

Rohmer M, Knani M, Simonin P, Sutter B and Sahm H (1993) Isoprenoid biosynthesis in bacteria: a novel pathway for early steps leading to isopentenyl diphosphate. Biochem J 295:517–524

Rosenstiel TN, Fisher AJ, Fall R and Monson RK (2002) Differential accumulation of dimethylallyl diphosphate in leaves and needles of isoprene- and methylbutenol-emitting and nonemitting species. Plant Physiol 129:1276–1284

Ruzicka L (1938) Die architektur der polyterpene. Angew Chem 51:5–11

Ruzicka L, Eschenmoser A and Heusser H (1953) The isoprene rule and the biogenesis of isoprenoid compounds. Experientia 9:357–396

Sarijeva G, Knapp M and Lichtenthaler HK (2007) Differences in photosynthetic activity, chlorophyll and carotenoid levels, and in chlorophyll fluorescence parameters in green sun and shade leaves of *Ginkgo* and *Fagus*. J Plant Physiol 164:950–955

Schade GW, Goldstein AH, Gray DW and Lerdau MT (2000) Canopy and leaf level 2-methyl-3-buten-2-ol fluxes from a ponderosa pine plantation. Atmos Environ 34:3535–3544

Schindler C, Reith P and Lichtenthaler HK (1994) Differential levels of carotenoids and decrease of zeaxanthin cycle performance during leaf development in a green and an aurea variety of tobacco. J Plant Physiol 143:500–507

Schnarrenberger C, Jacobshagen S, Müller B and Krüger I (1990) Evolution of isozymes of sugar phosphate metabolism in green algae. In: Ogita ZI and Market CL (eds) Isozymes. Structure, Function and Use in Biology and Medicine. Wiley Liss, New York, pp. 743–764

Schnitzler J-P, Graus M, Kreuzwieser J, Heizmann U, Rennenberg H, Wisthaler A and Hansel A (2004) Contribution of different carbon sources to isoprene biosynthesis in poplar leaves. Plant Physiol 135:152–160

Schuhr CA, Radykewicz T, Sagner S, Latzel C, Zenk MH, Arigoni D, Bacher A, Rohdich F and Eisenreich W (2003) Quantitative assessment of crosstalk between the two isoprenoid biosynthesis pathways in plants by NMR spectroscopy. Phytochem Rev 2:3–16

Schulze-Siebert D and Schulze G (1987) ß-carotene synthesis in isolated chloroplasts. Plant Physiol 84:1233–1237

Schulze-Siebert D, Heinecke D, Scharf H and Schulze G (1984) Pyruvate-derived amino acids in spinach chloroplasts. Plant Physiol 76:465–471

Schwarz MK (1994) Terpenbiosynthese in *Ginkgo biloba*. Ph.D. thesis, Eidgen Techn Hochschule. Zürich, Switzerland

Schwender J, Lichtenthaler HK, Seemann M and Rohmer M (1995) Biosynthesis of isoprenoid chains of chlorophylls and plastoquinone in *Scenedesmus* by a novel pathway. In: Mathis P (ed) Photosynthesis: From Light to Biosphere. Kluwer, Amsterdam, pp. 1001–1004

Schwender J, Seeman M, Lichtenthaler HK and Rohmer M (1996) Biosynthesis of isoprenoids (carotenoids, sterols, prenyl side-chains of chlorophyll and plastoquinone) via a novel pyruvate/glycero-aldehyde-3-phosphate non-mevalonate pathway in the green alga *Scenedesmus*. Biochem J 316:73–80

Schwender J, Zeidler J, Gröner R, Müller C, Focke M, Braun S, Lichtenthaler FW and Lichtenthaler HK (1997) Incorporation of 1-deoxy-D-xylulose into isoprene and phytol by higher plants and algae. FEBS Lett 414:129–134

Schwender J, Müller C, Zeidler J and Lichtenthaler HK (1999) Cloning and heterologous expression of a cDNA encoding 1-deoxy-D-xylulose-5-phosphate reductoisomerase of *Arabidopsis thaliana*. FEBS Lett 455:140–144

Schwender J, Gemünden C and Lichtenthaler HK (2001) Chlorophyta exclusively use the 1-deoxyxylulose 5-phosphate/2-C-methylerythritol 4-phosphate pathway for the biosynthesis of isoprenoids. Planta 212:416–423

Seemann M, Campos N, Rodríguez-Concepción M, Hoeffler J-F, Grosdemange-Billiard C, Boronat A and Rohmer M (2002) Isoprenoid biosynthesis via the methylerythritol phosphate pathway: accumulation of 2-*C*-methyl-D-erythritol 2,4-cyclodiphosphate in a *gcpE* deficient mutant of *Escherichia coli*. Tetrahedron Lett 43:775–778

Sharkey TD (1996) Isoprene emission by plants and animals. Endeavour 20:74–78

Sharkey TD and Yeh S (2001) Isoprene emission from plants. Annu Rev Plant Physiol Plant Mol Biol 52:407–436

Sharkey TD, Yeh S, Wiberley AE, Falbel TG, Gong D and Fernandez DE (2005) Evolution of the isoprene biosynthetic pathway in kudzu. Plant Physiol 137:700–712

Silver GM and Fall R (1995) Characterization of aspen isoprene synthase, an enzyme responsible for leaf isoprene emission to the atmosphere. J Biol Chem 270:13010–13016

Sisler EC and Klein WH (1963) The effect of age and various chemicals on the lag phase of chlorophyll synthesis in dark grown bean seedlings. Physiol Plant 16:315–322

Stumpf PK (1984) Fatty acid biosynthesis in higher plants. In: Numa S (ed) Fatty Acid Metabolism and Its Regulation. Elsevier Science Publishers BV, Amsterdam, pp. 155–179

Thornber JP (1975) Chlorophyll-proteins: light-harvesting and reaction center components of plants. Annu Rev Plant Physiol 26:127–158

Van den Hoek C, Mann DG and Jahn HM (1995) Algae, an Introduction to Phycology. Cambridge University Press, Cambridge

Wallach O (1885) Zur Kenntnis der Terpene und der ätherischen Öle. Justus Liebigs Ann Chem 227:277–302

Wild A, Höpfner M, Rühle W and Richter M (1986) Changes in the stochiometry of photosystem II components as an adaptive response to high-light and low-light conditions during growth. Z Naturforsch C 41:597–603

Wildermuth MC and Fall R (1996) Light-dependent isoprene emission (Characterization of a thylakoid-bound isoprene synthase in *Salix discolor* chloroplasts). Plant Physiol 112:171–182

Wildermuth MC and Fall R (1998) Biochemical characterization of stromal and thylakoid-bound isoforms of isoprene synthase in willow leaves. Plant Physiol 116:1111–1123

Wilson RJM (2002) Progress with parasite plastids. J Mol Biol 319:257–274

Wolf DE, Hoffmann CH, Aldrich PE, Skeggs HR, Wright LD and Folkers K (1956) β-Hydroxy-β-methyl-δ-dihydroxy-β-methylvaleric acid (divalonic acid), a new biological factor. J Am Chem Soc 78:4499

Wolfertz M, Sharkey TD, Boland W and Kuhnemann F (2004) Rapid regulation of the methylerythritol 4-phosphate pathway during isoprene synthesis. Plant Physiol 135:1939–1945

Zeidler J (2001) Precursors and inhibitors of isoprene synthesis in plants (in German). Karlsruhe Contr Plant Physiol 38:1–157

Zeidler J and Lichtenthaler HK (2001) Biosynthesis of 2-methyl-3-buten-2-ol emitted from needles of *Pinus ponderosa* via the non-mevalonate DOXP/MEP pathway of isoprenoid formation. Planta 213:323–326

Zeidler JG, Lichtenthaler HK, May HU and Lichtenthaler FW (1997) Is isoprene emitted by plants synthesized via the novel isopentenylpyrophosphate pathway? Z Naturforsch 52c:15–23

Zeidler JG, Schwender J, Müller C, Wiesner J, Weidemeyer C, Beck E, Jomaa H and Lichtenthaler HK (1998) Inhibition of the non-mevalonate 1-deoxy-D-xylulose-5-phosphate pathway of plant isoprenoid biosynthesis by fosmidomycin. Z Naturforsch 53c:980–986

Zeidler J, Schwender J, Müller C and Lichtenthaler HK (2000) The isoprenoid biosynthesis of plants as test-system for drugs against malaria and pathogenic bacteria. Biochem Soc Trans 28:798–800

Zhang H, Sharifi MR and Nobel PS (1995) Photosynthetic characteristics of sun versus shade plants of *Encelia farinosa* as affected by photosynthetic photon flux density. Intercellular CO_2 concentration, leaf water potential, and leaf temperature. Austr J Plant Physiol 22:833–841

Zhou WX and Nes WD (2000) Stereochemistry of hydrogen introduction at C-25 in ergosterol synthesized by the mevalonate-independent pathway. Tetrahedron Lett 41:2791–2795

Chapter 8

The Methylerythritol 4-Phosphate Pathway: Regulatory Role in Plastid Isoprenoid Biosynthesis

Albert Boronat

Faculty of Biology, Department of Biochemistry and Molecular Biology, University of Barcelona, 08028-Barcelona, Spain

Summary

Plastids synthesize a variety of isoprenoid compounds that are essential for plant growth and development, like carotenoids, gibberellins, abscisic acid and the side chain of chlorophylls, plastoquinone and tocopherols. Some plastidial isoprenoids are also relevant for their biotechnological interest. The recently elucidated methylerythritol 4-phosphate (MEP) pathway provides the isopentenyl diphosphate and dimethylallyl diphosphate required for the synthesis of plastidial isoprenoids. At present, very little is known about the regulatory mechanisms underlying the control of the metabolic flux through the MEP pathway and its crosstalk with the cytosolic mevalonate (MVA) pathway. Here I summarize recent aspects related to the regulatory role of the MEP pathway in the synthesis of plastidial isoprenoids and the use of reverse genetic strategies to study the crosstalk between the MVA and the MEP pathways.

Author for Correspondence, e-mail: aboronat@bq.ub.es

Constantin A. Rebeiz et al. (eds.), The Chloroplast: Basics and Applications, pp. 119–126.

I Introduction

Isoprenoids (also known as terpenoids) are ubiquitous compounds found in all living organisms. Some isoprenoids play essential roles in particular cell functions. Sterols contribute to eukaryotic membrane architecture, acyclic polyprenoids are found in the side chain of ubiquinone, plastoquinone and chlorophylls, and carotenoids play a role in photosynthetic organisms. Although the role of other isoprenoids is less evident, like that of the vast array of plant secondary metabolites, some are known to play key roles in the adaptative responses to different environmental challenges (Croteau et al., 2000).

In spite of the remarkable diversity of structure and function, all isoprenoids are derived from a basic five-carbon unit, isopentenyl diphosphate (IPP), and its isomer dimethylallyl diphosphate (DMAPP). Addition of IPP units to DMAPP, catalyzed by prenyltransferases, leads to the synthesis of geranyl diphosphate (GPP, C10), farnesyl diphosphate (FPP, C15) and geranylgeranyl diphosphate (GGPP, C20), which are the starting points of downstream pathways for the synthesis of the different groups of isoprenoid end products (Fig. 1). For many years, it was accepted that IPP was synthesized from acetyl-CoA through the well-known acetate/mevalonate (MVA) pathway (McGarvey and Croteau, 1995). However, an alternative MVA-independent pathway for the biosynthesis of IPP and DMAPP was identified a few years ago in photosynthetic bacteria, algae and plants (Lichtenthaler, 1999; Lichtenthaler, 2010). This novel pathway, currently known as the methylerythritol 4-phosphate (MEP) pathway, is widely distributed in nature and is present in most eubacteria, apicomplexan protozoa (like the malaria parasite *Plasmodium falciparum,* as shown by Jomaa et al., 1999), green algae and higher plants (Eisenreich et al., 2004: Lichtenthaler, 2010).

Abbreviations: DXS – 1-deoxy-D-xylulose 5-phosphate synthase; DXR – 1-deoxy-D-xylulose 5-phosphate reductoisomerase; HDR – 1-hydroxy-2-methyl-2-(E)-butenyl diphosphate reductase; HMGR – 3-hydroxy-3-metylglutaryl coenzyme A reductase; MEP – 2-C-methyl-D-erythritol 4-phosphate; MEV – mevinolin; MVA – mevalonate; GGPP – geranylgeranyl diphosphate; FSM – fosmidomycin

The first reaction of the MEP pathway is catalyzed by the enzyme 1-deoxy-D-xylulose 5-phosphate (DXP) synthase (DXS) and involves the condensation of (hydroxyethyl) thiamin derived from pyruvate with the C1 aldehyde group of D-glyceraldehyde 3-phosphate to produce DXP. In the second step, an intramolecular rearrangement and reduction of DXP by the enzyme DXP reductoisomerase (DXR) yields 2C-methyl-D-erythritol 4-phosphate (MEP), currently considered the first committed precursor of plastidic isoprenoids. MEP is then converted to 2-C-methyl-D-erythritol 2,4-cyclodiphosphate (MEcPP) by the sequential action of the enzymes 4-diphosphocytidyl-2C-methyl-D-erythritol (CDP-ME) synthase (CMS), CDP-ME kinase (CMK) and MEcPP synthase (MCS). The opening reduction of the MEcPP ring catalyzed by the enzyme 1-hydroxy-2-methyl-2-(E)-butenyl diphosphate (HMBPP) synthase (HDS) forms HMBPP. Finally, the enzyme HMBPP reductase (HDR) catalyzes the simultaneous formation of IPP and DMAPP, which can be interconverted by the action of IPP isomerase (Fig. 1) (reviewed in Rodríguez-Concepción and Boronat, 2002; Eisenreich et al., 2004).

In plants, the MVA and the MEP pathways coexist although they are localized in different cell compartments (Lichtenthaler et al., 1997). The MVA pathway is localized in the cytosol and provides the IPP required for the synthesis of sterols, sesquiterpenes, dolichols cytokinins, brassinosteroids and the prenyl moiety used for protein prenylation (Fig. 1). MVA-derived IPP is also used for the synthesis of ubiquinone in the mitochondria (Disch et al., 1998). The MEP pathway is compartmentalized in the plastids and provides IPP and DMAPP for the synthesis of isoprene, monoterpenes, carotenoids, abscisic acid, gibberellins and the side chain of chlorophylls, tochopherols, plastoquinone and phylloquinone. MEP pathway enzymes are encoded by nuclear genes and imported to plastids. The exchange of IPP or derived prenyl phosphates between the MVA and MEP pathway has been reported in different plant species (see below) and represents and additional level of complexity in the organization and regulation of plant isoprenoid biosynthesis.

In addition to their biological relevance in growth and development many plant isoprenoids

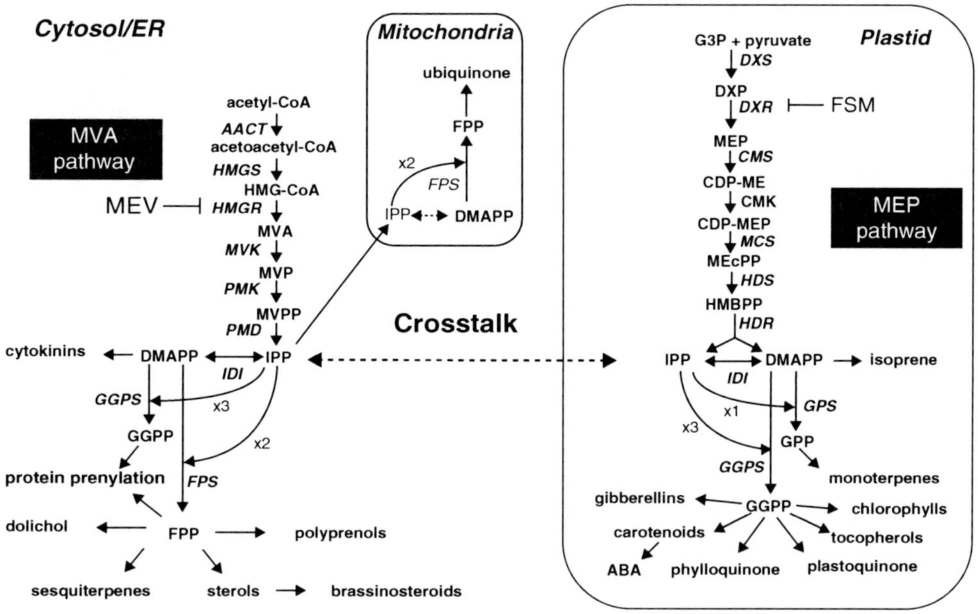

Fig. 1. Organization and compartmentalization of isoprenoid biosynthesis in plants. HMG-CoA, 3-hydroxy-3-metylglutaryl coenzyme A; MVA, mevalonate; MVP, 5-phosphomevalonate; MVPP, 5-diphosphomevalonate; G3P, D-glyceraldehyde 3-phosphate; DXP, 1-deoxy-D-xylulose 5-phosphate; MEP, 2C-metyl-D-erythritol 4-phosphate; CDP-ME, 4-diphosphocytidyl-2C-methyl-D-erythritol; CDP-MEP, 4-diphosphocytidyl-2C-methyl-D-erythritol 2-phosphate; ME-cPP, 2C-methyl-D-erythritol 2,4-ciclodiphosphate; HBMPP, 1-hydroxy-2-metyl-2-(E)-butenyl 4-diphosphate; IPP, isopentenyl diphosphate; DMAPP, dimethylallyl diphosphate; GPP, geranyl diphosphate; FPP, farnesyl diphosphate; GGPP, geranylgeranyl diphosphate; ABA, abscisic acid. Enzymes are indicated in italic: *AACT*, acetoacetyl-CoA thiolase; *HMGS*, HMG-CoA synthase; *HMGR*, HMG-CoA reductase; *MVK*, MVA kinase; *PMK*, MVP kinase; *PMD*, MVPP decarboxylase; *IDI*, IPP isomerase; *GPS*, GPP synthase; *FPS*, FPP synthase; *GGPS*, GGPP synthase; *DXS*, DXP synthase; *DXR*, DXP reductoisomerase; *CMS*, CDP-ME synthase; *CMK*, CDP-ME kinase; *MCS*, MEcPP synthase; *HDS*, HMBPP synthase; *HDR*, HMBPP reductase. The steps inhibited by mevinolin (MEV) and fosmidomycin (FSM) are indicated.

have biotechnological interest in different fields. This is the case of artemisinin, used as antimalarial drug, and taxol, a potent antitumoral agent. Other isoprenoids are relevant in human health and nutrition like beta carotene, tocopherols and phytosterols. Others have industrial interest as polymers (rubber), fragrances or agrochemicals. Because of this, metabolic engineering approaches have been undertaken to increase or modify isoprenoid contents in plants. Although in some cases the obtained results were successful in other cases led to unpredicted results, thus reflecting, at least in part, the lack of knowledge about the mechanisms underlying the regulation of isoprenoid biosynthesis in plants. Here I present an update on the currrent knowledge about the recently elucidated MEP pathway and its role in the regulation of plastid isoprenoid biosynthesis.

II Regulatory Role of the MEP Pathway in Plastid Isoprenoid Biosynthesis

It is now well established that the flux of precursors through the MVA pathway is limiting for the synthesis of sterols (the main isoprenoid product derived from cytosolic IPP) and that the synthesis of MVA from HMG-CoA, catalyzed by the enzyme HMG-CoA reductase (HMGR) represents a key regulatory step. In plants, as in other eukaryotic organisms, HMGR is a highly regulated enzyme and its activity is under the control of a variety of mechanisms operating at transcriptional and post-transcriptional level (Stermer et al., 1994; Bach et al., 1999). The first indication about the limiting role of HMGR in sterol biosynthesis was reported in a tobacco mutant line overproducing sterols (Gondet et al., 1992). This observation was confirmed by using transgenic plants engineered to overproduce HMGR. The heterologous expression of N-terminal truncated forms of HMGR lacking the membrane binding domain in transgenic tobacco plants resulted in increased sterol levels in leaves (Chappell et al., 1995) and seeds (Harker et al., 2003). More recently, it has also been reported that transgenic tomato plants constitutively expressing the full length Arabidopsis HMGR1 isoform show increased levels of sterols in leaves and fruits (Enfissi et al., 2005).

In contrast to the MVA pathway, where a single enzyme was shown to play the major regulatory role in the synthesis of cytosolic IPP, the flux of intermediates through the MEP pathway to produce IPP and DMAPP in the plastid seems to be shared by several enzymes. Recent results derived from studies using transgenic plants have revealed that at least three enzymes (DXS, DXR and HDR) can modulate the flux of precursors through the MEP pathway.

A first indication about the regulatory role of DXS came from the observation that DXS transcript levels correlate with the production of particular plastidic isoprenoids in different plant systems, like Arabidopsis seedlings (Estévez et al., 2000), pepper fruit (Bouvier et al., 1998), tomato fruit (Lois et al., 2000), peppermint oil gland secretory cells (Lange et al., 1998), *Catharanthus roseus* cells (Veau et al., 2000; van der Fits and Memelink, 2000) and maize mycorrhizal roots (Walter et al., 2000, 2002). However, the confirmation of the role of DXS in controlling the synthesis of plastidic isoprenoids was obtained from the analysis of transgenic plants in which DXS activity was up-regulated. Early work from Estévez et al. (2001) showed that the overexpression of DXS in Arabidopsis resulted in increased levels of various plastidic isoprenoids such as chlorophylls, tocopherols, carotenoids, abscisic acid and gibberellins. The fruit specific expression of bacterial DXS in tomato led to a significant increase in the accumulation of carotenoids, although in this case tocopherol levels remained unchanged (Enfissi et al., 2005). Two recent reports have shown that the up-regulation of DXS enhanced the production of monoterpenes in essential oil of transgenic spike lavender (Muñoz-Bertomeu et al., 2006) and increased the levels of carotenoids in potato tubers (Morris et al., 2006).

The regulatory role of DXR in the synthesis of plastididal isoprenoids is less clear. Thus, in a way similar to DXS, a clear positive correlation between DXR transcript levels and the synthesis of particular plastidic isoprenoids has been reported in Arabidopsis seedlings in response to light (Carretero-Paulet et al., 2002), arbuscule development in maize roots (Hans et al., 2004), *Catharanthus roseus* cells producing monoterpene indole alkaloids (Veau et al., 2000) and *Camptotheca acuminata* callus cultures elicited with methyl jasmonate (Yao et al., 2008). In contrast, this correlation has not been observed during carotenoid biosynthesis in tomato fruit ripening (Rodríguez-Concepción et al., 2001) and the diurnal emission of volatiles in snapdragon flowers (Dudareva et al., 2005; Nagegowda et al., 2010). However, the up-regulation of DXR led to increased production of monoterpene essential oils in transgenic peppermint plants (Mahmoud and Croteau, 2001) and enhanced levels of carotenoids and chlorophylls in transgenic Arabidopsis plants (Carretero-Paulet et al., 2006). These results suggested that the reaction catalyzed by DXR may be limiting for isoprenoid biosynthesis but depends on the cell type, tissue or plant species.

The last step of the MEP pathway, catalyzed by the enzyme HDR, also seems to represent an important regulatory step in the synthesis of plastidic isoprenoids. In a similar way to DXS, a correlation between HDR transcript levels and the accumulation of plastidial isoprenoids was observed during tomato fruit ripening and Arabidopsis seedling deetiolation (Botella-Pavía et al., 2004). In addition, transgenic Arabidopsis plants constitutively expressing tomato HDR exhibited enhanced levels of carotenoids similar to those reported in transgenic plants overexpressing DXS (Botella-Pavía et al., 2004). However, further work is needed to confirm whether HDR also plays a similar regulatory role in other plant species.

The analysis of Arabidopsis transgenic plants overexpressing DXS, DXR and HDR together with taxadiene synthase (TXS) has provided new insights into the limiting role of the MEP pathway in the synthesis of plastidial isoprenoids. TXS catalyzes the conversion of GGPP to taxadiene, the first committed precursor in the synthesis of paclitaxel (also known as taxol), a potent anticancer drug produced by the yew tree. Transgenic Arabidopsis plants constitutively expressing TXS from *Taxus baccata* were able to synthesize taxadiene, although at relatively low levels

(Besumbes et al., 2004). To evaluate whether an enhanced production of isoprenoid precursors through the MEP pathway resulted in higher rates of taxadiene biosynthesis, TXS overexpressing plants were crossed with plants overexpressing DXS, DXR or HDR (Botella-Pavía et al., 2004; Carretero-Paulet et al., 2006). The hemizygous DXS-TXS, DXR-TXS and HDR-TXS plants showed a 6.5-, 3,5- and 13-fold increase in the accumulation of taxadiene, respectively, compared with the control hemizygous plants harboring the TXS transgene alone. These results reinforce the previous observations about the limiting role of the MEP pathway in the synthesis of plastididial isoprenoids and highlighted the major regulatory role of DXS and HDR.

Considering that taxadiene and carotenoids are both synthesized from the same precursor (GGPP), the high increase of taxadiene production in the DXS-TXS, DXR-TXS and HDR-TXS transgenic plants is in sharp contrast with the relatively low increase in the carotenoid levels (Botella-Pavía et al., 2004; Carretero-Paulet et al., 2006). These results confirm previous reports indicating that other enzymes in the downstream pathways also catalyze key regulatory steps in the synthesis and accumulation of particular plastidial isoprenoids.

III Crosstalk Between the MVA and the MEP Pathways

In spite of the different subcellular compartmentalization of the MVA and the MEP pathways in plants, the exchange of IPP/DMAPP or derived prenyl diphosphates between the two compartments is well documented (Schwender et al., 2001; Hemmerlin et al., 2003; Laule et al., 2003; Hampel et al., 2005). However, very little is known about the regulatory mechanisms involved.

The crosstalk between the MVA and the MEP pathway has been reported in Arabidopsis (Kasahara et al., 2002; Nagata et al., 2002; Laule et al., 2003). Although a limited exchange of prenyl diphosphates has been shown to occur, its rate is not sufficient to rescue the block of one of the pathways with precursors synthesized by the other pathway. As a consequence, the specific block of either the MVA or the MEP pathway results in a developmental block at the seedling stage. A strategy based on the identification and characterization of Arabidopsis mutants able to overcome either the inhibition of the MVA pathway or the MEP pathway has been considered to uncover mechanisms potentially resulting in the activation of the exchange of isoprenoid precursors between the cytosol and the plastid.

One such Arabidopsis mutant (*rim1)*, resistant to both mevinolin (MEV), an inhibitor of HMGR, and fosmidomycin (FSM), and inhibitor of DXR, was shown to be defective in phytochrome B, the major light photoreceptor. The MEV-resistant phenotype of the *rim1* mutant can be explained by the increased levels of HMGR activity resulting from an activation of gene expression. While the MEV-resistant phenotype of the *rim1* mutant was caused by increased HMGR activity, no evidence was found for an up-regulation of the MEP pathway that could explain the resistance to FSM. Most likely, the FSM-resistant phenotype of *rim1* was derived from an enhanced intake of MVA-derived isoprenoid precursors by the plastid (Rodríguez-Concepción et al., 2004).

In the *rif10* mutant, resistant to FSM, the level of flux-controlling enzymes of the MEP pathway is post-transcriptionally up-regulated. This phenotype is linked to a decreased accumulation of chlorophylls and carotenoids, resulting in plants that are paler and smaller than the wild type. The *rif10* mutant is impaired in plastid RNA processing due to a mutation in the At3g03710 gene encoding the chloroplast-targeted exoribonuclease polyribonucleotide phosphorylase. FSM resistance and other *rif10*-like phenotypes were also observed in wild-type Arabidopsis, tomato and rice seedlings grown in the presence of sublethal concentrations of chloramphenicol (an inhibitor of protein synthesis in plastids). By contrast, treatment with norflurazon (an inhibitor of carotenoid biosynthesis causing a similar pale phenotype) did not result in FSM resistance. Together, the results support the notion that plastome-encoded proteins are involved in negatively regulating the post-transcriptional accumulation of MEP pathway enzymes in chloroplasts (Sauret-Güeto et al., 2006).

Recently, an Arabidopsis mutant (*loi1*) resistant to both lovastatin (an inhibitor of HMGR) and clomazone (an inhibitor of DXS) has been reported by Kobayashi et al. (2007). The lovastatin-resistant phenotype of *loi1* was derived from the post-transcriptional up-regulation of HMGR.

The *loi1* mutant is affected in the *LOI1* gene, which encodes a novel pentatricopeptide repeat (PPR) protein. PPR proteins are thought to regulate the expression of genes encoded by plastid and mitochondrial genomes. The LOI1 protein is assumed to be localized in mitochondria and has the ability to bind single-stranded nucleic acids, suggesting than the post-transcriptional regulation of mitochondrial RNA may be also involved in isoprenoid biosynthesis in both the MVA and MEP pathways (Kobayashi et al., 2007).

Taken together, these results suggest that a variety of mechanisms unrelated to the MVA or the MEP pathway may affect the homeostasis of isoprenoid biosynthesis in plant cells.

IV Perspectives for Metabolic Engineering of Plastid Isoprenoids

The recent elucidation of the MEP pathway and the crosstalk between the MVA and the MEP pathways has led to reconsider many aspects concerning the organization and regulation of plant isoprenoid biosynthesis. Recent work derived from the molecular analysis of genes and enzymes of the MEP pathway is shedding light on the regulatory role of the MEP pathway in the synthesis of plastidic isoprenoids. Considering the biotechnological interest of many plastidic isoprenoids, metabolic engineering approaches aimed at increasing their production in plant cells must take into account the following aspects, that: (1) the flux of intermediates through the MEP pathway is limiting for the synthesis of the structural precursors of plastidial isoprenoids, (2) the flux of intermediates through the MEP pathway in some plants is shared by at least three enzymes (DXS, DXR and HDR), and (3) a variety of mechanisms not directly related with the MVA and the MEP pathway may affect the homeostasis of isoprenoid biosynthesis in plant cells.

Acknowledgements

Work from my group has been funded by grants BIO2006-003704 from the Spanish Ministerio de Educación y Ciencia and 2005SGR00914 from the Generalitat de Catalunya

References

Bach TJ, Boronat A, Campos N, Ferrer A and Vollack KU (1999) Mevalonate biosynthesis in plants. Crit Rev Biochem Mol Biol 34: 107–122

Besumbes O, Sauret-Gueto S, Phillips MA, Imperial S, Rodriguez-Concepcion M and Boronat A (2004) Metabolic engineering of isoprenoid biosynthesis in Arabidopsis for the production of taxadiene, the first committed precursor of Taxol. Biotechnol Bioeng 88: 168–175

Botella-Pavía P, Besumbes O, Phillips MA, Carretero-Paulet L, Boronat, A and Rodríguez-Concepción M (2004) Regulation of carotenoid biosynthesis in plants: evidence for a key role of hydroxymethylbutenyl diphosphate reductase in controlling the supply of plastidial isoprenoid precursors. Plant J 40: 188–199

Bouvier F, d'Harlingue A, Suire C, Backhaus RA and Camara B (1998) Dedicated roles of plastid transketolases during the early onset of isoprenoid biogenesis in pepper fruits. Plant Physiol 117: 1423–1431

Carretero-Paulet L, Ahumada I, Cunillera N, Rodriguez-Concepcion M, Ferrer A, Boronat A and Campos N (2002) Expression and molecular analysis of the Arabidopsis DXR gene encoding 1-deoxy-D-xylulose 5-phosphate reductoisomerase, the first committed enzyme of the 2-C-methyl-D-erythritol 4-phosphate pathway. Plant Physiol 129: 1581–1591

Carretero-Paulet L, Cairó A, Botella-Pavía P, Besumbes O, Campos N, Boronat A and Rodríguez-Concepción M (2006) Enhanced flux through the methylerythritol phosphate pathway in Arabidopsis plants overexpressing deoxyxylulose 5-phosphate reductoisomerase. Plant Mol Biol 62: 683–695

Chappell J, Wolf F, Proulx J, Cuellar R and Saunders C (1995) Is the reaction catalyzed by 3-hydroxy-3-methylglutaryl coenzyme A reductase a rate-limiting step for isoprenoid biosynthesis in plants?. Plant Physiol 109: 1337–1343

Croteau R, Kutchan T and Lewis N (2000). Natural products (secondary metabolites). In: Buchanan B, Gruissem W and Jones R (eds) Biochemistry and Molecular Biology of Plants. American Society of Plant Biologists, Rockville, MD, pp. 1250–1268

Disch A, Hemmerlin A, Bach TJ and Rohmer M (1998) Mevalonate-derived isopentenyl diphosphate is the biosynthetic precursor of ubiquinone prenyl side chain in tobacco BY-2 cells. Biochem J 331: 615–621

Dudareva N, Andersson S, Orlova I, Gatto N, Reichelt M, Rhodes D, Boland W and Gershenzon J (2005) The non-mevalonate pathway supports both monoterpene and sesquiterpene formation in snapdragon flowers. Proc Natl Acad Sci USA 102: 933–938

Eisenreich W, Bacher A, Arigoni D and Rohdich F (2004) Biosynthesis of isoprenoids via the non-mevalonate pathway. Cell Mol Life Sci 61: 1401–1426

Enfissi EMA, Fraser PD, Lois LM, Boronat A, Schuch W and Bramley PM (2005) Metabolic engineering of the mevalonate and non-mevalonate isopentenyl diphosphate-forming pathways for the production of health-promoting isoprenoids in tomato. Plant Biotech J 3: 17–27

Estévez JM, Cantero A, Romero C, Kawaide H, Jiménez LF, Kuzuyama, T, Seto H, Kamiya Y and León P (2000) Analysis of the expression of CLA1, a gene that encodes the 1-deoxyxylulose 5-phosphate synthase of the 2-C-methyl-D-erythritol-4-phosphate pathway in Arabidopsis. Plant Physiol 124: 95–103

Estévez JM, Cantero A, Reindl A, Reichler S and León P (2001) 1-Deoxy-D-xylulose-5-phosphate synthase, a limiting enzyme for plastidic isoprenoid biosynthesis in plants. J Biol Chem 276: 22901–22909

Gondet L, Weber T, Maillot-Vernier P, Benveniste P and Bach TJ (1992) Regulatory role of microsomal 3-hydroxy-3-methylglutaryl-coenzyme A reductase in a tobacco mutant that overproduces sterols. Biochem Biophys Res Commun 186: 888–893

Hampel D, Mosandl A and Wüst M (2005) Biosynthesis of mono- and sesquiterpenes in carrot roots and leaves (*Daucus carota* L.): metabolic cross talk of cytosolic mevalonate and plastidial methylerythritol phosphate pathways. Phytochemistry 66: 305–311

Hans J, Hause B, Strack D and Walter MH (2004) Cloning, characterization, and immunolocalization of a mycorrhiza-inducible 1-deoxy-d-xylulose 5-phosphate reductoisomerase in arbuscule-containing cells of maize. Plant Physiol 134: 614–624

Harker M, Holmberg N, Clayton JC, Gibbard CL, Wallace AD, Rawlins S, Hellyer SA, Lanot A and Safford R (2003) Enhancement of seed phytosterol levels by expression of an N-terminal truncated Hevea brasiliensis (rubber tree) 3-hydroxy-3-methylglutaryl-CoA reductase. Plant Biotechnol J 1: 113–121

Hemmerlin A, Hoeffler JF, Meyer O, Tritsch D, Kagan IA, Grosdemange-Billiard C, Rohmer M and Bach TJ (2003) Crosstalk between the cytosolic mevalonate and the plastidial methylerythritol phosphate pathways in tobacco bright yellow-2 cells. J Biol Chem 278: 26666–26676

Jomaa H, Wiesner J, Sanderbrand S, Altincicek B, Weidemeyer C, Hintz M, Türbachova I, Eberl M, Zeidler J, Lichtenthaler HK, Soldati D and Beck E (1999) Inhibitors of the nonmevalonate pathway of isoprenoid biosynthesis as antimalarial drugs. Science 285: 1573–1576

Kasahara H, Hanada A, Kuzuyama T, Takagi M, Kamiya Y and Yamaguchi S (2002) Contribution of the mevalonate and methylerythritol phosphate pathways to the biosynthesis of gibberellins in Arabidopsis. J Biol Chem 277: 45188–45194

Kobayashi K, Suzuki M, Tang J, Nagata N, Ohyama K, Seki H, Kiuchi R, Kaneko Y, Nakazawa M, Matsui M, Matsumoto S, Yoshida S and Muranaka T (2007) LOVASTATIN INSENSITIVE 1, a novel pentatricopeptide repeat protein, is a potential regulatory factor of isoprenoid biosynthesis in Arabidopsis. Plant Cell Physiol 48: 322–331

Lange BM, Wildung MR, McCaskill D and Croteau R (1998) A family of transketolases that directs isoprenoid biosynthesis via a mevalonate-independent pathway. Proc Natl Acad Sci USA 95: 2100–2104

Laule O, Furholz A, Chang HS, Zhu T, Wang X, Heifetz PB, Gruissem W and Lange M (2003) Crosstalk between cytosolic and plastidial pathways of isoprenoid biosynthesis in *Arabidopsis thaliana*. Proc Natl Acad Sci USA 100: 6866–6871

Lichtenthaler HK, Rohmer M and Schwender J (1997) Two independent biochemical pathways for isopentenyl diphosphate (IPP) and isoprenoid biosynthesis in higher plants. Physiol Plant 101: 643–652

Lichtenthaler HK (1999) The 1-deoxy-D-xylulose-5-phosphate pathway of isoprenoid biosynthesis in plants. Annu Rev Plant Physiol Plant Mol Biol 50: 47–65

Lichtenthaler HK (2010) The non-mevalonate DOXP/MEP pathway of chloroplast isoprenoid and pigment biosynthesis. In: Rebeiz CA, Benning C, Daniel H, Hoober K, Lichtenthaler HK, Portis A, Tripathy B (eds) The Chloroplast: Basics and Applications. Springer, The Netherlands, pp. 93–118

Lois LM, Rodríguez-Concepción M, Gallego F, Campos N and Boronat A (2000) Carotenoid biosynthesis during tomato fruit development: regulatory role of 1-deoxy-D-xylulose 5phosphate synthase. Plant J 22: 503–513

Mahmoud SS and Croteau R (2001) Metabolic engineering of essential oil yield and composition in mint by altering expression of deoxyxylulose phosphate reductoisomerase and menthofuran synthase. Proc Natl Acad Sci USA 98: 8915–8920

McGarvey DJ and Croteau R (1995) Terpenoid metabolism. Plant Cell 7: 1015–1026

Morris WL, Ducreux LJ, Hedden P, Millam S and Taylor MA (2006) Overexpression of a bacterial 1-deoxy-D-xylulose 5-phosphate synthase gene in potato tubers perturbs the isoprenoid metabolic network: implications for the control of the tuber life cycle. J Exp Bot 57: 3007–3018

Muñoz-Bertomeu J, Arrillaga I, Ros R and Segura J (2006) Up-regulation of 1-deoxy-D-xylulose-5-phosphate synthase enhances production of essential oils in transgenic spike lavender. Plant Physiol 142: 890–900

Nagata N, Suzuki M, Yoshida S and Muranaka T (2002) Mevalonic acid partially restores chloroplast and etioplast development in Arabidopsis lacking the non-mevalonate pathwa. Planta 216: 345–350

Nagegowda DA, Rhodes D and Dudareva N (2010) The role of the methyl-erythritol-phosphate pathway in rhythmic emission of volatiles. In: Rebeiz CA, Benning C, Daniel H, Hoober K, Lichtenthaler HK, Portis A, Tripathy B (eds) The Chloroplast: Basics and Applications. Springer, The Netherlands, pp. 139–153

Rodríguez-Concepción M, Ahumada I, Diez-Juez E, Sauret-Gueto S, Lois LM, Gallego F, Carretero-Paulet L, Campos N and Boronat A (2001) 1-Deoxy-D-xylulose 5-phosphate reductoisomerase and plastid isoprenoid biosynthesis during tomato fruit ripening. Plant J 27: 213–222

Rodríguez-Concepción M and Boronat A (2002) Elucidation of the methylerythritol phosphate pathway for isoprenoid biosynthesis in bacteria and plastids. A metabolic milestone achieved through genomics. Plant Physiol 130: 1079–1089

Rodríguez-Concepción M, Forés O, Martínez-García JF, González V, Phillips MA, Ferrer A and Boronat A (2004) Distinct light-mediated pathways regulate the biosynthesis and exchange of isoprenoid precursors during Arabidopsis seedling development. Plant Cell 16: 144–156

Sauret-Güeto S, Botella-Pavia P, Flores-Pérez U, Martinez-García JF, San Roman C, León P, Boronat A and Rodríguez-Concepción M (2006) Plastid cues posttranscriptionally regulate the accumulation of key enzymes of the methylerythritol phosphate pathway in Arabidopsis. Plant Physiol 141: 75–84

Schwender J, Gemünden C and Lichtenthaler HK (2001) Chlorophyta exclusively use the 1-deoxyxylulose 5-phosphate/2-C-methylerythritol 4-phosphate pathway for the biosynthesis of isoprenoids. Planta 212: 416–423

Stermer BA, Bianchini GM and Korth KL (1994) Regulation of HMG-CoA reductase activity in plants. J Lipid Res 35: 1133–1140

Van der Fits L and Memelink J (2000) ORCA3, a jasmonate-responsive transcriptional regulator of plant primary and secondary metabolism. Science 289: 295–297

Veau B, Courtois M, Oudin A, Chenieux JC, Rideau M and Clastre M (2000) Cloning and expression of cDNAs encoding two enzymes of the MEP pathway in *Catharanthus roseus*. Biochim Biophys Acta 1517: 159–163

Walter MH, Fester T and Strack D (2000) Arbuscular mycorrhizal fungi induce the non-mevalonate methylerythritol phosphate pathway of isoprenoid biosynthesis correlated with accumulation of the 'yellow pigment' and other apocarotenoids. Plant J 21: 571–578

Walter MH, Hans J and Strack D (2002) Two distantly related genes encoding 1-deoxy-d-xylulose 5-phosphate synthases: differential regulation in shoots and apocarotenoid-accumulating mycorrhizal roots. Plant J 31: 243–254

Yao H, Gong Y, Zuo K, Ling H, Qiu C, Zhang F, Wang Y, Pi Y, Liu X, Sun X and Tang K (2008) Molecular cloning, expression profiling and functional analysis of a DXR gene encoding 1-deoxy-d-xylulose 5-phosphate reductoisomerase from *Camptotheca acuminata*. J Plant Physiol 165: 203–213

Chapter 9

The Role of Plastids in Protein Geranylgeranylation in Tobacco BY-2 Cells

Esther Gerber, Andréa Hemmerlin, and Thomas J. Bach*
Institut de Biologie Moléculaire des Plantes, Centre National de la Recherche Scientifique (Unité Propre de Recherche 2357 conventionée à l'Université de Strasbourg), Département Reseaux Metaboliques, Institut de Botanique, 28 rue Goethe, F-67083 Strasbourg, France

Dring N. Crowell
Department of Biological Sciences, Idaho State University, 650 Memorial Drive, Pocatello, ID 83209-8007

Michel Rohmer
Université de Strasbourg/CNRS, Institut de Chimie (Unité Mixte de Recherche UMR 7177), 4 rue Blaise Pascal, F-67070 Strasbourg, France (MR)

Summary

Isoprenylation of proteins, a fundamental process in eukaryotic cells, is characterized by the formation of a thioether bond between the cysteine residue of a carboxyterminal CaaX motif and a C_{15} (farnesyl) or a C_{20} (geranylgeranyl) isoprenyl group. Using a rice calmodulin (*Oriza sativa* CaM61) fragment with a motif for geranylgeranylation, we developed an experimental system allowing us to visualize the localization of isoprenylated proteins in tobacco bright yellow-2 (TBY-2) cells.

---

*Author for Correspondence, e-mail: Thomas.Bach@ibmpcnrs.unistra.fr

Constantin A. Rebeiz et al. (eds.), The Chloroplast: Basics and Applications, pp. 127–137.

Our results clearly demonstrate that the geranylgeranylated proteins are mainly associated with the plasma membrane. Abbrogation of the isoprenylation reaction by inhibitors or mutations in the CaaX box triggers a change in the intracellular targeting of the chimeric proteins, leading to their accumulation in the nucleus instead of at the plasma membrane.

By using specific inhibitors (mevinolin and fosmidomycin) of both isoprenoid precursor pathways occurring in higher plants, we provide evidence for the essential role played by the plastidial methylerythritol phosphate (MEP) pathway in the synthesis of geranylgeranyl diphosphate needed for the covalent modification of such proteins and their association with the plasma membrane in TBY-2 cells. Interestingly, low μM concentrations of exogenous geranylgeraniol were capable of complementing inhibition of the MEP pathway. Such chemical complementation experiments allow for distinguishing between different levels of inhibition, for instance inhibition of precursor biosynthesis versus inhibition of prenyltransferases.

This experimental system can be used to evaluate new herbicides and drugs, which could interact with the MEP pathway or the geranylgeranylation of proteins after having entered the cells and after their potential conversion into an active compound by cellular metabolic processes. It might also be useful in demonstrating toxic effects of specific inhibitors and in measuring biosynthetic fluxes.

I Introduction

With some 50,000 entities described to date (Bach, 1995, Sacchettini and Poulter, 1997; Croteau et al., 2000; Chang et al., 2007), isoprenoids represent the most diverse group of primary and secondary metabolites in plants. These compounds all originate from a branched five-carbon unit, the so-called active isoprene, isopentenyl diphosphate (IPP) and its isomer dimethylallyl diphosphate (DMAPP). Many primary and secondary isoprenoid plant products act as essential nutrients in the human diet (e.g., pro-vitamin A, vitamins E and K), flavors and fragrances (e.g., mono-, sesqui- and diterpenes), pharmaceuticals (e.g., steroids), food colors and dyes (e.g., shikonin), cancer chemotherapeutic drugs (e.g., taxol), or anti-malarial agents (e.g., artemisinin). In higher plants, it was revealed that two pathways are involved in the biosynthesis of the active isoprene unit. Over the course of evolution, plants have maintained the well-known eukaryotic mevalonic acid (MVA) pathway (Bochar et al., 1999) in the cytosol, also called the classical pathway, and acquired the more recently discovered prokaryotic 2-*C*-methyl-D-erythritol 4-phosphate (MEP) or alternative pathway (Rohmer et al., 1993, 1996; Rohmer, 1999) from the endosymbiotic ancestor of plastids; for more details see Chapter 7 (Lichtenthaler, 2010) and 8 (Boronat, 2010) of this book. Under normal physiological conditions, cytoplasmic isoprenoids, i.e., sterols, or the side chain of mitochondrial ubiquinone are synthesized from MVA-derived IPP (Disch et al., 1998), whereas plastidial isoprenoids are derived from simultaneously formed DMAPP and IPP (Hoeffler et al., 2002), which are synthesized via the MEP pathway (Schwarz, 1994; Arigoni et al., 1997; Lichtenthaler et al., 1997a, b; Lichtenthaler, 1999; Eisenreich et al., 1998; Rohmer, 1999; Rodríguez-Concepción and Boronat, 2002; Kuzuyama and Seto, 2003). There is ample evidence that both IPP and DMAPP biosynthetic pathways cooperate under certain conditions, in the transport of intermediates across membranes, which remain to be clearly identified. To address this and other open questions, we designed a plant model based on isoprenylation-dependent subcellular localization, of green fluorescent protein. This approach is useful for visualize metabolic fluxes originating either from MEP or MVA-derived pools of intermediates.

Abbreviations: BD – basic domain; DMAPP – dimethylallyl diphosphate; DX – deoxyxylulose; Fol – farnesol; FPP – farnesyl diphosphate; Ftase – FPP-specific prenyltransferase; GFP – green fluorescent protein; GFP-BD-CVIL – GFP, to which the basic domain of rice calmodulin CaM61 bearing a carboxyterminal isoprenylation motif has been fused; GFP-BD-SVIL – the same, but with the isoprenylation motif disrupted by site-directed mutagenesis; GGol – geranylgeraniol; GGPP – geranylgeranyl diphosphate; GGTase – I GGPP-specific prenyltransferase type I; IPP – isopentenyl diphosphate; MEP – methylerythritol phosphate; MVA – mevalonic acid; TBY-2 – tobacco Bright Yellow-2

II Protein Isoprenylation in Plants

A The Chemical Modification of a C-Terminal Cysteine

This lipophilic modification, through formation of a chemically stable thioether bond between a carboxyterminal cysteine residue and a C_{15} (farnesol) or a C_{20} (geranylgeraniol) isoprenol residue, is typical of eukaryotic organisms (Clarke, 1992; Sinensky, 2000), including plants (for literature see Crowell, 2000; Gutkowska et al., 2004). The sole requirement for isoprenylation of many proteins is the presence of a CaaX tetrapeptide motif at the carboxyterminal end, with aa representing two aliphatic amino acyl residues. If the X is leucine, the protein is recognized predominantly by a geranylgeranyl diphosphate (GGPP)-specific protein prenyltransferase (GGTase type I). If the X is cysteine, serine, methionine, glutamine, or alanine, the protein is recognized by a farnesyl diphosphate (FPP)-specific prenyltransferase (FTase) (Caplin et al., 1998). Both of these prenyltransferases share a common α subunit, but possess different β subunits, which provide substrate specificity, although this is not absolute in vitro (Caplin et al., 1998). As in animal and yeast cells, the prenylation of plant proteins by FTase or GGTase I is followed by proteolytic removal of the carboxyterminal tripeptide aaX (Bracha et al., 2002; Cadiñanos et al., 2003), after which the carboxyterminal isoprenylated cysteine residue is carboxymethylated (Crowell et al., 1998; Rodríguez-Concepción et al., 2000; Crowell and Kennedy, 2001). The latter step is reversible and might therefore play a role in the insertion of the modified protein into the membrane (Crowell, 2000).

B Functions of Protein Prenylation in Plants

The covalent isoprenylation of a considerable number of proteins, among those many small GTP binding proteins, is important for their function in eukaryotic cells (Clarke, 1992; Zhang and Casey, 1996). Research on protein isoprenylation in plants has been driven by findings that implicate isoprenylated proteins in hormone signaling (Crowell and Randall, 1996; Cutler et al., 1996; Pei et al., 1998; Burnett et al., 2000; Ziegelhoffer et al., 2000). In addition, the chaperone ANJ1 (Zhu et al., 1993), which is involved in protection against heat stress through regulation of protein folding, and various GTP binding proteins implicated in the regulation of plant development are isoprenylated (Lin et al., 1996; Yalovsky et al., 2000; Bonetta et al., 2000; Li et al., 2001; Galichet and Gruissem, 2003; Running et al., 2004; Bloch et al., 2005). Homeostasis of essential heavy metals may also depend on the isoprenylation of proteins (Dykema et al., 1999). Moreover, inhibition studies suggested a role for protein isoprenylation in the regulation of the cell cycle (Qian et al., 1996; Hemmerlin and Bach, 1998; Hemmerlin et al., 2000). A rather extensive list of plant proteins shown to be isoprenylated in vitro and/or in vivo has been compiled (Crowell, 2000).

C Isoprenylation of Proteins in Tobacco BY-2 Cells

[^{14}C]MVA-labeling of proteins in mevinolin-treated TBY-2 cells was demonstrated some time ago (Randall et al., 1993; Morehead et al., 1995). Complete inhibition of endogenous MVA synthesis by mevinolin was necessary to obtain visible bands by fluorography of [^{14}C]MVA-labeled proteins on SDS-polyacrylamide gels. Inhibition studies using the protein farnesyltransferase (PFT) inhibitor chaetomellic acid suggested a role for protein isoprenylation in TBY-2 cell cycle regulation (Hemmerlin et al., 2000). The inhibitor was toxic, and at low micromolar concentrations induced cell death in about 35% of the cells, compared to 20% in mevinolin-treated cells. Furthermore, the cells were arrested at the transition from G2 to M phase (Hemmerlin et al., 2000), suggesting a requirement for one or more isoprenylated (farnesylated?) proteins to pass the G2-M control point.

D Origin of the Prenyl Residue Used for Protein Modification

1 A Double Origin of Prenyl Diphosphates

It has commonly been assumed that the prenyl diphosphates used as substrates for the cytosolic protein prenyl transferases are generated via the MVA pathway. In early studies using spinach seedlings fed with [^{3}H]MVA, the label was found to be associated with nuclear and mitochondrial proteins (Shipton et al., 1995). However, what was considered to be a pure mitochondrial preparation might have been contaminated by nuclei, in view of the practically identical tritium labeling pattern. A later study with plastids isolated from [^{3}H]MVA-fed spinach leaves reported the labeling of membrane fractions considered to represent mainly the outer envelopes of chloroplasts, and to a lesser extent inner envelope membranes and thylakoids (Parmryd et al., 1997). However, labeling with MVA (Shipton et al., 1995) does not prove that the cytosolic MVA pathway is solely responsible for the delivery of isoprenyl residues, although incorporation of exogenous [^{3}H]Fol and [^{3}H]GGol was observed (Parmryd et al., 1999), with only partially overlapping patterns.

To our knowledge, and also stated in a recent review (Crowell, 2000), the extent to which plastidial MVA-independent isoprenoid biosynthesis contributes to the delivery of isoprenyl residues for protein isoprenylation was an open question. It was therefore necessary to demonstrate whether a MEP-derived precursor can be used to label proteins or not. In plant cells only exogenous 1-deoxy-D-xylulose (DX), the dephosphorylated first product of the MEP pathway, can be integrated into the plastid isoprenoid biosynthetic pathway (Hemmerlin et al., 2006b). In a previous study using tobacco TBY-2 cells, it was demonstrated that enzymatically synthesized, [^{14}C]-labeled DX was incorporated into low MW proteins (Hemmerlin et al., 2003a). These proteins migrated in SDS polyacrylamide gels in a range where GTP-binding proteins from the same cell line have already been identified by GTP blot analysis (Hemmerlin et al., 1999), and also in a range where proteins became labeled by exogenous GGol (Parmryd et al., 1999). As a consequence, the question arises, in which compartments are the prenyl diphosphate substrates really built up? FPP should represent the common precursor of non-plastidial isoprenoids and even should not occur in plastids, in contrast to C_5, C_{10} and C_{20} prenyl diphosphates. However, it is clear that some exchange of prenyl diphosphates occurs between the cytosol and plastids (*cf* Schwarz, 1994; Adam et al., 1999; Kasahara et al., 2002; Hemmerlin et al., 2003a; Laule et al., 2003; see also Chapter 13 in this volume and literature cited therein (Nagegowda et al., 2010)). However, these interactions have mainly been shown under stress conditions used for the elicitation of defense reactions (Piel et al., 1998; Jux et al., 2001), or after treatment with specific inhibitors (Kasahara et al., 2002; Hemmerlin et al., 2003a; Laule et al., 2003). Thus, the extent to which translocation of isoprenoid intermediates across membranes contributes to differentially localized pathways, remains open. With this in mind, we have expanded our previous studies on isoprenoid cross-talk (Hemmerlin et al., 2003a) with the use of correspondingly transformed TBY-2 cells suited for both microscopic and biosynthetic studies.

2 Construction of a Tool to Test the Origin of Geranylgeranyl Residues in Prenylated Proteins

(a) State of the Art

Although the cloning of an array of Arabidopsis genes encoding various isoprenyl diphosphate synthases was reported (Okada et al., 2000), the substrate specificity was not always unequivocally shown. This might be due to a lack in chain-length specificity of prenyltransferases, especially under in vitro conditions, which might lead to reactions that most likely do not occur in living cells. While it is clear that in yeast (Schafer and Rine, 1992) and animal cells, which rely exclusively on the cytosolic and putatively peroxisomal (Breitling and Krisans, 2002; Kovacs et al., 2002) MVA pathway for protein geranygeranylation (Schafer and Rine, 1992), a functional cytoplasmic GGPP synthase must exist, the situation in higher plant cells need not be the same.

(b) Tobacco BY-2 Cell Suspensions as a Suitable Tool

Tobacco (*Nicotiana tabacum*) cv. Bright Yellow-2 (TBY-2) root-derived suspension cells are characterized by their pale-yellow color, and a rapid cell division. They do not agglomerate into calli as other plant cell suspensions often do, but can form a string of a few cells, which is very convenient for microscopic observations. Furthermore, they contain all organelles found in other cell types, with a considerable number of mitochondria (Nagata et al., 1992). In particular, the cells contain functional proplastids, which can evolve into leucoplasts or amyloplasts under specific conditions (Nagata et al., 1992;

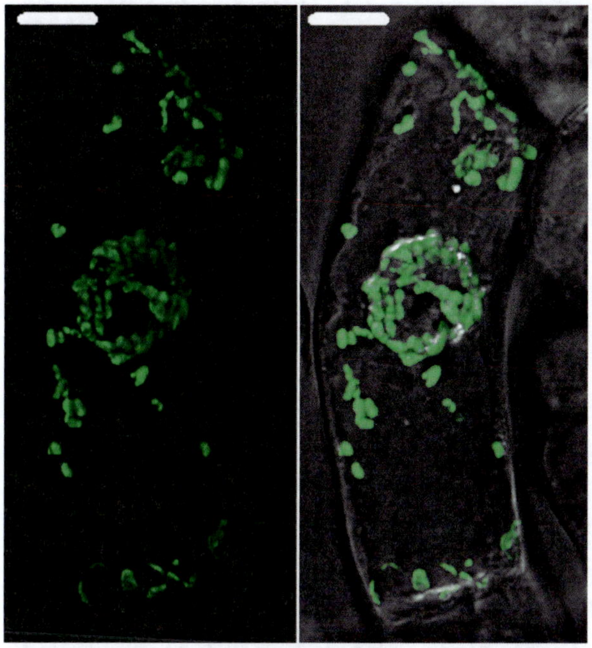

Fig. 1. Tobacco BY-2 plastids visualized by expression of a green fluorescent protein (GFP) fusion protein. TBY-2 cells were transiently transformed by bombardment with tungsten particles carrying a construct coding for the transient peptide of chrysanthemyl diphosphate synthase from *Artemisia tridentata* (Hemmerlin et al., 2003b) fused to the N-terminal side of GFP, the gene being expressed under the control of the constitutive cauliflower mosaic virus 35S promoter. The image represents a projection of 50 pictures acquired through different confocal levels of the cell and shows a strong agglomeration of the plastids around the nucleus. The *white bar* represents 10 µm. Left panel: Fluorescence image; right panel: The same combined with transmission image.

Miyazawa et al., 2002), but never into chloroplasts. These plastids can for instance be visualized by confocal microscopy after transient expression of specific plastid-targeted peptides (Hemmerlin et al., 2003b) fused to the N-terminal side of GFP (Fig. 1).

(c) Description of the System and Results

In order to examine the origin of GGPP used for protein prenylation, we have extended our studies by using transformed TBY-2 cells expressing GFP engineered for isoprenylation, which can be analyzed by confocal microscopy. This approach was partially based on earlier observations of a differential localized protein (Rodríguez-Concepción et al., 1999). A further aim was to develop a more general test for inhibitors that might potentially interfere with protein isoprenylation, either through inhibition of protein prenyltransferases or through interference with isoprenoid precursor biosynthesis. Moreover, such an experimental approach should pave the way for intelligent bioengineering of isoprenoid pathways, avoiding potential interference with regulatory processes involving isoprenylated proteins such as hormonal signaling and cell division.

The unusual carboxyterminal extension of a rice (*Oryza sativa*) calmodulin (OsCaM61) that includes a domain with several basic aminoacyl residues (BD) and a CVIL geranylgeranylation motif was fused to green fluorescent protein (GFP) such that it replaced the calmodulin functional moiety, yielding GFP-BD-CVIL. Furthermore, the corresponding fusion construct was placed under the control of a gradually inducible promoter (Aoyama and Chua, 1997), which is essential to perform the above-mentioned measurements of pools and flux rates in stably transformed TBY-2 cells. That the isoprenylated chimeric protein translocates to the plasma membrane and does not reside in the rudimentary cell wall of TBY-2 cells, was demonstrated by mannitol-induced osmotic plasmolysis and co-localization with the specific marker FM4-64™ from Molecular Probes (Bolte et al., 2004) (Fig. 2a–d). Further, the viability of transformed TBY-2 cells was not affected by expression of this engineered GFP, as indicated by labeling with the red fluorochrome MitoTracker™ (Molecular Probes), which signals the differential electronegative potential in active mitochondria (Fig. 2e–h). The microscopic techniques used have been described in detail (Hemmerlin et al., 2004, 2006a, b).

We made use of inhibitors like mevinolin, affecting MVA biosynthesis (Alberts et al., 1980) and fosmidomycin, specifically inhibiting the MEP synthase (Shigy, 1989; Kuzuyama et al., 1998) to examine the differential contributions of the cytoplasmic MVA and plastidial MEP pathways to the GGPP pool. Without inhibitor treatment, GFP-BD-CVIL accumulated almost exclusively at the plasma membrane (Fig. 3a, b). While mevinolin treatment did not exert any noticeable effect on the localization of the GFP fusion protein (Fig. 3c, d), inhibition by fosmidomycin led to the accumulation of modified GFP in the nucleus instead of at the plasma membrane, and more importantly in the nucleolus (Fig. 3e–f). This was observed with a mutant form of the GFP fusion protein bearing an SVIL motif, which prevents isoprenylation

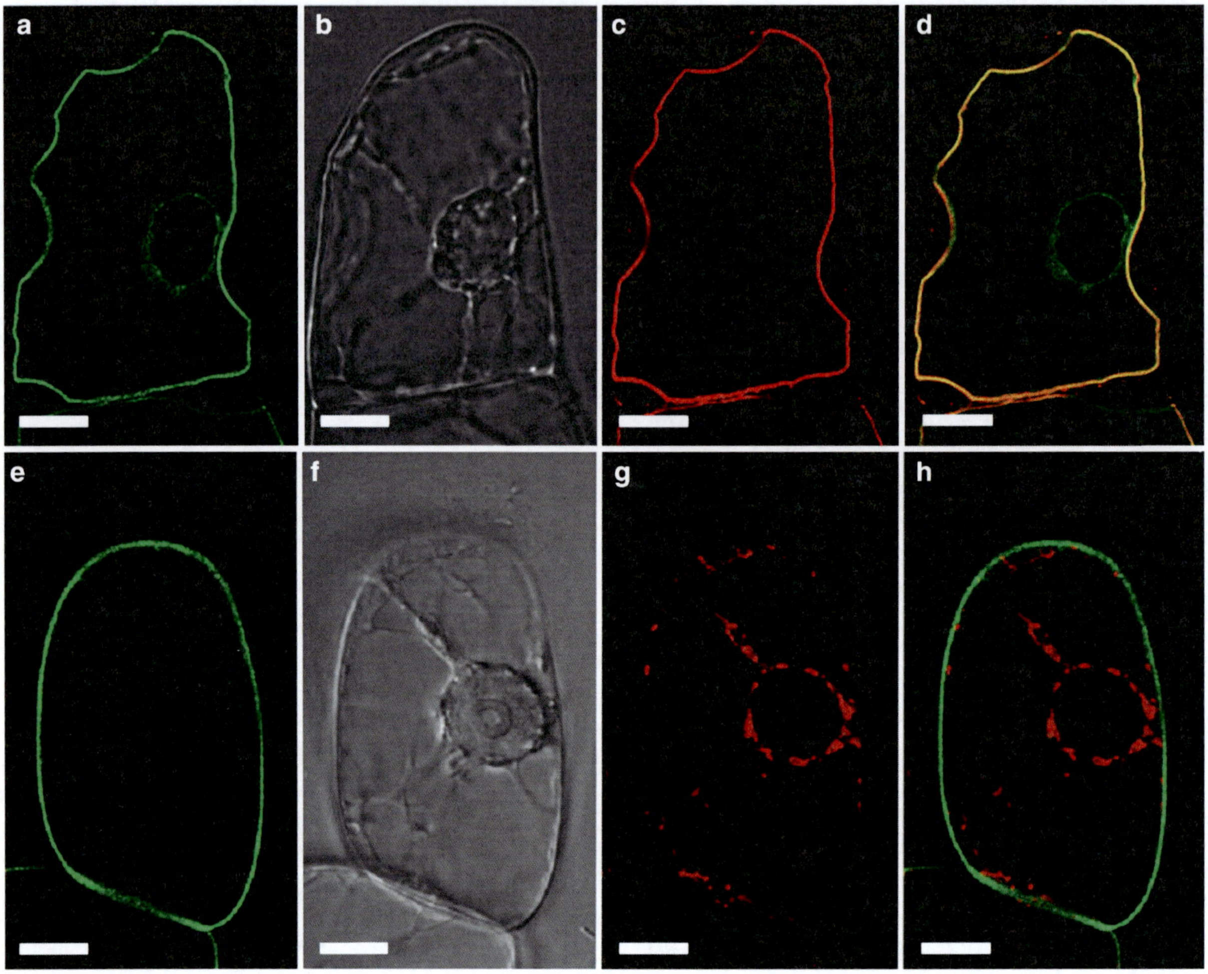

Fig. 2. Co-localization of chimeric GFP with the plasma membrane and viability of TBY-2 cells. GFP was fused at its carboxy-terminus to the C-terminal part (basic domain, BD) of rice calmodulin OsCaM51, bearing a CVIL geranylgeranylation motif (yielding GFP-BD-CVIL). Its expression was induced by 10 μM dexamethasone before treatment for 15 h with the plasma membrane-specific fluorescence marker FM4-64™ (6.7 μg/ml), and D-mannitol (0.23 M), or without mannitol but in the presence of MitoTracker (0.1 μM) that labels mitochondria in viable cells. The images show single optical sections through the middle of cells indicating the localization of the nucleus. (**a–d**): Plasmolysis of TBY-2 cells expressing GFP-BD-CVIL and co-localization with FM4-64™. The cells were exposed for 10 min to 0.23 M D-mannitol before the addition of fluorescent dye, followed by confocal analysis. (**a**) green channel; (**b**) Nomarski (differential interference contrast); (**c**) red channel detecting the localization of the marker; (**d**) overlay of A + C, indicating clear co-localization with the periphery of the cell. (**e–h**) TBY-2 cells expressing GFP-BD-CVIL were treated with MitoTracker™ immediately before microscopic inspection. (**e**) Green channel; (**f**) Nomarski (differential interference contrast); (**g**) red channel detecting labeled mitochondria; (**h**) merged image E + G. *White bars* correspond to 10 μm.

Fig. 3. (continued) 5 μM mevinolin, resulting in practically complete localization of GFP-BD-CVIL to the plasma membrane, as in the untreated control. (**d**) Phase contrast image of the same cells. (**e**) Transformed TBY-2 cells were treated with 40 mM fosmidomycin, resulting in nuclear localization of GFP-BD-CVIL. (**f**) Phase contrast image of the same cell. (**g**) Expression of a chimeric protein that cannot be geranylgeranylated due to replacement of the cysteinyl residue in the prenyltransferase recognition motif by serine (GFP-BD-SVIL). Note the almost complete absence of fluorescence from the cell's periphery and the concentration in the nucleolus. (**h**) Phase contrast image of the same cell, indicating the localization of the nucleus and especially of the nucleolus. (**i**) Expression of GFP-BD-CVIL in TBY-2 cells treated with 40 μM fosmidomycin and 20 μM geranylgeraniol, which results in complete complementation of the fosmidomycin inhibition. (**j**) Phase contrast image of the same cell. (**k**) Expression of GFP-BD-CVIL in stably transformed TBY-2 cells treated with 40 μM fosmidomycin and 20 μM farnesol. Note that the majority of fluorescence remains localized to the nucleus. (**l**) Phase contrast image of the same cell. *White bars* correspond to 10 μm.

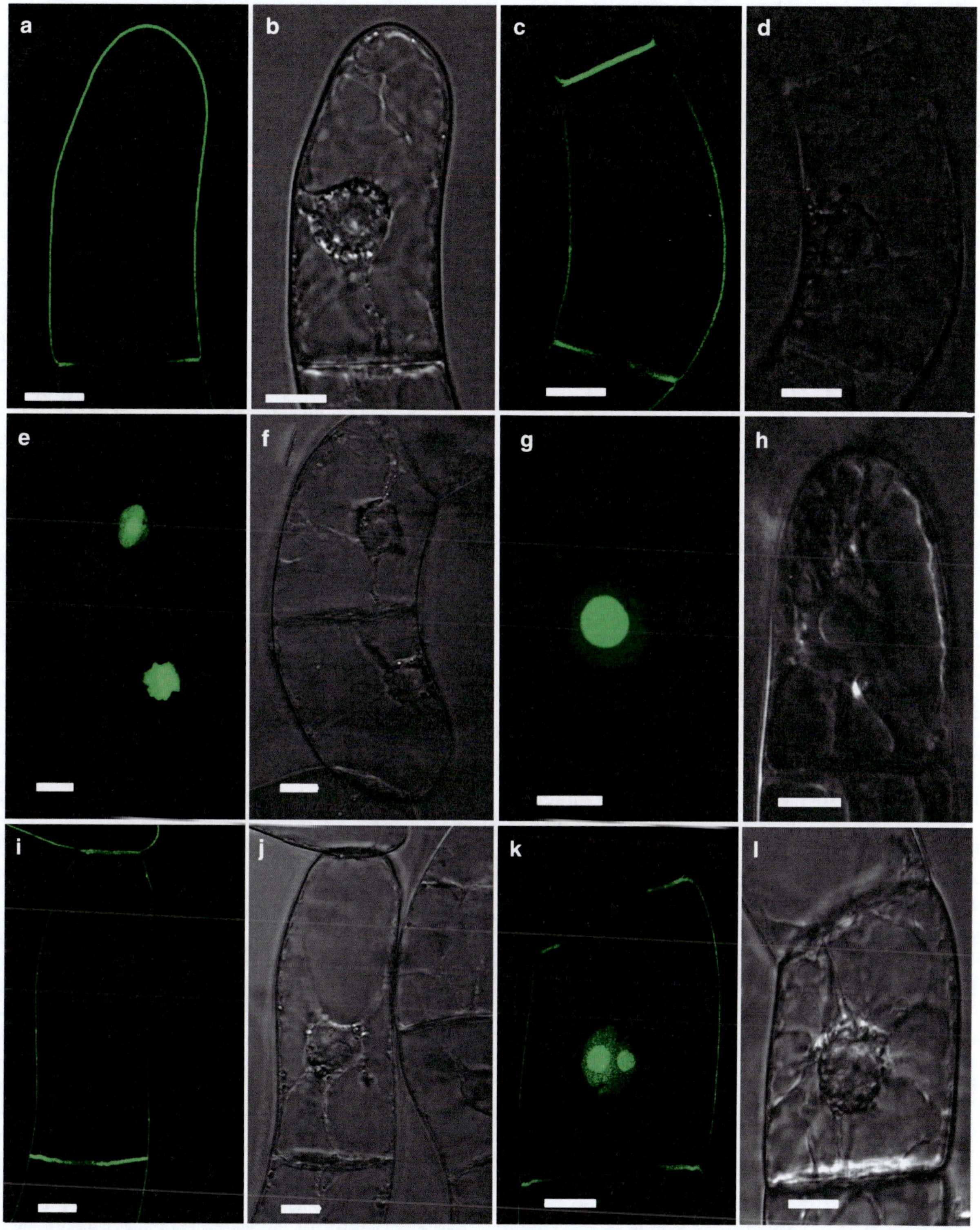

Fig. 3. Differential contributions of the cytoplasmic MVA and plastidial MEP pathways to the GGPP pool. Stably transformed TBY-2 cells in stationary phase were diluted into fresh medium and treated under standard conditions (3 h of pre-incubation followed by induction with 10 μM dexamethasone and by another 15 h of continuous incubation before confocal microscopic inspection). (**a**) GFP, at its carboxyterminus fused to the C-terminal part of rice calmodulin OsCaM51, bearing a CVIL geranylgeranylation motif (GFP-BD-CVIL). The resulting fluorescence is clearly localized to the periphery of the cell. (**b**) Transmission image of the same cell, indicating structural details, especially the position of the nucleus. (**c**) Transformed TBY-2 cells were treated with

(Fig. 3g, h). Thus, despite the plasma membrane localization of modified, geranylgeranylated GFP, our observations provided evidence for an essential role of the plastid-localized MEP pathway in this process. Interestingly, very low concentrations (>3 μM) of exogenous GGol compensated for fosmidomycin-induced miss-localization of chimeric GFP in the nucleus (Fig. 3i, j), suggesting a double phosphorylation system able of converting it to its diphosphate form (Thai et al., 1999), which is then used for modification of engineered GFP. Partial complementation was also observed at considerably higher concentrations of Fol (Fig. 3k, l), but is has to be kept in mind that Fol becomes toxic to TBY-2 cells, depending on the cell suspension density (Hemmerlin and Bach, 2000). In follow-up experiments, we demonstrated that such a chemical complementation by GGol does not work in the presence of a protein prenyltransferase inhibitor. Thus by a simple in vivo approach, an immediate distinction between MEP pathway and prenyl transferase inhibitors is possible.

III Conclusion and Perspectives

In conclusion, transformed, intact TBY-2 cells expressing prenylated GFP fusion proteins under optimized conditions will be useful as a versatile tool for rapid identification of toxic effects exerted by specific inhibitors, measurement of biosynthetic flux rates, and testing of new herbicides and pharmaceuticals (antibacterial and antiparasitic drugs) that might either interfere with the MEP pathway or with protein geranylgeranylation. Studies with the bleaching herbicide oxo-clomazone and chemical complementation experiments with isoprenols and MVA support this prediction (Gerber et al., 2009). Further work is underway to establish unequivocally the biosynthetic origin of all isoprene units in geranylgeranylated GFP in transgenicTBY-2 cells, amongst others by mass spectral analyses of tagged and purified GFP-BD-CVIL.

Acknowledgements

Our work is supported by the CNRS and the Agence Nationale de la Recherche (AH, MR, TJB), by the Université Louis Pasteur (MR) and by the Institut Universitaire de France (MR). We also thank the CNRS, the Université Louis Pasteur, the Région Alsace, and the Association pour la Recherche sur le Cancer for the support of the Inter-Institute confocal microscopy equipment at the IBMP in Strasbourg. We wish to thank Prof. N.-H. Chua for the pTA plasmid, Dr. A.W. Alberts for a generous gift of mevinolin and Dr. R.J. Eilers for a sample of fosmidomycin. We express our thanks to technical staff members at the IBMP's central services.

References

Adam KP, Thiel R and Zapp J (1999) Incorporation of 1-[1-^{13}C]Deoxy-D-xylulose in chamomile sesquiterpenes. Arch Biochem Biophys 369: 127–132

Alberts AW, Chen J, Kuron G, Hunt, V, Huff J, Hoffman C, Rothrock J, Lopez M, Joshua H, Harris E, Patchett A, Monagan R, Curries S, Stapley E, Albers-Schönberg G, Hensens O, Hirshfield J, Hogsteen K, Liesch J and Springer J (1980) Mevinolin, a highly potent competitive inhibitor of hydroxymethylglutaryl-coenzyme A reductase and a cholesterol-lowering agent. Proc Natl Acad Sci USA 77: 3957–3961

Aoyama T and Chua NH (1997) A glucocorticoid-mediated transcriptional induction system in transgenic plants. Plant J 11: 605–612

Arigoni D, Sagner S, Latzel C, Eisenreich W, Bacher A and Zenk MH (1997) Terpenoid biosynthesis from 1-deoxy-D-xylulose in higher plants by intramolecular skeletal rearrangement. Proc Natl Acad Sci USA 94: 10600–10605

Bach TJ (1995) Some new aspects of isoprenoid biosynthesis in plants – a review. Lipids 30: 191–202

Bloch D, Lavy M, Efrat Y, Efroni I, Bracha-Drori K, Abu-Abied M, Sadot E and Yalovsky S (2005) Ectopic expression of an activated RAC in Arabidopsis disrupts membrane cycling. Mol Biol Cell 16: 1913–1927

Bochar DA, Friesen JA, Stauffacher CV and Rodwell VW (1999) Biosynthesis of mevalonic acid from acetyl-CoA. In: Cane DE (ed) Comprehensive Natural Product Chemistry, Isoprenoids Including Steroids and Carotenoids, Vol 2. Pergamon Press, Tarrytown, New York, pp. 15–44

Bolte S, Talbot C, Boutté Y, Catrice O, Read ND and Satiat-Jeunemaître B (2004) FM-dyes as experimental probes for dissecting vesicle trafficking in living plant cells. J Microsc 214: 159–173

Bonetta D, Bayliss P, Sun S, Sage T and McCourt P (2000) Farnesylation is involved in meristem organization in Arabidopsis. Planta 211: 182–190

Boronat A (2010) The methylerythritol 4-phosphate pathway: regulatory role in plastid isoprenoid biosynthesis. In: Rebeiz CA, Benning C, Daniel H, Hoober K, Lichtenthaler HK, Portis A, Tripathy B (eds) The Chloroplast: Basics and Applications. Springer, Netherlands, pp. 119–126

Bracha K, Lavy M and Yalovsky S. (2002) The Arabidopsis AtSTE24 is a CAAX protease with broad substrate specificity. J Biol Chem 277: 29856–29864

Breitling R and Krisans SK (2002) A second gene for peroxisomal HMG-CoA reductase? A genomic reassessment. J Lipid Res 43: 2031–2036

Burnett EC, Desikan R, Moser RC and Neill SJ (2000) ABA activation of an MBP kinase in *Pisum sativum* epidermal peels correlates with stomatal responses to ABA. J Exp Bot 51: 197–205

Cadinaños J, Varela I, Mandel DA, Schmidt WK, Diaz-Perales A, Lopez-Otin C and Freije JM. (2003) AtFACE-2, a functional prenylated protein protease from *Arabidopsis thaliana* related to mammalian Ras-converting enzymes. J Biol Chem 278: 42091–42097

Caplin BE, Ohya Y and Marshall MS (1998) Amino acid residues that define both the isoprenoid and CAAX preferences of the *Saccharomyces cerevisiae* protein farnesyltransferase. Creating the perfect farnesyltransferase. J Biol Chem 273: 9472–9479

Chang MJ, Eachus RA, Trieu W, Ro DK and Keasling JD (2007) Engineering Escherichia coli for production of functionalized terpenoids using plant P450s. Nat Chem Biol 2007 5: 274–277

Clarke S. (1992) Protein isoprenylation and methylation at a carboxyterminal cysteine residue. Annu Rev Biochem 61: 355–386

Croteau R, Kutchan TM and Lewis NG (2000) Natural products (secondary metabolites). In: Buchanan BB, Gruissem W, Jones RL (eds) Biochemistry and Molecular Biology of Plants, Chapter 24. American Society of Plant Physiologists, Rockville, MD, pp. 1250–1318

Crowell DN (2000) Functional implications of protein isoprenylation in plants. Progr Lipid Res 39: 393–408

Crowell DN and Randall SK (1996) Protein farnesylation and phytohormone signal transduction. Trends Plant Sci 1: 407–408

Crowell DN, Sen SE and Randall SK (1998) Prenylcysteine alpha-carboxyl methyltransferase in suspension-cultured tobacco cells. Plant Physiol 118: 11–123

Crowell DN and Kennedy M (2001) Identification and functional expression in yeast of a prenylcysteine alpha-carboxyl methyltransferase gene from *Arabidopsis thaliana*. Plant Mol Biol 45: 469–476

Cutler S, Ghassemian M, Bonetta D, Cooney S and McCourt P (1996) A protein farnesyl transferase involved in abscisic acid signal transduction in Arabidopsis. Science 273: 1239–1241

Disch A, Hemmerlin A, Bach TJ and Rohmer M (1998) Mevalonate-derived isopentenyl diphosphate is the biosynthetic precursor of ubiquinone prenyl side chain in tobacco BY-2 cells. Biochem J 331: 615–621

Dykema PE, Sipes PR, Marie A, Biermann BJ, Crowell DN and Randall SK (1999) A new class of proteins capable of binding transition metals. Plant Mol Biol 41: 139–150

Eisenreich W, Schwarz M, Cartayrade A, Arigoni D, Zenk MH and Bacher A (1998) The deoxyxylulose phosphate pathway of terpenoid biosynthesis in plants and microorganisms. Chem Biol 5: R22–R233

Galichet A and Gruissem W (2003) Protein farnesylation in plants: old friends and new targets. Curr Opin Plant Biol 6: 530–535

Gerber E, Hemmerlin A, Hartmann, M, Hartmann MA, Heintz D, Mutterer J, Rodriguez-Concepcion M, Boronat M, Rohmer M, Crowell DN and Bach TJ (2009) The plastidial 2C-methyl-D-drythriol 4-phosphate pathway provides the isoprenyl moiety for protein geranylgeranylation in tobacco BY-2 cells. Plant Cell 21: 285–300

Gutkowska M, Bienkowski T, Hung VS, Wanke M, Hertel J, Danikiewicz W and Swiezewska E (2004) Proteins are polyisoprenylated in *Arabidopsis thaliana*. Biochem Biophys Res Commun 322: 998–1004

Hemmerlin A and Bach TJ (1998) Effects of mevinolin on cell cycle progression and viability of tobacco BY-2 cells. Plant J 14: 65–74

Hemmerlin A and Bach TJ (2000) Farnesol-induced cell death and stimulation of HMG-CoA reductase activity in tobacco BY-2 cells. Plant Physiol 123: 1257–1268

Hemmerlin A, Brown SC and Bach TJ (1999) Function of mevalonate in tobacco cell proliferation. Acta Bot Gall 146: 85–100

Hemmerlin A, Fischt I and Bach TJ (2000) Differential interaction of branch-specific inhibitors of isoprenoid biosynthesis with cell cycle progression in tobacco BY-2 cells. Physiol Plant 110: 343–350

Hemmerlin A, Hoeffler J-F, Meyer O, Tritsch D, Kagan IA, Grosdemange-Billiard C, Rohmer M and Bach TJ (2003a) Cross-talk between the cytosolic mevalonate and the plastidial methylerythritol phosphate pathways in tobacco Bright Yellow-2 cells. J Biol Chem 278: 26666–26676

Hemmerlin A, Rivera SB, Erickson HK and Poulter CD (2003b) Enzymes encoded by the farnesyl diphosphate synthase gene family in the Big Sagebrush *Artemisia tridentata* ssp. *spiciformis*. J Biol Chem 278: 32132–32140

Hemmerlin A, Gerber E, Feldtrauer J-F, Wentzinger L, Hartmann M-A, Tritsch D, Hoeffler J-F, Rohmer M and Bach TJ (2004) A review of tobacco BY-2 cells as an excellent system to study the biosynthesis and function of sterols and other isoprenoids. Lipids 39: 723–735

Hemmerlin A, Reents R, Mutterer J, Feldtrauer J-F, Waldmann H and Bach TJ (2006a) Monitoring farnesol-induced toxicity in tobacco BY-2 cells: application of a fluorescent analog. Arch Biochem Biophys 448: 93–103

Hemmerlin A, Tritsch D, Hartmann M, Pacaud K, Hoeffler J-F, van Dorsselaer A, Rohmer M and Bach TJ (2006b) A cytosolic *Arabidopsis* D-xylulose kinase catalyzes the phosphorylation of 1-deoxy-D-xylulose into a precursor of the plastidial isoprenoid pathway. Plant Physiol 142: 441–457

Hoeffler J-F, Hemmerlin A, Grosdemange-Billiard C, Bach TJ and Rohmer M (2002) Isoprenoid biosynthesis in higher plants and in *Escherichia coli*: on the branching in the methylerythritol phosphate pathway and independent biosynthesis of isopentenyl diphosphate and dimethylallyl diphosphate. Biochem J 366: 573–583

Jux A, Gleixner G and Boland W (2001) Classification of terpenoids according to the methylerythritolphosphate or the mevalonate pathway with natural $^{12}C/^{13}C$ isotope ratios: dynamic allocation of resources in induced plants. Angew Chem Int Ed 40: 2091–2093

Kasahara H, Hanada A, Kuzuyama T, Takagi M, Kamiya Y and Yamaguchi S (2002) Contribution of the mevalonate and methylerythritol phosphate pathways to the biosynthesis of gibberellins in Arabidopsis. J Biol Chem 277: 45188–45194

Kovacs WJ, Olivier LM and Krisans SK (2002) Central role of peroxisomes in isoprenoid biosynthesis. Progr Lipid Res 41: 369–391

Kuzuyama T and Seto H (2003) Diversity of the biosynthesis of the isoprene units. Nat Prod Rep 20: 171–183

Kuzuyama T, Shimizu T, Takahashi S and Seto H (1998) Fosmidomycin, a specific inhibitor of 1-deoxy-D-xylulose 5-phosphate reductoisomerase in the nonmevalonate pathway for terpenoid biosynthesis. Tetrahedron Lett 39: 7913–7916

Laule O, Fürholz A, Chang H-S, Zhu T, Wang X, Heifetz PB, Gruissem W and Lange BM (2003) Crosstalk between cytosolic and plastidial pathways of isoprenoid biosynthesis in *Arabidopsis thaliana*. Proc Natl Acad Sci USA 100: 6866–6871

Li H, Shen JJ, Zheng ZL, Lin Y and Yang Z (2001) The Rop GTPase switch controls multiple developmental processes in Arabidopsis. Plant Physiol 126: 670–684

Lichtenthaler HK (1999) The 1-deoxy-D-xylulose-5-phosphate pathway of isoprenoid biosynthesis in plants. Annu Rev Plant Physiol Plant Mol Biol 50: 47–65

Lichtenthaler HK, Schwender J, Disch A and Rohmer M (1997a) Biosynthesis of isoprenoids in higher plant chloroplasts proceeds via a mevalonate-independent pathway. FEBS Lett 400: 271–274

Lichtenthaler HK, Rohmer M and Schwender J (1997b) Two independent biochemical pathways for isopentenyl diphosphate (IPP) and isoprenoid biosynthesis in higher plants. Physiol Plant 101: 643–652

Lichtenthaler HK (2010) The non-mevalonate DOXP/MEP pathway of chloroplast isoprenoid and pigment biosynthesis. In: Rebeiz CA, Benning C, Daniel H, Hoober K, Lichtenthaler HK, Portis A, Tripathy B (eds) The Chloroplast: Basics and Applications. Springer, Netherlands, pp. 93–118

Lin Y, Wang Y, Zhu JK and Yang Z. (1996) Localization of a Rho GTPase implies a role in tip growth and movement of the generative cell in pollen tubes. Plant Cell 8: 293–303

Miyazawa Y, Kato H, Muranaka T and Yoshida S (2002) Amyloplast formation in cultured tobacco BY-2 cells requires a high cytokinin content. Plant Cell Physiol 43: 1534–1541

Morehead TA, Biermann BJ, Crowell DN and Randall SK (1995) Changes in protein isoprenylation during the growth of suspension-cultured tobacco cells. Plant Physiol 109: 277–284

Nagata T, Nemoto Y and Hasezawa S (1992) Tobacco BY-2 cell line as the "HeLa" cell in cell biology of higher plants. Int Rev Cytol 132: 1–30

Nagegowda DA, Rhodes D and Dudareva N (2010) The role of the methyl-erythritol-phosphate pathway in rhythmic emission of volatiles. In: Rebeiz CA, Benning C, Daniel H, Green B, Hoober K, Lichtenthaler HK, Portis A, Tripathy B (eds) The Chloroplast: Basics and Applications. Springer, Netherlands, pp. 139–153

Okada K, Saito T, Nakagawa T, Kawamukai M and Kamiya Y (2000) Five geranylgeranyl diphosphate synthases expressed in different organs are localized into three subcellular compartments in Arabidopsis. Plant Physiol 122: 1045–1056

Parmryd I, Andersson B and Dallner G. (1999) Protein prenylation in spinach chloroplasts. Proc Natl Acad Sci USA 96: 10074–10079

Parmryd I, Shipton CA, Andersson B and Dallner G (1997) Protein prenylation in spinach – tissue specificity and greening-induced changes. Arch Biochem Biophys 339: 73–78

Pei ZM, Ghassemian M, Kwak CM, McCourt P and Schroeder JI (1998) Role of farnesyltransferase in ABA regulation of guard cell anion channels and plant water loss. Science 282: 287–290

Piel J, Donath J, Bandemer K and Boland W (1998) Mevalonate-independent biosynthesis of terpenoid volatiles in plants: induced and constitutive emission of volatiles. Angew Chem Int Ed 37: 2478–2481

Qian D, Zhou D, Ju R, Cramer CL and Yang Z (1996) Protein farnesyltransferase in plants: molecular characterization and involvement in cell cycle control. Plant Cell 8: 2381–2394

Randall SK, Marshall MS and Crowell DN (1993) Protein isoprenylation in suspension-cultured tobacco cells. Plant Cell 5: 433–442

Rodríguez-Concepción M, Yalovsky S, Zik M, Fromm H and Gruissem W. (1999) The prenylation status of a novel plant calmodulin directs plasma membrane or nuclear localization of the protein. EMBO J 18: 1996–2007

Rodríguez-Concepción M, Toledo-Ortiz G, Yalovsky S, Caldelari D and Gruissem W (2000) Carboxyl-methylation of prenylated calmodulin CaM53 is required for efficient plasma membrane targeting of the protein. Plant J 24: 775–784

Rodríguez-Concepción M and Boronat A (2002) Elucidation of the methylerythritol phosphate pathway for isoprenoid biosynthesis in bacteria and plastids. A metabolic milestone achieved through genomics. Plant Physiol 130: 1079–1089

Rohmer M (1999) The discovery of a mevalonate-independent pathway for isoprenoid biosynthesis in bacteria, algae and higher plants. Nat Prod Rep 16: 565–574

Rohmer M, Knani M, Simonin P, Sutter B and Sahm H. (1993) Isoprenoid biosynthesis in bacteria: a novel pathway for the early steps leading to isopentenyl diphosphate. Biochem J 295: 517–524

Rohmer M, Seemann M, Horbach S, Bringer-Meyer S and Sahm H (1996) Glyceraldehyde 3-phosphate and pyruvate as precursors of isoprenic units in an alternative non-mevalonate pathway for terpenoid biosynthesis. J Am Chem Soc 118: 2564–2566

Running MP, Lavy M, Sternberg H, Galichet A, Gruissem W, Hake S, Ori N and Yalovsky S (2004) Enlarged meristems and delayed growth in plp mutants result from lack of CaaX prenyltransferases. Proc Natl Acad Sci U S A 101: 7815–7820

Sacchettini JC and Poulter CD (1997) Creating isoprenoid diversity. Science 277: 1788–1789

Schafer WR and Rine J (1992) Protein prenylation: genes, enzymes, targets, and functions. Annu Rev Genet 26: 209–237

Schwarz, M. (1994) Terpen-Biosynthese in *Ginkgo biloba*: Eine überraschende Geschichte. Ph.D. thesis 10951, Eidgenössische Technische Hochschule, Zürich, Switzerland

Shigy Y (1989) Inhibition of bacterial isoprenoid synthesis by fosmidomycin, a phosphonic acid-containing antibiotic. J Antimicrob Chemother 24: 131–145

Shipton CA, Parmryd I, Swiezewska E, Andersson B and Dallner G (1995) Isoprenylation of plant proteins in vivo. Isoprenylated proteins are abundant in the mitochondria and nuclei of spinach. J Biol Chem 270: 566–572

Sinensky M (2000) Functional aspects of polyisoprenoid protein substituents: roles in protein-protein interaction and trafficking. Biochim Biophys Acta 1529: 203–209

Thai L, Rush JS, Maul JE, Devarenne T, Rodgers DL, Chappell J and Waechter CJ (1999) Farnesol is utilized for isoprenoid biosynthesis in plant cells via farnesyl pyrophosphate formed by successive monophosphorylation reactions. Proc Natl Acad Sci U S A 96: 13080–13085

Yalovsky S, Kulukian A, Rodríguez-Concepción M, Young CA and Gruissem W (2000) Functional requirement of plant farnesyltransferase during development in Arabidopsis. Plant Cell 12: 1267–1278

Zhang FL and Casey PJ (1996) Protein prenylation: molecular mechanisms and functional consequences. Annu Rev Biochem 65: 241–269

Zhu JK, Bressan RA and Hasegawa PM (1993) Isoprenylation of the plant molecular chaperone ANJ1 facilitates membrane association and function at high temperature. Proc Natl Acad Sci USA 90: 8557–8561

Ziegelhoffer EC, Medrano LJ and Meyerowitz EM (2000) Cloning of the Arabidopsis WIGGUM gene identifies a role for farnesylation in meristem development. Proc Natl Acad Sci USA 97: 7633–7638

Chapter 10

The Role of the Methyl-Erythritol-Phosphate (MEP) Pathway in Rhythmic Emission of Volatiles

Dinesh A. Nagegowda, David Rhodes, and Natalia Dudareva*
Department of Horticulture and Landscape Architecture, Purdue University, West Lafayette, IN 47907, USA

Summary

Plants release an array of terpenoid compounds including hemiterpenes, mono- and sesquiterpenes, irregular terpenes and some diterpenes throughout their life cycles. These secondary metabolites play crucial roles in pollinator attraction, defense, communication and interaction with the surrounding environment. Release of these compounds from flowers and undamaged and herbivore attacked leaves follows a rhythmic profile, which is induced by illumination and often controlled by a circadian clock. In plants two distinct biochemical pathways localized in different subcellular compartments, the cytosolic mevalonic acid (MVA) pathway and plastidial methyl-erythritol-phosphate (MEP) pathway, are responsible for the biosynthesis of basic carbon building blocks for terpenoid compounds. Mounting evidence suggests that the flux through the MEP pathway changes rhythmically over a daily light/dark cycle peaking during the day. In this chapter we discuss the contribution of the MEP pathway to the rhythmic emission of terpenoids released from different plant tissues and the regulatory steps controlling the flux through this pathway.

*Author for Correspondence, e-mail: dudareva@purdue.edu

Constantin A. Rebeiz et al. (eds.), The Chloroplast: Basics and Applications, pp. 139–153.

I Introduction

Volatile terpenoids represent more than 2% of >20,000 different terpene molecules known to date (Knudsen and Gershenzon, 2006). Because of the physicochemical requirements for volatility, this group of natural products is restricted to low-molecular weight compounds (<300 Da) with low boiling points and high vapor pressure at ambient temperatures. All terpenoids are derived from the fusion of C_5 basic isoprene blocks and classified according to the number of C_5 units in their basic skeleton. The volatile fraction of terpenoids is represented by a simple five-carbon compound isoprene (C_5) as well as by monoterpenes (C_{10}), comprising the most abundant volatile constituents, followed by sesquiterpenes (C_{15}), irregular terpenes (C_8–C_{18}), and some diterpenes (C_{20}) (Knudsen and Gershenzon, 2006). These volatiles are produced in a wide range of plant species and released from flowers, fruits, and vegetative tissues into the atmosphere and from roots into the soil. The primary functions of volatile terpenoids are to defend plants against herbivores and pathogens or to provide a reproductive advantage by attracting pollinators and seed dispersers (Gershenzon and Dudareva, 2007; Dudareva et al., 2006). Within a species the blend of emitted terpenoids differ quantitatively and qualitatively with some compounds in common (reviewed in Dudareva et al., 2006). Moreover, in some species the release of volatile terpenoids from flowers and undamaged and herbivore-attacked leaves displays a rhythmic pattern throughout the photoperiod (Dudareva et al., 2006).

In plants, two distinct biochemical pathways localized in different subcellular compartments are responsible for the biosynthesis of the universal five carbon building blocks, isopentenyl diphosphate (IPP) and its isomer dimethylallyl diphosphate (DMAPP) (Chappell, 1995; Lichtenthaler et al., 1997b; Lange et al., 2000) (Fig. 1). In the cytosol, IPP is synthesized from three molecules of acetyl-CoA by the classical mevalonic acid (MVA) pathway (Qureshi and Porter, 1981; Newman and Chappell, 1999), while in plastids, it is derived from pyruvate and glyceraldehyde-3-phosphate via the methyl-erythritol-phosphate (MEP) pathway (Lichtenthaler et al., 1997a, b; Eisenreich et al., 1998; Lichtenthaler, 1999; Rohmer, 1999). The cytosolic MVA pathway and the plastidic MEP pathway are reviewed in Chapter 7 of this book (Lichtenthaler, 2010).

DMAPP generated from the MEP pathway in plastids is used for isoprene formation (Schwender et al., 1997; Zeidler et al., 1997) via isoprene synthases (Silver and Fall, 1995; Schnitzler et al., 1996; Miller et al., 2001). It is also used by methylbutenol synthase to produce the hemiterpene methylbutenol emitted by ponderosa pines (Zeidler and Lichtenthaler, 2001). In both compartments, IPP and DMAPP undergo condensation catalyzed by short-chain prenyltransferases to form precursors for monoterpene and diterpene biosynthesis in plastids, and sesquiterpenes in the cytosol (Dudareva et al., 2004). Although subcellular compartmentation of the MVA and MEP pathways in plants allows both pathways to operate independently and contribute to sesquiterpene, and to monoterpene and diterpene formation, respectively, the biosynthesis of certain monoterpenes and sesquiterpenes in some plant species occurs via the cooperation of both

Abbreviations: aa – amino acid; ALA – alamethicin; CaMV – cauliflower mosaic virus; CCD – carotenoid cleavage dioxygenase; CDP – ME, 4-diphosphocytidyl-2-*C*-methyl-D-erythritol; CDP-ME2P – 4-diphosphocytidyl-2-*C*-methyl-D-erythritol 2-phosphate; CMK – 4-diphosphocytidyl-2-*C*-methyl-D-erythritol kinase; CMS – 4-diphosphocytidyl-2-*C*-methyl-D-erythritol synthase; DMAPP – dimethylallyl diphosphate; DOX – 1-deoxy-D-xylulose; DXP – 1-deoxy-D-xylulose 5-phosphate; DXR – 1-deoxy-D-xylulose 5-phosphate reductoisomerase; DXS – 1-deoxy-D-xylulose 5-phosphate synthase; FaNES – *Fragaria ananassa* nerolidol synthase; FPP – farnesyl diphosphate; FTC – Forest tent caterpillar; EST – expressed sequence tag; GA-3P – glyceraldehyde-3-phosphate; GGPP – geranyl geranyl diphosphate; GGPPS – geranyl geranyl diphosphate synthase; GPP – geranyl diphosphate; GPPS – geranyl diphosphate synthase; GPPS – SSU geranyl diphosphate synthase small subunit; HDR – 1-hydroxy-2-methyl-2-(*E*)-butenyl 4-diphosphate reductase; HDS – 1-hydroxy-2-methyl-2-(*E*)-butenyl 4-diphosphate synthase; HMBPP – 1-hydroxy-2-methyl-2-(*E*)-butenyl 4-diphosphate; HMG-CoA – 3-hydroxy-3-methylglutaryl-CoA; HMGR – 3-hydroxy-3-methylglutaryl-CoA reductase; IDI – isopentenyl diphosphate isomerase; IPP – isopentenyl diphosphate; ISPS – isoprene synthase; MCS – 2-*C*-methyl-D-erythritol 2,4-cyclodiphosphate synthase; ME-2,4cPP – 2-*C*-methyl-D-erythritol 2,4-cyclodiphosphate; MEP – 2-*C*-methyl-D-erythritol 4-phosphate; MeJA – methyl jasmonate; MVA – mevalonic acid; MVL – mevalolactone; RACE – rapid amplification of cDNA ends; RT-PCR – reverse transcriptase-polymerase chain reaction; TLC – thin layer chromatography; TPS – terpene synthase;

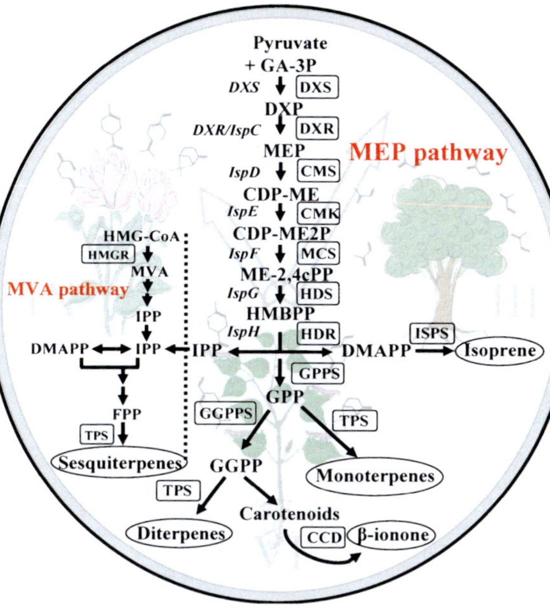

Fig. 1. MEP pathway and rhythmic emission of terpenoids in plants. CCD, carotenoid cleavage dioxygenase; CMK/IspE, 4-diphosphocytidyl-2-C-methyl-D-erythritol kinase; CMS/IspD, 4-diphosphocytidyl-2-*C*-methyl-D-erythritol synthase; CDP-ME, 4-diphosphocytidyl-2-*C*-methyl-D-erythritol; CDP-ME2P, 4-diphosphocytidyl-2-*C*-methyl-D-erythritol 2-phosphate; DMAPP, dimethylallyl diphosphate; DXP, 1-deoxy-D-xylulose 5-phosphate; DXR, DXP reductoisomerase; DXS, DXP synthase; FPP, farnesyl diphosphate; GA-3P, glyceraldehyde-3-phosphate; GGPP, geranyl geranyl diphosphate; GGPPS, GGPP synthase; GPP, geranyl diphosphate; GPPS, GPP synthase; HDS/IspG, 1-hydroxy-2-methyl-2-(*E*)-butenyl 4-diphosphate synthase; HDR/IspH, 1-hydroxy-2-methyl-2-(*E*)-butenyl 4-diphosphate reductase; HMBPP, 1-hydroxy-2-methyl-2-(*E*)-butenyl 4-diphosphate; HMG-CoA, 3-hydroxy-3-methylglutaryl-CoA; HMGR, 3-hydroxy-3-methylglutaryl-CoA reductase; IPP, isopentenyl diphosphate; ISPS, isoprene synthase; MCS/IspF, 2-*C*-methyl-D-erythritol 2,4-cyclodiphosphate synthase; ME-2,4cPP, 2-*C*-methyl-D-erythritol 2,4-cyclodiphosphate; MEP, 2-*C*-methyl-D-erythritol 4-phosphate; TPS, terpene synthase. Names of the enzymes are *boxed* and the corresponding genes in MEP pathway are *italicized* on the *left*. Volatile compounds are indicated within the *ovals*.

pathways (Adam et al., 1999; Piel et al., 1998; Dudareva et al., 2005). Although all the genes and enzymes responsible for each step in these pathways have been isolated and functionally characterized, to date very little is known about the complex regulatory mechanisms controlling the flux through these pathways and their cooperation in the biosynthesis of volatile terpenoids and their emission. In this chapter we discuss the contribution of the MEP pathway to the biosynthesis of terpene volatiles and the regulation involved in their rhythmic emission.

II The MEP Pathway and Rhythmic Emission of Floral Volatiles

Terpenoids represent the largest class of floral volatiles and include such well known and widely distributed constituents of floral scents as the monoterpenes, linalool, limonene, myrcene, ocimene, geraniol, and the sesquiterpenes, farnesene, nerolidol, caryophyllene, and germacrene. Emission of these compounds from flowers of many plant species occurs at similar levels over a daily light/dark cycle (e.g. *Clarkia breweri* (Pichersky et al., 1994)), while in some flowers their emission exhibits distinct diurnal or nocturnal patterns. Nocturnal monoterpene emission was observed in five *Nicotiana* species, *N. rustica*, *N. alata*, *N. forgetiana*, *N. bonariensis*, and *N. langsdorffii* (Raguso et al., 2003) in contrast to diurnal oscillations in emission of terpenoids from *Rosa hybrida* L. cv Honesty (Helsper et al., 1998), *Antirrhinum majus* (Dudareva et al., 2003), *Petunia hybrida* (Simkin et al., 2004), and *Arabidopsis thaliana* (Aharoni et al., 2003) (Table 1).

Moreover, within a single scent bouquet emission of various monoterpene compounds can follow different profiles. In *Hesperis matronalis* flowers, for example, 1,8-cineole is released primarily during the day, *(E)*-β-ocimene mainly near dusk whereas the levels of linalool emission remain constant over a daily light/dark cycle (Nielsen et al., 1995).

The rhythmic release of floral volatiles can be induced by illumination or darkness, or can be controlled by a circadian clock. Nocturnal oscillations in emission of volatiles were reported to be controlled by a circadian clock in contrast to diurnal rhythmicity in volatile emission which is controlled by irradiation levels. However, a circadian nature of diurnal rhythmicity in emission of terpenoids was recently shown in rose and snapdragon flowers (Helsper et al., 1998; Dudareva et al., 2003). Moreover, snapdragon flowers which emit three monoterpenes, myrcene, *(E)*-β-ocimene, and linalool and a sesquiterpene, nerolidol, were used as a model system to investigate the contribution of both IPP biosynthetic pathways to the regulation of rhythmic emission of terpenoids (Dudareva et al., 2005).

Table 1. Terpenoid compounds showing rhythmic emission in plants.

Major terpenoid compound(s)	Plant species	Source	Reference
Constitutively emitted terpenoids			
Myrcene and (*E*)-β-ocimene	*Antirrhinum majus*	Flower	Dudareva et al., 2003
Linalool and nerolidol	*Antirrhinum majus*	Flower	Dudareva et al., 2003
Myrcene, limonene, (*E*)-β-ocimene, β-caryophyllene, thujopsene and α-humulene	*Arabidopsis thaliana*	Flower	Aharoni et al., 2003
β-Pinene	*Artemisia annua*	Leaf	Lu et al., 2002
1,8 Cineole and (*E*)-β-ocimene	*Hesperis matronalis*	Flower	Nielsen et al., 1995
1,8 Cineole, limonene, myrcene, sabinene	*Nicotiana* spp.[a]	Flower	Raguso et al., 2003
β-ionone	*Petunia hybrida*	Leaf and flower	Simkin et al., 2004
trans-β-ocimene, linalool and 1,8-cineole	*Pinus pinea*	Leaf	Staudt et al., 1997
α-Pinene, linalool and β-pinene + sabinene	*Pistacia lentiscus*	Leaf	Hansen et al., 1997
Geraniol, citronellol, nerol, *E*-citral, *Z*-citral, methylgeranylate	*Rosa hybrida*	Flower	Helsper et al., 1998
1,8-Cineol, α-pinene and β-pinene	*Rosmarinus oficinalis*	Leaf	Hansen et al., 1997
Induced terpenoids			
4,8,12-Trimethyltrideca-1,3,7,11-tetraene (TMTT)	*Capsicum annuum*	Leaf	Kunert et al., 2002
(*E*)-β-ocimene, (*E*)-4,8- dimethyl 1,3,7-nonatriene (DMNT), (*E*)-β-farnesene and (*E,E*)-α-farnesene	*Gossypium hirsutum*	Leaf	Loughrin et al., 1994
(*E*)-β-ocimene and DMNT	*Gossypium hirsutum*	Leaf	Loughrin et al., 1997
DMNT and TMTT	*Phaseolus lunatus*	Leaf	Kunert et al., 2002
Linalool, 1,8-cineole, and (*E*)-β-farnesene, terpinen-4-ol	*Picea abies*	Leaf	Martin et al., 2003
(-)-Linalool	*Picea sitchensis*	Leaf	Miller et al., 2005
(−)-Germacrene D	*Populus trichocarpa*	Leaf	Arimura et al., 2004a
Isoprene			
Isoprene	*Elaeis guineensis*	Leaf	Wilkinson et al., 2006
	Eucalyptus spp	Leaf	
	Ficus bengalensis	Leaf	
	Ficus religiosa	Leaf	Padhy and Varshney, 2005
	Mangifera indica	Leaf	
	Melia azedarach	Leaf	
	Populus × *canescens*	Leaf	Loivamäki et al., 2007
	Populus deltoides	Leaf	Funk et al., 2003
	Quercus alba	Leaf	Geron et al., 2000
	Quercus rubra	Leaf	Funk et al., 2003
	Syzygium jambolanum	Leaf	Padhy and Varshney, 2005
	Ulex europaeus	Leaf	Cao et al., 1997

[a]Nocturnal emission

Treatment of cut snapdragon flowers with pathway-specific inhibitors (fosmidomycin for the MEP pathway and mevinolin for the MVA pathway) revealed that fosmidomycin inhibits emission of both monoterpenes and the sesquiterpene nerolidol, while mevinolin has virtually no effect on the amount of emitted compounds. These results suggested that the MVA pathway does not contribute to nerolidol formation and that both monoterpene and sesquiterpene biosynthesis in snapdragon flowers relies on the plastidial supply of IPP precursors via the MEP pathway. Consistent with these results, exogenously supplied stable isotope-labeled 1-deoxy-[5,5-$^2H_2$]-D-xylulose ([$^2H_2$]-DOX), a specific precursor of the MEP pathway, was incorporated into both monoterpenes and nerolidol but the level of labeling of these compounds was greater at night than during the day. The oscillations in the level of labeling of these compounds could be the result of the rhythmic operation of the endogenous MEP pathway with greater flux during the light period thus reducing the incorporation of exogenous substrate.

The biosynthesis of plastidial IPP precursors is directly linked to photosynthesis. The two immediate precursors of the MEP pathway, pyruvate and glyceraldehyde 3-phosphate, are derived from the

Calvin cycle, suggesting that the rhythmic changes in the flux through this pathway could be induced either by light or controlled by an endogenous clock. The exposure of snapdragon flowers to continuous darkness for 3 days revealed a persistence in the oscillations in emission and in the levels of labeling of monoterpenes and nerolidol, indicating that the flux through the MEP pathway follows a diurnal rhythm which is controlled by a circadian clock (Dudareva et al., 2005). Moreover, the elimination of the contribution of the MEP pathway to nerolidol biosynthesis by the inhibitor fosmidomycin in the presence of exogenous deuterium labeled [2,2-$^2H_2$]-mevalolactone ([$^2H_2$]-MVL) resulted in the elimination of rhythmicity in nerolidol emission and the level of its labeling. Thus, the diurnal fluctuations in terpenoid emission in snapdragon flowers are likely the result of rhythmicity in the flux through the MEP pathway.

Headspace analysis of terpenoids released from Arabidopsis flowers revealed that nearly all monoterpenes (e.g., myrcene, limonene, and *(E)*-β-ocimene) and sesquiterpenes (β-caryophyllene, thujopsene, and α-humulene) also exhibit a clear diurnal pattern in their emission (Aharoni et al., 2003) (Table 1). Moreover, the overexpression in Arabidopsis of a strawberry linalool/nerolidol synthase (*FaNES1*) targeted to plastids under the control of a constitutive *35S CaMV* promoter resulted in production of linalool, whose emission followed diurnal oscillations as well. These results suggest that the diurnal emission of compounds synthesized by endogenous and introduced terpene synthases may be determined by the availability of precursors, the biosynthesis of which could occur rhythmically within the biosynthetic pathways. Other factors like glycosidase or glycosyltransferase activities as well as the light activation of the *35S CaMV* promoter (Schnurr and Guerra, 2000) in the case of the introduced linalool, can not be excluded (Aharoni et al., 2003). Although the contribution of plastidial MEP and cytosolic MVA pathways to the biosynthesis of terpenoid precursors was not analyzed in Arabidopsis, the existence of crosstalk between these two pathways and the trafficking of isoprenoid intermediates from the plastid to the cytosol has been demonstrated in Arabidopsis seedlings (Laule et al., 2003).

The initial step of the MEP pathway includes the condensation of pyruvate and glyceraldehyde 3-phosphate with the formation of 1-deoxy-D-xylulose-5-phosphate (DXP) in a reaction catalyzed by the transketolase DXP synthase (DXS) (Fig. 1). The produced DXP is then used in plants as a precursor for IPP biosynthesis as well as for the biosynthesis of the cofactors, thiamin pyrophosphate and pyridoxal phosphate (Julliard and Douce, 1991; Himmeldirk et al., 1996). The first step specific for IPP production is catalyzed by DXP reductoisomerase (DXR) and involves the conversion of DXP to methylerythritol phosphate (Fig. 1). While DXS is thought to be an important rate-controlling step of the MEP pathway (Lois et al., 2000; Estévez et al., 2001), DXR may also serve as a significant control point of the metabolic flux through the pathway since it catalyzes the first committed step of the MEP pathway towards terpenoid biosynthesis (Takahashi et al., 1998; Mahmoud and Croteau, 2001; Carretero-Paulet et al., 2006). The post-transcriptional regulation of the MEP pathway is described in chapter 8 of this book (Boronat, 2010). Analysis of *DXS* and *DXR* expression in snapdragon flowers over a daily light/dark cycle revealed that only *DXS* transcripts show a rhythmic pattern which peaks during the light period and strongly correlates with the pattern of diurnal monoterpene and nerolidol emissions (Dudareva et al., 2005) (Table 2). These results suggest that transcriptional regulation of *DXS* expression determines the rhythmic profile of the flux through the MEP pathway in snapdragon flowers. The lack of diurnal oscillations in *DXR* transcript levels could be due to a minor role of DXR in the regulation of the MEP pathway in snapdragon flowers or post-transcriptional regulation of DXR activity. Additionally, the generic *DXR* probe used in these experiments could have recognized more than one possible *DXR* isoform, and so masked the correlation of expression of a specific *DXR* isogene with monoterpene and sesquiterpene emission (Dudareva et al., 2005).

The fact that exogenously supplied [$^2H_2$]-DOX did not affect the total amount of emitted terpenoids and did not eliminate rhythmicity in their emission provides evidence that some additional regulatory mechanisms also take place downstream of DXS. While the expression of other genes in the MEP pathway downstream of DXR was not analyzed in snapdragon flowers, expression analysis of geranyl diphosphate synthase (GPPS), which catalyzes

Table 2. MEP pathway and downstream genes showing rhythmic expression.

Gene	Plant species	Reference
1-Deoxy-D-xylulose 5-phosphate synthase (*DXS*)	*Antirrhinum majus*	Dudareva et al., 2005
1-Deoxy-D-xylulose 5-phosphate reductoisomerase (*DXR*)	*Populus trichocarpa*	Arimura et al., 2004a
1-Deoxy-D-xylulose 5-phosphate reductoisomerase (*DXR*)	*Artemisia annua*	Lu et al., 2002
1-Deoxy-D-xylulose 5-phosphate reductoisomerase (*DXR/IspC*)		
4-Diphosphocytidyl-2-*C*-methyl-D-erythritol synthase (*CMS or IspD*)		
4-Diphosphocytidyl-2-*C*-methyl-D-erythritol kinase (*CMK/IspE*)		
2-*C*-methyl-D-erythritol 2,4-cyclodiphosphate synthase (*MCS/IspF*)	*Arabidopsis thaliana*	Hsieh and Goodman, 2005
1-Hydroxy-2-methyl-2-(*E*)-butenyl 4-Diphosphate synthase (*HDS/IspG*)		
1-Hydroxy-2-methyl-2-(*E*)-butenyl 4-diphosphate reductase (*HDR/IspH*)		
Isoprene synthase (*ISPS*)	*Populus* × *canescens*	Loivamäki et al., 2007
Pinene synthase (*QH6*)	*Artemisia annua*	Lu et al., 2002
Myrcene and (*E*)-β-ocimene synthases	*Antirrhinum majus*	Dudareva et al., 2003
(−)-Germacrene D synthase	*Populus trichocarpa*	Arimura et al., 2004a
Carotenoid cleavage dioxygenase (*PhCCD1*)	*Petunia hybrida*	Simkin et al., 2004

a single condensation of IPP and DMAPP to form GPP (Fig. 1), the substrate for monoterpene biosynthesis, and the monoterpene synthases responsible for myrcene and *(E)*-β-ocimene formation was performed in petal tissue during a daily light/dark cycle. In snapdragon, only a small subunit (GPPS.SSU) of a heterodimeric GPP synthase controls the rate of GPP production (Tholl et al., 2004). Levels of transcripts for *GPPS.SSU* and monoterpene synthases exhibited similar weak diurnal oscillations which are retained under continuous dark, suggesting that their cyclic expression is under circadian control (Dudareva et al., 2003; Tholl et al., 2004). Although the expression of *DXS, GPPS.SSU* and monoterpene synthases all positively correlated with monoterpene emission, the molecular mechanisms involved in the regulation of the flux towards terpenoids still remain unknown.

In Arabidopsis, the high levels of *DXS* and *DXR* gene expression were also found in the inflorescences, which was consistent with the high emission of terpenoids from this part of the plant (Carretero-Paulet et al., 2002). The expression of the *DXR* gene closely paralleled that of *DXS*, but exhibited a slightly more restricted pattern. Its expression begins later than *DXS* in emerging inflorescences, suggesting that in contrast to snapdragon, DXR instead of DXS might be limiting for the onset of isoprenoid biosynthesis in Arabidopsis flowers (Carretero-Paulet et al., 2002) (Table 2). Consistent with diurnal emission of terpenoids, the expression of both genes as well as the rest of the MEP pathway genes was significantly induced by light (Carretero-Paulet et al., 2002; Hsieh and Goodman, 2005). The only exception is the 1-hydroxy-2-methyl-2-*(E)*-butenyl 4-diphosphate reductase (HDR) which catalyzes the last step of the MEP pathway and is expressed constitutively regardless of light/dark conditions (Hsieh and Goodman, 2005). Although the expression of the MEP pathway genes was not analyzed in the inflorescences over a daily light/dark cycle, their expression in 2-week-old Arabidopsis seedlings exhibited diurnal oscillations and was significantly suppressed during the transition from light to dark. Moreover, *DXS* and *DXR* expression followed two different diurnal profiles with *DXS* peaking earlier in the light cycle (Hsieh and Goodman, 2005). Whether a circadian clock is involved in the regulation of expression of the MEP pathway genes as well as the flux through the MEP pathway in Arabidopsis still remains to be determined.

Experiments with pathway-specific precursors and inhibitors in snapdragon flowers revealed that the endogenous MVA pathway does not contribute to nerolidol formation and is blocked before the formation of mevalonic acid. The fact that $[^2H_2]$-MVL supplied to snapdragon flowers was efficiently incorporated into nerolidol and led to an increase of its emission during both day and night indicated that the later enzymes of the MVA pathway were active. 3-Hydroxy-3-methylglutaryl-coenzyme A reductase (HMGR, EC 1.1.1.34), which catalyzes the NADPH-dependent reduction of 3-hydroxy-3-methylglutaryl-coenzyme A (HMG-CoA) to mevalonic acid, is the key regulatory enzyme of the MVA pathway and has been extensively studied in some plant species. *Arabidopsis thaliana*, for example, contains two differentially expressed *HMGR* genes which encode three isoforms (Lumbreras et al., 1995). In contrast to Arabidopsis, three *HMGR* genes have been identified in snapdragon. To isolate these genes we used two different approaches, RT-PCR in combination with 5′ and 3′ rapid amplification of cDNA ends (RACE) and cDNA library screening. Since a search of 11,600 nonreduntant expressed sequence tags (ESTs) from a normalized snapdragon cDNA library constructed from mRNAs isolated from different vegetative and floral organs revealed no clones with homology to the HMGRs, degenerate primers were designed in the HMGR conserved domain and used in RT-PCR experiments with total RNA isolated from upper and lower lobes of snapdragon flowers. RT-PCR resulted in a fragment of ~450 nucleotides which showed homology to known HMGRs. 5′ and 3′ RACE was used to recover the corresponding full-length clone. The obtained full-length cDNA clone (designated as *AmHMGR1*) is 2,195 nucleotides in size and encodes an open reading frame of 1,758 nucleotides, corresponding to a protein of 586 aa (Fig. 2). It has 174 and 263 nucleotides in its 5′ and 3′ untranslated regions, respectively.

For functional characterization of the protein encoded by *AmHMGR1* cDNA, the coding region was subcloned into the pET-28a expression vector and expressed in *Escherichia coli* Rosetta cells. The HMGR activity of the isopropylthio-β-galactoside-induced bacterial crude extracts and recombinant His-tag purified protein was determined by measuring the ^{14}C-mevalonate formation from the ^{14}C labeled substrate 3-hydroxy-3-methylglutaryl-CoA. Thin layer chromatography (TLC) separation of the reaction products showed a single band with an R_f value corresponding to that of mevalonic acid, thus confirming the HMGR activity of the enzyme (Fig. 3).

Since HMGRs are known to belong to a multigene family (Bach et al., 1999), the *AmHMGR1* cDNA was used as a probe to screen a snapdragon petal specific cDNA library to identify other *HMGR* genes, if any. This screening resulted in two truncated clones with sequence homology to AmHMGR1. Full length cDNAs for each clone were obtained by RACE amplification and designated as *AmHMGR2* and *AmHMGR3*. *AmHMGR2* and *AmHMGR3* cDNAs were 2,080 and 2,071 nucleotides in size and encoded proteins of 548 aa and 555 aa, respectively. Sequence alignment and phylogenetic analysis of the AmHMGR deduced amino acid sequences revealed that AmHMGR2 and AmHMGR3 are closely related with 85% aa identity and both share about 70–72% identity with AmHMGR1 (Figs. 2 and 4). All three AmHMGRs exhibited high aa sequence identity to HMGRs from *Andrographis paniculata* (73–81%), *A. thaliana* (73–75%) and *Hevea brasiliensis* (76–77%) (Figs. 2 and 4).

To determine the contribution of these genes to nerolidol biosynthesis and emission, their expression was analyzed by semi-quantitative RT-PCR in different floral tissues of 5 day-old flowers, a developmental stage with high levels of terpenoid emission (Dudareva et al., 2003). The highest level of expression of all three genes was found in stamens (Fig. 5). *AmHMGR1* and *AmHMGR2* were also highly expressed in tubes and pistils of snapdragon flowers, while low levels of expression of all three genes were observed in ovaries and upper and lower petal lobes, the tissues primarily involved in the formation and emission of terpenoid volatiles (Dudareva et al., 2003). In green tissues, *AmHMGR2* was expressed at a low level in leaves, while low levels of both *AmHMGR1* and *AmHMGR3* expression were found in sepals (Fig. 5). Low levels of expression of all three *AmHMGRs* in scent producing parts of the flower (upper and lower petal lobes) suggest that the MVA pathway contributes little if any to nerolidol biosynthesis in snapdragon flowers. This statement is also supported by our earlier feeding experiments with [$^2H_2$]-MVL in the presence of fosmidomycin (Dudareva et al., 2005) where in the absence of the MEP pathway [$^2H_2$]-MVL feeding led to almost complete deuterium labeling of nerolidol.

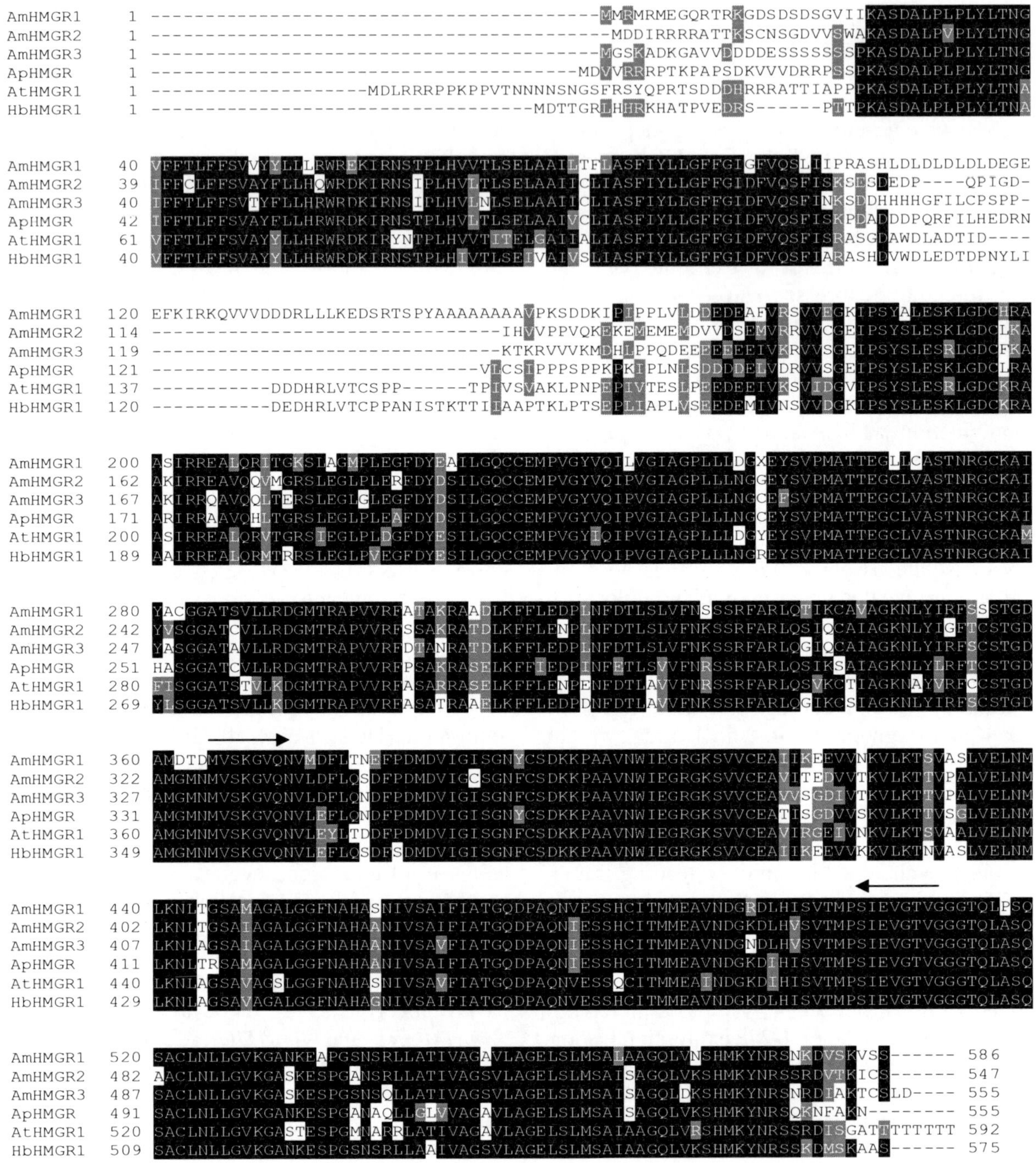

Fig. 2. Alignment of deduced AmHMGR amino acid sequences to other plant HMGRs. Alignment of AmHMGR predicted amino acid sequences with HMGRs from *Andrographis paniculata* (AAP14352), *Arabidopsis thaliana* HMGR1 (P14891) and *Hevea brasiliensis* (AAQ63055). Alignment was performed using ClustalW and *shaded* using the BoxShade Version 3.21 software program (Human Genome Sequencing Center, Houston, TX). Residues *shaded in black* indicate conserved residues (identical in at least four out of six sequences shown), and residues *shaded in gray* are similar in at least two of six sequences shown. *Dashes* indicate gaps that have been inserted for optimal alignment. The positions of degenerate forward and reverse primers used for obtaining ~450 bp of *AmHMGR1* are indicated by *arrows*.

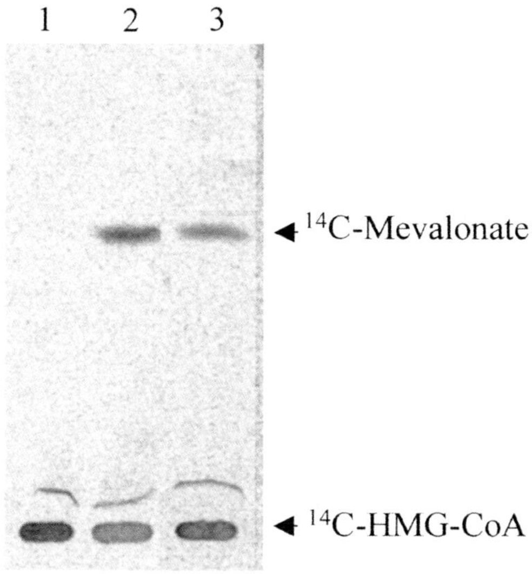

Fig. 3. Analysis of HMGR activity of a recombinant *E. coli* expressed AmHMGR1. Reactions were carried out using crude cell extracts from pET28-Rosetta (control, lane 1), pET28-HMGR1-Rosetta (lane 2) and Ni-NTA purified fraction from pET28-HMGR1-Rosetta cells (lane 3) in the presence of ^{14}C-3-hydroxy-3-methylglutaryl-CoA. Product identification was performed by thin layer chromatography. The R_f value of the product formed by AmHMGR1 corresponds to that of mevalonate.

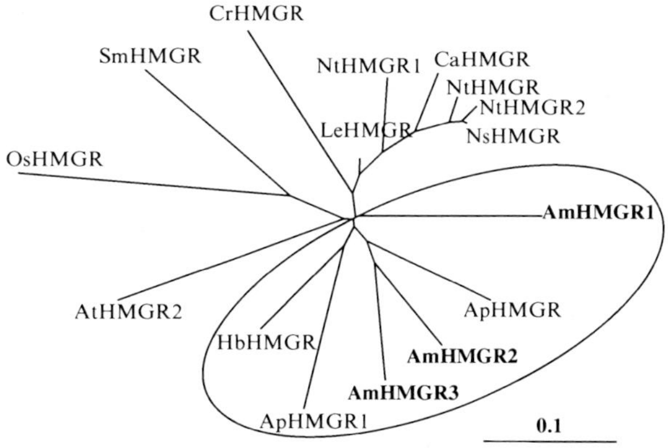

Fig. 4. Phylogenetic tree illustrating the relatedness of AmHMGR proteins to other plant HMGRs. The unrooted neighbor joining tree was created using ClustalX and TreeView for visualization. AmHMGR1, AmHMGR2 and AmHMGR3, *Antirrhinum majus* HMGR1 (EF666139), HMRG2 (EF666140) and HMGR3 (EF666141); ApHMGR, *Andrographis paniculata* HMGR (AAP14352); AtHMGR1 and AtHMGR2, *Arabidopsis thaliana* HMGR1 (P14891) and HMGR2 (P43256); CaHMGR, *Capsicum annuum* HMGR (Q9XEL8); CrHMGR, *Catharanthus roseus* HMGR (AAT52222); HbHMGR, *Hevea brasiliensis* HMGR (AAQ63055); LeHMGR, *Lycopersicon esculentum* HMGR (AAL16927); NtHMGR, NtHMGR1 and NtHMGR2, *Nicotiana tabacum* HMGR (AAL54879), HMGR1 (AAB87727) and HMGR2 (AAL54878); NsHMGR, *Nicotiana sylvestris* HMGR (Q01559); OsHMGR, *Oryza sativa* HMGR (AAA21720); SmHMGR, *Solanum melongena* HMGR (AAQ12265).

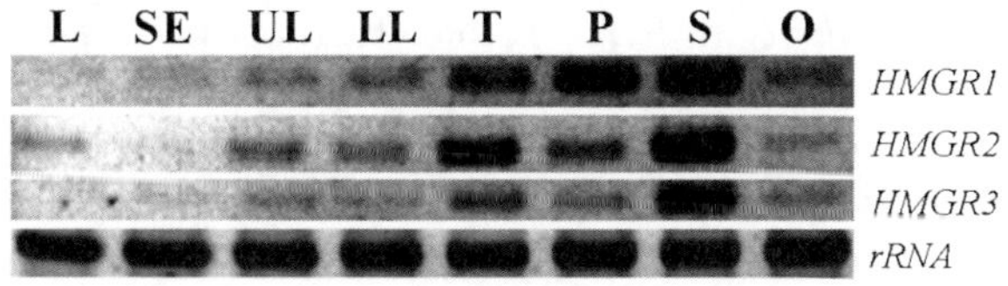

Fig. 5. Tissue specific *AmHMGR* mRNA expression in snapdragon flowers. Semiquantitative quantitative RT-PCR analysis of *AmHMGR* in young leaves (L), upper (UL) and lower (LL) petal lobes, tubes (T), pistils (P), stamens (S), ovaries (O), and sepals (SE) of 5-day-old snapdragon flowers is shown. One μg of total RNA from each tissue was used for cDNA preparation and RT-PCR was performed for 25 cycles.

III The MEP Pathway and Rhythmic Emission of Leaf Volatiles

Rhythmic release of terpenoids is not only limited to flowers, but is also found in vegetative tissue. An evergreen holm oak *Quercus ilex* widespread in the Mediterranean forests emits 14 terpenes, out of which α-pinene is the most abundant (Loreto et al., 1996; Staudt and Bertin, 1998). Emission of these compounds is light-dependent and three distinct classes were identified based on their responses to light induction: a rapidly induced class including α-pinene, a more slowly induced class including *cis*-β-ocimene, and the most slowly induced class with 3-methyl-3-buten-1-ol as a representative (Loreto et al., 1996). Diurnal oscillations in the emission of terpenoids with peak emission during the day were also reported in Norway spruce (*Picea abies* L.) (Martin et al., 2003), Stone pine (*Pinus pinea* L.) (Staudt et al., 1997), rosemary (*Rosmarinus officinalis* L.), pistachio (*Pistacia lentiscus* L.) (Hansen et al., 1997), and Chinese wormwood (*Artemisia annua*) (Lu et al., 2002) (Table 1). Although the involvement of both the MVA and MEP pathways in the regulation of diurnal terpenoid emission in these species was not investigated, the expression of DXR and β-pinene synthase was analyzed in *A. annua* during a daily light/dark cycle (Lu et al., 2002). The levels of mRNA transcripts for both genes peaked shortly after noon showing a diurnal rhythm in their expression. Moreover, diurnal oscillations in the expression of β-pinene synthase were retained in plants exposed to constant light or constant dark, suggesting the involvement of a circadian clock in the regulation of its expression (Lu et al., 2002) (Table 2).

IV The MEP Pathway and Rhythmic Emission of Herbivore-Induced Plant Volatiles

Terpenoids, monoterpenes, sesquiterpenes and homoterpenes, are often released from plant vegetative tissues in response to damage and herbivore attack. These emitted volatiles can directly repel (De Moraes et al., 2001; Kessler and Baldwin, 2001) or intoxicate (Vancanneyt et al., 2001) microbe and animals, or attract natural predators of attacking herbivores, indirectly protecting the signaling plant from further damage via tritrophic interactions (Mercke et al., 2004; Arimura et al., 2004b; Degen et al., 2004). The monoterpenes linalool, *(E)*-β-ocimene and α-pinene, as well as the sesquiterpenes germacrene D, caryophyllene and α- and β-farnesene are the compounds most often released from herbivore-damaged tissues. In most cases, the release of herbivore-induced terpenoids follows diurnal cycles with a peak of emission during the day (Table 1). Examples include emission from grape leaves (*Vitis labrúsca* L.) infested by Japanese beetles (*Popillia japonica*) (Loughrin et al., 1997), cotton plants (*Gossypium hirsutum*) injured by beet armyworms (*Spodoptera exigua* Hübner) (Loughrin et al., 1994), poplar trees (*Populus trichocarpa* × *deltoides*) infested with forest tent caterpillars (FTC) (*Malacosoma disstria*) (Arimura et al., 2004a), Sitka spruce (*Picea sitchensis*) attacked by white pine weevils (*Pissodes strobi*) (Miller et al., 2005), and lima beans (*Phaseolus lunatus*) infested with the Egyptian cotton leafworms (*Spodoptera littoralis*) (Arimura et al., 2005) (Table 1). Moreover, induced diurnal emission of mono-and sesquiterpenoids was found in Sitka spruce and Norway spruce after exposure of intact plants to methyl jasmonate (MeJA), an elicitor simulating insect or pathogen attack, and in lima bean treated with alamethicin (ALA), an ion channel-forming fungal elicitor (Miller et al., 2005; Martin et al., 2003; Kunert et al., 2002).

Similar to the situation with the constitutive emission of volatile terpenoids from vegetative tissues, the contribution of the MVA and MEP pathways to the regulation of rhythmic emission of herbivore-induced volatiles was not investigated. The only example includes hybrid poplar where the expression of (−)-germacrene D synthase and one *DXR* gene (*PtdDXR1*) was analyzed in local FTC-infested and systemic leaves (Arimura et al., 2004a). While the FTC-induced expression of (−)-germacrene D synthase exhibited an obvious diurnal profile that peaked during the light period and closely matched the actual pattern of FTC-induced volatile release, FTC did not affect the abundance of *PtdDXR1* transcripts which displayed only slight diurnal fluctuations (Arimura et al., 2004a) (Table 2).

V The MEP Pathway and Rhythmic Emission of Isoprene

Vegetative tissues of many plant species including mosses, ferns, gymnosperms and angiosperms release a highly volatile five-carbon terpene isoprene into the atmosphere (Kesselmeier and Staudt, 1999; Sharkey and Yeh, 2001). Although the biological function of isoprene is still unclear, this hemiterpene may act to increase the tolerance of photosynthesis to high temperatures by stabilizing the thylakoid membranes (Sharkey et al., 2001; Peñuelas et al., 2005; Velikova and Loreto, 2005), protect plants against extensive light (Peñuelas and Munne-Bosch, 2005), serve as an antioxidant by quenching reactive oxygen species (Loreto and Velikova, 2001; Affek and Yakir, 2002) or as an overflow valve for carbon and energy excess (Rosenstiel et al., 2002; Magel et al., 2006). Isoprene emission displays a clear diurnal pattern as was found in many trees including gorse *Ulex europaeus* (Cao et al., 1997), some oak species *Quercus alba*, *Q, rubra*, *Q. robur* (Geron et al., 2000; Funk et al., 2003; Brüggemann and Schnitzler, 2002a), eastern cottonwood *Populus deltoides* (Funk et al., 2003), eucalyptus *Eucalyptus* sp., banyan *Ficus bengalenis*, peepul *Ficus religiosa*, mango *Mangifera indica*, chinaberry *Melia azedarach*, jambolan *Syzygium jambolanum* (Padhy and Varshney, 2005), poplar *Populus* spp. (Mayrhofer et al., 2005), and oil palm *Elaeis guineensis* (Wilkinson et al., 2006) (Table 1). These diurnal oscillations in isoprene emission can be induced by light or controlled by a circadian clock. Circadian control of isoprene biosynthesis was shown in oil palm and grey poplar (Wilkinson et al., 2006; Loivamäki et al., 2007) raising the issue of its widespread reach in the plant kingdom.

Isoprene is synthesized from DMAPP in chloroplasts in a reaction catalyzed by isoprene synthase (ISPS) (Fig. 1) (Silver and Fall, 1995; Schnitzler et al., 1996; Sharkey et al., 2005; Sasaki et al., 2005) and shares the MEP pathway with monoterpenes, diterpenes and carotenoids as tetraterpenes. Within the plastid, isoprene synthase competes with GPP synthase for DMAPP utilization and has much higher K_m values for DMAPP (in the millimolar range) (Wildermuth and Fall, 1998) when compared with that of GPP synthases (in the micromolar range) (Tholl et al., 2001; 2004; Burke and Croteau, 2002). Rhythmic emission of isoprene can be regulated by the availability of DMAPP substrate and/or by the activity of isoprene synthase. Although the regulatory mechanisms controlling rhythmic isoprene emission are not completely understood, analysis of *PcISPS* and *PcDXR* gene expression in shoot cultures of grey poplar over a daily light/dark cycle revealed diurnal oscillations which were retained in continuous light only for *PcISPS* indicating that its expression is controlled by a circadian clock while the *PcDXR* expression is light-dependent (Loivamäki et al., 2007) (Table 2). However, light was found to be a trigger of *PcISPS* gene expression as well, since a two-fold increase in the *PcISPS* transcript levels was observed under constant light conditions. Consistent with the observed results circadian regulatory elements and putative light elements were found in the promoter region of PcISPS gene (Wilkinson et al., 2006; Loivamäki et al., 2007). Despite the circadian rhythm in *PcISPS* expression, the levels of PcISPS protein and its activity did not display diurnal fluctuations suggesting that the availability of DMAPP might be an important factor controlling circadian changes in isoprene emission (Loivamäki et al., 2007). Indeed, leaf DMAPP levels and isoprene emission were closely coordinated and showed similar diurnal variations (Mayrhofer et al., 2005). Light-dependent DMAPP production with highest levels from predawn to midday was also found in all isoprene and methylbutenol (a C_5 terpenoid similar to isoprene) emitting and nonemitting species (Brüggemann and Schnitzler, 2002a; Rosenstiel et al., 2002; Magel et al., 2006). Although the capacity to emit isoprene was clearly associated with elevated DMAPP levels (Rosenstiel et al., 2002), diurnal variations in cellular DMAPP levels may be a general characteristic of plant metabolism.

The last step in the MEP pathway is catalyzed by the IspH protein which converts 1-hydroxy-2-methyl-2-(*E*)-butenyl 4-diphospate into IPP and DMAPP (Fig. 1) at a 5:1 ratio (Adam et al., 2002). This IPP:DMAPP ratio within the cell is adjusted by isopentenyl diphosphate isomerase (IDI) which catalyzes the isomerization of IPP to DMAPP and might play a regulatory role in determining DMAPP levels. Analysis of IDI activity in oak leaves revealed that it is always higher than ISPS activity and that both activities can fully account for the observed isoprene emission. However, like ISPS, IDI activity does not display diurnal oscillations and cannot be responsible for diurnal variations in DMAPP levels and isoprene emission (Brüggemann and Schnitzler, 2002b), which can thus be attributed to circadian diurnal changes in the flux through the MEP pathway.

Feeding experiments with dideuterated deoxyxylulose (DOX-d_2) were performed to understand the regulatory mechanisms controlling the flux through the MEP pathway in *Eucalyptus globulus* (Wolfertz et al., 2004). The exogenous DOX-d_2 displaced the endogenous sources of carbon for isoprene biosynthesis but did not lead to an increase in isoprene emission suggesting that the DXS activity is negatively feedback modulated by the intermediates of the MEP pathway downstream from deoxyxylulose 5-phosphate (Wolfertz et al., 2004).

DMAPP is the last precursor of isoprene and can be formed not only from the plastidial MEP pathway but also from extrachloroplastic courses of carbon via the cytosolic MVA pathway. The contribution of different sources of DMAPP to isoprene emission was analyzed by comparing the labeling patterns of DMAPP and emitted isoprene in mature leaves of *Populus nigra* and *Phragmites australis* exposed to $^{13}CO_2$ (Loreto et al., 2004). A rapid, high level of ^{13}C labeling of emitted isoprene (90% in 15 min) along with a partial DMAPP labeling (28–36%) indicates that the labeled DMAPP represents a chloroplastic DMAPP which contributes to isoprene emission. Pretreatment of leaves with fosmidomycin resulted in residual isoprene emission and a very low ^{13}C labeling of the DMAPP pool suggesting that the residual isoprene is formed from the extra-chloroplastic sources and that at the very least, the MEP and MVA pathways are not cross-linked following inhibition of the plastidial pathway (Loreto et al., 2004; Nogués et al., 2006).

VI Conclusions

The crucial role of volatile terpenoids in the plant life cycle highlights the importance of understanding their biosynthesis and regulation of their formation and emission. Terpenoid emission from flowers and undamaged and herbivore attacked leaves often exhibit rhythmic patterns. However, reports concerning rhythmicity in terpenoids concentrate mainly on in planta chemical composition and emission profiles leaving the regulatory mechanisms of rhythmic emission still unknown. The recent discovery of the MEP pathway revealed that IPP and DMAPP could be synthesized not only in the cytosol but also in plastids and the exchange of intermediates between subcellular compartments could exist, thus adding an additional level of complexity to the investigation of the regulation of the flux towards volatile terpenoids. Despite the discovery of all the genes of the MEP pathway (see Fig. 1) the contribution of these genes to the regulation of the rhythmic emission of terpenoids in plants still remains to be determined. Further investigations of the MEP pathway enzymes and direct measurements of their levels and activities will provide new insights into the complex regulatory network of isoprenoid biosynthesis in plants. The determination of IPP and DMAPP pools in different cellular compartments in combination with feeding experiments with pathway specific precursors and inhibitors will uncover the contribution of each of the IPP biosynthetic pathways to the rhythmic emission of terpenoids. The integration of metabolic profiling with transcriptomic and proteomic datasets will help to elucidate the regulatory aspects of the isoprenoid network in plants. At present it is not known at what level the circadian clock controls terpenoid emission. Thus, the discovery of principles underlying circadian clocks and potential connections between circadian oscillations in gene expression and oscillations in metabolic activity are expected to yield important new insights into the role of the endogenous biological clock in the regulation of rhythmic emission of terpenoids.

Acknowledgements

This work was supported by grants from the National Science Foundation (Grant No. MCB-0615700) (N.D., D.R.), the Fred Gloeckner Foundation (N.D.) and the German Academic Exchange Service (N.D.).

References

Adam P, Hecht S, Eisenreich W, Kaiser J, Grawert T, Arigoni D, Bacher A and Rohdich F (2002) Biosynthesis of terpenes: studies on 1-hydroxy-2-methyl-2-(E)-butenyl 4-diphosphate reductase. Proc Natl Acad Sci USA 99: 12108–12113

Adam KP, Thiel R and Zapp J (1999) Incorporation of 1-[1-(13)C]Deoxy-D-xylulose in chamomile sesquiterpenes. Arch Biochem Biophys 369: 127–132

Affek HP and Yakir D (2002) Protection by isoprene against singlet oxygen in leaves. Plant Physiol 129: 269–277

Aharoni A, Giri AP, Deuerlein S, Griepink F, de Kogel WJ, Verstappen FWA, Verhoeven HA, Jongsma MA, Schwab W and Bouwmeester HJ (2003) Terpenoid metabolism in wild-type and transgenic Arabidopsis plants. Plant Cell 15: 2866–2884

Arimura G, Huber DPW and Bohlmann J (2004a) Forest tent caterpillars (*Malacosoma disstria*) induce local and systemic diurnal emissions of terpenoid volatiles in hybrid poplar (*Populus trichocarpa* x *deltoides*): cDNA cloning, functional characterization, and patterns of gene expression of (–)-germacrene D synthase, PtdTPS1. Plant J 37: 603–616

Arimura G, Kost C and Boland W (2005) Herbivore-induced, indirect plant defences. Biochim Biophys Acta 1734: 91–111

Arimura G, Ozawa R, Kugimiya S, Takabayashi J and Bohlmann J (2004b) Herbivore-induced defense response in a model legume: Two-spotted spider mites, *Tetranychus urticae*, induce emission of (*E*)-β-ocimene and transcript accumulation of (*E*)-β-ocimene synthase in *Lotus japonicus*. Plant Physiol 135: 1976–1983

Bach TJ, Boronat A, Campos N, Ferrer A and Vollack KU (1999) Mevalonate biosynthesis in plants. Crit Rev Biochem Mol Biol 34: 107–122

Boronat A (2010) The methylerythritol 4-phosphate pathway: regulatory role in plastid isoprenoid biosynthesis.

In: Rebeiz CA, Benning C, Daniel H, Hoober K, Lichtenthaler HK, Portis A, Tripathy B (eds) The Chloroplast: Basics and Application. Springer, Dordrecht, Netherlands, pp. 119–126

Brüggemann N and Schnitzler JP (2002a) Diurnal variation of dimethylallyl diphosphate concentrations in oak (*Quercus robur* L.) leaves. Physiol Plantarum 115: 190–196

Brüggemann N and Schnitzler JP (2002b) Relationship of isopentenyl diphosphate (IDP) isomerase activity to isoprene emission of oak leaves. Tree Physiol 22: 1011–1018

Burke C and Croteau R (2002) Interactions with the small subunit of geranyl diphosphate synthase modifies the chain length specificity of geranylgeranyl diphosphate synthase to produce geranyl diphosphate. J Biol Chem 277: 3141–3149

Cao X-L, Boissard C, Juan AJ and Hewitt CN (1997) Biogenic emissions of volatile organic compounds from gorse (*Ulex europaues*): diurnal emission fluxes at Kelling Heath, England. J Geophys Res 102: 18903–18915

Carretero-Paulet L, Ahumada I, Cunillera N, Rodriguez-Conception M, Ferrer A, Boronat A and Campos N (2002) Expression and molecular analysis of the *Arabidopsis* DXR gene encoding 1-deoxy-D-xylulose 5-phosphate reductoisomerase, the first committed enzyme of the 2-C-methyl-D-erythritol 4-phosphate pathway. Plant Physiol 129: 1581—1591

Carretero-Paulet L, Cairo A, Botella-Pavia P, Besumbes O, Campos N, Boronat A, and Rodriguez-Concepcion M (2006) Enhanced flux through the methylerythritol 4-phosphate pathway in Arabidopsis plants overexpressing deoxyxylulose 5-phosphate reductoisomerase. Plant Mol Biol 62: 683–695

Chappell J (1995) Biochemistry and molecular biology of the isoprenoid biosynthetic pathway in plants. Ann Rev Plant Physiol Plant Mol Biol 46: 521–547

De Moraes CM, Mescheer MC and Tumlinson JH (2001) Caterpillar-induced nocturnal plant volatiles repel nonspecific females. Nature 410: 577–580

Degen T, Dillmann C, Marion-Poll F and Turlings TCJ (2004) High genetic variability of herbivore-induced volatile emission within a broad range of maize inbred lines. Plant Physiol 135: 1928–1938

Dudareva N, Andersson S, Orlova I, Gatto N, Reichelt M, Rhodes D, Boland W and Gershenzon J (2005) The nonmevalonate pathway supports both monoterpene and sesquiterpene formation in snapdragon flowers. Proc Natl Acad Sci U S A 102: 933–938

Dudareva N, Martin D, Kish CM, Kolosova N, Gorenstein N, Faldt J, Miller B and Bohlman J (2003) (*E*)-β-Ocimene and myrcene synthase genes of floral scent biosynthesis in snapdragon: Function and expression of three terpene synthase genes of a new TPS-subfamily. Plant Cell 15: 1227–1241

Dudareva N, Negre F, Nagegowda DA and Orlova I (2006) Plant volatiles: recent advances and future perspectives. Crit Rev Plant Sci 25: 417–440

Dudareva N, Pichersky E and Gershenzon J (2004) Biochemistry of plant volatiles. Plant Physiol 135: 1893–1902

Eisenreich W, Schwarz M, Cartayrade A, Arigoni D, Zenk MH and Bacher A (1998) The deoxyxylulose phosphate pathway of terpenoid biosynthesis in plants and microorganisms. Chem Biol 5: R221–R233

Estévez JM, Cantero A, Reindl A, Reichler S and León P (2001) 1-Deoxy-D-xylulose-5-phosphate synthase, a limiting enzyme for plastidic isoprenoid biosynthesis in plants. J Biol Chem 276: 22901–22909

Funk JL, Jones CG, Baker CJ, Fuller HM, Giardina CP and Lerdau MT (2003) Diurnal variation in the basal emission rate of isoprene. Ecol Appl 13: 169–178

Gershenzon J and Dudareva N (2007) The function of terpene natural products in the natural world. Nature Chem Biol 3: 408–414

Geron C, Guenther A, Sharkey T and Arnts RR (2000) Temporal variability in basal isoprene emission factor. Tree Physiol 20: 799–805

Hansen U, Van Eijk J, Bertin N, Staudt M, Kotzias D, Seufert G, Fugit J-L, Torres L, Cecinato A, Brancaleoni E, Ciccioli, P and Bomboi, T (1997) Biogenic emissions and CO_2 gas exchange investigated on four Mediterranean shrubs. Atmos Environ 31: 157–166

Helsper JPF, Davies JA, Bouwmeester HJ, Krol AF and van Kampen MH (1998) Circadian rhythmicity in emission of volatile compounds by flowers of *Rosa hybrida* L. cv. Honesty. Planta 207: 88–95

Himmeldirk K, Kennedy IA, Hill RE, Sayer BG and Spenser ID (1996) Biosynthesis of vitamins B1 and B6 in *Escherichia coli*: concurrent incorporation of 1-deoxy-D-xylulose into thiamin and pyridoxol. J Chem Soc Chem Commun 1187–1188

Hsieh MH and Goodman HM (2005) The Arabidopsis IspH homolog is involved in the plastid nonmevalonate pathway of isoprenoid biosynthesis. Plant Physiol 138: 641–653

Julliard JH and Douce R (1991) Biosynthesis of the thiazole moiety of thiamine (vitamin B1) in higher plant chlroplasts. Proc Natl Acad Sci USA 88: 2042–2045

Kesselmeier J and Staudt M (1999) Biogenic volatile organic compounds (VOC): an overview on emission, physiology and ecology. J Atm Chem 33: 23–88

Kessler A and Baldwin IT (2001) Defensive function of herbivore-induced plant volatile emissions in nature. Science 291: 2141–2144

Knudsen JT and Gershenzon J (2006) The chemistry diversity of floral scent. In: Dudareva N and Pichersky E (eds) Biology of Floral Scent. CRC Press, Boca Raton, FL, pp. 27–52

Kunert M, Biedermann A, Koch T and Boland W (2002) Ultrafast sampling and analysis of plant volatiles by a hand-held miniaturized GC with pre-concentration unit: kinetic and quantitative aspects of plant volatile production. J Sep Sci 25: 677–684

Lange BM, Rujan T, Martin W and Croteau R (2000) Isoprenoid biosynthesis: the evolution of two ancient and distinct pathways across genomes. Proc Natl Acad Sci USA 97: 13172–13177

Laule O, Furholz A, Chang HS, Zhu T, Wang X, Heifetz PB, Gruissem W and Lange BM (2003) Crosstalk between cytosolic and plastidial pathways of isoprenoid biosynthesis in *Arabidopsis thaliana*. Proc Natl Acad Sci USA 100: 6866–6871

Lichtenthaler HK (1999) The 1-deoxy-D-xylulose-5-phosphate pathway of isoprenoid biosynthesis in plants. Annu Rev Plant Physiol Plant Mol Biol 50: 47–66

Lichtenthaler HK (2010) The non-mevalonate DOXP/MEP pathway of chloroplast isoprenoid and pigment biosynthesis. In: Rebeiz CA, Benning C, Daniel H, Hoober K, Lichtenthaler HK, Portis A, Tripathy B (eds) The Chloroplast: Basics and Application, Chapter 7. Springer, Netherlands, pp. 95–118

Lichtenthaler HK, Schwender J, Disch A and Rohmer M (1997a) Biosynthesis of isoprenoids in higher plant chloroplasts proceeds via a mevalonate independent pathway. FEBS Lett 400: 271–274

Lichtenthaler HK, Rohmer M and Schwender J (1997b) Two independent biochemical pathways for isopentenyl diphosphate (IPP) and isoprenoid biosynthesis in higher plants. Physiol Plant 101: 643–652

Lois LM, Rodriguez-Concepcion M, Gallego F, Campos N and Boronat A (2000) Carotenoid biosynthesis during tomato fruit development: regulatory role of 1-deoxy-D-xylulose 5-phosphate synthase. Plant J 22: 503–513

Loivamäki M, Louis S, Cinege G, Zimmer I, Fischbach RJ and Schnitzler JP (2007) Circadian rhythms of isoprene biosynthesis in grey poplar leaves. Plant Physiol 143: 540–551

Loreto F, Ciccioli P, Brancaleoni E, Cecinato A, Frattoni M and Sharkey T (1996) Different sources of reduced carbon contribute to form three classes of terpenoid emitted by *Quercus ilex* L. leaves. Proc Natl Acad Sci U S A 93: 9966–9969

Loreto F and Velikova V (2001) Isoprene produced by leaves protects the photosynthetic apparatus against ozone damage, quenches ozone products, and reduces lipid peroxidation of cellular membranes. Plant Physiol 127: 1781–1787

Loreto F, Pinelli P, Brancaleoni E and Ciccioli P (2004) ^{13}C labeling reveals chloroplastic and extra-chloroplastic pools of dimethylallyl pyrophosphate and their contribution to isoprene formation. Plant Physiol 135: 1903–1907

Loughrin JH, Manukian A, Heath RR, Turlings TCJ and Tumlinson JH (1994) Diurnal cycle of emission of induced volatile terpenoids herbivore-injured cotton plants. Proc Natl Acad Sci U S A 91: 11836–11840

Loughrin JH, Potter DA, Hamilton-Kemp TR and Byers ME (1997) Diurnal emission of volatile compounds by Japanese Beetle-damaged grape leaves. Phytochemistry 45: 919–923

Lu S, Xu R, Jia JW, Pang JH, Matsuda SPT and Chen XY (2002) Cloning and functional characterization of a beta-pinene synthase from *Artemisia annua* that shows a circadian pattern of expression. Plant Physiol 130: 1335–1348

Lumbreras V, Campos N and Boronat A (1995) The use of an alternative promoter in the *Arabidopsis thaliana HMG1* gene generates an mRNA that encodes a novel 3-hydroxy-3 methylglutaryl coenzymeA reductase isoform with an extended N-terminal region. Plant J 8: 541–549

Magel E, Mayrhofer S, Muller A, Zimmer I, Hampp R and Schnitzler J-P (2006) Photosynthesis and substrate supply for isoprene biosynthesis in poplar leaves. Atmos Environ 40: S138–S151

Mahmoud SS and Croteau RB (2001) Metabolic engineering of essential oil yield and composition in mint by altering expression of deoxyxylulose phosphate reductoisomerase and menthofuran synthase. Proc Natl Acad Sci USA 98: 8915–8920

Martin D, Gershenzon J and Bohlmann J (2003) Induction of volatile terpene biosynthesis and diurnal emission by methyl jasmonate in foliage of Norway spruce. Plant Physiol 132: 1586–1599

Mayrhofer S, Teuber M, Zimmer I, Louis S, Fischbach RJ and Schnitzler J-P (2005) Diurnal and seasonal variation of isoprene biosynthesis-related genes in grey poplar leaves. Plant Physiol 139: 474–484

Mercke P, Kappers IF, Verstappen FWA, Vorst O, Dicke M and Bouwmeester HJ (2004) Combined transcript and metabolite analysis reveals genes involved in spider mite induced volatile formation in cucumber plants. Plant Physiol 135: 2012–2024

Miller B, Madilao LL, Ralph S and Bohlmann J (2005) Insect-induced conifer defense: White pine weevil and methyl jasmonate induce traumatic resinosis, de novo formed volatile emissions, and accumulation of terpenoid synthase and putative octadecanoid pathway transcript in Sitka spruce. Plant Physiol 137: 369–382

Miller B, Oschinski C and Zimmer W (2001) First isolation of an isoprene synthase gene from poplar and successful expression of the gene in *Escherichia coli*. Planta 213: 483–487

Newman JD and Chappell J (1999) Isoprenoid biosynthesis in plants: Carbon partitioning within the cytoplasmic pathway. Crit Rev Biochem Mol Biol 34: 95–106

Nielsen JK, Jakobsen HB, Friis P, Hansen K, Moller J and Olsen CE (1995) Asynchronous rhythms in the emission of volatiles from *Hesperis matronalis* flowers. Phytochemistry 38: 847–851

Nogués I, Brilli F and Loreto F (2006) Dimethylallyl diphosphate and geranyl diphosphate pools of plant species characterized by different isoprenoid emissions.Plant Physiol 141: 721–730

Padhy PK and Varshney CK (2005) Isoprene emission from tropical tree species Environ Pollut 135: 101–109

Peñuelas J, Llusia J, Asensio D and Munne-Bosch S (2005) Linking isoprene with plant thermotolerance, antioxidants and monoterpene emissions. Plant Cell Environ 28: 278–286

Peñuelas J and Munne-Bosch S (2005) Isoprenoids: an evolutionary pool for photoprotection. Trends Plant Sci 10: 166–169

Pichersky E, Raguso RA, Lewinsohn E and Croteau R (1994) Floral scent production in *Clarkia* (Onagraceae). I. Localization and developmental modulation of monoterpene emission and linalool synthase activity. Plant Physiol 106: 1533–1540

Piel J, Donath J, Bandemer K and Boland W (1998) Mevalonate-independent biosynthesis of terpenoid volatiles in plants: Induced and constitutive emission of volatiles. Angew Chem Int Ed 37: 2478–2481

Qureshi N and Porter W (1981) Conversion of acetyl-coenzyme A to isopentenyl pyrophosphate. In: Porter JW and Spurgeon SL (eds) Biosynthesis of Isoprenoid Compounds. Wiley, New York, pp. 47–94

Raguso RA, Levin RA, Foose SE, Holmberg MW and McDade LA (2003) Fragrance chemistry, nocturnal rhythms and pollination "syndromes" in *Nicotiana*. Phytochemistry 63: 265–284

Rohmer M (1999) The discovery of a mevalonate-independent pathway for isoprenoid biosynthesis in bacteria, algae and higher plants. Nat Prod Rep 16: 565–574

Rosenstiel TN, Fisher AJ, Fall R and Monson RK (2002) Differential accumulation of dimethylallyl diphosphate in leaves and needles of isoprene and methylbutenol-emitting and nonemitting species. Plant Physiol 129: 1276–1284

Sasaki K, Ohara K and Yazaki K (2005) Gene expression and characterization of isoprene synthase from *Populus alba*. FEBS Lett 579: 2514–2518

Schnitzler J-P, Arenz R, Steinbrecher R and Lehning A (1996) Characterization of an isoprene synthase from leaves of *Quercus petraea* (Mattuschka). Liebl Bot Acta 109: 216–221

Schnurr JA and Guerra DJ (2000) The *CaMV-35S* promoter is sensitive to shortened photoperiod in transgenic tobacco. Plant Cell Rep 19: 279–282

Schwender J, Zeidler J, Gröner R, Müller C, Focke M, Braun S, Lichtenthaler FW and Lichtenthaler HK (1997) Incorporation of 1-deoxy-D-xylulose into isoprene and phytol by higher plants and algae. FEBS Lett 414: 129–134

Sharkey TD, Chen XY and Yeh S (2001) Isoprene increases thermotolerance of fosmidomycin-fed leaves. Plant Physiol 125: 2001–2006

Sharkey TD and Yeh S (2001) Isoprene emission from plants. Annu Rev Plant Physiol Plant Mol Biol 52: 407–436

Sharkey TD, Yeh S, Wiberley AE, Falbel TG, Gong DM and Fernandez DE (2005) Evolution of the isoprene biosynthetic pathway in Kudzu. Plant Physiol 137: 700–712

Silver GM and Fall R (1995) Characterization of aspen isoprene synthase, an enzyme responsible for leaf isoprene emission to the atmosphere. J Biol Chem 270: 13010–13016

Simkin AJ, Underwood BA, Auldridge M, Loucas HM, Shibuya K, Schmelz E, Clark DG and Klee HJ (2004) Circadian regulation of the PhCCD1 carotenoid cleavage dioxygenase controls emission of beta-ionone, a fragrance volatile of petunia flowers. Plant Physiol 136: 3504–3514

Staudt M and Bertin N (1998) Light and temperature dependence of the emission of cyclic and acyclic monoterpenes from holm oak (*Quercus ilex* L.) leaves. Plant Cell Environ 21: 385–395

Staudt M, Bertin N, Hansen U, Seufert G, Ciccioli P, Foster P, Frenzel B and Fugit J-L (1997) Seasonal and diurnal patterns of monoterpene emissions from *Pinus pinea* (L.) under field. Atmos Environ 31: 145–156

Takahashi S, Kuzuyama T, Watanabe H and Seto H (1998) A 1-deoxy-D-xylulose 5-phosphate reductoisomerase catalyzing the formation of 2-C-methyl-D-erythritol 4-phosphate in an alternative nonmevalonate pathway for terpenoid synthesis. Proc Natl Acad Sci USA 95: 9879–9884

Tholl D, Croteau R and Gershenzon J (2001) Partial purification and characterization of the short-chain prenyltransferases, geranyl diphospate synthase and farnesyl diphosphate synthase, from *Abies grandis* (grand fir). Arch Biochem Biophys 386: 233–242

Tholl D, Kish CM, Orlova I, Sherman D, Gershenzon J, Pichersky E and Dudareva N. (2004) Formation of monoterpenes in *Antirrhinum majus* and *Clarkia breweri* flowers involves heterodimeric geranyl diphosphate synthases. Plant Cell 16: 977–992

Vancanneyt G, Sanz C, Farmaki T, Paneque M, Ortego F, Castanera P and Sanchez-Serrano JJ (2001) Hydroperoxide lyase depletion in transgenic potato plants leads to an increase in aphid performance. Proc Natl Acad Sci USA 98: 8139–8144

Velikova V and Loreto L (2005) On the relationship between isoprene emission and thermotolerance in *Phragmites australis* leaves exposed to high temperatures and during the recovery from a heat stress. Plant Cell Environ 28: 318–327

Wildermuth MC and Fall R (1998) Biochemical characterization of stromal and thylakoid-bound isoforms of isoprene synthase in willow leaves. Plant Physiol 116: 1111–1121

Wilkinson MJ, Owen SM, Possell M, Hartwell J, Gould P, Hall A, Vickers C and Hewitt CN (2006) Circadian control of isoprene emissions from oil palm (*Elaeis guineensis*). Plant J 47: 960–968

Wolfertz M, Sharkey TD, Boland W and Kühnemann F (2004) Rapid regulation of the methylerythritol 4-phosphate pathway during isoprene synthesis. Plant Physiol 135: 1939–1945

Zeidler JG, Lichtenthaler HK, May HU and Lichtenthaler FW (1997) Is isoprene emitted by plants synthesized via the novel isopentenylpyrophosphate pathway? Z Naturforsch 52c: 15–23

Zeidler J and Lichtenthaler HK (2001) Biosynthesis of 2-methyl-3-buten-2-ol emitted from needles of *Pinus ponderosa* via the non-mevalonate DOXP/MEP pathway of isoprenoid formation. Planta 213: 323–326

Chapter 11

Tocochromanols: Biological Function and Recent Advances to Engineer Plastidial Biochemistry for Enhanced Oil Seed Vitamin E Levels

Henry E. Valentin*
Monsanto Co., Calgene Campus, 1920 Fifth Street, Davis, CA 95616, USA

Qungang Qi
Monsanto Co., 700 Chesterfield Parkway West, Chesterfield, MO 63017, USA

Summary

Tocochromanols (tocopherols and tocotrienols) are important lipophilic antioxidants for animals and humans. Their biological activity is expressed as vitamin E activity. This review describes some recent findings about tocochromanol function and their biosynthesis in plants, summarizes the current state of the art of tocochromanol pathway engineering, and compares different strategies to engineer the vitamin E content in cyanobacteria and plants with a focus on oilseed as target tissues. Limitations in our understanding of the tocochromanol biosynthetic pathway are discussed.

*Author for Correspondence, e-mail: henry.e.valentin@monsanto.com

Constantin A. Rebeiz et al. (eds.), The Chloroplast: Basics and Applications, pp. 155–169.

I Introduction

Tocochromanols (α-, β-, γ-, and δ-tocopherol, and α-, β-, γ-, and δ-tocotrienol) (Fig. 1) represent a class of lipophilic antioxidants that are only synthesized by oxygen evolving photosynthetic organisms, including plants, algae, and cyanobacteria. Green tissues of vascular plants and algae contain in their chloroplasts primarily α-tocopherol as regular constituents with some traces of γ-tocopherol (Lichtenthaler, 1969a, 1977). Tocotrienols can occur in leaves of some plants, but α-tocopherol remains the major compound. Tocotrienols are distinguished from tocopherols by the presence of three *trans* double bonds in the isoprenoid derived side chain and are most commonly found besides α-tocopherol in monocotyledoneous plants, especially in oil seeds. The Greek prefix refers to the grade of methylation of the aromatic head group with α-tocopherol and α-tocotrienol representing the highest methylated tocochromanols and δ-tocopherol as well as δ-tocotrienol having the lowest grade of methylation. The beneficial biological activity of tocochromanols on animals and humans was discovered nearly a century ago (Evans and Bishop, 1922) and is commonly expressed as vitamin E activity (IUPC-IUP, 1982). It is most commonly determined using the rat fetal resorption assay, by dialuric acid induced hemolysis, or by monitoring the onset of nutritional muscular dystrophy (Weiser et al., 1986; Sheppard et al., 1993; Mino et al., 1988; Chow, 2001).

It is believed that the beneficial affects of tocochromanols are derived from their radical scavenging activity in lipophilic environments, resulting in the stabilization of polyunsaturated fatty acids in membrane lipids. More recent evidence suggests that tocopherols also play a role in intracellular signaling influencing seed germination, seedling development, and stress responses (Maeda and DellaPenna, 2007; Munné-Bosch and Alegre, 2002). In animals, all tocochromanols are taken up with similar rates (Gloor et al, 1966; Traber and Arai, 1999). Although γ-tocopherol is superior to α-tocopherol in detoxifying electrophiles in lipophilic environments (Cooney et al., 1993) it exhibits only 10–20% of the vitamin E activity of α-tocopherol. This difference in activity is presumably due to the preferred retention and distribution of α-tocopherol throughout the mammalian body (Hosomi et al., 1997; Burton et al., 1998; Traber and Sies, 1996).

Recent research has suggested that in lipophilic environments tocotrienols are more effective in protecting against lipid peroxidation than α-tocopherol (Theriault et al., 1999). Nevertheless, tocotrienols appear to have only about one third or less of the vtamin E activity of tocopherols in animals and humans (Chow, 2001; Hunter and Cahoon, 2007). As a result of this discrepancy between the radical scavenging activity of the tocopherol isoforms in lipophilic environments, and their vitamin E activity in animals and humans, other functions such as signal transduction have been discussed to explain the unexpected high vitamin E activity of α-tocopherol (Brigelius-Flohé and Davies, 2007; Brigelius-Flohé et al., 2002; Rimbach et al., 2002 and references cited therein). Recent results obtained with mutants in mice, corn, and potatoes appear to link tocopherols with sugar metabolism. α-Tocopherol binding protein knockout mice exhibited serum α-tocoperol levels reduced to one-third of WT-levels. These mice did not exhibit any signs of vitamin E deficiency, but had serum glucose levels reduced under all tested feeding conditions and intraperitoneal injected glucose was metabolized significantly faster than in WT-mice (Birringer et al., 2007). Plant mutants harboring a knockout mutation in tocopherol cyclase resulting in loss of tocochromanol accumulation exhibited a deficiency in sucrose export from leaf source tissues (Provencher et al., 2001; Kumar et al., 2005; Hofius et al., 2004). If these data point towards a role of tocopherols in sugar metabolism, or if the observed phenotypes are just coincidental consequences of altered membrane fluidity or membrane integrity remains to be seen in future experiments.

Abbreviations: ChlP – chlorophyllase; GGH – geranylgeranyl-reductase; GGDP – geranylgeranyl-diphosphate; HGA – homogentisic acid; HGGT – homogentisate geranylgeranyltransferase; HPPD – p-hydroxyphenyl pyruvate dioxygenase; HST – homogentisate solanelsyltransferase; 2M6PBQ – 2-methyl-6-phytylbenzoquinol; 2M6SBQ – 2-methyl-6-solanesylbenzoquinol; PDP – phytoldiphosphate; pHPPA – p-hydroxyphenylpyruvate; SDP – solanesyldiphosphate; TyrA – bifunctional chorismate mutase/prephenate dehydrogenase; VTE1 – tocopherol cyclase; VTE2 – homogentisate phytyltransferase; VTE3 – 2-methyl-6-phytylbenzoquinol methyltransferase; VTE4 – γ-tocopherol methyltransferase; WT – wild type

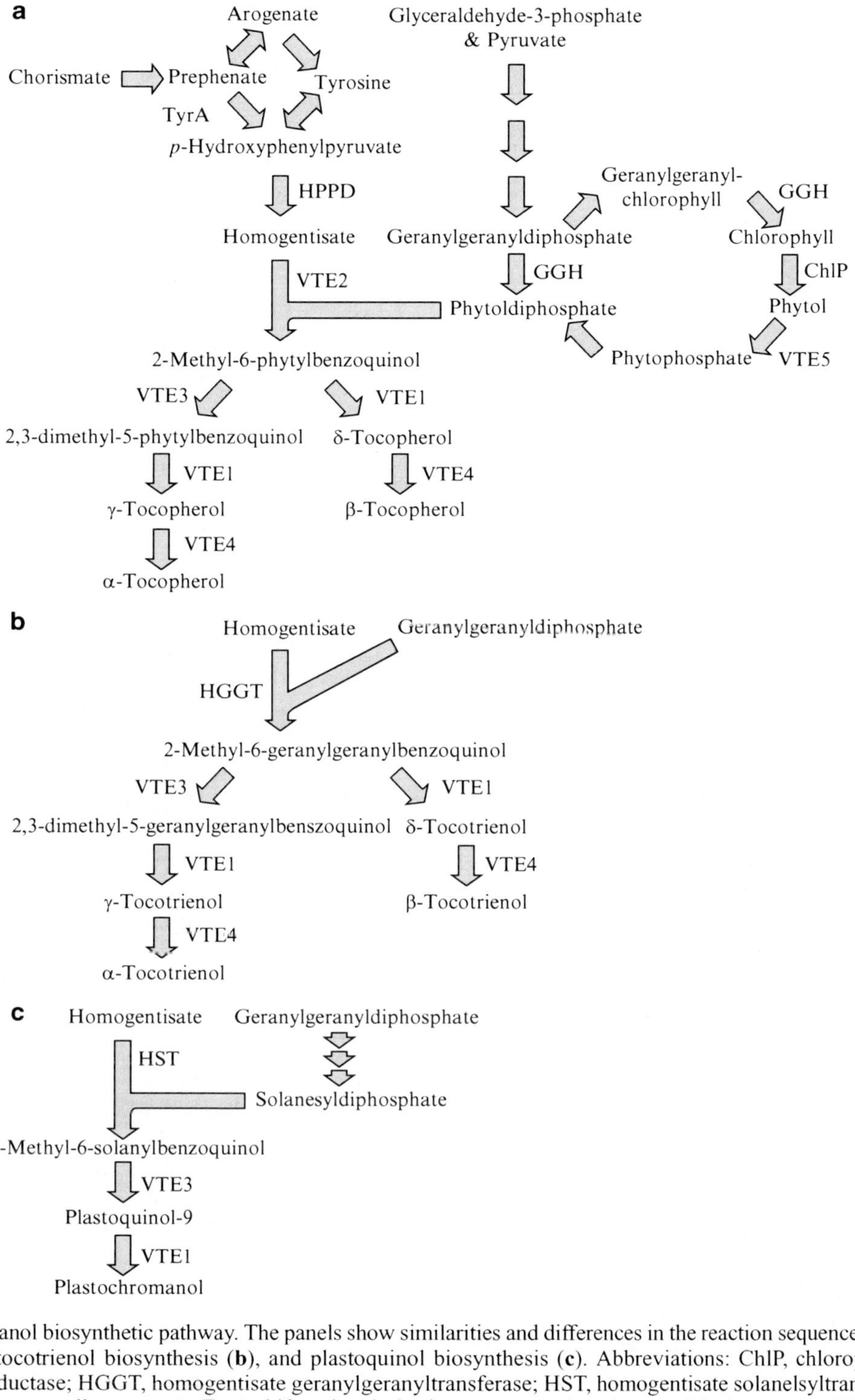

Fig. 1. Tocochromanol biosynthetic pathway. The panels show similarities and differences in the reaction sequences for tocopherol biosynthesis (**a**), tocotrienol biosynthesis (**b**), and plastoquinol biosynthesis (**c**). Abbreviations: ChlP, chlorophyllase; GGH, geranylgeranyl-reductase; HGGT, homogentisate geranylgeranyltransferase; HST, homogentisate solanelsyltransferase; HPPD, p-hyroxyphenylpyruvate dioxygenase; TyrA, bifunctional chorismate mutase/prephenate dehydrogenase; VTE1, tocopherol cyclase; VTE2, homogentisate phytyltransferase; VTE3, 2-methyl-6-phytylbenzoquinol methyltransferase; VTE4, γ-tocopherol methyltransferase.

The vitamin E activity of 1 mg chemically synthesized all racemic α-tocopherol acetate is defined as 1 international unit (IU). Naturally occurring sterioisomeric *2R,4′R,8′R*-α-tocopherol is 1.5-fold more active than synthetic α-tocopherol acetate (Chow, 2001; Sheppard et al., 1993). The National Institute of Health (NIH) currently suggests a recommended daily allowance (RDA) of 15 mg isomeric pure α-tocopherol, or 22.4 IU for humans (http://ods.od.nih.gov/factsheets/VitaminE.asp). This amount of vitamin E activity is typically found in the average Western diet. Accordingly, pathogenicity as a result of vitamin E deficiency is rare in western cultures and limited to individuals with metabolic impairments, such as deficiencies in fat absorption due to the inability to secrete bile acids (Triantafillidis et al., 1998), deficiencies in fat metabolism, individuals with rare genetic abnormalities in the α-tocopherol transfer protein (Cavalier et al., 1998), and prematurely born infants with birth weights of less than 1,500 g (Farrell and Roberts, 1994).

Vitamin E supplementation at therapeutic doses (100–1,000 IU) has been associated with cancer reduction, improved immune response, and cardiovascular health benefits (Buring and Hennekens, 1997; Tangney, 1997; Bramley et al., 2000). However, some more recent epidemiologic studies failed to confirm those cardiovascular (Yusuf et al., 2000; Waters et al., 2002) and cancer reducing health benefits (Graham et al., 1992; Wu et al., 2002). Tocotrienols have been reported to have health benefits that are distinct from the beneficial effects of tocopherols, such as hypocholesterolemic health benefits (Qureshi and Qureshi, 1993; Qureshi et al., 1995; Brown and Goodman, 1998).

Large vitamin E volumes are required in the animal feed industry where high doses of vitamin E are applied to improve the quality and shelf life of meat (Sanders et al., 1997), as well as in pharmaceutical and cosmetic applications (Edwards, 2001). The majority of Vitamin E requirements for the animal feed market and cosmetic applications are satisfied with synthetic α-tocopherol acetate produced from isophytol and trimethylhydroquinone. These tocopherol precursors originate from fossil fuels. With fossil fuel resources declining, and the demand for tocopherol applications increasing due to further growth in world population, in addition to changes in food consumption with preference towards increased meat consumption, production cost for chemically derived tocopherols are expected to increase rather than to decline in the foreseeable future.

Natural tocopherols are synthesized by plants and by some blue-green photosynthetic bacteria (Cyanobacteria) which possess an oxygenic photosynthesis with two photosynthetic light reactions (Lichtenthaler, 1969a). The level of α-tocopherol in green plant tissues is relatively low. During chlorophyll breakdown and chromoplast formation in ripening green fruits, the synthesis of α-tocopherol continues in parallel to the accumulation of secondary carotenoids with the result that red or orange-red fruits, e.g. of bell pepper, have relatively good α-tocopherol levels (Lichtenthaler, 1969b, 1977). Biotechnologically engineered crops have the potential to serve as an alternative vitamin E source with production cost substantially below chemically synthesized tocopherols. The highest naturally occurring tocopherol concentrations are found in oil seeds of various plants and in chloroplasts of sun-exposed green leaves (sun leaves) of various trees such as beech (*Fagus sylvatica*) or in green leaves of several year old fig trees (*Ficus elastica*). In these sun-exposed or several year old green leaves α-tocopherol is accumulated in high amounts in the large osmiophilic plastoglobuli of the chloroplast stroma (Lichtenthaler, 2007) and its level can reach or exceed the level of chlorophyll *b* (Lichtenthaler, 2007). On a dry weight basis the α-tocopherol level of sun leaves reaches similar levels as in oil seeds (see also Table 1). However, the isolation of α-tocopherol from leaves is economically not feasible.

Thus, the best source for natural tocopherols are the seeds of oilseed crops. For these crops processing capabilities for oil extraction and in some cases for tocochromanol enrichment (e.g. soybean and oil palm) are well established. The vast majority of commercially sold natural tocopherol is produced from soybean seeds. Due to its limited availability and consequently its premium price, it is predominantly sold for human applications. As a result, current processing technology for natural tocopherols is designed to enrich the tocopherol fraction during the soybean oil refining process where tocopherols are collected with the deodorizer distillate. The deodorizer distillate is subjected to additional purification steps to obtain

Table 1. Literature reports about maximum tocochromanol levels achieved through metabolic engineering compared to wild-type tocochromanol levels, and the highest natural tocochromanol source, wheat germ oil.

Host organism/crop	Tissue	Tocopherol level (ng/mg fresh tissue)	Fold increase	Max. tocopherol level (ng/mg fresh tissue)	Tocotrienol fraction (%)	Trans genes	Reference
Synechocystis sp. *PCC6803*	na	133	16.5	2,200	nd	HPPD, tyrA, VTE2, GGH	Karunanandaa et al., 2005
A. thaliana	Leaf	40–65[a]	15	700–900	85	*HGGT*	Cahoon et al., 2003
A. thaliana	Leaf	8.4	4.4	38.2	nd	*VTE2*	Collakova and DellaPenna, 2003a
A. thaliana	Seed	349	1.4	478	nd	*VTE2*	Collakova and DellaPenna, 2003a
A. thaliana	Seed	527	1.8	926	nd	*VTE2*	Savidge et al., 2002
A. thaliana	Seed	541	5	2,710	62	*HPPD, tyrA, VTE2,*	Karunanandaa et al., 2005
Nicotiana tabacum	Leaf	55[a]	10	552[a]	90	*HPPD, PDH*	Rippert et al., 2004
Canola	Seed	227	3.7	829	46	*HPPD, tyrA, VTE2,*	Karunanandaa et al., 2005
Glycine max	Seed	326	14.7	4,800	94	*HPPD, tyrA, VTE2, GGH*	Karunanandaa et al., 2005
	Seed	350	11.4	3,000–4,000	Unknown	*HGGT*	E.B. Cahoon, 2008, personal communication
Zea mays	Seed	50–70	4–6	250–334	74	*HGGT*	Cahoon et al., 2003
Wheat germ oil	na	2,778	na	na	na	na	Slover, 1971

Tocochromanol contents are reported as levels referred to fresh sample weight, unless indicated otherwise. The number of transgenes referres to pathway genes excluding the selectable marker. Abbreviations: [a], tocopherol content referred to dry cell matter; *GGH*, geranylgeranyldiphosphate dehydrogenase; *HGGT*, homogentisate geranylgeranyl transferase; *PDH*, prephenate dehydrogenase; *HPPD*, *p*-hydroxyphenylpyruvate dioxygenase; na, not applicable; nd, below the limit of detection, or not reported; *tyrA*, bacterial bifunctional prephenate dehydrogenase; *VTE2*, homogentisate phytyltransferase.

tocopherol rich oil. For these reasons, oilseed, and particularly soybean represent the best target for the biotechnological cost efficient production of tocopherols. As indicated above, α-tocopherol has the highest vitamin E activity among all tocopherol isoforms making it the preferred target for engineering transgenic plants to produce vitamin E. Other applications may benefit from slightly modified strategies. For example, stabilization of biodiesel may be most efficiently accomplished by increasing γ-tocopherol levels in the oil as γ-tocopherol is a more efficient radical scavenger in lipophilic environments.

II Tocochromanol Biosynthesis and Regulation

Within chloroplasts tocopherols are bound to the chloroplast envelope, the photochemically active thylakoids and large amounts are found in the osmiophilic plastoglobuli of the plastid stroma which increase in number and size with increasing age and α-tocopherol content of leaves (Lichtenthaler, 2007 and literature cited therein). The enzymatic reactions in tocochromanol biosynthesis were identified approximately 20 years ago (Schultz-Siebert et al., 1987; Soll et al., 1980). In plants, tocochromanol biosynthesis takes place in the plastids, and all enzymes with the exception of the tocopherol cyclase are associated with the plastid envelope (Soll et al., 1980, 1985; Vidi et al., 2006). The tocopherol cyclase appears to be bound to the plastoglobules, which represent plastidic lipid reservoirs for excess plastid lipids, predominantly for α-tocopherol and plastoquinone-9 (Lichtenthaler, 2007), that are surrounded by a lipoprotein matrix that is associated with plastidial thylakoid membranes (Jotham et al., 2007; Vidi et al., 2006). Little is known about the localization of the tocopherol biosynthetic enzymes in cyanobacteria.

Homogentisic acid (HGA) and an isoprenoid diphosphate such as phytoldiphosphate (PDP) or geranylgeranyldiphosphate (GGDP) serve as substrates for tocopherol and tocotrienol biosynthesis, respectively. In plants HGA is derived from tyrosine. A transaminase converts tyrosine to *p*-hydroxyphenylpyruvate (pHPPA), which is then converted to HGA through the action of the *p*-hydroxyphenylpyruvate dioxygenase (HPPD) (Fig. 1a). In bacteria and yeast pHPPA is synthesized from chorismate and prephenate, respectively. The bacterial TyrA enzyme represents a bifunctional chorismate mutase, prephenate dehydrogenase. pHPPA is subsequently converted to HGA catalyzed by HPPD. PDP is derived from the plastidial isoprenoid biosynthesis, the Desoxy-d-xylulose diphosphate/Methylerythritol phosphate pathway (DOXP/MEP pathway) which has been discovered recently by a joint action of the groups of Rohmer and Lichtenthaler (reviewed with further details by Lichtenthaler, 1999, 2007, 2010; Rohmer, 2003; Boronat, 2010). In addition, PDP can also derive from a Salvage pathway using phytol coming from turnover and breakdown of chlorophyll (Ischebeck et al., 2006; Valentin et al., 2006; Rise et al., 1989; Fig. 1a).

The first committed reaction in tocochromanol biosynthesis is the prenylation of HGA with PDP or GGDP resulting in the formation of 2-methyl-6-phytylbenzoquinol (2M6PBQ) or 2-methyl-6-geranylgeranylbenzoquinol (2M6GGB), respectively. The prenylation with PDP is typically catalyzed by the homogentisate phytyltransferase encoded by *slr1736* in *Synechocystis* (*Syn-vte2*), or in Arabidopsis by the *VITAMIN E2* gene (*At-VTE2*) (Collakova and DellaPenna, 2001; Schledz et al., 2001; Savidge et al., 2002). While in vivo results and in vitro data suggest that VTE2 can catalyze the prenylation with GGDP as well (Rippert et al., 2004; Karunanandaa et al., 2005; Sadre et al., 2006), a different class of homogentisate prenyltransferase enzyme with apparent substrate preference for GGDP has been isolated from barley (Cahoon et al., 2003). This type of enzyme has been designated homogentisate geranylgeranyl transferase (HGGT) and appears to be responsible for tocotrienol biosynthesis in monocotyledoneous plants. Recently a third type of homogentisate prenyltrasferase from *Arabidopsis thaliana* and *Chlamydomonas reinhardtii* was discovered and functionally characterized (Sadre et al., 2006; Venkatesh et al., 2006). These enzymes exhibited a strong substrate preference for solanesyldiphosphate (SDP) under certain in vitro assay conditions resulting in the formation of 2-methyl-6-solanesylbenzoquinol (2M6SBQ), a plastoquinol-9 precursor (Sadre et al., 2006). Accordingly, these enzymes are referred to as homogentisate solanesyltransferase (HST). Surprisingly, overexpression of *HST* in Arabidopsis resulted in increased tocopherol levels as well as increased plastoquinol-9 (Sadre et al., 2006; Venkatesh et al., 2006). However, Arabidopsis seed harboring seed-specific and constitutive *HST* antisense constructs did not show a reduction in tocopherol levels (Sadre et al., 2006). This appears consistent with a substrate preference of HST for 2M6SBQ.

The metabolic fate of the VTE2 reaction product appears to be determined by the relative enzymatic activities of the tocopherol cyclase (VTE1), and the 2M6PBQ-methyltransferase (VTE3). It has been demonstrated recently that overexpression of *VTE3* in soybean seed is sufficient to direct nearly all tocopherols into the pool of α-, and γ-tocopherols (van Eenennaam et al., 2003a, b). Non-transgenic soybean contain on average 30–35% β-, and δ-tocopherol. Interestingly, VTE3 appears to have developed independently in plants and green photosynthetic bacteria, as suggested by their low level of sequence similarity (Cheng et al., 2003; van Eenennaam et al., 2003b). Independent characterization of VTE3 in two different laboratories revealed recently that this enzyme does not only catalyze the methylation of the tocopherol precursors 2M6PBQ (Fig. 1a) and 2M6GGBQ (Fig. 1b), but also methylates the plastoquinol precursor 2M6SBQ (Fig. 1c) (Motohashi et al., 2003; Cheng et al., 2003). The essential function of VTE3 in two independent metabolic pathways could increase the challenge of tocopherol pathway engineering, as altered tocopherol intermediate pools as a result of enhanced tocopherol flux could lead to substrate competition between 2M6PBQ, or 2,2M6GGBQ and the plastoquinol precursor 2M6SBQ.

The VTE3 reaction product, 2,3-dimethyl-5-phytylbenzoquinol (2,3DM5PBQ), serves as substrate for VTE1 to form γ-tocopherol (Fig. 1b). This enzyme also utilizes 2M6PBQ and plastoquinol-9 as substrate, resulting in the formation of δ-tocopherol and plastochromanol, respectively (Kumar et al., 2005). The function of VTE1 in tocopherol biosynthesis was discovered with the help of mutational analysis in *Synechocystis* and *Arabidopsis* (Subramaniam et al., 2001; Porfirova et al., 2002; Sattler et al., 2003). Interestingly, the coding sequence of corn *VTE1* had been discovered in a sucrose transport deficient corn mutant previously (Provencher et al., 2001). However, the link between phenotype and tocochromanol biosynthesis was not made at the time and continues to be elusive. Additional experiments may be required to determine whether the observed sugar transport deficient phenotype is a result of the lack to tocochromanols, a lack of plastoquinol and derivatives, or whether it is derived from a yet to be identified function of VTE1.

The final reaction to convert γ-tocopherol to α-tocopherol is catalyzed by the γ-methyltransferase encoded by *VTE4* (Fig. 1b). By analogy to the methylation of γ-tocopherol, VTE4 can also methylate δ-tocopherol to form β-tocopherol. The *VTE4*-gene was also discovered through bioinformatics analysis and confirmed by mutational analysis in *Synechocystis* (Shintani and DellaPenna, 1998). For tocotrienol synthesis, it appears that 2M6GGBQ undergoes the same subsequent methylation and cyclization reactions as 2M6PBQ during synthesis of the corresponding tocopherols to produce α-, β-, γ- and δ-tocotrienols (Munné-Bosch and Alegre, 2002; Soll and Schultz, 1980; d'Harlingue and Camara, 1985; Stocker et al., 1993; Cheng et al., 2003).

III Tocochromanol Pathway Engineering for Enhancement of Vitamin E

The vitamin E activity of tocochromanols varies with the number and position of methyl groups on the aromatic head group, and the degree of unsaturation of the isoprenoid side chain. The highest vitamin E activity is provided by α-tocopherol which exhibits the highest degree of methylation on the aromatic head group, and has a fully saturated isoprenoid side chain. Oil seeds, which are among the richest natural tocochromanol sources, frequently contain a substantial fraction of their tocochromanols in the biologically less active form of γ-tocopherol and the seed of monocotyledone plants contain large amounts of tocotrienols which appear to have less vitamin E activity than α-tocopherol as well (Chow, 2001). As a result, vitamin E activity can be enhanced via three principally different strategies, (i) the conversion of all tocochromanols into α-tocopherol, or (ii) by increasing the concentration of all tocopherol isoforms, and (iii) by a targeted increase of tocotrienols. The first strategy is predominantly applicable to crop seed, since leaves and cyanobacteria produce mainly α-tocopherol. In plants, optimization of seed tocopherol composition has been employed successfully in Arabidopsis (Shintani and DellaPenna, 1998; van Eenennaam et al., 2003a), canola, and soybean (van Eenennaam et al., 2003a, b). The second strategy has been utilized to increase the vitamin E content of *Synechocystis*, Arabidopsis, tobacco, canola, and soybean seed (Savidge et al., 2002; Collakova and DellaPenna, 2003b; Rippert et al., 2004; Karunanandaa et al., 2005). A focused effort to increase tocotrienol levels (strategy iii) in Arabidopsis, corn and soybean has been also pursued by Cahoon and collaborators (Cahoon et al., 2003, http://www.imba.missouri.edu/report/2005_1_2.php).

IV Optimized Tocochromanol Composition

Arabidopsis and canola seed contain predominantly γ-tocopherol, which has only 10% of the vitamin E activity of α-tocopherol. Transgenic expression of the Arabidopsis *VTE4* (*At-VTE4*) in Arabidopsis and canola seed resulted in the formation of 95% and 100% α-tocopherol, respectively (Shintani and DellaPenna, 1998; van Eenennaam et al., 2003a). This correlates with a tenfold and a twofold increase in vitamin E activity in Arabidopsis and canola, respectively. Total tocopherol contents in transgenic and non-transgenic control seed were equivalent.

Soybean seed contain a more diverse tocopherol composition, consisting of 10–20% α-tocopherol, 2–5% β-tocopherol, 60–70% γ-tocopherol, and 20–30% δ-tocopherol. Expression of *At-VTE4* resulted in conversion of all γ-tocopherol to α-tocopherol, and the simultaneous complete conversion of δ-tocopherol to β-tocopherol (Van Eenennaam et al., 2003b). This result was consistent with utilization of δ-tocopherol and γ-tocopherol as substrates for VTE4 as shown by in vitro enzyme assays (Shintani and DellaPenna, 1998). Indeed in vitro enzyme assays suggested that neither β-tocopherol nor δ tocopherol can be utilized by VTE3 (Cheng et al., 2003). It was not predictable therefore whether transgenic expression of VTE3 would divert the metabolic flux from δ-, and β-tocopherol towards γ-, and α-tocopherol.

However, transgenic expression of *At-VTE3* in soybean seed resulted in nearly complete conversion of all tocopherols to γ-, and α-tocopherol (van Eenennaam et al., 2003b). These results indicate an important role of VTE3 in controlling the flux from 2M6PBQ to γ-, and α-tocopherol. Combined expression of *At-VTE3* and *At-VTE4* in soybean seed resulted in the conversion of >95% of tocopherols to α-tocopherol. In soybean this correlates with a fivefold increase in seed vitamin E activity (van Eenennaam et al., 2003b).

An alternative potentially more efficient strategy to engineer tocochromanol composition by using designer transcription factors has been explored by van Eenennaam et al. (2004). If such a transcription factor could be designed to recognize conserved DNA sequences in the 5′-regions of both methyltransferases, the expression of a single transcription factor gene could potentially be sufficient to obtain a similar result as through the transgenic expression of VTE3 and VTE4 described above. A transcription factor that would recognize conserved sequences in the 5′-regions of the two methyltransferases and other genes that are involved in the biosynthesis of tocochromanol precursors could potentially even combine multiple strategies to increase the vitamin E content. However, the use of zinc finger transcription factors cited above provided only incremental increases in α-tocopherol, requiring additional optimization to convert all tocochromanols to α-tocopherol (van Eenennaam et al., 2004).

Since the reactions in tocotrienol biosynthesis are believed to be equivalent to the reactions found in tocopherol biosynthesis (Fig. 1a and b) it can be assumed that β-, γ-, and δ-tocotrienol can be converted to α-tocotrienol by overexpressing VTE4 and potentially VTE3. However a conversion of tocotrienols to the more biologically active tocopherols through a direct reduction of the isoprenoid side chain is not described in the literature.

V Enhancement of Total Tocochromanol Content

Although most of the biochemical reactions leading to the formation of the eight main tocochromanols have been discovered some 20 years ago, the regulatory mechanism of the overall pathway remains poorly understood. It has been hypothesized that the first committed reaction, catalyzed by VTE2 (Fig. 1a) or HGGT (Fig. 1b), plays a regulatory role in tocochromanol biosynthesis. Indeed, transgenic expression of *Arabidopsis thaliana VTE2* (*At-VTE2*) or *HGGT* in *Arabidopsis* and other host plants resulted in substantially increased tocochromanol levels (Cahoon et al., 2003; Karunanandaa et al., 2005). *Arabidopsis* and soybean seed tocopherol levels increased up to 1.8- and 1.4-fold for the best performing *At-VTE2* expressing events, respectively (Savidge et al., 2002; Collakova and DellaPenna, 2003a; Karunanandaa et al., 2005). In *Arabidopsis* leaf, this effect was even more pronounced, and total leaf tocopherol levels increased up to 4.4-fold (Collakova and DellaPenna, 2003a). Interestingly the effect of transgenic *At-VTE2* expression in Arabidopsis leaf was still surpassed by the impact of environmental factors, such as high light, and nutrient deficiency, which resulted in up to 18-fold increased leaf tocopherol levels (Collakova and DellaPenna, 2003b). This observation might indicate either that the potential of *At-VTE2* overexpression was not utilized to its full potential under non-stressed conditions, or that other reactions, or factors contribute to tocopherol flux regulation.

Feeding experiments using soybean suspension cultures reported by Karunanandaa et al. (2005) suggest that tocopherol biosynthesis is further limited by the availability of its precursors HGA and PDP. A limitation in tocopherol flux, particularly by HGA is also supported by data obtained through transgenic expression of HPPD, and through combined expression of *HPPD* and a prephenate dehydrogenase (*tyrA*) (Tsegaye et al., 2002; Rippert et al., 2004; Karunanandaa et al., 2005). In the photosynthetic model organism *Synechocystis* sp. PCC 6803 expression of *HPPD* resulted in a brown coloration of the culture supernatant, indicative of HGA excretion, and a sevenfold increase in total tocochromanol levels (Karunanandaa et al., 2005; Qi et al., 2005). In *Arabidopsis* seed, transgenic expression of *HPPD* increased seed tocopherol levels by up to 0.28-fold only (Tsegaye et al., 2002; Karunanandaa et al., 2005). The relatively low impact of *HPPD* expression in Arabidopsis and other plants is likely due to the fact that plants synthesize HGA via arogenate (Fig. 1). Bacteria in contrast synthesize HGA directly from prephenate via pHPPA, catalyzed by the bifunctional prephenate dehydrogenase (TyrA) (Fig. 1). This pathway is much shorter than the reaction sequence leading to HGA biosynthesis in plants and does not include tyrosine as an intermediate. As a result the tyrosine feedback regulation, which regulates tyrosine biosynthesis, and may lead to limiting pHPPA biosynthesis in plants, does not affect the bacterial pathway for HGA biosynthesis. Therefore, the bacterial pathway is much more efficient at increasing the HGA pools when compared to the plant tyrosine-derived pathway. Thus it is not surprising that combined expression of *tyrA* and *HPPD* in plants resulted in a dark seed phenotype that correlated with 60- and 800-fold increased HGA-levels in mature *Arabidopsis* and soybean seed, respectively (Karunanandaa et al., 2005). The high HGA seed phenotype correlated with an average seed tocopherol increase of 1.8-, and 2.6-fold in *Arabidopsis* and soybean, respectively (Karunanandaa et al., 2005). In tobacco leaf total tocopherol increases of up to tenfold were observed with a similar gene combination (Rippert et al., 2004). Interestingly, Karunanandaa et al. and Rippert et al. found that combined expression of HPPD and a prephenate dehydrogenase, or a bifunctional prephenate dehydrogenase led to the accumulation of substantial tocotrienol levels, suggesting that PDP was limiting tocopherol biosynthesis in high HGA plants (Rippert et al., 2004; Karunanandaa et al., 2005).

It is generally accepted that PDP for tocopherol biosynthesis is derived from GGDP which is catalyzed by the geranylgeranyl reductase (GGH, Fig. 1a). This hypothesis is derived from labeling experiments using chloroplast preparations (Soll et al., 1983), and from enzyme assays performed with recombinant *E. coli* expressed Arabidopsis *GGH* (*At-GGH*) (Keller et al., 1998). However, transgenic expression of *At-GGH* in high HGA soybean co-expressing *HPPD*, *tyrA*, and *VTE2* have so far failed to divert the metabolic flux from tocotrienol formation to tocopherol formation (Karunanandaa et al., 2005).

More recent data obtained from an *Arabidopsis* mutant screen (Norris et al., 2004; Valentin et al., 2006), and biochemical data obtained with extracts from Arabidopsis seedlings (Ischebeck et al., 2006) indicate that a substantial portion of phytol for tocopherol biosynthesis may be recycled from chlorophyll degradation, or chlorophyll turnover. The *Arabidopsis* phytol kinase mutant *(vte5-1)* described by Valentin et al. (2006) accumulated only 20% of Wt seed tocopherol levels, suggesting that in non-mutant Arabidopsis seed, at least 80% of phytol used for tocopherol biosynthesis, passes through a free phytol intermediate (Norris et al., 2004). Interestingly, the presence of strong phytol kinase activity in spinach leaf chloroplast preparations was reported already 30 years ago (Soll et al., 1980). These data are supported by a recent publication of Ischebeck et al. (2006), who identified two distinct kinase activities, a CTP dependent phytol kinase, and a phytolphosphate kinase that utilizes CTP, GTP, UTP, and ATP as co-substrates. However, a free phytol intermediate as major tocopherol precursor appears to contrast the previously generally accepted belief that PDP for tocopherol and chlorophyll biosynthesis is provided through direct reduction of GGDP catalyzed by GGH (Schultz et al., 1985; Grusak and DellaPenna, 1999; Hofius and Sonnewald, 2003). The commonly accepted model of a direct reduction of GGDP by GGH appears to be supported by the observation that GGH knockout mutants and antisense experiments performed in tobacco and *Synechocystis* clearly demonstrate, that GGH is essential for tocopherol biosynthesis in bacteria and plants (Shpilyov et al., 2005; Graßes et al., 2001). These apparently contradicting data might be explained by a GGH substrate preference for geranylgeranyl chlorophyll rather than GGDP, or through a preferred localization of GGH in the plastidial stroma. In 1983 Soll et al. reported the presence of two independent pathways for the reduction of the geranylgeranyl moiety in spinach chloroplasts (Soll et al., 1983). One pathway for the direct reduction of GGDP to PDP was associated with the plastid envelope membrane. The second pathway, a reduction of geranylgeranyl chlorophyll, was localized in the thylakoid fraction. With these data in mind, the discovery of a phytol kinase as an essential gene for biosynthesis of at least 80% of seed tocopherols in Arabidopsis might merely suggest that 80% of phytol for tocopherol biosynthesis is formed in the thylakoids via geranylgeranyl chlorophyll, and up to 20% of phytol, or rather PDP is formed directly via reduction of GGDP at the inner plastidial envelope membrane. In agreement with this hypothesis, bioinformatics and in vitro enzyme assays has revealed that VTE5-1 is a membrane enzyme that is associated with the chloroplast envelope (Ischebeck et al., 2006; Valentin et al., 2006). The inner plastidial membrane is also thought to be the localization of chlorophyllase (Matile et al., 1997), and the remaining tocopherol pathway enzymes (Soll et al., 1980).

The identification of a phytol kinase as pathway enzyme for tocopherol biosynthesis in *Arabidopsis* seed opens an interesting new perspective on the biochemical reaction sequences leading to chlorophyll, tocopherol and phylloquinone K1. Additional experiments will be required to validate the relevance of these enzymes and their impact on phytol metabolism.

Enzymes downstream of VTE2 such as the two methyltransferases VTE3 and VTE4, and the tocopherol cyclase VTE1, were thought to have no or only a marginal impact on total tocopherol levels. However, two recent studies provide evidence that under high flux conditions, VTE3 becomes rate limiting. Collakova and DellaPenna concluded on the bases of quantitative mRNA analysis of tocopherol related genes in stressed non-transgenic and transgenic Arabidopsis plants that under high flux conditions the methylation of 2M6PBQ can become rate limiting (Collakova and DellaPenna, 2003b). Karunanandaa et al. came to similar conclusions based on results obtained with transgenic high tocopherol soybean seed (Karunanandaa et al., 2005). These authors demonstrated that the total tocopherol content in transgenic high tocopherol soybean seed expressing *E. coli tyrA*, *Arabidopsis HPPD*, and *At-VTE2*, increased even further upon crossing these lines with soybean lines expressing Arabidopsis-*VTE3* and Arabidopsis-*VTE4*.

Constitutive expression of the tocopherol cyclase *VTE1* has provided evidence that leaf tocopherol levels may be limited by VTE1 (Kanwischer et al., 2005). However, these data have not been confirmed in the scientific literature.

VI Enhancement of Tocotrienol Biosynthesis

A strategy to specifically enhance tocotrienol levels has been implemented via the transgenic expression of barley HGGT. This enzyme is distinguished from VTE2 by its preference for GGDP as co-substrate in vivo. Expression of this gene in Arabidopsis leaf and corn seed resulted in substantial tocotrienol accumulation (Table 1) (Cahoon et al., 2003). In transgenic soybean expressing HGGT, total tocochoromanol levels of approximately 2,500 ppm with 85% being tocotrienols have been reported online (http://www.imba.missouri.edu/report/2005_1_2.php). More recently tocochromanol levels of 3,000–4,000 ppm have been reported (Table 1) (E.B. Cahoon, 2008, personal communication). This is certainly the highest tocochromanol increase obtained through the expression of a single gene, and very near the range reported by Karunanandaa et al (Karunanandaa et al., 2005). Additional experiments need to explore if *HGGT* expression in combination with strategies that increase the HGA-levels can increase total tocochromanol levels well beyond currently described levels. A potential shortfall of this strategy is that tocotrienols have substantially lower vitamin E activity than tocopherols (Valentin and Qi, 2005). In addition, a vitamin E product containing tocotrienols rather than tocopherols may require additional studies to demonstrate equivalent bioefficacy. On the other hand, tocotrienols appear to provide health benefits that are distinct from tocopherols. In fact, tocotrienols extracted from palm oil are currently commercially sold as human nutritional supplements. And some agricultural biotech companies appear to continue working on tocotrienol enriched grains (Cahoon et al., 2008).

VII Conclusions and Outlook

During the past decade, research developments, as described in this chapter, document significant progress toward our understanding of genetic, molecular, and biochemical aspects of tocochromanol biosynthesis in plants, and developing biotech crop seed with enhanced vitamin E. However, to increase seed tocopherol levels the activity of multiple enzymes may need to be enhanced. Alternatively, expression levels of regulatory genes such as kinases or transcription factors might need to be manipulated to alter the activity of multiple pathway genes to enhance flux through the entire tocopherol pathway. The unresolved issue, in this respect, concerns the degree to which all genes in biosynthetic pathways are coordinately controlled. In addition, analysis and selection of genetic natural variations in seed vitamin E have also been initiated for the identification of high tocopherol species and quantitative vitamin E loci affecting tocopherol content and composition in seeds (Gilliland et al., 2006; García-Moreno et al., 2006; Ujiie et al., 2005; Rocheford et al., 2002). The integration of conventional breeding, marker-assisted breeding, and transgenic efforts would accelerate the development of high tocochromanol crop products.

Three different biotechnological strategies have been employed to increase the vitamin E content of seed and two of these strategies, expression of *VTE3* or *HGGT*, have been successfully employed in leaf tissue and cyanobacterial cells. In seed, which frequently contain γ-, and δ-tocopherol, expression of one or two methyltransferases, *VTE3* and *VTE4*, is sufficient to convert nearly all tocopherols to α-tocopherol. The concept of using designed transcription factors to activate internal methyltransferases for the same task has been demonstrated, but needs further refinement. Further vitamin E enhancements in seed and in leaves can be obtained only by increasing the total tocochromanol level. Preferably a strategy to increase total tocochromanols is combined with a strategy to optimize tocochromanol composition, such a strategy would not only support conversion of all tocochromanols into the most active form for animals and humans, but would also further enhance total tocochromanol flux (Karunanandaa et al., 2005). Substantial tocochromanol increases have been achieved expressing a HGGT (Table 1). However, total tocochromanol levels were still lower than in experiments that combined VTE2-expression with expression of other tocopherol pathway enzymes that provide tocopherol precursors.

The alternative strategy to increase total tocopherols by increasing the supply of tocopherol precursors such as HGA, in some cases combined with expression of *VTE2*, has been very successful, as well, resulting in tocochromanol levels up to 4,800 ppm in soybean seed. Unfortunately, these high HGA and tocochromanol seeds exhibited adverse morphology, and difficulty in germination (Karunanandaa et al., 2005). Interestingly, tocotrienol levels in these soybean seed accounted for up to 94% of total tocochromanols. This was even higher than the tocotrienol levels seen in corn seed, soybean seed or Arabidopsis leaf expressing HGGT, which contained up to 85% tocotrienols. This appears to suggest that under extreme conditions when HGA is available in excess and PDP is limiting while GGDP continues to be available, VTE2 can produce substantial levels of tocotrienol precursors. Nevertheless, all efforts to increase tocochromanol levels point towards a limitation in PDP biosynthesis, while GGDP appears to be available in substantial amounts.

The discovery of a phytol kinase as an essential enzyme for the biosynthesis of 80% of tocopherols in wild-type Arabidopsis seed is not in agreement with the established view of the tocopherol biosynthetic pathway. In contrast it suggests that the vast majority of phytol for tocopherol biosynthesis is obtained from chlorophyll catabolism (Fig. 1a). Alternatively, phytol may be synthesized by a completely different pathway, as recently suggested for *Euglena gracilis* (Kim et al., 2004). A thorough understanding of the biochemical reactions leading to the formation of PDP will be essential to provide sufficient PDP for increased tocopherol biosynthesis. Microbial model systems such as *Synechocystis* may allow dissection of some of the remaining questions in tocopherol biosynthesis. The ability of microbes to grow heterotrophically might allow the characterization of tocopherol biosynthesis in chlorophyll free mutants in order to evaluate the in vivo relevance of a direct reduction of GGDP to PDP by GGH for tocopherol and vitamin K biosynthesis. The identification of a phytolphosphate kinase in Arabidopsis seedlings is consistent with a salvage pathway as a key metabolic pathway for the biosynthesis of the majority of tocopherols in bacteria and plants (Fig. 1a). While plants appear to harbor a phytol kinase paralog, no such gene has not been found in *Synechocystis* or other photosynthetic cyanobacteria (Valentin et al., 2006). Furthermore, characterization of the regulation and the identification of rate limiting reactions of the tocopherol biosynthetic pathway have just been initiated (Savidge et al., 2002; Collakova and DellaPenna, 2001, 2003a, b). Finally, while some literature suggests that tocopherols can be modified to their respective fatty acid esters in vivo (Roshchin et al., 1986) the subsequent metabolic fate of tocopherols appears completely unknown, and it remains to be tested whether tocopherols are catabolized in photosynthetic plants and cyanobacteria.

The feasibility of developing commercial processes for the production of tocopherols based on transgenic bacteria or plants may depend on such factors as the final level of α-tocopherol in the target tissue, the value of an isomerically pure α-tocopherol versus a racemic α-tocopherol mixture as it is obtained from chemical synthesis, the ability of tocotrienols to substitute for the metabolic function of tocopherols in animals, the future price of development of petrochemical resources and chemical synthesis, as well as developments in processing technology, and public acceptance. Although a large number of products originating from transgenic sources have been introduced to the markets during the last 2 decades, non-medical transgenic products are still viewed with intense scrutiny particularly by European consumers. Vitamin E enhanced oil with the prospect of improved nutrition might provide the direct consumer benefit which became so obvious for transgenic products with medical applications and was incremental for their public acceptance, more than a decade ago.

In addition, for plants it will be essential to avoid poor agronomic performance such as reduced seed vigor, or yield. Although cyanobacteria are much more accessible to metabolic engineering than plants due to shorter generation times, and a substantially larger genetic toolbox, few examples from the current patent literature focus on this prospect and those examples appear to be limited to the production of tocopherol precursors such as phytol (Millis et al., 1999).

References

IUPAC-IUP Joint Commission on Biochemical Nomenclature (1982) Nomenclature of tocopherols and related compounds. Recommendation 1981. Eur J Biochem 123: 473–475

Birringer M, Kuhlow D, Pfluger PT, Landes N, Schulz TJ, Glaubitz M, Florian S, Pfeiffer A, Schuelke M, Brigelius-Flohé R and Ristow M (2007) Improved glucose metabolism in mice lacking a-tocopherol transfer protein. Eur J Nutr 46: 397–405

Boronat A (2010) The methylerythritol 4-phosphate pathway: regulatory role in plastid isoprenoid biosynthesis. In: Rebeiz CA, Benning C, Daniel H, Hoober K, Lichtenthaler HK, Portis A, Tripathy B (eds) The Chloroplast: Basics and Applications, Chapter 8. Springer, The Netherlands, pp. 119–126

Bramley PM, Elmadfa I, Kafatos A, Kelly FJ, Manios Y, Roxborough HE, Schuch W, Sheehy PJA and Wagner K-H (2000) Vitamin E. J Sci Food Agric 80: 913–938.

Brigelius-Flohé R and Davies KJA (2007) Is vitamin E an antioxidant, a regulator of signal transduction and gene expression, or a 'junk' food? Comments on the two accompanying papers: "Molecular mechanism of a-tocopherol action" by A. Azzi and "Vitamin E, antioxidant and nothing more" by M. Traber and J. Atkinson. Free Radic Biol Med 43: 2–3

Brigelius-Flohé R, Kelly FJ, Salonen JK, Neuzil J, Zingg JM and Azzi A (2002) The European perspective on vitamin E: current knowledge and future research. Am J Clin Nutr 76: 703–716.

Brown DJ and Goodman J (1998) A review of vitamins A, C, and E and their relationship to cardiovascular disease. Clin Excell Nurse Pract.2: 10–22

Buring JE and Hennekens CH (1997) Antioxidant vitamins and cardiovascular disease. Nutr Rev 55: S53–S60

Burton GW, Traber MG and Acuff RV (1998) Human plasma and tissue α-tocopherol concentrations in response to supplementation with deuterated natural and synthetic vitamin E. Am J Clin Nutr 67: 669–684

Cahoon EB, Hall SH, Ripp KG, Ganzke TS, Hitz WD and Coughlan SJ (2003) Metabolic redesign of vitamin E biosynthesis in plants for tocotrienol production and increased antioxidant content. Nat Biotech 21: 1082–1087

Cahoon EB, Coughlan SJ, Cahoon RE and Butler KH (2008) Compositions and methods for altering tocotrienol content. US20080005812A1.

Cavalier L, Ouahchi K, Kayden H, Donato S, Reutenaucer L, Mandel JL and Koenig M (1998) Ataxia with isolated vitamin E deficiency: heterogeneity of mutations and phenotypic variability in a large number of families. Am J Hum Genet 62: 301–310.

Cheng Z, Sattler S, Maeda H, Sakuragi Y, Bryant DA and DellaPenna D (2003) Highly divergent methyltransferases catalyze a conserved reaction in tocopherol and plastoquinone synthesis in cyanobacteria and photosynthetic eukaryotes. Plant Cell 15: 2343–2356

Chow CK (2001) Vitamin E. In: Rucker RB, Suttie JW, McCormick DB, Machlin LJ (eds) Handbook of Vitamins, 3rd ed. Marcel Dekker Inc., New York, pp. 165–197

Collakova E and DellaPenna D (2001) Isolation and functional analysis of homogentisate phytyltransferase from *Synechocystis* sp. PCC 6803 and Arabidopsis. Plant Physiol 127: 1113–1124

Collakova E and DellaPenna D (2003a) Homogentisate phytyltransferase activity is limiting for tocopherol biosynthesis in Arabidopsis. Plant Phys 131: 632–642

Collakova E and DellaPenna D (2003b) The role of homogentisate phytyltransferase and other tocopherol pathway enzymes in regulation of tocopherol biosynthesis during abiotic stress. Plant Physiol 133: 930–940

Cooney RW, France AA, Harwood PJ, Hatch-Pigott V, Custer LJ and Mordan LJ (1993) γ-Tocopherol detoxification of nitrogen dioxide: superiority to α-tocopherol. Proc Natl Acad Sci USA 90: 1771–1775

d'Harlingue A and Camara B (1985) Plastid enzymes of terpenoid biosynthesis. J Biol Chem 260: 15200–15203

Edwards H (2001) Vitamin E: an important anti-oxidant in the skin? In: Edwards H (ed) Retinoids and Lipid-Soluble Vitamins in Clinical Practice, Vol 17. New York, NY, pp. 43–47

Evans HM and Bishop KS (1922) On the existence of a hitherto unrecognized dietary factor essential for reproduction. Science 56: 650–651.

Farrell P and Roberts R (1994) Vitamin E. In: Shils M, Olson JA, Shike M (eds) Modern Nutrition in Health and Disease, 8th ed. Lea & Febiger, Philadelphia, PA, pp. 326–341

García-Moreno MJ, Vera-Ruiz EM, Fernández-Martínez JM, Velasco L and Pérez-Vich B (2006) Genetic and molecular analysis of high gama-tocopherol content in sunflower. Crop Sci 46: 2015–2021

Gilliland LU, Magallanes-Lundback M, Hemming C, Supplee A, Koornneef M, nie Bentsink L and DellaPenna D (2006) Genetic basis natural variation in seed vitamin E levels in *Arabidopsis thaliana*. PNAS 103: 18834–18841

Gloor U, Wuersch J, Schwiter U and Wiss O (1966) Resorption, retention, distribution, and metabolism of dl-α-tocopheramine, dl-N-methyl-γ-tocopheramine, and γ-tocopherol in relation to dl-α-tocopherol in rats. Helv Chim Acta 49: 2303–1312

Graham S, Sielezny M, Marshall J, Priore R, Freudenheim J, Brasure J, Haughey B, Nasca P and Zdeb M (1992) Diet in the epidemiology of Postmenopausal Breast Cancer in the New York State Cohort. Am J Epidemiol 136: 3127–3137

Graßes T, Grimm B, Koroleva O and Jahns P (2001) Loss of α-tocopherol in tobacco plants with decreased geranylgeranyl reductase activity does not modify photosynthesis in optimal growth conditions but increases sensitivity to high-light stress. Planta 213: 620–628

Grusak MA and DellaPenna D (1999) Improving the nutrient composition of plants to enhance human nutrition and health. Ann Rev Plant Physiol Plant Mol Biol 50: 133–161

Hofius D and Sonnewald U (2003) Vitamin E biosynthesis: biochemistry meets cell biology. Trends Plant Sci 8: 6–8

Hofius E, Hajirezaei MR, Geiger M, Tschiersch H, Melzer M and Sonnewald U (2004) RNAi-mediated tocopherol deficiency imparis photoassimilate export in transgenic potato plants. Plant Physiol 135: 1256–1268.

Hosomi A, Arita M, Sato Y, Kiyose C, Ueda T, Igarashi O, Arai H and Inoue K (1997) Affinity for α-tocopherol transfer protein as a determinant of the biological activities of vitamin E analogs. FEBS Lett 409: 105–108

Hunter SC and Cahoon EB (2007) Enhancing vitamin E in oliseeds: unraveling tocopherol and tocotrienol biosynthesis. Lipids 42: 97–108.

Ischebeck T, Zbierzak AM, Kanwischer M and Doermann P (2006) A salvage pathway for phytol metabolism in Arabidopsis. J Biol Chem 281: 2470–2477.

Jotham RA II, Frost E, Vidi PA, Kessler F and Staehelin LA (2007) Plastoglobules are lipoprotein subcompartments of the chloroplast that are permanently coupled to thylakoid membranes and contain biosynthetic enzymes. Plant Cell 18: 1693–1703

Kanwischer M, Porfirova S, Bergmueller E and Doermann P (2005) Alterations in tocopherol cyclase activity in transgenic and mutant plants of Arabidopsis affect tocopherol content, tocopherol composition, and oxidative stress. Plant Physiol 137: 713–723

Karunanandaa B, Qi Q, Hao M, Baszis S, Jensen P, Wong Y-HH, Jiang J, Venkatramesh M, Gruys KJ, Moshiri F, Post-Beittenmiller D, Weiss JD and Valentin HE (2005) Metabolically engineered oilseed crops with enhanced seed tocopherol. Metab Eng 7: 384–400

Keller P, Bouvier F, d'Harlingue A and Camara B (1998) Metabolic compartmentation of plastid prenyllipid biosynthesis. Evidence for the involvement of a multifunctional geranylgeranyl reductase. Eur J Biochem 251: 413–417

Kim D, Filtz MR and Proteau PJ (2004) The methylerythritol phosphate pathway contributes to carotenoid but not phytol biosynthesis in *Euglena gracilis*. J Nat Prod 67: 1067–1069

Kumar R, Raclaru M, Schuessler T, Gruber J, Sadre R, Luehs W, Zarhloul KM, Friedt W, Enders D, Frentzen M and Weier D (2005) Characterization of plant tocopherol cyclases and their overexpression in transgenic *Brassica napus* seeds. FEBS Lett 579: 1357–1364

Lichtenthaler HK (1969a) Localization and functional concentrations of lipoquinones in chloroplasts. In: Metzner H (ed) Progress in Photosynthesis Research, Vol. I. Laupp, Tübingen, Germany, pp. 304–314

Lichtenthaler HK (1969b) Zur Synthese der lipophilen Plastidenchinone und Sekundär-carotinoide während der Chromoplastenentwicklung. Ber Dtsch Bot Ges 82: 483–497

Lichtenthaler HK (1977) Regulation of prenylquinone biosynthesis in higher plants. In: Tevini M, Lichtenthaler HK (eds) Lipids and Lipid Polymers in Higher Plants. Springer-Verlag, Berlin, pp. 29–44

Lichtenthaler HK (1999) The 1-deoxy-D-xylulose-5-phosphate pathway of isoprenoid biosynthesis in plants. Annu Rev Plant Physiol Plant Mol Biol 50: 47–65

Lichtenthaler HK (2007) Biosynthesis, accumulation and emission of carotenoids, α-tocopherol, plastoquinone, and isoprene in leaves under high photosynthetic irradiance. Photosynth Res 92: 163–179

Lichtenthaler HK (2010) The DOXP/MEP pathway of chloroplast isoprenoid and pigment biosynthesis. In: Rebeiz CA, Benning C, Daniel H, Hoober K, Lichtenthaler HK, Portis A, Tripathy B (eds) The Chloroplast: Basics and Applications. Springer, Dordrecht, Netherlands, pp. 89–93

Maeda H and DellaPenna D (2007) Tocopherol functions in photosynthetic organisms. Curr Opin Plant Biol 10. 260–265

Matile P, Schellenbert M and Vicente F (1997) Localization of chlorophyllase in the chloroplast envelope. Planta 201: 96–99

Millis JR, Saucy GG, Maurina-Brunker J, McMullin TW and Hyatt JA (1999) Method of vitamin production. European patent application EP1095001A4

Mino M, Tamai H, Yasuda K, Yamada C, Igarashi O, Hayashi M, Hirahara F, Katsui G and Kijima S (1988) Biopotencies of tocopherol analogs as determined by dialuric acid induced hemolysis in rats. Bitamin 62: 241–246

Motohashi R, Ito T, Kobayashi M, Taji T, Nagata N, Asami T, Yoshida S, Yamaguchi-Shinozaki K and Shinozaki K (2003) Functional analysis of the 37 kDa inner envelope membrane polypeptide in chloroplast biogenesis using a *Ds*-tagged Arabidopsis pale-green mutant. Plant J. 34: 719–731

Munné-Bosch S and Alegre L (2002) The function of tocopherols and tocotrienols in plants. Crit Rev Plant Sci 21: 31–57

Nomenclature rules for vitamin E (972.31) (1990) Official methods of analysis. Assoc Off Anal Chem, 15th ed. 109: 8–15

Norris S, Lincoln K, Abad M, Scott M, Eilers R, Hartsuyker K, Kindle K, Hirshberg J, Karunanandaa B, Moshiri F, Stein JC, Valentin HE and Venkatesh TV (2004) Tocopherol biosynthesis related genes and uses thereof. International patent application WO 2004013312 A2

Porfirova S, Bergmüller E, Tropf S, Lemke R and Dörmann P (2002) Isolation of an Arabidopsis mutant lacking vitamin E and identification of a cyclase essential for all tocopherol biosynthesis. PNAS 99: 12495–12500

Provencher LM, Miao L, Sinha N and Lucas WJ (2001) *Sucrose Defective Export1* encodes a novel protein implicated in chloroplast-to-nucleus signaling. Plant Cell 13: 1127–1141

Qi Q, Hao M, Ng WO, Slater S, Baszis S, Weiss DJ and Valentin HE (2005). Application of the *Synechococcus nirA* promoter to establish an inducible expression system for engineering the *Synechocystis* tocopherol pathway. Appl Environ Microbiol 70: 5678–5684

Qureshi N and Qureshi AA (1993) Tocotrienols: novel hypocholesterolemic agents with antioxidant properties. In: Packer L, Fuchs J (eds) Vitamin E in Health and Disease. Marcel Dekker, New York, pp. 247–267

Qureshi AA, Bradlow BA, Brace L, Manganello J, Peterson DM, Pearce BC, Wright JK, Gapor A and Elson CE (1995) Response of hypercholesterolemic subjects to administration of tocotrienols. Lipids 30: 1171–1177

Rimbach G, Minihane AM, Majewicz J, Fischer A, Pallauf J, Virgli F and Weinberg PD (2002) Regulation of cell signaling by vitamin E. Proc Nutr Soc 61: 415–425

Rippert P, Scimemi C, Dubald M and Matringe M (2004) Engineering plant shikimate pathway for production of tocotrienol and improving herbicide resistance. Plant Physiol 134: 92–100

Rise M, Cojocaru M, Gottlieb, EH and Goldschmidt EF (1989) Accumulation of α-tocopherol in senescing organs as related to chlorophyll degradation. Plant Physiol 89:1028–1030

Rocheford TR, Wong JC, Egesel CO and Lambert RJ (2002) Enhancement of vitamin E levels in corn. Am Coll Nutr 21: 191S–198S

Rohmer M (2003) Mevalonate-independent methylerythritol phosphate pathway for isoprenoid biosynthesis. Elucidation and distribution. Pure Appl Chem 75: 375–387

Roshchin V, Fragina A and Solov'ev V (1986) Polyprenols and tocopherols from needles of *Picea abies* (L.) Karst Rast Resur 22: 530–537

Sadre R, Gruber J and Frentzen M (2006) Characterization of homogentisate prenyltransferases involved in plastoquinone-9 and tocochromanol biosynthesis. FEBS Lett 580: 5357–5362

Sanders SK, Morgan JB, Wulf DM, Tatum JD, Williams SN and Smith GC (1997) Vitamin E supplementation of cattle and shelf-life of beef for the Japanese market. J Anim Sci 75: 2634–2640

Sattler SE, Cahoon EB, Coughlan SJ and DellaPenna D (2003) Characterization of tocopherol cyclases from higher plants and cyanobacteria. Evolutionary implications for tocopherol synthesis and function. Plant Phys 132: 2184–2195

Savidge B, Weiss JD, Wong Y-HH, Lassner MW, Mitsky TA, Shewmaker CK, Post-Beittenmiller D and Valentin HE (2002) Isolation and characterization of tocopherol prenyltransferase genes from *Synechocystis* PCC 6803 and Arabidopsis. Plant Physiol 129: 321–33

Schledz M, Seidler A, Beyer P and Neuhaus G (2001) A novel phytyltransferase from *Synechocystis* sp. PCC 6803 involved in tocopherol biosynthesis. FEBS Lett 499: 15–20

Schultz G, Soll J, Fiedler E and Schulze-Siebert D (1985) Synthesis of prenylquinones in chloroplasts. Plant Physiol. 64: 123–129

Schultz-Siebert D, Homeyer U, Soll, J and Schultz G (1987) Synthesis of plastoquinone-9, α-tocopherol and phylloquinone (vitamin K1) and its integration in chloroplast carbon metabolism of higher plants. In: Stumpf P, Mudd JB, Nes WD (eds) The Metabolism, Structure and Function of Plant Lipids. Plenum Press, New York, pp. 29–36

Sheppard AJ, Pennington JA and Weihrauch JL (1993) Analysis and distribution of vitamin E in vegetable oils and foods. In: Packer L, Fuchs J (eds) Vitamin E in Health and Disease. Marcel Dekker, New York, pp. 9–31

Shintani D and DellaPenna D (1998) Elevating the vitamin E content of plants through metabolic engineering. Science 282: 2098–2100

Shpilyov AV, Zinchenko VV, Shestakov SV, Grimm B and Lokstein H (2005) Inactivation of the geranylgeranyl reductase (ChlP) gene in the cyanobacterium *Synechocystis* sp. PCC 6803. Biochim Biophys Acta 1706: 195–203

Slover HT (1971) Tocopherols in foods and fats. Lipids 6: 291–296

Soll J and Schultz G (1980) 2-Methyl-6-phytylquinol and 2,3-dimethyl-5-phytylquinol as precursors of tocopherol synthesis in spinach chloroplasts. Phytochemistry 19: 215–218

Soll J, Kemmerling M and Schultz G (1980) Tocopherol and plastoquinone synthesis in spinach chloroplasts subfractions. Arch Biochem Biophys 204: 544–550

Soll J, Schultz G, Rüdiger W and Benz J (1983) Hydrogenation of geranylgeraniol. Plant Physiol 71: 849–854

Soll J, Schultz G, Joyard J, Douce R and Block MA (1985) Localization and synthesis of prenylquinones in isolated outer and inner envelop membranes from spinach chloroplasts. Arch Biochem Biophys 238: 290–299

Stocker A, Ruttimann A and Woggon WD (1993) Identification of the tocopherol cyclase in the blue-green algae *Anabaena variabilis* Kutzing (cyanobacteria). Helv Chim Acta 76: 1729–1738

Subramaniam S, Slater S, Karberg K, Chen R, Valentin HE and Wong Y-HH (2001) Nucleic acid sequences to proteins involved in tocopherol synthesis. International patent application WO 01/79472

Tangney CC (1997) Vitamin E and cardiovascular disease. Nutr Today 32: 13–22

Theriault A, Chao JT, Wang Q, Gapor A and Adeli K (1999) Tocotrienol: a review of its therapeutic potential. Clin Biochem 32: 309–319

Traber MG and Arai H (1999) Molecular mechanisms of vitamin E transport. Annu Rev Nutr 19: 343–355

Traber MG and Sies H (1996) Vitamin E and humans: Demand and delivery. Ann Rev Nutr 16: 321–347

Triantafillidis JK, Kottaras G, Sgourous S, Cheracakis P, Driva G, Konstantellou E, Parasi A, Choremi H and Samouilidou E (1998) A beta-lipoproteinemia: clinical and laboratory features therapeutic manipulations, and follow-up study of three members of a Greek family. J Clin Gagroenterol 26: 207–211

Tsegaye Y, Shintani DK and DellaPenna D (2002) Overexpression of the enzyme *p*-hydroxyphenylpyruvate dioxygenase in Arabidopsis and its relation to tocopherol biosynthesis. Plant Physiol Biochem 40: 913–920

Ujiie A, Yamada T, Fujimoto K, Endo Y and Kitamura K (2005) Identification of soybean varieties with high α-tocopherol content breeding. Science 55: 123–125

Valentin HE and Qi Q (2005) Biotechnological production and application of vitamin E: current state and prospects. Appl Microbiol Biotechnol 68: 436–444

Valentin HE, Lincoln K, Moshiri F, Jensen PK, Qi Q, Venkatesh TV, Karunanandaa B, Baszis S, Norris SR, Savidge B, Gruys KJ and Last RL (2006) The Arabidopsis vitamin E pathway gene 5-1 mutant reveals a critical role for phytol kinase in seed tocopherol biosynthesis. Plant Cell 18: 212–224

Van Eenennaam A, Valentin HE, Karunanandaa B, Hao M, Aasen E and Levering C (2003a) Methyltransferase genes and uses of thereof. WO 0/3016482

Van Eenennaam AL, Lincoln K, Durrett TP, Valentin HE, Shewmaker CK, Thorne GM, Jiang J, Baszis SR, Levering CK, Aasen ED, Hao M, Stein JC, Norris SR and Last RL (2003b) Engineering vitamin E content: from Arabidopsis mutant to soy oil. Plant Cell 15: 3007–3019

Van Eenennaam AL, Li G, Venkatramesh M, Levering C, Gong X, Jamieson AC, Brebar EJ, Shewmaker CK and Casey C (2004) Elevation of seed α-tocopherol levels using plant-based transcription factors targeted to an endogenous locus. Metab Eng 6: 101–108

Venkatesh TV, Karunanandaa B, Free DL, Rottnek JM, Baszis SR and Valentin HE (2006) Identification and characterization of an Arabidopsis homogentisate phytyltransferase paralog. Planta 223: 1134–1144

Vidi PA, Kanwischer M, Baginsky S, Austin JR, Csucs G, Dörmann P, Kessler F and Bréhélin C (2006) Tocopherol cyclase (VTE1) localization and vitamin E accumulation in chloroplast plastoglobule lipoprotein particles. J Biol Chem 281: 1125–11234

Waters DD, Alderman EL, Hsia J, Howard BV, Cobb FR, Rogers WJ, Ouyang P, Thompson P, Tardif JC, Higginson L, Bittner V, Steffes M, Gordon DJ, Proschan M, Younes N and Verter JI (2002) Effects of hormone replacement therapy and antioxidant vitamin supplements on coronary atherosclerosis in postmenopausal women: a randomized controlled trial. J Am Med Assoc 288: 2432–2440

Weiser H, Vecchi M and Schlachter M (1986) Stereoisomers of α-tocopherol equivalents of all-rac-, 2-ambo- and RRR- α-tocopherol evaluated by simultaneous determination of resorption-gestation, myopathy and liver storage capacity in rats. Int J Vit Nutr Res 56: 45–56

Wu K, Willett WC, Chan JM, Fuchs CS, Colditz GA, Rimm EB and Giovannucci EL (2002) A prospective study on supplemental vitamin E intake and risk of colon cancer in women and men. Cancer Epidemiol Biomark Prev 11: 1298–1304

Yusuf S, Dagenais G, Pogue J, Bosch J and Sleight P (2000) Vitamin E supplementation and cardiovascular events in high-risk patients. The Heart Outcomes Prevention Eval uation Study Investigators. N Engl J Med 342: 154–160

Chapter 12

The Anionic Chloroplast Membrane Lipids: Phosphatidylglycerol and Sulfoquinovosyldiacylglycerol

Christoph Benning
Department of Biochemistry and Molecular Biology, Michigan State University, East Lansing, MI 48824, USA

Summary

Four polar lipids provide the basic building blocks for the membranes of chloroplasts of all plants and cyanobacteria: the two galactolipids mono-and digalactosyldiacylglycerol, the sulfolipid sulfoquinovosyldiacylglycerol, and the phospholipid phosphatidylglycerol. All four lipids are glycerolipids with a diacylglycerol backbone, but only one is a phosphoglycerolipid. While the two galactolipids are not charged at biological pH, the sulfolipid and phospholipid carry a negative charge. The focus here is on the two chloroplast anionic lipids. Enzymes involved in their assembly are well described and are localized in the chloroplasts. Mutants affecting the biosynthesis of one or both anionic lipids of the model plant *Arabidopsis thaliana*, the unicellular algae *Chlamydomonas reinhardtii* or different cyanobacteria are available providing tools for the functional analysis of these lipids. This genetic analysis has shown that sulfolipid can substitute to some extent for phosphatidylglycerol under phosphate limiting growth conditions, when plants replace phospholipids with non-phosphorous glycolipids. However, phosphatidylglycerol mutants have a much more severe phenotype than sulfolipid mutants suggesting specific roles for phosphatidylglycerol in the photosynthetic membrane that cannot be filled by the sulfolipid. Double mutant analysis suggests that the net negative charge of the photosynthetic membrane is critical for proper function.

Author for Correspondence, e-mail: benning@msu.edu

Constantin A. Rebeiz et al. (eds.), The Chloroplast: Basics and Applications, pp. 171–183.

I Introduction

The chloroplast contains one of the most extensive membrane systems found in nature, the photosynthetic membranes, which are organized into thylakoids. Inserted into the photosynthetic membranes are the pigment protein complexes of light harvesting antennae and photosynthetic reaction centers as well as other protein complexes required for photosynthetic electron transport. These protein complexes are embedded in a matrix of polar glycerolipids, the thylakoid membrane lipids. The glycerolipids of the photosynthetic membranes in cyanobacteria and plants fall into four classes: The two non-charged galactoglycerolipids mono- and digalactosyldiacylglycerol, and the two anionic lipids phosphatidylglycerol (PtdGro) and sulfoquinovosyldiacylglycerol (SQDG). The biosynthesis and function of the two galactolipids is discussed in Chapters 13 and 14. Here, I am focusing on the biosynthesis and function of the two anionic lipids with emphasis on *Arabidopsis thaliana* as a genetic model, which has greatly contributed to our current understanding of lipid metabolism and function in plants. For additional information on the function of the two anionic chloroplast lipids in photosynthesis the reader is referred to a recent comparison by Frentzen (2004).

The basic structure of the two anionic lipids PtdGro and SQDG of the photosynthetic membrane is shown in Fig. 1a. Attached to the sn-3 position of a glycerol backbone is the headgroup determining the respective lipid class. For PtdGro the headgroup is a glycerol-3-phosphate linked in a phosphodiester bond. For SQDG it is a 6-deoxy-6-sulfoglucose (sulfoquinovose) linked to sn-3 in α-configuration at the anomeric (C1) carbon of the sulfoquinovose. As such, PtdGro is a phosphoglycerolipid, and SQDG a non-phosphorous glycoglycerolipid. Different molecular species of the two glycerolipids are typically found in plants, which are distinguished by their acyl substituents at the sn-1 and sn-2 positions of the glycerol backbone. This leads to a large number of similar, but not identical lipid molecules, a fact that contributes to the complexity of the lipid composition of thylakoid membranes. Acyl groups typically found in the two anionic lipids are shown in Fig. 1b. It should be noted that PtdGro and SQDG are rich in saturated fatty acids (Browse and Somerville, 1994), particularly palmitic acid (16:0; total carbon number: number of double bonds). Palmitic acid is a minor acyl group in the two galactolipids. More strikingly, PtdGro contains an acyl group specifically in the sn-2 position of the glycerol backbone not fond in other lipids, *trans* Δ^3 hexadecenoate (Murata, 1983; Dubacq and Tremolieres, 1983). The possible roles of these distinct acyl compositions of PtdGro and SQDG will be discussed below.

The assembly of the two anionic thylakoid lipids depends on enzymes located inside the plastid. While the enzymes and genes involved in the biosynthesis of SQDG are known, some of the enzymes involved in plastid PtdGro biosynthesis have not yet been experimentally verified at the molecular level. In cases where the respective gene is not yet known, comparative genomics provides possible candidates (Beisson et al., 2003; Lykidis, 2007). The two lipids are also distinct with regard to the origin of their diacylglycerol moiety. While it seems clear that a subfraction of the plastid glycoglycerolipids including SQDG in Arabidopsis and many other plants are imported from the endoplasmic reticulum (ER), PtdGro seems to be primarily assembled from plastid-derived diacylglycerol moieties (Browse et al., 1986). This two-pathway-hypothesis for plastid glycoglycerolipid biosynthesis was originally proposed by Roughan and colleagues (Roughan et al., 1980; Roughan and Slack, 1982). Its molecular and biochemical basis, in particular the underlying lipid trafficking

Abbreviations: ACP – acyl carrier protein; Acyl-ACP – acyl-acyl carrier protein; APR – adenosylphosphosulfate reductase; APS – adenosylphosphosulfate; ATP – adenosine triphosphate; ATS – ATP-sulfurylase; ATS1 – acyl-ACP:glycerol 3-phosphate acyltransferase; ATS2 – acyl-ACP:lysophosphatidic acid acyltransferase; CDP-DAG – CDP-diacylglycerol; CDS – CDP-DAG synthetase; DAG – diacylglycerol; ER – endoplasmic reticulum; FAS – fatty acid synthase; Fd-GOGAT – ferredoxin-dependent glutamate synthase; GSH – reduced glutathione; GSSG – oxidized glutathione; iE – inner envelope; oE – outer envelope; pCDS – plastid CDP-diacylglycerol synthetase; PGP1 – phosphatidylglycerol phosphate synthase; pGPAT – plastid glycerolphosphate acyltransferase; PGPP – phosphatidylglycerol phosphate phosphatase; pLPAAT – plastid lysophosphatidic acid acyltransferase; PPi – orthophosphate; PtdGro – phosphatidylglycerol; PtdGroP – phosphatidylglycerol phosphate; PtdOH – phosphatidic acid; SQD1 – UDP-SQ synthase; SQD2 – SQDG synthase; SQDG – sulfoquinovosyldiacylglycerol; SO_3^- – sulfite; Thy thylakoid membrane; TLC – thin-layer chromatography; UDP-Glc – UDP-glucose; UDP-SQ – UDP-sulfoquinovose.

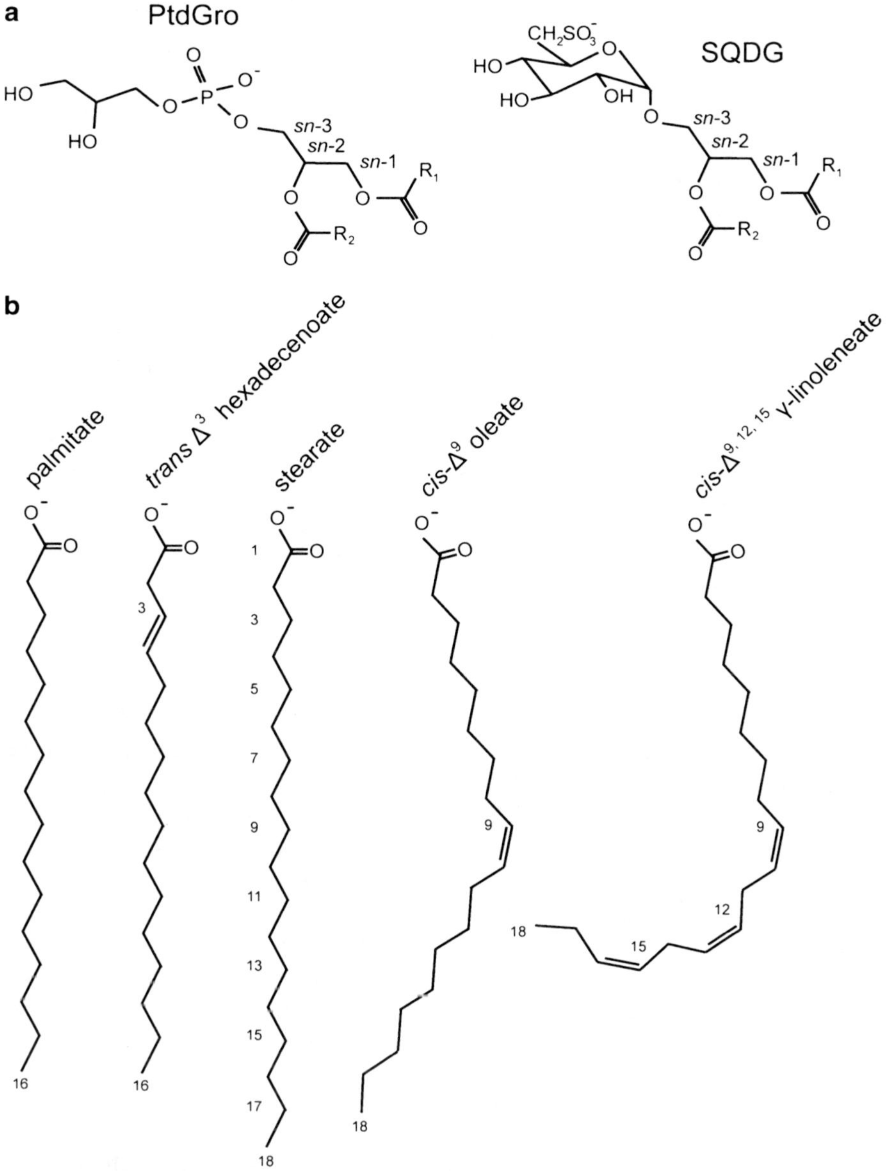

Fig. 1. Basic structures of phosphatidylglycerol (PtdGro) and sulfoquinovosyl-diacylglycerol (SQDG) (**a**) and fatty acyl substituents in these lipids (**b**). The carbons (*sn*-1, 2, 3) of the glycerol backbone in the two lipids are indicated in (**a**). Fatty acyl substituents are marked as R-groups in the two structures. The carbon numbering of the acyl groups is indicated in (**b**).

phenomena are not yet fully understood (Benning et al., 2006). However, lipids derived from the plastid or the ER pathways can be distinguished based on their acyl composition (Heinz and Roughan, 1983), a fact that permits the calculation of fluxes through the two pathways (Browse et al., 1986).

Genetic mutants of Arabidopsis deficient in either of the two anionic lipids or both, as well as cyanobacterial mutants and mutants of the unicellular algae *Chlamydomonas reinhardtii* are available. Their phenotypic analysis has provided insights into the role of these anionic lipids in photosynthetic membranes. Of particular note is the hypothesis that the non-phosphorous glycoglycerolipid SQDG can substitute for the phosphoglycerolipid PtdGro under phosphate-limiting conditions (Benning et al., 1993; Essigmann et al., 1998; Yu et al., 2002; Yu and Benning, 2003). This hypothesis will be discussed in detail below.

II Biosynthesis of Plastidic Phosphatidylglycerol

Analyzing lipids from *Scenedesmus*, Benson and Maruo (1958) discovered PtdGro. In plants, at least three parallel sets of enzymes involved in PtdGro biosynthesis are associated with the ER, the inner mitochondrial membrane and the inner plastid membrane, respectively (Kinney, 1993; Frentzen, 2004). The proposed pathway for the biosynthesis of plastid PtdGro in Arabidopsis and presumably other seed plants is shown in Fig. 2. At least two enzymes are exclusively committed to the biosynthesis of PtdGro, plastid PtdGro-phosphate synthase and PtdGro-phosphate phosphatase. In Arabidopsis, two PtdGro synthases have been identified (Müller and Frentzen, 2001) encoded by *PGP1* and *PGP2*. Of these *PGP1* encodes a dually targeted enzyme associated with the plastid and the mitochondrion (Babiychuk et al., 2003) while PGP2 appears to be associated with the ER. Mutants of Arabidopsis carrying a mutation in the *pgp1* gene are available (Xu et al., 2002; Hagio et al., 2002; Babiychuk et al., 2003). A gene encoding PtdGro-phosphate phosphatase in plants or any other eukaryote has not yet been positively identified, but candidate genes have been proposed based on comparative genomics (Lykidis, 2007).

Glycerol 3-phosphate and CDP-diacylglycerol (CDP-DAG) provide the substrate for the plastid PtdGro-phosphate synthase (Fig. 2). CDP-DAG is probably produced from phosphatidic acid (PtdOH) and CTP catalyzed by a plastidic CDP-DAG synthetase (CDS). Two cDNAs encoding plant eukaryotic-type CDS have been characterized (Kopka et al., 1997). Although the localization of these CDS proteins has not been experimentally verified, it seems likely that they are associated with the ER. The Arabidopsis genome contains a total of five predicted CDS encoding genes (Lykidis, 2007). Four of these cluster together with ER-associated isoforms, and one is similar to cyanobacterial CDS (Lykidis, 2007). This latter form seems to be a likely candidate for the plant plastid-targeted CDS involved in PtdGro biosynthesis. Because none of the other glycerolipids inside the plastids require CDP-DAG, but diacylglycerol (DAG), the plastid CDS might be solely dedicated to PtdGro biosynthesis. Its substrate specificity for particular PtdOH molecular species could also account to some degree for the PtdGro characteristic molecular species composition.

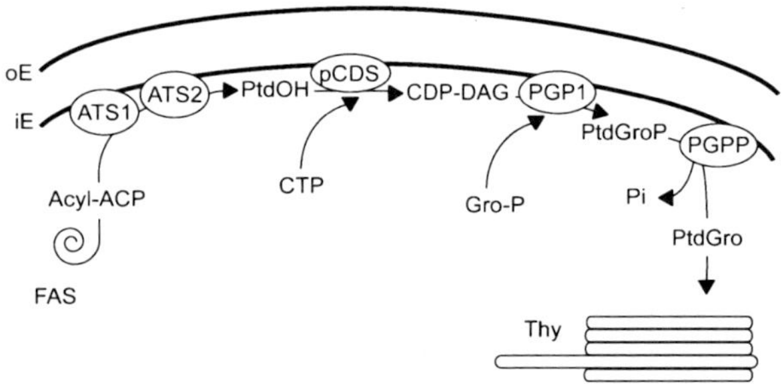

Fig. 2. Schematic representation of the proposed pathway for phosphatidylglycerol biosynthesis in plastids. The outer (oE) and inner (iE) envelope membranes, and thylakoid membranes (Thy) are indicated. Enzymes are: ATS1, acyl-ACP: glycerol 3-phosphate acyltransferase; ATS2, acyl-ACP: lysophosphatidic acid acyltransferase; FAS, fatty acid synthase; pCDS, plastid CDP-diacylglycerol synthetase; PGP1, phosphatidylglycerol phosphate synthase; and PGPP, phosphatidylglycerol phosphate phosphatase. Metabolites are: Acyl-ACP, acyl-acyl carrier protein; CDP-DAG, CDP-diacylglycerol; PtdGro, phosphatidylglycerol; PtdGroP, phosphatidylglycerol phosphate; PtdOH, phosphatidic acid.

The biosynthesis of PtdOH in plastids requires the action of two acyltransferases, acyl-acyl carrier protein (ACP): glycerol 3-phopshate acyltransferase (pGPAT) and acyl-ACP: lyso-PtdOH acyltransferase (pLPAAT). The cDNA encoding the squash and Arabidopsis pGPAT have been described (Ishizaki et al., 1988; Nishida et al., 1993). The Arabidopsis pGPAT is encoded by a gene designated *ATS1* (formerly *ACT1*; Fig. 2) and a number of mutant alleles are available (Kunst et al., 1988; Xu et al., 2006). Likewise pLPAAT in Arabidopsis is encoded by *ATS2* (Fig. 2) for which mutants are available as well (Kim and Huang, 2004; Yu et al., 2004). Contrary to the *ats1* mutations, the available *ats2* mutations are embryo-lethal and the respective mutants cannot be biochemically studied.

With regard to PtdGro biosynthesis in plastids it is important to note that while in different Arabidopsis *ats1* mutant alleles the biosynthesis of the predominant galactoglycerolipids by the plastid pathway is blocked and almost entirely depends on the import of the DAG moiety from the ER, PtdGro biosynthesis is only mildly affected (Kunst et al., 1988). This prompted a detailed investigation of the different *ats1* alleles to determine whether those available were leaky, or whether an ATS1-independent mechanism for the biosynthesis of the CDP-DAG precursor of PtdGro needed to be invoked. The conclusion from this analysis was that even the most severe allele of *ats1* still produced a mutant ATS1 protein with residual activity, presumably sufficient to produce PtdGro (Xu et al., 2006). Furthermore, it was concluded that PtdOH formed in the plastid is preferentially channeled into CDP-DAG and PtdGro biosynthesis, and that in the *ats1* mutant galactoglycerolipids can be fully derived from ER-imported DAG moieties in Arabidopsis (see also Chapter 13). It should be noted that our understanding of the mechanism for the transfer of lipids from the ER to the plastid is still rudimentary (Awai et al., 2006; Benning et al., 2006). For example, the identity of the lipid species transported from the ER to the plastid is unknown, as is the detailed mechanism by which its DAG moiety is incorporated into ER-derived thylakoid lipids.

The observation that plant plastid PtdGro biosynthesis is somewhat different from the biosynthesis of the thylakoid glycoglycerolipids is further supported by the presence of an unusual fatty acid in PtdGro, *trans* Δ^3-hexadecenoate (Murata, 1983; Dubacq and Tremolieres, 1983), which is exclusively present in the sn-2 position of the PtdGro glycerol back bone. One of the first lipid mutants of Arabidopsis described, *fad4* (JB60), is deficient in molecular species of PtdGro containing this fatty acids (Browse et al., 1985). Whether the *FAD4* gene encodes a PtdGro-specific fatty acid desaturase has to await mapping of the *fad4* locus and identification and analysis of the respective protein.

III Biosynthesis of Sulfoquinovosyldiacylglycerol

The sulfolipid SQDG was discovered by Benson and coworkers in the late 1950s (Benson et al., 1959) and the reader may also consult a recent, more detailed review on the pathway of SQDG biosynthesis (Benning et al., 2008). Genetics provided the access to the enzymes, and the respective genes were first identified using a mutant approach in the purple bacterium *Rhodobacter sphaeroides* (Benning and Somerville, 1992a, b; Rossak et al., 1995, 1997). Conserved in all SQDG containing organisms is the SqdB protein as it is called in bacteria, or the SQD1 protein in seed plants, which resembles sugar nucleotide-modifying enzymes (Benning and Somerville, 1992a; Güler et al., 1996; Essigmann et al., 1998). It is the enzyme catalyzing the first reaction of the now experimentally established sugar nucleotide pathway for SQDG biosynthesis, which is shown in Fig. 3. The SQD1 enzyme from Arabidopsis is localized in the plastid (Essigmann et al., 1998) and catalyzes the formation of UDP-sulfoquinovose (UDP-SQ) from UDP-glucose and sulfite (Sanda et al., 2001). A reaction mechanism involving the transient formation of an UDP-4-keto-5,6-glucoseene intermediate to which sulfite is added at carbon 6 was proposed based on (a) chemical synthesis of sulfoquinovose from glucose and sulfite (Lehmann and Benson, 1964), (b) similarity to the mechanism of other sugar nucleotide-modifying enzymes (Pugh et al., 1995), (c) modeling of the SQD1 protein and (d) demonstration of the presence of a tightly bound catalytic NAD^+ (Essigmann et al., 1999).

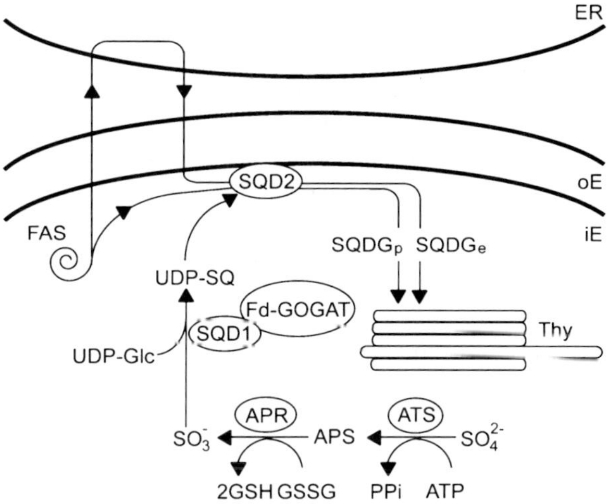

Fig. 3. Sulfolipid biosynthesis in plant chloroplasts. The inner envelope membrane (iE), the outer envelope membrane (oE), the endoplasmic reticulum (ER), and the thylakoid membranes (Thy) are shown. Enzymes are: APR, adenosylphosphosulfate reductase; ATS, ATP-sulfurylase, FAS, fatty acid synthase; Fd-GOGAT, ferredoxin-dependent glutamate synthase; SQD1, UDP-SQ synthase; and SQD2, SQDG synthase. Sulfolipid (SQDG) can be derived from the plastid pathway of thylakoid lipid biosynthesis ($SQDG_p$), or the ER pathway of thylakoid lipid biosynthesis ($SQDG_e$). Substrates and intermediates are: APS, adenosylphosphosulfate; ATP, adenosine triphosphate; GSSG, oxidized glutathione; GSH, reduced glutathione; PPi, orthophosphate; SO_3^-, sulfite; SO_4^{2-}, sulfate; UDP-Glc; UDP-glucose; UDP-SQ, UDP-sulfoquinovose (Reproduced from Benning et al., 2008, with kind permission of Springer Science and Business Media).

The mechanism was further corroborated by determining the crystal structure of SQD1 (Mulichak et al., 1999).

The native SQD1 protein from spinach is more active than the recombinant SQD1 protein from Arabidopsis. It purified as a large 250 kDa complex containing ferredoxin-dependent glutamate synthase (Shimojima and Benning, 2003; Shimojima et al., 2005). It was speculated that the FMN-cofactor of the glutamate synthase channels sulfite to SQD1 (Shimojima et al., 2005), in order to alleviate the toxic effects of free sulfite, which is harmful to the cell even at low concentrations. Sulfite is derived from the sulfate assimilation pathway in chloroplasts, where it is formed by reduction of adenosine 5′-phosphosulfate (APS) by the action of APS reductase (Leustek et al., 2000; Kopriva, 2006). The origin of the UDP-glucose substrate of SQD1 in the chloroplasts remains unclear (Benning, 2007).

The second enzyme directly involved in SQDG biosynthesis in plants is the SQDG synthase, designated SQD2 in Arabidopsis (Fig. 3). This enzyme is similar to glycosyltransferases and presumably uses UDP-sulfoquinovose and DAG as substrates to produce UDP and SQDG at the inner envelope of the chloroplast. Genetic evidence in Arabidopsis is available as a mutant of Arabidopsis disrupted in SQD2 completely lacks SQDG (Yu et al., 2002). Although the enzymatic activity has not yet been directly demonstrated for recombinant SQD2 from Arabidopsis, the corresponding native enzyme of spinach was characterized at the biochemical level using synthetic UDP-SQDG as a substrate (Heinz et al., 1989; Seifert and Heinz, 1992). Interestingly, *Rb. sphaeroides* uses a different, unrelated glycosyltransferase encoded by *sqdD* to produce SQDG from UDP-sulfoquinovose and a not yet identified acceptor molecule (Rossak et al., 1995). However, the SQDG synthase in cyanobacteria seems to be an ortholog of SQD2 of plants (Güler et al., 2000). In fact, the cyanobacterial ortholog of *SQD2*, designated *sqdX*, was first identified in *Synechococcus* sp. PCC7942 as a putative glycosyltransferase encoded by a gene neighboring the respective cyanobacterial *SQD1* ortholog, *sqdB*, which is responsible for the biosynthesis of UDP-sulfoquinovose (Güler et al., 2000). That SQD1 and SQD2 are minimally required for SQDG biosynthesis was demonstrated by co-expression of the respective Arabidopsis cDNAs in *E. coli.* This approach led to the biosynthesis of SQDG in this bacterium, which normally lacks sulfolipid (Yu et al., 2002).

IV Functions of Plastid Phosphatidylglycerol

The essential role of PtdGro in photosynthesis was recently discussed in detail by Wada and Murata (2007). In principle, three independent lines of evidence provide a clear indication that PtdGro plays an intricate role in the function of photosynthetic membrane complexes: First, PtdGro is found in recent crystal structures of photosynthetic complexes; Second, removal of PtdGro from isolated photosynthetic complexes affects their function in vitro; And third, genetic mutants of different mutants lacking PtdGro demonstrate the in vivo role of PtdGro. In addition, transgenic plants with altered acyl composition of plastid PtdGro were useful in studying the role of specific molecular species of PtdGro in chilling sensitivity of plants (Nishida and Murata, 1996).

Crystal structures of photosystem I (PSI) of the cyanobacterium *Thermosynechococcus elongatus* showed three PtGro molecules per subunit, with one found in symmetrical arrangement with a molecule of monogalactosyldiacylglycerol (Jordan et al., 2001). This specific arrangement seems to suggest specific roles for the proper function of this complex. Likewise, the photosystem II (PSII) structure of the same cyanobacterium contained one PtdGro molecule per subunit (Loll et al., 2005, 2007). Crystal structures of the spinach light harvesting complex II (LHCII) revealed one PtdGro and one digalactosyldiacylglycerol molecule per monomer and it was suggested that LHCII trimer formation was dependent on the presence of PtdGro (Liu et al., 2004). On the contrary, the crystal structure of the cytochrome b_6/f complex from *C. reinhardtii* lacked a specific PtdGro but contained SQDG (Stroebel et al., 2003).

Many of these findings of specific PtdGro association with photosynthetic complexes were preceded by the functional analysis of isolated complexes following phospholipase treatment to remove phospholipids. For example, trimerization of LHCII was affected following phospholipase treatment (Nussberger et al., 1993). Specific amino acids in the N-terminus of LHCII were found to be critical for trimerization and the interaction of LHCII with PtdGro (Hobe et al., 1994, 1995). Electron transport was impaired when thylakoid membranes were treated with phospholipases that removed much of the PtdGro (Jordan et al., 1983; Droppa et al., 1995).

One of the earliest described PtdGro mutants was affected in the acyl composition of PtdGro in *C. reinhardtii* (Seras et al., 1989). This mutant lacked *trans* Δ^3 hexadecenoate in the sn-2 position of PtdGro and showed a pleiotropic photosynthetic phenotype explained by the decreased synthesis of two PSII core subunits, D1 and apoCP47 (Pineau et al., 2004). However, the equivalent *fad4* mutant of Arabidopsis showed little or no defect in photosynthesis (Browse et al., 1985; McCourt et al., 1985) suggesting a more subtle function of the respective molecular species of PtdGro in organisms other than *C. reinhardtii*. This conclusion was also supported by experiments on transgenic plants targeting a bacterial DAG kinase to the envelopes of transgenic tobacco plants (Fritz et al., 2007). Much of the plastid PtdGro in these transgenic lines had an acyl composition consistent with its origin from the ER-pathway. While the transgenic plants were slower growing, photosynthesis was little impaired. Again, this experiment showed that the acyl composition of plastid PtdGro might not be as critical for photosynthesis as the experiments with *C. reinhardtii* mentioned above implied.

However, chilling sensitivity in seed plants is directly affected by the acyl composition of PtdGro (Nishida and Murata, 1996). Murata and co-workers expressed cDNAs encoding pGPATs (ATS1 orthologs, see Fig. 3) from chilling sensitive and chilling resistant squash and Arabidopsis, respectively, in tobacco (Murata et al., 1992; Moon et al., 1995). Due to the different substrate specificity of the two enzymes, the unsaturation of the acyl groups of PtdGro was altered in the two sets of transgenic lines leading to increased chilling tolerance for the Arabidopsis cDNA expressing transgenics and decreased for the squash expressing transgenics. This effect correlated with a change in the critical temperature of rigid lipid domain formation (Szalontai et al., 2003). In a similar approach an *Escherichia coli* GPAT encoding gene was introduced into Arabidopsis (Wolter et al., 1992), which typically is chilling resistant. As a consequence, the abundance of molecular species of PtdGro with saturated and *trans*-monounsaturated acyl groups increased leading to chilling sensitivity of the transgenic plants.

Mutants disrupted in the *cdsA* gene (Sato et al., 2000b) or the *pgsA* gene (Hagio et al., 2000) of the cyanobacterium *Synechocystis* sp. PCC6803 required PtdGro supplementation for growth suggesting an essential role for this lipid. Adjusting the concentration of PtdGro in the medium enabled studies, which correlated the abundance of this lipid with photosynthetic competence, or assembly and maintenance of the photosynthetic machinery. Initially, PSII activity was found to be affected at the level of the electron acceptor Q_B (Gombos et al., 2002). Furthermore, D1 PSII reaction center protein turnover was impaired under high-light growth conditions in the mutant suggesting a requirement of PtdGro for the maintenance of PSII (Sakurai et al., 2003). In subsequent studies it became clear that PSI assembly also required the presence of PtdGro (Domonkos et al., 2004; Sato et al., 2004). While these results are very intriguing, it has to be kept in mind that all results were obtained for one bacterial species, *Synechocystis* sp. PCC6803. As will be discussed below for SQDG mutants, lipid defects can have different physiological effects in different species.

In general, results observed for *Synechocystis* sp. PCC6803 described above were supported by the analysis of *pgp1* mutants of Arabidopsis. Mutants carrying a leaky *pgp1-1* allele showed 30% reduction in PtdGro content, reduced amounts of chlorophyll and impaired photosynthesis (Xu et al., 2002). Loss-of-function alleles of *pgp1* resulted in pale white plants completely lacking chlorophyll, which were not viable on soil but could be maintained when supplied with sugar on agar-solidified medium (Hagio et al., 2002; Babiychuk et al., 2003). The complete lack of chloroplast development in these mutants is a clear indication of the essential role of PtdGro in plants, suggesting that plastid PtdGro has specific functions that cannot be substituted by other lipids under the conditions tested.

V Functions of Sulfoquinovosyldiacylglycerol

Initially, a role for SQDG as a proton conducting pathway in photosynthetic membrane was hypothesized based on theoretical considerations (Haines, 1983). The presence of SQDG in most photosynthetic, but absence from most non-photosynthetic organisms (Benning, 2007) also suggested a possible role in photosynthesis which was corroborated by in vitro studies. For example, photosynthetic electron transport in isolated thylakoid membranes was inhibited by SQDG-specific antibodies (Radunz and Schmid, 1992). In addition, the biochemical detection of SQDG in isolated photosynthetic complexes (Menke et al., 1976; Gounaris and Barber, 1985; Pick et al., 1985; Sigrist et al., 1988) and the presence of SQDG in the cytochrome b_6f complex of *C. reinhardtii* (Stroebel et al., 2003) provided circumstantial evidence for a role of SQDG in photosynthesis. A specific requirement of SQDG for the reconstitution of chloroplast ATP synthase did suggest a functional interaction between SQDG and protein complexes of the photosynthetic membrane (Pick et al., 1987). This combined evidence led to the assumption that SQDG might be essential for photosynthesis in obligate photoautotrophic organisms (Barber and Gounaris, 1986).

Because there were reasonable concerns that SQDG might be essential for obligate photoautotrophic organisms, mutants deficient in SQDG biosynthesis were first isolated in the facultative photoheterotrophic purple bacterium *Rb. sphaeroides* (Benning and Somerville, 1992a, b). A *Rb. sphaeroides* mutant disrupted in *sqdB*, which completely lacked SQDG, was able to grow normally on a complete medium and showed no impairment in photosynthetic electron transport (Benning et al., 1993). However, when this mutant was grown under phosphate-limiting conditions, growth slowed sooner than for the corresponding wild type. Comparison of polar lipids in phosphate-deprived and phosphate-replete wild-type cells of *Rb. sphaeroides* showed that phospholipids were replaced by non-phosphorous glycolipids such as SQDG or betaine lipid as cells became phosphate starved (Benning et al., 1995). Because the *sqdB* mutant did not reduce PtdGro content to the same extent as wild type following phosphate deprivation, it was postulated that SQDG can substitute PtdGro, at least to some extent, under these conditions (Benning et al., 1993).

Similarly, an *sqdB* mutant of the cyanobacterium *Synechococcus* sp. PCC7942 lacking SQDG showed growth impairment under phosphate-depleted conditions, but grew like wild type on phosphate-replete medium (Güler et al., 1996). Closer examination by ultra-fast fluorescence spectroscopy, showed however that electron transfer rates within PSII were altered in the mutant without measurable effects on oxygen evolution rates (Güler et al., 1996). In contrast, when the *sqdB* gene was disrupted in *Syenchocystis* sp. PCC6803, supplementation of SQDG in the medium was required for growth and photosynthesis was affected even under normal growth conditions (Aoki et al., 2004). This fact should be considered prior to the generalization of functional data based on mutant analysis in a single species (see above for discussion of PtdGro function).

Analysis of SQDG-deficient mutants of the unicellular algae *C. reinhardtii*, also showed impairment of the photosynthetic apparatus and growth impairment under normal growth conditions (Sato et al., 1995, 2003; Minoda et al., 2002, 2003). This phenotype was more severe under phosphate-limited growth conditions (Riekhof et al., 2003). Growth of an SQDG-deficient Arabidopsis mutant disrupted in *SQD2* was not impaired unless the plants were starved for phosphate (Yu et al., 2002) similar to the observations made in *Rb. sphaeroides* and *Synechcococcus* sp. PCC7492 described above. No detailed analysis of photosynthesis in this SQDG-deficient mutant of Arabidopsis is available at this time.

In non-photoysnthetic bacteria that contain SQDG such as *Sinorhizobium melilotii* (Cedergren and Hollingsworth, 1994), SQDG-deficiency had only subtle effects on growth, even under phosphate-limited conditions and did not affect the ability of *Sr. melilotii* (Weissenmayer et al., 2000). Thus, SQDG-deficiency has varying effects in different organisms, but unlike PtdGro it seems clear that SQDG is not essential for photosynthesis. Rather, SQDG seems to be an excellent anionic lipid in thylakoid membranes and other cell membranes in non-photosynthetic bacteria. Within limits, it can substitute for PtdGro, but cannot assume the roles of PtdGro found in association with specific photosynthetic complexes as described above.

VI The Importance of Anionic Lipids in Chloroplasts

It is interesting to note that thylakoid membranes of plastids contain the neutral but polar galactoglycerolipids and anionic lipids SQDG and PtdGro, but no zwitter-ionic lipids or positively charged lipids. The varying effects in SQDG-deficient mutants of different species and the stronger effects of PtdGro deficiency on photosynthetic competence even in a mutant with slightly reduced PtdGro content (Xu et al., 2002), suggests that anionic lipids in general are necessary for the proper functioning of photosynthetic membranes. After all, protons are transferred and charge is separated across the thylakoid membrane during photosynthetic electron transport and it has been proposed that anionic lipid headgroups act as a proton-conducting pathway along the surface of the membrane (Haines, 1983). The question, whether SQDG-deficiency causes growth impairment in some organisms but not in others was experimentally tested in Arabidopsis by creating a synthetic mutant by combining the *sqd2* (Yu et al., 2002) mutant allele causing complete loss of SQDG and the weak *pgp1-1* allele (Xu et al., 2002) into one Arabidopsis line. The resulting double mutant showed impaired growth, a reduction in overall anionic lipids, and impaired photosynthesis (Yu and Benning, 2003). Essentially, the effects of the two mutations were additive. Thus the conclusion was that anionic charge in the headgroups of thylakoid membrane lipids is essential for the proper functioning of the photosynthetic membrane. This is also evident from experiments done under phosphate-limiting growth conditions, when in all organisms tested, SQDG amounts increased while those of PtdGro decreased, but the total amount of anionic lipids remained constant (Benning et al., 1993; Güler et al., 1996; Essigmann et al., 1998; Sato et al., 2000a; Yu et al., 2002; Riekhof et al., 2003; Yu and Benning, 2003). Thus, based on these and other data discussed by others (Sato et al., 2000a; Frentzen, 2004), there is a general consensus that anionic lipids play an important role in the proper functioning of photosynthetic membranes and that SQDG and PtdGro can substitute for each other to maintain a constant level of anionic lipids. This conclusion does not preclude the possibility that the two lipids have very specific functions in association with complexes of the photosynthetic apparatus. At this time, those more specific functions are more evident for PtdGro than for SQDG.

VII Future Perspectives

While much has been learned from biochemical and genetic analysis with regard to the biosynthesis and function of the anionic lipids PtdGro and SQDG in chloroplasts, some interesting questions remain to be answered. At this time, there are obvious gaps in our knowledge about the biosynthesis of PtdGro in plastids. Not all enzymes involved in its biosynthesis, for example FAD4 introducing the *tans*-double bond into palmitic acid of PtdGro, have been identified. Moreover, it is not entirely clear whether the proposed pathway of PtdGro biosynthesis (Fig. 2) in plastids is actually correct. Based on labeling studies and its fatty acid composition PtdGro is the only one of the four chloroplast lipids which apparently does not derive from the ER-pathway. However, even a very severe reduction of the plastid pathway by nearly eliminating the activity of ATS1 does not abolish PtdGro formation. Could there be an alternative mechanism for PtdGro biosynthesis in plastids. There are numerous plants in which the galactoglycerolipids are exclusively derived from the ER-pathway (Mongrand et al., 1998). Yet PtdGro even in these plants still has a fatty acid composition suggesting that it is assembled in the plastid. It has been proposed that plastid PtdOH phosphatase activity is low in these plants (Heinz and Roughan, 1983), but molecular evidence for this hypothesis is still missing. Thus, PtdGro biosynthesis in plastids has either its own separate set of enzymes for the assembly of PtdOH in the plastid, or the general plastid lipid synthesis machinery preferentially channels PtdOH into PtdGro possibly due to limiting PtdOH phosphatase activity involved in the production of plastid-derived DAG for galactoglycerolipid biosynthesis.

There are also remaining questions regarding the biosynthesis of SQDG (Benning, 2007). The formation of UDP-sulfoquinovose catalyzed by SQD1 is dependent on UDP-glucose. While the presence of ADP-glucose in plastids is well documented, a source for UDP-glucose in plastids remains to be identified. Furthermore, a functional significance for the observed interaction between SQD1 and glutamate synthase (Shimojima et al., 2005) needs to be verified. Furthermore, the SQD2 protein has not yet been produced in recombinant form and awaits direct biochemical analysis.

Finally, the availability of lipid mutants in plants algae and bacteria have greatly contributed to our current understanding of lipid function. However, more synthetic mutants should be generated and a more detailed analysis of photosynthesis might shed additional light on the roles of these anionic lipids in plants. As more crystal structures of protein complexes of the thylakoid membrane become available, new specific associations of SQDG and PtdGro with different protein complexes might be discovered that will lead to testable hypotheses regarding their specific functions. Examples for the SQDG/PtdGro substitution hypothesis under phosphate limitation in natural environments are still rare. One example is the discovery of the prevalence of SQDG over PtdGro in marine picocyanobacteria in the North-Pacific gyre (Van Mooy et al., 2006). As more lipids of photosynthetic organisms grown in natural environments are analyzed, correlations between environmental factors and lipid composition will become apparent. The relevance of the presence of anionic lipids in thylakoid membranes of the chloroplast seems clear, but it is till a long way from a full understanding of the precise reasons for the requirement of these lipids in photosynthetic membranes.

Acknowledgements

Work on membrane lipid biosynthesis and the regulation of thylakoid lipid biosynthesis in the Benning lab has been supported in part by grants from the U.S. National Science Foundation and the U.S. Department of Energy.

Note: (1) The *FAD4* gene of Arabidopsis encoding the missing fatty acid transdesaturase or a component thereof was recently isolated (Gao et al., 2009); and (2) The gene for a plastid UDP-glucose pyrophosphorylase, essential for sulfolipid biosynthesis, was recently identified in *Arabidopsis* (Okazaki et al., 2009).

References

Aoki M, Sato N, Meguro A and Tsuzuki M (2004) Differing involvement of sulfoquinovosyl diacylglycerol in photosystem II in two species of unicellular cyanobacteria. Eur J Biochem 271: 685–693

Awai K, Xu C, Lu B and Benning C (2006) Lipid trafficking between the endoplasmic reticulum and the chloroplast. Biochem Soc Trans 34: 395–398

Babiychuk E, Müller F, Eubel H, Braun HP, Frentzen M and Kushnir S (2003) Arabidopsis phosphatidylglycerophosphate synthase 1 is essential for chloroplast differentiation, but is dispensable for mitochondrial function. Plant J 33: 899–909

Barber J and Gounaris K (1986) What role does sulfolipid play within the thylakoid membrane? Photosynth Res 9: 239–249

Beisson F, Koo AJ, Ruuska S, Schwender J, Pollard M, Thelen JJ, Paddock T, Salas JJ, Savage L, Milcamps A, Mhaske VB, Cho Y and Ohlrogge JB (2003) Arabidopsis genes involved in acyl lipid metabolism. A 2003 census of the candidates, a study of the distribution of expressed sequence tags in organs, and a web-based database. Plant Physiol 132: 681–697

Benning C (2007) Questions remaining in sulfolipid biosynthesis: a historical perspective. Photosynth Res 92: 199–203

Benning C and Somerville CR (1992a) Identification of an operon involved in sulfolipid biosynthesis in *Rhodobacter sphaeroides*. J Bacteriol 174: 6479–6487

Benning C and Somerville CR (1992b) Isolation and genetic complementation of a sulfolipid-deficient mutant of *Rhodobacter sphaeroides*. J Bacteriol 174: 2352–2360

Benning C, Beatty JT, Prince RC and Somerville CR (1993) The sulfolipid sulfoquinovosyldiacylglycerol is not required for photosynthetic electron transport in *Rhodobacter sphaeroides* but enhances growth under phosphate limitation. Proc Natl Acad Sci USA 90: 1561–1565

Benning C, Huang ZH and Gage DA (1995) Accumulation of a novel glycolipid and a betaine lipid in cells of *Rhodobacter sphaeroides* grown under phosphate limitation. Arch Biochem Biophys 317: 103–111

Benning C, Xu C and Awai K (2006) Non-vesicular and vesicular lipid trafficking involving plastids. Curr Opin Plant Biol 9: 241–247

Benning C, Garavito MR and Shimojima M (2008) Sulfolipid biosynthesis and function in plants. In: Hell R, Dahl C, Knaff D, Leustek T (eds) Sulfur Metabolism in Phototrophic Organism, Series Govindjee, ed., Advances in Photosynthesis and Respiration, Chapter 10. Springer, Dordrecht, Netherlands, pp. 189–204

Benson AA and Maruo B (1958) Plant phospholipids. Identification of the phosphatidyl glycerols. Bichim Biophys Acta 27: 189–195

Benson AA, Daniel H and Wiser R (1959) A sulfolipid in plants. Proc Natl Acad Sci USA 45: 1582–1587

Browse J and Somerville CR (1994) Glycerolipids. In: Meyerowitz EM and Somerville CR (eds) Arabidopsis. Cold Spring Harbor Press, Cold Spring Harbor, New York, pp. 881–912

Browse J, McCourt P and Somerville C (1985) A mutant of Arabidopsis lacking a chloroplast-specific lipid. Science 227: 763–765

Browse J, Warwick N, Somerville CR and Slack CR (1986) Fluxes through the prokaryotic and eukaryotic pathways of lipid synthesis in the "16:3" plant *Arabidopsis thaliana*. Biochem J 235: 25–31

Cedergren RA and Hollingsworth RI (1994) Occurrence of sulfoquinovosyl diacylglycerol in some members of the family *Rhizobiaceae*. J Lipid Res 35: 1452–1461

Domonkos I, Malec P, Sallai A, Kovacs L, Itoh K, Shen G, Ughy B, Bogos B, Sakurai I, Kis M, Strzalka K, Wada H,

Itoh S, Farkas T and Gombos Z (2004) Phosphatidylglycerol is essential for oligomerization of photosystem I reaction center. Plant Physiol 134: 1471–1478

Droppa M, Horvath G, Hideg E and Frakas T (1995) The role of phospholipids in regulating photosynthetic electron transport activities: treatment of thylakoids with phospholipase C. Photosynthes Res 46: 287–293

Dubacq JP and Tremolieres A (1983) Occurrence and function of phosphatidylglycerol containing delta-3-trans-hexadecenoic acid in photosynthetic lamellae. Physiol Veg 21: 293–312

Essigmann B, Güler S, Narang RA, Linke D and Benning C (1998) Phosphate availability affects the thylakoid lipid composition and the expression of *SQD1*, a gene required for sulfolipid biosynthesis in *Arabidopsis thaliana*. Proc Natl Acad Sci USA 95: 1950–1955

Essigmann B, Hespenheide BM, Kuhn LA and Benning C (1999) Prediction of the active-site structure and NAD^+ binding in SQD1, a protein essential for sulfolipid biosynthesis in *Arabidopsis*. Arch Biochem Biophys 369: 30–41

Frentzen M (2004) Phosphatidylglycerol and sulfoquinovosyldiacylglycerol: anionic membrane lipids and phosphate regulation. Curr Opin Plant Biol 7: 270–276

Fritz M, Lokstein H, Hackenberg D, Welti R, Roth M, Zahringer U, Fulda M, Hellmeyer W, Ott C, Wolter FP and Heinz E (2007) Channeling of eukaryotic diacylglycerol into the biosynthesis of plastidial phosphatidylglycerol. J Biol Chem 282: 4613–4625

Gao J, Ajjawi I, Manoli III A, Sawin A, Xu C, Froehlich JE, Last RL and Benning C (2009) FATTY ACID DESATURASE4 of *Arabidopsis* encodes a protein distinct from characterized desaturases. Plant J 60: 832–829

Gombos Z, Varkonyi Z, Hagio M, Iwaki M, Kovacs L, Masamoto K, Itoh S and Wada H (2002) Phosphatidylglycerol requirement for the function of electron acceptor plastoquinone Q(B) in the photosystem II reaction center. Biochemistry 41: 3796–3802

Gounaris K and Barber J (1985) Isolation and characterisation of a photosytem II reaction center lipoprotein complex. FEBS Lett 188: 68–72

Güler S, Seeliger A, Härtel H, Renger G and Benning C (1996) A null mutant of *Synechococcus* sp. PCC7942 deficient in the sulfolipid sulfoquinovosyl diacylglycerol. J Biol Chem 271: 7501–7507

Güler S, Essigmann B and Benning C (2000) A cyanobacterial gene, *sqdX*, required for biosynthesis of the sulfolipid sulfoquinovosyldiacylglycerol. J Bacteriol 182: 543–545

Hagio M, Gombos Z, Varkonyi Z, Masamoto K, Sato N, Tsuzuki M and Wada H (2000) Direct evidence for requirement of phosphatidylglycerol in photosystem II of photosynthesis. Plant Physiol 124: 795–804

Hagio M, Sakurai I, Sato S, Kato T, Tabata S and Wada H (2002) Phosphatidylglycerol is essential for the development of thylakoid membranes in *Arabidopsis thaliana*. Plant Cell Physiol 43: 1456–1464

Haines TH (1983) Anionic lipid headgroups as a proton-conducting pathway along the surface of membranes: a hypothesis. Proc Natl Acad Sci USA 80: 160–164

Heinz E and Roughan G (1983) Similarities and differences in lipid metabolism of chloroplasts isolated from 18:3 and 16:3 plants. Plant Physiol 72: 273–279

Heinz E, Schmidt H, Hoch M, Jung KH, Binder H and Schmidt RR (1989) Synthesis of different nucleoside 5'-diphospho-sulfoquinovoses and their use for studies on sulfolipid biosynthesis in chloroplasts. Eur J Biochem 184: 445–453

Hobe S, Prytulla S, Kuhlbrandt W and Paulsen H (1994) Trimerization and crystallization of reconstituted light-harvesting chlorophyll a/b complex. EMBO J 13: 3423–3429

Hobe S, Forster R, Klingler J and Paulsen H (1995) N-proximal sequence motif in light-harvesting chlorophyll a/b-binding protein is essential for the trimerization of light-harvesting chlorophyll a/b complex. Biochemistry 34: 10224–10228

Ishizaki O, Nishida I, Agata K, Eguchi G and Murata N (1988) Cloning and nucleotide sequence of cDNA for the plastid glycerol-3-phosphate acyltransferase from squash. FEBS Lett 238: 424–430

Jordan B, Chow W-S and Baker A (1983) The role of phospholipids in the molecular organisation of pea chloroplasts: effect of phospholipid depletion on photosynthetic activities. Bichim Biophys Acta 725: 77–86

Jordan P, Fromme P, Witt H, Klukas O, Saenger W and Krauss N (2001) Three-dimensional structure of cyanobacterial photosytem I at 2.5 A resolution. Nature 411: 909–917

Kim HU and Huang AH (2004) Plastid lysophosphatidyl acyltransferase is essential for embryo development in Arabidopsis. Plant Physiol 134: 1206–1216

Kinney AJ (1993) Phospholipid head groups. In: Moore TS (ed) Lipid Metabolism in Plants. CRC Press, Boca Raton, FL, pp. 259–284

Kopka J, Ludewig M and Muller-Rober B (1997) Complementary DNAs encoding eukaryotic-type cytidine-5'-diphosphate- diacylglycerol synthases of two plant species. Plant Physiol 113: 997–1002

Kopriva S (2006) Regulation of sulfate assimilation in Arabidopsis and beyond. Ann Bot (Lond) 97: 479–495

Kunst L, Browse J and Somerville C (1988) Altered regulation of lipid biosynthesis in a mutant of *Arabidopsis* deficient in chloroplast glycerol 3-phosphate acyltransferase activity. Proc Natl Acad Sci USA 85: 4143–4147

Lehmann J and Benson AA (1964) The plant sulfolipid. IX. Sulfosugar syntheses from methyl hexoseenides. J Am Chem Soc 86: 4469–4472

Leustek T, Martin MN, Bick JA and Davies JP (2000) Pathways and regulation of sulfur metabolism revealed through molecular and genetic studies. Annu Rev Plant Physiol Plant Mol Biol 51: 141–165

Liu Z, Yan H, Wang K, Kuang T, Zhang J, Gui L, An X and Chang W (2004) Crystal structure of spinach major light-harvesting complex at 2.72 A resolution. Nature 428: 287–292

Loll B, Kern J, Saenger W, Zouni A and Biesiadka J (2005) Towards complete cofactor arrangement in the 3.0 A resolution structure of photosystem II. Nature 438: 1040–1044

Loll B, Kern J, Saenger W, Zouni A and Biesiadka J (2007) Lipids in photosystem II: interactions with protein and cofactors. Biochim Biophys Acta 1767: 509–519

Lykidis A (2007) Comparative genomics and evolution of eukaryotic phospholipid biosynthesis. Prog Lipid Res 46: 171–199

McCourt P, Browse J, Watson J, Arntzen CJ and Somerville CR (1985) Analysis of photosynthetic antenna function in a mutant of *Arabidopsis thaliana* (L.) lacking trans-hexadecenoic acid. Plant Physiol 78: 853–858

Menke W, Radunz A, Schmid GH, Koenig F and Hirtz RD (1976) Intermolecular interactions of polypeptides and lipids in the thylakoid membrane. Z Naturforsch [C] 31: 436–444

Minoda A, Sato N, Nozaki H, Okada K, Takahashi H, Sonoike K and Tsuzuki M (2002) Role of sulfoquinovosyl diacylglycerol for the maintenance of photosystem II in *Chlamydomonas reinhardtii*. Eur J Biochem 269: 2353–2358

Minoda A, Sonoike K, Okada K, Sato N and Tsuzuki M (2003) Decrease in the efficiency of the electron donation to tyrosine Z of photosystem II in an SQDG-deficient mutant of Chlamydomonas. FEBS Lett 553: 109–112

Mongrand S, Besoule J-J, Cabantous F and Cassagne C (1998) The C16:3/C18:3 fatty acid balance in photosynthetic tissues from 468 plant species. Phytochemistry 49: 1049–1064

Moon BY, Higashi S, Gombos Z and Murata N (1995) Unsaturation of the membrane lipids of chloroplasts stabilizes the photosynthetic machinery against low-temperature photoinhibition in transgenic tobacco plants. Proc Natl Acad Sci USA 92: 6219–6223

Mulichak AM, Theisen MJ, Essigmann B, Benning C and Garavito RM (1999) Crystal structure of SQD1, an enzyme involved in the biosynthesis of the plant sulfolipid headgroup donor UDP-sulfoquinovose. Proc Natl Acad Sci USA 96: 13097–13102

Müller F and Frentzen M (2001) Phosphatidylglycerophosphate synthases from *Arabidopsis thaliana*. FEBS Lett 509: 298–302

Murata N (1983) Molecular species composition of phosphatidyglycerols from chilling-sensitive and chilling-resistant plants. Plant Cell Physiol 24: 476–479

Murata N, Ishizaki-Nishizawa O, Higashi S, Hayashi H, Tasaka Y and Nishida I (1992) Genetically engineered alteration in the chilling sensitivity of plants. Nature 356: 710–713

Nishida I and Murata N (1996) Chilling sensitivity in plants and cyanobacteria: the crucial contribution of memrbane lipids. Ann Rev Plant Physiol 47: 568

Nishida I, Tasaka Y, Shiraishi H and Murata N (1993) The gene and the RNA for the precursor to the plastid-located glycerol 3-phosphate acyltransferase of *Arabidopsis thaliana*. Plant Mol Biol 21: 267–277

Nussberger S, Dorr K, Wang DN and Kuhlbrandt W (1993) Lipid-protein interactions in crystals of plant light-harvesting complex. J Mol Biol 234: 347–356

Okazaki Y, Shimojima M, Sawada Y, Toyooka K, Narisawa T, Mochida K, Tanaka H, Matsuda F, Hirai A, Hirai MY, Ohta H and Saito K (2009) A chloroplastic UDP-glucose pyrophosphorylase from *Arabidopsis* is the committed enzyme for the first step of sulfolipid biosynthesis. Plant Cell 21: 892–909

Pick U, Gounaris K, Weiss M and Barber J (1985) Tightly bound sulfolipids in chloroplast CF0-CF1. Biochim Biophys Acta 808: 415–420

Pick U, Weiss M, Gounaris K and Barber J (1987) The role of different thylakoid glycolipids in the function of reconstituted chloroplast ATP synthase. Biochim Biophys Acta 891: 28–39

Pineau B, Girard-Bascou J, Eberhard S, Choquet Y, Tremolieres A, Gerard-Hirne C, Bennardo-Connan A, Decottignies P, Gillet S and Wollman FA (2004) A single mutation that causes phosphatidylglycerol deficiency impairs synthesis of photosystem II cores in *Chlamydomonas reinhardtii*. Eur J Biochem 271: 329–338

Pugh CE, Roy AB, Hawkes T and Harwood JL (1995) A new pathway for the synthesis of the plant sulpholipid, sulphoquinovosyldiacylglycerol. Biochem J 309: 513–519

Radunz A and Schmid GH (1992) Binding of lipids onto polypeptides of the thylakoid membrane I. Galactolipids and sulpholipid as prosthetic groups of core peptides of the photosystem II complex. Z Naturforsch [C] 47: 406–415

Riekhof WR, Ruckle ME, Lydic TA, Sears BB and Benning C (2003) The sulfolipids 2'-O-acyl-sulfoquinovosyldiacylglycerol and sulfoquinovosyldiacylglycerol are absent from a *Chlamydomonas reinhardtii* mutant deleted in SQD1. Plant Physiol 133: 864–874

Rossak M, Tietje C, Heinz E and Benning C (1995) Accumulation of UDP-sulfoquinovose in a sulfolipid-deficient mutant of *Rhodobacter sphaeroides*. J Biol Chem 270: 25792–25797

Rossak M, Schäfer A, Xu N, Gage DA and Benning C (1997) Accumulation of sulfoquinovosyl-1-O-dihydroxyacetone in a sulfolipid- deficient mutant of *Rhodobacter sphaeroides* inactivated in *sqdC*. Arch Biochem Biophys 340: 219–230

Roughan PG and Slack CR (1982) Cellular organization of glycerolipid metabolism. Ann Rev Plant Physiol 33: 97–132

Roughan PG, Holland R and Slack CR (1980) The role of chloroplasts and microsomal fractions in polar-lipid synthesis from [1-^{14}C]acetate by cell-free preparations from spinach (*Spinacia oleracea*) leaves. Biochem J 188: 17–24

Sakurai I, Hagio M, Gombos Z, Tyystjarvi T, Paakkarinen V, Aro EM and Wada H (2003) Requirement of phosphatidylglycerol for maintenance of photosynthetic machinery. Plant Physiol 133: 1376–1384

Sanda S, Leustek T, Theisen M, Garavito M and Benning C (2001) Recombinant Arabidopsis SQD1 converts UDP-glucose and sulfite to the sulfolipid head precursor UDP-sulfoquinovose *in vitro*. J Biol Chem 276: 3941–3946

Sato N, Tsuzuki M, Matsuda Y, Ehara T, Osafune T and Kawaguchi A (1995) Isolation and characterization of mutants affected in lipid metabolism of *Chlamydomonas reinhardtii*. Eur J Biochem 230: 987–993

Sato N, Hagio M, Wada H and Tsuzuki M (2000a) Environmental effects on acidic lipids of thylakoid membranes. Biochem Soc Trans 28: 912–914

Sato N, Hagio M, Wada H and Tsuzuki M (2000b) Requirement of phosphatidylglycerol for photosynthetic function in thylakoid membranes. Proc Natl Acad Sci USA 97: 10655–10660

Sato N, Aoki M, Maru Y, Sonoike K, Minoda A and Tsuzuki M (2003) Involvement of sulfoquinovosyl diacylglycerol in the structural integrity and heat-tolerance of photosystem II. Planta 217. 245–251

Sato N, Suda K and Tsuzuki M (2004) Responsibility of phosphatidylglycerol for biogenesis of the PSI complex. Biochim Biophys Acta 1658: 235–243

Seifert U and Heinz E (1992) Enzymatic characteristics of UDP-sulfoquinovose:diacylglycerol sulfoquinovosyltransferase from chloroplast envelopes. Bot Acta 105: 197–205

Seras M, Garnier J, Tremolieres A and Guyon D (1989) Lipid biosynthesis in cells of the wild-type and of 2 photosynthesis mutants of *Chlamydomonas reinhardtii*. Plant Physiol Biochem 27: 393–399

Shimojima M and Benning C (2003) Native uridine 5'-diphosphate-sulfoquinovose synthase, SQD1, from spinach purifies as a 250-kDa complex. Arch Biochem Biophys 413: 123–130

Shimojima M, Hoffmann-Benning S, Garavito RM and Benning C (2005) Ferredoxin-dependent glutamate synthase moonlights in plant sulfolipid biosynthesis by forming a complex with SQD1. Arch Biochem Biophys 436: 206–214

Sigrist M, Zwillenberg C, Giroud CH, Eicehnberger W and Boschetti A (1988) Sulfolipid associated with the light harvesting complex associated with photosytem II apoproteins of *Chlamydomonas reinhardtii*. Plant Sci 58: 15–23

Stroebel D, Choquet Y, Popot JL and Picot D (2003) An atypical haem in the cytochrome b(6)f complex. Nature 426: 413–418

Szalontai B, Kota Z, Nonaka H and Murata N (2003) Structural consequences of genetically engineered saturation of the fatty acids of phosphatidylglycerol in tobacco thylakoid membranes. An FTIR study. Biochemistry 42: 4292–4299

Van Mooy BA, Rocap G, Fredricks HF, Evans CT and Devol AH (2006) Sulfolipids dramatically decrease phosphorus demand by picocyanobacteria in oligotrophic marine environments. Proc Natl Acad Sci USA 103: 8607–8612

Wada H and Murata N (2007) The essential role of phosphatidylglycerol in photosynthesis. Photosynth Res 92: 205–215

Weissenmayer B, Geiger O and Benning C (2000) Disruption of a gene essential for sulfoquinovosyldiacylglycerol biosynthesis in *Sinorhizobium meliloti* has no detectable effect on root nodule symbiosis. Mol Plant Microbe Interact 13: 666–672

Wolter FP, Schmidt R and Heinz E (1992) Chilling sensitivity of *Arabidopsis thaliana* with genetically engineered membrane lipids. EMBO J 11: 4685–4692

Xu C, Härtel H, Wada H, Hagio M, Yu B, Eakin C and Benning C (2002) The *pgp1* locus of Arabidopsis encodes a phosphatidylglycerol synthase with impaired activity. Plant Physiol 129: 594–604

Xu C, Cornish AJ, Froehlich JE and Benning C (2006) Phosphatidylglycerol biosynthesis in chloroplasts of Arabidopsis mutants deficient in acyl-ACP glycerol-3-phosphate acyltransferase. Plant J 47: 296–309

Yu B and Benning C (2003) Anionic lipids are required for chloroplast structure and function in Arabidopsis. Plant J 36: 762–770

Yu B, Xu C and Benning C (2002) Arabidopsis disrupted in SQD2 encoding sulfolipid synthase is impaired in phosphate-limited growth. Proc Natl Acad Sci USA 99: 5732–5737

Yu B, Wakao S, Fan J and Benning C (2004) Loss of plastidic lysophosphatidic acid acyltransferase causes embryo-lethality in Arabidopsis. Plant Cell Physiol 45: 503–510

Chapter 13

Biosynthesis and Function of Monogalactosyldiacylglycerol (MGDG), the Signature Lipid of Chloroplasts

Yuki Nakamura
Yong Loo Lin School of Medicine, Centre for Life Sciences, Department of Biochemistry, National University of Singapore, 28 Medical Drive, Singapore 117456, Republic of Singapore

Mie Shimojima and Hiroyuki Ohta*
Center for Biological Resources and Informatics, Tokyo Institute of Technology, 4259-B-65 Nagatsuta-cho, Midori-ku, Yokohama 226-8501, Japan
Research Center for the Evolving Earth and Planets, Tokyo Institute of Technology, 4259-B-65 Nagatsuta-cho, Midori-ku, Yokohama 226-8501, Japan

Koichi Kobayashi
Graduate School of Bioscience and Biotechnology, Tokyo Institute of Technology, 4259-B-65 Nagatsuta-cho, Midori-ku, Yokohama 226-8501, Japan

*Author for Correspondence, e-mail: ohta.h.ab@m.titech.ac.jp

Constantin A. Rebeiz et al. (eds.), The Chloroplast: Basics and Applications, pp. 185–202.

Summary

Monogalactosyldiacylglycerol (MGDG) is a predominant membrane constituent of chloroplasts and other plastids, which occupies more than 50% of total photosynthetic membrane lipids. Therefore, the biosynthesis of MGDG has been considered to be crucial for plant photosynthetic function. Since the success in the identification of MGDG synthase from cucumber, significant progress was made in this decade to unravel the nature of MGDG synthase. These include not only basic studies with regard to biochemical features, regulatory systems, in vivo function, but also emerging new focus such as tools for searching the evolutionary origin of chloroplasts, design of herbicides and functions of novel class of oxylipin-containing galactolipids. In this article, we review advances made in this decade to understand a broad array of function and biosynthetic pathways of MGDG.

I Introduction

Monogalactosyldiacylglycerol (MGDG) is a glycoglycerolipid which has one galactose moiety bound to the *sn*-3 position of glycerol to form a hydrophilic head group and two acyl chains at both the *sn*-1 and *sn*-2 positions of the glycerol backbone. MGDG is a predominant membrane lipid in seed plants: it constitutes more than 50% of the total chloroplast membrane lipid (Douce and Joyard, 1980). Therefore, it is said that MGDG is the most abundant polar lipid on earth. In eukaryotes, the galactolipid is found exclusively in photosynthetic organisms, such as seed plants, fungi, green mosses and algae with a few exceptions of a certain non-green protista such as Apicomplexan which have a plastid called apicoplasts containing MGDG and other galactolipids in their membranes (Maréchal et al., 2002). Indeed, galactolipids are a rare lipid class in animals. MGDG is synthesized in the envelope of plastids by galactosylation of *sn*-1,2-diacylglycerol (DAG) from uridine-diphosphate-galactose (UDP-Gal). The enzyme catalyzing this reaction step is called MGDG synthase (MGD). It was intensively studied, because it was considered to be a key enzyme in the formation of photosynthetic membranes. However, it was only a decade ago when the isolation of the gene encoding MGD was reported (Shimojima et al., 1997). Since then, significant progress has been made to elucidate the biochemical properties, in vivo function and regulation of MGD. In this review, we aim to summarize the research progress made after the isolation of the MGD gene in seed plants. For information before gene identification, readers are advised to refer to the following well-documented reviews (Joyard and Douce, 1987; Joyard et al., 1998).

Abbreviations: BA – 6-benzyladenine; DAG – *sn*-1,2 diacylglycerol; DGDG – digalactosyldiacylglycerol; DTT – dithiothreitol; GL2 – GLABRA2; GT – glycosyltransferase; IAA – indole-3-acetic acid; LPP – lipid phosphate phosphatase; MGD – monogalactosyldiacylglycerol synthase; MGDG – monogalactosyldiacylglycerol; MGDG-O 12-oxophytodienoic acid-containing monogalactosyldiacylglycerol; MGlcDG – monoglucosyldiacylglycerol; NEM – *N*-ethylmaleimide; NPA – *N*-(1-naphthyl) phthalamic acid; NPC – non-specific phospholipase C; OPDA – 12-oxophytodienoic acid; PA – phosphatidic acid; PAP – phosphatidic acid phosphatase; PC – phosphatidylcholine; PCIB – *p*-chlorophenoxyisobutyric acid; PE – phosphatidylethanolamine; PG – phosphatidylglycerol; Phi – phosphite; Pi – phosphate; PI – phosphatidylinositol; PLC – phospholipase C; PLD – phospholipase D; SQDG – sulfoquinovosyldiacylglycerol; TAG – triacylglycerol; TIBA – 2,3,5-triiodobenzole; UDP-Gal – uridine diphosphate-galactose; UDP-Glc – uridine diphosphate-glucose; WT – wild type;

II Identification of MGDG Synthase in Seed Plants

In seed plants, MGDG is synthesized from DAG and UDP-Gal by MGDG synthase (UDP-galactose: 1,2-*sn*-diacylglycerol 3-*β*-D-galactosyltransferase, EC 2.4.1.46). Due to its extremely low abundance and tight association with cellular membranes, the purification of native MGDG synthase from plants was very challenging (Slabas, 1997). Initial efforts were made by two groups for more than 10 years. They purified approximately 20 kDa protein from the envelope fraction of spinach chloroplasts and tentatively suggested it was MGDG synthase (Maréchal et al., 1991; Teucher and Heinz, 1991). However, the data were not conclusive leaving doubts whether this was indeed MGDG synthase since the amount of purified protein was too low to determine the amino acid sequences. Using a solubilized microsome fraction of cucumber cotyledons, Shimojima and co-workers purified MGD to >11,000-fold and isolated a protein with approximate molecular mass of 47 kDa (Ohta et al. 1995b; Shimojima et al., 1997). Successful amino acid sequencing determined the full-length of MGDG synthase as a homolog of *E. coli* MurG, a glycosyltransferase involved in formation of peptidoglycan. In vitro expression of recombinant protein showed significant activity to produce MGDG from DAG and UDP-Gal, which let us to stipulate that this protein encodes a functional MGD, thus concluding the long search for the MGD purification (Shimojima et al., 1997). The encoded protein for cucumber MGDG synthase (CsMGD) had a transit peptide that is targeted to the chloroplast. Later on, a screening of a spinach cDNA library by CsMGD cDNA identified a spinach MGDG synthase (SoMGD1) gene encoding a protein with a molecular mass of 45 kDa in mature form and sequence similarity of 73.4% to CsMGD (Miège et al., 1999). The enzyme activity of SoMGD1 was confirmed by in vitro reconstitution and activity assay of mature protein in *E. coli*, suggesting the MGD protein is widespread among seed plants.

III Biochemical Properties of MGDG Synthase

A Enzymatic Features of MGDG Synthase

Even before the isolation of the MGDG synthase in 1997, the enzymatic features were well-characterized by several groups particularly for spinach MGDG synthase (Joyard and Douce, 1987; Joyard et al., 1998). However, successful gene isolation of MGDG synthase by Shimojima and co-workers allowed us to study detailed enzymological features of MGDG synthase by using the recombinant protein.

Various factors are reported to change the activity of MGD. Addition of dithiothreitol (DTT) activates SoMGD (elimination of DTT from the assay mixture reduced MGD activity by 85%) and CsMGD (Miège et al., 1999; Yamaryo et al., 2006). On the other hand, a thiol-blocking reagent *N*-ethylmaleimide (NEM) and *o*-phenanthroline severely inhibit the activity. Recent study by Yamaryo and co-workers showed in vitro regulation of CsMGD by redox state. Since CsMGD requires a reductant such as DTT for maintaining its activity and addition of NEM to inhibit it, it was thought that the thiol groups of cysteine residues in MGD are involved in the regulation of activity (Yamaryo et al., 2006). When recombinant CsMGD was oxidized by $CuCl_2$, the activity was diminished. However, reduction by DTT treatment recovered the activity and this change was confirmed to be reversible. This recovery was observed with thioredoxin (Trx), a possible reductant although it remains to be elucidated whether CsMGD and Trx co-localize in vivo. In vitro reconstitution of Trx-mediated transmission of redox potential showed that oxidized MGD recovered its activity by reduction of its thiol groups by Trx which was reduced by NADPH and NADPH-dependent thioredoxin reductase (NTR). The activation of MGD by lipid molecules was suggested by several reports. An early study reported phosphatidylglycerol (PG) as a potent activation factor (Covès et al., 1988). However, Ohta and co-workers showed in cucumber that activation by phosphatidic acid (PA) is the most reasonable, considering the in vivo concentration of each lipid class (60 μM PA showed 30-fold increase in MGD activity) (Ohta et al., 1995).

B Subcellular Localization of MGDG Synthase

Spinach MGD synthase was reported to be located in the inner envelope membrane of chloroplasts (Dorne et al., 1982; Block et al., 1983) whereas in the case of Pea (*Pisum sativa*) the activity was located in the outer envelope membranes (Cline and Keegstra, 1983). After the MGD gene was isolated in 1997, localization studies with MGD were done using antibodies. In spinach, the result of western blot analyses with anti-SoMGD and in vitro import analyses suggested that SoMGD is a monotopic protein associated with the inner envelope, so that a fraction of the protein could be removed by 0.1 M Na_2CO_3, (Miège et al., 1999). The localization was then investigated for three isoforms of Arabidopsis MGD (Awai et al., 2001; Xu et al., 2005) (described in detail below). MGD1 was shown to be localized in the outer leaflet of the inner envelope (Xu et al., 2005), whereas MGD2 and MGD3 were exclusively localized at the outer envelope (Awai et al., 2001). These results correlated with the existence of putative transit peptides: AtMGD1 and SoMGD1 having a demonstrated transit peptide, whereas AtMGD2 and AtMGD3 having no corresponding target sequences.

C Three-Dimensional Structure of MGDG Synthase

The 3D structure of MGDG synthase is hard to establish since it is a membrane protein. However, MurG, a bacterial glycosyltransferase which catalyzes the transfer of *N*-acetyl-glucosamine onto lipid 1, shows sequence similarity to MGD monomers. Using the established 3D structure of *E. coli* MurG as a template, a folding model for spinach MGD was computed (Botté et al., 2005). The predicted monomer structure was a double Rossmann fold. The possible binding site for UDP-Gal was suggested in the cleft which separates the two Rossmann folds. As for a DAG-binding site, they combined the obtained model with phylogenetic and biochemical information, and suggested the possible site in a specific region of the N-terminal domain. Furthermore, they suggested that certain hydrophobic amino acids of the N-terminal domain are involved in the association of MGD with membranes (Botté et al., 2005).

D Two Types of MGDG Synthase in Arabidopsis

In Arabidopsis, three isoforms of MGD were isolated and named AtMGD1, AtMGD2 and AtMGD3 (Table 1) (Awai et al., 2001). These isoforms are categorized into two subgroup, either Type A (AtMGD1) or Type B (AtMGD2 and AtMGD3). AtMGD1 shows more than 80% amino acid identity to CsMGD and SoMGD, and possesses a transit peptide. By contrast, AtMGD2 and AtMGD3 show relatively lower amino acid identities to AtMGD1 (about 60%) and no transit peptides were annotated. Immunochemical studies as well as in vitro transport experiments with intact Pea chloroplasts showed that AtMGD1 is localized to the inner envelope whereas AtMGD2 and AtMGD3 were both in the outer envelope (Awai et al., 2001). Furthermore, a recent report determined the topological orientation of AtMGD1 concluding that this protein faces the inter-membrane space of both envelopes (Xu et al., 2005). Enzyme activity assays of recombinant AtMGD expressed in *E. coli* showed significant activity for the incorporation of galactose into DAG to form MGDG (Awai et al., 2001). However, there was a difference in substrate preference for DAG molecular species; while Type A equally utilized both 18:2/18:2-DAG and 18:1/16:0-DAG, Type B preferred 18:2/18:2-DAG to 18:1/16:0-DAG, suggesting that Type B is more specialized for DAG derived from the eukaryotic pathway, assembling the DAG backbone at the ER (Roughan and Slack, 1982). This result is in contrast to SoMGD which has broad substrate selectivity for DAG molecular species (affinity; 18:2/18:2-DAG>18:1/18:1-DAG>18:1/16:0-DAG) (Miège et al., 1999). Interestingly, this activity matched well with that of native MGD in an isolated envelope fractions (Maréchal et al., 1995), suggesting that SoMGD is capable of accepting both prokaryotic and eukaryotic DAG.

Table 1. Comparison of two types of MGD in Arabidopsis.

	Isoform	Transit peptide	DAG preference	Suborganelle localization	Expressed organs	Regulation factors
Type A	MGD1	Yes	Prokaryotic/eukaryotic	Inner envelope	Ubiquitously expressed	Light/cytokinin
Type B	MGD2	No	Eukaryotic	Outer envelope	Non-photosynthetic organs	Auxin/Pi signal
	MGD3	No	Eukaryotic	Outer envelope	Non-photosynthetic organs	Auxin/Pi signal

Expression studies also suggest functional differences between Type A and Type B. RT-PCR showed that Type B genes were highly expressed in certain non-photosynthetic organs (AtMGD2; flower, AtMGD3; root) whereas Type A was expressed in all organs analyzed. Furthermore, AtMGD3 was highly expressed in seedlings and its level decreased as they grew (Awai et al., 2001). A more detailed expression study using GUS reporter assays showed that expression of MGD1::GUS was detected widely in all green tissues, whereas the staining in transformants harboring MGD2::GUS or MGD3::GUS was observed only in restricted parts (e.g. leaf tips) (Kobayashi et al., 2004). However, the type B::GUS transformants showed intense staining in growing pollen tubes, which suggests possible involvement of Type B MGD in galactolipid synthesis in pollen tube formation. In seedlings, GUS expression of MGD3 was detected in addition to MGD1, which is in good agreement with the result obtained by RT-PCR. Although roots seem unlikely the primary organ for MGD expression, intense GUS staining of Type B appears when plants are subject to phosphate starvation (described in detail in Section IV) (Awai et al., 2001; Kobayashi et al., 2004).

E MGDG Synthesis in Non-photosynthetic Organs

In contrast to photosynthetic organs in which MGDG is involved in the assembly of photosynthetic membranes, non-photosynthetic organs such as flowers and roots were long ignored in the context of galactolipid biosynthesis although early studies showed existence of considerable amounts of galactolipids in such organs (Kleinig and Liedvogel, 1978; Camara et al., 1983). The identification of two types of MGD in Arabidopsis and their differential expression pattern (Awai et al., 2001; Kobayashi et al., 2004) suggested that overall galactolipid biosynthesis and function of MGD might be distinct in non-photosynthetic organs. Above all, the intense staining observed in anthers and elongating pollen tubes of Type B::GUS transformants imply the importance of MGDG synthesis during the reproductive process. Basic characterization of galactolipid biosynthesis in flowers was conducted with *Petunia hybrida*, and results showed several distinct features (Nakamura et al., 2003). In flowers, digalactosyldiacylglycerol (DGDG) exceeded MGDG as the major galactolipid, although the ratio of MGDG to DGDG is nearly 2:1 in photosynthetic organs presumably for proper formation of photosynthetic membranes (Murphy, 1982). In addition, MGDG synthesis activity increased during flower development: the highest activity was observed in mature pistils, almost twice as high as in leaves. These results suggest that biosynthesis of MGDG is significant in flowers and has as of now an unknown function. Although no further analyses were reported to date, mutant analyses on type B MGD in Arabidopsis may give us clues for a better understanding of this enzyme class.

By contrast, galactolipid especially MGDG was very low in roots of Arabidopsis. However, when it comes to phosphate starvation, a rapid increase in DGDG (but not MGDG) occurs (Härtel et al., 2000). This increase is primarily due to activation of type B MGD (Awai et al., 2001; Kobayashi et al., 2004) and DGD1/DGD2 (Kelly and Dörmann, 2002; Kelly et al., 2003) enzymes involved in the subsequent galactosylation to yield DGDG. Although details of DGDG will be described elsewhere in this book, it should be noted that the increased DGDG substitutes for the major portion of decreased phospholipids in root plasma membranes of oat (Andersson et al., 2003) and mitochondria of Arabidopsis (Jouhet et al., 2004).

IV Function and Regulation of MGDG Synthase

A Regulation of Type A MGDG Synthase

Chloroplasts do not exist in seeds but are formed from its precursors, proplastids, during the process of germination. The most prominent morphological change that takes place within a plastid is the formation of the thylakoid membrane. Considering that thylakoid membrane lipids mostly consist of galactolipids, one could consider the formation of the thylakoid membrane as active galactolipid synthesis in view of lipid formation. Indeed, a rapid increase in MGDG synthase activity occurs upon illumination on dark-germinated cucumber seedlings (Ohta et al., 1995a). Although the activity increased tenfold during the first 4 days of darkness after germination, illumination of those 4-day-old seedlings increased the activity by threefold in the following 42 h. At this time, accumulation of galactolipids was also increased by six to seven times that before illumination. Here, one intriguing point to focus on is how MGD can accomplish the regulation of the rapid increase of MGDG that occurs during germination and greening, despite its low amount (>0.1% of total envelope protein). The most important regulatory factor for MGDG synthesis is light. Expression of Type A MGD is induced by light, which triggers an increase of MGDG synthase activity followed by the accumulation of MGDG and DGDG. Detailed analyses of CsMGD in cucumber seedlings showed profound evidence of regulation of MGDG synthase by light and phytohormones (Yamaryo et al., 2003). Concomitant analysis of gene expression and product accumulation of MGDG synthesis that occurs by illuminating dark-germinated cucumber seedlings revealed that induction of *MGD* expression was followed by the subsequent increase in the amount of enzyme and MGDG production. When detached cotyledons are subject to illumination, expression as well as activity of MGD was less induced. However, accumulation of MGDG was not affected, suggesting that light is not merely an inducer of MGD expression but is involved in MGDG production in a distinct way. On the other hand, cytokinin treatment induced the mRNA level of MGD in darkness. Since cytokinin is known to be transported from roots or hypocotyl in the seedlings, excision of cotyledons may interrupts cytokinin signaling and thus reduces gene expression. This suggests that gene expression is regulated by light via cytokinin. However, production of MGDG was not fully recovered by cytokinin treatment of dark-grown seedlings. Simultaneous treatment of illumination and cytokinin to detached cotyledons favorably recovered MGDG production, suggesting that light and cytokinin cooperatively regulate MGDG synthesis. The effect of illumination was further studied and results showed that red light (600–700 nm), but not far-red light, is required for MGDG production whereas illumination of far-red light is sufficient to induce gene expression of CsMGD. Since the cucumber *lh* mutant (a Phy B disruptant) showed no changes in MGDG synthesis at least in cucumber, Phy A is likely to be involved in the photoreception for MGDG synthesis.

B Regulation of Type B MGDG Synthase

Although the expression of type B MGD is restricted to certain parts of organs (MGD2; flowers, MGD3; roots) under normal growth condition, it is rapidly induced when plants suffer phosphate (Pi) limitation particularly in roots (Awai et al., 2001; Kobayashi et al., 2004). Phosphate is an essential macronutrient for plant growth, development and reproduction (Raghothama, 1999). It plays a decisive roles not only in signaling cascades but is also a constitutive component of nucleic acids and phospholipids. Although Pi is mostly assimilated by the root surface, it often becomes deficient in soil since a major portion of Pi forms insoluble salts such as calcium phosphate or aluminium phosphate that are unavailable for assimilation by roots. When plants suffer Pi limitation, highly integrated systems are activated for assimilation of external Pi and internal phosphorus storage. Root architecture is modified to increase root hair formation presumably for the sake of efficient Pi uptake by enlarging the absorptive root surface areas (Lynch, 1995). Another primary adaptation mechanism to Pi starvation is the alteration of membrane lipid composition; a replacement of phospholipids by a non-phosphorus galactolipid DGDG. Indeed, this increased DGDG is localized outside the plastids (Härtel et al., 2000) and DGDG becomes a component of the plasma membrane lipid (Andersson et al., 2003). In Arabidopsis, several genes for galactolipid biosynthesis (MGD2/3, DGD1/2) are known to be activated (Awai et al., 2001, Kelly and Dörmann, 2002, Kelly et al., 2003). Since no increase in MGDG was observed during Pi starvation, the type B MGD induced here is considered to function in the substrate supply to the DGDG production.

As mentioned above, Pi starvation induces formation of lateral roots and root hairs. This architectural modification resembles that of morphological changes triggered by auxin action. Moreover, spatial analysis of type B MGD::GUS activity during Pi starvation showed intense GUS staining in epithem cells of hydathodes, the base of trichomes, stipules and lateral root branches, similar to the staining pattern of auxin-derived GUS activity of the auxin-responsive element::GUS (DR5::GUS) construct (Avsian-Kretchmer et al., 2002; Aloni et al., 2003; Kobayashi et al., 2006). Indeed, Pi-starved roots are reported to increase sensitivity to auxin. When Pi-starved roots were treated with auxin inhibitors, 2,3,5-triiodobenzole (TIBA; an inhibitor of auxin transport) and *p*-chlorophenoxyisobutyric acid (PCIB, an inhibitor of auxin effects), the GUS staining of type B MGD was diminished. A similar effect was observed with *N*-(1-naphthyl) phthalamic acid (NPA), an inhibitor of auxin transport, suggesting that expression of type B MGD in Pi-starved roots requires polar transport of auxin from the shoot (Kobayashi et al., 2006). In contrast to auxin that promotes formation of lateral roots and root hair, cytokinin is known to affect root elongation and lateral root formation. In addition, cytokinin is reported to be inhibitory during Pi starvation. When Pi-starved plants were treated with 6-benzyladenine (BA), a synthetic cytokinin, the expression of type B MGD was strongly repressed in roots but not in shoots. Moreover, a simultaneous treatment of BA and indole-3-acetic acid (IAA), a representative auxin, recovered the expression, suggesting that the effects of auxin and cytokinin are antagonistic in the expression of type B MGD during Pi starvation. Such transcriptional regulation of type B MGD by auxin and cytokinin was confirmed also by membrane lipid alteration; accumulation of DGDG during Pi starvation was cancelled by the treatment of BA and TIBA in roots but not in shoots. Furthermore, analyses of the *axr4-2aux1-7* double mutant, which is resistant to auxin and shows a great reduction in lateral root formation and gravitropism (Hobbie and Estelle, 1995; Yamamoto and Yamamoto, 1999) showed decreases in the expression of type B MGD as well as DGDG accumulation during Pi starvation, suggesting that auxin signaling may be involved in Pi-starved roots in vivo. Although activation of type B MGD synthases requires auxin during Pi starvation, exogenous auxin treatment itself could not induce the expression of these genes under normal growth condition, suggesting that the Pi starvation-inducible activation requires an additional signal(s) other than auxin (Kobayashi et al., 2006). Phosphite (Phi) is an inactive analog of the Pi anion, which mimics Pi in signaling pathways, thereby suppressing various Pi-starvation inducible responses (Ticconi et al., 2001; Varadarajan et al., 2002). Because Phi is not an available source of phosphorus for plants under Pi-starved condition, plants still suffer Pi starvation even in the presence of Phi. Addition of Phi in Pi-starved plants cancelled the expression level of type B MGD synthase and accumulation of DGDG. However, IAA treatment of this –Pi/+Phi plants did not induce the expression of type B, indicating that suppression of Type B by Phi treatment was independent of auxin (Kobayashi et al., 2006). Under –Pi/+Phi conditions, plants would be in a state of 'pseudo Pi sufficiency', even though they were starved for available Pi, because Phi mimics Pi signaling. Therefore, these data suggest that type B MGD gene expression does not depend on the intracellular Pi availability but rather is regulated by a Pi-sensing signaling system. A recent report suggests that expression of MGD3 by Pi starvation is mainly dependent on cell division activity that determines the magnitude of the Pi starvation responses (Lai et al., 2007).

C In Vivo Function of MGDG Synthase by Mutant Analyses

Indirect evidence about the function of MGDG synthesis is based on a fact that appears to be correlated to photosynthesis. However, an in vivo function of MGD synthase was directly tested for the first time when Jarvis and co-workers reported a leaky mutant in which a T-DNA was inserted in the upstream region of MGD1 (Jarvis et al., 2000). This mutant showed phenotypes both in chloroplast morphology and membrane lipid metabolism. In mutant chloroplasts, the organelle size, thylakoid membranes, and size and number of starch grains were decreased. However, these morphological changes were not observed in etioplasts, suggesting that the phenotype may be associated with photosynthetic function. As for lipid metabolism, the content of MGDG was decreased by nearly 50% of the wild typte (WT) and three-fourth of MGDG synthase activity disappeared. These results suggest that MGD1 may be involved in the bulk of MGDG synthesis and photosynthesis in chloroplasts. However, there was no decrease in DGDG content as well as DGDG biosynthetic activity, suggesting that suppression of MGD1 does not affect the substrate supply to DGDG synthesis, i.e. rather MGD2/3 would be involved in this step. These observations gave us profound insight into the in vivo function of Type A MGD, although we still wondered whether B MGDG synthase could compensate

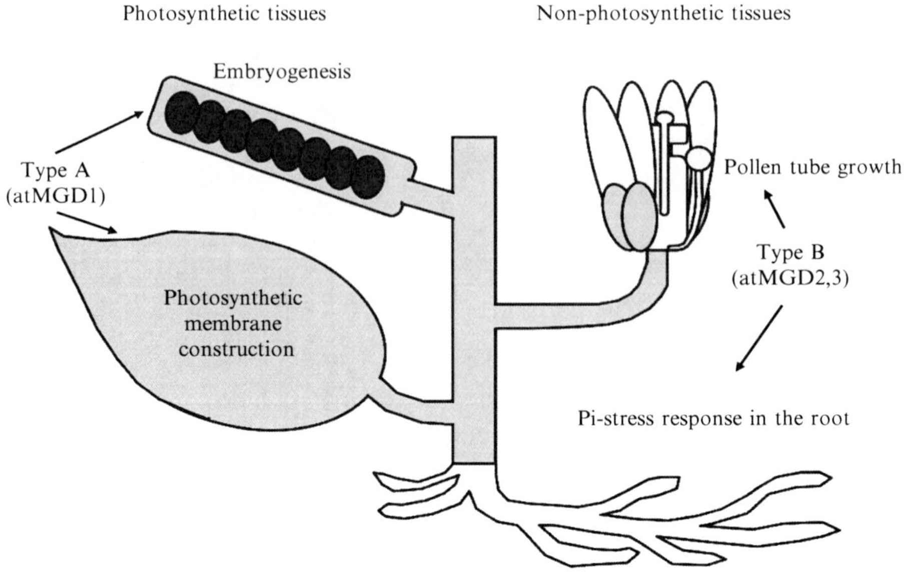

Fig. 1. Physiological functions of *type A* and *type B MGDs* for plant growth. *Type A MGD1* are expressed widely in photosynthetic tissues. By contrast, Type B MGD2 and 3 showed intense expression particularly in non-photosynthetic tissues during Pi starvation and pollen tube growth.

for Type A function and what it would be like if MGD1 was completely eliminated. Moreover, it was still uncertain whether MGDG is essential for chloroplast function. To address this question, we recently isolated and characterized a null mutant of MGD1, designated *mgd1-2*, obtained from an available T-DNA tagged mutant line (Kobayashi et al., 2007). When the siliques of *mgd1/MGD1* heterologous mutant were examined, about 25% of the embryos were aberrant, suggesting that these seeds may be *mgd1-2/mgd1-2* homozygous with a likely embryonic lethal phenotype. Interestingly, however, when the wrinkled seed were planted, they germinated. The phenotype of *mgd1-2* was intriguing; the mutant plant was very small and white compared to WT. Lipid analyses of *mgd1-2* revealed that plastidic membrane lipids, such as MGDG, DGDG, and sulfoquinovosyldiacylglycerol (SQDG), were hardly detectable, presumably due to the deficiency in chloroplasts formation. Indeed, an observation of mutant chloroplasts by electron microscopy showed a disrupted structure of thylakoid membranes. Thus, results of analyses with *mgd1* (referred to as *mgd1-1* hereafter) and *mgd1-2* indicated that MGD1 is responsible for the bulk of MGDG synthesis and cannot be replaced by Type B MGDG synthases. Therefore, its function is indispensable for photosynthetic ability and formation of proper chloroplasts.

As for type B MGD (MGD2 and MGD3), preliminary results of mutant analyses showed no significant impact on lipid metabolism in photosynthetic organs, although analyses in non-photosynthetic organs are still ongoing. Since Awai and co-workers showed that MGD2 and MGD3 were expressed in different organs (Awai et al., 2001; Kobayashi et al., 2004), it is of great interest to find out whether these two isozymes are functionally redundant or have distinct functions each other (Fig. 1).

V Substrate Supply Systems for MGDG Synthesis

The substrates for MGDG synthesis are DAG and UDP-Gal. Here, our focus is on the supply of DAG for MGDG biosynthesis. DAG is supplied by multiple metabolic pathways, and production of DAG itself presents a complex problem as this lipid class serves as substrate for galactolipids, phospholipids and even triacylglycerol (TAG). For the enzymes involved in the biosynthesis of DAG and its precursors, readers are referred to

Joyard et al. (1998). The biosynthesis of MGDG occurs exclusively in plastids (Douce and Joyard, 1980). However, DAG is known to be supplied either within plastids (prokaryotic pathway) or by way of the ER (eukaryotic pathway) (Roughan and Slack, 1982). Early studies suggested that in both pathways, DAG is derived from different lipid precursors. In the prokaryotic pathway, DAG is produced from phosphatidic acid (PA) by PA phosphatase (PAP) (Joyard and Douce, 1977). On the other hand, eukaryotic DAG is generally assumed to be derived from PC synthesized in the ER (PC hypothesis). This hypothesis is based on in situ pulse chase studies, where radiolabel transiently allocated to the cellular pool of PC being incorporated mainly into MGDG (Roughan, 1970; Heinz, 1977; Douce and Joyard, 1980; Roughan and Slack, 1982). This hypothesis involves (1) the synthesis of PC, (2) its transfer (or that of DAG) to chloroplasts, and (3) the integration of the DAG backbone into MGDG and DGDG. Although the transport manner and the transported lipid have not been identified yet, lysoPC (*A. porrum*; Mongrand et al., 1997, 2000), DAG (*B. napus*; Williams et al., 2000) and PA (*A. thaliana*; Xu et al., 2005; Awai et al., 2006b) have been proposed recently as the possible transported lipid. An early study showed the possibility that PC could be the transported lipid: when spinach chloroplasts were pre-incubated with [^{14}C]-containing liposomes and phospholipid transfer proteins prior to incubation with phospholipase C (PLC) and UDP-Gal, formation of [^{14}C]MGDG was observed (Oursel et al., 1987). Recently, Jouhet and coworkers reported with suspension culture of *A. pseudoplatanus* that, during phosphate starved conditions which activate galactolipid biosynthesis, a transient increase of PC was followed by its rapid decrease and concomitant increase in DAG (Jouhet et al., 2003). Furthermore, the fatty acid composition of PC and DAG were very similar, suggesting that DAG is supplied by the hydrolysis of PC. Another report by Andersson and coworkers who studied the conditions that would allow the conversion of chloroplast-localized PC into the precursor for MGDG synthesis showed that cytosolic phospholipase D (PLD) activity that resides in the >100kDa fraction is involved in the substrate supply for MGDG synthesis (Andersson et al., 2004). Since PLD reaction gives rise to PA, subsequent reaction by PAP should be coupled to the reaction. This idea of the combinatory two-step reaction seems somewhat irrational, but the involvement of activity for direct hydrolysis by a PC-hydrolyzing PLC (PC-PLC) is not clear even though the activity itself was reported in several studies (Kates, 1955; Strauss et al., 1976; Chrastil and Parrish, 1987; Rouet-Mayer et al., 1995; Scherer et al., 2002). However, significant progress has been made in recent 3 years with Arabidopsis for identifying and characterizing key enzymes involved in this mechanism. As mentioned earlier, Arabidopsis has the machinery for MGDG synthesis both at the outer and inner envelope membrane of chloroplast. Because these two pathways are separated from each other (Benning and Ohta, 2005), the DAG supply for these two pathways could rely on independent enzymes/mechanisms (Fig. 2).

A DAG Supply to the Outer Envelope

Outer envelope-localized MGDG synthesis does not contribute to the bulk of MGDG synthesis under normal growth condition (Benning and Ohta, 2005). Therefore, the outer envelope-specific flux is often masked by the inner envelope-localized pathway, and therefore it is difficult to estimate the contribution of the outer envelope system (Benning and Ohta, 2005). It was reported that phosphate starvation activates specifically the outer envelope-localized pathway. Our group made use of this condition to identify which metabolic pathway is activated for DAG supply (Nakamura et al., 2005). Hydrolysis of PC to yield DAG is shown to be highly induced upon Pi starvation, and moreover, inhibition of PLD activity by *n*-BuOH showed no significant decrease in the activity, suggesting that the DAG supply from PC is mediated by one-step hydrolysis of PC-PLC. Based on homology search using known bacterial PC-PLCs, six putative PC-PLCs were detected in Arabidopsis and designated non-specific phospholipase C (NPC). Expression studies revealed that one of them, *NPC4*, was strongly induced upon Pi starvation. In vitro enzyme assays of recombinant NPC4 showed significant PLC activity for PC and phosphatidylethanolamine (PE) but not phosphatidylinositol (PI) or PA. Its subcellular localization using a specific anti- NPC4 antibody was confirmed to be in the plasma membrane. Two independent T-DNA lines giving rise to *npc4* knock

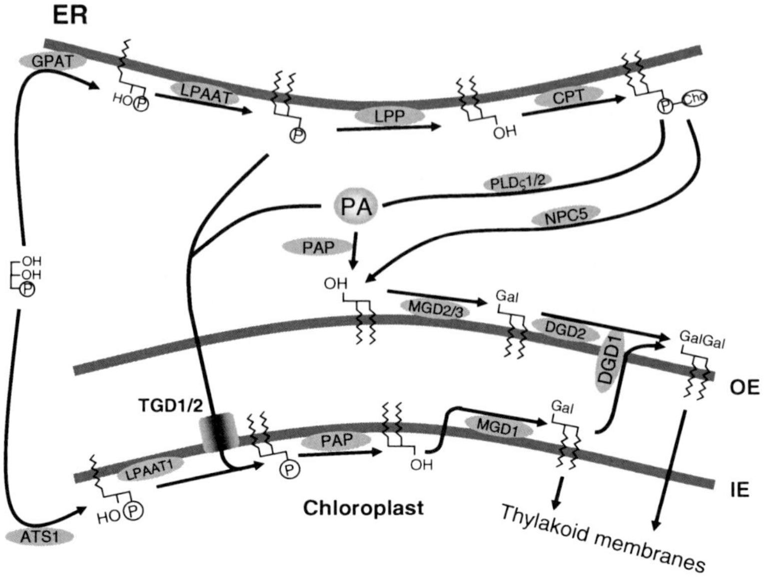

Fig. 2. A proposed model for two DAG supply systems for galactolipid biosynthesis in Arabidopsis. ATS1; glycerol-3-phosphate acyl-ACP acyltransferase 1, CPT; Diacylglycerol CDP-Choline Choline phosphotransferase, DAG, diacylglycerol; DGDG, digalactosyldiacylglycerol; DGD, DGDG synthase; G3P, glycerol-3-phosphate; GPAT, G3P acyltransferase; LPAAT, lysophosphatidic acid acyltransferase; MGDG, monogalactosyldiacylglycerol; MGD, MGDG synthase; NPC, non-specific phospholipase C; PA, phosphatidic acid; PLD, phospholipase D. IE; inner envelope, OE; outer envelope.

out mutants (*npc4-1* and *npc4-2*) almost cancelled the induction of total PC-PLC activity, suggesting that the induced PC-PLC activity upon Pi starvation is largely attributable to NPC4. However, disruption of NPC4 did not affect lipid composition, which left the main contributor for DAG supply from PC an open question. In an independent study, the disruption of NPC5 affected the accumulation of galactolipid DGDG upon Pi starvation (Gaude et al., 2008). Detailed analyses of NPC5 showed that it had several features distinct from NPC4 although they share more than 80% amino acid identity, and expression level at Pi starvation. Enzymatic activity both in vitro and in vivo were much lower than that of NPC4, and in contrast to the plasma membrane-associated NPC4, NPC5 was primarily a soluble protein. The decrease of DGDG accumulation in Pi-starved *npc5* mutants was observed both in shoots and roots. Particularly in shoots, the amount of DGDG accumulated upon Pi starvation decreased by ~50%, suggesting that nearly half the Pi starvation-induced DGDG accumulation was attributable to NPC5.

Apart from the significant contribution of PLC to DGDG accumulation, recent reports have shown the involvement of PLD for DGDG synthesis during Pi starvation. Arabidopsis has 12 isoforms of PLD, designated PLDα(3), β(2), γ(3), δ, ε and ζ(2) (Wang, 2005). Among them, only PLDζ(2) is a PH-PX type of PLD that is structurally similar to mammalian PLD (Qin and Wang, 2002). Gene expression of *PLDζ2* was reported to be induced by Pi starvation, and deletion of *PLDζ2* resulted in a decrease in DGDG accumulation of 60% in roots, and showed a premature change in root architecture upon Pi starvation (Cruz-Ramírez et al., 2006). Furthermore, possible involvement of PLDζ1 was shown in addition to PLDζ2 by analyzing a *pldζ1pldζ2* double knock out mutant; the co-deletion of both *PLDζ1* and *PLDζ2* revealed a further decrease in DGDG accumulation in comparison with the parental single disruptants (Li et al., 2006a, b). Mostly, PLDs have been studied as enzymes producing a signaling molecule, because PA is a second messenger in plants (Wang, 2005). The physiological function of PLDζ1 first reported by Ohashi and coworkers was that *PLDζ1* is a target of GLABRA2 (GL2), a transcriptional factor involved in root hair formation. An altered expression level of

PLDζ1 using an inducible gene expression system resulted in abnormal root hair morphology (Ohashi et al., 2003). In the case of the *pldζ1pldζ2* double mutant, a decrease in primary root elongation and increase in lateral root length were observed following Pi starvation (Li et al., 2006a). Since PA content in *pldζ1pldζ2* was decreased only under Pi starvation, this morphological change might be due to the altered PA signal in addition to the decreased DGDG content. Considering that no significant changes in the DGDG content were observed in shoots of these mutants, it appears that multiple systems are involved in DAG supply during Pi starvation; NPC is dominant in shoots while both NPC and PLD cooperatively play roles in roots. This idea is also supported by the analysis of Pi-starved oat roots. In a purified plasma membrane fraction of Pi-starved oat roots, activity of PLD and a subsequent DAG-producing reaction by PAP were dominant although an NPC4-like protein was detected by the anti-AtNPC4 antibody (Andersson et al., 2005).

B DAG Supply to the Inner Envelope

Supply of DAG to the inner envelope of chloroplasts is considered to be mediated by PAP (Joyard and Douce, 1977). The activity is tightly associated with the inner envelope since Block et al. (1983) demonstrated that only the inner envelope membrane contains the PAP activity, whereas both the inner and outer envelopes were capable of synthesizing PA (Block et al., 1983). The biochemical properties of envelope PAP is strikingly different from similar enzymes described in various cell fractions of animals or yeast (Bishop and Bell 1985; Carman and Henry 1989). While its extraplastidic activity in plant tissues resembles that of animal or yeast with regard to optimal pH and cation sensitivity, the activity found in the envelope fraction showed clear differences (Joyard and Douce 1987; Stymne and Stobart, 1987). First, the pH optimum for the envelope enzyme is alkaline (~9.0) and Mg^{2+} is a powerful inhibitor with 5 mM of Mg^{2+} inhibiting the activity by 75% (Joyard and Douce, 1979). For animal or yeast PAPs, the pH optimum varied from neutral to slightly acidic and no inhibitory effect by Mg^{2+} was observed.

For the in vivo regulation of envelope PAP activity, DAG, which is a reaction product of PAP, was reported to be a competitive inhibitor of the activity (Malherbe et al., 1992). Using isolated intact chloroplasts, it was shown that the activity reached its maximum when the DAG level was low, and decreased as the DAG level increased. Therefore, steady state activity of PAP may monitor in vivo the DAG/PA molar ratio. Such feedback inhibition of PAP could cause PA accumulation which can be utilized for phosphatidylglycerol biosynthesis.

Another interesting feature of envelope PAP is a clear difference in activity between 16:3 and 18:3 plants (Heinz and Roughan, 1983; Frentzen et al., 1983; Gardiner and Roughan, 1983). In contrast to 16:3 plants, in which the envelope-localized Kornberg-Pricer pathway contributes significantly to the galactolipid biosynthesis, the chloroplasts from 18:3 plants possess rather low PAP activity (Heinz and Roughan, 1983). This activity was too low to sustain the full rate of galactolipid yield. This difference might explain the reason why 18:3 plants contain small amounts of galactolipid and sulfolipid with C16 fatty acid in the *sn*-2 position.

Despite the importance of this enzyme in membrane lipid metabolism, identification of relevant protein has not yet been reported. In Arabidopsis, there are four isoforms of lipid phosphate phosphatase (LPP) which are homologs of animal and yeast LPPs. Biochemical studies showed that LPP1 and LPP2 have PAP activity in vitro (Pierrugues et al., 2001). However, the pH optimum was not alkaline and Mg^{2+} addition was not inhibitory to the activity of both isoforms. Moreover, neither of these four isoforms is predicted to be chloroplast localized. Because chloroplast PAP activity is exclusively detected in the inner envelope, the precursor protein of chloroplast PAP should have a transit peptide. This suggests that they are unlikely to be chloroplast PAPs. Indeed, expression of LPP1 was induced by UV-B and various stress treatments, and LPP2 was shown to be involved in ABA signaling downstream of ABI3 (Katagiri et al., 2005). Therefore, these isoforms are regarded as attenuators of the PA signal rather than enzymes supplying DAG for galactolipid biosynthesis in chloroplasts. In an attempt to unravel the phylogenetic origin of chloroplast PAP, our group recently found a novel subfamily of LPPs with prokaryotic origin (Nakamura et al., 2007). Because cyanobacteria, a potential origin of chloroplast, have no mammalian or yeast LPP homologs in their genome sequence (although they have PAP activity), we hypothesized that both cyanobacterial LPP and animal/yeast LPP originate from a common ancestral LPP ortholog. In view of this hypothesis, we found a putative LPP ortholog in a more primitive organism *Chlorobium tepidum*, and indeed found its homolog in cyanobacteria *Synechocystis* sp. PCC6803 and *Anabaena* sp. PCC7120. Interestingly, there are five homologs of the cyanobacterial LPP in Arabidopsis (not identical to known LPP1-4) and three of them were localized in chloroplasts (Nakamura et al., 2007). As far as we know, these are the most plausible candidates for chloroplastic PAP involved in DAG supply for MGDG synthesis.

It was generally assumed that chloroplastic PAP provides only prokaryotic DAG molecular species, with its substrate, PA. However, isolation of a novel class of mutants, *trigalactosyldiacylglycerol 1* (*tgd1*) and *tgd2*, suggest that part of eukaryotic PA may be available to chloroplast-localized lipid metabolism. The *tgd1-1* mutant, isolated as a chemicially-induced mutant accumulating trigalactosyldiacylglycerol (TGDG), showed a significant disruption in the biosynthesis of ER-derived thylakoid lipids (Xu et al., 2003). The mutation was found in a permease-like protein as a possible component of a multipartite ABC transporter similar to those found in bacteria (Xu et al., 2003). The localization of TGD1 is most likely in the inner envelope of chloroplasts (Xu et al., 2003, 2005), and one of the intriguing phenotypes of *tgd1-1* exhibited reduced incorporation of PA into chloroplast, suggesting that the TGD1 protein is part of a multi-component ABC-type PA transporter (Xu et al., 2005; Benning et al., 2006). This idea was supported by the isolation and characterization of the *tgd2-1* mutant in which import of ER-derived lipid into chloroplasts was disrupted (Awai et al., 2006b). The TGD2 is a PA-binding protein which possesses a predicted mycobacterial cell entry domain. TGD2 is also an inner membrane-tethered protein facing the outer envelope membranes (Awai et al., 2006b). Although it is still unknown whether TGD1 and TGD2 interact in vivo, presumed bacterial orthologs in Gram-negative bacteria are typically organized in transcriptional unit, suggesting that TGD1 and TGD2 may be involved in the transport of ER-derived PA into chloroplasts. Further studies are expected to isolate yet unidentified component(s) and characterize how the PA transporter delivers substrate for chloroplastic PAP.

VI MGDG Synthesis in Photoautotrophic Prokaryotes

Although MGDG and other galactolipids are rare lipids in animals, photoautotrophic prokaryotes and certain non-green bacteria have considerable amounts of MGDG as a membrane constituent. The chloroplast lipid composition in seed plants is very similar to that of cyanobacteria (Wada and Murata, 1998). However, intriguing is that the biosynthesis for MGDG is different between chloroplasts and cyanobacteria. In cyanobacteria, MGDG is synthesized in a two-step pathway: first the DAG backbone incorporates UDP-Glucose (UDP-Glc) to form monoglucosyldiacylglycerol (MGlcDG), which then is converted to MGDG by Glucose 5-epimerase (Fig. 3) (Sato and Murata, 1982a, b; Sato, 1994; Hölzl et al., 2005; Awai et al., 2006a). Since cyanobacteria such as *Synechocystis* sp. PCC6803 and *Anabaena* sp. PCC7120 have no MGDG synthase homolog, isolation of the

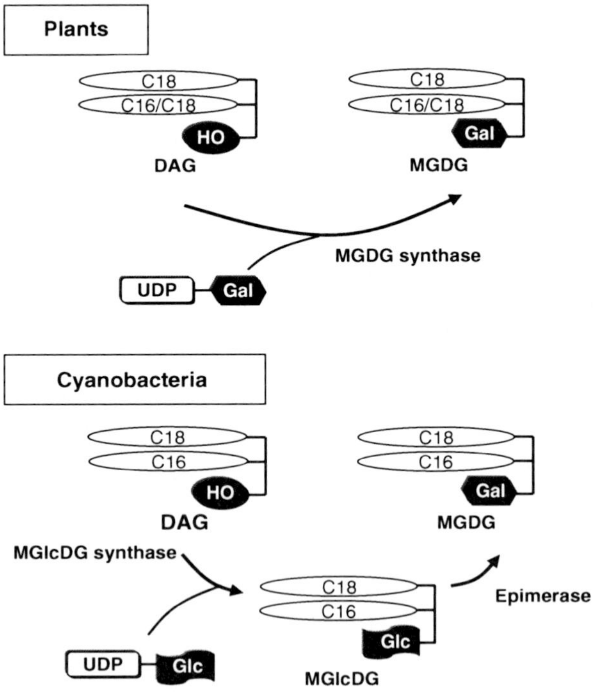

Fig. 3. MGDG synthesis in seed plants and cyanobacteria. DAG, diacylglycerol; Glc, glucose; Gal, galactose.

relevant gene for MGlcDG synthase was unsuccessful. Recently, Awai and co-workers reported that a gene, which belongs to a glycosyltransferase family (GT family) different from *MGD* in seed plants, is responsible for MGlcDG synthesis in Synechocystis and Anabaena (Awai et al., 2006a). By employing a comparative genomic analysis approach, candidate genes were selected based on the following criteria: (1) Glycosyltransferase (GT) motifs should exist in the amino acid sequence of the candidate gene product. (2) The protein should not have significant similarity to previously characterized GT. (3) The enzyme should be conserved between *Synechocystis* and *Anabaena* but not in Arabidopsis. From 3,100 and 5,300 genes annotated in the genome of *Synechocystis* and *Anabaena*, respectively, 67 genes were picked due to the existence of the GT motif. Among them, 21 were unknown and four of them were highly conserved only in cyanobacterial species. The recombinant proteins for these four candidates were subjected to an in vitro enzyme assay, and one of them (sll1377 in *Synechocystis* and all4933 in *Anabaena*) turned out to be the MGlcDG synthase (Awai et al., 2006a). Unlike the MGD of seed plants and as a member of the GT28 family, the MGlcDG synthases were categorized in the GT2 family as typical D…DxD and QxxRW motifs are found (Charnock et al., 2001). The anomeric configuration of MGDG from cyanobacteria and MGlcDG produced in MGlcDG synthase-expressing *E coli* was shown to be *β*, which is in agreement with other members of the GT2 family which produces *β*-linked glycolipids. Enzyme characterization of MGlcDG synthase showed that the activity was dependent on Mg^{2+}: the activity was reduced to ~10% without Mg^{2+} or upon chelation of existing Mg^{2+} by EDTA. The preferable acyl species of DAG was 1-oleoyl-2-palmitoyl rather than 1,2-dipalmitoyl. These features are mostly in agreement with the in vivo activity of cyanobacterial MGlcDG synthase (Sato and Murata 1982a), suggesting that the identified gene product is responsible for the bulk of MGlcDG synthase activity. The fact that no homozygous knock out mutant of MGlcDG was successfully isolated (Awai et al., 2006b) implies that this step is essential for cyanobacterial viability. Now that the gene for MGlcDG synthase is isolated, an open question is why the biosynthetic pathway for MGDG is not conserved between the cyanobacteria and chloroplasts of seed plants. When available genome resources were examined, plant-type MGDG synthase was conserved in Arabidopsis, rice (*Oryza sativa*), maize (*Zea mays*), soybean (*Glycine max*), tobacco (*Nicotiana tabacum*) and cucumber (*Cucumis sativums*), moss (*Physcomitrella patens*), green algae (*Chlamydomonas reinharditii*) and red algae (*Cyanidioschyzon merolae*). On the other hand, MGlcDG synthase is found widely in cyanobacteria and some primitive algae, such as *Cyanidioschyzon merolae* and *Chlamydomonas reinharditii*. This fact leads to the presumption that the plant-type MGDG synthase was acquired soon after the emergence of eukaryotic photoautotrophs. However, the story may be far more complicated as we recently found that *Gloeobacter violaceus* PCC7421, that is known to branch off at the earliest stage in the cyanobacterial clade, had both MGDG synthase and MGlcDG synthase homologs (Shimojima M and Ohta H, unpublished observation). Therefore, our current hypothesis as to the evolution of galactolipid synthesis is as follows; in ancient cyanobacteria, both MGD and MGlcDG synthase existed. As cyanobacteria evolved, they conserved MGlcDG synthase rather than MGD. On the other hand, endosymbiosis introduced both types into host cells which then were encoded in the nuclear genome of primitive photosynthetic eukaryotes (e.g. Cyanidioschyzon). For eukaryotes, MGDG synthase was selected instead of MGlcDG synthase. Although our knowledge is too tenuous to discuss such an evolution, our current focus is to characterize the primitive cyanobacterium *Gloeobacter* as well as to discover the most ancient photosynthetic eukaryote which still possess both cyanobacterial type and plant-type MGDG biosynthesis. Such evidence will extend our hypothesis and suggest that MGDG synthesis would be a powerful tool to trace the evolution of chloroplasts.

VII Future Perspectives

After the first isolation of MGDG synthase in cucumber (Shimojima et al., 1997), significant progress has been made in this decade to unravel the physiological functions of MGDG. However, as evidence accumulates and our understanding grows, our scientific curiosity goes to emerging new questions.

The first question is the function of MGDG in plant morphogenesis and life cycle. Since MGDG is a predominant component of plastids, its function is likely to be related to whatever role plastids play. As suggested by the analysis of *mgd1-2* (Kobayashi et al., 2007), formation of MGDG is critical for photosynthetic function. In addition, specific expression of type B MGD in non-photosynthetic organs suggests a possible role of MGD other than photosynthesis. Indeed, a distinct lipid-containing plastid called elaioplast is formed in anthers (Wu et al., 1997), and it is possible that MGD may play a role in such organelles.

Second, we are interested in the regulation of MGDG synthesis. The regulation would be controlled both by upstream regulatory factor(s) of MGD and mechanisms for supplying substrate for MGDG synthesis (i.e. DAG). Results of expression studies strongly suggest that both type of MGD are under different regulatory control. At this time we do not know the nature of possible mediators which link light/cytokinin to type A MGD expression, and auxin/Pi signals to type B MGD expression. To fill the gap, our future studies should include forward genetic approaches to isolate mutants with defects in the expression of MGDs. On the substrate supply side, we have not yet identified all major sources for DAG.

Third, novel functions of MGDG other than as a major component of plastid membranes is of particular interest. Recently, a new class of MGDG was found which contains 12-oxo-phytodienoic acid (OPDA) as acyl moiety (Stelmach et al 2001; Hisamatsu et al., 2003, 2005). OPDA is a precursor for jasmonic acid biosynthesis and serves as a signal transducer in seed plants by triggering various genes involved in plant defense (Stinzi et al., 2001; Taki et al., 2005). The OPDA-containing MGDG, or MGDG-O, was first identified in wounded Arabidopsis (Stelmach et al., 2001). The correct chemical structure of MGDG-O was determined by NMR analyses to be *sn-1-O*-(12-oxophytodienoyl)-*sn2-O*-(hexadecatrienoyl)-monogalactosyl diglyceride. This lipid was localized in the plastids and its production was induced by wounding which suggests that the formation of MGDG-O may relate to wounding-induced signal transduction (Stelmach et al., 2001). Later, Andersson and co-workers isolated a related lipid structure, Aradisopside E, which contains two OPDA and one dinor-OPDA in a monogalactosyl diglycerol backbone (Andersson et al., 2006). When a bacterial avirulence protein, AvrRpm1, was induced in plants, the Aradisopside E accumulated as much as 7–8% of the total lipid content. Interestingly, Arabidopside E inhibited the growth of a bacterial pathogen in vitro, suggesting that such OPDA-containing MGDG may protect plants against pathogen attack (Andersson et al., 2006). Indeed, an independent study revealed that MGDG is not the only lipid that contains OPDA. A lipidomics analyses detected considerable amounts of OPDA-containing DGDG and PG in Arabidopsis. Furthermore, abundance of some of these compounds increased by 200–1,000-fold within 15 min after wounding (Buseman et al., 2006). A series of reports mentioned above also supports the notion that OPDA-containing plastid lipids may serve for functional plant defenses. It would be interesting to the underlying mechanism which form such structures. Isolation of key enzymes involved in the metabolism of OPDA-containing lipids may not only elucidate the metabolic flow but may also give us a clue to understand the production of free OPDA and its role in signal transduction.

In addition to the above-mentioned basics, various application of MGDG metabolism have been initiated. An attempt to design specific inhibitors of MGD has been made to produce effective herbicides (Botté et al., 2005). Since MGDG is widely found in plants and is indispensable membrane lipid, selective and effective inhibition of MGDG synthesis may cause lethal effects in weeds.

Acknowledgements

This work was supported in part by a Grand-in-aid for Scientific Research on Priority Areas Nos.17051009 and 18056007 from MEXT of Japan. Y.N, and K.K were supported by a doctoral fellowship from Japan Society for the Promotion of Science (JSPS).

References

Aloni R, Schwalm K, Kanghans M and Ullrich CI (2003) Gradual shift in sites of free-auxin production during leaf-primordium development and their role in vascular differentiation and leaf morphogenesis in Arabidopsis. Planta 216:841–853

Andersson MX, Stridh MH, Larsson KE, Liljenberg C and Sandelius AS (2003) Phosphate-deficient oat replaces a major portion of the plasma membrane phospholipids with the galactolipid digalactosyldiacylglycerol. FEBS Lett 537:128–132

Andersson MX, Kjellberg JM and Sandelius AS (2004) The involvement of cytosolic lipases in converting phosphatidylcholine to substrate for galactolipid synthesis in the chloroplast envelope. Biochim Biophys Acta 1684:46–53

Andersson MX, Larsson KE, Tjellström H, Liljenberg C and Sandelius AS (2005) Phosphate-limited oat. J Biol Chem 280(30):27578–27586

Andersson MX, Hamberg M, Kourtchenko O, Brunnström Å, McPhail KL, Gerwick WH, Göbel C, Feussner I and Ellerström M (2006) Oxylipin profiling of the hypersensitive response in *Arabidopsis thaliana*. J Biol Chem 281(42):31528–31537

Avsian-Kretchmer O, Cheng JC, Chen L, Moctezuma E and Sung ZR (2002) Indole acetic acid distribution coincides with vascular differentiation pattern during Arabidopsis leaf ontogeny. Plant Physiol 130:199–209

Awai K, Maréchal E, Block MA, Brun D, Masuda T, Shimada H, Takamiya K, Ohta H and Joyard J (2001) Two types of MGDG synthase genes, found widely in both 16:3 and 18:3 plants, differentially mediate galactolipid synthesis in photosynthetic and nonphotosynthetic tissues in *Arabidopsis thaliana*. Proc Natl Acad Sci USA 98(19):10960–10965

Awai K, Kakimoto T, Awai C, Kaneko T, Nakamura Y, Takamiya K, Wada H and Ohta H (2006a) Comparative genomic analysis revealed a gene for monoglucosyldiacylglycerol synthase, and enzyme for photosynthetic membrane lipid synthesis in cyanobacteria. Plant Physiol 141:1120–1127

Awai K, Xu C, Tamot B and Benning C (2006b) A phosphatidic acid-binding protein of the chloroplast inner envelope involved in lipid trafficking. Proc Natl Acad Sci USA 103(28):10817–10822

Bishop WR and Bell RM (1985) Assembly of the endoplasmic reticulum phospholipid bilayer: the phosphatidylcholipe transporter. Cell 42:51–60

Block MA, Dorne A-J, Joyard J and Douce R (1983) The phosphatidic acid phosphatase of the chloroplast envelope is located on the inner envelope membrane. FEBS Lett 164:111–115

Benning C and Ohta H (2005) Three enzyme systems for galactoglycerolipid biosynthesis are coordinately regulated in plants. J Biol Chem 280(4):2397–2400

Benning C, Xu C and Awai K (2006) Non-vesicular and vesicular lipid trafficking involving plastids. Curr Opin Plant Biol 9:1–7

Botté C, Jeanneau C, Snajdrova L, Bastien O, Imberty A, Breton C and Maréchal E (2005) Molecular modeling and site-directed mutagenesis of plant chloroplast monogalactosyldiacylglycerol syntghase reveal critical residues for activity. J Biol Chem 280(41):34691–34701

Buseman CM, Tamura P, Sparks AA, Baughman EJ, Maatta S, Zhao J, Roth MR, Esch SW, Shah J, Williams TD and Welti R (2006) Wounding stimulates the accumulation of glycerolipids containing oxophytodienoic acid and dinoroxophytodienoic aicd in Arabidopsis leaves. Plant Physiol 142:2839

Camara B, Bardat F, Dobgo O, Brangeon J and Monéger R (1983) Terpenoid metabolism in plastids. Isolation and biochemical characteristics of capsicum annuum chromoplasts. Plant Physiol 3:94–99

Carman GM and Henry SA (1989) Phospholipid biosynthesis in yeast. Annu Rev Biochem 58:635–669

Charnock SJ, Henrissat B and Davies GJ (2001) Three-dimensional structures of UDP-sugar glycosyltransferase illuminate the biosynthesis of plant polysaccharides. Plant Physiol 125:527–531

Chrastil J and Parrish FW (1987) Phospholipase C and D in Rice Grains. J Agric Food Chem 35:624--627

Cline K and Keegstra K (1983) Galactosyltransferases involved in galactolipid biosynthesis are located in the outer membrane of pea chloroplast envelopes. Plant Physiol 71:366–372

Covès J, Joyard J and Douce R (1988) Lipid requirement and kinetic studies of solubilized UDP-galactose:diacylglycerol galactosyltransferase activity from spinach chloroplast envelope membranes. Proc Natl Acad Sci USA 85:4966–4970

Cruz-Ramírez A, Oropeza-Aburto A, Razo-Hernández F, Ramírez-Chávez E and Herrera-Estrella R (2006) Phospholipase DZ2 plays an important role in extraplastidic galactolipid biosynthesis and phosphate recycling in Arabidopsis roots. Proc Natl Acad Sci USA 103(17):6765–6770

Dörmann P and Benning C (2002) Galactolipids rule in seed plants. Trends Plant Sci 7:112–118

Dorne A-J, Block MA, Joyard J and Douce R (1982) The Galactolipid:galactolipid galactosyltransferase is located on the outer membrane of the chloroplast envelope. FEBS Lett 145:30–34

Douce R and Joyard J (1980) Lipids: structure and function. In: Stumpf PK (ed) The Biochemistry of Plants Vol.4. Academic, New York, pp. 321–362

Frentzen M, Heinz E, McKeon TA and Stumpf PK (1983) Specificities and selectivities of glycerol-3-phosphate acyltransferase and monoacylglycerol-3-phosphate acyltransferase from pea and spinach chloroplasts. Eur J Biochem 129:629–636

Gardiner SE and Roughan PG (1983) Relationship between fatty acyl-composition of diacylgalactosylglycerol and turnover of chloroplast phosphatidate. Biochem J 210:949–952

Gaude N, Nakamura Y, Scheible W-R, Ohta H and Dörmann P (2008) Phospholipase C5 (NPC5) is involved in galactolipid accumulation during phosphate limitation in leaves of Arabidopsis. Plant J 56:28–39

Härtel H, Dörmann P and Benning C (2000) DGD1-independent biosynthesis of extraplastidic galactolipids after phosphate starvation in Arabidopsis. Proc Natl Acad Sci USA 97:10649–10654

Heinz E (1977) Enzymatic reactions in galactolipid biosynthesis. In: Tevini M and Licthenthaler HK (eds) Lipids and Lipid Polymers. Springer, Berlin, pp. 102–120

Heinz E and Roughan PG (1983) Similarities and differences in lipid metabolism of chloroplasts isolated from 18:3 and 16:3 plants. Plant Physiol 72:273–279

Hisamatsu Y, Goto N, Hasegawa K and Shigemori H (2003) Arabidopsides A and B, two new oxylipins from *Arabidopsis thaliana*. Tetrahedron Lett 44:5553–5556

Hisamastu Y, Goto N, Sejiguchi M, Hasegawa K and Shigemori H (2005) Oxylipins Arabidopsides C and D from *Arabidopsis thaliana*. J Nat Prod 68:600–603

Hobbie L and Estelle M (1995) The axr4 auxin-resistant mutants of *Arabidopsis thaliana* define a gene important for root gravitropism and lateral root initiation. Plant J 7:211–220

Hölzl G, Zähringer U, Warnecke D and Heinz E (2005) Glycoengineering of cyanobacterial thylakoid membrane for future studyes on the role of glycolipid in photosynthesis. Plant Cell Physiol 46:1766–1778

Jarvis P, Dörmann P, Peto CA, Lutes J, Benning C and Chory J (2000) Galactolipid deficiency and abnormal chloroplast development in the *Arabidopsis MGD synthase 1* Mutant. Proc Natl Acad Sci USA 97:8175–8179

Jouhet J, Maréchal E, Bligny R, Joyard J and Block MA (2003) Transient increase of phosphatidylcholine in plant cells in response to phosphate deprivation. FEBS Lett 544:63–68

Jouhet J, Maréchal E, Baldan B, Bligny R, Joyard J and Block, MA (2004) Phosphate deprivation induces transfer of DGDG galactolipid from chloroplast to mitochondria. J Cell Biol 167:863–874

Joyard J and Douce R (1977) Site of synthesis of phosphatidic acid and diacylglycerol in spinach chloroplast. Biochim Biophys Acta 486(2):273–85

Joyard J and Douce R (1979) Characterization of phosphatidate phosphohydrolase activity associated with chloroplast envelope membranes. FEBS Lett 102:147–150

Joyard J and Douce R (1987) Galactolipid Synthesis. In: Stumpf PK (ed) The Biochemistry of Plants, Vol 9. Academic, New York, pp. 215–274

Joyard J, Maréchal E, Miège C, Block MA, Dorne A-J and Douce R (1998) Structure distribution and biosynthesis of glycerolipids from higher plant chloroplasts. In: Siegenthaler PA and Murata N (eds) Lipids in Photosynthesis: Structure, Function and Genetics. Kluwer, The Netherlands, pp. 21–52

Katagiri T, Ishiyama K, Kato T, Tabata S, Kobayashi M and Shinozaki, K (2005) An important role of phosphatidic acid in ABA signaling during germination in *Arabidopsis thaliana*. Plant J 43:107–117

Kates M (1955) Hydrolysis of lecithin by plant plastid enzymes. Can J Biochem Physiol 33:575–589

Kelly AA and Dörmann P (2002) DGD2, an Arabidopsis gene encoding a udp-galactose-dependent digalactosyldiacylglycerol synthas is expressed during growth under phosphate-limiting conditions. J Biol Chem 277(2):1166–1173

Kelly AA, Froelich JE and Dörmann P (2003) Disruption of two digalactosyldiacylglycerol synthases genes DGD1 and DGD2 in Arabidopsis reveals the existence of an additional enzyme of galactolipid synthesis. Plant Cell 15:2694–2706

Kleinig H and Liedvogel B (1978) Fatty acid synthesis by isolated chromoplasts from the daffodil. Eur J Biochem 83:499–505

Kobayashi K, Awai K, Takamiy K and Ohta H (2004) Arabidopsis type B monogalactosyldiacylglycerol synthase genes are expressed during pollen tube growth and induced by phosphate starvation. Plant Physiol 134:640–648

Kobayashi K, Masuda T, Takamiya K and Ohta H (2006) Membrane lipid alteration during phosphate starvation in regulated by phosphate signaling and auxin/cytokinin cross-talk. Plant J 47:238–248

Kobayashi K, Kondo M, Fukuda A, Nishimura M and Ohta H (2007) Galactolipid synthesis in chloroplast inner envelope is essential for proper thylakoid biogenesis, photosynthesis, and embryogenesis. Proc Natl Acad Sci USA 104:17216–17221

Kocsis MG and Weselake RJ (1996) Phosphatidate phosphatases of mammals, yeast, and higher plants. Lipids 31(8):785–802

Lai F, Thacker J, Li Y and Doerner P (2007) Cell division activity determines the magnitude of phosphate starvation responses in Arabidopsis. Plant J 50:545–556

Li M, Qin C, Welti R and Wang X (2006a) Double knockouts of phospholipase Dζ1 and Dζ2 in Arabidopsis affect root elongation during phosphate-limited growth but do not affect root hair patterning. Plant Physiol 140: 761–770

Li M, Welti R and Wang X (2006b) Quantitative profiling of Arabidopsis polar glycerolipids in response to phosphorus starvation. Roles of phospholipase Dζ1 and Dζ2 in phosphatidylcholine hydrolysis and digalactosyldiacylglycerol accumulation in phosphorus-starved plants. Plant Physiol 142:750–761

Lynch JP (1995) Root architecture and plant productivity. Plant Physiol 109:7–13

Malherbe A, Block MA, Joyard J and Douce R (1992) Feedback inhibition of phosphatidate phosphatase from spinach chloroplast envelope membranes by diacylglycerol. J Biol Chem 267:23546–23553

Maréchal E, Miège C, Block MA, Douce R and Joyard J (1995) The catalytic site of MGDG synthase from spinach chloroplast envelope: a biochemical study of the structure and metal content. J Biol Chem 270:5714–5722

Maréchal E, Block MA, Joyard J and Douce R (1991) Purification of UDP-galactose:1,2-diacylglycerol galactosyltransferase from spinach chloroplast envelope membranes. C R Acad Sci Paris, t 313 Série III:521–528

Maréchal E, Azzouz N, Macedo CS, Block MA, Feagin JE, Schwarz RT and Joyard J (2002) Synthesis of chloroplast galactolipids in apicomplexan parasites. Eukaryot Cell 1(4):653–656

Miège C, Maréchal E, Shimojima M, Awai K, Block MA, Ohta H, Takamiya K, Douce R and Joyard J (1999) Biochemical and topological properties of type a MGDG synthase, a spinach chloroplast envelope enzyme catalyzing the synthesis of both pokaryotic and eukaryotic MGDG. Eur J Biochem 265:990–1001

Mongrand S, Bessoule J-J and Cassagne C (1997) A re-examination in vivo of the phosphatidylcholine-galactolipid metabolic relationship during plant lipid biosynthesis. Biochem J 327:853–858

Mongrand S, Cassagne C and Bessoule J-J (2000) Import of lyso-phosphatidylcholine into chloroplasts likely at the origin of eukaryotic plastidial lipids. Plant Physiol 122:845–852

Murphy DJ (1982) The importance of non-planar bilayer regions in photosynthetic membranes and their stabilisation by galactolipids. FEBS Lett 150:19–26

Nakamura Y, Arimitsu H, Yamaryo Y, Awai K, Masuda T, Shimada H, Takamiya K and Ohta H (2003) Digalactosyldiacylglycerol is a major glycolipid in floral organs of petunia hybrida. Lipids 38:1107–1112

Nakamura Y, Awai K, Masuda T, Yoshioka Y, Takamiya K and Ohta H (2005) A novel phosphatidylcholine-hydrolyzing phospholipase C induced by phosphate starvation in Arabidopsis. J Biol Chem 280:7469–7476

Nakamura Y, Tsuchiya M and Ohta H (2007) Plastidic phosphatidic acid phosphatases identified in a distinct subfamily of lipid phosphate phosphatases with prokaryotic origin. J Biol Chem 282:29013–29021

Ohashi Y, Oka A, Rodrigues-Pousada R, Possenti M, Ruberti I, Morelli G and Aoyama T (2003) Modulation of phospholipid signaling by GLABRA2 in root-hair pattern formation. Science 300:1427–1430

Ohta H, Shimojima M, Ookata K, Masuda T, Shioi Y and Takamiya K (1995a) A close relationship between increases in galactosyltransferase activity and the accumulation of galactolipids during plastid development in cucumber seedlings. Plant Cell Physiol 36(6):1115–1120

Ohta H, Shimojima M, Arai T, Masuda T, Shioi Y and Takamiya K (1995b) UDP-galactose :diacylglycerol galactosyltransferase in cucumber seedlings: purification of the enzyme and the activation by phosphatidic acid. In: Kader JC and Mazliak P (eds) Plant Lipid Metabolism. Kluwer, The Netherlands, pp. 152–155

Oursel A, Escoffier JC, Kader JP, Dubacq JP and Trémolières (1987) Last step in the cooperative pathway for galactolipid synthesis in spinach leaves: formation of monogalactosyldiacylglycerol with C18 polyunsaturated acyl group at both carbon atoms of the glycerol. FEBS Lett 219:393–399

Pierrugues O, Brutesco C, Oshiro J, Gouy M, Deveaux Y, Carman GM, Thuriaux P and Kazmaier M (2001) Lipid phosphate phosphatases in Arabidopsis. J Biol Chem 276:20300–20308

Qin C and Wang X (2002) The Arabidopsis phospholipase D family. Characterization of a calcium-independent and phosphatidylcholine-sensitive PLDζ1 with distinct regulatory domains. Plant Physiol 128:1057–1068

Raghothama KG (1999) Phosphate acquisition. Annu Rev Plant Physiol Plant Mol Biol 50:665–693

Rouet-Mayer MA, Valentova O, Simond-Cote E, Daussan J and Thévenot C (1995) Critical analysis of phospholipid hydrolyzing activities in ripening tomato fruits. Study by spectrofluorimetry and high-performance liquid chromatography. Lipids 30(8):739–746

Roughan PG (1970) Turnover of the glycerolipids of pumpkin leaves. The importance of phosphatidylcholine. Biochem J 117:1–8

Roughan PG and Slack CR (1982) Cellular organization of glycerolipid metabolism. Annu Rev Plant Physiol 33:97–132

Sato N (1994) Effect of Exogenous glucose on the accumulation of monoglucosyl diacylglycerol in the cyanobacterium synechosystis PCC6803. Plant Physiol Biochem 32:121–126

Sato N and Murata N (1982a) Lipid biosynthesis in the blue-green alga, *Anabaena variabilis* I, lipid classes. Biochim Biophys Acta 710:271–278

Sato N and Murata N (1982b) Lipid biosynthesis in the blue-green alga, *Anabaena variabilis* III, UDPglucose:diacylglycerol glucosyltransferase activity *in vitro*. Plant Cell Physiol 23:1115–1120

Scherer GFE, Paul RU, André H and Martinec J (2002) Down-regulation by elicitors of phosphatidylcholine-hydrolyzing phospholipase C and up-regulation of phospholipase A in plant cells. Biochem Biophys Res Commun 293:766–770

Shimojima M, Ohta H, Iwamatsu A, Masuda T, Shioi Y and Takamiya K (1997) Cloning of the gene for monogalactosyldiacylglycerol synthase and its evolutionary origin. Proc Natl Acad Sci USA 94:333–337

Slabas T (1997) Galactolipid biosynthesis genes and endosymbiosis. Trends Plant Sci 2(5):161–162

Stelmach BA, Müller A, Hennig P, Gebhardt S, Schubert-Zsilavecz M and Weiler EW (2001) A novel class of oxylipins, *sn*1-*O*-(12-Oxophytodienoyl)-*sn*2-*O*-(hexadecatrienoyl)-monogalactosyl diglyceride, from *Arabidopsis thaliana*. J Biol Chem 276(16):12832–12838

Stinzi A, Weber H, Reymond P, Browse J and Farmer EE (2001) Plant defense in the absence of jasmonic acid: the role of cyclopentenones. Proc Natl Acad Sci USA 98:12837–12842

Strauss H, Leibovitz-Ben Gershon Z and Heller M (1976) Enzymatic hydrolysis of 1-monoacyl-sn-glycerol-3-phosphorylcholine (1-lysolecithin) by phospholipases from peanut seeds. Lipids 11:442–448

Stymne S and Stobart AK (1987) Triacylglycerol biosynthesis. In: Stumpf PK (ed) The Biochemistry of Plants, Vol 9, Lipids: Structure and Function. Academic, New York, pp. 175–214

Taki N, Sasaki-Sekimoto Y, Obayashi T, Kikuta A, Kobayashi K, Ainai T, Yagi K, Sakurai N, Suzuki H, Masuda T, Takamiya K, Shibata D, Kobayashi Y and Ohta H (2005) 12-Oxo-phytodienoic acid triggers expression of a distinct set of genes and plays a role in wound-induced gene expression in Arabidopsis. Plant Pysiol 136:1268–1283

Teucher T and Heinz E (1991) Purification of UDP-galactose:diacylglycerol galactosyltransferase from chloroplast envelopes of spinach (*Spinacia oleracea* L.). Planta 184:319–326

Ticconi CA, Delatorre CA and Abel S (2001) Attenuation of phosphate starvation responses by phosphite in Arabidopsis. Plant Physiol 127:963–972

Varadarajan DK, Karthikeyan AS, Matilda PD and Ragothama KG (2002) Phosphite, an analog of phosphate, suppresses the coordinated expression of genes under phosphate starvation. Plant Physiol 129:1232–1240

Wada H and Murata N (1998) Membrane lipids in cyanobacteria. In: Siegenthaler PA and Murata N (eds) Lipids in Photosynthesis: Structure, Function and Genetics. Kluwer, The Netherlands, pp. 65–81

Wang X (2005) Regulatory functions of phospholipase D and phosphatidic acid in plant growth, development, and stress responses. Plant Physiol 139:566–573

Williams JP, Imperial V, Khan MU and Hodson JN (2000) The role of phosphatidylcholine in fatty acid exchange and desaturation in *Brassica napus* L. leaves. Biochem J 349:127–133

Wu SSH, Platt KA, Ratnayake C, Wang T-W, Ting JTL and Huang AHC (1997) Isolation and characterization of neutral-lipid-containing organelles and globuli-filled plastids from *Brassica napus* tapetum. Proc Natl Acad Sci USA 94:12711–12716

Xu C, Fan J, Riekhof W, Froehlich JE and Benning C (2003) A permiase-like protein involved in ER to thylakoid lipid transfer in Arabidopsis. EMBO J 22:2370–2379

Xu C, Fan J, Froehlich JE, Awai K and Benning C (2005) Mutation of the TGD1 chloroplast envelope protein affects phosphatidate metabolism in Arabidopsis. Plant Cell 17:3094–3110

Yamamoto M and Yamamoto KT (1999) Effects of neutral and synthetic auxins on the gravitropic growth habit of roots in two auxin-resistant mutants of Arabidopsis, axr1 and axr4: evidence for defects in the auxin influx mechanism of axr4. J Plant Res 112:391–396

Yamaryo Y, Kanai D, Awai K, Shimojima M, Masuda T, Shimada H, Takamiya K and Ohta H (2003) Light and cytokinin play a co-operative role in MGDG synthesis in greening cucumber cotyledons. Plant Cell Physiol 44:844–855

Yamaryo Y, Motohashi K, Takamiya K, Hisabori T and Ohta H (2006) In vitro reconstitution of monogalactosyldiacylglycerol (MGDG) synthase regulation by thioredoxin. FEBS Lett 580:4086–4090

Chapter 14

Synthesis and Function of the Galactolipid Digalactosyldiacylglycerol

Peter Dörmann

Institute of Molecular Physiology and Biotechnology of Plants, Karlrobert-Kreiten-Strasse 13, 53115 Bonn, Germany

Summary

Seed plants contain two galactolipids, monogalactosyldiacylglycerol (MGDG) and digalactosyldiacylglycerol (DGDG) in their chloroplast membranes. DGDG is synthesized from UDP-galactose and MGDG by DGDG synthases in the envelope membranes of plastids. Galactolipids were identified in the X-ray structures of photosynthetic complexes. Deficiency of DGDG as observed in the *dgd1* mutant of *Arabidopsis* has severe consequences for the efficiency of the photosynthetic machinery. The amount of DGDG increases when plants are grown under phosphate limitation. Under these conditions, DGDG serves as a surrogate for phospholipids thus saving phosphate for more important cellular processes. During senescence, chlorophyll and MGDG are degraded resulting in the release of free phytol and free fatty acids. At the same time, fatty acid phytyl esters accumulate in the plastoglobulies of chloroplasts. These esters serve as sink for the deposition of phytol and fatty acids which otherwise would destabilize the membrane bilayer structure.

Author for Correspondence, e-mail: doermann@uni-bonn.de

Constantin A. Rebeiz et al. (eds.), The Chloroplast: Basics and Applications, pp. 203–211.

I Introduction

Chloroplasts of seed plants are characterized by the presence of a unique set of lipids that are absent from other organisms including animals and fungi (Benson, 1971; Joyard et al., 1998). The two galactolipids monogalactosyldiacylglycerol (MGDG) and digalactosyldiacylglycerol (DGDG), the sulfolipid sulfoquinovosyldiacylglycerol (SQDG) and the phospholipid phosphatidylglycerol (PG) represent the most abundant glycerolipids in plastids. The lipid composition of chloroplasts resembles that of other organisms performing oxidative photosynthesis, in particular that of cyanobacteria. The high conservation of the lipid composition between cyanobacteria and seed plants has to be seen in view of the endosymbiont hypothesis which implies that the plastidial progenitor of chloroplasts, a former cyanobacterium, was engulfed by an ancestral eukaryote giving rise to the first plant cell. Thus, along with oxygenic photosynthesis, a number of cyanobacterium-like processes and biosynthetic pathways were introduced into the plant cell. Plastids of today's seed plants are capable of producing a large variety of cyanobacterium-derived compounds (e.g., galactolipids, essential fatty acids, essential amino acids, etc.) that are not only crucial for the plant, but also for human nutrition.

The fact that the lipid composition of cyanobacteria and chloroplasts is highly similar suggests that the genes involved in their synthesis might also be related. However, after isolating the galactolipid synthesis genes from plants (Shimojima et al., 1997; Dörmann et al., 1999) and comparing their sequence with the *Synechocystis* genome, it became clear that cyanobacteria must contain a different set of galactolipid synthesis genes. Therefore, after establishment of the progenitor chloroplast, the galactolipid synthase genes in plants have apparently evolved independently. In line with this finding, the pathway of galactolipid synthesis in cyanobacteria was found to be different from plants, and the sequence of the MGDG synthase from plants is divergent from that of the monoglucosyldiacylglycerol synthase involved in galactolipid synthesis in cyanobacteria (Awai et al., 2006).

The strict correlation of the occurrence of the two galactolipids MGDG and DGDG with the presence of oxygenic photosynthesis suggested that these lipids are highly important to sustain a maximal efficiency of the photosynthetic light reactions (Benson et al., 1958; Benson, 1971). Indeed, recent studies confirmed that the galactolipids MGDG and DGDG, as well as the two other thylakoid lipids PG and SQDG are of crucial importance for photosynthesis in plants and cyanobacteria.

A second function of galactolipids was discovered in the past years. In contrast to the ubiquitously distributed phospholipids, galactolipids are phosphate-free. Phosphate is one of the most important nutrients and oftentimes limiting for plant growth (Raghothama, 1999). About one third of organically bound phosphorus is fixed in the phospholipid pool of plant membranes, thus rendering the membranes one of the largest sinks for phosphate (Poirier et al., 1991). Seed plants developed a number of mechanisms to adapt to conditions of phosphate limitation with which they are frequently confronted in natural habitats (Daram et al., 1999; Versaw and Harrison, 2002; Wu et al., 2003; Ticconi and Abel, 2004). One such adaptive process includes the replacement of a proportion of the phospholipids with the galactolipid DGDG in the cellular membranes.

II Structure and Occurrence of Digalactosyldiacylglycerol

Galactolipids are phosphorus-free glycoglycerolipids containing one, two or several galactose residues in their polar head group. In contrast to MGDG which carries one galactose in β-glycosidic linkage, the second galactose in DGDG is bound in α-anomeric configuration (Carter et al., 1956). Therefore, the structure of DGDG can be depicted as 1,2-diacyl-3-(α-D-galactopyranosyl-1→6-β-D-galactopyranosyl)-*sn*-glycerol (Fig. 1). In addition to MGDG and DGDG, additional galactolipids were identified in plants carrying three or more galactose residues (oligogalactolipids; Benson et al., 1958). However, these oligogalactolipids where only identified in low amounts in

Abbreviations: DGDG – digalactosyldiacylglycerol; GGGT – galactolipid:galactolipid galactosyltransferase; LCHII – light harvesting complex II; MGDG – monogalactosyldiacylglycerol; PC – phosphatidylglycerol; PE – phosphatidylethanolamine; PG – phosphatidylglycerol; PI – phosphatidylinositol; PSI, PSII – photosystem I, II; SQDG – sulfoquinovosyldiacylglycerol

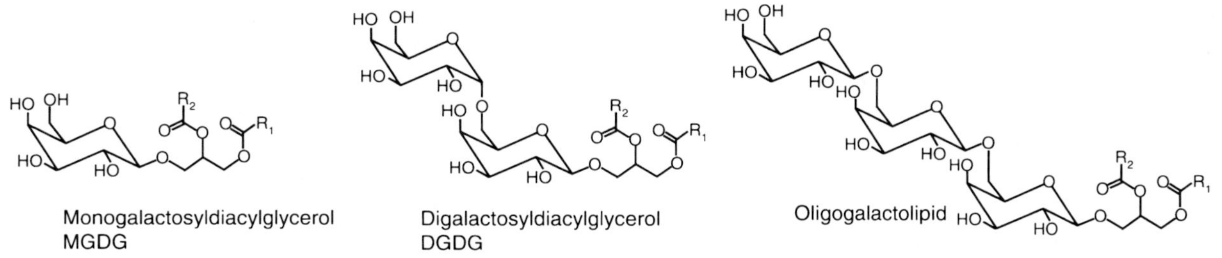

Fig. 1. Structure of galactolipids in seed plants. Seed plants and cyanobacteria contain two galactolipids, MGDG and DGDG. The galactose residue of MGDG and the inner galactose of DGDG are bound to diacylglycerol in β-glycosidic linkage. The outermost galactose moiety of DGDG is linked to the C6 position of the inner galactose in α-Adzuki linkage. Furthermore, some plant species contain low amounts of oligogalactolipids (here trigalactosyldiacylglycerol from Adzuki beans with all-β anomeric structure is shown).

certain organs of a limited number of plant species. For example, Adzuki beans (*Vigna angularis*) where shown to accumulate oligogalactolipids with all-β-anomeric linkages (Kojima et al., 1990). Similarly, an *Arabidopsis* mutant, *tgd1*, affected in chloroplast lipid trafficking, contains high amounts of all-β-oligogalactolipids in the leaves (Xu et al., 2003). In addition, oligogalactolipids accumulate in large amounts in vitro in isolated envelope membranes of chloroplasts indicating that the capability of oligogalactolipid synthesis is inherent to seed plants, but is strongly suppressed in leaves under normal growth conditions (Dorne et al., 1982; Cline and Keegstra, 1983).

III Synthesis of Digalactosyldiacylglycerol and Oligogalactolipids

DGDG synthesis is localized to the envelope membranes of plastids (Douce, 1974). Two DGDG synthases were identified in Arabidopsis, DGD1 and DGD2, and these two enzymes are localized in the outer envelope membrane (Dörmann et al., 1999; Froehlich et al., 2001; Kelly and Dörmann, 2002) (Fig. 2). DGD1 is responsible for the synthesis of the predominant fraction of DGDG in leaves, while expression of DGD2 is low under normal growth conditions (Kelly et al., 2003). Under phosphate limiting conditions, DGD2 expression is induced (Kelly et al., 2003). The DGDG synthases DGD1 and DGD2 attach the second galactose moiety to MGDG in α-glycosidic linkage using UDP-galactose as galactose-donor. DGDG produced in the outer envelope is subsequently redistributed to the thylakoids and, during phosphate deprivation, to extraplastidial membranes of the cell (Fig. 2).

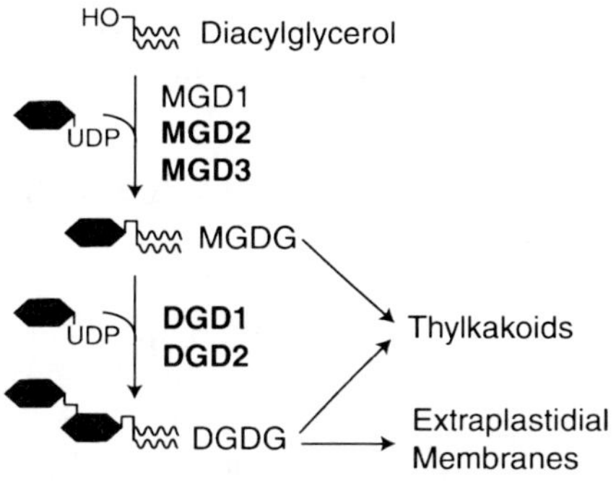

Fig. 2. Galactolipid synthesis. Galactolipids in Arabidopsis are synthesized from diacylglycerol and UDP-galactose in two steps by the MGDG synthases (MGD1, MGD2, MGD3) and DGDG synthases (DGD1, DGD2). Under normal growth conditions, MGDG and DGDG accumulate in the thylakoid membranes of chloroplasts. During phosphate deprivation, expression of MGD2, MGD3, DGD1 and DGD2 is induced (indicated in *bold*). Under these conditions, the amount of DGDG increases, and this lipid is transported to the thylakoids and to extraplastidial membranes where it replaces phospholipids. Galactose is indicated as a *black hexagon*, and fatty acids as *waved lines*.

The pathway for the synthesis of oligogalactolipids in plants remains an unsolved question. Previously, a galactolipid-dependent galactosyltransferase (galactolipid:galactolipid galactosyltransferase) activity was described in plants that was claimed to be responsible for the synthesis of DGDG and oligogalactolipids (van Besouw and Wintermans, 1978; Heemskerk et al., 1990). While the synthesis of DGDG in leaves can be attributed to DGDG synthases (DGD1, DGD2; Kelly et al., 2003), the role of the galactolipid:galactolipid galactosyltransferase in oligogalactolipid synthesis is still unclear. The galactose in oligogalactolipids of the *tgd1* mutant and of Adzuki beans shows β-glycosidic linkage, indicating that the

DGDG synthases which form α-glycosidic bonds cannot be involved in their synthesis. Additional evidence for this hypothesis comes from the fact that oligogalactolipid synthesis is still functional in the Arabidopsis *dgd1 dgd2* and *tgd1 dgd1* double mutants (Kelly et al., 2003; Xu et al., 2003). Furthermore, since none of the MGDG synthases (MGD1, MGD2, MGD3) shows processive activity (i.e. sequenctial glycosylation activity), an additional, hitherto unknown protein must exist in *Arabidopsis* that is capable of oligogalactolipid synthesis (Awai et al., 2001). However, the function of oligogalactolipids, the identity of the oligogalactolipid synthase and its role in chloroplast lipid synthesis remain enigmatic.

IV Function of Digalactosyldiacylglycerol in Photosynthesis

The lipids of the thylakoids establish the matrix of the bilayer membrane into which the pigment protein complexes of the photosynthetic machinery are embedded. The fact that some lipids were found to be closely bound to specific photosynthetic complexes indicated that they might play a functional role in photosynthesis (Murata et al., 1990; Nußberger et al., 1993; Sakurai et al., 2006). Crystallization and structural analysis of the photosynthetic complexes revealed that they contain lipid molecules closely bound to the polypeptide chains of the PS I and II and of the cytochrome b_6/f complex, indicating that these lipids have to be considered as cofactors in analogy to chlorophylls and carotenoids (Table 1; for reviews see: Fyfe et al., 2005; Jones, 2007). For example, DGDG and the phospholipid PG were identified within the trimers of LHC II (Liu et al., 2004; Standfuss et al., 2005). Three DGDG and three PG molecules localize to one LHCII trimer, and the presence of lipids, in particular PG, was shown to support LHCII trimerization in vitro (Nußberger et al., 1993). Furthermore, a "belt" of lipids surrounding the core of PSII was identified in PSII crystal structures (Loll et al., 2005). All of the four thylakoid lipids (MGDG, DGDG, SQDG and PG) were identified in PSII, and it is believed that the lipid belt serves as a lubricating shell around the PSII core proteins supporting the replacement of D1 protein damaged during photooxidation. The in vivo role of DGDG in photosynthesis was studied using *Arabidopsis* plants carrying mutations in the DGDG synthase genes DGD1 or DGD2. The *dgd1* mutant of *Arabidopsis* contains only one tenth of the WT amount of DGDG, and the synthesis of this residual amount of DGDG can be attributed to the activity of the second DGDG synthase, DGD2 (Dörmann et al., 1995; Kelly et al., 2003). A range of photosynthesis-related parameters was shown to be affected in the *dgd1* mutant, including a change in stability or activity of PSII, PSI and LHCII. The PSII quantum yield is reduced in *dgd1*, and this effect is even more pronounced when plants are exposed to high light (Dörmann et al., 1995). PSI stability is affected because PSI subunits are more susceptible to removal by chaotropic reagents in *dgd1* as compared to wild type (Guo et al., 2005). Furthermore, PSI subunit stoichiometry and P700 oxidation level are affected in *dgd1* (Ivanov et al., 2006). Interestingly, the PSI-derived limitations in photosynthesis observed for *dgd1* can be recovered by growth at low temperatures (Hendrickson et al., 2006). The content of photosynthetic pigments is altered in *dgd1*, i.e. the chlorophyll a/b ratio is decreased and the ratio of peripheral antenna complexes of PSII to the PSII core complex is reduced (Härtel et al., 1997). The electron transfer from Q_A to Q_B in PSII and the activity of the water oxidizing complex

Table 1. Lipids identified in x-ray structures of photosynthetic complexes.

Complex	Species	Lipids detected	Reference
PSI	*Thermosynechococcus elongatus*	3 PG, 1 MGDG	Jordan et al., 2001
PSII	*Thermosynechococcus elongatus*	6 MGDG, 4 DGDG, 1 PG, 3 SQDG	Loll et al., 2005
bacterial reaction center	*Rhodobacter sphaeroides*	1 cardiolipin, 1 PC, 1 GlcGalDG	Camara-Artigas et al., 2002
LHCII (trimer)	*Spinacia oleracea*	3 DGDG, 3 PG	Liu et al., 2004
LHCII (trimer)	*Pisum sativum*	3 DGDG, 3 PG	Standfuss et al., 2005
LHC	*Amphidinium carterae*	2 DGDG	Hofmann et al., 1996
Cytochrome b_6/f	*Chlamydomonas reinhardtii*	2 MGDG, 1 SQDG	Stroebel et al., 2003

are compromised in *dgd1* (Reifarth et al., 1997). The content of xanthophyll cycle pigments is also altered, because *dgd1* shows a higher zeaxanthin content even after prolonged dark exposure (Härtel et al., 1998). Excitation with light flashes above the saturation level results in elevated carotenoid triplet formation in *dgd1* (Steffen et al., 2005). The alterations of photosynthesis are more severe in the *dgd1 dgd2* double mutant indicating that a residual pool of DGDG in *dgd1* is still closely associated with and stabilizes the photosynthetic complexes (Steffen et al., 2005). The change in electron flow through PSII as detected in *dgd1* can be explained by a decreased DGDG content in the PSII core (see Loll et al., 2005). Furthermore, LCHII trimer formation is reduced as demonstrated by green gel electrophoresis of thylakoid preparations from *dgd1* (Dörmann et al., 1995). This effect can be attributed to the occurrence of DGDG in LCHII (Liu et al., 2004). Interestingly, the addition of decylmaltoside to *dgd1* thylakoid preparations rescues the in vitro destabilisation of LCHII trimers, indicating that amphipatic diglycosyl molecules (e.g. DGDG or decylmaltoside) can bind to LCHII trimers and support trimerization in vitro.

The strong reduction of DGDG content in the *dgd1* mutant is accompanied by changes of the biophysical and structural properties of the thylakoid bilayer membranes. Therefore, specific effects of the individual lipid molecules with regard to the binding and stabilization of membrane-associated protein complexes might be obscured by the changes in overall membrane architecture. In an alternative, transgenic engineering approach to study the specific functions of DGDG in growth and photosynthesis, DGDG was replaced with a foreign, bacterial lipid that contains a mixed sugar composition in its head group (Hölzl et al., 2006). To this end, a bacterial β-glucosyltransferase (βGlcT) involved in glycolipid synthesis was introduced into the *dgd1* mutant of Arabidopsis. The ectopic expression of βGlcT resulted in the accumulation of β-glucosyl-β-galactosyldiacylglycerol (βGlcβGalDG) in the chloroplast membranes, accompanied with the complementation of the growth retardation of the *dgd1* mutant (Hölzl et al., 2006). Therefore, similar to DGDG, accumulation of βGlcβGalDG supports the establishment of functional thylakoid bilayer membranes in the chloroplast. However, chlorophyll fluorescence measurements revealed that the PSII quantum yield of βGlcGalDG-accumulating *dgd1* mutant plants was not fully restored to wild type levels. Furthermore, under high light conditions, the decrease in PSII quantum yield of βGlcβGalDG-accumulating *dgd1* plants was not complemented at all. Therefore, DGDG represents the superior lipid for supporting maximal efficiency of photosynthesis, in particular under high light conditions, suggesting that galactose, but not glucose is the preferred sugar in the head group of thylakoid glycolipid (Hölzl et al., 2006). The decrease of in vitro LHCII trimerization state as observed by green gel analysis of *dgd1* thylakoid preparations (Dörmann et al., 1995) was restored in βGlcβGalDG-containing *dgd1* plants (Hölzl et al., 2006). The fact that βGlcβGalDG can stabilize LHCII trimers in a similar way as DGDG indicates that the decrease in PSII efficiency observed by chlorophyll fluorescence measurements is not associated with changes in LHCII trimerization status. However, the replacement of DGDG with βGlcβGalDG in the core of PSII per se (Loll et al., 2005) might be an explanation for the decrease in PSII quantum yield as observed for βGlcβGalDG-containing *dgd1* plants.

V Digalactosyldiacylglycerol as Surrogate for Phospholipids

Phosphorus represents one of the most important nutrients and oftentimes limits plant growth in natural environments (Raghothama, 1999). Since a large proportion of organically bound phosphate resides in the phospholipids of biological membranes (Poirier et al., 1991), plants have developed an adaptive strategy of replacing phospholipids with non-phosphorus glycolipids during conditions of phosphate limitation. Thus, phosphate deprivation results in the decrease of phospholipids with a concomitant increase in DGDG and SQDG while the amount of MGDG remains more or less unchanged (Essigmann et al., 1998; Härtel et al., 2000). Along with the increase in DGDG synthesis, expression of the galactolipid synthase genes MGD2, MGD3, DGD1 and DGD2 is up-regulated (Awai et al., 2001; Kelly and Dörmann, 2002; Kelly et al., 2003). The increase in galactolipid synthesis and MGD2 and MGD3 expression during phosphate-deprivation depends on auxin and cytokinin cross-talk (Kobayashi et al., 2006). During phosphate deficiency, DGDG does not only accumulate in plastids, but its content also increases in extraplastidial membranes, e.g. plasma membrane and mitochondria (Härtel et al., 2000; Andersson et al., 2003; Jouhet et al., 2004) (Fig. 2). The effect of DGDG accumulation in thylakoid membranes and in extraplastidial membranes on the functional integrity of membrane-linked processes such as photosynthesis or membrane transport has not been characterized.

VI Changes in Galactolipid Content During Stress and Senescence

Similar to other lipids in biological membranes, galactolipids are subject to constant metabolic turnover. Thus, galactosidase and galactolipase activities in the chloroplasts trigger in the release of galactose, diacylglycerol, lyso-lipids and fatty acids from galactolipids. Galactolipases were first described in plant tissues by Sastry and Kates (1964) and Helmsing (1969), but the identity of the respective genes remains elusive. On the other hand, galactolipids are constantly synthesized de novo from diacylglycerol and UDP-galactose by action of MGDG and DGDG synthases. The regulation of galactolipid turnover during normal and stress conditions, as well as the fate of galactolipid degradation products has been poorly studied. Nitrogen deprivation and senescence result in the disintegration of thylakoid membranes accompanied by chlorophyll and lipid breakdown. Furthermore, plastoglobulies, small spherical lipid-containing droplets in the stroma of plastids, increase in size during stress suggesting that these suborganellar structures adsorb part of the chlorophyll and lipid degradation products (e.g. Gaude et al., 2007). During nitrogen deprivation, the amounts of chlorophyll and MGDG are strongly decreased, while the DGDG content is slightly increased (Gaude et al., 2007). Degradation of chlorophyll and MGDG leads to the release of free phytol and fatty acids into the thylakoid membrane (Fig. 3). Due to their detergent-like characteristics, phytol and fatty acids destabilize the bilayer membrane. Interestingly, fatty acid phytyl esters accumulate in the plastoglobulies of chloroplasts under stress or senescence, suggesting that they represent a sink for phytol and free fatty acids released during chlorophyll and galactolipid breakdown (Ischebeck et al., 2006; Gaude et al., 2007). However, our knowledge on the physiological and biochemical processes involved in galactolipid turnover is still very limited, and the identity of most of the genes involved remains unknown.

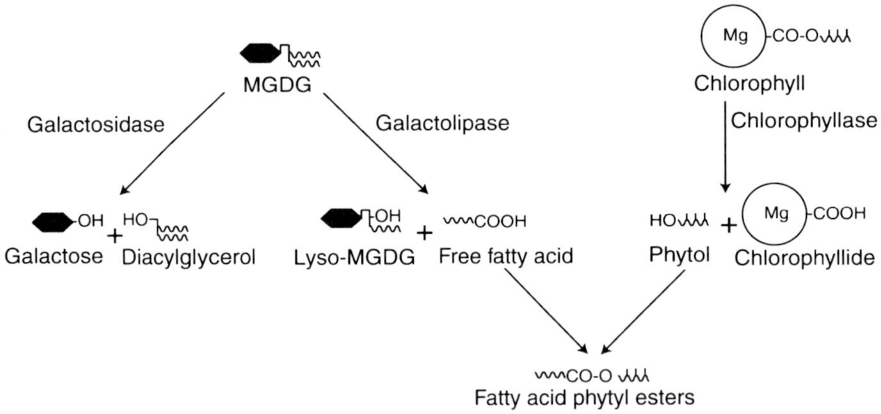

Fig. 3. Lipid degradation during stress or senescence. Under chlorotic stress conditions such as nitrogen deprivation or senescence, the galactolipid MGDG can be hydrolyzed by galactosidases or galactolipases resulting in the release of galactose/diacylglycerol or lyso-MGDG/free fatty acid, respectively. Chlorophyll is broken down by chlorophyllase to phytol and esters chlorophyllide. A fraction of free fatty acids and phytol released during stress is used for the synthesis of fatty acid phytyl esters which are deposited in the plastoglobulies of chloroplasts. Galactose is indicated as a *black hexagon*, and fatty acids as *waved lines*. The tetrapyrrole ring system of chlorophyll is depicted as a *circle with Mg in the center*. A *waved line with short extensions* indicates the phytyl moiety.

VII Conclusions

Recent years have seen a tremendous progress in our understanding of the synthesis of galactolipids in plants, of their function in photosynthesis and during phosphate deprivation in plastidial and extraplastidial membranes. The identification of MGDG and DGDG synthase genes and of galactolipid deficient plant mutants has provided the tools to study the regulation of synthesis and the in vivo roles of galactolipids. In addition to MGDG and DGDG, plants contain oligogalactolipids in low amounts, but their function and biosynthetic pathway remains unknown. Crystallization and structural analysis of the photosynthetic complexes confirmed that galactolipids have critical functions in photosynthesis. Replacement of DGDG with the alternative glycolipid βGlcβGalDG provided the means to uncover specific functions of the galactolipid in photosynthesis. However, additional genetic manipulations are required to replace the entire galactolipid matrix in thylakoids with glucolipids. DGDG accumulation in plant membranes represents an important adaptive mechanism to phosphate deprivation. However, the homeostasis of lipid content in membranes during normal and stress conditions, i.e. the regulation of metabolic turnover and de novo synthesis, remains unclear. Thus, future studies will have to focus on the characterization of the pathways and regulative factors involved in galactolipid homeostasis in plant membranes during normal growth and stress conditions.

References

Andersson MX, Stridh MH, Larsson KE, Liljenberg C and Sandelius AS (2003) Phosphate-deficient oat replaces a major portion of the plasma membrane phospholipids with the galactolipid digalactosyldiacylglycerol. FEBS Lett 537: 128–132

Awai K, Maréchal E, Block MA, Brun D, Masuda T, Shimada H, Takamiya Ki, Ohta H and Joyard J (2001) Two types of MGDG synthase genes found widely in both 16:3 and 18:3 plants differentially mediate galactolipid syntheses in photosynthetic and nonphotosynthetic tissues in *Arabidopsis thaliana*. Proc Natl Acad Sci USA 98: 10960–10965

Awai K, Kakimoto T, Awai C, Kaneko T, Nakamura Y, Takamiya K-i, Wada H and Ohta H (2006) Comparative genomic analysis revealed a gene for monoglucosyldiacylglycerol synthase, an enzyme for photosynthetic membrane lipid synthesis in cyanobacteria. Plant Physiol 141: 1120–1127

Benson AA (1971) Lipids of chloroplasts. In: Gibbs M (ed) Structure and Function of Chloroplasts. Springer-Verlag, Berlin, pp. 129–148

Benson AA, Wiser R, Ferrari RA and Miller JA (1958) Photosynthesis of galactolipids. J Am Chem Soc 80: 4740

Camara-Artigas A, Brune D and Allen JP (2002) Interactions between lipids and bacterial reaction centres determined by protein crystallography. Proc Natl Acad Sci USA 99: 11055–11060

Carter HE, McCluer RH and Slifer ED (1956) Lipids of wheat flour. I. Characterization of galactosylglycerol components. J Am Chem Soc 78: 3735–3738

Cline K and Keegstra K (1983) Galactosyltransferases involved in galactolipid biosynthesis are located in the outer membrane of pea chloroplast envelopes. Plant Physiol 71: 366–372

Daram P, Brunner S, Rausch C, Steiner C, Amrhein N and Bucher M (1999) Pht2;1 encodes a low-affinity phosphate transporter from Arabidopsis. Plant Cell 11: 2153–2166

Dörmann P, Hoffmann-Benning S, Balbo I and Benning C (1995) Isolation and characterization of an Arabidopsis mutant deficient in the thylakoid lipid digalactosyl diacylglycerol. Plant Cell 7: 1801–1810

Dörmann P, Balbo I and Benning C (1999) *Arabidopsis* galactolipid biosynthesis and lipid trafficking mediated by DGD1. Science 284: 2181–2184

Dorne A-J, Block MA, Joyard J and Douce R (1982) The galactolipid:galactolipid galactosyltransferase is located on the outer membrane of the chloroplast envelope. FEBS Lett. 145: 30–34

Douce R (1974) Site of biosynthesis of galactolipids in spinach chloroplasts. Science 183: 852–853

Essigmann B, Gueler, S, Narang RA, Linke D and Benning C (1998) Phosphate availability affects the thylakoid lipid composition and the expression of *SQD1*, a gene required for sulfolipid biosynthesis in *Arabidopsis thaliana*. Proc Natl Acad Sci USA 95: 1950–1955

Froehlich JE, Benning C and Dörmann P (2001) The digalactosyldiacylglycerol (DGDG) synthase DGD1 is inserted into the outer envelope membrane of chloroplasts in a manner independent of the general import pathway and does not depend on direct interaction with monogalactosyldiacylglycerol synthase for DGDG biosynthesis. J Biol Chem 276: 31806–31812

Fyfe PK, Hughes AV, Heathcote P and Jones MR (2005) Proteins, chlorophylls and lipids: x-ray analysis of a three-way relationship. Trends Plant Sci 10: 275–282

Gaude N, Bréhélin C, Tischendorf G, Kessler F and Dörmann P (2007) Nitrogen deficiency in *Arabidopsis* affects galactolipid composition and gene expression and results in accumulation of fatty acid phytyl esters. Plant J 49: 729–739

Guo J, Zhang Z, Bi Y, Yang W, Xu Y and Zhang L (2005) Decreased stability of photosystem I in *dgd1* mutant of *Arabidopsis thaliana*. FEBS Lett 579: 3619–3624

Härtel H, Lokstein H, Dörmann P, Grimm B and Benning C (1997) Changes in the composition of the photosynthetic apparatus in the galactolipid-deficient *dgd1* mutant of *Arabidopsis thaliana*. Plant Physiol 115: 1175–1184

Härtel H, Lokstein H, Dörmann P, Trethewey RN and Benning C (1998) Photosynthetic light utilization and xanthophyll cycle activity in the galactolipid deficient *dgd1* mutant of *Arabidopsis thaliana*. Plant Physiol Biochem 36: 407–417

Härtel H, Dörmann P and Benning C (2000) DGD1-independent biosynthesis of extraplastidic galactolipids following phosphate deprivation in Arabidopsis. Proc Natl Acad Sci USA 97: 10649–10654

Heemskerk JWM, Storz T, Schmidt RR and Heinz E (1990) Biosynthesis of digalactosyldiacylglycerol in plastids from 16:3 and 18:3 Plants. Plant Physiol 93: 1286–1294

Helmsing PJ (1969) Purification and properties of galactolipase. Biochim Biophys Acta 189: 95–105

Hendrickson L, Vlčková A, Selstam E, Huner N, Öquist G and Hurry V (2006) Cold acclimation of the Arabidopsis *dgd1* mutant results in recovery from photosystem I-limited photosynthesis. FEBS Lett 580: 4959–4968

Hofmann E, Wrench PM, Sharples FP, Hiller RG, Wilte W and Diederichs K (1996) Structural basis of light harvesting by carotenoids: peridinin-chlorophyll-protein from *Amphidinium carterae*. Science 272: 1788–1744

Hölzl G, Witt S, Kelly AA, Zähringer U, Warnecke D, Dörmann P and Heinz E (2006) Functional differences between galactolipids and glucolipids revealed in photosynthesis of higher plants. Proc Natl Acad Sci USA 103: 7512–7517

Ischebeck T, Zbierzak AM, Kanwischer M and Dörmann P (2006) A salvage pathway for phytol metabolism in *Arabidopsis*. J Biol Chem 281: 2470–2477

Ivanov AG, Hendrickson L, Krol M, Selstam E, Öquist G, Hurry V and Huner NPA (2006) Digalactosyldiacylglycerol deficiency impairs the capacity for photosynthetic intersystem electron transport and state transitions in *Arabidopsis thaliana* due to photosystem I acceptor-side limitations. Plant Cell Physiol 47: 1146–1157

Jones MR (2007) Lipids in photosynthetic reaction centres: structural roles and functional holes. Prog Lipid Res 46: 56–87

Jordan P, Fromme P, Witt HT, Klukas O, Saenger W and Krauß N (2001) Three-dimensional structure of cyanobacterial photosystem I at 2.5 Å resolution. Nature 411: 909–917

Jouhet J, Maréchal E, Baldan B, Bligny R, Joyard J and Block MA (2004) Phosphate deprivation induces transfer of DGDG galactolipid from chloroplast to mitochondria. J Cell Biol 167: 863–874

Joyard J, Maréchal E, Miège C, Block MA, Dorne A-J and Douce R (1998) Structure, distribution and biosynthesis of glycerolipids from higher plant chloroplasts. In: Siegenthaler P-A, Murata N (eds) Lipids in Photosynthesis: Structure, Function and Genetics. Kluwer, Netherlands, pp. 21–52

Kelly AA and Dörmann P (2002) *DGD2*, an *Arabidopsis* gene encoding a UDP-galactose dependent digalactosyldiacylglycerol synthase is expressed during growth under phosphate limiting conditions. J Biol Chem 277: 1166–1173

Kelly AA, Froehlich JE and Dörmann P (2003) Disruption of the two digalactosyldiacylglycerol synthase genes *DGD1* and *DGD2* in Arabidopsis reveals the existence of an additional enzyme of galactolipid synthesis. Plant Cell 15: 2694–2706

Kobayashi K, Masuda T, Takamiya K and Ohta H (2006) Membrane lipid alteration during phosphate starvation is regulated by phosphate signalling and auxin/cytokinin cross-talk. Plant J 47: 238–248

Kojima M, Seki K, Ohnishi M, Ito S and Fujino Y (1990) Structure of novel glyceroglycolipids in Adzuki bean (*Vigna angularis*) seeds. Biochem Cell Biol 68: 59–64

Liu Z, Yan H, Wang K, Kuang T, Zhang J, Gui L, An X and Chang W (2004) Crystal structure of spinach major lightharvesting complex at 2.72 Å resolution. Nature 428: 287–292

Loll B, Kern J, Saenger W, Zouni A and Biesiadka J (2005) Towards complete cofactor arrangement in the 3.0 Å resolution structure of photosystem II. Nature 438: 1040–1044

Murata N, Higashi S-I and Fujimura Y (1990) Glycerolipids in various preparations of photosystem II from spinach chloroplasts. Biochim Biophys Acta 1019: 261–268

Nußberger S, Dörr K, Wang N and Kühlbrandt W (1993) Lipid-protein interactions in crystals of plant light-harvesting complex. J Mol Biol 234: 347–356

Poirier Y, Thoma S, Somerville C and Schiefelbein J (1991) A mutant of Arabidopsis deficient in xylem loading of phosphate. Plant Physiol 97: 1087–1093

Raghothama KG (1999) Phosphate acquisition. Annu Rev Plant Physiol Plant Mol Biol 50: 665–693

Reifarth F, Christen G, Seeliger AG, Dörmann P, Benning C and Renger G (1997) Modification of the water oxidizing complex in leaves of the *dgd1* mutant of *Arabidopsis thaliana* deficient in the galactolipid digalactosyldiacylglycerol. Biochemistry 36: 11769–11776

Sakurai I, Shen JR, Leng J, Ohashi S, Kobayashi M and Wada H (2006) Lipids in oxygen-evolving photosystem II complexes of cyanobacteria and higher plants. J Biochem (Tokyo) 140: 201–209

Sastry PS and Kates M (1964) Hydrolysis of monogalactosyl and digalactosyl diglycerides by specific enzymes in runner-bean leaves. Biochemistry 3: 1280–1287

Shimojima M, Ohta H, Iwamatsu A, Masuda T, Shioi Y and Takamiya K-i (1997) Cloning of the gene for monogalactosyldiacylglycerol synthase and its evolutionary origin. Proc Natl Acad Sci USA 94: 333–337

Standfuss J, Terwisscha van Scheltinga AC, Lamborghini M and Kühlbrandt W (2005) Mechanisms of photoprotection and nonphotochemical quenching in pea light-harvesting complex at 2.5 Å resolution. EMBO J 24: 919–928

Steffen R, Kelly AA, Huyer J, Dörmann P and Renger G (2005) Investigations on the reaction pattern of photosystem II in leaves from *Arabidopsis thaliana* wild type plants and mutants with genetically modified lipid content. Biochemistry 44: 3134–3142

Stroebel D, Choquet Y, Popot J-L and Picot D (2003) An atypical haem in the cytochrome *b6f* complex. Nature 426: 413–418

Ticconi CA and Abel S (2004) Short on phosphate: plant surveillance and countermeasures. Trends Plant Sci 9: 548–555

van Besouw A and Wintermans JF (1978) Galactolipid formation in chloroplast envelopes. I. Evidence for two mechanisms in galactosylation. Biochim Biophys Acta 529: 44–53

Versaw WK and Harrison MJ (2002) A chloroplast phosphate transporter PHT2;1 influences allocation of phosphate within the plant and phosphate-starvation responses. Plant Cell 14: 1751–1766

Wu P, Ma L, Hou X, Wang M, Wu Y, Liu F and Deng XW (2003) Phosphate starvation triggers distinct alterations of genome expression in Arabidopsis roots and leaves. Plant Physiol 132: 1260–1271

Xu C, Fan J, Riekhof W, Froehlich JE and Benning C (2003) A permease-like protein involved in ER to thylakoid lipid transfer in *Arabidopsis*. EMBO J 22: 2370–2379

Chapter 15

The Chemistry and Biology of Light-Harvesting Complex II and Thylakoid Biogenesis: *raison d'etre* of Chlorophylls *b* and *c*

J. Kenneth Hoober* and Laura L. Eggink
School of Life Sciences and Center for the Study of Early Events in Photosynthesis, Arizona State University, Tempe, AZ 85287-4501, USA

Min Chen and Anthony W.D. Larkum
School of Biological Sciences, University of Sydney, NSW 2006, Australia

Summary

Accumulation of light-harvesting complexes (LHCs) in plants is minimal in mutant strains that lack chlorophyll (Chl) *b*. Experiments with the model chlorophyte alga, *Chlamydomonas reinhardtii*, and isolated chloroplasts demonstrated that LHC apoproteins (LHCPs) are not imported into the chloroplast at a significant rate unless Chl *b* is synthesized. Conversion of chlorophyllide (Chlide) *a* to Chlide *b* occurs by oxidation of the 7-methyl group to the 7-formyl group by Chlide *a* oxygenase, which is located on the chloroplast envelope inner membrane. These results led to the hypothesis that Chl *b* has properties that allow interactions with the apoproteins that are distinct from those with Chl *a* and that Chl *b* regulates retention of LHCPs in the plastid, with subsequent assembly of LHCs in the envelope inner membrane. A consequence of introduction of the electronegative formyl group is withdrawal of electron density toward the periphery of the tetrapyrrole macrocycle, along the X axis of the molecule, which reduces electron density around the central Mg atom and thus increases exposure of the positive charge of the Mg.

*Author for Correspondence, e-mail: khoober@asu.edu

Constantin A. Rebeiz et al. (eds.), The Chloroplast: Basics and Applications, pp. 213–229.

The enhanced Lewis acid 'hardness' of the metal in Chl *b* promotes formation of coordination bonds with 'hard' Lewis bases such as peptide bond carbonyl groups in LHCPs, which are ligands that are relatively unfavorable for coordination with Chl *a*. This concept is supported by comparison with Chl *c* and Chl *d*. In Chl *c*, withdrawal of electron density on the X axis is achieved by extension of conjugation of the ring π system through the double-bond of the *trans*-acrylate side-chain on C17 to the unesterified, electronegative carboxyl group and by introduction of electronegative groups on C7 and C8. The C3-vinyl group of Chl *a* is oxidized to the electronegative formyl group in Chl *d*, which withdrawals electron density along the Y axis of the molecule and extends the Q_y transition dipole moment. Chl *c* serves the same role in chromophyte algae as Chl *b* does in plants, whereas Chl *d* substitutes for Chl *a* in the cyanobacterium *Acaryochloris marina*.

I Introduction

Most photosynthetic organisms, from cyanobacteria to plants, contain chlorophyll (Chl) *a* as the principal Chl in photosynthetic reaction centers. Plants and green algae also contain Chl *b* as a secondary Chl, with a Chl *a*:*b* ratio of 2.5–4. Chromophyte algae, such as the dinoflagellates, heterokonts, haptophytes and cryptophytes, contain Chl *c* rather than Chl *b* as the secondary Chl (see Fig. 1 for structures). These types of Chl reside within thylakoid membranes in chloroplasts. Whereas the core complexes of photosystem (PS) I and PS II contain Chl *a*, light-harvesting complexes (LHCs) contain nearly equal amounts of the secondary Chls along with Chl *a*. Essentially all Chl *b* and Chl *c* are found in the light-harvesting complexes (LHCs) (Green and Durnford, 1996; Durnford et al., 1999; De Martino et al., 2000; Goss et al., 2000).

Although the chloroplast displays a variety of features among these organisms, its monophyletic origin by endosymbiosis of an ancient cyanobacterium has received increasingly strong support (Palmer, 2003; Bhattacharya and Medlin, 2004; Rodriguez-Ezpeleta et al., 2005). In addition, Tomitani et al. (1999) provided evidence that the gene for chlorophyllide *a* oxygenase (CAO), the enzyme that catalyzes conversion of chlorophyllide (Chlide) *a* to Chlide *b*, also has a common origin. The evolutionary relationship of CAO in plants with the enzyme in the cyanobacterial prochlorophytes suggests that the original ancestor of plastids contained Chl *b* and that modern cyanobacteria, along with the red algae, lost this ability. A secondary endosymbiotic event in which a red alga was engulfed by a non-photosynthetic eukaryotic host gave rise, after divergence, to the chromophyte algae (Bachvaroff et al., 2005a; Shalchian-Tabrizi et al., 2006). As with the primary endosymbiotic event, the secondary event apparently also occurred one time. Synthesis of Chl *c* in chromophytes probably evolved to solve the problem in LHC assembly that was achieved in chlorophytes with Chl *b*.

Because of their location in LHCs, the function of Chls *b* and *c* has traditionally been ascribed as an enhancement of light absorption because of their slightly shifted absorption spectra relative to Chl *a*. Eggink et al. (2001, 2004) and Chen et al. (2005a) proposed that although Chls *b* and *c* have spectral characteristics somewhat different from those of Chl *a*, the primary role of Chl *b* (and Chl *c*) is instead on *assembly* of LHCs. The altered spectral properties are indeed consequences of structural changes in the molecule. But the structural changes also influence the Lewis acid properties of the central Mg atom and thereby its interaction with LHC apoproteins (LHCPs) (Belanger and Rebeiz, 1984; Tamiaki et al., 1998; Noy et al., 2000). Therefore, synthesis of Chl *b* increases the absorptive cross-section of photosystems by supporting assembly of LHCs. The focus of this chapter is on the properties of Chl *b* (and Chl *c*) that provide this role in chloroplast development, a role that cannot be provided by Chl *a*. Our discussion of the properties of the Chls also includes Chl *d*, which is the most recently described species of Chl and the major type in the cyanobacterium

Abbreviations: Asn (N) – asparagine; BChl – bacteriochlorophyll; CAO – chlorophyllide *a* oxygenase; Chl – chlorophyll; Chlide – chlorophyllide; D – Debye; EPR – electron paramagnetic resonance; Gln (Q) – glutamine; Glu (E) – glutamic acid; Gly (G) – glycine; His (H) – histidine; Ile (I) – isoleucine; Leu (L) – leucine; LHC – light-harvesting complex; LHCP – light-harvesting complex apoprotein; PChlide – protochlorophyllide; Phe (F) – phenylalanine; Pro (P) – proline; Ser (S) – serine; Trp (W) – tryptophan; Val (V) – valine

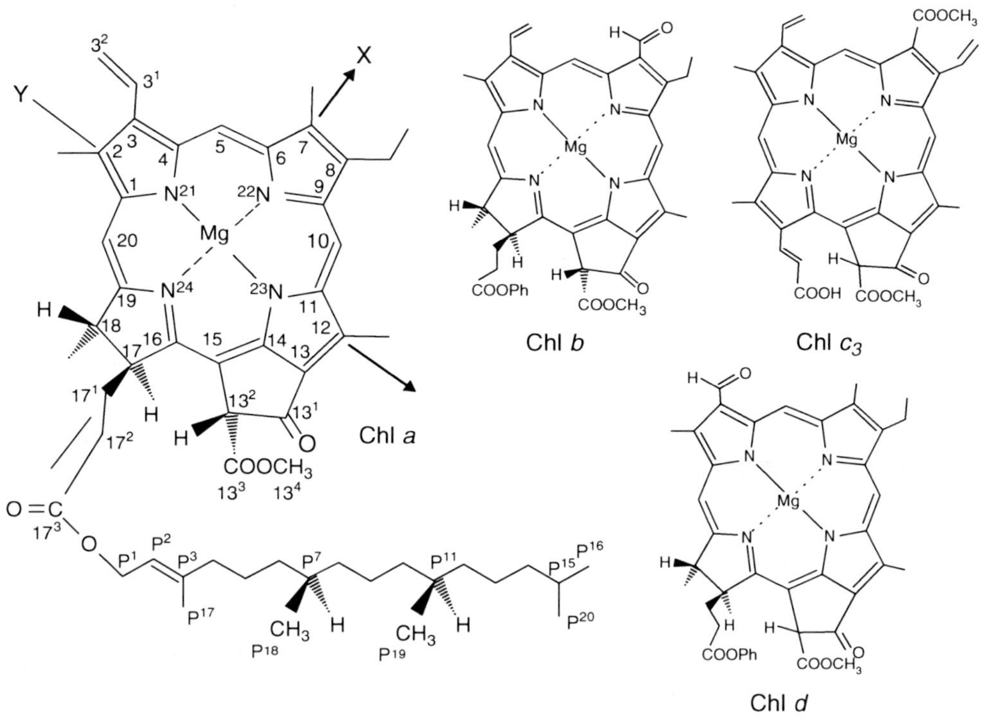

Fig. 1. Structures of the major chlorophylls. Chl *a* contains a 7-methyl group, which in Chl *b* is oxidized to a formyl group. Chls *a* and *b* are otherwise the same, including esterification of the C17 propionyl group with the isoprenoid alcohol, phytol (Ph = phytol). The predominant mono-vinyl forms of the Chls are illustrated. The two geometrical axes of the molecular framework are shown on the structure of Chl *a*. The *functional* electronic transition, Q_y, of Chl *a* is 70° clockwise from the X axis (Fragata et al., 1988; Sundholm, 2003; Cai et al., 2006). The Q_y axis of Chl *b* is 61° clockwise from the X axis (Simonetto et al., 1999). The Q_x transitions are within a few degrees of the X axis. The C17–C18 double-bond in Chl *c* is not reduced, and an additional double bond is introduced into the sidechain to produce the C17[1] *trans*-acrylate group, while the carboxyl group remains unesterified. Chl *c* species differ at positions 7 and 8; c_1: 7 = $-CH_3$, 8 = $-C_2H_5$; c_2: 7 = $-CH_3$, 8 = $-C_2H_3$; c_3: 7 = $-COOCH_3$, 8 = $-C_2H_3$. Chl *d* contains a formyl group on position 3, formed by oxidation of the 3-vinyl group.

Acaryochloris marina (Miyashita et al., 1997; Akiyama et al., 2002).

A Chlorophyll a

Conversion of 3,8-divinyl-Chl *a* to 3-monovinyl-Chl *a* by reduction of the 8-vinyl to an ethyl group is the final step in Chl *a* biosynthesis and yields the predominant form of Chl *a* (Belanger and Rebeiz, 1984; Kolossov and Rebeiz, 2001; Nagata et al., 2005). As a result, Chl *a* has electron-donating methyl and ethyl groups at positions 7 and 8, respectively (Fig. 1). Along with reduction of the C17-C18 double bond to a single bond, which converts the porphyrin precursor protochlorophyllide (Pchlide) *a* to the chlorin ring system, these groups impose an electron density, from opposite sides of the molecule along the X axis, on the pyrrole nitrogens, which partially shields the positive charge of the central Mg atom. In addition, the 3-vinyl and 13[1]-keto groups exert weak electron withdrawing effects on opposite ends of the Y axis.

The geometrical coordinates of the molecular framework for Chl *a* are shown in Fig. 1. The primary X axis transects the molecule from the position of C17 to C7. The Y axis transects the molecule from C2 to C12. The experimentally determined, functional Q_y transition-moment direction is a vector 70° clockwise from the X axis (Fragata et al., 1988; Simonetto et al., 1999; Sundholm, 2003; Cai et al., 2006). This displacement from the geometrical Y axis places the functional Q_y vector from near C1 to near C11. The long-wavelength Q_y absorption maximum of Chl *a* is at 665 nm (in methanol). The functional Q_x vector is within a few degrees of the geometrical X axis.

B Chlorophyll b

Synthesis of Chl *b* involves incorporation of the electronegative oxygen atom to generate the 7-formyl group. The oxygen provides a significant pull on electrons away from the core of the molecule along the X axis, which weakens the Q_y dipole strength of the molecule from 5.33 D for Chl *a* to 4.41 D in Chl *b* (at a refractive index for the environment of 1.35) (Knox and Spring, 2003; Linnanto et al., 2006). Because the dipole strength determines the magnitude of the absorption coefficients, among other properties, the Q_y absorption coefficient of Chl *b* is only 50 to 62% (depending on solvent) of the Q_y absorption coefficient of Chl *a* and is shifted to higher energy, with a maximum at 652 nm (in methanol). The Q_y transition-moment direction is displaced further from the Y axis and is at an angle of only 61° clockwise from the X axis (Simonetto et al., 1999), described as a transect from near C20 to near C10.

Intuitively, the Q_x transition moment of Chl *b* should be stronger than that of Chl *a*. Computational analysis of molecular orbitals suggest that the Q_x oscillator strength of Chl *b* is two-fold greater than for Chl *a,* with a maximum at 538 nm (Linnanto and Korppi-Tommola, 2004). The magnetic circular dichroism spectrum of Chl *b* contains a weak negative transition at 540 nm and an intense negative transition at 625 nm (Frackowiak et al., 1987). From the argument that a stronger molecular dipole is consistent with a longer wavelength absorption maximum, the strong longer-wavelength negative transition in the magnetic circular dichroism spectrum of Chl *b* may correspond to the Q_x transition.

Molecular orbital calculations suggest that, with weaker Q_y and stronger Q_x transitions, the electronic distribution in the Chl *b* molecule is essentially circular (Hoff and Amesz, 1991; Linnanto and Korppi-Tommola, 2004). The withdrawal of electron density from the pyrrole nitrogen atoms lowers their pK values by about two pH units (Phillips, 1963; Smith, 1975). The lessened electron density around the pyrrole nitrogens results in less shielding of the Mg atom of Chl *b* and allows it to more strongly express its positive point charge. Molecular orbital calculations give the Mg in Chl an atomic charge with a value of +1.0 ± 0.3 (Linnanto and Korppi-Tommola, 2004).

C Chlorophyll c

Chl *c* is synthesized from PChlide, the precursor of Chlide *a*, not by typical reduction of the double bond between C17 and C18 but by introduction of an additional double bond in the sidechain to form the 17^1-*trans*-acrylate group (Fig. 1). Thus Chl *c* retains the porphyrin ring system. The acrylate carboxyl group usually remains unesterified, in contrast to other Chls, and as a result, conjugation of the ring π system is extended to the carboxyl group. Two of the three major forms of Chl *c* also have electron withdrawing groups on C7 and C8. Chl c_2 has an 8-vinyl group and Chl c_3 has, in addition, a 7-methylcarboxylate ester (Porra, 1997) (Chl c_3 is illustrated in Fig. 1). The 17^1-*trans*-acrylate sidechain, with the electronegative carboxyl group, and the groups on C7 and C8 exert a strong electronegative pull at each end of the X axis, which results in lower electron density around the Mg atom (Dougherty et al., 1970). Absorbance spectra of the series (c_1, c_2 and c_3) suggest that the functional Q_x transition moment increasingly dominates the long-wavelength absorption peaks (Jeffrey and Wright, 1987; Helfrich et al., 2003), further reduces the absorption coefficient and blue-shifts the Q_y absorbance maximum to 630 nm for Chl c_1 (in acetone). Conversion of the propionate sidechain of Pchlide *a* to the acrylate sidechain of Chl *c* inhibits the ability of NADPH:Pchlide oxidoreductase to reduce the C17–C18 double bond as occurs in Chl *a* synthesis (Helfrich et al., 2003), which suggests that sidechain desaturation occurs prior to potential interaction of Pchlide *a* with the oxidoreductase.

D Chlorophyll d

Chl *d* is synthesized by oxidation of the 3-vinyl group of Chl *a* to a formyl group, whose electron-withdrawing character extends the Q_y vector but should also cause the transition direction to align more closely with the molecular Y axis. As a result, the lowest energy, Q_y absorption band shifts from 665 nm (in methanol) for Chl *a* to a longer wavelength (lower energy) maximum of 697 nm for Chl *d* and increases the dipole strength, which is proportional to the molar absorption coefficient. Chl *d* is the predominant Chl in *Acaryochloris marina* (Chl *d*/Chl *a* ratio *ca.* 30:1) in which it provides the primary electron donor in PS I (Hu et al., 1998) and PS II (Chen et al., 2005b). Kobayashi et al. (2007) found that the redox potential of Chl *d* in vitro was +0.88 V, intermediate between Chl *a* (+0.81 V) and Chl *b* (+0.93 V).

II Coordination Chemistry of Chlorophyll and Ligands

Steps in the synthesis of Chl *a* from protoporphyrin IX include (a) insertion of the Mg ion by magnesium chelatase, (b) esterification of the 13^3-carboxyl group to produce Mg-protoporphyrin IX methyl ester, and (c) formation of the fifth isocyclic ring, which generates the electron-withdrawing 13^1-carbonyl group on the Y axis of the final precursor, PChlide *a*. A water molecule is likely retained during the chelatase reaction to provide the axial ligand for the Mg ion. Reduction of the C17–C18 double bond in PChlide *a* to a single bond and reduction of the 8-vinyl group to the ethyl group, which may occur in either order, provide electron-donating groups on the X axis. The result is substantial shielding of the effective positive charge on the Mg ion by the electron cloud of the π bond system and a weakening of the bond between the negative end of the dipole in the water ligand.

The following discussion is an attempt to understand how the structural differences between Chls *a* and *b* result in different preferences for coordinating ligands and draws extensively on work by Hartwich et al. (1998) and Noy et al. (1998, 2000) on the effects of metal substitution in bacteriochlorophyll (BChl) *a*. Substituting the central Mg in BChl *a* with other metals (e.g., Ni) of greater electronegativity strongly influences the Q_x but not the Q_y transition energies of the tetrapyrrole (Hartwich et al., 1998). In Chls *a* and *b* the metal is the same, but the argument can be applied in reverse, in which substitution of peripheral groups on the X axis alters the environment of the central Mg ion and thus its effective electronegativity. We suggest that this change in the properties of the metal is a major factor that determines ligand partners in Chl–protein complexes.

The Mg ion in the center of the tetrapyrrole is coordinated within the plane of the macrocycle by the four nitrogens of the pyrrole rings. Two axial positions are available for coordination with exogenous ligands, although under biological conditions the Mg is usually five-coordinated (Pascal et al., 2002). These unfilled orbitals define the Lewis acid properties of the Mg ion in Chl. An axial coordination bond will form when a Lewis base, which has non-bonded electrons, shares a pair with Mg. Two key factors, in addition to the local environment within the membrane, seem to influence ligand preferences by Chls *a* and *b*. First is the *availability of electrons* in the Lewis base. The second critical factor is the *strength of its dipole*. Although both factors are involved in formation of coordination bonds with the chlorophylls, Chl *a* is attracted to ligands rich in electrons whereas Chl *b* is attracted to ligands that maintain a strongly negative end of a fixed dipole.

The availability of non-bonded electrons correlates with the pK of the group when H^+ is the Lewis acid. Water has a pK_a of −1.74, an indication that protonation occurs with difficulty because non-bonded electrons are strongly attracted to the electronegative oxygen atom and not readily available for bonding. The fixed dipole moment of water is 1.85 D in the gas phase and 2.70 D in the fully H-bonded state (dielectric constant = 81) (see Hoober et al., 2007, for a review). In a membrane environment, with a dielectric constant of 2–4, the dipole moment of water is likely close to the value in the gas phase. The dielectric constant rises from 5 to 10 through an interface zone at the surface of a membrane (Tanizaki and Feig, 2005), and thus mass action of water becomes a significant factor in bonding in this zone. In contrast, an electron pair on a nitrogen atom of the imidazole group of His readily binds H^+, which is reflected in its pK value of 6–7. Although the dipole moment of imidazole is 3.67 D in the gas phase (Spackman, 1992), its aromatic character allows the dipole to reorient in the presence of other dipoles (Epstein et al., 1982).

Measurements of electronegative equalization with Ni-BChl *a* indicated that a partial charge (ca. 0.3*e*) is transferred from imidazole to the metal ion upon coordination (Noy et al., 2000). Mg is a less electronegative metal than Ni (Noy et al., 1998), and consequently in Chls the charge transfer may be somewhat less. Transfer of a partial charge of this magnitude to the metal would not substantially change the charge of the metal. This transfer of electron density from a ligand such as the imidazole to the metal would, however, induce a partial positive charge on the ligand and potentially cause repulsion by the positive charge on the metal. The electron cloud that encloses Mg in Chl *a* may accommodate a slight positive charge on the ligand and allow bond formation. Coordination of imidazole with Chl *b* may be less favorable because the positive charge on Mg is more strongly expressed. Furthermore, the functionally more electronegative Mg ion in Chl *b* may induce a greater charge transfer from imidazole to the metal than with Chl *a*, which would increase the tendency to repel the imidazole group. The less-shielding electron cloud in Chl *b* would not override this outcome, which may be a factor in the paucity of imidazole as a ligand of Chl *b*.

Instead, the "hard" Mg in Chl *b* is found to be coordinated with the negative end of 'hard' Lewis bases that contain an electronegative, oxygen-induced, fixed dipole. As judged by their pK values, electrons are not readily available in water molecules or carbonyl groups of amides in proteins (Hoober et al., 2007). Electron density is unlikely to be transferred from these groups to the metal in Chl *b*, because oxygen is more electronegative than Mg. Coordination bonds with these ligands are then characterized primarily by electrostatic interactions with the negative end – the oxygen atom – of the dipole. Chl *b* binds more strongly to water than does Chl *a* (Ballschmitter et al., 1969) and apparently with sufficient force to prevent weakly dipolar groups from forming coordination bonds by displacing the water molecule.

Although electron rich and with a full negative charge, the resonance structure of a carboxyl group lowers its pK to 3–5 and its dipole moment to a relatively low value of 1.52 D (Dudev et al., 1999). In LHCs the carboxyl group of Glu forms an ion pair with the guanidinium group of Arg (Liu et al., 2004; Standfuss et al., 2005). The resulting charge-compensation reduces repulsion by the tetrapyrrole electron cloud, which allows a carboxyl oxygen to be a favorable ligand for Chl *a*. The ion pair is occasionally also a ligand of Chl *b*. The dipole moments of amide groups of Gln and Asn are 3.5–3.9 D (Hoober et al., 2007), but sidechain amides in native LHCs are not commonly ligands of Chl *b*. In reconstituted complexes, binding to some amide ligands is mixed (Bassi et al., 1999; Remelli et al., 1999), which suggests a competition between Chls *a* and *b* for these ligands. The dipole moment of a peptide bond amide is 4.2 D (Gunner et al., 2000) and is sufficiently strong to displace water from Chl *b*, as indicated by two critical ligands of Chl *b* in LHCII (Liu et al., 2004; Standfuss et al., 2005). Calculations with the model compound, N-methylacetamide, indicated that interaction with a positive charge can increase its dipole to 6 D or more (Whitfield et al., 2006), which would lead to a strong electrostatic bond with Chl *b*. H-bonding of the 7-formyl oxygen on Chl *b* with other Chl molecules and amino acid sidechains in LHCs would further strengthen its electron-withdrawing property and thereby influence the electrostatic nature of the bond between Mg and its ligand.

III Binding of Chlorophyll to Proteins

A sequence motif –Glu-x-x-His-x-Arg– (–ExxHxR–) occurs in membrane-spanning helix-1 of LHCPs, the first of the Chl-binding domains to enter the envelope membranes during import into the chloroplast. This motif has been conserved throughout evolution from small, single-membrane-spanning polypeptides in cyanobacteria to all LHCPs and related proteins in plants (Dolganov et al., 1995; Durnford et al., 1999; Funk and Vermaas, 1999). In a few of these proteins, His is replaced by Asn. A similar motif, with Asn as a conserved substitution for His, is located in helix-3 of LHCPs as –Glu-x-x-Asn-x-Arg– (–ExxNxR–) (Jansson, 1999). The crystal structure of LHCII revealed four molecules of Chl *a* bound to these motifs. Two are bound to Glu/Arg pairs formed by complementary alignment of the helices during folding of the protein. A Chl *a* is bound to the imidazole group of His in helix-1 and another to the sidechain amide of the Asn in helix-3 (Liu et al., 2004; Standfuss et al., 2005).

Molecular modeling of the amino acid sequence in helix-1 indicated that the motif, when the protein is in an extended conformation during import, should provide two ligands for Chl *a,* one provided by an internal loop formed by an ion-pair between the carboxyl group of Glu and the nearby guanidinium group of Arg and the second by the imidazole ring of His (Eggink and Hoober, 2000; Chen and Cai, 2007). In the sequence of LHCPs in plants, the most common amino acid next to Arg in the motif in helix-1 is Trp, the only fluorescent amino acid in proteins. Therefore, a simple experiment was designed to test binding of Chl to this motif. A 16-mer peptide was synthesized that contained the native sequence around these ligands, –EVIHSRW–, or the inverse –WRSHIVE–. Trp (W) next to Arg (R) provided the ability to assay binding by Förster resonance energy transfer from Trp to Chl. The peptides were mixed with Chl in detergent micelles that mimic a membrane environment, and binding was measured by fluorescence emission from Chl *a* at 675 nm upon excitation of Trp at 280 nm. The results showed that indeed the peptide bound two molecules of Chl *a* (Eggink and Hoober, 2000).

This experimental design was extended to examine whether other Chl species also bound to the peptide. The data showed that, in addition to Chl *a,* Chl *d* readily bound the peptide with essentially the same affinity (Fig. 2). However, Chls *b* and *c* did not bind effectively. Computational modeling of these interactions, by calculating the theoretical ΔH of formation of the complexes with the peptide in the gas phase, suggested that complex formation is thermodynamically very favorable with Chls *a* and *d* but unfavorable with Chls *b* and *c* (Table 1). The modeling predicted that association of Chl *b* in particular is strongly influenced by the dielectric constant and is more favorable when sufficient water is available for Chl *b* and the ligands to be bridged by a water molecule. These theoretical ΔH values correlate well with the empirical data shown in Fig. 2.

The requirement of Chl for folding of LHCPs was demonstrated during reconstitution studies (Paulsen et al., 1993; Reinsberg, et al., 2001). In these experiments, Chl *a* bound with rapid

Table 1. Heat of formation of Chl–peptide complexes calculated with the Parametric Method 5 for association of each of the species of Chl with a 16-mer maquette of helix-1 of Lhcb1 described in Fig. 2. The more negative the value, the more thermodynamically stable the complex (Adapted from Chen and Cai, 2007).

	Heat of formation (kcal/mol)	
Species	In vacuum	In water
Chl *a*	−706	−870
Chl *b*	770	−160
Chl c_1	1,001	372
Chl c_2	1,115	310
Chl c_3	877	84
Chl *d*	−806	−949
BChl *a*	−892	−1,047

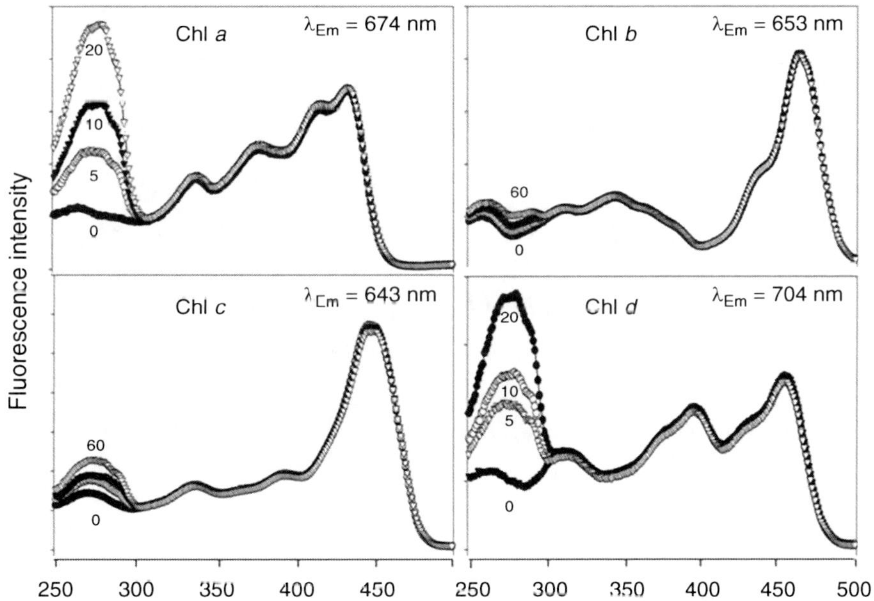

Fig. 2. Association of a 16-mer maquette of helix-1 in Lhcb1 that contains a Chl-binding motif (Eggink and Hoober, 2000). Binding was assayed by Förster resonance energy transfer from a Trp residue, adjacent to Arg in the peptide, to Chl. Interaction was expressed as emergence of a peak in the excitation spectrum at 280 nm, the absorption maximum of Trp. Each assay contained 100 nM of Chl; peptide was added to the numbers (in μM) under the peak at 280 nm in each panel (From Chen et al., 2005a).

kinetics to the protein, assayed by energy transfer from a chromophore attached to the protein, while most of the Chl *b* entered the complex more slowly (Horn and Paulsen, 2004; Horn et al., 2007). These kinetic experiments did not exclude the possibility that a Chl *b* molecule also bound rapidly. A key interaction is likely the coordination of Chl *b* with the peptide backbone carbonyl of Tyr24, which in the crystal structure of LHCII is a ligand of Chl *b*. As described below, this interaction seems to be essential for efficient import of LHCPs into the chloroplast.

IV Chlorophyll Assignments in Light Harvesting Complex II (LHCII)

In Chl *b-less* mutants of higher plants, only a few of the apoproteins for LHCI and LHCII accumulate in the organelle in vivo (Król et al., 1995; Bossmann et al., 1997; Espineda et al., 1999). The positions of Chl *b* in LHCII may provide clues to this requirement of Chl *b* for integration of the apoprotein into the plastid envelope. Resolution of the crystal structure of LHCII revealed the identity of each of the 14 Chl molecules and their ligands in the complex. The complex had essentially the same structure whether isolated from spinach (Liu et al., 2004) or pea (Standfuss et al., 2005). The complex contains eight Chl *a*, six Chl *b* and three integral xanthophyll moieties associated with a single polypeptide of 232 amino acids, about 25 kDa in mass. Figure 3 illustrates the binding sites and ligands for the Chls. Chl *a* interacts with His64, His212, the carboxyl oxygen of two charge-compensated Glu/Arg ion pairs that are contributed by helices 1 and 3, the amide sidechains of Asn183 and Gln197, an oxygen atom of the phosphoryl group of the bound phosphatidyl glycerol, and a water molecule.

Two Chl *b* molecules are coordinated with the oxygen atom of peptide bonds, one (*b*9) at Tyr24 and the second (*b*14) at Val119. These peptide bond carbonyl groups are available for coordination because formation of an intra-helix H-bond is precluded by the presence of Pro residues one helical turn distant. Chl *b*14, bonded to the peptide carbonyl of Val119, is located near the membrane surface and the local dielectric constant

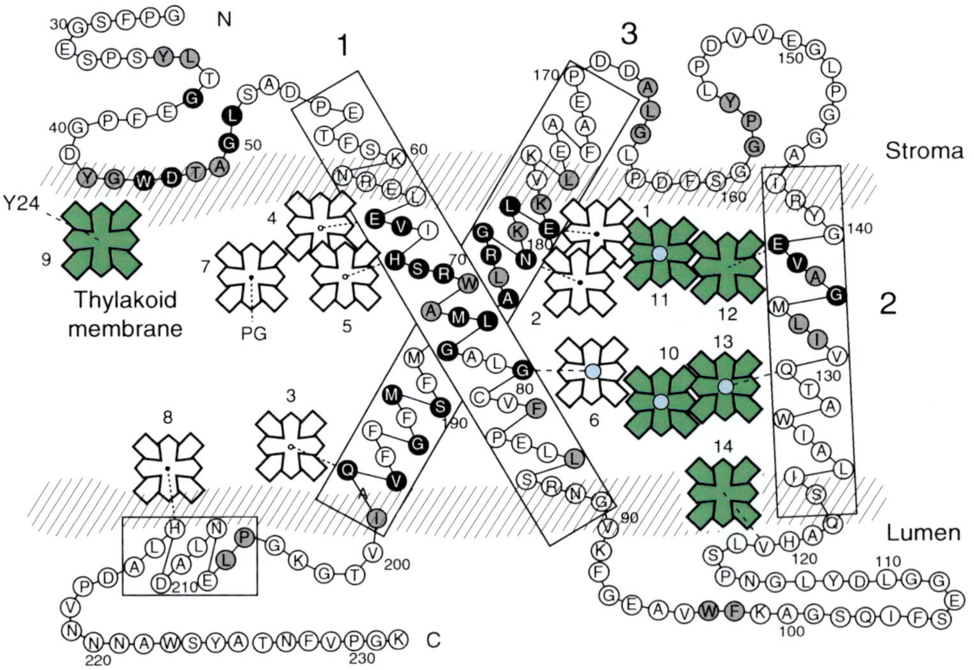

Fig. 3. Location of each of the Chls in LHCII. Chl *b* is indicated by *closed* (green) symbols, Chl *a* by *open* symbols. A water ligand is indicated by a *blue dot* in a symbol. The Chls are numbered according to Standfuss et al. (2005) with Chl *a* numbered 1–8 and Chl *b* 9–14. Note that the protein contains only three His residues, and Chl *b* is coordinated with the backbone carbonyl of Val119 rather than Chl *a* to the imidazole of His120 (Adapted from Green and Durnford, 1996).

may be too high for favorable bond formation between Chl *a* and His120. Chl *b*12 is bound to the sidechain carboxyl group of Glu139, which is sufficiently near Arg142 for charge compensation. Interestingly, three of the six molecules of Chl *b* (*b*10, *b*11 and *b*13) in LHCII retain a water molecule as the ligand and thus the Mg atom interacts with the protein and other Chls via H-bonded water bridges. Chl *b* provides stability to LHCs through these relatively strong electrostatic bonds with ligands that are unfavorable for coordination with Chl *a*.

An available peptide carbonyl group is present in Gly78 because of Pro at position 82. Interestingly, this position (*a*6) is filled with water-ligated Chl *a* rather than an expected Chl *b*. The adjacent, bulky sidechains of Leu77 and Phe81 may hinder Chl *b* from approaching sufficiently near to coordinate directly with this carbonyl group. This suggestion is supported by a result from reconstitution experiments with a mutant Lhcb1 in which Gly78 was substituted with Phe. When reconstitution was accomplished with a mixture of nearly equal amounts of Chls *a* and *b*, the additional steric hindrance by the bulk of Phe resulted in the loss of Chl *b* as compared with reconstitution with wild-type Lhcb1 (Rogl and Kühlbrandt, 1999). Thus, this site may have been filled with Chl *b* during in vitro reconstitution experiments but with Chl *a* during assembly in vivo. This position is perhaps filled with Chl *a* early during assembly in vivo by mass action from a higher concentration of Chl *a*.

Of particular interest in regard to assembly of the complex in vivo are the Chls associated with helix-1, which is the first membrane-spanning domain to interact with the chloroplast envelope during import into the chloroplast after synthesis in the cytosol. A mutant Lhcb1 with Ala substituted for His in the motif EVIHSR was not imported into isolated chloroplasts (Kohorn, 1990), which suggested that binding of Chl *a* to this motif is required for complete import. Reinbothe et al. (2006) found that LHCPs are also not imported into chloroplasts isolated from a Chl *b-less* mutant strain of *Arabidopsis*. In addition, LHCPs are not imported into the chloroplast at a detectable rate in vivo in the absence of Chl synthesis in the mutant strain *cbn*1-113*y* of *C. reinhardtii*. In these cells, LHCPs were synthesized at nearly the same rate in the dark as in the light (Fig. 4a) but were not detected immunochemically in the chloroplast (Fig. 4b). Instead, the proteins accumulated in the cytosol and vacuoles in the dark (Park and Hoober, 1997). Evidence obtained with in vitro and in vivo systems indicates that the primary role of Chl *b* is to facilitate assembly of stable Chl-protein complexes (Król et al., 1995; Preiss and Thornber, 1995; Bossmann et al., 1997; Hoober

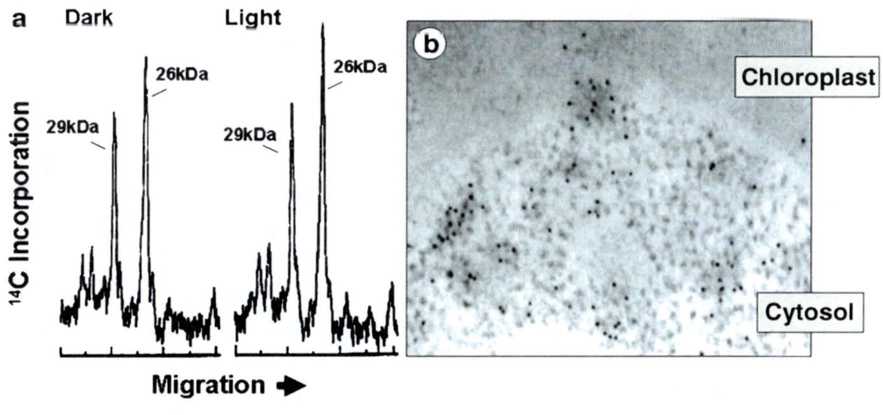

Fig. 4. Accumulation of the major LHCPs in *C. reinhardtii cbn*1-113*y*, a Chl *b-less* mutant strain. (**a**) Cells were grown in the dark, and then the temperature was raised to 38°C. After 1.5 h pre-incubation a portion of the cell suspension was transferred to the light. Cells remaining in the dark and those in the light were labeled 15 min with [^{14}C]arginine. A radioautogram of the gel was prepared after electrophoresis of proteins in a membrane fraction of the cells and then was scanned to estimate the rate of synthesis of the major LHCPs, designated as 26 and 29 kDa polypeptides (From Hoober et al., 1990). (**b**) Immunoelectron microscopic detection of LHCPs in the *b-less* strain of *C. reinhardtii* after incubation for 1.5 h at 38°C in the dark. A section of the cell was incubated with antibodies raised against the 29 kDa major LHCP, and bound antibodies were detected with protein A-gold. Gold particles were located predominantly over the cytosol and vacuoles, with few detected over the chloroplast (From Park and Hoober, 1997).

and Eggink, 1999). We interpret these data to suggest that, without Chl *b* molecules bound to helix-1, import into the chloroplast is aborted. The two peptide bond carbonyl ligands to Chls *b*9 and *b*14 are at the ends of helix-1, at Tyr24 and Val119 (see Fig. 3), which likely provide an anchor in the membrane for the N-terminal domain of the protein. A high rate of Chl *a* synthesis can compensate for the absence of Chl *b*, as indicated by the normal levels of LHCPs in thylakoid membranes in Chl *b-less* mutants of *Chlamydomonas* grown in the light on acetate as a carbon source (Michel et al., 1983; Hoober et al., 1990; Chunaev et al., 1991; Park and Hoober, 1997).

Thus, Chl *b* appears to enhance import and accumulation of LHCPs in chloroplasts under normal conditions by assisting in anchoring helix-1 in the chloroplast envelope during assembly of LHCs. Stability of the initial assembly intermediates is essential for the remainder of the protein to enter the chloroplast envelop membranes, interact with additional Chls and xanthophylls, and complete folding of the mature complex. These results point to the envelope as the site of (at least initial) interaction of LHCPs with Chl. Very soon after assembly of LHCIIs, they combine with reaction centers and the nascent photosynthetic unit is transferred to thylakoid membranes (Hoober, 2006). Consequently, stable LHCs accumulate and the capacity of the plant to harvest light energy is greatly increased by expansion of the antenna.

V Cellular Location of Chlorophyll *b* Synthesis and LHCII Assembly

Observations of LHCII assembly can be correlated with the sub-chloroplast location of Chl *b* synthesis. Conditions were established in which dark-grown, degreened cells of *C. reinhardtii y*1 were induced to synthesize Chl at a linear rate when cells were exposed to light at 38°C. This system thus allowed examination of assembly of LHCs and formation of thylakoid membranes within the first minutes of greening (Hoober et al., 1991). The linear accumulation of LHCPs correlated with the immediate and nearly linear accumulation of Chl *b* (Maloney et al., 1989). Immunoelectron microscopy showed that antibodies against LHCP extensively decorated thylakoid membranes in control, green cells in *C. reinhardtii*. During chloroplast development, LHCPs were initially detected in the chloroplast along the envelope (White et al., 1996).

Whereas these results were obtained with rapidly greening algal cells at 38°C, the initial localization of LHCPs was also examined with cells greening at 25°C. Chl accumulation was dramatically slowed by treatment with high concentrations of chloramphenicol, an inhibitor of chloroplast protein synthesis. As shown in Fig. 5b, LHCPs were detected only along the chloroplast envelope in cells treated with chloramphenicol, in contrast to the rich decoration by immunogold labeling of internal thylakoid membranes in control cells that were greening normally (Fig. 5a).

The proteins accumulated also at a similar rate in electron-dense granules in small vacuoles in the cytosol (White et al., 1996). The granules were composed largely of polyphosphate (Komine et al., 2000). At the higher temperature, the granules were strongly immunolabled (Komine et al., 1996, 2000). Thus, LHCPs that were made in excess of the chloroplast's capacity to assemble LHCs were not transported into the chloroplast stroma. Instead, excess LHCPs were shunted to vacuoles for degradation (Park et al., 1999). In a mutant strain of *C. reinhardtii* designated MC9, which is deficient in assembly of LHCs, LHCPs accumulated in vacuoles at a rate nearly tenfold greater than in wild-type (Eggink et al., 1999). These results suggest that import of LHCPs is regulated by the availability of Chl.

Regardless of whether most of the proteins accumulated in or outside of the chloroplast, electrophoretic analysis indicated that all detectable proteins were mature-sized. Models of the initiation of import after synthesis of the LHCP precursor in the cytosol indicate that as the N-terminal transit sequence enters the stroma most of the protein remains in the cytosol. Thus processing to remove the transit sequence may occur before the protein is completely imported. Chl *b* is possibly required at this stage to prevent the proteins from escaping back into the cytosol. These results provide indirect evidence that Chl *b* is synthesized in the envelope, where it immediately interacts with LHCPs. Chlide *a* oxygenase (CAO) has been localized immunochemically to the inner membrane of the envelope (Eggink et al., 2004; Reinbothe et al., 2006) and apparently obtains its substrate within the envelope (Reinbothe and Reinbothe, 1996; Joyard et al., 1998).

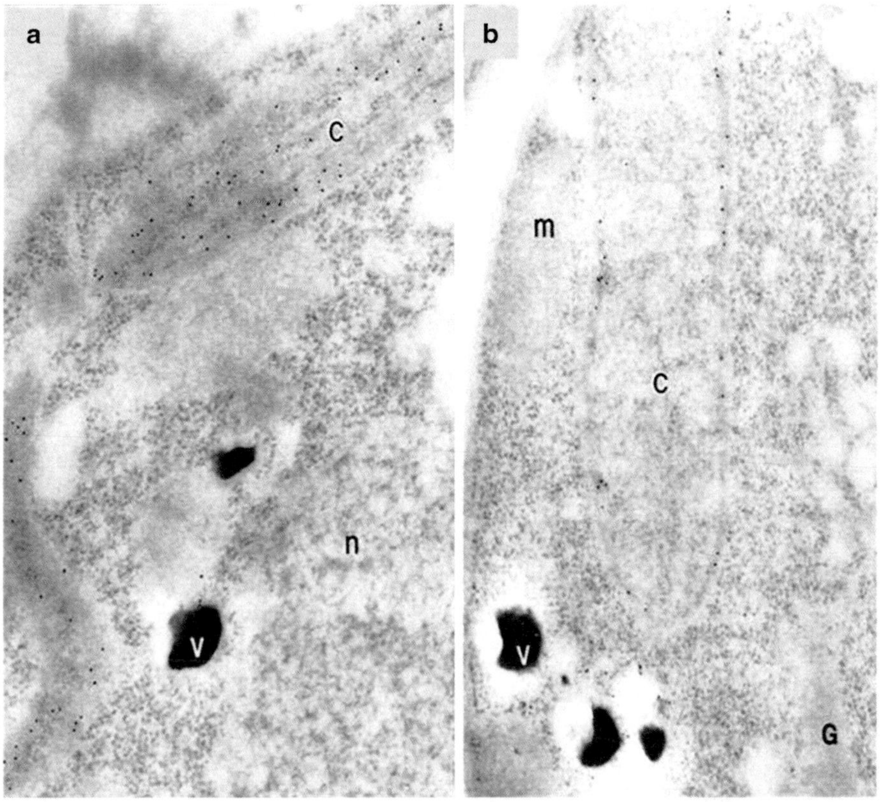

Fig. 5. Immunoelectron microscopic detection of LHCPs in the chloroplast of greening *C. reinhardtii y1* cells. Dark-grown cells were exposed to light at 25°C for 6 h and then processed for microscopy. Sections were incubated with antibodies raised against LHCP and then with gold-conjugated protein A. (**a**) Gold particles were abundant over thylakoid profiles within the chloroplast in control cells. (**b**) Cells were exposed to light in the presence of 200 μg ml^{-1} chloramphenicol, an inhibitor of chloroplast protein synthesis that severely inhibited Chl synthesis (Hoober, 1972). Gold particles were abundant over the chloroplast envelope but few were detected in the interior of the organelle. The amounts of LHCPs in the two samples were approximately the same (From Eggink et al., 2001).

The chlorin is immediately esterified with geranylgeranyl diphosphate (Maloney et al., 1989). More direct evidence for localization of these activities is given below.

Kinetic analyses of the assembly of LHCs, and their connection to an energy transducing apparatus that trapped light energy absorbed by LHCs, showed that functional membranes were assembled within minutes after exposure to light (White et al., 1996). Activities of PS I and PS II appeared with no lag under these conditions, and the fluorescence induction kinetics of the initial membranes were characteristic of a state 2 electron transport system, in which PS I and PS II had not achieved the segregation normally found in mature membranes (White and Hoober, 1994). These results indicate that complete and functional photosynthetic membranes are assembled rapidly, within the time expected for synthesis of a LHCP in the cytosol, import into the chloroplast and assembly into a complex with Chl. These results are consistent with the multibranched pathway-sublocation model of biogenesis proposed by Rebeiz et al. (1999, 2003). Differentiation from small, state 2-type vesicles into mature, state 1-type membranes required a much longer time.

Examination of the ultrastructure of rapidly greening cells revealed an abundance of vesicles emanating from the inner membrane (Hoober et al., 1991). The correlation of the appearance of vesicles, the immunological detection of LHCPs on the envelope, the kinetics of assembly of LHCs, and the immediate trapping of energy absorbed by the antenna complexes provided convincing support for the chloroplast envelope as the site of assembly of the membrane (Hoober, 2006). Although these observations have been made most definitively with studies with the model alga *C. reinhardtii*, consistent data have been obtained with plants (see Hoober and Eggink, 1999; Hoober and Arygroudi-Akoyunoglou, 2004, for reviews). Mutant strains of the plant *Arabidopsis* (Kroll et al., 2001; Aseeva et al., 2004; Vothknecht and Soll, 2006) lack an activity necessary for generation of vesicles from the inner envelope membrane, which is necessary for transfer of material to the interior of the chloroplast for expansion of the thylakoid membrane system.

VI Chlorophyllide *a* Oxygenase

Synthesis of Chl *b* from Chl *a*, whether before or after esterification, is catalyzed by a membrane-bound activity designated Chlide *a* oxygenase (CAO) (Tanaka et al., 1998; Espineda et al., 1999). An antiserum against this protein was obtained in rabbits after expression of *Arabidopsis CAO* cDNA in *Escherichia coli*. Resolution of membranes from *C. reinhardtii y1* on a sucrose gradient followed by an immunoblot after electrophoresis of fractions revealed CAO on membranes with the density of the envelope inner membrane (Fig. 6). This localization was confirmed by an analysis of outer and inner membranes of chloroplasts from *Arabidopsis* and barley by Reinbothe et al. (2006), with the active site facing the outer membrane. These results provide conclusive evidence for location of CAO on the chloroplast envelope, where it provides Chl(ide) *b* for import of LHCPs and assembly of LHCs.

Proteins were immunoprecipitated from detergent-solubilized membranes from primary leaves of *Arabidopsis* with antibodies against CAO and examined by EPR spectroscopy to determine whether mononuclear iron and Rieske iron-sulfur centers, predicted by the amino acid sequence (Tanaka et al., 1998; Espineda et al., 1999), were present. A signal indicative of the predicted high-spin mononuclear ferric iron was readily detected at g = 4.3. The spectrum of a typical Rieske iron-sulfur complex was not detected, but a spectral feature was observed at g = 2.057. Most interestingly, a remarkably stable radical signal was

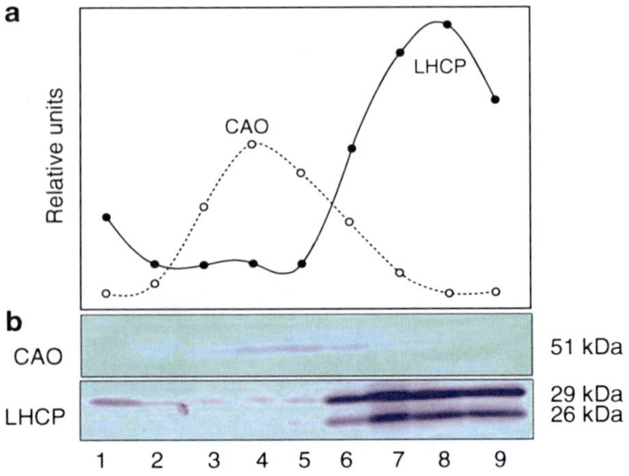

Fig. 6. Cells of *C. reinhardtii y1* were grown 3 days in the dark to partially deplete the cells of thylakoid membranes. Cells were broken and membranes were resolved on a 0.4–1.5 M continuous sucrose gradient. Fractions were collected, the proteins in each were separated by gel electrophoresis, and an immunoblot of the gel was prepared with antibodies against either CAO or LHCP. As shown in (**b**), CAO was detected in the middle of the gradient (fractions 3–6), the buoyant position of the inner membrane of the chloroplast envelope. Thylakoid membranes were recovered in fractions 6–9. Panel (**a**) illustrates results of a densitometric scan of the bands shown in (**b**) (From Eggink et al., 2004).

detected at g = 2.0042, a value consistent with a Tyr radical (Fig. 7a). A similar EPR spectrum was obtained for proteins immunoprecipitated from membranes of *Chlamydomonas* (Fig. 7b). To determine whether the EPR signals were indeed from the enzyme, *CAO* cDNA was expressed in *E. coli*. The protein was recovered in the membrane fraction after centrifugation of broken cells. The EPR spectrum of the recombinant protein, shown in Fig. 8, was similar to that obtained for the native enzyme by immunoprecipitation. Moreover, addition of Chl *a* to the recombinant enzyme quenched the radical signal. Expression of *CAO* cDNA in reticulocyte lysates produced a highly active enzyme that in vitro converted approximately 80% of added Chlide *a* to Chlide *b* (Reinbothe et al., 2006).

The spectrum for the native enzyme (Fig. 7) was nearly identical to the EPR spectrum obtained by Jäger-Vottero et al. (1997) of purified chloroplast envelope membranes. These investigators also detected an unusual g = 2.057 signal, which was not observed at higher temperatures, a characteristic of iron–sulfur centers. The similarities in these spectra, and the ability

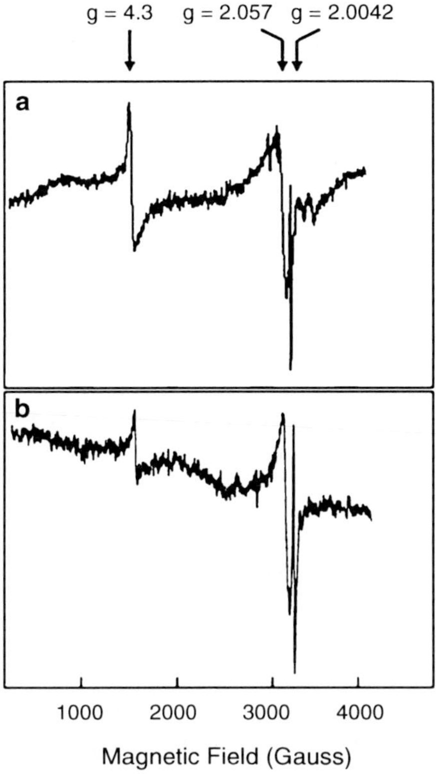

Fig. 7. EPR spectra of (**a**) CAO immunoprecipitated from an extract of leaves of *Arabidopsis* and (**b**) an extract of cells of *C. reinhardtii* obtained at X-band (9.4 GHz) at 7.5 K (From Eggink et al., 2004).

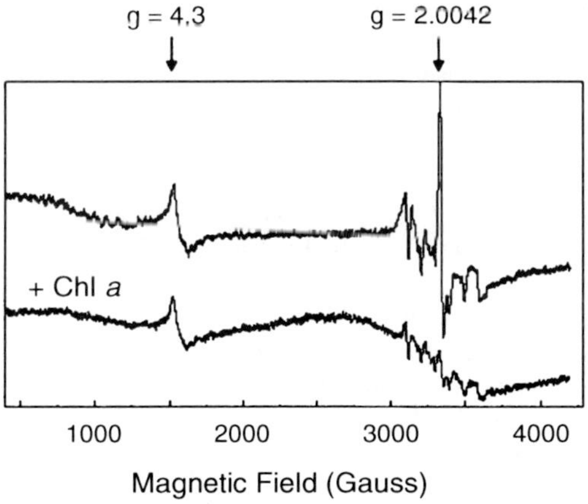

Fig. 8. EPR spectra of a membrane pellet fraction of *E. coli* cells induced to express cDNA for *Arabidopsis* CAO. The spectra were obtained at 125 K, a temperature at which the g = 2.057 signal was not observed. The signals between 3,200 and 3,700 Gauss were caused by adventitious Mn^{2+}, which could be removed by a wash with chelating agents. When Chl *a* was added to the sample, the stable radical signal disappeared (From Eggink et al., 2004).

to quench the radical upon addition of Chl *a*, suggested that the EPR spectrum of the envelope membranes was contributed largely by CAO. An experiment to identify the position of the radical relative to a trypsin-cleavable site in CAO indicated a Tyr residue near the C-terminus (Tyr422 in *C. reinhardtii* and Tyr518 in *Arabidopsis*) (Eggink et al., 2004). It is unclear at this stage of the work whether the radical is involved in the reaction, although the ability of Chl *a* to quench the radical supports this conclusion. Possible mechanisms for this reaction include abstraction of an electron or hydrogen from Chl *a*, analogous to the mechanisms of action of lipoxygenase and ribonucleotide reductase, the latter of which also contains a stable Tyr radical (Que and Ho, 1996; Yun et al., 2002). The redox potential of a Tyr radical (+0.93 V) is sufficiently high to oxidize Chl *a* ($E_{ox}^{\,1}$ = +0.81 V, or near +0.5 V when paired with another polyaromatic molecule) (Watanabe and Kobayashi, 1991). A radical form of the chlorin substrate could then react with molecular oxygen. Alternatively, the phenolic group of Tyr may contribute an electron during formation of an activated iron-oxygen complex that generates the 7-hydroxymethyl Chl intermediate, as described by Oster et al. (2000). These possible reaction pathways were proposed by Porra et al. (1994, 1997), who determined that the oxygen atom in the 7-formyl group of Chl *b* is derived from molecular oxygen. Our studies provide an approach toward an analysis of the mechanism of this reaction.

VII Conclusions

The most fundamental biological process is photosynthesis in plants, which supports all living organisms. The key process in photosynthesis is activation of Chl as a reducing agent by the absorption of light energy. The efficiency of this process is dramatically enhanced by the light-harvesting antenna. Thus, the enzyme-catalyzed reaction that oxidizes Chlide *a* to Chlide *b*, which is required for assembly of the light-harvesting complexes, is among those high in importance.

Our evidence that Chl *b* synthesis is required to retain LHCPs in the chloroplast leads to two principal conclusions. First, the last steps of Chl synthesis, including that of Chl *b*, occur in the envelope membranes, at least during the early stages of chloroplast development. Interaction of the proteins with Chl and assembly of a functional energy transducing membrane also is initiated in the envelope. Secondly, an understanding of the assembly of LHCs must include the chemistry of coordination bond formation between Chls and the proteins. These interactions are apparently necessary for correct folding of the protein, and they provide an important factor, along with the local concentrations of the Chls, that establishes the specificity of occupancy of binding sites for Chl in LHCs.

An understanding of the factors that enhance assembly and stability of light-harvesting complexes can possibly lead to engineering more robust organisms for a number of useful, 'green' processes. Algae provide a potential 'scrubber' of CO_2 in power plant effluents. Conditions in photobioreactors that allow growth of algal cultures to achieve high cell densities gradually become light limited, and under these conditions production may become greater with organisms that have an expanded light-harvesting antenna.

It may also emerge, based on our data, that oxidation of Chlide *a* to Chlide *b* occurs by a radical mechanism, which would be sensitive to radical scavengers and anti-oxidants that are abundant in foods and nutritional supplements. Such chemicals in the environment could substantially deter growth of efficient, photosynthetic organisms. Much detailed work remains to be done on all aspects of the proposals made in this article.

Acknowledgement

MC thanks the Australian Research Council (DP045233 and DP0665169) for financial support. We thank Zheng-Li Cai for computational calculations and molecular modeling.

References

Akiyama M, Miyashita H, Kise H, Watanabe T, Mimuro M, Miyachi S and Kobayashi M (2002) Quest for minor but key chlorophyll molecules in photosynthetic reaction centers – unusual pigment composition in the reaction centers of the chlorophyll *d*-dominated cyanobacterium *Acaryochloris marina*. Photosynth Res 74: 97–107

Aseeva E, Ossenbühl F, Eichacker LA, Wanner G, Soll J and Vothknecht U (2004) Complex formation of Vipp1 depends on its α-helical PspA-like domain. J Biol Chem 279: 35535–35541

Bachvaroff TR, Puerta MVS and Delwiche CF (2005) Chlorophyll *c*-containing plastid relationships based on analyses of a multigene data set with all four chromalveolate lineages. Mol Biol Evol 22: 1772–1782

Ballschmitter K, Cotton TM and Katz JJ (1969) Chlorophyll-water interactions. Hydration, dehydration and hydrates of chlorophyll. Biochim Biophys Acta 180: 347–359

Bassi R, Croce R, Cugini D and Sandonà D (1999) Mutational analysis of a higher plant antenna protein provides identification of chromophores bound into multiple sites. Proc Natl Acad Sci USA 96: 10056–10061

Belanger FC and Rebeiz CA (1984) Chlorophyll biogenesis. Spectroscopic study of net spectral shifts induced by axial coordination in metalated tetrapyrrole. Spectrochimica Acta 40A: 807–827

Bhattacharya D and Medlin LK (2004) Dating algal origin using molecular clock methods. Protist 155: 9–10

Bossmann B, Knoetzel J and Jansson S (1997) Screening of *chlorina* mutants of barley (*Hordeum vulgare L.*) with antibodies against light-harvesting proteins of PSI and PSII: absence of specific antenna proteins. Photosynth Res 52: 127–136

Cai ZL, Crossley JC, Reimers JR, Kobayashi R and Amos RD (2006) Density functional theory for charge transfer: the nature of N-bands of porphyrins and chlorophylls revealed through CAM-B3LYP, CASPT2, and SAC-CI calculations. J Phys Chem B 110: 15624–15632

Chen M and Cai ZL (2007) Theoretical study on the thermodynamic properties of chlorophyll *d*-peptides coordinating ligand. Biochim Biophys Acta 1767(6): 603–609

Chen M, Eggink LL, Hoober JK and Larkin AWD (2005a) Influence of structure on binding of chlorophylls to peptide ligands. J Am Chem Soc 127: 2052–2053

Chen M, Telfer A, Lin S, Pascal A, Larkum AWD, Barber J and Blankenship RE (2005b) The nature of the photosystem II reaction centre in the chlorophyll *d*-containing prokaryote, *Acaryochloris marina*. Photochem Photobiol Sci 4: 1060–1064

Chunaev AS, Mirnaya ON, Maslov VG and Boschetti A (1991) Chlorophyll *b*- and loroxanthin-deficient mutants of *Chlamydomonas reinhardtii*. Photosynthetica 25: 291–301

De Martino A, Douady D, Quinet-Szely M, Rousseau B, Crépineau F, Apt K and Caron L (2000) The light-harvesting antenna of brown algae. Highly homologous proteins encoded by a multigene family. Eur J Biochem 267: 5540–5549

Dolganov NAM, Bhaya D and Grossman AR (1995) Cyanobacterial protein with similarity to the chlorophyll *a/b* proteins of higher plants: evolution and regulation. Proc Natl Acad Sci USA 92: 636–640

Dougherty RC, Strain HH, Svec WA, Uphaus RA and Katz JJ (1970) The structure, properties and distribution of chlorophyll *c*. J Am Chem Soc 92: 2826–2833

Dudev T, Cowan JA and Lim C (1999) Competitive binding in magnesium coordination chemistry: water versus ligands of biological interest. J Am Chem Soc 121: 7665–7673

Durnford DG, Deane JA, Tan S, McFadden GI, Gantt E and Green BR (1999) A plylogenetic assessment of the eukaryotic light-harvesting proteins, with implications for plastid evolution. J Mol Evol 48: 59–68

Eggink LL and Hoober JK (2000) Chlorophyll binding to peptide maquettes containing a retention motif. J Biol Chem 275: 9087–9090

Eggink LL, Park HS and Hoober JK (1999) The role of the envelope in assembly of light-harvesting complexes in the chloroplast: distribution of LHCP between chloroplast and vacuoles during chloroplast development in *Chlamydomonas reinhardtii.* In: Argyroudi-Akoyunoglou JH and Senger H (eds) The Chloroplast: From Molecular Biology to Biotechnology. Kluwer, Dordrecht, The Netherlands, pp. 161—166

Eggink LL, Park HS and Hoober JK (2001) The role of chlorophyll *b* in photosynthesis: hypothesis. BMC Plant Biol 1: 2

Eggink LL, LoBrutto R, Brune DC, Brusslan J, Yamasato A, Tanaka A and Hoober JK (2004) Synthesis of chlorophyll *b*: localization of chlorophyllide *a* oxygenase and discovery of a stable radical in the catalytic subunit. BMC Plant Biol 4: 5

Epstein J, Ruble JR and Craven BM (1982) The charge density in imidazole by X-ray diffraction at 103 and 293 K. Acta Cryst B38: 140–149

Espineda CE, Linford AS, Devine D and Brusslan JA (1999) The *AtCAO* gene, encoding chlorophyll *a* oxygenase, is required for chlorophyll *b* synthesis in *Arabidopsis thaliana.* Proc Natl Acad Sci USA 96: 10507–10511

Frackowiak D, Bauman D, Manikowski H, Browett WR and Stillman MJ (1987) Circular dichroism and magnetic circular dichroism spectra of chlorophylls *a* and *b* in nematic liquid crystals. Biophys Chem 28: 101–114

Fragata M, Nordén B and Kurucsev T (1988) Linear dichroism (250-700 nm) of chlorophyll *a* and pheophytin *a* oriented in a lamellar phase of glycerylmonooctanoate/H_2O. Characterization of electronic transitions. Photochem Photobiol 47: 133–143

Funk C and Vermaas W (1999) A cyanobacterial gene family coding for single-helix proteins resembling part of the light-harvesting proteins from higher plants. Biochemistry 38: 0307–9404

Goss R, Wilhelm C and Garab G (2000) Organization of the pigment molecules in the chlorophyll *a/b/c* containing alga *Mantoniella squamata* (Prasinophyceae) studied by means of absorption, circular and linear dichroism spectroscopy. Biochim Biophys Acta 1457: 190–199

Green BR and Durnford DG (1996) The chlorophyll-carotenoid proteins of oxygenic photosynthesis. Annu Rev Plant Physiol Plant Mol Biol 47: 685–714

Gunner MR, Saleh MA, Cross E, ud-Doula A and Wise M (2000) Backbone dipoles generate positive potentials in all proteins: origins and implications of the effect. Biophys J 78: 1126–1144

Hartwich G, Fiedor L, Simonin I, Cmiel E, Schäfer W, Noy D, Scherz A and Scheer H (1998) Metal-substituted bacteriochlorophylls. 1. Preparation and influence of metal and coordination on spectra. J Am Chem Soc 120: 3675–3683

Helfrich M, Bommer B, Oster U, Klement H, Mayer K, Larkum AWD and Rüdiger W (2003) Chlorophylls of the *c* family: absolute configuration and inhibition of NADPH:protochlorophyllide oxidoreductase. Biochim Biophys Acta 1605: 97–103

Hoff AJ and Amesz J (1991) Visible absorption spectroscopy of chlorophylls. In: Scheer H (ed) Chlorophylls. CRC Press, Boca Raton, FL, pp. 723–738

Hoober JK (1972) A major polypeptide of chloroplast membranes in *Chlamydomonas reinhardtii.* J Cell Biol 52: 84–96

Hoober JK (2006) Chloroplast development: whence and whither. In: Wise RR and Hoober JK (eds) The Structure and Function of Plastids, Advances in Photosynthesis and Respiration, Vol 23. Springer, Dordrecht, The Netherlands, pp. 27–51

Hoober JK and Arygroudi-Akoyunoglou JH (2004) Assembly of light-harvesting complexes of photosystem II and the role of chlorophyll *b*. In: Papageorgiou G and Govindjee (eds) Chlorophyll a Fluorescence: The Signature of Photosynthetic Efficiency and Green Plant Productivity, Advances in Photosynthesis and Respiration, Vol 19. Springer, Dordrecht, The Netherlands, pp. 679–712

Hoober JK and Eggink LL (1999) Assembly of light-harvesting complex II and biogenesis of thylakoid membranes in chloroplasts. Photosynth Res 61: 197–215

Hoober JK, Maloney MA, Asbury LR and Marks DB (1990) Accumulation of chlorophyll *a/b*-binding polypeptides in *Chlamydomonas reinhardtii y*-1 in the light or dark at 38°C. Plant Physiol 92: 419–426

Hoober JK, Boyd CO and Paavola LG (1991) Origin of thylakoid membranes in *Chlamydomonas reinhardtii y*-1 at 38°C. Plant Physiol 96: 1321–1328

Hoober JK, Eggink LL and Chen M (2007) Chlorophylls, ligands and assembly of light-harvesting complexes in chloroplasts. Photosynth Res 94: 387–400

Horn R and Paulsen H (2004) Early steps in the assembly of light-harvesting chlorophyll *a/b* complex – time-resolved fluorescence measurements. J Biol Chem 279: 44400–44406

Horn R, Grundmann G and Paulsen H (2007) Consecutive binding of chlorophylls *a* and *b* during the assembly *in vitro* of light-harvesting chlorophyll-*a/b* protein (LHCIIb). J Mol Biol 366: 1045–1054

Hu Q, Miyashita H, Iwasaki I, Kurano N, Miyachi S, Iwaki M and Itoh S (1998) A photosystem I reaction center driven by chlorophyll *d* in oxygenic photosynthesis. Proc Natl Acad Sci USA 95: 13319–13323

Jäger-Vottero P, Dorne A-J, Jordanov J, Douce R and Joyard J (1997) Redox chains in chloroplast envelope membranes: spectroscopic evidence for the presence of electron carriers, including iron-sulfur centers. Proc Natl Acad Sci USA 94: 1597–1602

Jansson S (1999) A guide to the *Lhc* genes and their relatives in *Arabidopsis*. Trends Plant Sci 4: 236–240

Jeffrey SW and Wright SW (1987) A new spectrally distinct component in preparations of chlorophyll *c* from the micro-alga *Emiliania huxleyi* (Prymnesiophyceae). Biochim Biophys Acta 894: 180–188

Joyard J, Teyssier E, Miège C, Berny-Seigneurin D, Maréchal E, Block MA, Dorne A-J, Rolland N, Ajlani G and Douce R (1998) The biochemical machinery of plastid envelope membranes. Plant Physiol 118: 715–723

Knox RS and Spring BQ (2003) Dipole strengths in the chlorophylls. Photochem Photobiol 77: 497–501

Kobayashi M, Ohashi S, Iwamoto K, Shiraiwa Y, Kato Y and Watanabe T (2007) Redox potential of chlorophyll *d in vitro*. Biochim Biophys Acta 1767: 596–602

Kohorn BD (1990) Replacement of histidines of light harvesting chlorophyll *a/b* binding protein II disrupts chlorophyll-protein complex assembly. Plant Physiol 93: 339–342

Kolossov VL and Rebeiz CA (2001) Chloroplast biogenesis 84: solubilization and partial purification of membrane-bound [4-vinyl] chlorophyllide *a* reductase from etiolated barley leaves. Anal Biochem 295: 214–219

Komine Y, Park H, Wolfe GR and Hoober JK (1996) Secretory granules in the cytoplasm of a wall-less mutant of *Chlamydomonas reinhardtii* containing processed light-harveting complex apoproteins and HSP70. J Photochem Photobiol B36: 301–306

Komine Y, Eggink LL, Park H and Hoober JK (2000) Vacuolar granules in *Chlamydomonas reinhardtii*: polyphosphate and a 70-kDa polypeptide as major components. Planta 210: 897–905

Kroll D, Meierhoff K, Bechtold N, Kinoshita M, Westphal S, Vothknecht UC, Soll J and Westhoff P (2001) *VIPP1*, a nuclear gene of *Arabidopsis thaliana* essential for thylakoid membrane formation. Proc Natl Acad Sci USA 98: 4238–4242

Król M, Spangfort MD, Huner NPA, Öquist G, Gustafsson P and Jansson S (1995) Chlorophyll *a/b*-binding proteins, pigment conversions, and early light-induced proteins in a chlorophyll *b*-less barley mutant. Plant Physiol 107: 873–883

Linnanto J and Korppi-Tommola J (2004) Semiempirical PM5 molecular orbital study on chlorophylls and bacteriochlorophylls: comparison of semiempirical, ab initio, and density functional results. J Comput Chem 25: 123–137

Linnanto J, Martiskainen J, Lehtovuori V, Ihalainen J, Kananavicius R, Barbato R and Korppi-Tommola J (2006) Excitation energy transfer in the LHC-II trimer: a model based on the new 2.72 Å structure. Photosynth Res 87: 267–279

Liu Z, Yan H, Wang K, Kuang T, Zhang J, Gui L, An X and Chang W (2004) Crystal structure of spinach major light-harvesting complex at 2.72 Å resolution. Nature 428: 287–292

Maloney MA, Hoober JK and Marks DB (1989) Kinetics of chlorophyll accumulation and formation of chlorophyll-protein complexes during greening of *Chlamydomonas rinhardtii y*1 at 38°C. Plant Physiol 91: 1100–1106

Michel HP, Tellenbach M and Boschetti A (1983) A chlorophyll *b*-less mutant of *Chlamydomonas reinhardtii* lacking in the light-harvesting chlorophyll *a/b*-protein complex but not in its apoproteins. Biochim Biophys Acta 725: 417–424

Miyashita H, Adachi K, Kurano N, Ikemoto H, Chihara M and Miyachi S (1997) Pigment composition of a novel oxygenic photosynthetic prokaryote containing chlorophyll *d* as the major chlorophyll. Plant Cell Physiol 38: 274–281

Nagata N, Tanaka R, Satoh S and Tanaka A (2005) Identification of a vinyl reductase gene for chlorophyll synthesis in *Arabidopsis thaliana* and implications for the evolution of prochlorococcus species. Plant Cell 17: 233–240

Noy D, Fiedor L, Hartwich G, Scheer H and Scherz A (1998) Metal-substituted bacteriochlorophylls. 2. Changes in redox potentials and electronic transition energies are dominated by intramolecular electrostatic interaction. J Am Chem Soc 120: 3684–3693

Noy D, Yerushalmi R, Brumfeld V, Ashur I, Scheer H, Baldridge KK and Scherz A (2000) Optical absorption and computational studies of [Ni]-bacteriochlorophyll-*a*. New insight into charge distribution between metal and ligands. J Am Chem Soc 122: 3937–3944

Oster U, Tanaka R, Tanaka A and Rüdiger W (2000) Cloning and functinal expression of the gene encoding the key enzyme for chlorophyll *b* synthesis (CAO) from *Arabidopsis thaliana*. Plant J 21: 305–310

Palmer JD (2003) The symbiotic birth and spread of plastids: how many times and whodunit? J Phycol 39: 4–11

Park H and Hoober JK (1997) Chlorophyll synthesis modulates retention of apoproteins of light-harvesting complex II by the chloroplast in *Chlamydomonas reinhardtii*. Physiol Plant 101: 135–142

Park H, Eggink LL, Roberson RW and Hoober JK (1999) Transfer of proteins from the chloroplast to vacuoles in *Chlamydomonas reinhardtii* (Chlorophyta). J Phycol 35: 528–538

Pascal A, Caffarri S, Croce R, Sandonà D, Bassi R and Robert B (2002) A structural investigation of the central chlorophyll *a* binding sites in the minor photosystem II antenna protein, Lhcb4. Biochemistry 41: 2305–2310

Paulsen H, Finkenzeller B and Kühlein N (1993) Pigments induce folding of light-harvesting chlorophyll *a/b* binding protein. Eur J Biochem 215: 809–816

Phillips JN (1963) Physico-chemical properties of porphyrins. In: Florkin M and Stotz EH (eds) Comprehensive Biochemistry, Vol 9. Elsevier, Amsterdam, pp. 34–72

Porra RJ (1997) Recent progress in porphyrin and chlorophyll biosynthesis. Photochem Photobiol 65: 492–516

Porra RJ, Schäfer W, Cmiel E, Katheder I and Scheer H (1994) The derivation of the formyl-group of chlorophyll *b* in higher plants from molecular oxygen. Eur J Biochem 219: 671–679

Preiss S and Thornber JP (1995) Stability of the apoproteins of light-harvesting complex I and II during biogenesis of thylakoids in the chlorophyll *b*-less barley mutant *chlorina f2*. Plant Physiol 107: 709–717

Que L Jr and Ho RYN (1996) Dioxygen activation by enzymes with mononuclear non-heme iron active sites. Chem Rev 96: 2607–2624

Rebeiz CA, Ioannides IM, Kolossov V and Kopetz KJ (1999) Chloroplast biogenesis 80. Proposal of a unified multibranched chlorophyll *a/b* biosynthetic pathway. Photosynthetica 36: 117–128

Rebeiz CA, Kolossov VL, Briskin D and Gawienowski M (2003) Chloroplast biogenesis: Chlorophyll biosynthetic heterogeneity, multiple biosynthetic routes and biological spin-offs. In: Nalwa HS (ed) Handbook of Photochemistry and Photobiology, Vol 4. American Scientific Publishers, Los Angeles, CA, pp. 183–248

Reinbothe S and Reinbothe C (1996) The regulation of enzymes involved in chlorophyll biosynthesis. Eur J Biochem 237: 323–343

Reinbothe C, Bartsch S, Eggink LL, Hoober JK, Brusslan J, Andrade-Paz R, Monnet J and Reinbothe S (2006) A role for chlorophyllide *a* oxygenase in the regulated import and stabilization of light-harvesting chlorophyll *a/b* proteins. Proc Natl Acad Sci USA 103: 4777–4782

Reinsberg D, Ottmann K, Booth PJ and Paulsen H (2001) Effects of chlorophyll *a*, chlorophyll *b*, and xanthophylls on the *in vitro* assembly kinetics of the major light-harvesting chlorophyll *a/b* complex, LHCIIb. J Mol Biol 308: 59–67

Remelli R, Varotto C, Sandonà D, Croce R and Bassi R (1999) Chlorophyll binding to monomeric light-harvesting complex: a mutational analysis of chromophore-binding residues. J Biol Chem 274: 33510–33521

Rodriguez-Ezpeleta N, Brinkmann, H, Burey SC, Roure B, Burger G, Löffelhardt W, Bohnert HJ, Philippe H and Lang BF (2005) Monophyly of primary photosynthetic eukaryotes: green plants, red algae and glaucophytes. Curr Biol 15: 1325–1330

Rogl H and Kühlbrandt W (1999) Mutant trimers of light-harvesting complex II exhibit altered pigment content and spectroscopic features. Biochemistry 38: 2281–2287

Shalchian-Tabrizi K, Skånseng M, Ronquist F, Klaveness D, Bachvaroff TR, Delwiche CF, Botnen A, Tengs T and Jakobsen KS (2006) Heterotachy processes in Rhodophyte-derived secondhand plastid genes: implications for addressing the origin and evolution of dinoflagellate plastids. Mol Biol Evol 23: 1504–1515

Simonetto R, Crimi M, Sandonà D, Croce R, Cinque G, Breton J and Bassi R (1999) Orientation of chlorophyll transition moments in the higher-plant light-harvesting complex CP29. Biochemistry 38: 12974–12983

Smith KM (1975) General features of the structure and chemistry of porphyrin compounds. In: Smith KM (ed) Porphyrins and Metalloporphyrins. Elsevier Scientific, Amsterdam, pp. 1–28

Spackman MA (1992) Molecular electric moments from x-ray diffraction data. Chem Rev 92: 1769–1797

Standfuss J, Terwisscha van Scheltinga AC, Lamborghini M and Kühlbrandt W (2005) Mechanisms of photoprotection and nonphotochemical quenching in pea light-harvesting complex at 2.5 Å resolution. EMBO J 24: 919–928

Sundholm D (2003) A density-functional-theory study of bacteriochlorophyll *b*. Phys Chem Chem Phys 5: 4265–4271

Tamiaki H, Yagai S and Miyatake T (1998) Synthetic zinc tetrapyrroles complex with pyridine as a single axial ligand. Bioorg Med Chem 6: 2171–2178

Tanaka A, Ito H, Tanaka R, Tanaka N, Yoshida K and Okada K (1998) Chlorophyll *a* oxygenase (CAO) is involved in chlorophyll *b* formation from chlorophyll *a*. Proc Natl Acad Sci USA 95: 12719–12723

Tanizaki S and Feig M (2005) A generalized Born formalism for heterogeneous dielectric environments: application to the implicit modeling of biological membranes. J Chem Phys 122: 124706

Tomitani A, Okada K, Miyashita H, Matthijs HCP, Ohno T and Tanaka A (1999) Chlorophyll *b* and phycobilins in the common ancestor of cyanobacteria and chloroplasts. Nature 400: 159–162

Vothknecht UC and Soll J (2006) Protein import into chloroplasts: who, when, and how? In: Wise RR and Hoober JK (eds) The Structure and Function of Plastids, Advances in Photosynthesis and Respiration, Vol 23. Springer, Dordrecht, The Netherlands, pp. 53–74

Watanabe T and Kobayashi M (1991) Electrochemistry of chlorophylls. In: Scheer H (ed) Chlorophylls. CRC Press, Boca Raton, FL, pp. 287–315

White RA and Hoober JK (1994) Biogenesis of thylakoid membranes in *Chlamydomonas reinhardtii y*1: a kinetic study of initial greening. Plant Physiol 106: 583–590

White RA, Wolfe GR, Komine Y and Hoober JK (1996) Localization of light-harvesting complex apoproteins in the chloroplast and cytoplasm during greening of *Chlamydomonas reinhardtii* at 38°C. Photosynth Res 47: 267–280

Whitfield TW, Crain J and Martyna GJ (2006) Structural properties of liquid *N*-methylacetamide via *ab initio*, path integral, and classical molecular dynamics. J Chem Phys 124: 094503

Yun D, Krebs C, Gupta GP, Iwig DF, Huynh BH and Bollinger JM Jr (2002) Facile electron transfer during formation of cluster X and kinetic competence of X for tyrosyl radical production in protein R2 of ribonucleotide reductase from mouse. Biochemistry 41: 981–990

Chapter 16

Folding and Pigment Binding of Light-Harvesting Chlorophyll *a/b* Protein (LHCIIb)

Harald Paulsen* and Christoph Dockter
Institut für Allgemeine Botanik, Universität Mainz, Müllerweg 6, D-55099 Mainz, Germany

Aleksei Volkov
Max-Planck-Institut für Polymerforschung, Ackermannweg 10, D-55128 Mainz, Germany

Gunnar Jeschke
Institut für Physikalische Chemie, Universität Konstanz, 78475 Konstanz, Germany

Summary

The major light-harvesting chlorophyll *a/b* protein (LHCIIb) is one of the most abundant proteins of the chloroplast in green plants. It contains roughly half of the chlorophylls involved in photosynthesis, and exhibits an unusual ability to self-organize in vitro. Simply mixing the apoprotein, native or recombinant, with its pigments, chlorophyll *a*, chlorophyll *b*, and xanthophylls, in detergent solution, suffices to trigger protein folding and the assembly of about 18 pigments in their correct binding sites. A study of the mechanism of this self-organization seems worthwhile since (1) our knowledge about membrane protein folding is scarce compared to what we know about the folding of water-soluble proteins, (2) the mechanism of LHCIIb formation in vitro may give useful clues about the so-far unknown pathway of its assembly in the chloroplast, and (3) a thorough understanding of the process may facilitate the application of recombinant LHCIIb in hybrid constructs such as photovoltaic devices or the construction of potentially useful proteins or other polymers that spontaneously bind other dyes at a similarly high density.

During the assembly of recombinant LHCIIb, the formation of protein secondary structure is triggered by the binding of pigments. Chlorophylls are bound in two apparent kinetic phases. A faster one in the range of tens of seconds reflects the binding of chlorophyll *a* along with xanthophylls.

*Author for Correspondence, e-mail: Paulsen@mail.uni-mainz.de

Constantin A. Rebeiz et al. (eds.), The Chloroplast: Basics and Applications, pp. 231–244.

During the slower step in the range of minutes, mostly chlorophyll *b* is bound. The intermediate complex lacking chlorophyll *b* is unstable and susceptible to proteases. The resulting two-step model of LHCIIb formation is able to explain why LHCIIb in vivo exhibits an apparently constant chlorophyll *a:b* ratio although several binding sites have been found to be accessible to both of the two chlorophylls in vitro and in plants over-producing chlorophyll *b*. Time-resolved electron paramagnetic resonance (EPR) techniques have been established to assess protein folding beyond secondary structure formation during LHCIIb assembly, and first results of EPR-monitored kinetics are shown.

I Introduction

Chloroplasts contain the presumably most abundant membrane protein on Earth, the major light-harvesting chlorophyll (Chl) *a/b* protein (LHCIIb) of photosystem (PS) II. Its main function is to enlarge the absorption cross section of PSII (and, in some instances, PSI as well) in green plants, thus increasing the efficiency of photosynthesis at ambient light intensities. Additionally, LHCIIb serves to balance the flow of excitation energy between the two photosystems in a process called state transition (Bellafiore et al., 2005), and it can rapidly switch from light harvesting to the dissipation of light energy to prevent damage to PSII due to over-excitation (Niyogi et al., 2005; Szabó et al., 2005); this function is known as non-photochemical quenching (NPQ). State transition and NPQ are short-term responses to rapidly changing light intensities; the long-term adaptation of the photosynthetic apparatus to different illumination intensities experienced by the plant is by changes in the size of the light-harvesting antenna of the photosynthetic apparatus and specifically the number of LHCIIb per PSII (Tanaka and Melis, 1997; Masuda et al., 2003). For some biotechnological applications of plant photosynthesis, such as solar hydrogen production by green algae, it is advantageous to reduce the light-harvesting capacity of the photosynthetic apparatus (Melis et al., 2000); therefore, transgenic *Chlamydomonas reinhardttii* strains have been constructed that contain less LHCIIb than the wildtype (Kruse et al., 2005). Thus, both for understanding the regulation of photosynthesis in plants and for some biotechnological applications of photosynthesis in plants with bioengineered chloroplasts, it is essential to understand how the size of the light-harvesting antenna is regulated. This again requires a thorough grasp of individual processes that may contribute to this regulation, i.e. the biosynthesis of the light-harvesting proteins and pigments and their assembly into light-harvesting complexes. The present chapter describes some work aimed to unravel the assembly process of the major light-harvesting complex LHCIIb.

Monomeric LHCIIb complexes contain an apoprotein of 25–27 kDa (Lhcb1, 2, or 3) that non-covalently binds 14 Chl molecules (eight Chl *a* and six Chl *b*) and four carotenoids (two luteins, one neoxanthin, and one violaxanthin). In the thylakoid, most or all of LHCIIb is organized in trimeric complexes, presumably homo and heterotrimers of the three different apoproteins (Jackowski et al., 2001). The structure of the trimeric complex is known in near-atomic detail (Liu et al., 2004; Standfuss et al., 2005). Less detailed is our knowledge about the biogenesis of LHCIIb. The apoprotein is coded for in the nucleus, synthesized by cytoplasmic ribosomes in its precursor form, and imported into the chloroplast where it is inserted into the thylakoid via the signal recognition particle pathway (Li et al., 1995; Schünemann, 2004). How the protein is assembled with its pigments is not known, and neither do we know how the pigments travel from the sites of their biosynthesis to the site of LHCIIb assembly. Several models of this process are presented and evaluated in Chapter 1 of this book. Assembly is thought to take place in the thylakoid; this process can be reproduced in vitro by inserting Lhcb1 into isolated thylakoid membranes (Cline, 1986; Kuttkat et al., 1995) or etioplast membranes complemented with Chls (Kuttkat et al., 1997). However, LHCIIb may also assemble in the envelope membrane as indicated mostly by work with *Chlamydomonas* (Hoober and Eggink, 1999). Both Chl *a* and Chl *b* as well

Abbreviations: CAO–chlorophyllide *a* oxygenase; CD–circular dichroism; Chl–chlorophyll; cpSRP–chloroplast signal recognition particle; EPR–electron paramagnetic resonance; LHCI, LHCII–light-harvesting Chl *a/b* complex of PSI and PSII, respectively; LHCIIb–major subunit of LHCII with apoproteins Lhcb1-3; NPQ–non-photochemical quenching; PSI, PSII photosystems I and II, respectively; s–seconds; SDS–sodium dodecylsulfate; $\tau 1$, $\tau 2$–reaction times 1,2 = inverted rate constants k1, k2 in kinetic analyses

as carotenoids must be present for accumulating LHCIIb, since in vascular plants no unpigmented or even partially pigmented apoprotein is accumulated when pigments, specifically Chl *b*, are missing (Darr et al., 1986; Harrison et al., 1993; Krol et al., 1995; Bossmann et al., 1997; Kuttkat et al., 1997). Presumably, full pigmentation is necessary to stabilize the apoprotein against degradation by proteases in the thylakoid (Bellemare et al., 1982). Alternatively, if LHCIIb assembly takes place in the envelope, it is thought that in the absence of pigments the protein is retracted from the envelope membrane and transported to the vacuole (Hoober and Eggink, 1999).

LHCIIb behaves as a self-organizing complex in vitro. If the unfolded apoprotein is mixed with pigments in detergent solution under renaturing conditions, it spontaneously folds and binds the correct complement of pigments to form structurally virtually authentic LHCIIb (Plumley and Schmidt, 1987; Paulsen et al., 1990; Hobe et al., 1994). Pigment binding triggers the folding of the protein (Paulsen et al., 1993), and the minimum selection of pigments required for producing a stable pigment–protein complex is Chl *b* and lutein (Hobe et al., 2003). The simplest experimental procedure for reconstituting LHCIIb in vitro is to mix a solution of its apoprotein in the ionic detergent sodium dodecylsulfate (SDS) with a pigment solution in a non-ionic detergent. The mixing can be done very quickly, for instance by using a stopped-flow apparatus, which has enabled us to do time-resolved measurements of the folding process in the time range of seconds to minutes.

II Time-Resolved Measurements of LHCIIb Assembly In Vitro

A Fluorescence as a Monitor for LHCIIb Assembly

A useful monitor for the formation of functional LHCIIb is the establishment of excitation energy transfer from Chl *b* to Chl *a*. If Chl *b* is excited before complex formation occurs, it will emit much of its excitation energy as Chl *b* fluorescence (Fig. 1). As soon as Chl *b* is assembled into LHCIIb, it will efficiently transfer its excitation energy to Chl *a*; this intra-LHCIIb energy transfer from pigment to pigment is one of the essential functions of this pigment–protein complex. Consequently, emission

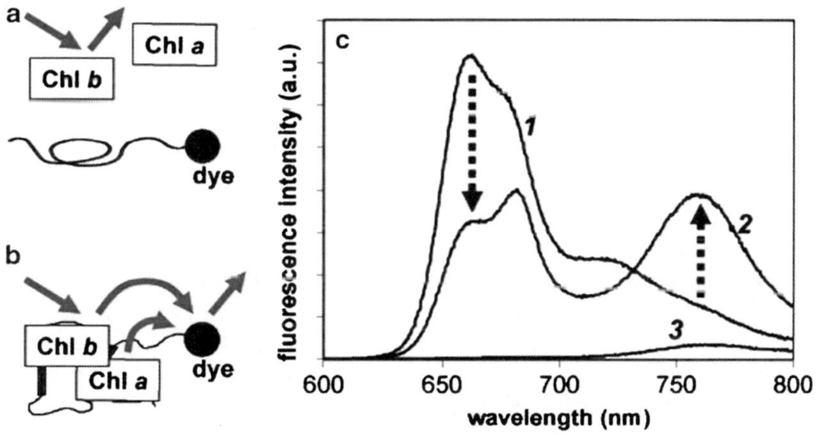

Fig. 1. Detection of LHCIIb assembly *via* energy transfer from Chls to an acceptor dye. (**a**) Sketch of the situation prior to assembly. Chl *b* is excited (*downward arrow*) and emits most of the excitation energy via fluorescence (*upward arrow*). The dye-labeled protein (*curved line*) does not fluoresce. (**b**) Sketch of the situation after assembly. Protein-bound Chl *b* transfers most of its excitation energy to the acceptor dye, either directly or via Chl *a* (*bent arrows*), stimulating the dye to emit fluorescence. (**c**) Steady-state fluorescence emission spectra upon excitation at 470 nm (excitation predominantly of Chl *b*) of (1) the assay mixture containing LHCIIb apoprotein ΔC49, a mutant version lacking 49 C-terminal amino acids and unable to form stable pigment–protein complexes, (2) the LHCIIb apoprotein, labeled with the acceptor dye DY731 in the stromal loop, upon reconstitution with pigments (corresponding to B), and (3) ΔC49 at the same concentration and under the same condition as in spectrum (1) but in the absence of pigments. The dotted arrows indicate the most prominent changes in fluorescence emission intensity upon the establishment of Chl *b*-dye energy transfer during LHCIIb assembly, i.e., the quenched donor (Chl *b*) fluorescence at 660 nm and the sensitized acceptor (dye) fluorescence at 759 nm. Panel C is adapted from Horn et al., 2007.

will no longer originate from Chl *b* but (mostly) from Chl *a*. In other words, if Chl *b* is continuously excited and fluorescence emission is observed as the LHCIIb pigment-protein complex forms, then Chl *b* fluorescence (maximum at around 660 nm) will be expected to decrease, and Chl *a* fluorescence (maximum at around 680 nm) will rise. This is exactly what was observed (Booth and Paulsen, 1996). The traces of the decreasing Chl *b* and the increasing Chl *a* signal could both be fitted with the sum of two exponentials with time constants of 10–50 s and 1 to several minutes, respectively. These two kinetically resolved phases may indicate two different pigment binding steps during the assembly of LHCIIb; however, they may also reflect higher-order kinetics that can be approximated by two exponentials. To distinguish between these possibilities, the energy transfer from Chl *b* to Chl *a* as a monitor for LHCIIb formation has its drawbacks. It requires that both Chl *a* and Chl *b* be bound to the protein. If a first event in the assembly process would be the binding of either Chl *a* or Chl *b* to the protein, this step would evade detection.

Therefore, the kinetic analysis of LHCIIb assembly in vitro was repeated by using a different monitor, the energy transfer from Chls to an acceptor dye covalently attached to the protein (Horn et al., 2007). The infrared-absorbing hemicyanine dye DY731, made by Dyomics, Jena, Germany, with excitation and emission maxima of 736 and 759 nm, respectively, has an absorption band extending far enough into the visible range to make it an efficient excitation energy acceptor for both Chl *a* and Chl *b*. The distances at which the strongly distance-dependent Förster energy transfer is 50% (Förster critical distance R_0) are estimated to be >50 Å for both Chls. The distances, as estimated from the LHCIIb crystal structure (Liu et al., 2004), between the dye attached to the stromal loop of the LHCIIb apoprotein and all Chl molecules in the LHCIIb crystal structure are less than 50 Å, so all Chls should be able to transfer their excitation energy to the acceptor dye. Consistently, purified DY731-labelled LHCIIb exhibits an energy transfer efficiency from the Chls to the dye of >90% (not shown). Earlier experiments had shown that attaching a similar dye to one of various positions in the LHCIIb apoprotein, including the N- and C-terminal domains as well as the stromal and lumenal loops, did not change the kinetics of LHCIIb assembly (Horn and Paulsen, 2004); therefore, the dye attached does not interfere with protein folding or pigment binding.

In Fig. 1, the fluorescence emission spectrum of a reconstitution assay,with dye-labeled LHCIIb protein (spectrum 2 in panel C) is compared with that of the assay mixture containing dye-labeled control protein ΔC49, a derivative of the LHCIIb apoprotein unable to bind pigments because of a 49-amino acid C-terminal deletion (spectrum 1). Excitation at 470 nm was near the absorption maximum of Chl *b*. Upon excitation at this wavelength in the absence of Chls, the fluorescent dye exhibits very little fluorescence at 759 nm as seen in spectrum 3. Therefore, the substantial dye emission at 759 nm in spectrum 2 is due to excitation energy transfer from Chl (panel B). Consistently, Chl *b* emission in spectum 2 is much lower compared to that in spectrum *1*, the control experiment in which the Chls do not bind to the dye-labeled protein and, therefore, cannot engage in energy transfer to the dye (panel A). Chl emission in spectrum *2* is not expected to be fully quenched despite efficient energy transfer in the reconstituted pigment–protein complexes because (i) a stoichiometric excess of pigments over protein is used in the reconstitution assay and less than 100% of the protein molecules become reconstituted with pigments, and (ii) only 90% of the protein is labeled. Thus there are always some reconstituted LHCIIb molecules in which Chl *b* can transfer its excitation energy only as far as Chl *a*. One way to analyze the time course of LHCIIb formation is to monitor the rise in acceptor dye fluorescence at 759 nm (dashed upward arrow in panel C). This is shown in Fig. 2.

The kinetic trace of the rising acceptor dye fluorescence during the assembly of LHCIIb (trace 1, P + ABC in Fig. 2) can be fitted by two exponentials with time constants of about 40 s ($\tau 1$) and 4 min ($\tau 2$, see Table 1); both phases contribute about equally to the total signal change. Thus, this monitor shows the same two apparent phases of the assembly process as the ones seen by using the rise in Chl *b*-to-Chl *a* energy transfer as a monitor (see above).

Since both Chls are able to transfer their excitation energy to the acceptor dye, this setup allows to study the assembly of incomplete LHCIIb with only one of the Chl species. If Chl *b* is omitted from the reaction, the slower phase $\tau 2$ completely disappears (trace 2, P + AC). Conversely, when

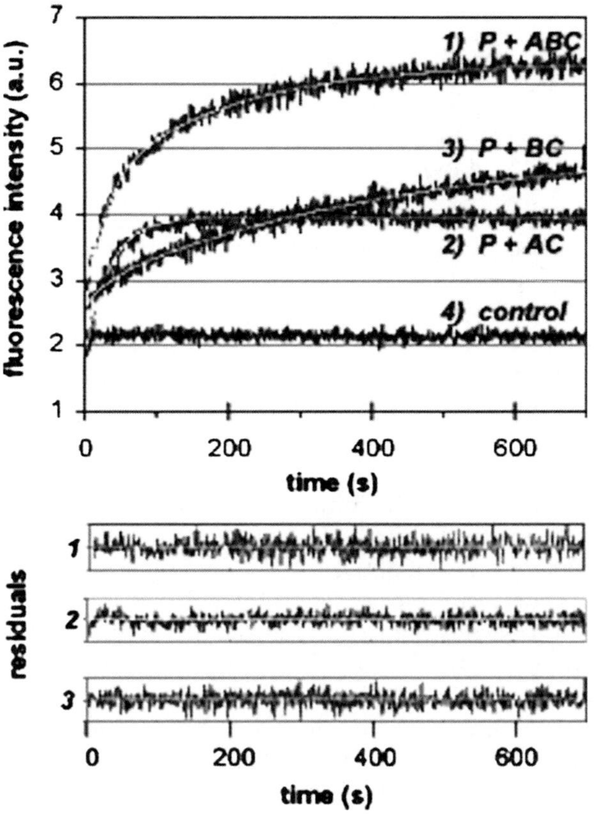

Fig. 2. Assembly of LHCIIb apoprotein with pigments as monitored by the rise of Chl-stimulated acceptor dye fluorescence at 759 nm. Reactions were started by rapidly mixing (1) LHCIIb apoprotein plus Chl *a*, Chl *b*, and carotenoids (P + ABC), (2) the protein plus Chl *a* and carotenoids (P + AC), (3) the protein plus Chl *b* and carotenoids (P + BC), and (4) as in (1) but with the LHCIIb apoprotein replaced by the non-functional mutant protein ΔC49. The white lines in traces (1)–(3) represent the bi-exponential (1 and 3) or mono-exponential (2) fits given in Table 1. Residuals, i.e., data points in the kinetic trace minus the according value in the fit curve, of these fits are given in the lower panels. The upper and lower limits of each of these panels correspond to ±0.3 fluorescence units in the upper panel (Reprinted from Horn et al., 2007 with permission from Elsevier).

Chl *a* is omitted, the slower phase contributes 80% of the total signal change (trace 3, P + BC, and Table 1). As a control, the same experiment was performed with the dye-labeled non-functional protein ΔC49 (trace 4, control) which is unable to bind Chls and, therefore, does not give rise to any Chl-sensitized dye fluorescence (Fig. 1). This observation assigns the slower phase in LHCIIb assembly exclusively to Chl *b* binding and the faster phase mostly to Chl *a* binding.

The faster binding kinetics of Chl *a* compared to that of (most of) Chl *b* suggests that an intermediate is formed with carotenoids and Chl *a* bound to the protein but no Chl *b*. In fact, earlier experiments have shown that, in the absence of Chl *b*, Chl *a* binds to the LHCIIb apoprotein; however, this complex is quite labile and dissociates upon mere dilution (Horn and Paulsen, 2004). In fact, LHCIIb was successfully assembled in a "double-jump" experiment, in which Chl *a*, along with carotenoids, was bound to the protein in the first step, and then in the second step, Chl *b* was added to this mixture. The kinetic traces obtained by monitoring Chl binding *via* the rise in sensitized acceptor dye fluorescence were very similar to the ones shown in Fig. 2, with the first step exhibiting only the fast kinetics τ1, and the second step showing predominantly the slower kinetics τ2 (Horn et al., 2007). A very similar picture was obtained when the kinetics of LHCIIb assembly was measured by using a different monitor, the circular dichroism (CD) signal at 492 nm (Horn et al., 2007). This strong negative CD band contains contributions of several LHCIIb-bound pigments, including Chl *a*, Chl *b*, and neoxanthin (Simonetto et al., 1999). Taken together, these data clearly show that the two apparent kinetic phases observed in LHCIIb assembly reflect two distinguishable pigment binding events rather than only one with higher-order kinetics. Moreover, the slower phase is exclusively assigned to Chl *b* and the faster phase mostly to Chl *a* binding.

B A Two-step Model of Pigment Binding

The different binding kinetics of Chl *a* and Chl *b* during LHCIIb assembly implies that Chl *a* binding precedes that of (most of) Chl *b* unless a large excess amount of Chl *b* is present. A two-step model depicting sequential binding of Chl *a* and Chl *b* is given in Fig. 3. A surplus of Chl *b* is clearly absent when LHCIIb is formed in vivo where the availability of Chl *b* limits the amount of Chl *a*/*b* complexes assembled (Shimada et al., 1990; Harrison and Melis, 1992; Preiss and Thornber, 1995). Mutant plants lacking Chl *b* do not accumulate LHCIIb (Darr et al., 1986; Bossmann et al., 1997), presumably because the protein needs to be stabilized by Chl *b* in order to be resistant to degradation by proteases (Apel and Kloppstech, 1980). As an exception, the Chl

Fig. 3. Two step scheme of LHCIIb assembly. The protein (*thin line*) already containing some α-helical segments (*short sections of thick line*) binds Chl *a* and carotenoids in a faster step (~10 s–1 min) and then binds Chl *b* (and possibly additional Chl *a*) in a slower second step (several minutes) (Reprinted from Horn et al., 2007 with permission from Elsevier).

b-deficient *C. reinhardtii* strain *pg-113* accumulates LHCIIb apoprotein when grown in a medium containing acetate (Michel et al., 1983) possibly because under this heterotrophic condition Chl *a* is synthesized at a higher rate than during photosynthetic growth and therefore is able to stabilize the LHCIIb apoprotein (see chapter 19). Consistently, the intermediate complex observed during LHCIIb assembly in vitro, containing Chl *a* but no Chl *b*, is structurally labile and sensitive toward proteases (Horn and Paulsen, 2004). Only the binding of Chl *b* during the second, slower step in Fig. 3 stabilizes the complex against dissociation and renders it resistant towards proteases such as trypsin and thermolysin except an N-terminal segment (Bellemare et al., 1982; Cline, 1988; Kuttkat et al., 1995, 1997; Horn and Paulsen, 2004).

The molecular basis for Chl *a* and Chl *b* binding into LHCIIb with different kinetics is unclear. From comparing the Chl *a* and Chl *b* binding sites in the LHCIIb crystal structure (Liu et al., 2004; Standfuss et al., 2005) one difference is that most of the Chl *b* binding sites not only provide ligands for the central Mg in the pigment (in three cases this ligand is a protein-bound water molecule) but also provide hydrogen bond partners to its C7-formyl group. Chl *a* contains a C7 methyl group instead which in most cases is accommodated by a non-polar region of the binding pocket. Possibly, the establishment of this non-directional hydrophobic interaction is less time-consuming than the formation of the hydrogen bond of the formyl group in Chl *b*. On the other hand, some but not all Chl *a* molecules undergo additional hydrogen bonding via their $C13^1$ keto groups (located in the isocyclic ring of the chlorophyll molecule) which also would be expected to slow down their binding into the complex. The group of Hoober reported that Chl *a* but not Chl *b* binds to synthetic peptides containing the sequence Glu-Ile-Val-His-Ser-Arg found in LHCIIb and proposed that this results from different Lewis acid strengths of the central Mg in the two Chls (Eggink and Hoober, 2000; Hoober and Eggink, 2001; Eggink et al., 2001). The formyl group in the periphery of Chl *b* increases the Lewis acid strength of its Mg compared to that of Chl *a*, and tightens the binding of a water molecule as a fifth ligand; consequently, a soft Lewis base such as histidine, a typical protein ligand for Chl *a*, may not be strong enough to replace the tightly bound water, so that harder Lewis bases such as carbonyl groups or protein-bound water molecules are required to bind Chl *b* to the complex (Chen et al., 2005). It is possible that the replacement or exchange of tightly bound water in Chl *b* is the time-consuming step that slows down the binding of this pigment into LHCIIb.

On the other hand, the kinetic analysis of LHCIIb assembly shows that part of Chl *b* binds to the complex as fast as Chl *a* does. Therefore, it seems more likely that the two different Chl binding kinetics observed are connected with different binding sites rather than with the Chl molecules themselves. Earlier titration experiments of the Chl *b* binding sites in LHCIIb with different amounts of Chl *a* and Chl *b* have shown that five Chl *b* binding sites are exclusively filled with Chl *b* and are not accessible to Chl *a*. An additional site exhibits a slight preference for Chl *b* over Chl *a* whereas the other binding sites more or less prefer Chl *a* but can also bind Chl *b* (Hobe et al., 2003). It should be noted that recombinant LHCIIb reconstituted under relatively stringent conditions contains only five Chl *b* and seven Chl *a* molecules per apoprotein whereas in the crystal structure of native LHCIIb, six Chl *b* and eight Chl *a* molecules are observed (Liu et al., 2004; Standfuss et al., 2005). If the binding sites accepting both Chl *a* and Chl *b* (termed Chl *a* sites in what follows) are the fast binding sites and the exclusively Chl *b* binding sites are slow, this would explain why the binding of Chl *a*, which is unable to bind to the Chl *b* binding sites, exclusively follows the faster kinetics whereas Chl *b* which can bind to either binding site, shows both the faster and the predominant slower kinetics The binding of Chl *b* to its cognate binding sites occurs with high affinity, rendering it virtually irreversible. In the absence of Chl *b*, Chl *a* binds to the Chl *a* binding sites in a reversible fashion. Presumably, the same is true for Chl *b* binding to Chl *a* binding sites as long as the Chl *b* binding sites are not yet filled. Only when the Chl *b* binding sites are occupied by Chl *b*, the entire complex, including the Chl *a* sites, is stabilized, possibly via pigment–pigment interactions or by a conformational change of the protein.

In LHCIIb reconstituted in vitro, Chl *b* can be trapped in the non-selective Chl *a* sites by lowering the Chl *a:b* ratio in the reconstitution mixture, leading to LHCIIb versions with a very low Chl *a* content (Kleima et al., 1999) or even exclusively containing Chl *b* (Hobe et al., 2003). On the other hand, LHCIIb in plants consistently contains Chls *a* and *b* at a ratio close to 1.3, and neither of the LHCIIb crystal structures (Liu et al., 2004; Standfuss et al., 2005) gives any indication of a mixed occupation of binding sites. The model depicted in Fig. 4 is able to explain this even if the non-selective character of Chl *a* binding sites seen in reconstitution experiments in vitro (Remelli et al., 1999; Hobe et al., 2003) is also true in vivo. If Chl *b* is the limiting component for LHCIIb assembly, then its limited availability will be sufficient to direct this pigment to the Chl *b* binding sites since in the Chl *a* binding sites it is outcompeted by Chl *a* present in excess amounts.

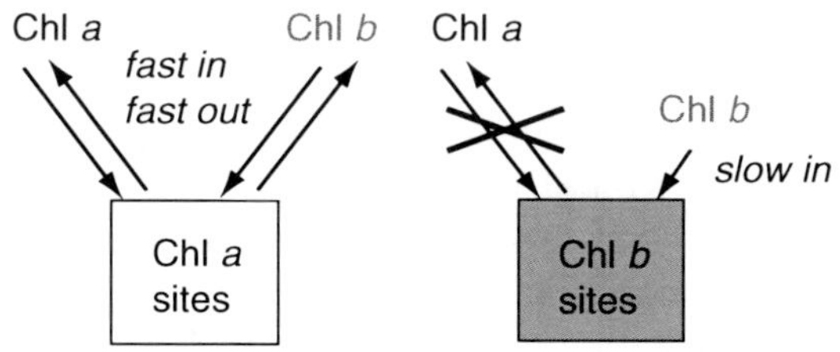

Fig. 4. Model of initial Chl binding during the assembly of LHCIIb. Chl *a* sites can either bind Chl *a* or Chl *b* in a fast, reversible fashion. Chl *b* binding sites bind only Chl *b* in a slow and virtually irreversible fashion. Once the Chl *b* binding sites are filled, both Chl *a* and Chl *b* binding becomes irreversible (Reprinted from Horn et al., 2007 with permission from Elsevier).

Thus, the potential non-selectivity of the Chl *a* binding sites is masked without requiring any further regulatory mechanism or the assumption of a principle difference in the binding behaviour of native and recombinant LHCIIb apoprotein (Rogl and Kühlbrandt, 1999).

Apparently, Chl *a* binding sites not only in LHCIIb but also in other Chl–protein complexes possess the ability to bind Chl *b* if its amount is increased. An *Arabidopsis* mutant accumulating very high amounts of Chl *b* (Chl *a:b* ratio of 1.1 instead of 3.4 in the wildtype) due to the expression of a prokaryotic chlorophyllide *a* oxygenase (CAO), the enzyme responsible for converting Chl *a* into Chl *b*, not only contains LHCIIb with an increased Chl *b* content (Chl *a:b* of 0.8 instead of 1.4 in the wildtype) but also has 40% of the Chl *a* in the core antenna complexes replaced with Chl *b* (Hirashima et al., 2006). Even in cyanobacteria that do not synthesize Chl *b*, the expression of a plant protein for CAO led to the functional incorporation of Chl *b* into PSI and PSII (Satoh et al., 2001; Xu et al., 2001).

It has repeatedly been proposed in the literature that Chl *a* is converted to Chl *b* as it is bound to LHCIIb or another Chl *a/b* binding protein (Porra et al., 1994; Plumley and Schmidt, 1995; Hoober and Eggink, 2001; Xu et al., 2001). This would make the Chl *a*-containing intermediate in the assembly pathway towards LHCIIb the substrate of the Chl *b*-synthesizing enzyme CAO (Fig. 5), a condition that has not yet been tested experimentally. A direct interaction between the precursor of the LHCIIb apoprotein and CAO has been reported to take place in the envelope membrane during import of the protein into isolated chloroplasts from *Arabidopsis*. This interaction has been proposed to direct Chl *b* (or chlorophyllide *b*) to its proper destinations in the Chl *a/b*

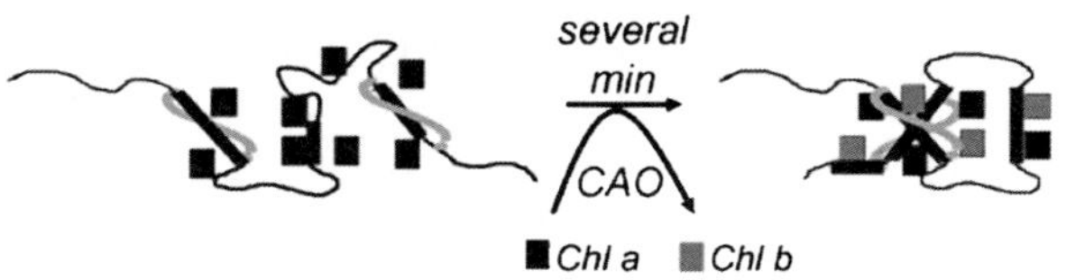

Fig. 5. Alternative second step in the two-step scheme of LHCIIb assembly. Once the intermediate protein complex containing Chl *a* and carotenoids has formed (first step in Fig. 3), some of the protein-bound Chl *a* molecules serve as substrates for chlorophyllide (or Chl) *a* oxygenase (CAO) and become converted to Chl *b* molecules.

proteins (Hoober and Eggink, 2001; Hirashima et al., 2006). If the model depicted in Fig. 4 can be extended to other Chl-binding proteins, it would provide a simpler explanation. It implies that, similar to the situation described above for LHCIIb, many or all of the Chl *a* binding sites are in principle able to bind Chl *b* as well but in fact become filled with Chl *a* since Chl *b* is present only in small amounts during the assembly of the complexes. This requires the activity of CAO to be controlled such that Chl *b* synthesis never exceeds the amount needed to fill proper Chl *b* binding sites in Chl *a/b* proteins. A possibility for such a control has already been described. The N-proximal domain in plant CAO renders the enzyme unstable in response to the presence of Chl *b* (Yamasato et al., 2005). Those pigments that are irreversibly bound to Chl *b* binding sites are not likely to be active in this regulation. On the other hand, Chl *b* molecules that are reversibly bound to Chl *a* binding sites (Fig. 4) are available for other interactions such as the destabilization of CAO. Thus, CAO would be downregulated as soon as all available proper Chl *b* binding sites are occupied, ensuring that the (non-selective) Chl *a* binding sites are filled with Chl *a* only.

It should be noted that the considerations presented above about the mechanism of pigment binding into LHCIIb are independent of where LHCIIb assembly actually takes place. It is generally thought that in higher plants the LHCIIb apoprotein in its precursor form is imported into chloroplasts and then, upon processing, inserted into the thylakoid where presumably protein folding and the binding of pigments takes place (Payan and Cline, 1991). The chloroplast signal recognition particle (cpSRP) has been identified as the carrier protein transporting the hydrophobic LHCIIb apoprotein across the aqueous stroma compartment (Li et al., 1995), and other components of the cpSRP pathway have been shown to be involved in the insertion of the LHCIIb apoprotein into the thylakoid (Schünemann, 2004). In an alternative scenario, originally based on observations in *C. reinhardtii*, the inner envelope membrane is the site of LHCIIb assembly (Hoober and Eggink, 1999). The LHCIIb apoprotein interacts with pigments already as it inserts into the envelope membrane during import into the chloroplast, and only if pigment binding takes place, the protein is retained by the chloroplast (Park and Hoober, 1997; Reinbothe et al., 2006); otherwise it will be retracted and sequestered to the vacuole (Wolfe et al., 1997; Park et al., 1999). Regardless of whether LHCIIb assembly takes place in the envelope or in the thylakoid, the molecular mechanism of this process is unknown. In particular it is unclear how the occurrence of uncoupled Chls is avoided which are potentially harmful as they can give rise to highly reactive singlet oxygen.

C Protein Folding During LHCIIb Assembly

The LHCIIb apoprotein is thought to be folded as it is complexed with pigments, either in the envelope or in the thylakoid membrane (see above). For the folding of α-helical membrane proteins in general, a two-step mechanism has been proposed in which the trans-membrane α helices are formed in a fast first step and then associate in their proper positions relative to one another in a second, slower step (Popot and Engelman, 2000; Engelman et al., 2003). The analysis of LHCIIb assembly in vitro showed that in this case α helix formation is dependent on cooperative pigment binding and therefore is relatively slow (Paulsen et al., 1993; Horn and Paulsen, 2002). Reconstitution of LHCIIb with pigments can be achieved with the protein initially dissolved in SDS where part of the α-helical structure is pre-formed before the onset of assembly. In this case, the completion of α helix formation is triggered by pigment binding. However, LHCIIb assembly in vitro can also be achieved starting with the completely unfolded apoprotein in guanidinium hydrochloride. In this case, the entire secondary structure formation is dependent on pigment binding (Yang et al., 2003).

The kinetics of secondary protein structure formation during LHCIIb assembly can be measured by monitoring the circular dichroism (CD) in the UV domain where the polypeptide chain absorbs and where α helix, β sheet, and random-coil protein structures give rise to characteristic CD signals. Time-resolved CD measurements during the assembly of LHCIIb revealed that the formation of α helices follows the same kinetics as the binding of Chls to the protein, leading to the conclusion that pigment binding and secondary structure formation are tightly coupled processes (Horn and Paulsen, 2002). The CD monitor gives no information, however, on the formation of tertiary structure, so we do not know in which time window of the assembly process the α helices assemble into their native arrangement, and what the contributions are of the hydrophilic loops and the N- and C-terminal domains. The latter question is particularly interesting, as the hydrophilic protein domains are involved in a number of functionally important properties of LHCIIb. According to the crystal structure, the loop domains are positioned close to the membrane surface to produce a flat surface of the LHCIIb molecule which may be a prerequisite for efficient stacking of the thylakoid membrane (Kühlbrandt et al., 1994). The hydrophilic N-terminal domain contains the phosphorylation site that is involved in state transitions, i.e., the balanced distribution of LHCIIb between PSII and PSI (Bellafiore et al., 2005), and the trimerization motif involved in the oligomerization of LHCIIb monomers into trimeric complexes (Hobe et al., 1995). The amino acid sequence of the lumenal loop has been shown to have a significant impact on the stability of LHCIIb (Mick et al., 2004a, b).

A possible means to gather information about the positioning and the dynamic behaviour of individual protein domains is to measure intramolecular distances. This has frequently been done by employing Förster energy transfer between fluorescent labels. One drawback of using Förster energy transfer for studying tertiary structure dynamics in LHCIIb is that, in order to avoid spectral interference with the intrinsic pigments, fluorescent labels in the infrared spectral domain need to be used which usually are quite large and therefore are likely to interfere with the folding behavior of the protein.

This complication is avoided when molecular distances are measured via the electron paramagnetic resonance (EPR) of spin-labelled proteins (Hubbell and Altenbach, 1994; Rabenstein and Shin, 1995; Hubbell et al., 2000; Steinhoff, 2002). The spin labels used for this technique, for instance the proxyl label (Fig. 6), are no larger than amino acid side chains and therefore less likely to disturb the native protein structure. By using pulse EPR such as the four-pulse double electron-electron resonance technique, distances between spin labels of up to 50 Å can be measured (Pannier et al., 2000).

Recombinant versions of LHCIIb containing two cysteins in various protein domains have been

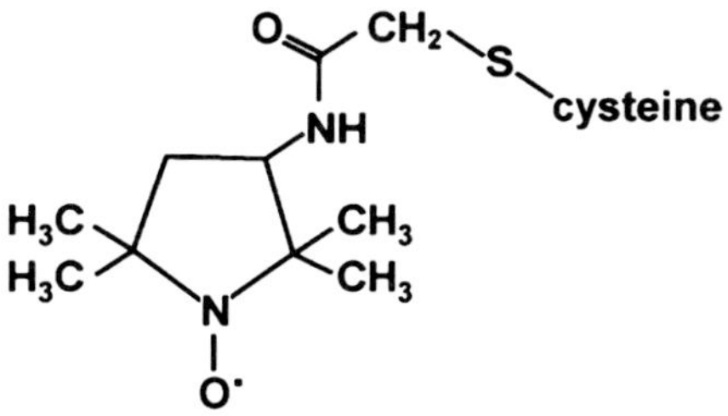

Fig. 6. Structure of the proxyl spin label 3-N-acetylamino-2,2,5,5-tetramethylpyrrolidine-1-oxyl coupled to the sulfhydryl group of a cysteine.

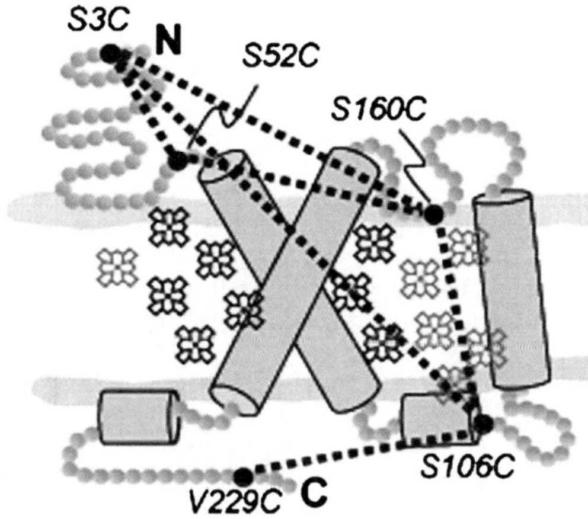

Fig. 7. Structural model of LHCIIb based on crystal structure analyses (Liu et al., 2004; Standfuss et al., 2005) with intramolecular distances measured via EPR. The labeled *black dots* indicate amino acids (S and V for serine and valine, respectively) with their position numbers, replaced in pairs by cysteines (C). The distances between these positions are indicated by *dotted lines*. N and C, amino and carboxy terminus, respectively, of the polypeptide (Adapted from Jeschke et al., 2005).

spin-labelled specifically at the cysteine sulfhydryl groups (Jeschke et al., 2005). These proteins, complexed with pigments in vitro, have then been used to measure distance distributions between the labeled protein domains (Fig. 7). An example for two such distance distributions is given in Fig. 8. In the upper panel, the distance distribution has been measured between a position near the C terminus and another one in the lumenal loop. The result shows a relatively broad distribution between 2.5 and 5.5 nm with the maximum coinciding quite well with the distance predicted from the crystal structure (Liu et al., 2004). In the lower panel, the two spin labels were attached to two amino acids in the N-terminal hydrophilic protein domain, one

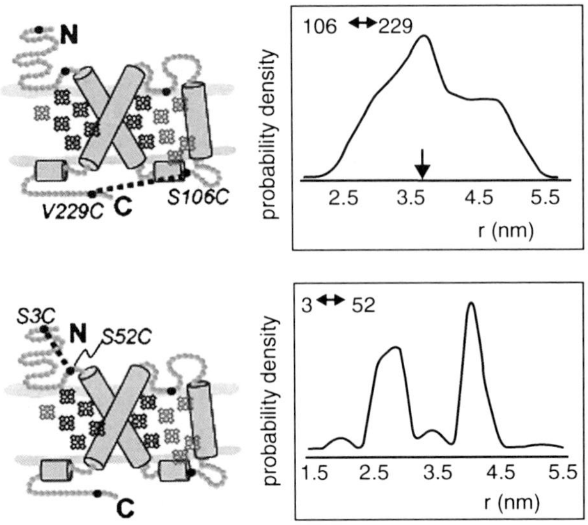

Fig. 8. Two EPR measurements of intramolecular distances in spin-labeled LHCIIb. The *left-hand sketches* are the according extracts from Fig. 7. Distance (r) distributions are given for spin label pairs in positions 106 and 229 (lumenal loop and near C terminus, respectively, *upper right-hand panel*) and in positions 3 and 52 (N- and C-proximal end, respectively, of the N-terminal hydrophilic domain). The *arrow* indicates the distance taken from the crystal structure (Liu et al., 2004) between the midpoints of the amino acid side chains in the labeling positions 106 and 229 (Adapted from Jeschke et al., 2005).

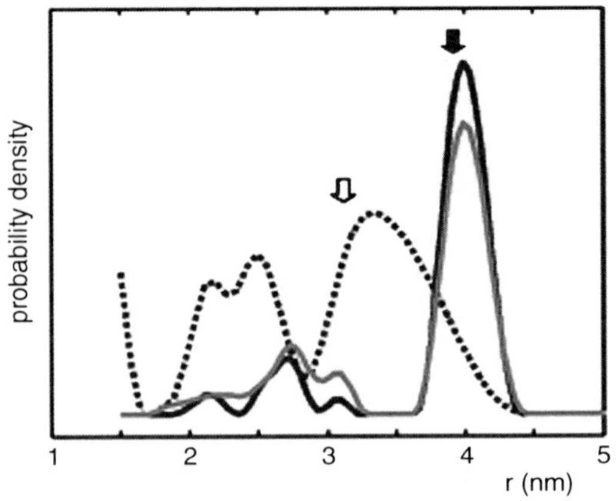

Fig. 9. Kinetics of protein folding during LHCIIb assembly as monitored by EPR. Distances distributions were measured by EPR (Jeschke et al., 2005) between spin labels in positions 106 (lumenal loop) and 160 (stromal loop) of the LHCIIb apoprotein before folding (*black dotted line*) or 22 s and 997 s (*grey* and *black solid line*, respectively) upon the onset of folding. The *filled arrow* marks the distance between the label positions taken from the LHCIIb crystal structure (Liu et al., 2004), and the *empty arrow* indicates the mean distance between the label positions as calculated from modeling the unfolded protein as a globular structure.

in position 3 near the very N terminus and the other position 52, near the N-proximal trans-membrane helix. In this case, the measured distance cannot be compared to that taken from the crystal structures since in the latter the N-terminal 9 amino acids have not been resolved. The distance distribution seen in the lower panel of Fig. 8 clearly is bimodal with peaks at 2.8 and 4.1 nm. This strongly suggests that the N-terminal domain can adopt two different structural states that exhibit different distances between the ends of the domain. A bimodal distance distribution is also seen in other distance measurements including the N terminus. Moreover, an apparently bimodal distance distribution is obtained when intermolecular distances are measured between LHCIIb trimers with each monomer carrying one label in position 3 near the N terminus (Jeschke et al., 2005). These observations may be the first direct indication of the existence of different conformational states of LHCIIb in aqueous solution. Conformational changes of LHCIIb have been hypothesized to be involved in its regulatory functions such as state transition (Nilsson et al., 1997; Zer et al., 2003) and NPQ (Horton et al., 2000; Pascal et al., 2005).

When the intramolecular distance distributions are measured via EPR during the assembly of LHCIIb using a freeze-quench technique, the folding of the LHCIIb apoprotein can be monitored. Figure 9 shows a preliminary result in which the distance distributions between labels in positions 106 and 160 in the lumenal and stromal loops, respectively, have been measured before and at intervals of 22 and 997 s after the onset of the assembly process. The mean distance calculated for the randomly coiled protein is 3.1 nm. As expected, before assembly a broad distribution with contributions at both shorter and longer distances is observed (dotted line). The maximum in the two distance distributions obtained during the assembly is at 4 nm, near the expected distance between the amino acid side chains in these positions in the fully folded protein, taken from the LHCIIb crystal structure (Liu et al., 2004) of 3.9 nm. These distance distributions still exhibit a minor contribution at shorter distances corresponding to randomly coiled protein. Looking at the change in the distance distribution occurring during the time span from 22 s (grey line) to

16 min (black solid line) into the assembly process, a rise is seen in the peak at 3.9 nm, presumably representing the fully folded protein, at the expense of the distance peaks at 3.1 nm and shorter distances that dominate the data of the unfolded protein. If these distance assignments are correct, this would mean that after 22 s of folding time, when much of $\tau 1$ but little of $\tau 2$ has passed, in most of the protein molecules the stromal and lumenal loops are already at their native distance from one another. Only a slight increase in the amount of protein with properly positioned loops is seen after an additional folding time of several minutes. It is conceivable that even after 997 s (16 min) of folding time when, according to the kinetic analysis, the assembly process should be complete, a fraction of the protein would not exhibit the "native" distance between the two loops since the yield in reconstituted protein under the conditions used here usually is >80% but always <100%. The preliminary conclusion from these data would be that much of the positioning of the loop domains in the LHCIIb apoprotein already occurs during the first apparent kinetic phase ($\tau 1$) of LHCII assembly. Clearly, these data need to be substantiated by many more measurements. However, it can be expected that the approach of monitoring distance changes between different protein positions during LHCIIb assembly should in principle be able to yield a complete three-dimensional picture of the protein folding pathway.

III Concluding Remarks

The capability to self-organize spontaneously in vitro is a remarkable property shared by many members of the Chl *a/b* protein family (Paulsen, 2006). The significance of this property for the assembly of Chl *a/b* proteins in vivo is difficult to estimate as long as we know so little about the molecular environment in which the process takes place. The LHCIIb apoprotein is inserted into the thylakoid membrane *via* the cpSRP pathway (see Section II.B) involving the putative membrane receptor ALB3 (Tu et al., 1999; Moore et al., 2003; Ossenbühl et al., 2004). Therefore, if LHCIIb assembly in fact takes place in the thylakoid, ALB3 becomes a potential candidate for being part of the molecular machinery involved. On the other hand, the spontaneous folding and pigment binding of the LHCIIb apoprotein in vitro may be of interest for a number of biotechnological applications. One possibility is the employment of recombinant LHCIIb or derivatives thereof for light-harvesting in charge-separating units, for instance in electrochemical photovoltaic cells (Wolf-Klein et al., 2002). Even if the direct use of chlorophyll-protein complexes for such applications turns out to be disadvantageous because of the inherent tendency of this pigment to undergo photooxidative reactions, the assembly of LHCIIb may still be worthwhile to study under the aspect of practical applications. The LHCIIb apoprotein folds to bind its pigments in such a way that the local Chl concentration in the trimeric complex exceeds 0.4 M (Standfuss et al., 2005). This is a concentration that cannot be achieved in any organic solvents. It arranges the Chls in such a way that they can rapidly exchange their excitation energy but do not form Chl aggregates which would efficiently dissipate that excitation energy. A thorough understanding of how this is achieved in LHCIIb may help to construct proteins (or other polymers) that self-organize to bind more stable pigments in a similarly organized way.

Acknowledgment

Work in the author's laboratory has been funded by the *Deutsche Forschungsgemeinschaft* (Pa 324/3-4, SFB 625 TP B10).

References

Apel K and Kloppstech K (1980) The effect of light in the biosynthesis of the light-harvesting chlorophyll *a/b* protein. Planta 150: 426–430

Bellafiore S, Barneche F, Peltier G and Rochaix JD (2005) State transitions and light adaptation require chloroplast thylakoid protein kinase STN7. Nature 433: 892–895

Bellemare G, Bartlett S and Chua NH (1982) Biosynthesis of chlorophyll *a/b*-binding polypeptides in wild-type and the *Chlorina* f2 mutant of barley. J Biol Chem 257: 7762–7776

Booth PJ and Paulsen H (1996) Assembly of light-harvesting chlorophyll *a/b* complex *in vitro*. Time-resolved fluorescence measurements. Biochemistry 35: 5103–5108

Bossmann B, Knötzel J and Jansson S (1997) Screening of chlorina mutants of barley (*Hordeum vulgare L.*) with antibodies against light harvesting proteins of PS I and

PS II: absence of specific antenna proteins. Photosynth Res 52: 127–136

Chen M, Eggink LL, Hoober JK and Larkum AWD (2005) Influence of structure on binding of chlorophylls to peptide ligands. J Am Chem Soc 127: 2052–2053

Cline K (1986) Import of proteins into chloroplasts. J Biol Chem 261: 14804–14810

Cline K (1988) Light-harvesting chlorophyll *a/b* protein. Membrane insertion, proteolytic processing, assembly into LHCII, and localization to appressed membranes. Plant Physiol 86: 1120–1126

Darr SC, Somerville SC and Arntzen CJ (1986) Monoclonal antibodies to the light-harvesting chlorophyll *a/b* protein complex of Photosystem II. J Cell Biol 103: 733–740

Eggink LL and Hoober JK (2000) Chlorophyll binding to peptide maquettes containing a retention motif. J Biol Chem 275: 9087–9090

Eggink LL, Park H and Hoober JK (2001) The role of chlorophyll *b* in photosynthesis: hypothesis. BMC Plant Biol 1: 2

Engelman DM, Chen Y, Chin CN, Curran AR, Dixon AM, Dupuy AD, Lee AS, Lehnert U, Matthews EE, Reshetnyak YK, Senes A and Popot JL (2003) Membrane protein folding: beyond the two stage model – minireview. FEBS Lett 555: 122—125

Harrison MA and Melis A (1992) Organization and stability of polypeptides associated with the chlorophyll *a-b* light-harvesting complex of photosystem II. Plant Cell Physiol 33: 627–637

Harrison MA, Nemson JA and Melis A (1993) Assembly and composition of the chlorophyll *a-b* light-harvesting complex of barley (*Hordeum vulgare* L) – Immunochemical analysis of chlorophyll *b*-less and chlorophyll *b*-deficient mutants. Photosynth Res 38: 141–151

Hirashima M, Satoh S, Tanaka R and Tanaka A (2006) Pigment shuffling in antenna systems achieved by expressing prokaryotic chlorophyllide *a* oxygenase in *Arabidopsis*. J Biol Chem 281: 15385–15393

Hobe S, Fey H, Rogl H and Paulsen H (2003) Determination of relative chlorophyll binding affinities in the major light-harvesting chlorophyll *a/b* complex. J Biol Chem 278: 5912–5919

Hobe S, Förster R, Klingler J and Paulsen H (1995) N-proximal sequence motif in light-harvesting chlorophyll *a/b*-binding protein is essential for the trimerization of light-harvesting chlorophyll *a/b* complex. Biochemistry 34: 10224–10228

Hobe S, Prytulla S, Kühlbrandt W and Paulsen H (1994) Trimerization and crystallization of reconstituted light-harvesting chlorophyll *a/b* complex. EMBO J 13: 3423–3429

Hoober JK and Eggink LL (1999) Assembly of light-harvesting complex II and biogenesis of thylakoid membranes in chloroplasts. Photosynth Res 61: 197–215

Hoober JK and Eggink LL (2001) A potential role of chlorophylls *b* and *c* in assembly of light-harvesting complexes. FEBS Lett 489: 1–3

Horn R, Grundmann G and Paulsen H (2007) Consecutive binding of chlorophylls *a* and *b* during the assembly *in vitro* of light-harvesting chlorophyll-*a/b* protein (LHCIIb). J Mol Biol 366: 1045–1054

Horn R and Paulsen H (2002) Folding *in vitro* of light-harvesting chlorophyll *a/b* protein is coupled with pigment binding. J Mol Biol 318: 547–556

Horn R and Paulsen H (2004) Early steps in the assembly of light-harvesting chlorophyll *a/b* complex. J Biol Chem 279: 44400–44406

Horton P, Ruban AV and Wentworth M (2000) Allosteric regulation of the light-harvesting system of photosystem II. Philos Trans R Soc Lond Ser B-Biol Sci 355: 1361–1370

Hubbell WL and Altenbach C (1994) Investigation of structure and dynamics in membrane proteins using site-directed spin labeling. Curr Opin Struct Biol 4: 566–573

Hubbell WL, Cafiso DS and Altenbach C (2000) Identifying conformational changes with site-directed spin labeling. Nat Struct Biol 7: 735–739

Jackowski G, Kacprzak K and Jansson S (2001) Identification of Lhcb1/Lhcb2/Lhcb3 heterotrimers of the main light-harvesting chlorophyll *a/b*-protein complex of Photosystem II (LHC II). Biochim Biophys Acta 1504: 340–345

Jeschke G, Bender A, Schweikardt T, Panek G, Decker H and Paulsen H (2005) Localization of the N-terminal domain in light-harvesting chlorophyll *a/b* protein (LHCIIb) by electron paramagnetic resonance (EPR) measurements. J Biol Chem 280: 18623–18630

Kleima FJ, Hobe S, Calkoen F, Urbanus ML, Peterman EJG, van Grondelle R, Paulsen H and van Amerongen H (1999) Decreasing the chlorophyll *a/b* ratio in reconstituted LHCII: structural and functional consequences. Biochemistry 38: 6587–6596

Krol M, Spangfort MD, Huner NPA, Oquist G, Gustafsson P and Jansson S (1995) Chlorophyll *a/b*-binding proteins, pigment conversions, and early light-induced proteins in a chlorophyll *b*-less barley mutant. Plant Physiol 107: 873–883

Kruse O, Rupprecht J, Bader KP, Thomas-Hall S, Schenk PM, Finazzi G and Hankamer B (2005) Improved photobiological H_2 production in engineered green algal cells. J Biol Chem 280: 34170–34177

Kühlbrandt W, Wang DN and Fujiyoshi Y (1994) Atomic model of plant light-harvesting complex by electron crystallography. Nature 367: 614–621

Kuttkat A, Edhofer I, Eichacker LA and Paulsen H (1997) Light harvesting chlorophyll *a/b* binding protein stably inserts into etioplast membranes supplemented with Zn pheophytin *a/b*. J Biol Chem 272: 20451–20455

Kuttkat A, Grimm R and Paulsen H (1995) Light-harvesting chlorophyll *a/b*-binding protein inserted into isolated thylakoids binds pigments and is assembled into trimeric light-harvesting complex. Plant Physiol 109: 1267–1276

Li X, Henry R, Yuan JG, Cline K and Hoffman NE (1995) A chloroplast homologue of the signal recognition particle subunit SRP54 is involved in the posttranslational integration of a protein into thylakoid membranes. Proc Natl Acad Sci USA 92: 3789–3793

Liu Z, Yan H, Wang K, Kuang T, Zhang J, Gul L, An X and Chang W (2004) Crystal structure of spinach major light-harvesting complex at 2.72 Å resolution. Nature 428: 287–292

Masuda T, Tanaka A and Melis A (2003) Chlorophyll antenna size adjustments by irradiance in *Dunaliella salina* involve coordinate regulation of chlorophyll *a* oxygenase (CAO) and Lhcb gene expression. Plant Mol Biol 51: 757–771

Melis A, Zhang LP, Forestier M, Ghirardi ML and Seibert M (2000) Sustained photobiological hydrogen gas production upon reversible inactivation of oxygen evolution in the green alga *Chlamydomonas reinhardtii*. Plant Physiol 122: 127–135

Michel H, Tellenbach M and Boschetti A (1983) A chlorophyll *b*-less mutant of *Chlamydomonas reinhardtii* lacking in the light-harvesting chlorophyll *a/b*-protein complex but not in its apoprotein. Biochim Biophys Acta 725: 417–424

Mick V, Eggert K, Heinemann B, Geister S and Paulsen H (2004a) Single amino acids in the lumenal loop domain influence the stability of the major light-harvesting chlorophyll *a/b* complex. Biochemistry 43: 5467–5473

Mick V, Geister S and Paulsen H (2004b) The folding state of the lumenal loop determines the thermal stability of light-harvesting chlorophyll *a/b* protein (LHCIIb). Biochemistry 43: 14704–14711

Moore M, Goforth RL, Mori H and Henry R (2003) Functional interaction of chloroplast SRP/FtsY with the ALB3 translocase in thylakoids: substrate not required. J Cell Biol 162: 1245–1254

Nilsson A, Stys D, Drakenberg T, Spangfort MD, Forsen S and Allen JF (1997) Phosphorylation controls the three dimensional structure of plant light harvesting complex II. J Biol Chem 272: 18350–18357

Niyogi KK, Li XP, Rosenberg V and Jung HS (2005) Is PsbS the site of non-photochemical quenching in photosynthesis? J Exp Bot 56: 375–382

Ossenbühl F, Göhre V, Meurer J, Krieger-Liszkay A, Rochaix JD and Eichacker LA (2004) Efficient assembly of photosystem II in *Chlamydomonas reinhardtii* requires Alb3.1p, a homolog of *Arabidopsis* ALBINO3. Plant Cell 16: 1790–1800

Pannier M, Veit S, Godt A, Jeschke G and Spiess HW (2000) Dead-time free measurement of dipole-dipole interactions between electron spins. J Magn Reson 142: 331–340

Park H, Eggink LL, Roberson RW and Hoober JK (1999) Transfer of proteins from the chloroplast to vacuoles in *Chlamydomonas reinhardtii* (Chlorophyta): A pathway for degradation. J Phycol 35: 528–538

Park H and Hoober JK (1997) Chlorophyll synthesis modulates retention of apoproteins of light-harvesting complex II by the chloroplast in *Chlamydomonas reinhardtii*. Physiol Plant 101: 135–142

Pascal AA, Liu ZF, Broess K, Van Oort B, Van Amerongen H, Wang C, Horton P, Robert B, Chang WR and Ruban A (2005) Molecular basis of photoprotection and control of photosynthetic light-harvesting. Nature 436: 134–137

Paulsen H (2006) Reconstitution and pigment exchange. In: Grimm B, Porra R, Rüdiger W and Scheer H (eds) Advances in Photosynthesis and Respiration. Kluwer, Dordrecht, The Netherlands, pp. 275–285

Paulsen H, Finkenzeller B and Kühlein N (1993) Pigments induce folding of light-harvesting chlorophyll *a/b*-binding protein. Eur J Biochem 215: 809–816

Paulsen H, Rümler U and Rüdiger W (1990) Reconstitution of pigment-containing complexes from light-harvesting chlorophyll *a/b*-binding protein overexpressed in *E. coli*. Planta 181: 204–211

Payan LA and Cline K (1991) A stromal protein factor maintains the solubility and insertion competence of an imported thylakoid membrane protein. J Cell Biol 112: 603–613

Plumley FG and Schmidt GW (1987) Reconstitution of chlorophyll *a/b* light-harvesting complexes: xanthophyll-dependent assembly and energy transfer. Proc Natl Acad Sci USA 84: 146–150

Plumley FG and Schmidt GW (1995) Light-harvesting chlorophyll *a/b* complexes: interdependent pigment synthesis and protein assembly. Plant Cell 7: 689–704

Popot JL and Engelman DM (2000) Helical membrane protein folding, stability, and evolution. Annu Rev Biochem 69: 881–922

Porra RJ, Schäfer W, Cmiel E, Katheder I and Scheer H (1994) The derivation of the formyl-group oxygen of chlorophyll *b* in higher plants from molecular oxygen – achievement of high enrichment of the 7-formyl-group oxygen from $^{18}O_2$ in greening maize leaves. Eur J Biochem 219: 671–679

Preiss S and Thornber JP (1995) Stability of the apoproteins of light-harvesting complex I and II during biogenesis of thylakoids in the chlorophyll *b*-less barley mutant *chlorina* f2. Plant Physiol 107: 709–717

Rabenstein MD and Shin K (1995) Determination of the distance between two spin labels attached to a macromolecule. Proc Natl Acad Sci USA 92: 8239–8243

Reinbothe C, Bartsch S, Eggink LL, Hoober JK, Brusslan J, Andrade-Paz R, Monnet J and Reinbothe S (2006) A role for chlorophyllide *a* oxygenase in the regulated import and stabilization of light-harvesting chlorophyll *a/b* proteins. Proc Natl Acad Sci USA. 103: 4777–4782

Remelli R, Varotto C, Sandonà D, Croce R and Bassi R (1999) Chlorophyll binding to monomeric light-harvesting complex - A mutation analysis of chromophore-binding residues. J Biol Chem 274: 33510–33521

Rogl H and Kühlbrandt W (1999) Mutant trimers of light-harvesting complex II exhibit altered pigment content and spectroscopic features. Biochemistry 38: 16214–16222

Satoh S, Ikeuchi M, Mimuro M and Tanaka A (2001) Chlorophyll *b* expressed in cyanobacteria functions as a light harvesting antenna in Photosystem I through flexibility of the proteins. J Biol Chem 276: 4293–4297

Schünemann D (2004) Structure and function of the chloroplast signal recognition particle. Curr Genet 44: 295–304

Shimada Y, Tanaka A, Tanaka Y, Takabe T, Takabe T and Tsuji H (1990) Formation of chlorophyll-protein complexes during greening. 1. Distribution of newly synthesized chlorophyll among apoproteins. Plant Cell Physiol 31: 639–647

Simonetto R, Crimi M, Sandonà D, Croce R, Cinque G, Breton J and Bassi R (1999) Orientation of chlorophyll transition moments in the higher-plant light-harvesting complex CP29. Biochemistry 38: 12974–12983

Standfuss R, van Scheltinga ACT, Lamborghini M and Kühlbrandt W (2005) Mechanisms of photoprotection and nonphotochemical quenching in pea light-harvesting complex at 2.5 Å resolution. EMBO J 24: 919–928

Steinhoff HJ (2002) Methods for study of protein dynamics and protein-protein interaction in protein-ubiquitination by electron paramagnetic resonance spectroscopy. Front Biosci 7: 97–110

Szabó I, Bergantino E and Giacometti GM (2005) Light and oxygenic photosynthesis: Energy dissipation as a protection mechanism against photo-oxidation. EMBO Rep 6: 629–634

Tanaka A and Melis A (1997) Irradiance dependent changes in the size and composition of the chlorophyll *a/b* light harvesting complex in the green alga *Dunaliella salina*. Plant Cell Physiol 38: 17–24

Tu CJ, Schünemann D and Hoffman NE (1999) Chloroplast FtsY, chloroplast signal recognition particle, and GTP are required to reconstitute the soluble phase of light-harvesting chlorophyll protein transport into thylakoid membranes. J Biol Chem 274: 27219–27224

Wolf-Klein H, Kohl C, Müllen K and Paulsen H (2002) Biomimetic model of a plant photosystem consisting of a recombinant light-harvesting complex and a terrylene dye. Angew. Chem Int Ed 41: 3380–3382

Wolfe GR, Park H, Sharp WP and Hoober JK (1997) Light harvesting complex apoproteins in cytoplasmic vacuoles in *Chlamydomonas reinhardtii* (chlorophyta). J Phycol 33: 377–386

Xu H, Vavilin D and Vermaas W (2001) Chlorophyll *b* can serve as the major pigment in functional photosystem II complexes of cyanobacteria. Proc Natl Acad Sci USA 98: 14168–14173

Yamasato A, Nagata N, Tanaka R and Tanaka A (2005) The N-terminal domain of chlorophyllide *a* oxygenase confers protein instability in response to chlorophyll *b* accumulation in *Arabidopsis*. Plant Cell 17: 1585–1597

Yang C, Horn R and Paulsen H (2003) The light-harvesting chlorophyll *a/b* complex can be reconstituted *in vitro* from its completely unfolded apoprotein. Biochemistry 42: 4527–4533

Zer H, Vink M, Shochat S, Herrmann RG, Andersson B and Ohad I (2003) Light affects the accessibility of the thylakoid light-harvesting complex II (LHCII) phosphorylation site to the membrane protein kinase(s). Biochemistry 42: 728–738

Chapter 17

The Plastid Genome as a Platform for the Expression of Microbial Resistance Genes

Tracey A. Ruhlman, Jeffrey W. Cary, and Kanniah Rajasekaran*
Agricultural Research Service, Southern Regional Research Center, USDA, 1100 Robert E. Lee Blvd., New Orleans, LA 70124, USA

Summary

Stagnation in absolute yield increases of major food and fiber crops around the world is raising awareness that breeding for quantitative yield characteristics will most likely be insufficient to meet the needs of the burgeoning global population. Additionally the attractive agronomic characteristics of high absolute

*Author for Correspondence, e-mail: rajah.rajasekaran@ars.usda.gov

Constantin A. Rebeiz et al. (eds.), The Chloroplast: Basics and Applications, pp. 245–262.

yield and disease resistance have been challenging to stack into new cultivars by classical breeding. Losses incurred along the production-consumption continuum are not considered in evaluations of yield stability. Reports from the Food and Agriculture Organization indicate that up to 25% of all food productivity is lost due to post-harvest variables including pre-harvest infestations which result in accumulation of toxic metabolites in storage. Annual losses caused by mycotoxigenic fungi, such as members of the genus *Aspergillus*, can potentially reach up to a billion dollars due to market rejection and animal health impacts. Aflatoxin B_1, the most prevalent of the toxins produced by a number of *Aspergillus* and *Emericella* species, is classified as a class 1 carcinogen. Around the globe effective management programs for the detection and elimination of aflatoxins presents a major challenge. Much attention has been focused on implementing strategies to prevent pre-harvest infestation by *Aspergillus flavus*, the major source of aflatoxin contamination in food and feed crops. Genetic improvement of susceptible crop species may enhance resistance to microbial pathogens and facilitate reduced pesticide load, yet the possibility for transmission of novel genes to wild relatives has hampered acceptance of GM crops in some markets. Chloroplast transformation presents an attractive alternative to nuclear transformation and offers the potential to ameliorate this and other environmental concerns. Most agronomically important species exhibit maternal inheritance of organellar genomes eliminating the threat of transgene escape through pollen. Additionally, gene silencing is absent due to site directed, single copy insertion by homologous recombination. Foreign proteins can accumulate to high levels and are retained within the chloroplast envelope protecting them from degradation by host cytoplasmic proteases. In this study, a bacterial chloroperoxidase gene (*cpo-p*) was transformed into the tobacco chloroplast genome to test its efficacy against several plant pathogens and *Aspergillus flavus*.

Keywords: aflatoxin • antimicrobial proteins • Aspergillus flavus • chloroperoxidase • cotton, transgenic • disease resistance • drought tolerance

I Introduction

The emergence of civilization is thought to have been facilitated by the domestication of land plants. For more than 10,000 years we have selected and bred plants to achieve higher yields enabling the expansion of the first agrarian settlements into towns supporting more than 1,000 people and eventually the establishment of urban centers. Ironically, as monoculture systems with a few selected varieties of crops were introduced to accommodate the rising demand for food, feed and fiber, once-fertile soils became depleted in the specific nutrients required by the crop. In addition, by planting large areas with a single species, we have inadvertently selected for the particular pathogens and other pests that thrive on these crops. Improved agronomic practice, marker-assisted breeding programs and use of genetically engineered crops have been successful in addressing these concerns in modern agriculture to a remarkable degree (Evans, 1998; Sainju et al., 2003; Wang et al., 2005)

The architects of the 'green revolution' envisioned that increased global carrying capacity would result from the development of new crop cultivars, the use of irrigation systems, and the application of chemical fertilizers and pesticides. Indeed food production increased over 1,000%

Abbreviations: AFB_1–*Aspergillus flavus* toxin B_1; AMP–Antimicrobial peptides; Bt–*Bacillus thuringiensis*; CGIAR–Consultative Group on International Agricultural Research; CaMV–cauliflower mosaic virus; CPO-P–chloroperoxidase from *Pseudomonas pyrrocinia*; FAO–Food and Agriculture Organization; FDA–Food and Drug Administration, USA; GFP–green fluorescent protein; HBV–Hepatitis B virus; HCV–Hepatitis C virus; IARC–International Agency for Research on Cancer; IR–inverted repeat; KSA–Kernel screening assay; LPS–lipopolysaccharides; PAMP–pathogen associated molecular pattern; PEP–plastid encoded RNA polymerase; PRR–pattern recognition receptors; Prrn–plastid ribosomal operon promoter; RBS–ribosome binding site; SRRC–Southern Regional Research Center; USDA–United States Department of Agriculture; UTR–untranslated region; WHO–World Health Organization

from 1960 to 1990 but not without consequences in terms of production cost, dependence on chemical inputs, top soil erosion and salinization due to heavy fertilizer use and the development of pesticide-resistant species. The burgeoning global population is predicted to reach nine billion over the next 30–35 years (International Database, World Population Information, 2007) highlighting the ever widening gap between production and demand. The stagnation in absolute yield increases in major staple crops around the world is raising awareness that breeding for quantitative yield characteristics will most likely be insufficient to meet the needs of the future.

II Yield and Resistance

The attractive agronomic characteristics of high absolute yield and disease resistance, though not mutually exclusive, have been challenging to stack into new cultivars by classical breeding for a number of reasons. In addition to the difficulties associated with introgression of quantitative traits dependent on multiple loci and the complex inheritance patterns of polyploid species, several important crops have demonstrated yield penalties when bred for resistance traits (Staub and Grumet, 1993; Nilsson et al., 1999; Riday and Brummer, 2002; Brown, 2003).

Plants present a dynamic array of defense mechanisms shaped by the nature of the challenges encountered in their environment. In particular plants employ effective defense strategies against microbial pathogens that bear some similarity to innate immune responses seen in animal hosts (Staskawicz et al., 2001). One common strategy is the ability to recognize conserved pathogen associated molecular patterns (PAMPs).

In plants, perception of PAMPs, such as chitin, a component of fungal cell walls, and lipopolysaccharides (LPS's) characteristic of pathogenic bacteria, is facilitated by transmembrane pattern recognition receptors (PRRs). Engagement of PRRs by cognate ligands elicits basal defense responses intended to suppress pathogen growth and advance (Silipo et al., 2005; Jones and Dangl, 2006). Indicative of the co-evolutionary arms race often analogized to describe host-pathogen interactions, successful phytopathogens utilize effector proteins (Chang et al., 2005) which promote virulence and colonization of the host through attenuation of the basal defense response (Kim et al., 2005; Grant et al., 2006). Plants in kind have evolved intracellular resistance (R) proteins specific for pathogen virulence factors (Tornero et al., 2002). R-genes tend to be highly polymorphic across populations such that a range of effectors may be recognized in a particular host or cultivar while conspecifics may lack a particular R-protein and remain susceptible to the virulence function of that effector (Grant et al., 1995; Fluhr and Kaplan-Levy, 2002). The presence of resistant germplasm suggests the potential to transfer these traits to susceptible cultivars. Given the highly specific interactions involved in R-gene activation and signaling, and the apparent antagonism observed between different pathways (Traw et al., 2003; Laurie-Berry et al., 2006) the usefulness of this approach for cultivar improvement in some crops may be limited. On the other hand, wide use of the naturally occurring mutant allele for the R-gene *Mlo* has resulted in durable resistance to powdery mildew, an important fungal pathogen of barley. Despite small yield penalties associated with *mlo* expression (Stuiver and Custers, 2001), and increased susceptibility to another, less significant fungal pathogen (Jarosch et al., 1999), this example demonstrates that the potential exists to identify and utilize strategies based on regulation or over-expression of R-gene products.

Ultimately, yield stability, the producer's ability to achieve consistent yields year after year, is of most interest to the farmer. The acceptance of a new hybrid cultivar depends on the balance between pre-harvest losses due to less significant pests with a crop's resistance to a major destructive pathogen. The concept of yield stability does not, however, extend to the consideration of losses incurred along the production-consumption continuum. Reports from the Food and Agriculture Organization (FAO) indicate that up to 25% of all food productivity is lost due to post-harvest variables including infestation with number of pest species (FAO, 1989). The recommendations of the Consultative Group on International Agricultural Research (CGIAR) Technical Advisory Committee (TAC) emphasized the importance of research programs aimed at developing strategies to reduce wastage (FAO, 1997). Many agricultural crops are susceptible to post-harvest contamination which can negatively impact the eventual value of the commodity (FAO, 1989). Infestations which occur in the field may go undetected and proliferate in storage with the potential to accumulate toxic metabolic products. Among others, members of the genus *Aspergillus* are notorious in this regard.

III *Aspergillus flavus*: Managing a Food and Feed Safety Threat

In developed nations management of aflatoxin (*Aspergillus flavus* toxin) contamination represents a major regulatory challenge (Stoloff, 1980; Wu, 2004; Pitt and Hocking, 2006). Although not a plant pathogen, *A. flavus* is of particular interest as it poses a significant threat in terms of food and feed safety. *A. flavus* infestation occurs on seeds of cotton, corn, peanut and tree nuts particularly in oilseed hybrids. Market samples of sorghum, groundnut, cassava and soybean from developing nations routinely test positive for aflatoxin contamination (Williams et al., 2004) and a recent report demonstrated that aflatoxigenic strains of *A. flavus* colonize post-harvest stores of *Citrus aurantifolia* (sour lime) leading to the accumulation of aflatoxin B_1 and cyclopiazonic acid (Bamba and Sumbali, 2005).

A Economic and Health Impacts

Of perennial concern between the latitudes of 40° north and south of the equator, and in years of drought, even crops outside of this range are threatened. Aflatoxins are a closely related group of compounds with small differences in chemical composition (Fig. 1). The most prevalent, and most potent, aflatoxin B_1 (AFB_1) is only tolerated at very low levels, on the ppb scale, while contaminated stores (>20 ppb) can only be used in certain animal feeds or are destroyed resulting in dramatic economic losses (Bennett and Klich, 2003). In the United States alone the FDA estimates that approximately $500 million in costs can be attributed to market rejection and animal health impacts due to aflatoxin contamination (Vardon et al., 2003).

If the toxin goes undetected and enters food system as livestock fodder, it is incorporated into the hosts' tissues and can be transmitted though milk. In developing nations where regulatory mechanisms may be lacking, early exposure results in stunting and underweight in children (Hall and Wild, 1994). Immunologic suppression and nutritional consequences such as loss of food conversion efficiency is observed in livestock animals subjected to chronic aflatoxin exposure (Bondy and Pestka, 2000). The World Health Organization (WHO) International Agency for Research on Cancer (IARC) identifies AFB_1 as a class 1 carcinogen primarily perceived as an agent promoting liver cancers (IARC, 2002). There appears to be a synergistic relationship between aflatoxin and the hepatitis B (HBV) and C virus (HCV) viruses in the occurrence of hepatomas. In areas where aflatoxin and HBV occur together hepatomas represent 64% of all reported cancers (Williams et al., 2004). Chronic, as well as sporadic, aflatoxin contamination in a variety of field crops and agricultural commodities worldwide has had a serious impact on the economics and food safety of these products (Jelinek et al., 1989; Henry et al., 2002). The number of deaths in Indonesia due to aflatoxin-induced liver cancer is estimated at 20,000 per year (reviewed in Pitt, 2000). The short-term toxicity of aflatoxins and the chronic exposure of humans to these compounds in foods leading to liver cancer or death have been well established (CDC, 2004; Azziz-Baumgartner et al., 2005).

B Approaches to Intervention

Considerable interest has been focused on identifying interventions to prevent aflatoxin accumulation in commodities. Inherent to a successful breeding program is a recognizable resistance phenotype, defined according to the symptomatic responses of the plants to pathogen/pest exposure. *A. flavus* is not a pathogen of plants and produces no visible symptom which could be monitored to produce a successful hybrid. A new rapid screening system for maize has been employed to evaluate host resistance to aflatoxin accumulation. The kernel-screening assay (KSA) developed at the United States Department of Agriculture (USDA) Southern Regional Research Center (SRRC) relies on consistent and reproducible fungal colonization of maize kernels and allows quantitative measurement of aflatoxin production in the laboratory (Brown et al., 1995; Brown et al., 1997).

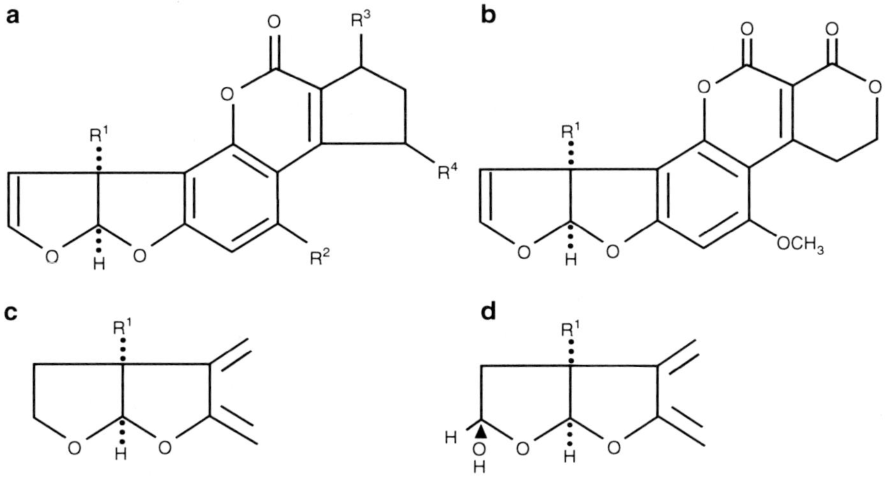

Aflatoxins	Structure	R^1	R^2	R^3	R^4
B_1	A	H	OCH_3	=O	H
M_1	A	OH	OCH_3	=O	H
P_1	A	H	OH	=O	H
Q_1	A	H	OCH_3	=O	OH
R_0	A	H	OCH_3	OH	H
R_0H_1	A	H	OCH_3	OH	OH
B_2	AC	H	OCH_3	=O	H
B_{2a}	AD	H	OCH_3	=O	H
M_2	AC	OH	OCH_3	=O	H
G_1	B	H	–	–	–
G_2	BC	H	–	–	–
G_{2a}	BD	H	–	–	–
GM	BC	OH	–	–	–

Fig. 1. Chemical structure of aflatoxins. (**a**) The B-type aflatoxins are characterized by a cyclopentane E-ring. These compounds have a blue fluorescence under long-wavelength ultraviolet light. (**b**) The G-type aflatoxins have a xanthone ring in place of the cyclopentane and fluoresce green under UV-light. (**c**) Aflatoxins of the B_2 and G_2 type have a saturated bis-furanyl ring. Only the bis-furan is shown. (**d**) Aflatoxins of the B_{1a} and G_{1a} type have a hydrated bis-furanyl structure.

The use of KSA to determine aflatoxin accumulation in seed from F1 and backcross lines found a range of resistance to accumulation, suggesting the potential to exploit heterosis for this trait in host resistance breeding strategies (Brown et al., 2001). The same group has used a β-glucuronidase (GUS)-expressing *A. flavus* to identify a maize line which had no apparent resistance to colonization, yet aflatoxin accumulation was low. This result raises the possibility of identifying kernel-specific factors which influence the rate of aflatoxin production (Brown et al., 2001). Rajasekaran et al. (2008) developed a screening procedure to monitor fungal growth, mode of entry, and colonization of cottonseeds and production of aflatoxins using a green fluorescent protein (GFP)-expressing an *Aspergillus flavus* strain. Elucidation of specific activities which limit production of aflatoxin despite colonization could lead to applications in biotechnology.

Studies in walnut have demonstrated the presence of seed-associated aflatoxin resistance variation in this important tree crop (Mahoney and Molyneux, 2004). Analysis of seed coats from several walnut species and cultivars demonstrated variability in gallic acid content that has been correlated with inhibition of aflatoxin production. However breeding programs in tree species can require considerable time to culminate. Thus Implementation of genetic engineering could provide plants expressing resistance phenotypes without great delays. As yet no genetic resistance to colonization or toxin production has been identified in the cotton germplasm emphasizing the need of genetic engineering. Current cotton improvement in terms of aflatoxin resistance is focused on the introduction of resistance proteins into the nuclear genome (Rajasekaran, 2004).

IV The Case for Transgenic Interventions

A Modifying the Nuclear Genome for Resistance

Crop species have been genetically engineered to resist viral pathogens (Fitch et al., 1992) and insect pests (Perlak et al., 1990), to tolerate drought (Shou et al., 2004) and herbicide treatment (Padgette et al., 1995), and to enhance nutritional value (Goto et al., 1999; Ye et al., 2000; Datta et al., 2003; Baisakh et al., 2006) through the incorporation of novel DNA into the nuclear genome.

The improvement of cotton through the introduction of foreign genes has been largely successful for agronomic traits such as insect (Perlak et al., 1990) and herbicide (Nida et al., 1996; Rajasekaran et al., 1996) resistance. In fact, for the 2007 season the USDA reports that genetically modified cotton comprised 87% of all upland cotton planted in the United States. Although not yet available for field application, transgenic cotton lines expressing foreign genes have demonstrated resistance to colonization by a number of more susceptible phytopathogens (Emani et al., 2003) as well as the more challenging *A. flavus* (Rajasekaran et al., 2005).

Expression of antimicrobial peptides (AMP) from a number of sources has been explored in the model plant *Nicotiana tabacum* (tobacco). Expression of AMPs from plants (De Bolle et al., 1996), magainin from *Xenopus laevis* (Zasloff, 1987; Zasloff et al., 1988), and cecropins originally isolated from the giant silk moth *Hyalophora cecropia* (Hultmark et al., 1980) has been explored as a means to enhance host resistance to fungal pathogens. More successful has been the use of synthetic peptide analogs which exhibit improved target specificity and are more resistant to degradation by host proteases (De Lucca et al., 1998; Cary et al., 2000). Furthermore these peptides are able to exert antifungal activity at lower concentrations than their natural predecessors (Cary et al., 2000).

Encouraged by results generated with tobacco transformants expressing the synthetic lytic peptide D4E1, Rajasekaran et al (2005) have recently demonstrated enhanced resistance to fungal colonization of cottonseed from transgenic plants expressing D4E1. Using a gfp-expressing *A. flavus* isolate to establish the level of infestation this group found colonization to be significantly reduced in seed from cotton plants (*Gossypium hirsutum* L.) transformed with the D4E1 gene fused to a strong, constitutive promoter (double 35S Cauliflower mosaic virus; CaMV). Transgenic expression of AMPs should confer resistance to a spectrum of microbial pathogens. In keeping with this, D4E1 plants were found to exhibit significant protection against colonization by *Fusarium verticillioides* (formerly *F. moniliforme*), *Verticillium dahliae*, and *Thielaviopsis basicola*.

The expression of AMP MSI-99, a magainin analog, has been first evaluated in tobacco followed by banana in an effort to enhance resistance to fungal pathogens. Transgenic expression resulted in resistance in both species (Chakrabarti et al., 2003). Like that of several characterized lytic AMPs, the antifungal activity of MSI-99 is thought to depend on the local concentration of peptide relative to the target membrane. When sufficient peptide has accumulated, pore formation is induced in the membrane leading to depolarization and loss of cellular integrity (De Lucca et al., 1998; Shai, 1999, 2002).

V Plastid Transformation

A An Alternative Approach to Biotechnological Improvement

In the interest of obtaining transgenic plants accumulating high levels of MSI-99, transformation of the tobacco plastid genome (plastome) has been explored (DeGray et al., 2001). Tobacco plants were generated which have the MSI-99 coding sequence fused to the strong constitutive ribosomal operon promoter (P*rrn*) stably integrated into the plastome. Because the integration site was situated within the inverted repeat (IR) region of the plastome templates for transcription of MSI-99 mRNA should be maximized as the foreign gene will be integrated into both repeats (Daniell et al., 2005b). Translation of the message was facilitated by the inclusion of a plastid preferred ribosome binding site (RBS) 5′ proximal to the start codon. Crude extracts from transplastomic tobacco plants expressing MSI-99 were found to inhibit growth of pre-germinated conidia of *V. dahliae*, *F. verticillioides*, as well as the more resistant *A. flavus*. Evaluation for *in planta* performance against *Psuedomonas syringae* pv *tabaci* and *Colletotrichum destructivum* showed that these lines were able to effectively inhibit colonization as evidenced by the lack of necrotic lesions at inoculation sites (Fig. 2).

Plastid genetic engineering offers an alternative platform to address current and future demands for improved food production. This technology has been implemented to confer desirable plant traits including insect resistance (McBride et al., 1995; De Cosa et al., 2001), herbicide resistance (Daniell et al., 1998; Iamtham and Day, 2000), salt tolerance (Kumar et al., 2004a), drought tolerance (Lee et al., 2003), disease resistance (DeGray et al., 2001), phytoremediation (Ruiz et al., 2003) and reversible male sterility (Ruiz and Daniell, 2005).

B Features of the Plastid Expression System

Indeed transformation of higher plant plastomes has enabled the accumulation of a diverse array of exogenous gene products (Daniell et al., 2005b; Daniell, 2006). A number of features of the plastid recombination and expression systems make this an attractive approach for crop improvement. In a mature leaf cell plastid DNA can comprise up to 20 % of the total cellular DNA content (Bendich, 1987). This is remarkable as in most angiosperms the plastome ranges in size from

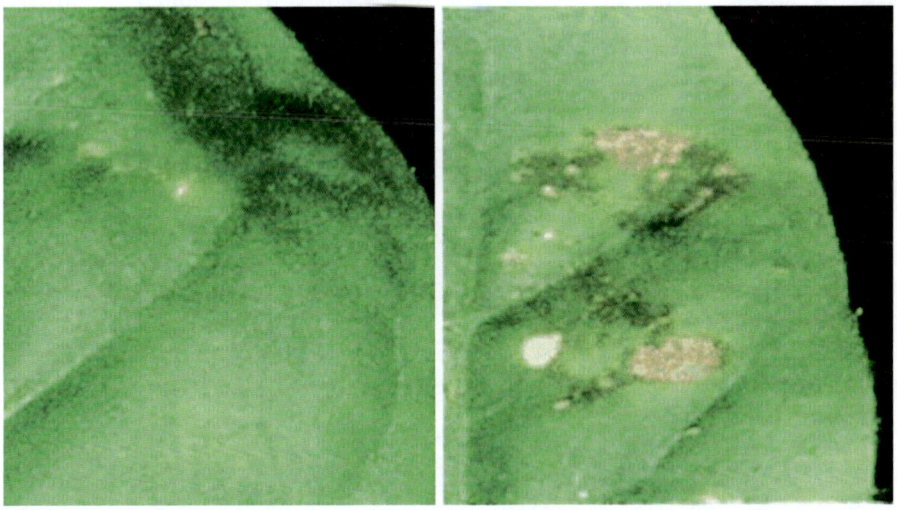

Fig. 2. In planta performance of transplastomic tobacco (*left*) expressing the MSI-99 antimicrobial peptide against *Colletotrichum destructivum*. The right frame shows susceptibility of wild type cultivar Petit Havana.

120 to 180 kb (Sugiura, 1992). The plastome persists in all plastid differentiation types and in the mesophyll cells of a mature leaf may be present in up to 10,000 copies (Maier et al., 2004).

1 The Plastome

(a) Integration of Foreign Sequences

Following the primary transformation event, where foreign genes are precisely integrated by homologous recombination, repeated rounds of selective regeneration drive the plastome toward homoplasmy. Because of the highly polyploid nature of the plastome, primary transformants will carry a mixed population of wild type and transformed plastome (transplastomic) copies, a state referred to as heteroplasmy. Southern analysis is carried out to assess the level of heteroplasmy in transplastomic lines. When hybridization with probes homologous to the plastome sequences flanking the integration site indicate the absence of untransformed (wild type) copies the plants are considered homoplasmic. In this state it follows that the foreign sequence present in the transplastome will be present at very high copy number, especially in photosynthesizing leaves (up to 20,000 copies in the IR region). Additionally the ability to specifically target single copy gene insertions by plastid transformation alleviates difficulties associated with nuclear transformation such as transgene silencing.

(b) Maternal Inheritance

Among the environmental concerns expressed over the use of transgenic plants in the field is the issue of transgene escape through pollen (Lu, 2003). Plastid transformants offer a remarkable degree of foreign gene containment (Ruf et al., 2007; Svab and Maliga, 2007). The plastome is maternally inherited in most species of agricultural interest (Daniell, 2002; Hagemann, 2004). In maternal inheritance systems, paternal transmission of plastids is impeded during either the first pollen mitosis via unequal plastid distribution, or during generative or sperm cell development via plastid degeneration (Birky, 2001). Therefore, the generative and sperm cells in mature pollen tend to be free of plastids. Confinement of transgenic plastids in maternal tissues abrogates the ability of recombinant sequences to disperse to weedy relatives, or nearby agricultural stands. An additional level of control is offered by the recently developed system for reversible male sterility via light regulated expression of the *phaA* gene of *Acinetobacter* sp. encoding β-ketothiolase in the chloroplast genome (Ruiz and Daniell, 2005).

C Moving Beyond the Model System

Although tobacco continues to serve as a model plant in plastid transformation experiments, transplastomic lines have been generated for several important crop species including tomato (Ruf et al., 2001; Wurbs et al., 2007), potato (Sidorov et al., 1999), soybean (Dufourmantel et al., 2004; Dufourmantel et al., 2005), cabbage (Liu et al., 2007), lettuce (Kanamoto et al., 2004; Lelivelt et al., 2005; Ruhlman et al., 2007), carrot (Kumar et al., 2004a) and cotton (Kumar et al., 2004b). The opportunity to transfer resistance traits to cotton through plastid transformation may lead to the introduction of transplastomic varieties which do not accumulate the toxic products of fungal metabolism such as aflatoxin in their seeds.

VI Identifying Candidate Genes for Aflatoxin Resistance

The identification and cloning of antimicrobial compounds from cellular organisms in combination with transgenic technologies offers the possibility to enhance disease resistance in agricultural crops. One aspect of the hypersensitive response triggered when higher plant cells are invaded by microbes such as fungi or bacteria is a respiratory burst generating reactive oxygen species such as hydrogen peroxide (Wojtaszek, 1997). A similar disease response occurring in animals goes one step further; the enzyme mediated generation of peracetic acid (AcOOH) from acetic acid (AcOH) and hypohalites ^{-}OX from hydrogen peroxide (Jacks et al., 2000).

$$H_2O_2 + X^- \rightarrow H_2O + {}^-OX$$

$$AcOH + H_2O_2 \rightarrow AcOOH + H_2O$$

Peracetic acid and hypohalites are much stronger antimicrobial agents than hydrogen peroxide but plants lack a haloperoxidase enzyme (van Pee, 1996).

A Chloroperoxidase

1 Antimicrobial Potential

A cDNA clone showing 38% sequence identity to a nonheme bromoperoxidase from *Streptomyces aureofaciens* was isolated from *Pseudomonas pyrrocinia*. Called chloroperoxidase (CPO-P) (Wiesner et al., 1986), this enzyme functions as a homodimer of ~32 kDa subunits which associate via a disulfide bridge between the single cysteine residues in each monomer. CPO-P is capable of catalyzing the conversion of indole to 7-chloroindole and was found to brominate the substrate monochlorodimedon in vitro (Wiesner et al., 1988). In vivo CPO-P is thought to be involved in the synthesis of pyrrolnitrin, a potent antifungal secondary metabolite (Wiesner et al., 1986, 1988; Burd and van Pee, 2003). CPO-P has been shown to have peroxidase and hydrolase activities (van Pee, 1996) however it appears that the oxidation of alkyl acids with hydrogen peroxide (Eq. 1) was not responsible for its antimicrobial effect when expressed in transgenic plants (Jacks et al., 2000; Rajasekaran et al., 2000).

2 Expression of CPO-P in Transgenic Plants

The chimeric *cpo-p* construct utilized the 35S CaMV promoter fused to the CPO-P coding region. It was introduced into the tobacco nuclear genome via *Agrobacterium*-mediated transformation to investigate its potential to confer disease resistance in higher plants (Rajasekaran et al., 2000). The CPO-P lines demonstrated significantly reduced anthracnose severity *in planta* caused by *C. destructivum*. Incubation with crude leaf extracts from the CPO-P transformants was effective in preventing colony establishment by pre-germinated conidia of several pathogenic fungi as well as the aflatoxin-producing saprophyte *A. flavus*. In careful in vitro analysis of purified CPO-P the enzyme was not found to affect the lethality of the putative physiological substrate H_2O_2 in transgenic plants, leading the authors to propose that the hydrolase activity may be responsible for enhanced resistance observed is these lines (Jacks et al., 2000, 2002). CPO-P exhibits hydrolase activity with 3,4-dihydrocoumarin (Kataoka et al., 2000), a degradation intermediate to salicylic acid, an established mediator of plant defense responses (Dong, 1998; Durrant and Dong, 2004).

VII An Environmentally Benign Approach

The Food and Feed Safety (FFS) division of the USDA-SRRC is interested in the translation of transgenic technologies into field applications which will benefit producers and deliver optimal food safety for consumers of agricultural products. In the furtherance of this mandate, we sought to generate plastid transformants for CPO-P. The expression of CPO-P in transgenic plants has shown promise as a means to enhance resistance traits. The high level of transgene containment offered by transplastomic strategies could assuage regulatory and public concern regarding the escape of transgenes thereby facilitating implementation in the field. Two lines of tobacco plastid transformants were generated using the pLD-CtV shuttle vector (Daniell et al., 1998; Ruhlman, 2005).

A Plastid Transformation Vector

The pLD-CtV vector carries a multiple cloning site (MCS) and all necessary elements for cloning, selection and tobacco plastid transformation by homologous recombination. Sequences include endogenous *trnI* and *trnA* to facilitate integration by homologous recombination in the intergenic spacer region of the ribosomal operon (*rrn*) of tobacco plastids. In addition to the *Escherichia coli* origin of replication, pLD-CtV carries a copy of the plastid origin of replication, *oriA* within the sequence of *trnI* (Kunnimalaiyaan and Nielsen, 1997; Kunnimalaiyaan et al., 1997). Transcription of integrated transgenes by the plastid encoded RNA polymerase (PEP) is enhanced by inclusion of an engineered *rrn* promoter region. The vector carries two selectable markers: amp^r (β-lactamase) for selection in *E. coli* and *aadA* (aminoglycoside-3′-adenyltransferase) for selection of transplastomic tobacco cells. The tobacco native full length *psbA* 3′ UTR is included to impart transcript stability.

The *rrn* promoter situated upstream of the *aadA* gene in pLD is sufficient to transcribe discistronic mRNA for the selectable marker and the downstream gene of interest. Once stably integrated into the plastid ribosomal operon, northern hybridization demonstrates the presence of various larger polycistronic mRNAs which carry the foreign sequence. These transcripts are the result of post-transcriptional processing of the ~72 kb primary transcript encoding the entire operon (Hartley, 1979) generated by the endogenous operon promoter.

B Determinants of Foreign Gene Expression in Plastids

Transcript availability is not the rate limiting factor for protein accumulation in mature leaf chloroplasts. Transcript stability, maturation and subsequent translation into protein products are influenced by several features of the mRNA. Native plastid mRNAs include 5′ and 3′ untranslated regions (UTRs), both of which confer on the molecule distinctive elements necessary for the eventual production of plastid proteins from mono- or polycistronic transcription units (Monde et al., 2000). Accumulation of polycistronic mRNAs as well as their efficient translation leading to high levels of foreign protein has been established (Jeong et al., 2004; Quesada-Vargas et al., 2005).

Native and heterologous elements have been successfully employed for the regulation of foreign protein expression in chloroplast transformation experiments. The ability to drive protein accumulation in the plastid system has lead to the implementation of the *psbA* 5′ UTR in many transformation experiments where high levels of foreign product is desired (De Cosa et al., 2001; Fernandez SanMillan et al., 2003; Leelavathi and Reddy, 2003; Dhingra et al., 2004; Molina et al., 2004; Watson et al., 2004; Chebolu and Daniell, 2007; Ruhlman et al., 2007).

1 The psbA 5′ UTR

(a) The Potential of *psbA* 5′ UTR Stems From Its Endogenous Role in Plastids

Photosystem II core protein D1 is a polytopic thylakoid membrane constituent with five membrane-spanning helices encoded by the plastid *psbA* gene. Expression of D1 is predominantly regulated at the level of translation and requires the participation of specific nuclear encoded factors imported into the plastids post-translationally from the cytoplasm. Photosystem II is highly susceptible to excessive light and the primary target of the damage is D1. If the core protein is not efficiently removed and replaced the result is impairment of electron transport, known as photoinhibition. It is this cycle of damage and replacement that makes the 5′UTR of *psbA* so attractive as a tool to enhance the level of foreign protein accumulation in transplastomic lines.

(b) Translational Control Is Highly Regulated and Dependent on Imported Trans-acting Protein Factors

In depth analyses have detailed the structure of the *psbA* 5′UTR which contains several representative translational motifs. There are three Shine Delgarno (SD) type ribosome binding sites (RBS) and a predicted stem loop region. RBS1 and RBS2 appear to act cooperatively, as deletion of both sites has a more severe effect than single deletions for either (Hirose and Sugiura, 1996). Detection of initiation complexes corresponding to ribosomes with initiator tRNA on RBS3 led to the proposition that the small subunit first binds transiently here then scans in the 5′→3′ direction to the legitimate AUG downstream. This scanning does not appear to be an absolutely required step as mutation or deletion of RBS3 reduced translation by less than 15% of wild type levels (Kim and Mullet, 1994). The stem loop structure found at −48 through −72 upstream of AUG possesses a predicted cleavage site at the 3′ end of the stem (Alexander *et al.*, 1998). Mutant lines for the stem loop structure demonstrated a twofold lower transcript abundance following a 17 nucleotide deletion at the 5′ end of the UTR. An eightfold depression in protein accumulation was observed resulting in an overall fourfold depression of translation efficiency. These results have been supported by a more recent study in which a series of deletion and single base mutants for the *psbA* 5′ stem loop were developed (Zou et al., 2003). Translation efficiency, as determined by GUS activity in relation to mRNA level, was depressed by 1.8- to sixfold in mutant lines over the lines with wild type *psbA* 5′ UTR.

(c) Light Regulation of Translation Via the *psbA* 5′ UTR

As many endogenous chloroplast proteins function in the photosynthetic unit it is only logical to predict that their expression will be regulated by light. Translation of *psbA* is stimulated by light to a greater extent than other mRNAs (Staub and Maliga, 1993, 1994; Eibl et al., 1999) reflecting the requirement for replacement of the PSII core subunit following photooxidative damage. Upregulation of D1 expression in response to light has been mainly investigated in etiolated seedling versus those grown in lighted growth chambers. Transplastomic lines chimeric for the *psbA* 5' region (including promoter) fused to the *uidA* (GUS) gene demonstrated 135- to 200-fold higher expression in light grown plants over those deprived of light. This was based on fluorogenic assay and illumination of dark grown seedlings which enhanced expression by 32- to 43-fold within 36 h. In these constructs the 3′ UTR was varied, with *rcbL* 3′ UTR resulting in the greatest enhancement. When constructs were employed with constant *psbA* 3′ regions and variable 5′ UTRs results were less dramatic, with light enhancement of GUS expression in the two- to fourfold range. Analysis of transcript abundance demonstrated only modest increases among transgenic lines exposed to light versus dark grown ones (Staub and Maliga, 1994). This indicates that for *psbA* enhancement of translation in the light lies in the 5′ leader sequence.

C The CPO-P Transplastomic Lines

Our constructs which carried the CPO-P coding region were expressed through different 5′ UTRs. One carried an AUG-proximal RBS (GGAGG), while the other carried the promoterless 5′ UTR of the tobacco native *psbA*.

1 Evaluating CPO-P Expression

(a) Protein Expression

By comparing the level of CPO-P accumulation in representatives of these two lines under standard diurnal cycles in the lab, we were unable to detect a significant difference between those which included the RBS and those with the *psbA* 5' UTR. Upon exposure to full sunlight, the superior performance of the *psbA* construct became evident. Indeed representatives of the *psbA-cpo-p* lines accumulated more foreign protein than their RBS counterparts, and this was reflected in a relative increase in the efficacy of crude plant extracts to inhibit the growth of pregerminated conidia of several phytopathogenic fungi and *A. flavus*.

(b) Analysis of Foreign Transcripts

Northern hybridization with the *cpo-p* probe revealed that in addition to the predicted dicistron and larger multicistronic species, in the *psbA* lines, a band with the correct electrophoretic mobility of the monocistron for *cpo-p* was detected (Fig. 3). This monocistron could only be the result of post-transcriptional processing, as the construct contained the 85 nucleotide 5′ UTR but no promoter. Most likely cleavage to the monocistronic form was facilitated by the stem loop element (−48 through −72). The stem loop structure is a common motif within plastid UTRs and for *psbA* it appears to be required for optimal expression possibly acting as a sort of antennae for complex formation by imported trans-acting factors. Following cleavage within the stem loop element, the protein complex can no longer form on the mRNA and activation of translation is abolished; formation of the complex prior to cleavage at this site protects the mRNA and translation can proceed (Alexander et al., 1998). In the absence of data to indicate which of the *cpo-p* mRNA species were associated with polysomes or complexed with protein factors we can only speculate as whether the monocistronic cleavage product was in fact a translatable message. On the other hand if the monocistron generated in the *psbA* lines was unstable it is unlikely that it would be detected at the correct size in northern blots. Rather we would expect to see various sized degradation products, or perhaps no signal at all.

(c) Continued Analysis

We are continuing to study the antimicrobial characteristics of CPO-P transplastomic lines through *in planta* pathogen challenge and hope the application of plastid biotechnology to problems such as aflatoxin contamination will usher in a new generation of elite lines with broad spectrum resistance to phytopathogens and which can be safely established in the field.

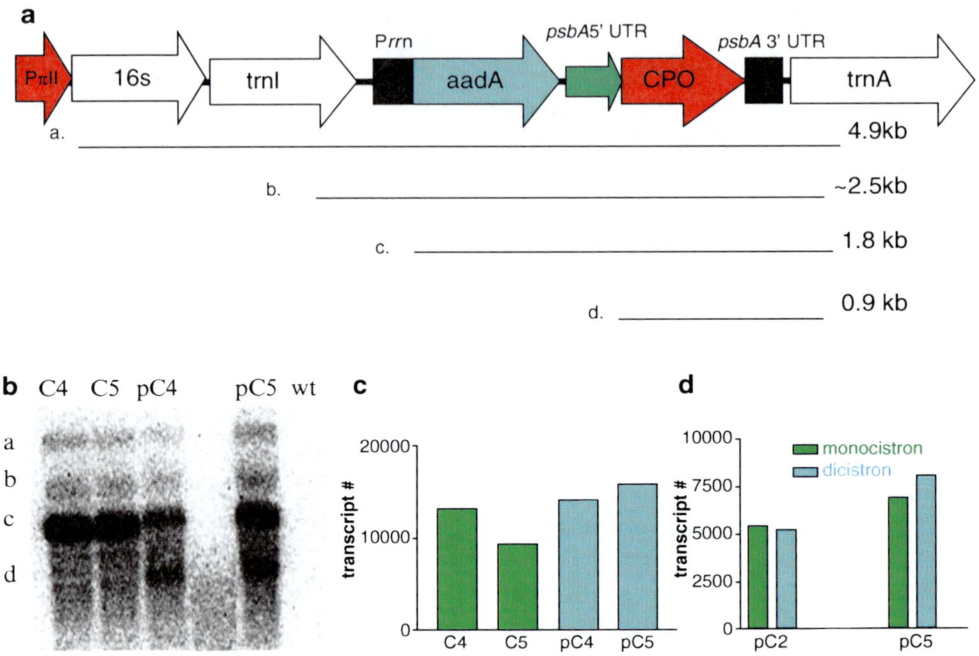

Fig. 3. Results of northern analysis of CPO lines. (**a**) Schematic representation of transcripts generated by chloroplast transformants indicating transcript sizes for *psbA* 5′UTR-CPO lines; CPO transcripts were found to be similar except that they were ~90 nucleotides shorter overall due to exclusion of *psbA* 5′ UTR. (**b**) Phosphorimage of transcripts hybridized to P^{32}-labeled, full length *cpo*. (**c**) Comparison of abundance between pooled (di- and monocistron) transcripts for *psbA* 5′UTR-CPO lines (pC2, pC4 and pC5) and *rbs* CPO lines (C4 and C5; dicistron) and (**d**) comparison of di- and monocistron abundance in *psbA* 5′UTR lines.

VIII Future Challenges: Control of Aflatoxin Contamination in Cottonseed

Important to the development of highly effective transplastomic approaches for control of *A. flavus* infection and aflatoxin contamination is an understanding of the role of different plastid genes during ovule development and seed filling. Strategies that seek to limit the potential for colonization in the field may also contribute to inhibition of post harvest toxin production. Genetic improvement of cotton via plastid transformation may prove to be an attainable goal if the research efforts of talented investigators are brought to bear on this challenging problem. Collaborative work should be initiated to bring expertise from molecular biology and cotton tissue culture to foster success in this undertaking.

A Taking a Direct Approach

For the most part seeds are quiescent and intuitively the approach that targets bioactive factors to the seed during development should have the greatest efficacy. The orchestration of expression patterns for maximal seed deposition of protein through the use of plastid endogenous regulatory elements or the localized overexpression of nuclear encoded proteins involved in seed development and metabolism could lead to plants with enhanced seed resistance to aflatoxin accumulation. Identifying and utilizing factors that either inhibit fungal colonization or subsequent aflatoxin production will be imperative to the development of such new strategies.

B Taking an Indirect Approach

Infestation and aflatoxin accumulation have been positively correlated to both water stress and mechanical damage from insect herbivores (Russell et al., 1976). Plastid biotechnology has demonstrated potential for the generation of plants able to resist drought through the introduction of genes that facilitate the accumulation of osmoprotectants.

1 Drought Tolerance

The trehalose phosphate synthase gene from the yeast *Saccharomyces cerevisiae* has been used to confer drought tolerance in transplastomic tobacco (Lee et al., 2003). Trehalose, a demonstrated osmoprotectant, accumulated in T1 plants up to 20 times that of wild type while pleiotropic effects were absent. Transplastomic seeds germinated on media containing 6% polyethylene glycol and the seedlings were able to recover from severe drying following rehydration. Elevated infestation of cotton seed has been demonstrated when drought stress occurs at anthesis (Klich and Chmielewski, 1985; Klich, 1987). Thus; enhancing drought tolerance could contribute to reduced preharvest colonization of bolls by *A. flavus* and in turn limit the occurrence of post harvest toxin accumulation.

2 Resistance to Herbivory

A primary entry route for the aflatoxin-producing fungus, *Aspergillus flavus*, into the cottonseed is through pink bollworm exit holes in unopened bolls. It has been shown that mycotoxin contamination was greatly reduced in Bt corn varieties due to reduction in insect damage (Munkvold et al., 1998, 1999; Dowd, 2000). By logical extension, preharvest aflatoxin contamination in Bt cotton would thus be greatly reduced due to minimization of exit holes caused by pink bollworms (Zipf and Rajasekaran, 2003). The ability to integrate and express foreign genes in operons in plastids has allowed the establishment of transplastomic tobacco plants with superior resistance to lepidopteron pests. Inclusion of the native chaperonin of the *Bacillus thuringiensis* (Bt) cry2Aa2 operon led to an abundance (about 46% of TSP) of Cry protein in transplastomic tobacco plants (De Cosa et al., 2001). An upstream sequence of the operon, orf2, encodes a protein involved in Cry folding into cuboidal crystals. In its crystalline form Cry is highly resistant to proteolytic activity (Crickmore and Ellar, 1992; Staples et al., 2001). Simultaneous integration of multiple foreign coding sequences such as the Cry operon in plastid genomes suggests the possibility of stacking desirable traits in a single transformation event.

C Generation of Transplastomic Cotton

Multi-gene engineering permitted the 'double gene single selection' system (Kumar et al., 2004b) that facilitated the generation of homoplasmic cotton transformants through the ability to apply selective pressure in green as well as non-green stages of development. As mentioned earlier current plantings of transgenic cotton represents more than 90% of the US production. The transgenic population comprises a single genetic background, that of *Gossypium hirsutum*, or upland cotton. Although cotton is generally an inbreeding plant, this genotype has the potential to hybridize with other *Gossypium* species in addition to non-transgenic *G. hirsutum* when the pollen is carried by wind. Maternal inheritance of the cotton plastome could assure transgene containment in the field. Also, with the ability to accumulate abundant foreign protein, expression of Bt in transplastomic cotton could abrogate the evolution of resistance in pest populations.

IX Conclusion

In recent years our fundamental understanding of host-microbe interaction has developed considerably. We have begun to tease out the genetic components that influence host resistance to microbial colonization. The use of advancing molecular technologies such as microarray expression profiling and proteomic tools are enabling the extension of our understanding to include factors that influence toxin biosyntheses by fungi such as *A. flavus*. Transgenic approaches such as plastid biotechnology are maturing beyond the introductory model systems into crop species and hold the promise to deliver efficacy in the field. By applying this expertise through creative collaborative projects aimed at controlling infestation and subsequent contamination of commodities we should be able to provide a higher level of safety to consumers of agricultural products in the future.

References

Alexander C, Faber N and Klaff P (1998) Characterization of protein-binding to the spinach chloroplast *psbA* mRNA 5' untranslated region. Nucleic Acids Res 26:2265–2272

Azziz-Baumgartner E, Lindblade K, Gieseker K, Rogers HS, Kieszak S, Njapau H, Schleicher R, McCoy LF, Misore A, DeCock K, Rubin C, Slutsker L and the Aflatoxin Investigative Group (2005) Case-control study of an acute aflatoxicosis outbreak, Kenya, 2004. Environ Health Perspect 113:1779–1783

Baisakh N, Rehana S, Rai M, Oliva N, Tan J, Mackill DJ, Khush GS, Datta K and Datta SK (2006) Marker-free transgenic (MFT) near-isogenic introgression lines (NIILs) of 'golden' indica rice (cv. IR64) with accumulation of provitamin A in the endosperm tissue. Plant Biotechnol J 4:467–475

Bamba R and Sumbali G (2005) Co-occurrence of aflatoxin B(1) and cyclopiazonic acid in sour lime (*Citrus aurantifolia* Swingle) during post-harvest pathogenesis by *Aspergillus flavus*. Mycopathologia 159: 407–411

Bendich AJ (1987) Why do chloroplasts and mitochondria contain so many copies of their genome? Bioessays 6:279–282

Bennett JW and Klich M (2003) Mycotoxins. Clin Microbiol Rev 16:497–516

Birky CW, Jr. (2001) The inheritance of genes in mitochondria and chloroplasts: laws, mechanisms, and models. Annu Rev Genet 35:125–148

Bondy GS and Pestka JJ (2000) Immunomodulation by fungal toxins. J Toxicol Environ Health B Crit Rev 3:109–143

Brown JKM (2003) A cost of disease resistance: paradigm or peculiarity? Trends Genet 19:667–671

Brown RL, Chen ZY, Menkir A, Cleveland TE, Cardwell K, Kling J and White DG (2001) Resistance to aflatoxin accumulation in kernels of maize inbreds selected for ear rot resistance in West and Central Africa. J Food Prot 64:396–400

Brown RL, Cleveland TE, Payne GA, Woloshuk CP, Campbell AW and White DG (1995) Determination of resistance to aflatoxin production in maize kernels and detection of fungal colonisation using an *Aspergillus flavus* transformant expressing *Escherichia coli* ß-glucuronidase. Phytopathology 85:983–989

Brown RL, Cleveland TE, Payne GA, Woloshuk CP and White DG (1997) Growth of an *Aspergillus flavus* transformant expressing *Escherichia coli* beta-glucuronidase in maize kernels resistant to aflatoxin production. J Food Prot 60:84–87

Burd VN and van Pee KH (2003) Halogenating enzymes in the biosynthesis of antibiotics. Biochemistry 68:1132–1135

Cary JW, Rajasekaran K, Jaynes JM and Cleveland TE (2000) Transgenic expression of a gene encoding a synthetic antimicrobial peptide results in inhibition of fungal growth in vitro and in planta. Plant Sci 154:171–181

CDC; Center for Disease Control, Atlanta, GA (2004) Outbreak of aflatoxin poisoning – Eastern and Central Provinces, Kenya, January–July 2004. MMWR 53:790–793

Chakrabarti A, Ganapathi TR, Mukherjee PK and Bapat VA (2003) MSI-99, a magainin analogue, imparts enhanced disease resistance in transgenic tobacco and banana. Planta 216:587–596

Chang JH, Urbach JM, Law TF, Arnold LW, Hu A, Gombar S, Grant SR, Ausubel FM and Dangl JL (2005) A high-throughput, near-saturating screen for type III effector genes from *Pseudomonas syringae*. Proc Natl Acad Sci USA 102:2549–2554

Chebolu S and Daniell H (2007) Stable expression of Gal/GalNAc lectin of *Entamoeba histolytica* in transgenic chloroplasts and immunogenicity in mice towards vaccine development for amoebiasis. Plant Biotechnol J 5:230–239

Crickmore N and Ellar DJ (1992) Involvement of a possible chaperonin in the efficient expression of a cloned CryIIA delta-endotoxin gene in *Bacillus thuringiensis*. Mol Microbiol 6:1533–1537

Daniell H (2002) Molecular strategies for gene containment in transgenic crops. Nat Biotechnol 20:581–586

Daniell H (2006) Production of biopharmaceuticals and vaccines in plants via the chloroplast genome. Biotechnol J 1:1071–1079

Daniell H, Datta R, Varma S, Gray S and Lee SB (1998) Containment of herbicide resistance through genetic engineering of the chloroplast genome. Nat Biotechnol 16:345–348

Daniell H, Kumar S and Dufourmantel N (2005a) Breakthrough in chloroplast genetic engineering of agronomically important crops. Trends Biotechnol 23:238–245

Daniell H, Ruiz ON and Dhingra A (2005b) Chloroplast genetic engineering to improve agronomic traits. Methods Mol Biol 286:111–138

Datta K, Baisakh N, Oliva N, Torrizo L, Abrigo E, Tan J, Rai M, Rehana S, Al-Babili S, Beyer P, Potrykus I and Datta SK (2003) Bioengineered 'golden' indica rice cultivars with ß-carotene metabolism in the endosperm with hygromycin and mannose selection systems. Plant Biotechnol J 1:81–90

De Bolle MF, Osborn RW, Goderis IJ, Noe L, Acland D, Hart CA, Torrekens S, Van Leuven F and Broekaert WF (1996) Antimicrobial peptides from *Mirabilis jalapa* and *Amaranthus caudatus*: expression, processing, localization and biological activity in transgenic tobacco. Plant Mol Biol 31:993–1008

De Cosa B, Moar W, Lee SB, Miller M and Daniell H (2001) Overexpression of the Bt cry2Aa2 operon in chloroplasts leads to formation of insecticidal crystals. Nat Biotechnol 19:71–74

De Lucca AJ, Bland JM, Grimm C, Jacks TJ, Cary JW, Jaynes JM and Cleveland TE and Walsh TJ (1998) Fungicidal properties, sterol binding, and proteolytic resistance of the synthetic peptide D4E1. Can J Microbiol 44: 514-520

DeGray G, Rajasekaran K, Smith F, Sanford J and Daniell H (2001) Expression of an antimicrobial peptide via the chloroplast genome to control phytopathogenic bacteria and fungi. Plant Physiol 127: 852–862

Dhingra A, Portis AR, Jr. and Daniell H (2004) Enhanced translation of a chloroplast-expressed RbcS gene restores small subunit levels and photosynthesis in nuclear RbcS antisense plants. Proc Natl Acad Sci USA 101:6315

Dong X (1998) SA, JA, ethylene, and disease resistance in plants. Curr Opin Plant Biol 1:316–323

Dowd P (2000) Indirect reduction of ear molds and associated mycotoxins in *Bacillus thuringiensis* corn under controlled and open field conditions: utility and limitations. J Econ Entomol 93:1669–1679

Dufourmantel N, Pelissier B, Garcon F, Peltier G, Ferullo JM and Tissot G (2004) Generation of fertile transplastomic soybean. Plant Mol Biol 55:479–489

Dufourmantel N, Tissot G, Goutorbe F, GarÃ§on F, Muhr C, Jansens S, Pelissier B, Peltier G and Dubald M (2005) Generation and analysis of soybean plastid transformants expressing *Bacillus thuringiensis* Cry1Ab protoxin. Plant Mol Biol 58:659–668

Durrant WE and Dong X (2004) Systemic acquired resistance. Annu Rev Phytopathol 42:185–209

Eibl C, Zou Z, Beck A, Kim M, Mullet J and Koop HU (1999) In vivo analysis of plastid *psbA*, *rbcL* and *rpl32* UTR elements by chloroplast transformation: tobacco plastid gene expression is controlled by modulation of transcript levels and translation efficiency. Plant J. 19:333–345

Emani C, Garcia JM, Lopata-Finch E, Pozo MJ, Uribe P, Kim DJ, Sunilkumar G, Cook DR, Kenerley CM and Rathore KS (2003) Enhanced fungal resistance in transgenic cotton expressing an endochitinase gene from *Trichoderma virens*. Plant Biotechnol J 1:321–336

Evans L (1998) Feeding the Ten Billion: Plants and Population Growth. Cambridge University Press, Cambridge, UK 247p

FAO (1989) Prevention of post harvest food losses – fruits, vegetables and root crops, a training manual. Food and Agriculture Organization of the United Nations, No. 17/2. ISBN 92-5-102766-8. Rome http://www.fao.org/docrep/T0073E00.htm. Accessed May 15, 2007

FAO (1997) Consultative Group on International Agricultural Research Technical Advisory Commitee Chairman's Report. Food and Agriculture Organization of the United Nations, No. SDR/TAC:IAR/98/6.1. Rome http://fao.org/docrep/W8427E/w8427e00.htm. Accessed May 15, 2008

Fernandez SanMillan A, Mingo-Castel A, Miller M and Daniell H (2003) A chloroplast transgenic approach to hyper-express and purify Human Serum Albumin, a protein highly susceptible to proteolytic degradation. Plant Biotechnol J 1:71–79

Fitch MMM, Manshardt RM, Gonsalves D, Slightom JL and Sanford JC (1992) Virus resistant papaya plants derived from tissues bombarded with the coat protein gene of papaya ringspot virus. Nat Biotech 10:1466–1472

Fluhr R and Kaplan-Levy RN (2002) Plant disease resistance: commonality and novelty in multicellular innate immunity. Curr Top Microbiol Immunol 270:23–46

Goto F, Yoshihara T, Shigemoto N, Toki S and Takaiwa F (1999) Iron fortification of rice seed by the soybean ferritin gene. Nat Biotechnol 17:282–286

Grant MR, Godiard L, Straube E, Ashfield T, Lewald J, Sattler A, Innes RW and Dangl JL (1995) Structure of the Arabidopsis RPM1 gene enabling dual specificity disease resistance. Science 269:843–846

Grant SR, Fisher EJ, Chang JH, Mole BM and Dangl JL (2006) Subterfuge and manipulation: type III effector proteins of phytopathogenic bacteria. Annu Rev Microbiol 60:425–449

Hall AJ and Wild CP (1994) Epidemiology of aflatoxin-related disease. In: Eaton DA, Groopman JD (eds) Human Health, Veterinary and Agricultural Significance.. Academic Press, San Diego, CA, pp. 233–258

Hagemann R (2004) The sexual inheritance of plant organelles. In: Daniell H, Chase C (eds) Molecular Biology and Biotechnology of Plant Organelles: Chloroplast and Mitochondria, Vol 1. Springer, Dordrecht, The Netherlands, pp. 93–114

Hartley MR (1979) The synthesis and origin of chloroplast low-molecular-weight ribosomal ribonucleic acid in spinach. Eur J Biochem 96:311–320

Henry SH, Bosch FX and Bowers JC (2002) Aflatoxin, hepatitis and worldwide liver cancer risks. Experiment Med Biol 504:229–233

Hirose T and Sugiura M (1996) Cis-acting elements and trans-acting factors for accurate translation of chloroplast *psbA* mRNAs: development of an in vitro translation system from tobacco chloroplasts. EMBO J 15:1687–1695

Hultmark D, Steiner H, Rasmuson T and Boman HG (1980) Insect immunity. Purification and properties of three inducible bactericidal proteins from hemolymph of immunized pupae of *Hyalophora cecropia*. Eur J Biochem 106:7–16

Iamtham S and Day A (2000) Removal of antibiotic resistance genes from transgenic tobacco plastids. Nat Biotechnol 18:1172–1176

IARC (2002) Some Traditional Herbal Medicines, Some Mycotoxins, Naphthalene and Styrene. IARC Monographs on the Evaluation of Carcinogenic Risks to Humans. International Agency on Cancer Research. World Health Organization, pp. 69–90

International Database, World Population Information (2007) United States Census Bureau. http://www.census.gov/ipc/www/idb/Accessed 15 May 2008

Jacks TJ, De Lucca AJ, Rajasekaran K, Stromberg K and van Pee K (2000) Antifungal and peroxidative activities of nonheme chloroperoxidase in relation to transgenic plant protection. J Agric Food Chem 48:4561–4564

Jacks TJ, Rajasekaran K, Stromberg KD, De Lucca AJ and van Pee KH (2002) Evaluation of peracid formation as the basis for resistance to infection in plants transformed with haloperoxidase. J Agric Food Chem 50:706–709

Jarosch B, Kogel K-H and Schaffrath U (1999) The Ambivalence of the barley Mlo Locus: Mutations conferring resistance against powdery mildew (*Blumeria graminis* f. sp. hordei) enhance susceptibility to the rice blast fungus *Magnaporthe grisea*. Mol Plant–Microbe Interact 12:508–514

Jelinek CF, Pohland AE and Wood GE (1989) Worldwide occurrence of mycotoxins in foods and feeds – an update. JAOAC 72:223–230

Jeong SW, Jeong WJ, Woo JW, Choi DW, Park YI and Liu JR (2004) Dicistronic expression of the green fluorescent protein and antibiotic resistance genes in the plastid for selection and tracking of plastid-transformed cells in tobacco. Plant Cell Rep 22:747–751

Jones JD and Dangl JL (2006) The plant immune system. Nature 444:323–329

Kanamoto H, Yamashita A, Okumura S, Hattori M and Tomizawa K (2004) The complete genome sequence of the *Lactuca sativa* (lettuce) chloroplast. Nippon Shokubutsu Seiri Gakki Nenkaioyobi Shinpojiumu Koen Yoshishu 45:112

Kataoka M, Honda K and Shimizu S (2000) 3,4-Dihydrocoumarin hydrolase with haloperoxidase activity from *Acinetobacter calcoaceticus* F46. Eur J Biochem 267:3–10

Kim J and Mullet JE (1994) Ribosome-binding sites on chloroplast *rbcL* and *psbA* mRNAs and light-induced initiation of D1 translation. Plant Mol Biol 25:437–448

Kim MG, da Cunha L, McFall AJ, Belkhadir Y, DebRoy S, Dangl JL and Mackey D (2005) Two *Pseudomonas syringae* type III effectors inhibit RIN4-regulated basal defense in Arabidopsis. Cell 121:749–759

Klich MA (1987) Relation of plant water potential at flowering to subsequent cottonseed infection by *Aspergillus flavus*. Phytopathology 77:739–746

Klich MA and Chmielewski MA (1985) Nectaries as entry sites for *Aspergillus flavus* in developing cotton bolls. Appl Environ Microbiol 50:602–604

Kumar S, Dhingra A and Daniell H (2004a) Plastid-expressed betaine aldehyde dehydrogenase gene in carrot cultured cells, roots, and leaves confers enhanced salt tolerance. Plant Physiol 136:2843–2854

Kumar S, Dhingra A and Daniell H (2004b) Stable transformation of the cotton plastid genome and maternal inheritance of transgenes. Plant Mol Biol 56:203–216

Kunnimalaiyaan M and Nielsen BL (1997) Fine mapping of replication origins (ori A and ori B) in *Nicotiana tabacum* chloroplast DNA. Nucleic Acids Res. 25:3681–3686

Kunnimalaiyaan M, Shi F and Nielsen BL (1997) Analysis of the tobacco chloroplast DNA replication origin (oriB) downstream of the 23 S rRNA gene. J Mol Biol 268:273–283

Laurie-Berry N, Joardar V, Street IH and Kunkel BN (2006) The *Arabidopsis thaliana* JASMONATE INSENSITIVE 1 gene is required for suppression of salicylic acid-dependent defenses during infection by *Pseudomonas syringae*. Mol Plant Microbe Interact 19:789–800

Lee SB, Kwon HB, Kwon SJ, Park SC, Jeong MJ, Han SE, Byun MO and Daniell H (2003) Accumulation of trehalose within transgenic chloroplasts confers drought tolerance. Mol Breed 11:1–13

Leelavathi S and Reddy VS (2003) Chloroplast expression of His-tagged GUS-fusions: a general strategy to overproduce and purify foreign proteins using transplastomic plants as bioreactors. Mol Breed 11:49–58

Lelivelt C, McCabe M, Newell C, de Snoo C, Dun K, Birch-Machin I, Gray J, Mills K and Nugent J (2005) Stable plastid transformation in lettuce (*Lactuca sativa* L.). Plant Mol Biol 58:763–774

Liu CW, Lin CC, Chen JJW and Tseng MJ (2007) Stable chloroplast transformation in cabbage (*Brassica oleracea* var. *capitata* L.) by particle bombardment. Plant Cell Rep 26:1733–1744

Lu BR (2003) Transgene containment by molecular means – is it possible and cost effective? Environ Biosafety Res 2:3–8

Mahoney N, Molyneux RJ (2004) Phytochemical inhibition of aflatoxigenicity in *Aspergillus flavus* by constituents of walnut (*Juglans regia*). J Agric Food Chem 52:1882–1889

Maier RM, Schmitz-Linneweber C, Daniell H and Chase C (2004) Plastid genomes. In: Daniell H and Chase C (eds) Molecular Biology and Biotechnology of Plant Organelles: Chloroplasts and Mitochondria, Vol 1. Springer, Dordrecht, The Netherlands, pp. 115–150

McBride KE, Svab Z, Schaaf DJ, Hogan PS, Stalker DM and Maliga P (1995) Amplification of a chimeric *Bacillus* gene in chloroplasts leads to an extraordinary level of an insecticidal protein in tobacco. Biotechnology (NY) 13:362–365

Molina A, Herva-Stubbs S, Daniell H, Mingo-Castel AM and Veramendi J (2004) High yield expression of a viral peptide animal vaccine in transgenic tobacco chloroplasts. Plant Biotechnol J 2:141–153

Monde RA, Schuster G and Stern DB (2000) Processing and degradation of chloroplast mRNA. Biochimie 82:573–582

Munkvold G, Starh HM, Logrieco A, Moretti A and Ritieni A (1998) Occurrence of fusaproliferin and beauvericin in *Fusarium*-contaminated livestock feed in Iowa. Appl Environ Microbiol 64:3923–3926

Munkvold G, Hellmich R and Rice L (1999) Comparison of fumonisin concentrations in kernels of transgenic Bt maize hybrids and non-transgenic hybrids. Plant Dis 83:130–138

Nida KL, Kolacz KH, Buehler RE, Deaton WR, Shuler WR, Armstrong TA, Taylor ML, Ebert CC, Rogan GJ, Padgette SR and Fuchs RL (1996) Glyphosate-tolerant cotton: genetic characterization and protein expression. J Agric and Food Chem 44:1960–1966

Nilsson NO, Hanson M, Panagopoulos AH, Tuvesson S, Ehlde M, Christiansson M, Rading IM, Rissler M and Kraft T (1999) QTL analysis of *Cercospora* leaf spot resistance in sugar beet. Plant Breed 118:327–334

Padgette SR, Kolacz KH, Delannay X, Re DB, Lavallee BJ, Tinius CN, Rhodes WK, Otero YI, Barry GF, Eichholtz DA, Peschke VM, Nida DL, Taylor NB and Kishore GM (1995) Development, identification, and characterization of a glyphosate-tolerant soybean line. Crop Sci 35:1451–1461

Perlak FJ, Deaton RW, Armstrong TA, Fuchs RL, Sims SR, Greenplate JT and Fischhoff DA (1990) Insect resistant cotton plants. Nat Biotech 8:939–943

Pitt JI (2000) Toxigenic fungi: which are important? Med Mycol 38(1):17–22

Pitt JI and Hocking AD (2006) Mycotoxins in Australia: biocontrol of aflatoxin in peanuts. Mycopathologia 162:233–243

Quesada-Vargas T, Ruiz ON and Daniell H (2005) Characterization of heterologous multigene operons in transgenic chloroplasts: transcription, processing, and translation. Plant Physiol 138:1746–1762

Rajasekaran K (2004) *Agrobacterium*-mediated genetic transformation of cotton. In: Curtis IS (ed) Transgenic Crops of the World – Essential Protocols. Kluwer, The Netherlands, pp. 243–254

Rajasekaran K, Cary J, Cotty P and Cleveland T (2008) Development of a GFP-Expressing *Aspergillus flavus* strain to study fungal invasion, colonization, and resistance in cottonseed. Mycopathologia 165:89–97

Rajasekaran K, Cary JW, Jacks TJ, Stromberg KD and Cleveland TE (2000) Inhibition of fungal growth in planta and in vitro by transgenic tobacco expressing a bacterial nonheme chloroperoxidase gene. Plant Cell Rep 19:333–338

Rajasekaran K, Cary JW, Jaynes JM and Cleveland TE (2005) Disease resistance conferred by the expression of a gene encoding a synthetic peptide in transgenic cotton (*Gossypium hirsutum* L.) plants. Plant Biotechnol J 3:545–554

Rajasekaran K, Grula JW, Hudspeth RL, Pofelis S and Anderson DM (1996) Herbicide-resistant Acala and Coker cottons transformed with a native gene encoding mutant forms of acetohydroxyacid synthase. Mol Breed 2:307–319

Riday H and Brummer CE (2002) Forage yield heterosis in alfalfa. Crop Sci 42:716–723

Ruf S, Hermann M, Berger IJ, Carrer H and Bock R (2001) Stable genetic transformation of tomato plastids and expression of a foreign protein in fruit. Nat Biotechnol 19:870–875

Ruf S, Karcher D and Bock R (2007) Determining the transgene containment level provided by chloroplast transformation. Proc Natl Acad Sci USA 104:6998–7002

Ruhlman T (2005) Chloroplast biotechnology in higher plants: Expressing microbial resistance genes in the plastid genome. Master Thesis, University of New Orleans, New Orleans, LA

Ruhlman T, Ahangari R, Devine A, Samsam M and Daniell H (2007) Expression of cholera toxin B-proinsulin fusion protein in lettuce and tobacco chloroplasts – oral administration protects against development of insulitis in non-obese diabetic mice. Plant Biotechnol J 5:495–510

Ruiz ON and Daniell H (2005) Engineering cytoplasmic male sterility via the chloroplast genome by expression of ß-ketothiolase. Plant Physiol 138:1232–1246

Ruiz ON, Hussein HS, Terry N and Daniell H (2003) Phytoremediation of organomercurial compounds via chloroplast genetic engineering. Plant Physiol 132:1344–1352

Russell TE, Watson TF and Ryan GF (1976) Field accumulation of aflatoxin in cottonseed as influenced by irrigation termination dates and pink bollworm infestation. Appl Environ Microbiol 31:711–713

Sainju UM, Whitehead WF and Singh BP (2003) Agricultural management practices to sustain crop yields and improve soil and environmental qualities. TSWJ 3:768–789

Shai Y (1999) Mechanism of the binding, insertion and destabilization of phospholipid bilayer membranes by alpha-helical antimicrobial and cell non-selective membrane-lytic peptides. Biochim Biophys Acta 1462:55–70

Shai Y (2002) From innate immunity to de-novo designed antimicrobial peptides. Curr Pharm Des 8:715–725

Shou H, Bordallo P and Wang K (2004) Expression of the *Nicotiana* protein kinase (NPK1) enhanced drought tolerance in transgenic maize. J Exp Bot 55:1013–1019

Sidorov VA, Kasten D, Pang SZ, Hajdukiewicz PT, Staub JM and Nehra NS (1999) Technical advance: stable chloroplast transformation in potato: use of green fluorescent protein as a plastid marker. Plant J 19:209–216

Silipo A, Molinaro A, Sturiale L, Dow JM, Erbs G, Lanzetta R, Newman MA and Parrilli M (2005) The elicitation of plant innate immunity by lipooligosaccharide of *Xanthomonas campestris*. J Biol Chem 280:33660–33668

Staples N, Ellar D and Crickmore N (2001) Cellular localization and characterization of the *Bacillus thuringiensis* Orf2 crystallization factor. Curr Microbiol 42:388–392

Staskawicz BJ, Mudgett MB, Dangl JL and Galan JE (2001) Common and contrasting themes of plant and animal diseases. Science 292:2285–2289

Staub J and Grumet R (1993) Selection for multiple disease resistance reduces cucumber yield potential. Euphytica 67:205–213

Staub JM and Maliga P (1993) Accumulation of D1 polypeptide in tobacco plastids is regulated via the untranslated region of the *psbA* mRNA. EMBO J. 12:601–606

Staub JM and Maliga P (1994) Translation of *psbA* mRNA is regulated by light via the 5'-untranslated region in tobacco plastids. Plant J 6.547–553

Stoloff L (1980) Aflatoxin control: past and present. J Assoc Off Anal Chem 63:1067–1073

Stuiver MH and Custers JH (2001) Engineering disease resistance in plants. Nature 411:865–868

Sugiura M (1992) The chloroplast genome. Plant Mol Biol 19:149–168

Svab Z and Maliga P (2007) Exceptional transmission of plastids and mitochondria from the transplastomic pollen parent and its impact on transgene containment. Proc Natl Acad Sci USA 104:7003–7008

Tornero P, Chao RA, Luthin WN, Goff SA and Dangl JL (2002) Large-scale structure-function analysis of the Arabidopsis RPM1 disease resistance protein. Plant Cell 14:435–450

Traw MB, Kim J, Enright S, Cipollini DF and Bergelson J (2003) Negative cross-talk between salicylate- and jasmonate-mediated pathways in the Wassilewskija ecotype of *Arabidopsis thaliana*. J Mol Ecol 12:1125–1135

van Pee KH (1996) Biosynthesis of halogenated metabolites by bacteria. Annu Rev Microbiol 50:375–399

Vardon P, McLaughlin C and Nardinelli C (2003) Potential economic costs of mycotoxins in the United States. In:

Mycotoxins: Risks in Plant, Animal, and Human Systems. CAST Task Force, Ames, IA

Wang Y, Xue Y and Li J (2005) Towards molecular breeding and improvement of rice in China. Trends Plant Sci 10:610–614

Watson J, Koya V, Leppla SH and Daniell H (2004) Expression of *Bacillus anthracis* protective antigen in transgenic chloroplasts of tobacco, a non-food/feed crop. Vaccine 22:4374–4384

Wiesner W, van Pee K-H and Lingens F (1986) Detection of a new chloroperoxidase in *Pseudomonas pyrrocinia*. FEBS Lett 209:321–324

Wiesner W, van Pee KH and Lingens F (1988) Purification and characterization of a novel bacterial non-heme chloroperoxidase from Pseudomonas pyrrocinia. J Biol Chem 263:13725

Williams JH, Phillips TD, Jolly PE, Stiles JK, Jolly CM and Aggarwal D (2004) Human aflatoxicosis in developing countries: a review of toxicology, exposure, potential health consequences, and interventions. Am J Clin Nutr 80:1106–1122

Wojtaszek P (1997) Oxidative burst: an early plant response to pathogen infection. Biochem J 322:681–692

Wu F (2004) Mycotoxin risk assessment for the purpose of setting international regulatory standards. Environ Sci Technol 38:4049–4055

Wurbs D, Ruf S and Bock R (2007) Contained metabolic engineering in tomatoes by expression of carotenoid biosynthesis genes from the plastid genome. Plant J 49:276–288

Ye X, Al Babili S, Kloti A, Zhang J, Lucca P, Beyer P and Potrykus I (2000) Engineering the provitamin A (beta-carotene) biosynthetic pathway into (carotenoid-free) rice endosperm. Science 287:303–305

Zasloff M (1987) Magainins, a class of antimicrobial peptides from *Xenopus* skin: isolation, characterization of two active forms, and partial cDNA sequence of a precursor. Proc Natl Acad Sci USA 84:5449–5453

Zasloff M, Martin B and Chen HC (1988) Antimicrobial activity of synthetic magainin peptides and several analogues. Proc Natl Acad Sci USA 85:910–913

Zipf A and Rajasekaran K (2003) Ecological impact of Bt cotton. J New Seed 5:115–135

Zou Z, Eibl C and Koop HU (2003) The stem-loop region of the tobacco *psbA* 5'UTR is an important determinant of mRNA stability and translation efficiency. Mol Genet Genomics 269:340–349

Chapter 18

Chloroplast Genetic Engineering: A Novel Technology for Agricultural Biotechnology and Bio-pharmaceutical Industry

N.D. Singh and H. Daniell*

Department of Molecular Biology and Microbiology, College of Medicine, University of Central Florida, Biomolecular Science Building, Room 336, Orlando, FL 32816-2364, USA

Summary

Chloroplast genetic engineering is becoming an attractive field in plant biotechnology. It offers several unique advantages, including high-level transgene expression, multi-gene engineering in a single transformation event and transgene containment by maternal inheritance, as well as lack of gene silencing, position and pleiotropic effects and undesirable foreign DNA. The hyper-expression of recombinant proteins within plastids offers a cost effective solution for using plants as bioreactors. Many transgenes have been stably integrated and expressed via the tobacco chloroplast genome to confer important agronomic traits including herbicide, insect, and disease resistance, drought and salt tolerance, and phytoremediation. Moreover, many vaccine antigens and biopharmaceutical proteins have been expressed at high levels via the chloroplast genome and their functionality has been evaluated using in vitro cell cultures as well as challenge with bacterial or viral pathogens using animal models. This technology has been extended to other crop species for the introduction of agronomic traits and production of low cost vaccine antigens and therapeutic proteins. Production of therapeutic proteins in chloroplasts eliminates the expensive fermentation technology. In addition, oral delivery of chloroplast-derived therapeutic proteins should eliminate

* Author for Correspondence, e-mail: daniell@mail.ucf.edu

Constantin A. Rebeiz et al. (eds.), The Chloroplast: Basics and Applications, pp. 263–284.

expensive purification steps, cold storage, cold transportation and delivery via sterile needles, thereby further decreasing their cost. Apart from the successful productions of vaccines and therapeutic proteins, chloroplast technology is also a good platform for the production of industrial products. Recently this technology has been extended for the production of enzymes used in ethanol production. In this chapter, we describe the overall concept of the chloroplast technology and the current status of chloroplast-derived vaccine antigens and biopharmaceutical proteins.

I Introduction

Chloroplasts are of prokaryotic origin. They are dynamic organelles within plant cells that house the photosynthetic apparatus. Apart from photosynthesis; other important metabolic activities take place within chloroplasts including the production of chlorophyll, starch, certain amino acids and lipids, some of the colorful pigments in flowers, vitamins and several key aspects of sulfur and nitrogen metabolism. Chloroplasts possess their own genome and a full complement of transcriptional and translational machinery to express their genes. In particular, the chloroplast contains protein synthesizing systems more similar to those of bacteria than those of eukaryotes. This is consistent with the hypothesis that these organelles had endosymbiotic origins (Gillham, 1994) and are the endosymbiotic remnants of a once free living cyanobacterial progenitor (Moreira et al., 2000). During the gradual integration of the acquired endosymbionts into the host cell's metabolism, the organellar genomes underwent a dramatic size reduction due to both massive gene loss and gene transfer to the nuclear genome (Martin et al., 1998). Therefore, the chloroplast of higher plants has retained a largely prokaryotic system of gene organization and expression, with the eukaryotic nuclear genome exerting significant regulatory control (Hager and Bock, 2000). Thus, chloroplasts are a class of plastids – organelles in plant cells – apparently derived from cyanobacterial ancestor that once lived symbiotically inside the plant cell. The number of plastids in each cell is variable, and each plastid contains multiple copies of its own genome, typically 50–100. Many plastid genomes have been sequenced. They resemble bacterial genomes in many respects, though features normally found in muticellular organisms, such as interrupted genes and RNA editing are also observed. The chloroplast genome codes for the transcription and translation machinery of the chloroplast along with numerous structural proteins. But the vast majority of the chloroplast proteins are encoded in the plant nucleus and proteins are imported into the chloroplast after their synthesis.

Genetic engineering offers the opportunity to modify the characteristics of plants by introduction of genes, from diverse organisms across their genetic barriers. Production of vaccine antigens and therapeutic proteins are very important to maintain health of our global community. The skyrocketing expense of current vaccines and therapeutic proteins can be attributed to their unnecessarily complex production and delivery – from the significant cost of fermentation systems to purification through the use of complex technologies and additional expenses associated with cold storage, transportation and sterile delivery. Therefore, the chloroplast transformation technology has been used as a high level protein expression system. Transgenes conferring agronomic traits, including herbicide, insect, and disease resistance, drought and salt tolerance, and phytoremediation have been stably integrated into chloroplast genomes of different crop species. Chloroplast transformation has been reported in different crop species including cauliflower (*Brassica oleracea*),

Abbreviations: AIAT – human alpha1-antitrypsin; BADH – betaine aldehyde dehydrogenase; BvCMO –choline monooxygenase; Cel6A – cellulase A; Cel6B – cellulase B; CSFV – classical swine fever virus; CRFK – Crandell feline kidney; EMC – encephalomyocarditis; EPSPS – 5-enolpyruvylshikimate-3-phosphate synthase; GFP – green fluorescence protein; GUS – β-glucuronidase; hCT-1 – Human cardiotrophin-1; HIV – Human immune deficiency virus; HPPD – bacterial 4-hydroxyphenylpyruvate dioxygenase; HSA – human serum albumin; hST – human somatotropin; IFN-α2β interferon alpha 2b; INF-γ – interferon gamma; IR – inverted repeat; LSC – large single copy region; PA – protective antigen; PBP – bioelastic protein-based polymers; PCR – polymerase chain reaction; PDS – particle delivery system; PEG – polyethylene glycol; PHB – polyhydroxy butyrate; pHBA – p-hydroxybenzoic acid; PMA – phenylmercuric acetate; PPT – phosphinothricin; Prrn – plastid ribosomal operon promoter; SSC – small single copy; TPS1 – trehalose phosphate synthase gene; UTR – untranslated region; VSV – CPE Vesicular stomatitis virus cytopathic effect

cabbage (*Brassica capitata*), oilseed rape (*Brassica napus*), lettuce (*Lactuca sativa*), petunia (*Petunia hybrid*), poplar (*Populus spp.*), potato (*Solanum tuberosum*), tobacco (*Nicotiana tabacum*), tomato (*Solanum lycopersicum*), carrot (*Daucus carota*), cotton (*Gossypium hirsutum*), soybean (*Glycine max*) and rice (*Oryza sativa*). Chloroplast-derived vaccine antigens against bacterial, viral and protozoan pathogens have been evaluated by immune responses, neutralizing antibodies, and pathogen or toxin challenge in suitable animal models. Moreover, human blood proteins including somatotropin, interferons, and insulin have been expressed in chloroplasts with successful functional evaluations. In addition, chloroplasts have been used as bioreactors for production of biopolymers, amino acids, and industrial enzymes.

II Genome and Organization

The chloroplast genome is typically a circular structure that consists of double stranded DNA of 120–220 kb arranged in monomeric or multimeric circles as well as in linear molecules (Palmer, 1985; Lilly et al., 2001). Most of the chloroplast genome has a highly conserved organization (Raubeson and Jansen, 2005), with two identical copies of a 20–30 kb inverted repeat region (IRA and IRB) which separate a large single copy (LSC) region and a small single copy (SSC) region (Fig. 1, Wakasugi et al., 2001). The majority of these genes code for proteins, typically involved in photosynthesis or gene expression, while the rest are transfer RNA or ribosomal RNA genes. About 40–50% of the chloroplast

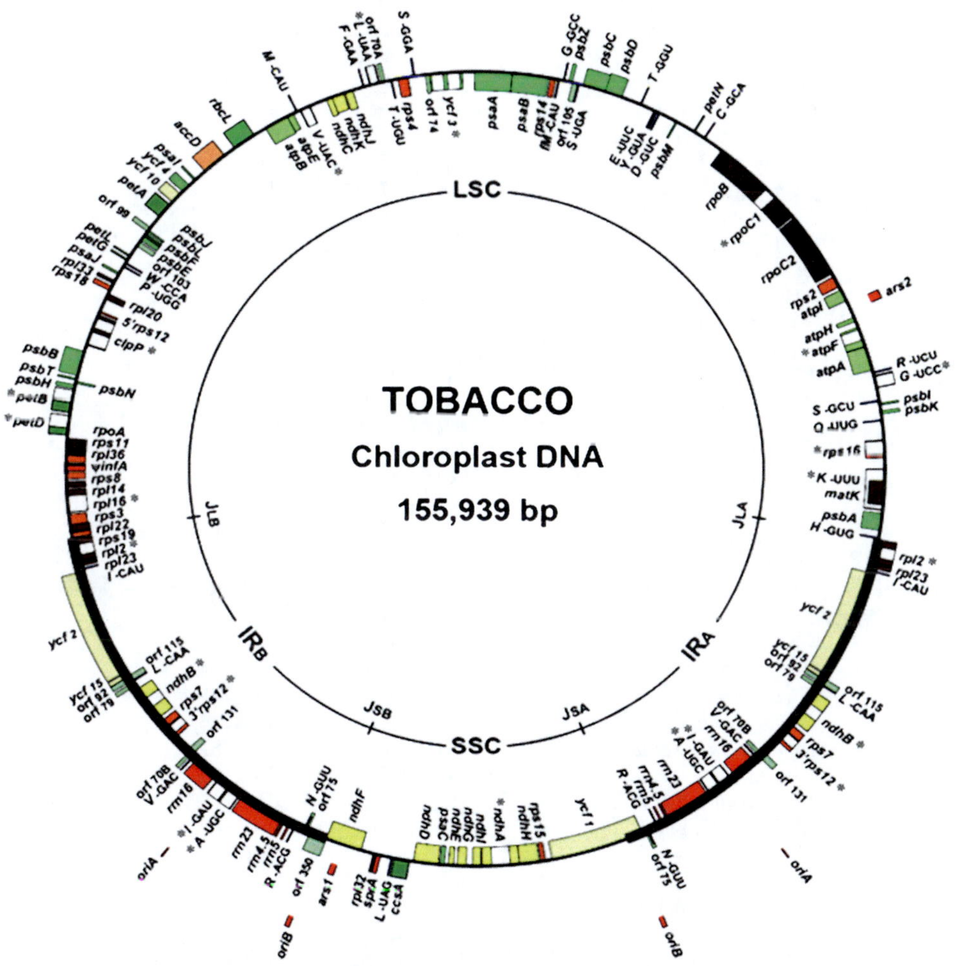

Fig. 1. Circular map of the tobacco chloroplast genome. The *thick lines* indicate the extent of the inverted repeats (IRa and IRb), which separate the genome into small (SSC) and large (LSC) single copy regions. Genes outside the map are transcribed in the counterclockwise direction and genes inside of the map are transcribed in the clockwise direction.

genome consists of non-coding spacer regions, introns and regulatory sequences. Most of the genes in plastid genomes are part of polycistronic transcriptional units, suggesting a bacterial origin (Mullet, 1993; Palmer, 1991). Plastid operons often have multiple promoters that enable a subset of genes to be transcribed within the operon (Kuroda and Maliga, 2002; Miyagi et al., 1998). Studies on organization and evolution of chloroplast genomes are advancing rapidly due to the availability of a number of completely sequenced genomes that have been published in the past few years (Verma et al., 2008) including several crop species (Table 1).

III Concept of Chloroplast Transformation

Chloroplast engineering is becoming an attractive and powerful tool in plant biotechnology for the improvement of crops. Chloroplast transformation is typically based on the delivery of DNA by

Table 1. List of the chloroplast genome sequence of different crop species.

Common name	Scientific name	Accession number	Genome size (bp)
Barley	*Hordeum vulgare*	NC_008590	136,462
Black cottonwood	*Populus trichocarpa*	NC_009143	157,033
Bread wheat	*Triticum aestivum*	NC_002762	134,545
Carrot	*Daucus carota*	NC_008325	155,911
Cassava	*Manihot esculenta*	NC_010433	161,453
Coffee	*Coffea arabica*	NC_008535	155,189
Cotton	*Gossypium barbadense*	NC_008641	160317
Cotton	*Gossypium hirsutum*	NC_007944	160301
Cucumber	*Cucumis sativus*	NC_007144	155,293
Eucalyptus	*Eucalyptus globulus*	NC_008115	160,286
Ginseng	*Panax ginseng*	NC_006290	156,318
Chickpea	*Cicer arietinum*	NC_011163	125319
Indian wild rice	*Oryza nivara*	NC_005973	134,494
Japanese black pine	*Pinus thunbergii*	NC_001631	119,707
Kidney bean	*Phaseolus vulgaris*	NC_009259	150,285
Korean pine	*Pinus koraiensis*	NC_004677	117,190
Lettuce	*Lactuca sativa*	NC_007578	152,765
Maize	*Zea mays*	NC_001666	140,384
Mulberry	*Morus indica*	NC_008359	158484
Orange	*Citrus sinensis*	NC_008334	160,129
Ornamental nightshade	*Solanum bulbocastanum*	NC_007943	155,371
Papaya	*Carica papaya*	NC_010323	160100
Potato	*Solanum tuberosum*	NC_008096	155,298
Indian rice	*Oryza sativa Indica Group*	NC_008155	134496
		NC_001320	134525
Sea Island cotton	*Gossypium barbadense*	NC_008641	160,316
Sorghum	*Sorghum bicolor*	NC_008602	140,754
Soybean	*Glycine max*	NC_007942	152,218
Spinach	*Spinacia oleracea*	NC_002202	150,725
Sugarcane	*Saccharum officinarum*	NC_006084	141,182
Sugarcane hybrid	*Saccharum hybrid*	NC_005878	141,182
Sugarcane	*Saccharum officinarum*	NC_006084	141,182
Sunflower	*Helianthus annuus*	NC_007977	151,104
Sweet orange	*Citrus sinensis*	NC_008334	160129
Tobacco	*Nicotiana tabacum*	NC_001879	155,943
Tomato	*Solanum lycopersicum*	NC_007898	155,461
Upland cotton	*Gossypium hirsutum*	NC_007944	160,301
Wheat	*Triticum aestivum*	NC_002762	134545
White poplar	*Populus alba*	NC_008235	156,505
Western balsam poplar	*Populus trichocarpa*	NC_009143	157,033
Wine grape	*Vitis vinifera*	NC_007957	160,928

the biolistic process (Daniell et al., 1990; Sanford et al., 1993) or occasionally by polyethylene glycol (PEG) treatment of protoplasts (Golds et al., 1993;O'Neill et al., 1993). This is followed by transgene integration into the chloroplast genome via homologous recombination facilitated by a RecA-type (Cerutti et al., 1992) system between the plastid-targeting sequences of the transformation vector and the targeted region of the plastid genome. To facilitate this recombination, chloroplast transformation vectors are designed with homologous flanking sequences on either side of the transgene cassette. Targeting sequences have no special properties other than the fact that they are homologous to the chosen target site and are generally about 1 kb in size. Both flanking sequences are essential for homologous recombination. Typically chloroplast transformation is accomplished by integration of the transgene into a few genome copies. This is followed by cell divisions under selection pressure to eliminate untransformed plastids, thereby achieving a homogeneous population of transformed plastid genomes. Transgenes have been stably integrated into several sites; however integration of transgenes into the IR region has the advantage of duplicating the introduced transgenes by the phenomenon of copy correction.

The concept of chloroplast genetic engineering was first conceived in the mid 1980s with the expression of bacterial and cyanobacterial genes in plastids in vivo (Daniell and McFadden 1987). Boyton et al. (1988) and Blowers et al. (1989) have demonstrated the stable complementation of partially deleted chloroplast gene in *Chlamydomonas reinhardtii*, a unicellular green algae with a single large chloroplast occupying approximately 60% of the cell volume. Then Daniell et al. (1990) demonstrated the first transient expression of foreign gene in chloroplasts of cultured tobacco cells, delivered by high-velocity DNA coated microprojectiles. This work was repeated in wheat leaves, calli and somatic embryos (Daniell et al., 1991). The *aad*A gene was first used as a selectable marker to transform the Chlamydomonas chloroplast genome (Goldschmidt-Clermont, 1991). Later, stable integration of the *aad*A selectable marker gene was achieved in the tobacco chloroplast genome (Svab and Maliga, 1993). Although transgenes were first integrated into transcriptionally silent spacer regions, integration of transgenes into transcriptionally active spacer regions offers several advantages including insertion of transgenes without 5′or 3′ untranslated regions (UTRs) or promoters. Many transgenes have been integrated in the transcriptionally active intergenic region between the *trn*I-*trn*A genes, within the rrn operon, located in the IR regions of the chloroplast genome (Grevich and Daniell, 2005). These integration sites carry the origin of replication (oriA), which facilitates the replication of introduced plasmid inside the chloroplast (Daniell et al., 1990), thereby increasing the template copy number and consequently enhancing the probability of transgene integration by homologous recombination. High levels of foreign protein expression (De Cosa et al., 2001) and homoplasmy in the first round of selection (Guda et al., 2000) was achieved when the transgenes were integrated into this site. Moreover, integration of transgenes between exons of *trn*A and *trn*I also facilitate correct processing of foreign transcripts because of the processing of introns present within both flanking regions. Using this technology transgenes have been integrated into the chloroplast genome of many crop species using different integration sites (Table 2).

IV Advantages of Plastid Transformation

One of the major environmental concerns in genetically modified (GM) crops is transgene containment. Lack of gene containment due to the pollen – mediated out-crosses of transgenes from nuclear transgenic plants to related crops or weeds have been a major concern (Daniell, 2002). Thus, the chloroplast transformation system is highly desirable in situations where outcrossing between crops and weeds or among crops is a concern (Daniell et al., 1998). In addition, the possibility of insects developing resistance to insecticidal proteins, due to the low levels of transgenes expression and toxicity of transgenic pollen to non-target insects, has raised environmental concerns for transgenic plants engineered for pest resistance (Daniell, 2002). To address these environmental concerns, several foreign proteins have been expressed via the chloroplast genome. Although pollen from plants shown to exhibit maternal plastid inheritance contains metabolically active plastids, the plastid DNA itself is lost during the process of pollen maturation

Table 2. Different crops which are successfully transformed through chloroplast genetic engineering.

Crop	Selectable marker/ gene of interest	Site of integration	Reference
Cabbage	*aad*A, *uid*A	*trn*V–*rrn*16S/*trn*I–*trn*A–*rrn*23S	Liu et al., 2007
Cabbage	*aad*A, *cry*1Ab	*trn*V–*rrn*16S/*trn*I–*trn*A–*rrn*23S	Liu et al., 2008
Cauliflower	*aad*A	*acc*D/*rbc*L	Nugent et al., 2006
Carrot	*aad*A, *badh*	*trn*I/*trn*A	Kumar et al., 2004a
Cotton	*aph*A6, *npt*II	*trn*I/*trn*A	Kumar et al., 2004b
Lettuce	*aad*A, *gfp*	*trn*I/*trn*A	Lelivelt et al., 2005
Lettuce	*aad*A, *gfp*	*acc*D/*rbc*L	Kanamoto et al., 2006
Lettuce	*aad*A, *CTB-Pins*	*trn*I/*trn*A	Ruhlman et al., 2007
Lettuce	*aad*A, *CTB-ama*1; *CTB-msp*1	*trn*I/*trn*A	Davoodi-Semiromi et al., 2010
Lettuce	*aad*A, *CTB-Pins, pag*	*trn*I/*trn*A	Ruhlman et al., 2010
Oilseed rape	*aad*A, *cry*1Aa10	*rps*7/*ndh*B	Hou et al., 2003
Oilseed *Brassicaceae*	*aad*A, *gfp*	*rrn*16 *trn*V/*rps*12/7	Skarjinskaia et al., 2003
Petunia	*aad*A, *uid*A	accD/rbcL	Zubkot et al., 2004
Poplar	*aad*A, *gfp*	*acc*D/*rbc*L	Okumura et al., 2006
Potato	*aad*A, gfp	*rrn*16 *trn*V/3′*rps*12/7	Sidorov et al., 1999
Potato	*aad*A, *gfp*	*acc*D/*rbc*L *rrn*16 *trn*V/3′*rps*12/7	Nguyen et al., 2005
Rice	*aad*A, *gfp*	*trn*I/*trn*A	Lee et al., 2006
Soybean	*aad*A	*trn*V/3′*rps*12/7	Dufourmantel et al., 2004
Soybean	*aad*A, *cry*1Ab	*trn*V/3′*rps*12/7	Dufourmantel et al., 2005
Soybean	*aad*A, *hppd*	*trn*V/3′*rps*12/7	Dufourmantel et al., 2007
Tomato	*aad*A	*trn*fM/*trn*G	Ruf et al., 2001
Tomato	*aad*A, *LCp*	*trn*fM/*trn*G	Wurbs et al., 2007

and hence is not transmitted to the next generation (Daniell, 2007). In chloroplast transformation, foreign genes are stably integrated into the spacer region of two functional genes through homologous recombination. This allows site specific integration and eliminates the concern of position effect frequently observed in nuclear transgenic plants (Daniell, 2002). Another advantage of chloroplast transformation is the absence of gene silencing, which is often a serious concern in nuclear transformation. There is no gene silencing in chloroplast transgenic lines at the transcriptional level, despite accumulation of transcripts 150-fold higher than in nuclear transgenic lines (Dhingra et al., 2004; Lee et al., 2003) or at the translational level despite accumulation of foreign proteins up to 46% of tsp (De Cosa et al., 2001). This high level of expression is due to polyploidy of the plastid genetic system, up to 10,000 copies of the chloroplast genome in each plant cell, resulting in the ability to sustain a high number of functional gene copies. Another major advantage of engineering the chloroplast genome resides in the expression of multiple transgenes as operons due to efficient translation of polycistronic messenger RNAs. This allows the engineering of multiple genes in a single transformation step and several heterologous operons have been expressed in transgenic chloroplasts and polycistrons are translated without processing into monocistrons (De Cosa et al., 2001; Quesada-Vargas et al., 2005). Moreover, foreign proteins synthesized in chloroplasts are properly folded with appropriate post-transcriptional modifications, including disulfide bonds (Arlen et al., 2007; Ruhlman et al., 2007; Staub et al., 2000) and lipid modifications (Glenz et al., 2006). Furthermore, all chloroplast transgenic lines express similar levels of foreign proteins within the range of physiological variations (Lee et al., 2003). Moreover, heterologous expression systems in prokaryotes for the expression of membrane proteins have been hampered by different synthesis, targeting, insertion and folding characteristics in their hosts. Adequate expression of membrane proteins continues to be a major challenge due the toxic effects which severely reduce cell growth product yields (Wagner et al., 2007; Wagner et al., 2006). More recently, a chloroplast transformation system was developed for the expression of membrane protein (Singh et al., 2008).

These results have significant implications for understanding membrane biogenesis and provide a first step in using chloroplast transformation as a means of expressing and accumulating high levels of native or foreign membrane proteins for structural studies or biomedical applications. Another major advantage of chloroplast technology is the elimination of fermentation, purification, cold chain, adjuvant or sterile injections which are normally used in routine vaccine productions. Bio-encapsulated vaccine antigens expressed in chloroplasts are protected in the stomach from acids and enzymes but are released to the immune system in the gut when plant cell walls are digested by bacteria that naturally colonize the gut.

V Chloroplast Transformation Vectors and Mode of Transgene Integration into Chloroplast Genome

The basic plastid transformation vector is comprised of flanking sequences and chloroplast specific expression cassettes (Fig. 2). Typically a basic chloroplast expression cassette consists of a strong plastid promoter, a selectable marker, 5′ and 3′ untranslated regions to enhance the levels of transcription and translation of the introduced foreign gene. The chloroplast promoter and regulatory elements can be amplified directly from the total cellular or chloroplast DNA using specific primers based on the available sequences of chloroplast genome. Commonly used 5′ regulatory sequences are derived from the *psb*A and *rbc*L genes whereas 3′s are from *psb*A, *rbc*L and *rps*16 genes. The majority of these regulatory sequences are derived from tobacco. However, endogenous regulatory sequences have also been used for chloroplast transformation (Kanamoto et al., 2006). Proper designing and understanding of chloroplast transformation vector with appropriate regulatory sequences play a key role in the efficiency of transformation and higher expression of the integrated transgenes (Verma and Daniell, 2007). Regulatory sequences are essential in the plastid transformation vector to compensate for degradation of the newly synthesized protein by chloroplast proteases. For example, the HSA protein expression was increased 500-fold resulting in the formation of protective inclusion bodies when the *HSA* coding sequence was regulated by the chloroplast *psb*A 5′and 3′ UTRs when compared with the ribosome binding site (Fernandez-San Millan et al., 2003). In another study, the insecticidal protein, Cry2Aa2 folded properly when the chaperone (encoded by the orf2 gene) was co-expressed, thereby protecting the protein from degradation and increasing the foreign protein accumulation over 128-fold, from 0.36% to 46.1% tsp (De Cosa et al., 2001).

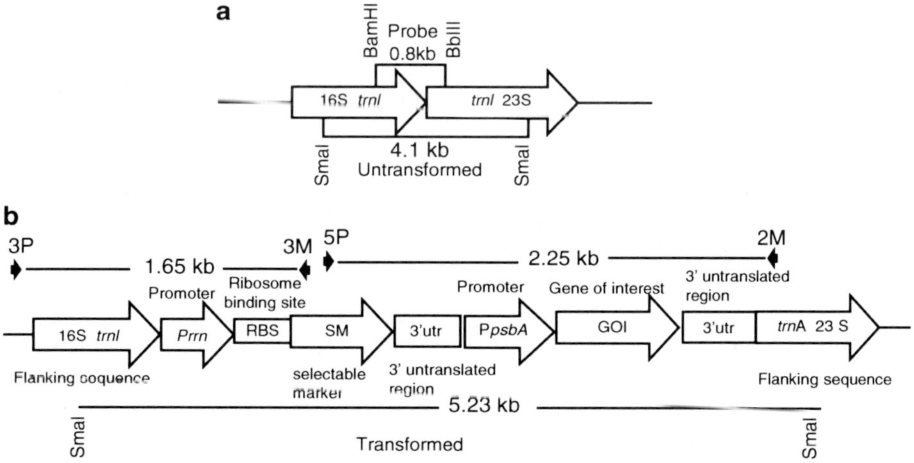

Fig. 2. Tobacco chloroplast transformation vector. (**a**) and (**b**) Schematic representation of the chloroplast flanking sequence used for homologous recombination, expression cassette with promoter, regulatory elements, selectable marker gene and gene of interest, primer annealing sites (3P/3M and 5P/2M) and expected products of Southern blots for untransformed and transgenic lines.

Once the chloroplast transformation cassette along with the proper regulatory sequences have been constructed, they are inserted into the spacer region between two functional genes, which serve as flanking sequences of the chloroplast genome, thus targeting the foreign genes to a precise location. Usually 1 kb each of homologous flanking sequence in either side of the chloroplast expression cassette is sufficient to facilitate homologous recombination. However, the frequency of homologous recombination has been enhanced by the use of long flanking sequences. Normally the PCR based technology is widely used to isolate these flanking sequences (Singh et al., 2009; Verma et al., 2008). Transgenes have been successfully integrated at different sites of the chloroplast genome using different regulatory sequences. Although transgenes were first integrated into transcriptionally silent spacer regions (Svab and Maliga, 1993), insertion of transgenes into transcriptionally active spacer regions offers exceptional advantages including insertion of transgenes without 5′or 3′UTRs or promoters (Verma and Daniell, 2007). This is the reason why a maximum number of transgene has been integrated between the transcriptionally active sites *trn*I and *trn*A. Moreover, this site is within the inverted repeats; therefore, 10,000 copies of the chloroplast genome per cell would double to 20,000 copies, thereby facilitating higher expression of foreign genes. Using this site of integration the highest accumulation of foreign proteins up to 46% of the total soluble protein have been obtained (De Cosa et al., 2001). Another unique property of this integration site is the presence of the chloroplast origin of replication (Kunnimalaiyaan and Nielsen, 1997; Lugo et al., 2004). This in turn may facilitate replication of foreign vectors within the chloroplasts (Daniell et al., 1990), enhance the probability of transgene integration and achieve homoplasmy even in the first round of selection (Guda et al., 2000). This was supported by the first successful Rubisco engineering obtained by integrating the *rbc*S gene at this site (Dhingra et al., 2004). All other earlier attempts of rubisco engineering at other integration sites in the chloroplast genome were only partially successful. Also integration of transgenes between exons of *trn*A and *trn*I facilitated correct processing of foreign transcripts. Thus, a number of transgenes have been integrated at various sites of the chloroplast genome of different crops.

VI Methods of Plastid Transformation and Recovery of Transplastomic Plants

Currently two methods have been used for DNA delivery into plastids; treatment of protoplast with polyethylene glycol (PEG-mediated) and the particle delivery system (PDS). However, PDS is widely adopted as it is less time consuming and simple. Normally plasmid DNA is coated with gold or tungsten particles and delivered into plant cells using particle bombardment (Daniell et al., 2005b; Singh et al., 2009). After several rounds of selection on appropriate medium containing selection agents (antibiotics), the putative transformants are screened (Fig. 3) via PCR analysis of the transgenic lines. Specific set of primers are designed to confirm the integration of transgenes into the chloroplast genome and to differentiate it from mutants or nuclear transgenic plants. Site specific chloroplast integration of the transgene cassette is determined by using a set of primers of which one anneals to the native chloroplast genome and the other anneals to the transgene cassette. No PCR amplification is observed in mutants and nuclear transgenic plants with these primers (Fig. 4a and b). Initially, few copies of the chloroplast genome incorporate the transgene cassette and after several round of selection, untransformed genome copies are replaced by transformed copies, thus leading to homoplasmy. Therefore, putative transformants undergo several round of selection until homoplasmy is achieved. Southern blot analysis then differentiates homoplasmic from heteroplasmic plants. Generally genomic DNA is isolated from the untransformed and transformed plants, digested with appropriate restriction endonuclease(s) and transferred to a membrane. This DNA bound membrane is probed with radiolabelled chloroplast flanking sequences used for homologous recombination (Fig. 2a). The transplastomic plants show a larger hybridizing fragment than the untransformed genome due to integration of the transgene cassette. Absence of untransformed fragment indicates that the transformant has reached homoplasmy (Fig. 4c). Then the confirmed homoplasmic plants are transferred to the green house. Expression of the foreign protein is confirmed by western blot analysis using appropriate antibodies (Fig. 4d). This is

Fig. 3. Selection of transplastomic plants. Transplastomic tobacco and lettuce shoots undergoing first (**a** and **d**), second (**b** and **e**), and third (**c** and **f**, rooting) rounds of selection, respectively.

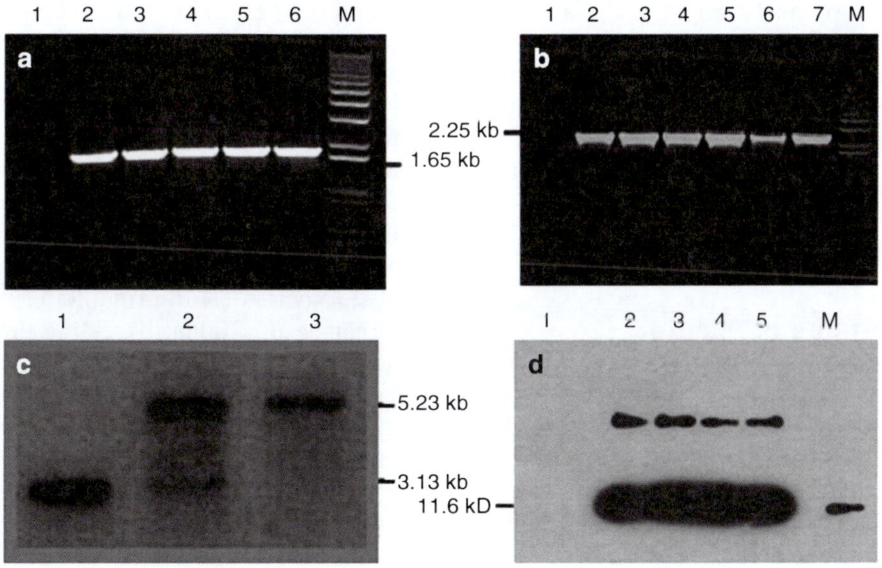

Fig. 4. Confirmation of transgene integration into the chloroplast genome and expression of foreign protein. (**a**) PCR products of 3P/3M primers. Lane 1, untransformed plant; lanes 2–5, transformed lines (1.6 kb); lane 6, positive control; lane M, 1kb DNA marker. (**b**) PCR product of 5P/2M primers. Lane 1, untransformed plant; lanes 2–6, transformed lines (3.2 kb). (**c**), Southern blot probed with radiolabeled flanking sequence showing homoplasmy. Lane 1, untransformed plant; lanes 2, heteroplasmic transplastomic plant; lane 3, homoplasmic transplastomic plant. (**d**) Western blot showing the expression of introduced gene. Lane 1, untransformed plant; lanes 2–5, different transplastomic lines; M, standard protein.

Fig. 5. Phenotype and maternal inheritance of transplastomic lines. (**a**) Mature transplastomic plant grown in the green house showing normal growth. (**b** and **c**) Untransformed (UT) and transgenic seeds of tobacco and lettuce germinated on medium containing selection agent, showing maternal inheritance.

followed by the assessment of effectiveness and the functionality of the expressed protein using animal models. Seeds from the homoplasmic plants are collected and checked for maternal inheritance by germination in a medium containing the selection agent. Transgenic seeds germinate and grow into green plants while untransformed plants bleach on the medium. The lack of transgene segregation suggests that transgenic lines are maternally inherited (Fig. 5).

VII Current Status of Plastid Transformation

Although several crop species are engineered by this technology, extensive work has been done in tobacco and many useful agronomic traits, therapeutic proteins, vaccine antigens etc. have been stably integrated and expressed in tobacco chloroplasts. A single tobacco plant is capable of generating a million seeds and yielding approximately 170 t of biomass per hectare (Cramer et al., 1999). Furthermore, tobacco is a non-food and non-feed self-pollinated crop which significantly minimize the transgene escape. However, the high content of nicotine and other toxic alkaloids has been a limiting factor for pharmaceutical applications that rely on oral delivery. Therefore, extension of this technology to other species is very important to overcome these concerns. Apart from the production of pharmaceutical proteins and vaccine antigens, this technology is highly useful for other economically important crops. Homoplasmic plants of soybean, carrot, and cotton have been regenerated via somatic embryogenesis after bombardment of embryogenic calli, and the use of species-specific plastid vectors (Dufourmantel et al., 2004, 2005; Kumar et al., 2004a, b). Some progress has also been made in improving the chloroplast transformation system of tomato. Using tomato, a plastid expression of a bacterial lycopene beta-cyclase gene resulted in herbicide resistance and triggerred conversion of lycopene, the main storage carotenoid of tomatoes, to betacarotene. That in turn, resulted in fourfold enhanced pro-vitamin A content of the fruits (Wurbs et al., 2007). A stable chloroplast transformation system has also been reported for cabbage (Liu et al., 2007, 2008), poplar (Okumura et al., 2006), and lettuce (Kanamoto et al., 2006; Lelivelt et al., 2005). Accumulation of a valuable therapeutic protein, the CTB-Pins fusion (Ruhlman et al., 2007), CTB-AMA1 and CTB-MSP1 (Davoodi-Semiromi et al., 2010) in the lettuce chloroplast genome has rendered this technology feasible for the production of therapeutic proteins and vaccine antigens. Also extensive work has been done in other crops with the successful integration of transgenes into the chloroplast genome at various sites (Table 2).

VIII Application of Chloroplast Technology for Agronomic Traits

One of the main reasons for the reduction and uncertainty in the production of agricultural crops is due to their susceptibility to insect pests and pathogens. These are controlled by application of chemically synthesized insecticides and fungicides. These pesticides are not only expensive but also affect the crop quality as well as pose a threat to the ecosystem. In this regard the biotechnological approach provides an opportunity to produce plant cultivars with agronomically desirable traits. In addition, engineering of insect resistant gene in plants are highly specific towards their target insects and are nonhazardous to humans and animals. Although many important crops have already been genetically modified via the nuclear genome, nuclear transgenic plants have many drawbacks including gene containment and the expression levels of the transgenes. Chloroplast genetic engineering eliminates many of these concerns due to gene containment and the high level expression of the introduced foreign genes. Several agronomically important genes have been introduced into plastid genome (Table 3). Genetically engineered tobacco plants expressing an insecticidal protein Cry2Aa2 have shown resistance against target insects. They were shown to correctly process a bacterial operon and express the Cry2Aa2 protein at high levels (~46.1% of total leaf protein) and resulted in the detection of cuboidal crystals using transmission electron microscopy (De Cosa et al., 2001). Even at lower level of Cry2Aa2 protein, chloroplast transgenic plants were able to kill insects that could withstand insecticidal protein concentrations 40,000 times higher than the normal level (Kota et al., 1999). This technology has also been extended to many other crops. Soybean plastids have been engineered with the *cry*1Ab gene which conferred high insecticidal activity against velvetbean caterpillar (Dufourmantel et al., 2005). More recently the *B. thuringiensis cry*1Ab gene was successfully transferred into the cabbage chloroplast genome with insecticidal effect on *Plutella xylostella* (Liu et al., 2008). Expression of *Bacillus thuringiensis cry*9Aa2 gene into tobacco chloroplast conferred resistance to the potato tuber moth (Chakrabarti et al., 2006).

Another problem in crop production resides in infestation by many plant diseases caused by different pathogens. The antimicrobial peptide MSI-99, an analog of magainin 2, was expressed in the tobacco chloroplast genome. In planta assays with the bacterial pathogen *P. syringae* pv *tabaci* and fungal pathogen, *Colletotrichum destructivum*,

Table 3. Agronomic traits expressed in plastid genome.

Agronomic traits	Gene	Promoter 5′/3′ UTRs	Literature cited
Insect resistance	*cry*1A(c)	P*rrn*/*rbcL*/*rps*16	McBride et al., 1995
	*cry*2Aa2	P*rrn*/ggagg (native)/*psb*A	Kota et al., 1999
	*cry*2Aa2 operon	Prrn/native 5′UTR/*psb*A	De Cosa et al., 2001
	*cry*1Aa10	Prrn/native 5′ UTR/*psb*A	Hou et al., 2003
	*cry*1Ab	P*rrn*/T7 gene10/*rbc*L	Dufourmantel et al., 2005
	*cry*9Aa2	Prrn/native 5′ UTR/*rbc*L	Chakrabarti et al., 2006
	*Cry*1Ab	Prrn/T*psb*A	Liu et al., 2008
Herbicide resistance	*aro*A (petunia)	Prrn/*ggagg*/*psb*A	Daniell et al., 1998
	bar	P*rrn*/*rbc*L/*psb*A	Iamtham and Day 2000
	hppd	*psb*A promoter/5′UTR/*rbc*L	Dufourmantel et al., 2007
Disease resistance	*MSI-99*	P*rrn*/*ggagg*/*psb*A	DeGray et al., 2001
Phytoremediation	*mer*A/*mer*B	P*rrn*/*ggagg*/*psb*A	Hussein et al., 2007; Ruiz et al., 2003
Stress tolerance	*tps1* (yeast)	P*rrn*/*ggagg*/*psb*A	Lee et al., 2003
	badh	P*rrn*/*ggagg*/*rps*16	Kumar et al., 2004a
	BvCMO	P*psb*A/T*psb*A P*rrn*/T*rps*	Zhang et al., 2008
Cytoplasmic male sterility	*pha*A	Prrn/*psb*A/*psb*A	Ruiz and Daniell 2005

transgenic plants showed no sign of necrosis even when 80,000 cells were used for inoculation, whereas a necrotic lesions was seen in un-transformed leaves even when a few cell were used for inoculation (DeGray et al., 2001). However, this phenomenon was not observed when the same gene was expressed via the nuclear genome, due to low expression levels (Chakrabarti et al., 2003; Li et al., 2001). These results clearly suggested that chloroplast can express antimicrobial peptides at high levels for the effective control of pathogens.

All commercial herbicide resistant transgenic lines were developed via nuclear transformation. One of the major drawbacks of nuclear engineered herbicide resistant plant is the possibility of transgene out-crossing to other related crops and weeds (Daniell, 1999, 2000). Therefore, the tobacco chloroplast genome was first engineered to express a 5-enolpyruvylshikimate-3-phosphate synthase (EPSPS) by integration of the *aro*A gene (Daniell et al., 1998). Chloroplast transgenic plants can resist up to 5 mM glyphosate which is tenfold greater than the lethal concentration. This finding opened up the possibility to engineer other herbicide resistant genes via the plastid genome. Later, another *Agrobacterium* strain CP4 EPSPS gene was expressed in tobacco plastids thereby resulting in a 250-fold more EPSPS enzyme than expressed in nuclear transformations (Ye et al., 2001). Furthermore, another herbicide resistant gene (bar gene) was successfully introduced into the tobacco chloroplast. This study showed that transgenic tobacco plants exhibited field-level tolerance to phosphinothricin (PPT), which was conferred by even lower levels of *bar* expression (Iamtham and Day, 2000; Lutz et al., 2001). The possibility of engineering herbicide resistant genes into plastids was also successful in another important crop. Bacterial 4-hydroxyphenylpyruvate dioxygenase (HPPD) from *P. fluorescens* was successfully expressed in soybean plastids, which showed stronger herbicide tolerance than nuclear transformants (Dufourmantel et al., 2007).

One of the major factors which adversely affect the crop production is abiotic stress. Trehalose phosphate synthase (*TPS1* gene) mediates the reaction that forms the osmoprotectant trehalose and it is thought to play a major role in protecting cells from damage caused by abiotic stress. Drought resistant tobacco plants were successfully developed by introducing the *TPS1* gene into the nuclear as well as in the chloroplast genome. Chloroplast transgenic plants showed 15–25% higher accumulation of trehalose than nuclear transgenic plants. Moreover, nuclear transgenic plants exhibited pleiotropic effect even at low concentrations. In contrast, despite the high accumulation of trehalose, chloroplast transgenic plants exhibited normal growth and showed drought tolerance by remaining green and healthy when germinated in media containing 3–6% polyethylene glycol (Lee et al., 2003).

Salinity has been categorized as a major agricultural problem in arid and semi-arid regions, in mismanaged land and in irrigation areas. Glycine betaine protects plant cells from salt stress by maintaining an osmotic balance with the environment (Robinson and Jones, 1986) and by stabilizing the quaternary structure of complex proteins (Papageorgiou and Murata, 1995). Thus, the betaine aldehyde dehydrogenase (*badh*) gene, which is responsible for glycine betaine synthesis was successfully expressed via the carrot plastid genome. The transgenic lines were able to survive a high concentration of NaCl (up to 400 mM), which is equivalent to halophytic conditions (Kumar et al., 2004a). More recently choline monooxygenase (BvCMO), the enzyme that catalyzes the conversion of choline to betaine aldehyde from beet was expressed in tobacco chloroplasts. Transplastomic plants exhibited tolerance to toxic levels of choline and exhibited salt/drought stress (Zhang et al., 2008).

Environmental pollution with heavy metals is a serious problem for plants, animals and man. Therefore, the development of plant based phytoremediation technologies for the clean-up of contaminated soils is of significant interest. Earlier attempts to genetically engineer plants with improved phytoremediation were based on nuclear transformation. However, improvements in phytoremediation were successfully accomplished via chloroplast technology. A native operon containing the *mer*A and *mer*B genes, which code for mercuric ion reductase (*mer*A) and organomercurial lyase (*mer*B), respectively, have been expressed via the tobacco chloroplast genome. Stable integration of the *mer*AB operon into the chloroplast genome resulted in high levels of tolerance for the organomercurial compound, phenylmercuric acetate (PMA) when grown in soil containing up to 400 μM PMA (Ruiz et al., 2003).

Further, transgenic tobacco plants engineered with bacterial *mer*A and *mer*B genes via the chloroplast genome were investigated to study the uptake and translocation of different forms of mercury (Hg) from roots to shoots, and their volatilization. Chloroplast-transgenic lines exhibited about 100-fold increase in the efficiency of Hg accumulation in shoots compared to untransformed plants. This was the first report of such high levels of Hg accumulation in green leaves or tissues. The combined expression of *mer*AB in the chloroplast genome enhanced the conversion of Hg^{2+} into Hg^0, which conferred tolerance by rapid volatilization and increased uptake of different forms of mercury, thus surpassing the concentrations found in the soil (Hussein et al., 2007).

Although naturally occurring cytoplasmic male sterility has been reported in some crop, such a system does not occur frequently in agricultural crops. Thus, the possibility of transgene escape has been a limiting factor for nuclear transformation. Moreover, the genetic segregation of nuclear transformation eventually dilutes male sterility. To address some of these concerns, cytoplasmic male sterility has been achieved by engineering the *pha*A gene that codes for β-ketothiolase via the chloroplast genome. The transgenic lines were normal except for the male sterility phenotype lacking pollen (Ruiz and Daniell, 2005). Further restoration of male fertility was observed under continuous illumination that increases acetyl-CoA carboxylase activity, thereby increasing the levels of plastidic fatty acid biosynthesis, which is especially needed for the formation of the exine pollen wall. This finding provides another possible method for transgene containment.

IX Chloroplast-Derived Vaccine Antigens

With the advent of modern molecular biology techniques, new strategies are developed for the production of subunit vaccines. These are vaccines comprised of proteins derived from pathogenic viruses, bacteria or parasites; in general the proteins are produced not by the pathogens themselves, but by expression of the gene encoding the protein in a "surrogate" organism. Unlike almost all other cell lines used for production of vaccines, components of plant cells have always been an important part of the normal human diet. Plants, therefore, offer significant new opportunities for making safe and effective oral vaccines. Plant-derived vaccines are not only cost effective but are also highly efficient. Also, orally delivered vaccines are simpler and may require less production and formulation regulations than injected vaccines. Bioencapsulation can protect the vaccine in the stomach and gradually release the antigen in the gut (Mor et al., 1998). Several vaccine antigens against bacterial (Arlen et al., 2008; Daniell et al., 2001a; Glenz et al., 2006; Koya et al., 2005; Tregoning et al., 2003, 2005; Davoodi-Semiromi et al., 2010), viral (Birch-Machin et al., 2004; Fernandez-San Millan et al., 2008; Lenzi et al., 2008; Molina et al., 2004, 2005) and protozoan (Chebolu and Daniell, 2007; Davoodi-Semiromi et al., 2010) pathogens have been expressed in transgenic chloroplasts (Table 4). For example, the currently available anthrax vaccine obtained from the cultures of *Bacillus anthracis* exhibits undesirable side effects. In an attempt to produce anthrax vaccine free of extraneous bacterial contaminants and in large quantities, PA was expressed in transgenic tobacco chloroplasts by inserting the *pag*A gene into the chloroplast genome (Koya et al., 2005; Watson et al., 2004). Cytotoxicity measurements in macrophage lysis assays showed that the chloroplast-derived PA was equivalent in potency to the PA produced in *B. anthracis*. Subcutaneously immunized mice with partially purified chloroplast-derived or *B. anthracis*-derived PA laced with adjuvant survived (100%) the challenge with lethal doses of toxin. These results demonstrated the immunogenic and immunoprotective properties of the chloroplast-derived anthrax vaccine antigen. It is possible to produce up to 360 million doses of fully functional anthrax vaccine in one acre of tobacco (Koya et al., 2005). Moreover, several chloroplast-derived vaccine antigens form correct disulfide bonds and assemble as heterodimers, pentamers or in other configurations with proper post-translational modifications (Kamarajugadda and Daniell, 2006; Verma and Daniell, 2007). Moreover, the immune responses of tobacco chloroplast-derived vaccines result in successful challenges with aerosol plague; there were ten billion spores in the spleen of control mice but none in immunized mice (Arlen et al., 2008). Comparison of subcutaneous injection (with adjuvant) and oral delivery of plant cells revealed that oral delivery offered greater protection and immunity (Arlen et al., 2008). Cervical cancer, the most prevalent cancer in women is caused by a certain type of a virus-human papillomavirus (HPV).

Table 4. Chloroplast-derived vaccine antigens.

Vaccines antigens	Gene	Functional evaluation	Reference
Cholera toxin B	*Ctx*B	GM1 ganglioside-binding assay. Long-term protection (50% mouse life span) against CT challenge in both oral (100%) and subcutaneously (89%) immunized mice; protection correlated with CTB-specific IgA and IgG1 titers in oral and IgG1 in subcutaneously immunized mice; increasing numbers of IL10$^+$ T-cell but not Foxp3$^+$ regulatory T-cells, suppression of IFN-γ and absence of IL-17 were observed in protected mice.	Daniell et al., 2001a; Davoodi-Semiromi et al., 2010
E. coli enterotoxin B	*LTB*	GM1 ganglioside-binding assay.	Kang et al., 2003
E. coli enterotoxin B	(*LTB*)	GM1 ganglioside-binding assay; oral immunization protected mice from CT challenge.	Rosales-Mendoza et al., 2009
Mutant of *E. coli* toxin	*LTK*63	GM1 ganglioside-binding assay.	Kang et al., 2004
Tetanus toxin	*Tet*C bacterial and synthetic	Mice developed immune response and survived the tetanus toxin challenge.	Tregoning et al., 2003,2005
Lyme disease	*Osp*A *Osp*A-T	Systemic immune response in mice. Protected mice against challenge with *Borrelia burgdorferi*.	Glenz et al., 2006
Anthrax protective	*Pag*	Macrophage lysis assay, systemic immune response, toxin neutralization assay, mice survived (100%) challenge with lethal doses of toxin.	Koya et al., 2005, Watson et al., 2004
Plaque	*CaF*1-*Lcr*V	Oral deliver offer greater protection and immunity than subcutaneous injection. Significant protection (88 %) was achieved in orally vaccinated mice than subcutaneous (33%) when challenged with *Y. pestis*.	Arlen et al., 2008
Canine parvovirus (*CPV*)	*CTB-2L21* *GFP-2L21*	GM1 ganglioside-binding assay. Produced high titers of anti2L21 Abs and able to recognize the VP2 protein. Rabbit sera were able to neutralize CPV in an in vitro infection assay	Molina et al., 2004, 2005
Rotavirus	*VP*6	Not reported	Birch-Machin et al., 2004
Hepatitis C	*NS*3	Not reported	Daniell et al., 2005a
Hepatitis E virus (HEV)	*HEV E2*	Antigenic response in mice.	Zhou et al., 2006
Human Papilloma-virus (HPV)	*L1*	Not reported	Lenzi et al., 2008
Human Papilloma-virus (HPV)	*L1*	Immunogenic response in mice after intraperitoneal injection, and neutralizing antibodies were detected.	Fernandez-San et al., 2008
Human immunode-ficiency virus type 1	*p24*	Not reported	McCabe et al., 2008
Swine fever virus (CSFV).	*CFSV E2*	Immunized mice elicited specific antibodies.	Shao et al., 2008
Ameobiasis	*Lec*A	Systemic immune response in mice.	Chebolu and Daniell 2007
Malaria	(*CTB-ama1* & *CTB-msp1*)	Sera of immunized mice completely inhibited proliferation of the malarial parasite and cross-reacted with the native parasite proteins/parasites in immunoblots and immuno-fluorescence studies, at the ring, trophozoite or schizont stage of the malarial parasite.	Davoodi-Semiromi et al., 2010

Although a vaccine based on virus-like particles has been developed and marketed, the high cost of the vaccine has always been a limiting factor. In order to replace the production system, the major structural protein of the HPV-16 capsid, L1 was expressed in the tobacco chloroplast genome (Fernandez-San Millan et al., 2008; Lenzi et al., 2008). It was shown that protein extracts from L1 transgenic plants were highly immunogenic and neutralizing antibodies were detected in intraperitonially injected mice (Fernandez-San Millan et al., 2008). Canine parvovirus (CPV) which infects dogs and other Canidae family members resulted into hemorrhagic gastroenteritis and mycocarditis. Antigens capable of eliciting a protective immune response against CPV, a linear antigenic peptide (2L21) from VP2 capsid protein of CPV was fused with CTB or GFP and was successfully expressed in tobacco chloroplasts (Molina et al., 2004, 2005). The correct expression of this epitope was verified by western blot using CPV-neutralizing monoclonal antibodies. Moreover, the fusion protein retained a pentamer form which is necessary for the GM1-ganglioside receptor binding, a characteristic of the CTB fusion protein. Immunized mice induced a high immunogenic response which recognized the VP2 protein from CPV. In vitro CPV neutralizing activity assays confirmed that the antibodies induced in rabbits by CTB-2L21 efficiently neutralized the CPV infection of CRFK cells. Another highly contagious pig disease, classical swine fever (CSF) is caused by classical swine fever virus (CSFV). The expression of a major neutralizing antigen for CSFV infection, the envelope glycoprotein *E2* gene in tobacco chloroplast is another example of an animal vaccine. Mice immunized with E2 expressed leaf extracts induced significant specific antibodies against CSFV (Shao et al., 2008). All these studies confirmed that the chloroplast genetic engineering system is most suitable for the production of vaccine antigens.

X Chloroplast-Derived Biopharmaceutical Proteins

Plants possess numerous advantages over traditional expression technologies for large scale production of recombinant proteins. Plant systems do not require prohibitively expensive fermenters. Also the post-translational modification and purification steps can be eliminated thereby reducing the production cost. Plant-derived biopharmaceuticals are less expensive to produce and store, are easily scaled up for mass production, and are safer than those derived from animals (Daniell et al., 2001b). Thus, several biopharmaceutical proteins have been expressed in plants through chloroplast engineering (Table 5). The successful expression of bioelastic

Table 5. Chloroplast derived biopharmaceutical therapeutic proteins.

Biopharmaceutical	Gene	Functional evaluation	Reference
Elastin-derived polymer	*EG121*	Not reported	Guda et al., 2000
Human somatotropin	*hST*	Growth response of Nb2 cell line in the presence of somatotropin. Rat lymphoma cell line Nb2 proliferated in proportion to the amount of somatotropin in the culture medium, until saturation is reached.	Staub et al., 2000
Human serum albumin	*hsa*	Not reported	Fernandez-San et al., 2003
Interferon gamma	*Gus-IFN-γ*	Protection of human lung carcinoma cells against infection by encephalomyocarditis virus.	Leelavathi and Reddy 2003
Interferon alpha 2b	*IFNa*2b	Transgenic IFN-a2b protected baby hamster kidney cells against cytopathic viral replication in vesicular stomatitis virus cytopathic effect assay, HeLa cells from HIV-1 entry and mice from a highly metastatic tumor line. Also, it increased the expression of major histocompatibility complex class I on splenocytes and the total number of natural killer cells	Arlen et al., 2007
Human Proinsulin	*CTB-Pins* (Pins-proinsulin)	GM1 binding assay, prevention of pancreatic insulitis and preservation of insulin-producing b-cells in cholera toxin B (CTB)-Pins treated NOD mice with lower blood or urine glucose levels. Increased expression of immunosuppressive cytokines such as interleukin-4 (IL-4) and IL-10 in the pancreas of CTB-Pins treated nonobese diabetic mice.	Ruhlman et al., 2007
Diabetes – Type 1	(*hGAD*65)	Immunoreactivity to diabetic sera.	Wang et al., 2008
Human alpha1-antitrypsin	(*A1AT*)	Fully active and bind to porcine pancreatic elastase.	Nadai et al., 2008
Human cardiotrophin-1	*hCT*-1	Not reported	Farran et al., 2008
Insulin-like growth factor	*IGF*-1n *IGF*-1s	Growth response in cultured HU-3 cells	Daniell and Ruiz 2009

protein-based polymers (PBP) into the chloroplast genome has opened the possibility for the production of biopharmaceutical protein in plants via chloroplast technology (Guda et al., 2000). PBP is a protein based polymer with many medical applications such as prevention of postsurgical adhesion and scars, wound coverings, artificial pericardia, tissue reconstruction and programmed drug delivery. Human somatotropin (hST) is utilized in the treatment of hypo pituitary dwarfism in children; Turner syndrome, chronic renal failure and HIV wasting syndrome; was successfully expressed via the tobacco chloroplast genome. Chloroplast-derived hST was properly disulfide bonded and fully functional (Staub et al., 2000). One of the most widely used intravenous proteins in many human therapies; human serum albumin (HSA) makes up ~60% of the protein found in blood serum. It has been used in multigram quantities to replace blood volume in trauma and various other clinical situations. However, the current demand is hardly met due to limited availability of human blood extraction and the potential of transmitting diseases as well as blood derived products. Therefore, an alternative way of producing this protein is needed. Earlier attempts of expressing HSA by nuclear transformations have resulted into 0.2% tsp in tubers and 0.02% in green leaves, which is below the level needed for cost effective production (Farran et al., 2002). In contrast, high levels of HSA (11.1% tsp) protein was expressed in tobacco chloroplast under the control of *psb*A regulatory sequences (Fernandez-San Millan et al., 2003).

Interferons are cytokines of the immune system which mediate a range of diverse functions including antiviral, antiproliferative, antitumor and immunomodulatory activities. They are used in various clinical treatments as potent enhancers of the immune system. Today, millions of people all over the world are infected by the hepatitis virus. The average annual cost of IFN-α2b treatment of hepatitis C infection is about $26,000, and is therefore unavailable to the majority of patients in developing countries. Therefore, interferon alpha 2b (IFN-α2b) was expressed in tobacco chloroplasts at high levels (20% tsp). It was shown that in vitro biological activity of the chloroplast derived IFN-α2b was similar to commercially produced PEG-Intron™. Chloroplast-derived IFN-α2b protected cells against cytopathic viral replication in the vesicular stomatitis virus cytopathic effect (VSV CPE) assay and inhibited early-stages of the human immune deficiency virus (HIV) infection. Finally, IFN-α2b purified from chloroplast transgenic lines (cpIFN-α2b) protected mice from a highly metastatic tumor line (Arlen et al., 2007).

Another class of cytokine, human interferon gamma (INF-γ) is used therapeutically to prevent viral proliferation in addition to several roles in immunoregulatory actions regarding pathogen bacteria. Attempts were made to express INF-γ through nuclear transformation but resulted in poor expression levels. However, in the tobacco chloroplast, INF-γ was expressed independently and with the fusion protein β-glucuronidase (GUS). Chloroplast transgenic lines produced INF-γ in the range of 6% tsp and offered complete protection against human lung carcinoma cell lines and against infection with encephalomyocarditis (Leelavathi and Reddy, 2003). A recently discovered cytokine with an excellent therapeutic potential, human cardiotrophin-1 (hCT-1) was expressed in the tobacco chloroplast genome (Farran et al., 2008). Serine-protease inhibitor, human alpha1-antitripsin (A1AT), which inhibits neutrophil elastase protects the pulmonary extracellular matrix from destruction (Brantly et al., 1988). Deficiency of A1AT is one of the most prevalent and potentially lethal hereditary diseases, resulting in lung problems such as emphysema or liver disorders (Wood and Stockley, 2007). Augmentation therapy was proposed for treatment, where a few grams of human alpha1-antitripsin (A1AT) were injected intravenously every week (Heresi and Stoller, 2008). The marketed product was purified from pooled human plasma. Since the supply of purified human A1AT is limited, an alternative option such as recombinant production systems is being considered (Karnaukhova et al., 2006). More recently, fully active human alpha1-antitripsin (A1AT) protein has been expressed in tobacco chloroplasts (Nadai et al., 2008).

XI Chloroplast-Derived Industrially Valuable Biomaterials

Apart from the successful production of vaccines and therapeutic proteins, chloroplast technology has also become a good platform for the produc-

Table 6. Chloroplast-derived industrially valuable biomaterials.

Biomaterial	Gene	Promoter 5′/3′UTRs	Reference
p-hydroxybenzoic acid	*ubi*C	P*rrn*/P*psb*A/*Tpsb*A	Viitanen et al., 2004
Monellin	*monellin*	P*rrn*/P*psb*A/TpsbA	Roh et al., 2006
Tryptophan	*ASA2*	P*rrn*/*rbc*L/rpL32 *rbc*L/*acc*D-ORF184	Zhang et al., 2001
Xylanase	*xyn*A	P*psb*A/T*psb*A	Leelavathi et al., 2003
Cellulase	*cel6A*, *cel6B*	P*psb*A/T*psb*A *rbc*L5′UTR/ggagg/T*rbc*L	Yu et al., 2007
Cellulase	*cel6A*	T7G105′UTR/T*psb*A	Gray et al., 2009
Cellulase, Pectinase, Xylanase	*cel*D, *pel*B, *pel*D, *xyn*2	P*rrn*/ggagg/P*psb*A/TpsbA	Verma et al., 2010
Polyhydroxybutyrate	*phb* operon	P*psb*A/T*psb*A	Lossl et al., 2003

tion of industrially useful biomaterials (Table 6). The p-hydroxybenzoic acid (pHBA), a monomer used in liquid crystal polymers is normally produced in small quantities in all plants. Chorismate pyruvate lyase encoded by the *ubi*C gene of *E. coli* that catalyzes the direct conversion of chorismate to pyruvate and pHBA. However, in chloroplasts, chorismate is converted to pHBA by ten consecutive enzymatic reactions, due to the lack of chorismate pyruvate lyase. Hyperexpression of the enzyme and accumulation of this polymer up to 25% of dry weight was achieved in tobacco chloroplast following the stable integration of the *ubi*C gene (Viitanen et al., 2004). The pathway for the synthesis of polyhydroxy butyrate (PHB), a polyester is produced by three bacterial enzymes and can be used in biodegradable plastics and elastomers. The polycistronic phb operon encoding this biosynthetic pathway was expressed in tobacco chloroplast (Lossl et al., 2003). In another attempt, the *xyn*A gene was transformed in tobacco chloroplast resulting into accumulation of thermostable xylanase (Leelavathi et al., 2003). Recently this chloroplast technology has been extended to expression systems for different enzymes responsible for the conversion of lignocellulosic biomass to glucose during the production of ethanol by fermentation (Verma et al., 2010). The conversion of cellulosic biomass into useful products involves three different classes of enzymes including endo-1,4-glucanase, exo-cellobiohydrolase and β-glucanase (Tomme et al., 1995). Two *Thermobifida fusca* thermostable cellulases, Cel6A and Cel6B have been expressed in tobacco chloroplasts, and enzyme assays have shown that both are active in hydrolyzing crystalline cellulose (Yu et al., 2007; Gray et al., 2009).

Epilogue

Inspite of many successful achievements in chloroplast transformation, extensive application of chloroplast technology to different plant species has been hampered by the lack of plastid transformation protocols for major crops. Some of the major obstacles are inadequate tissue culture and regeneration protocols, lack of appropriate selectable markers and low levels of transgene expression in non-green plastids. Leaf tissues are usually preferred for plastid transformation, because it is rapidly produced in large amounts and allows multiple successive rounds of selection and regeneration to achieve homoplasmic plants. Unfortunately this system is difficult for cereal crops where non-green tissue is used as explants. Other major obstacles include the difficulty of expressing transgenes in non-green plastids, in which gene expression and gene regulation systems are quite distinct from those of mature green chloroplasts (Verma and Daniell, 2007). Moreover, the number and size of proplastids present in non-green tissues are smaller than in fully developed chloroplasts of green leaf tissues. Therefore, no efficient protocols are available for the production of stable transplastomic plants in most of the cereal crop species.

Additionally, minimum availability of genome sequences is also one of the major obstacles to extend this technology to many crop species. It has been shown that the intergenic spacer regions and regulatory sequences contribute about 40–45% of the total chloroplast genome. Furthermore, spacer regions are not highly conserved (Saski et al., 2005). Moreover, the chloroplast transformation vectors utilize homologous flanking regions for recombination and insertion of foreign genes. Therefore, there is an urgent need to sequence chloroplast genomes to facilitate the transformation of crop species by making species-specific chloroplast vectors. It may also be necessary to develop new selection systems for monocot plastid transformation. However, production of transplastomic lines of carrot, cotton and soybean using non-green embryogenic cells containing proplastids has enabled this technology to move forward.

Acknowledgements

The investigations reported from the Daniell laboratory in this review were supported in part by grants from USDA 3611-21000-021-02S and NIH R01 GM 63879 to Henry Daniell.

References

Arlen PA, Falconer R, Cherukumilli S, Cole A, Cole AM, Oishi KK and Daniell H (2007) Field production and functional evaluation of chloroplast-derived interferon-alpha2b. Plant Biotechnol J 5:511–525

Arlen PA, Singleton M, Adamovicz JJ, Ding Y, Davoodi-Semiromi A and Daniell H (2008) Effective plague vaccination via oral delivery of plant cells expressing F1-V antigens in chloroplasts. Infect Immun 76:3640–3650

Birch-Machin I, Newell CA, Hibberd JM and Gray JC (2004) Accumulation of rotavirus VP6 protein in chloroplasts of transplastomic tobacco is limited by protein stability. Plant Biotechnol 2:261–270

Blowers AD, Bogorad L, Shark KB and Sanford JC (1989) Studies on Chlamydomonas chloroplast transformation: foreign DNA can be stably maintained in the chromosome. Plant Cell 1:123–132

Boynton JE, Gillham NW, Harris EH, Hosler JP, Johnson AM, Jones AR, Randolph-Anderson BL, Robertson D, Klein TM and Shark KB (1988) Chloroplast transformation in Chlamydomonas with high velocity microprojectiles. Science 240:1534–1538

Brantly M, Nukiwa T and Crystal RG (1988) Molecular basis of alpha-1-antitrypsin deficiency. Am J Med 84:13–31

Cerutti H, Osman M, Grandoni P and Jagendorf AT (1992) A homolog of *Escherichia coli* RecA protein in plastids of higher plants. Proc Natl Acad Sci USA 89:8068–8072

Chakrabarti A, Ganapathi TR, Mukherjee PK and Bapat VA (2003) MSI-99, a magainin analogue, imparts enhanced disease resistance in transgenic tobacco and banana. Planta 216:587–596

Chakrabarti SK, Lutz KA, Lertwiriyawong B, Svab Z and Maliga P (2006) Expression of the cry9Aa2 B.t. gene in tobacco chloroplasts confers resistance to potato tuber moth. Transgenic Res 15:481–488

Chebolu S and Daniell H (2007) Stable expression of Gal/GalNAc lectin of *Entamoeba histolytica* in transgenic chloroplasts and immunogenicity in mice towards vaccine development for amoebiasis. Plant Biotechnol J 5:230–239

Cramer CL, Boothe JG and Oishi KK (1999) Transgenic plants for therapeutic proteins: linking upstream and downstream strategies. Curr Top Microbiol Immunol 240:95-118

Daniell H (1999) Environmentally friendly approaches to genetic engineering. In Vitro Cell Dev Biol Plant 35:361–368

Daniell H (2000) Genetically modified crops: current concerns and solutions for next generation crops. Biotechnol Bioeng Rev 17:327–352

Daniell H (2002) Molecular strategies for gene containment in transgenic crops. Nat Biotechnol 20:581–586

Daniell H (2007) Transgene containment by maternal inheritance: effective or elusive? Proc Natl Acad Sci USA 104:6879–6880

Daniell H and McFadden BA (1987) Uptake and expression of bacterial and cyanobacterial genes by isolated cucumber etioplasts. Proc Natl Acad Sci USA 84:6349–6353

Daniell H, Krishna M and McFadden BF (1991) Transient expression of β-glucuronidase in different cellular compartments following biolistic delivery of foreign DNA into wheat leaves and calli. Plant Cell Rep 9:615–619

Daniell H, Ruiz ON and Dhingra A (2005b) Chloroplast genetic engineering to improve agronomic traits. Methods Mol Biol 286:111–138

Daniell H, Streatfield SJ and Wycoff K (2001b) Medical molecular farming: production of antibodies, biopharmaceuticals and edible vaccines in plants. Trends Plant Sci 6:219–226

Daniell H, Lee SB, Panchal T and Wiebe PO (2001a) Expression of the native cholera toxin B subunit gene and assembly as functional oligomers in transgenic tobacco chloroplasts. J Mol Biol 311:1001–1009

Daniell H, Ruiz,G, Denes B, Sandberg L and Langridge W (2009) Optimization of codon composition and regulatory elements for expression of human insulin like growth factor-1 in transgenic chloroplasts and evaluation of structural identity and function. BMC Biotechnol 9:33

Daniell H, Chebolu S, Kumar S, Singleton M and Falconer R (2005a) Chloroplast-derived vaccine antigens and other therapeutic proteins. Vaccine 23:1779–1783

Daniell H, Datta R, Varma S, Gray S and Lee SB (1998) Containment of herbicide resistance through genetic engineering of the chloroplast genome. Nat Biotechnol 16:345–348

Daniell H, Vivekananda J, Nielsen BL, Ye GN, Tewari KK and Sanford JC (1990) Transient foreign gene expression in chloroplasts of cultured tobacco cells after biolistic delivery of chloroplast vectors. Proc Natl Acad Sci USA 87:88–92

Davoodi-Semiromi A, Schreiber M, Nallapali S, Verma D, Singh ND, Banks RK, Chakrabarti D and Daniell H (2010)

Chloroplast-derived vaccine antigens confer dual immunity against cholera and malaria by oral or injectable delivery. Plant Biotech J 8:223–242

De Cosa B, Moar W, Lee SB, Miller M and Daniell H (2001) Overexpression of the Bt cry2Aa2 operon in chloroplasts leads to formation of insecticidal crystals. Nat Biotechnol 19:71–74

DeGray G, Rajasekaran K, Smith F, Sanford J and Daniell H (2001) Expression of an antimicrobial peptide via the chloroplast genome to control phytopathogenic bacteria and fungi. Plant Physiol 127:852–862

Dhingra A, Portis AR, Jr. and Daniell H (2004) Enhanced translation of a chloroplast-expressed RbcS gene restores small subunit levels and photosynthesis in nuclear RbcS antisense plants. Proc Natl Acad Sci USA 101:6315–6320

Dufourmantel N, Dubald M, Matringe M, Canard H, Garcon F, Job C, Kay E, Wisniewski JP, Ferullo JM, Pelissier B, Sailland A and Tissot G (2007) Generation and characterization of soybean and marker-free tobacco plastid transformants over-expressing a bacterial 4-hydroxyphenylpyruvate dioxygenase which provides strong herbicide tolerance. Plant Biotechnol J 5:118–133

Dufourmantel N, Pelissier B, Garcon F, Peltier G, Ferullo JM and Tissot G (2004) Generation of fertile transplastomic soybean. Plant Mol Biol 55:479–489

Dufourmantel N, Tissot G, Goutorbe F, Garcon F, Muhr C, Jansens S, Pelissier B, Peltier G and Dubald M (2005) Generation and analysis of soybean plastid transformants expressing *Bacillus thuringiensis* Cry1Ab protoxin. Plant Mol Biol 58:659–668

Farran I, Rio-Manterola F, Iniguez M, Garate S, Prieto J and Mingo-Castel AM (2008) High-density seedling expression system for the production of bioactive human cardiotrophin-1, a potential therapeutic cytokine, in transgenic tobacco chloroplasts. Plant Biotechnol 6:516–527

Farran I, Sanchez-Serrano JJ, Medina JF, Prieto J and Mingo-Castel AM (2002) Targeted expression of human serum albumin to potato tubers. Transgenic Res 11:337–346

Fernandez-San Millan A, Mingo-Castel A, Miller M and Daniell H (2003) A chloroplast transgenic approach to hyper-express and purify Human Serum Albumin, a protein highly susceptible to proteolytic degradation. Plant Biotechnol J 1:71–79

Fernandez-San Millan A, Ortigosa SM, Hervas-Stubbs S, Corral-Martinez P, Segui-Simarro JM, Gaetan J, Coursaget P and Veramendi J (2008) Human papillomavirus L1 protein expressed in tobacco chloroplasts self-assembles into virus-like particles that are highly immunogenic. Plant Biotechnol J 6:427–441

Gillham NW (1994) Organelle Genes and Genomes. Oxford University Press, NY

Glenz K, Bouchon B, Stehle T, Wallich R, Simon MM and Warzecha H (2006) Production of a recombinant bacterial lipoprotein in higher plant chloroplasts. Nat Biotechnol 24:76–77

Goldschmidt-Clermont M (1991) Transgenic expression of aminoglycoside adenine transferase in the chloroplast: a selectable marker for sitedirected transformation of *chlamydomonas*. Nucleic Acids Res 19: 4083–4089

Golds T, Maliga P and Koop HU (1993) Stable plastid transformation in PEG-treated protoplasts of *Nicotiana tabacum*. Nat Biotechnol 11:95–97

Gray BN, Ahner BA and Hanson MR (2009) High-level bacterial cellulase accumulation in chloroplast-transformed tobacco mediated by downstream box fusions. Biotechnol Bioeng 102:1045–1054

Grevich JJ and Daniell H (2005) Chloroplast genetic engineering: recent advances and future prespectives. Crit Rev plant Sci 24:83–107

Guda C, Lee SB and Daniell H (2000) Stable expression of biodegradable protein based polymer in tobacco chloroplasts. Plant Cell Report 19:257–262

Hager M and Bock R (2000) Enslaved bacteria as new hope for plant biotechnologists. Appl Microbiol Biotechnol 54:302–310

Heresi GA and Stoller JK (2008) Augmentation therapy in alpha-1 antitrypsin deficiency. Expert Opin Biol Ther 8:515–526

Hou BK, Zhou YH, Wan LH, Zhang ZL, Shen GF, Chen ZH and Hu ZM (2003) Chloroplast transformation in oilseed rape. Transgenic Res 12:111–114

Hussein HS, Ruiz ON, Terry N and Daniell H (2007) Phytoremediation of mercury and organomercurials in chloroplast transgenic plants: enhanced root uptake, translocation to shoots, and volatilization. Environ Sci Technol 41:8439–8446

Iamtham S and Day A (2000) Removal of antibiotic resistance genes from transgenic tobacco plastids. Nat Biotechnol 18:1172–1176

Kamarajugadda S and Daniell H (2006) Chloroplast-derived anthrax and other vaccine antigens: their immunogenic and immunoprotective properties. Expert Rev Vaccines 5:839–849

Kanamoto H, Yamashita A, Asao H, Okumura S, Takase H, Hattori M, Yokota A and Tomizawa K (2006) Efficient and stable transformation of *Lactuca sativa* L. cv. Cisco (lettuce) plastids. Transgenic Res 15:205–217

Kang TJ, Han SC, Jang MO, Kang KH, Jang YS and Yang MS (2004) Enhanced expression of B-subunit of *Escherichia coli* heat-labile enterotoxin in tobacco by optimization of coding sequence. Appl Biochem Biotechnol 117:175–187

Kang TJ, Loc NH, Jang MO, Jang YS, Kim YS, Seo JE and Yang MS (2003) Expression of the B subunit of E. coli heat-labile enterotoxin in the chloroplasts of plants and its characterization. Transgenic Res 12:683–691

Karnaukhova E, Ophir Y and Golding B (2006) Recombinant human alpha-1 proteinase inhibitor: towards therapeutic use. Amino Acids 30:317–332

Kota M, Daniell H, Varma S, Garczynski SF, Gould F and Moar WJ (1999) Overexpression of the *Bacillus thuringiensis*

(Bt) Cry2Aa2 protein in chloroplasts confers resistance to plants against susceptible and Bt-resistant insects. Proc Natl Acad Sci USA 96:1840–1845

Koya V, Moayeri M, Leppla SH and Daniell H (2005) Plant-based vaccine: mice immunized with chloroplast-derived anthrax protective antigen survive anthrax lethal toxin challenge. Infect Immun 73:8266–8274

Kumar S, Dhingra A and Daniell H (2004a) Plastid-expressed betaine aldehyde dehydrogenase gene in carrot cultured cells, roots, and leaves confers enhanced salt tolerance. Plant Physiol 136:2843–2854

Kumar S, Dhingra A and Daniell H (2004b) Stable transformation of the cotton plastid genome and maternal inheritance of transgenes. Plant Mol Biol 56:203–216

Kunnimalaiyaan M and Nielsen BL (1997) Fine mapping of replication origins (ori A and ori B) in *Nicotiana tabacum* chloroplast DNA. Nucleic Acids Res 25:3681–3686

Kuroda H and Maliga P (2002) Overexpression of the clpP 5'-untranslated region in a chimeric context causes a mutant phenotype, suggesting competition for a clpP-specific RNA maturation factor in tobacco chloroplasts. Plant Physiol 129:1600–1606

Lee SB, Kwon HB, Kwon SJ, Park SC, Jeong MJ, Han SE, Byun MO and Daniell H (2003) Accumulation of trehalose within transgenic chloroplasts confers drought tolerance. Mol Breed 11:1–13

Lee SM, Kang K, Chung H, Yoo SH, Xu XM, Lee SB, Cheong JJ, Daniell H and Kim M (2006) Plastid transformation in the monocotyledonous cereal crop, rice (*Oryza sativa*) and transmission of transgenes to their progeny. Mol Cells 21:401–410

Leelavathi S and Reddy V (2003) Chloroplast expression of His-tagged GUS-fusions: a general strategy to overproduce and purify foreign proteins using transplastomic plants as bioreactors. Mol Breed 11:49–58

Leelavathi S, Gupta N, Maiti S, Ghosh A and Reddy V (2003) Overproduction of an alkali- and thermo-stable xylanase in tobacco chloroplasts and efficient recovery of the enzyme. Mol Breed 11:59–67

Lelivelt CL, McCabe MS, Newell CA, Desnoo CB, van Dun KM, Birch-Machin I, Gray JC, Mills KH and Nugent JM (2005) Stable plastid transformation in lettuce (*Lactuca sativa* L.). Plant Mol Biol 58:763–774

Lenzi P, Scotti N, Alagna F, Tornesello ML, Pompa A, Vitale A, De Stradis A, Monti L, Grillo S, Buonaguro FM, Maliga P and Cardi T (2008) Translational fusion of chloroplast-expressed human papillomavirus type 16 L1 capsid protein enhances antigen accumulation in transplastomic tobacco. Transgenic Res 17:1091–1102

Li Q, Lawrence CB, Xing HY, Babbitt RA, Bass WT, Maiti IB and Everett NP (2001) Enhanced disease resistance conferred by expression of an antimicrobial magainin analog in transgenic tobacco. Planta 212:635–639

Lilly JW, Havey MJ, Jackson SA and Jiang J (2001) Cytogenomic analyses reveal the structural plasticity of the chloroplast genome in higher plants. Plant Cell 13:245–254

Liu CW, Lin CC, Chen JJ and Tseng MJ (2007) Stable chloroplast transformation in cabbage (*Brassica oleracea* L. var. capitata L.) by particle bombardment. Plant Cell Rep 26:1733–1744

Liu CW, Lin CC, Yiu JC, Chen JJ and Tseng MJ (2008) Expression of a *Bacillus thuringiensis* toxin (cry1Ab) gene in cabbage (*Brassica oleracea* L. var. capitata L.) chloroplasts confers high insecticidal efficacy against *Plutella xylostella*. Theor Appl Genet 117:75–88

Lossl A, Eibl C, Harloff HJ, Jung C and Koop HU (2003) Polyester synthesis in transplastomic tobacco (*Nicotiana tabacum* L.): significant contents of polyhydroxybutyrate are associated with growth reduction. Plant Cell Rep 21:891–899

Lugo SK, Kunnimalaiyaan M, Singh NK and Nielsen BL (2004) Required sequence elements for chloroplast DNA replication activity *in vitro* and in electroporated chloroplasts. Plant Sci 166:151–161

Lutz KA, Knapp JE and Maliga P (2001) Expression of bar in the plastid genome confers herbicide resistance. Plant Physiol 125:1585–1590

Martin W, Stoebe B, Goremykin V, Hapsmann S, Hasegawa M and Kowallik KV (1998) Gene transfer to the nucleus and the evolution of chloroplasts. Nature 393:162–165

McBride KE, Svab Z, Schaaf DJ, Hogan PS, Stalker DM and Maliga P (1995) Amplification of a chimeric Bacillus gene in chloroplasts leads to an extraordinary level of an insecticidal protein in tobacco. Biotechnology (N Y) 13:362–365

McCabe MS, Klaas M, Gonzalez-Rabade N, Poage M, Badillo-Corona JA, Zhou F, Karcher D, Bock R, Gray JC and Dix PJ (2008) Plastid transformation of high-biomass tobacco variety Maryland Mammoth for production of human immunodeficiency virus type 1 (HIV-1) p24 antigen. Plant Biotechnol J 9:914–929

Miyagi T, Kapoor S, Sugita M and Sugiura M (1998) Transcript analysis of the tobacco plastid operon rps2/atpI/H/F/A reveals the existence of a non-consensus type II (NCII) promoter upstream of the atpI coding sequence. Mol Gen Genet 257:299–307

Molina A, Hervas-Stubbs S, Daniell H, Mingo-Castel AM and Veramendi J (2004) High-yield expression of a viral peptide animal vaccine in transgenic tobacco chloroplasts. Plant Biotechnol J 2:141–153

Molina A, Veramendi J and Hervas-Stubbs S (2005) Induction of neutralizing antibodies by a tobacco chloroplast-derived vaccine based on a B cell epitope from canine parvovirus. Virology 342 (2):266–275

Mor TS, Gomez-Lim MA and Palmer KE (1998) Perspective: edible vaccines – a concept coming of age. Trends Microbiol 6:449–453

Moreira D, Le GH and Philippe H (2000) The origin of red algae and the evolution of chloroplasts. Nature 405:69–72

Mullet JE (1993) Dynamic regulation of chloroplast transcription. Plant Physiol 103:309–313

Nadai M, Bally J, Vitel M, Job C, Tissot G, Botterman J and Dubald M (2008) High-level expression of active human alpha1-antitrypsin in transgenic tobacco chloroplasts. Transgenic Res 18:173–183

Nguyen TT, Nugent G, Cardi T and Dix PJ (2005) Generation of homoplasmic plastid transformants of a commercial cultivar of potato (*Solanum tuberosum* L.). Plant Sci 168:1495–1500

Nugent GD, Coyne S, Nguyen TT,Kavanagh TT and Dix PJ (2006) Nuclear and plastid transformation of *Brassica oleracea* var. botrytis (cauliflower) using PEG-mediated uptake of DNA into protoplasts. Plant Sci 170:135–142

O'Neill C, Horvath GV, Horvath E, Dix PJ and Medgyesy P (1993) Chloroplast transformation in plants: polyethylene glycol (PEG) treatment of protoplasts is an alternative to biolistic delivery systems. Plant J 3:729–738

Okumura S, Sawada M, Park YW, Hayashi T, Shimamura M, Takase H and Tomizawa K (2006) Transformation of poplar (*Populus alba*) plastids and expression of foreign proteins in tree chloroplasts. Transgenic Res 15:637–646

Palmer JD (1991) Plastid chromosomes: atructure and evolution. In: Hermann RG (ed) The Molecular Biology of Plastids. Cell Culture and Somatic Cell Genetics of Plants. Springer, Vienna, pp. 5–53

Palmer JD (1985) Comparative organization of chloroplast genomes. Annu Rev Genet 19:325–354

Papageorgiou GC and Murata N (1995) The unusually strong stabilizing effects of glycinebeteaine on the structure and function in the oxygen-evolving photosystem II complex. Photosynthetica 44:243–252

Quesada-Vargas T, Ruiz ON and Daniell H (2005) Characterization of heterologous multigene operons in transgenic chloroplasts: transcription, processing, and translation. Plant Physiol 138:1746–1762

Raubeson LA and Jansen RK (2005) Chloroplast genomes of plants. In: Henry RJ (ed) Diversity and Evolution of Plants-Genotypic and Phenotypic Variation in Higher Plants. CABI, Cambridge, MA, pp. 45–68

Robinson SP and Jones GP (1986) Accumulation of glycinebetaine in chloroplasts provides osmotic adjustment during salt stress. Aust J Plant Physiol 13:659–668

Roh KH, Shin KS, Lee YH, Seo SC, Park HG, Daniell H and Lee SB (2006) Accumulation of sweet protein monellin is regulated by the psbA 5UTR in tobacco chloroplasts. J Plant Biol 49:34–43

Rosales-Mendoza S, Alpuche-Solis AG, Soria-Guerra RE, Moreno-Fierros L, Martinez-Gonzalez L, Herrera-Diaz A and Korban SS (2009) Expression of an *Escherichia coli* antigenic fusion protein comprising the heat labile toxin B subunit and the heat stable toxin, and its assembly as a functional oligomer in transplastomic tobacco plants. Plant J 2009 Jan; 57 (1):45–54. Epub 2008 Sep 30 57:45–54

Ruf S, Hermann M, Berger IJ, Carrer H and Bock R (2001) Stable genetic transformation of tomato plastids and expression of a foreign protein in fruit. Nat Biotechnol 19:870–875

Ruhlman T, Ahangari R, Devine A, Samsam M and Daniell H (2007) Expression of cholera toxin B-proinsulin fusion protein in lettuce and tobacco chloroplasts – oral administration protects against development of insulitis in non-obese diabetic mice. Plant Biotechnol J 5: 495–510

Ruhlman T, Verma D, Samson N and Daniell H (2010) The role of heterologous chloroplast sequence elements in transgene integration and expression. Plant Physiol 152:2088–2104

Ruiz ON and Daniell H (2005) Engineering cytoplasmic male sterility via the chloroplast genome by expression of {beta}-ketothiolase. Plant Physiol 138:1232–1246

Ruiz ON, Hussein HS, Terry N and Daniell H (2003) Phytoremediation of organomercurial compounds via chloroplast genetic engineering. Plant Physiol 132:1344–1352

Sanford JC, Smith FD and Russell JA (1993) Optimizing the biolistic process for different biological applications. Methods Enzymol 217:483–509

Saski C, Lee SB, Daniell H, Wood TC, Tomkins J, Kim HG and Jansen RK (2005) Complete chloroplast genome sequence of *Gycine max* and comparative analyses with other legume genomes. Plant Mol Biol 59:309–322

Shao HB, He DM, Qian KX, Shen GF and Su ZL (2008) The expression of classical swine fever virus structural protein E2 gene in tobacco chloroplasts for applying chloroplasts as bioreactors. C R Biol 331:179–184

Sidorov VA, Kasten D, Pang SZ, Hajdukiewicz PT, Staub JM and Nehra NS (1999) Technical advance: stable chloroplast transformation in potato: use of green fluorescent protein as a plastid marker. Plant J 19:209–216

Singh ND, Ding Y and Daniell H (2009) Chloroplast-derived vaccine antigens and biopharmaceuticals: protocols for expression, purification, or oral delivery and functional evaluation. Methods Mol Biol 483:163–192

Singh ND, Li M, Lee SB, Schnell D and Daniell H (2008) *Arabidopsis* Tic40 expression in tobacco chloroplasts results in massive proliferation of the inner envelope membrane and upregulation of associated proteins. The Plant Cell 20:3405–3417

Skarjinskaia M, Svab Z and Maliga P (2003) Plastid transformation in *Lesquerella fendleri*, an oilseed Brassicacea. Transgenic Res 12:115–122

Staub JM, Garcia B, Graves J, Hajdukiewicz PT, Hunter P, Nehra N, Paradkar V, Schlittler M, Carroll JA, Spatola L, Ward D, Ye G and Russell DA (2000) High-yield production of a human therapeutic protein in tobacco chloroplasts. Nat Biotechnol 18:333–338

Svab Z and Maliga P (1993) High-frequency plastid transformation in tobacco by selection for a chimeric aadA gene. Proc Natl Acad Sci USA 90:913–917

Tomme P, Warren RA and Gilkes NR (1995) Cellulose hydrolysis by bacteria and fungi. Adv Microb Physiol 37:1–81

Tregoning JS, Clare S, Bowe F, Edwards L, Fairweather N, Qazi O, Nixon PJ, Maliga P, Dougan G and Hussell T (2005) Protection against tetanus toxin using a plant-based vaccine. Eur J Immunol 35:1320–1326

Tregoning JS, Nixon P, Kuroda H, Svab Z, Clare S, Bowe F, Fairweather N, Ytterberg J, van Wijk KJ, Dougan G and Maliga P (2003) Expression of tetanus toxin fragment C in tobacco chloroplasts. Nucleic Acids Res 31:1174–1179

Verma D and Daniell H (2007) Chloroplast vector systems for biotechnology applications. Plant Physiol 145:1129–1143

Verma D, Samson NP, Koya V and Daniell H (2008) A protocol for expression of foreign genes in chloroplasts. Nat Protoc 3:739–758

Verma D, Kanagaraj A, Jin S, Singh ND, Kolattukudy PE and Daniell H (2010) Chloroplast-derived enzyme cocktails hydrolyse lignocellulosic biomass and release fermentable sugars. Plant Biotech J 8:332–350

Viitanen PV, Devine AL, Khan MS, Deuel DL, Van Dyk DE and Daniell H (2004) Metabolic engineering of the chloroplast genome using the *Echerichia coli* ubiC gene reveals that chorismate is a readily abundant plant precursor for p-hydroxybenzoic acid biosynthesis. Plant Physiol 136:4048–4060

Wagner S, Baars L, Ytterberg AJ, Klussmeier A, Wagner CS, Nord O, Nygren PA, van Wijk KJ and de Gier JW (2007) Consequences of membrane protein overexpression in *Escherichia coli*. Mol Cell Proteomics 6:1527–1550

Wagner S, Bader ML, Drew D and de Gier JW (2006) Rationalizing membrane protein overexpression. Trends Biotechnol 24:364–371

Wakasugi T, Tsudzuki T and Sugiura M (2001) The genomics of land plant chloroplasts: gene content and alteration of genomic information by RNA editing. Photosynth Res 70:107–118

Wang X, Brandsma M, Tremblay R, Maxwell D, Jevnikar AM, Huner N and Ma S (2008) A novel expression platform for the production of diabetes-associated autoantigen human glutamic acid decarboxylase (hGAD65). BMC Biotechnol 8:87

Watson J, Koya V, Leppla SH and Daniell H (2004) Expression of Bacillus anthracis protective antigen in transgenic chloroplasts of tobacco, a non-food/feed crop. Vaccine 22:4374–4384

Wood AM and Stockley RA (2007) Alpha one antitrypsin deficiency: from gene to treatment. Respiration 74:481–492

Wurbs D, Ruf S and Bock R (2007) Contained metabolic engineering in tomatoes by expression of carotenoid biosynthesis genes from the plastid genome. Plant J 49: 276–288

Ye GN, Hajdukiewicz PT, Broyles D, Rodriguez D, Xu CW, Nehra N and Staub JM (2001) Plastid-expressed 5-enolpyruvylshikimate-3-phosphate synthase genes provide high level glyphosate tolerance in tobacco. Plant J 25:261–270

Yu LX, Gray BN, Rutzke CJ, Walker LP, Wilson DB and Hanson MR (2007) Expression of thermostable microbial cellulases in the chloroplasts of nicotine-free tobacco. J Biotechnol 131:362–369

Zhang J, Tan W, Yang XH and Zhang HX (2008) Plastid-expressed choline monooxygenase gene improves salt and drought tolerance through accumulation of glycine betaine in tobacco. Plant Cell Rep 27:1113–1124

Zhang XH, Brotherton JE, Widholm JM and Portis AR, Jr. (2001) Targeting a nuclear anthranilate synthase alpha-subunit gene to the tobacco plastid genome results in enhanced tryptophan biosynthesis. Return of a gene to its pre-endosymbiotic origin. Plant Physiol 127:131–141

Zhou YX, Lee MY, Ng JM, Chye ML, Yip WK, Zee SY and Lam E (2006) A truncated hepatitis E virus ORF2 protein expressed in tobacco plastids is immunogenic in mice. World J Gastroenterol 12:306–312

Zubkot MK, Zubkot EI, van ZK, Meyer P and Day A (2004) Stable transformation of petunia plastids. Transgenic Res 13:523–530

Chapter 19

Engineering the Sunflower Rubisco Subunits into Tobacco Chloroplasts: New Considerations

Robert E. Sharwood and Spencer M. Whitney*
Molecular Plant Physiology, The Australian National University, GPO box 475, Canberra, ACT 2601, Australia

*Author for Correspondence, e-mail: spencer.whitney@anu.edu.au

Constantin A. Rebeiz et al. (eds.), The Chloroplast: Basics and Applications, pp. 285–306.

Summary

Many challenges confront Rubisco engineering in tobacco plastids. Here we use the *aad*A selectable marker gene flanked with *loxP* sites to select for transplastomic tobacco lines that replace the plastomic Rubisco large subunit gene (*rbc*L) with the comparable gene from sunflower (*rbc*L^S) or both *rbc*L^S and a synthetic gene (^{cm}*rbc*S) coding a sunflower Rubisco small subunit. Eight *rbc*L^S transformants (called t^{Rst}LA) and three *rbc*L^S-^{cm}*rbc*S transformants were obtained. During regeneration on selective medium recombination between the repeated 221-bp *psb*A 3'-sequences flanking ^{cm}*rbc*S elicited its excision in all transformants producing lines identical to t^{Rst}LA. Two marker-free (Δ*aad*A) transplastomic lines (t^{Rst}L) were produced by transiently introducing a CRE expressing T-DNA plasmid biolistically into t^{Rst}LA. The *rbc*L^S abundance in the t^{Rst}L increased five-fold relative to the t^{Rst}LA lines. However the level of hybrid L^sS^t Rubisco (comprising eight sunflower large and eight tobacco small subunits) only increased approximately 50% which improved the growth and phenotype of t^{Rst}L. The catalytic properties of L^sS^t were determined using whole leaf gas exchange measurements and verified previous findings that the hybrid enzyme is kinetically comparable to the native tobacco and sunflower Rubisco. Our results show L^sS^t production in t^{Rst}L is impeded post-translationally, either due to problems with the translational processing of *rbc*L^S or (and) from chaperone incompatibility problems that hinder productive folding of the sunflower large subunits or their assembly with tobacco small subunits. In the context of these findings we propose engineering strategies for addressing issues related to effectively engineering Rubisco subunits in tobacco plastids. We highlight how the development of a highly transformable *aad*A-free transplastomic tobacco line that only produces the smaller *Rhodospirillum rubrum* Rubisco might simplify some of these future engineering hurdles.

Abbreviations: 3-PGA – 3-phosphoglycerate; *aad*A – gene coding for spectinomycin resistance; carboxyarabinitol-P_2 – 2-carboxyarabinitol-1,5-bisphosphate; C_c – chloroplastic CO_2 partial pressure; C_i – intercellular CO_2 partial pressure; ^{cm}*rbc*S^S – plastome codon modified gene for the Rubisco small subunit from sunflower; ^{cm}*rbc*SH_6codon modified Rubisco small subunit gene with sequence coding a C-terminal 6x-histidine tag fusion; CRE – topoisomerase from P1 bacteriophage that catalyzes site-specific recombination of DNA; EPSP synthase – 5-enolpyruvylshikimate-3-phosphate synthase; Γ* – CO_2 compensation point in the absence of mitochondrial CO_2 release not associated with photorespiration; g_i – conductance for CO_2 transfer from intercellular airspaces to chloroplast stroma; K_c^{app} – Michaelis constant for CO_2; K_o – Michaelis constant for O_2; L^sS^t – Rubisco hexadecamer comprising eight sunflower large subunits and eight tobacco small subunits; MS – Murashige–Skoog salts; MS^{spec} – MS medium containing spectinomycin; pt^{Rst}LA – transforming plasmid directing the replacement of *rbc*L with *rbc*L^S, 221-bp of *psb*A 3'-untranslated sequence and an inversely oriented *lox*P-*aad*A-*lox*P gene; pt^{Rst}SLA – transforming plasmid directing the replacement of *rbc*L with *rbc*L^S, ^{cm}*rbc*S^S and an inversely oriented *aad*A gene; $pt^{Rst}SH_6$LA – transforming plasmid directing the replacement of *rbc*L with *rbc*L^S, ^{cm}*rbc*S^SH_6 and an inversely oriented *aad*A gene; *psb*A – gene that codes for the D1 protein of photosystem II; R_d – mitochondrial CO_2 release not associated with photorespiration; ribulose-P_2 – ribulose-1,5-bisphosphate; *rrn* – ribosomal RNA; Rubisco – ribulose-P_2 carboxylase/oxygenase; *rbc*L^C – cyanobacterial Rubisco large subunit gene; *rbc*L^S – sunflower Rubisco large subunit gene; *Rbc*S – nuclear gene for tobacco Rubisco small subunit; ^{cm}*rbc*S^SH_6 – codon modified sunflower Rubisco small subunit gene with sequence coding a C-terminal 6x-histidine tag fusion; $S_{c/o}$ – Rubisco relative specificity for CO_2 as opposed to O_2; T_0 – initial generation of vegetative tissue regenerated from transformed leaves; tob^{Rst} – tobacco with Rubisco comprising sunflower L- and tobacco S-subunits; t^{Rst}LA – homoplasmic tobacco plastome lines transformed with pt^{Rst}LA; t^{Rst}L – *aad* A-free (marker-less) t^{Rst}LA line; V_c^{max} – maximal Rubisco carboxylation rate; V_o^{max} – maximal Rubisco oxygenation rate

I Introduction

The advent of efficient methods for transforming the tobacco plastome has enabled development of experimental approaches to understanding gene regulation and protein expression in higher plant plastids (Koop et al., 2008). A useful foundation has already been laid but current knowledge falls far short of that required to predict whether a particular protein can be functionally produced in sufficient quantities in plastids. This is of particular relevance to ongoing efforts to modify photosynthetic CO_2-assimilation by targeted genetic manipulation of the most prominent stromal protein, Rubisco (ribulose 1,5-bisphosphate carboxylase/oxygenase; EC 4.1.1.39), whose catalytic large (L) subunit gene (*rbc*L) is coded within the large single copy region of plastid genomes (Shinozaki et al., 1986).

Rubisco plays a frequently rate-limiting role in higher plant photosynthesis and has a pervasive role on the efficiency by which plants use their resources of light, water and nitrogen (Andrews and Whitney, 2003; Zhu et al., 2004; Ghannoum et al., 2005; Raines, 2006; Parry et al., 2007). These limitations primarily stem from (a) the slow rate at which Rubisco can carboxylate its Calvin cycle 5-carbon substrate ribulose-P_2 (catalyzing ~one to four cycles per second under physiological conditions in tobacco) into two molecules of 3-phosphoglycerate (3 PGA) and (b) the inability of Rubiscos to distinguish against ribulose-P_2 oxygenation (Cleland et al., 1998). The addition of O_2, instead of CO_2, to ribulsoe-P_2 produces both 3-PGA and 2-phosphoglycolate with recycling of the latter to 3-PGA occurring via the photorespiratory pathway that consumes energy and results in net carbon loss (Wingler et al., 2000). Using site directed and random mutagenic tools (Spreitzer, 2003; Spreitzer et al., 2005; Greene et al., 2007), and by comparing Rubisco sequences from more than 20 different Rubisco crystal structures (Andersson and Taylor, 2003) an increasingly detailed understanding into Rubiscos conserved multi-step catalytic reaction has been gathered over the last 50 years (Portis and Parry, 2007). Mechanistic reasons for its slow catalysis (V_c^{max}) and variability in specificity for CO_2 over O_2 ($S_{c/o}$) remain unclear. However a comparison of the compiled kinetics for different Rubiscos provide evidence for mechanistic constraints that hypothesize that increased catalytic potency is curtailed as it comes at the expense of specificity, and vice versa (Tcherkez et al., 2006).

Rubisco engineering in higher plant plastids remains motivated by the finding that more efficient and specific forms of Rubisco have been characterized from a number of non-green algae. That in turn indicates that improving the kinetic prowess of higher plant Rubiscos is not immutable (Read and Tabita, 1994; Whitney et al., 2001). In comparison to higher plant hexadecameric form I Rubisco (comprising eight L and eight small (S) subunits) the Rubisco from non-green algae are structurally equivalent, have CO_2 affinities and turnover rates that can rival those of the higher-plant enzyme and have twofold higher CO_2/O_2 specificities. When transplanted into tobacco chloroplasts the non-green algal Rubisco L and S subunits can be abundantly expressed however they are unable to functionally assemble indicating that the folding and assembly requirements of non-green algal Rubiscos are not met by higher plant plastids (Whitney et al., 2001). In contrast, in tobacco plastids where the tobacco *rbc*L was replaced with the *rbc*L^C gene from the cyanobacterium *Synechococcus* PCC6301, no cyanobacterial L-subunits (soluble or insoluble) were detected despite producing appreciable levels of *rbc*L^C transcript (~10% of the abundant *rbc*L mRNA content in wild-type controls [Kanevski et al., 1999]). Comparable transformations that replaced *rbc*L with either the *rbc*M gene from *Rhodospirillum rubrum* or *rbc*L^S from sunflower were able to produce functional Rubisco, the latter producing a structurally stable hybrid hexadecamer (L^sS^t) comprising eight sunflower-L and eight nucleus encoded, cytosolic synthesized, tobacco-S subunits (Kanevski et al., 1999; Whitney and Andrews, 2001b). These transformants required CO_2-supplementation to grow autotrophically from seed. In tobacco-*rubrum* transformants, high CO_2 pressures were required due to the poor kinetic properties and low amount of produced homodimeric (L_2) *R. rubrum* form II Rubisco (Whitney and Andrews, 2003). In contrast the kinetic properties of the L^sS^t hexadecamer mimicked wild-type tobacco Rubisco but high CO_2 pressures were required to compensate for the scarcity of hybrid enzyme in the juvenile vegetative tissue of these transformants (Sharwood et al., 2008). With maturity the amount of leaf L^sS^t increased to levels suitable for supporting growth of the transformants to reproductive maturity in air without additional CO_2. While the production of *R. rubrum* Rubisco in tobacco chloroplasts is limited by translational processing of the *rbc*M message (Whitney and Andrews, 2003), the reason(s) for the limitation in L^sS^t production remain unresolved (Sharwood et al., 2008).

Prompted by a desire to more fully appreciate how to regulate the expression of foreign higher-plant form I Rubiscos in tobacco chloroplasts additional transplastomic tobacco lines producing variant levels of L^sS^t were generated to examine whether expression is limited at the transcriptional or post-transcriptional stage. As the sunflower and tobacco S subunits share only 77% sequence identity, a complementary sunflower S-subunit was also co-transplanted with its cognate L-subunit to examine to what extent the plastid synthesized sunflower S could outcompete the endogenous tobacco S subunits for assembly into hexadecamers.

II Transforming the Tobacco Plastome with Sunflower Rubisco Genes

A Replacing the Tobacco rbcL with Sunflower rbcL^S

The transforming plasmid pIK83 was used originally to derive transplastomic tobacco plants that replaced the tobacco *rbc*L coding sequence with the corresponding sunflower *rbc*L^S sequence (genbank accession number AF097517) while maintaining the tobacco *rbc*L promoter, 5′ untranslated region (UTR) and the first 26 nucleotides of conserved 5′ coding sequence (Kanevski et al., 1999). Seed from the transplastomic line Nt-pIK83-1 was obtained by grafting onto wild-type tobacco and its progeny referred to as tob^{Rst} (tobacco with Rubisco comprising sunflower L- and tobacco S-subunits) (Sharwood et al., 2008) (Fig. 1a). RNA blot analyses showed tob^{Rst} produced comparable levels of both monocistronic (*rbc*L^S) and dicistronic (*rbc*L^S-*aad*A) mRNAs whose overall abundance was reduced more than six-fold relative to the *rbc*L mRNA content in wild-type tobacco (Sharwood et al., 2008). It was found that the 174-bp of the *rbc*L 3′UTR (terminator) sequence maintained downstream of *rbc*L^S in tob^{Rst} was incomplete and appeared to perturb proper 3′-maturation of the *rbc*L^S message thus reducing the stability and abundance of the transcript (Sharwood et al., 2008). With this in mind we constructed the transforming plasmid, pt^{Rst}LA, which directed the identical replacement of *rbc*L with *rbc*L^S as in pIK83 but used 221-bp of 3′-untranslated sequence from the highly expressed tobacco *psb*A gene (that codes for the D1 protein of photosystem II) (Fig. 1a). Omission of the *rbc*L 3′UTR also removed sequence homology between *rbc*L^S and the *aad*A selectable marker gene (that codes for spectinomycin and streptomycin resistance) which previously allowed recombination with non-transformed plastome copies that lead to incorporation of only the *aad*A gene (Kanevski et al., 1999; Whitney et al., 1999). To ensure no read-through of *rbc*L^S the *aad*A gene was inserted in the opposite direction to *rbc*L^S and flanked by *loxP* sites to facilitate its excision later using CRE-recombinase (see Section IV.A).

B Co-transplanting rbcL^S and a Codon-Modified Sunflower ^{cm}rbcS Gene

1 A Need to Co-engineer Cognate L- and S-Subunits

If we are ultimately to transplant better Rubiscos into higher plant chloroplasts we need to extend our genetic manipulation accomplishments beyond *rbc*L and identify a means to efficiently engineer the foreign S-subunit into tobacco chloroplasts (Spreitzer, 2003). The work of Sharwood and co-workers (2008) suggests a limitation to hybrid L^sS^t production in tobacco chloroplasts might arise from incompatibilities between the tobacco S-subunits and sunflower L subunits that constrain their capacity to assemble into the stable L^sS^t hexadecamer (discussed in Section VIII.A.2). Transplanting in the sunflower S-subunits might therefore allow preference for its assembly with its cognate sunflower L-subunits. Previous attempts to relocate a native tobacco *Rbc*S gene to its pre-endosymbiotic origin in the plastome revealed very few plastome encoded S-subunits assembled into holoenzyme. This suggested either problems with their translation in the stroma or they competed poorly with the cytosol synthesized S-subunits for assembly (Whitney and Andrews, 2001a; Zhang et al., 2002). Only by inserting an *Rbc*S copy into a

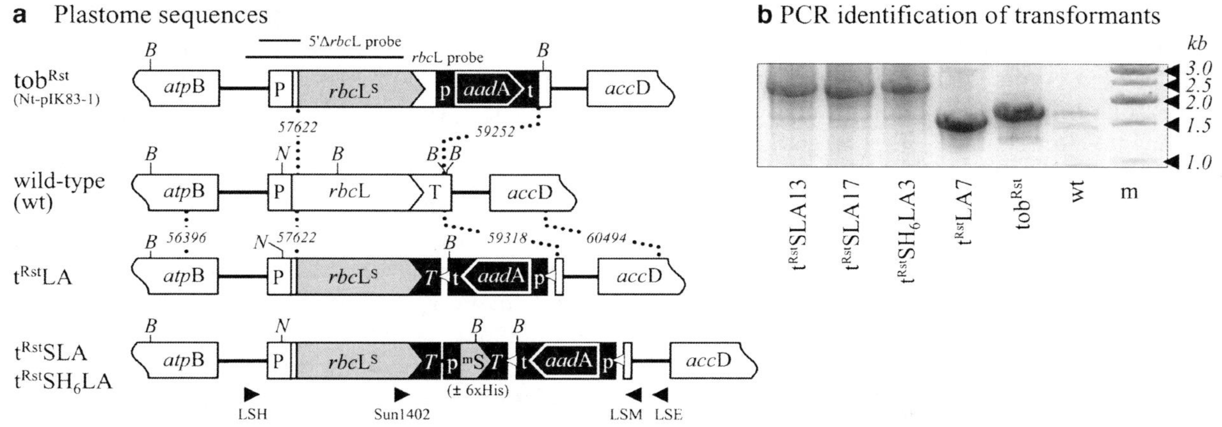

Fig. 1. Organization and analysis of transformed and wild-type tobacco plastomes. (**a**) The transforming plasmids ptRstLA, ptRstLSA and ptRstLSH$_6$A were used to transform the plastome of *Nicotinana tabacum* (L. cv Petit Havana) via the biolistic method (Svab and Maliga, 1993). Each plasmid contained plastome flanking sequence (see *dotted lines*, *numbers in italics* refer to tobacco plastome numbering; GenBank Z00044) that directed replacement of most of the tobacco *rbc*L and 3′UTR sequence (wt). In the t^{Rst}LA transformants the sequence was replaced with the *rbc*L gene from sunflower (*rbc*L^S) and 221-bp of the tobacco *psb*A 3′UTR sequence (*T*, complement of nucleotides 314–535 in Z00044) and an inversely oriented p-*aad*A-t gene cassette (p, 16S rDNA *rrn* promoter and T7 g10 5′UTR sequence; t, *rps*16 3′-untranslated region; (Svab and Maliga, 1993) containing flanking *loxP* sites (*white triangles*) cloned from plasmid p^{cm}trLA (Genbank AY827488). Transformed t^{Rst}LSA and t^{Rst}LSH$_6$A lines contain the same ^{S}*rbc*L-*T* and *aad*A sequence and an intervening codon modified sunflower *Rbc*S gene (^{cm}*rbc*S = ^{m}S, GenBank EU021288) that in t^{Rst}LSH$_6$A code a C-terminal 6x-histidine tag. The tob^{Rst} plant is the Nt-pIK83-1 line from Kanevski et al. (1999). The annealing positions of the 5′Δ*rbc*L and *rbc*L probes (Whitney and Andrews, 2003) and the primers LsH (5′- CTATGGAATTCGAACCTGAACTT -3′), Sun1402 (5′-GAGTTCCAGGCAATGGATAC-3′), LsM (5′-GACAATATACAGGATGGGTAG-3′) and LsE (Whitney and Andrews, 2001b) are shown. *acc*D, acetyl-CoA carboxylase *δ*-subunit gene; *atp*B, ATP synthase *β*-subunit gene; *B*, *Bam*HI; *N*, *Nco*I. (**b**) Plastome transformants identified by PCR using primers SunrbcL1402 (sequence unique to *rbc*L^S) and LsM (anneal downstream of transgene insertion site) using DNA extracted from tissue (see Whitney et al., 1999) after transfer to a second round of regeneration on spectinomycin (0.5 mg·ml^{-1}) containing selective medium (Svab and Maliga, 1993). The expected 2,429, 2,447, 1,650 and 1,810-bp fragments were amplified for the t^{Rst}LSA, t^{Rst}LSH$_6$A, t^{Rst}LA and tob^{Rst} transformants, respectively, and nothing for wt. m, 1 kb DNA ladder.

highly transcribed region of the plastome inverted repeat region in an anti-*Rbc*S tobacco line (where the *Rbc*S mRNA pool was reduced >80%) were sufficient levels of plastid synthesized S-subunits assembled into Rubisco (Dhingra et al., 2004). To examine whether subunit complementarity could circumvent limitations in the assembly of stromal synthesized small subunits, genes coding both sunflower subunits were co-introduced into the tobacco plastome using the transforming plasmids, pt^{Rst}SLA and pt^{Rst}SH$_6$LA (Fig. 1a). Both pt^{Rst}SLA and pt^{Rst}SH$_6$LA were derived from pt^{Rst}LA and directed the replacement of *rbc*L with *rbc*L^S and a synthetic sunflower S-subunit gene.

2 Altering the Codon Bias of a Sunflower RbcS^s Gene

To optimize the translatability of a sunflower S-subunit gene in chloroplasts a synthetic codon modified version of a nucleus encoded sunflower *Rbc*S^s mRNA (Genbank acc. No. Y00431, Waksman et al., 1987) was synthesized (^{cm}*rbc*S, Genbank acc. No. EU021288). The ^{cm}*rbc*S gene was assembled by splice overlap extension as described in Whitney and Sharwood (2008) and its codon bias matched to the abundantly expressed tobacco *rbc*L gene. Modifying the codon composition of other transgenes to match the adenine/thymine (A/T) nucleotide base preference (primarily at the third nucleotide position in the codon, Shimada and Sugiura, 1991) exhibited by tobacco plastome genes has improved translational processing of their mRNA's in plant plastids by as much as 2.5-fold (Maliga, 2003; Tregoning et al., 2003; Ye et al., 2003). In plasmid pt^{Rst}SLA the ^{cm}*rbc*S gene coded for the mature sunflower S-subunit while in pt^{Rst}SH$_6$LA additional sequence coding for a 6x-histidine (H$_6$)-tag was added in frame after the C-terminal tyrosine codon (^{cm}*rbc*SH$_6$).

The H_6-tag was incorporated to facilitate identification of the plastid synthesized sunflower SH_6 peptides using either immobilized metal affinity chromatography or immunoblot analysis with antibodies that recognize either H_6-tags or higher plant Rubisco S-subunits. As shown previously (Whitney and Andrews, 2001a) these analyses can quickly quantify the capacity of the plastid synthesized sunflower SH_6-subunits to out-compete the cytosolic synthesized tobacco S-subunits for assembly into Rubisco hexadecamers.

3 Using the T7g10 5′UTR to Regulate Sunflower S-Subunit Translation

To ensure that the ^{cm}*rbc*S^S and ^{cm}*rbc*S^SH_6 genes were highly transcribed they were equipped with the constitutive tobacco plastomic *rrn* (16S rDNA) promoter, the 63-bp T7 phage gene 10 (T7g10) 5′UTR sequence and the tobacco 221-bp *psb*A 3'UTR sequence. The T7g10 sequence has proven to be a versatile translational control sequence for enhancing foreign protein expression in tobacco plastids (Maliga, 2003). Examples include the sevenfold improvement in human somatropin accumulation when the *psb*A 5′UTR was replaced with the T7g10 sequence (Staub et al., 2000), the 200-fold improvement in the accumulation of the glyphosate resistance enzyme EPSP synthase after replacing the *rbc*L 5′UTR with the T7g10 sequence (Ye et al., 2003) and the two- and three-fold improvement in NPTII accumulation when the *rbc*L and *atp*B translational control regions were replaced with the T7g10 sequence, respectively (Kuroda and Maliga, 2001a, b).

C Transformation, Selection and Growth of the Transplastomic Lines

Five tobacco (*Nicotiana tabacum* L. cv Petit Havana (*N,N*)) leaves were transformed biolistically with each transforming plasmid and spectinomycin resistant plantlets regenerated on selective medium (0.6% (w/v) agar-solidified Murashige–Skoog (MS) salts (Murashige and Skoog, 1962) containing 3% (w/v) sucrose, 0.5 mg ml^{-1} spectinomycin and hormones (Svab and Maliga, 1993). Spectinomycin resistant ($spec^R$) plantlets were transferred onto fresh selective medium (first round of regeneration) and total DNA isolated from some of the material for PCR analysis using primers SunrbcL1402 (5'- GAGTTCCAGGCAATGGA-TAC -3', does not recognize tobacco *rbc*L) and LSM (5'- GACAATATACAGGATGGGTAG-3') (Fig. 1a and b). Eight bona fide t^{Rst}LA, two t^{Rst}SLA and one $t^{Rst}SH_6$LA plastome transformants were identified. Two t^{Rst}LA transformed lines (t^{Rst}LA5 and t^{Rst}LA7), both t^{Rst}SLA lines (t^{Rst}SLA13 and t^{Rst}SLA17) and the $t^{Rst}SH_6$LA3 line were regenerated twice more on selective medium until homoplasmic as determined by DNA blot analysis as described (Whitney et al., 1999). The plants were potted into 5 l pots of soil and grown to maturity in an air conditioned growth chamber as described (Whitney et al., 1999) with artificial illumination (~350 μmol quanta m^{-2} s^{-1}) and air supplemented with 0.5 or 1% (v/v) CO_2.

III Inadvertent Gene Excision by Recombination of Duplicated *psbA* 3′UTR Sequence

A Preferential Loss of Plastome Copies Containing ^{cm}*rbcS*^{S}

During regeneration on selective medium DNA blot analyses using the *rbc*L probe were performed to monitor the segregation status of the transformed and non-transformed (wild-type) plastome populations. After the second round of regeneration on selective medium both the t^{Rst}LA5 and t^{Rst}LA7 lines were already homoplasmic (only transformed plastomes present) as they showed the correct 3,451-bp *Bam*HI DNA fragment and no wild-type 2,126-bp and 1,238-bp DNA fragments (Fig. 2a). Curiously two large fragments were detected in *Bam*HI cut DNA from t^{Rst}SLA13, t^{Rst}SLA17 and $t^{Rst}SH_6$LA3 but no non-transformed plastome copies were evident. The upper fragments correlated with the 3,928 and 3,946 bp *Bam*HI DNA products expected for the t^{Rst}SLA and $t^{Rst}SH_6$LA transformants, respectively, while the smaller fragments were equivalent to the 3,451-bp DNA fragment from t^{Rst}LA7 and could not be explained as incomplete *Bam*HI digestion. After a further round of regeneration on selective medium the size of the amplified LSM/sun1402 PCR products from the t^{Rst}SLA and $t^{Rst}SH_6$LA transformants were identical to the 1,650-bp product amplified from t^{Rst}LA7 (Fig. 2b). Sequencing revealed that the ^{cm}*rbc*S^S and ^{cm}*rbc*S^SH_6 genes (along with their regulatory sequences) were absent and DNA blot analysis

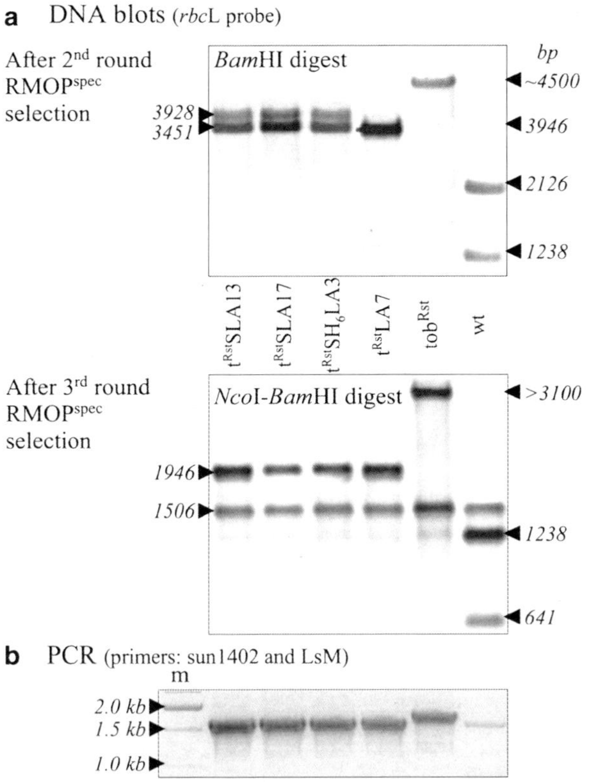

Fig. 2. DNA blots and PCR analysis of transformed and wild-type (wt) tobacco during regeneration on selective medium. (**a**) (Upper panel) Blot of the total DNA from tissue after two rounds of selection on spectinomycin (0.5 mg·ml^{-1}) containing selective medium (Svab and Maliga, 1993). The DNA was extracted, electrophoresed and blotted as described (Whitney et al., 1999). All the transformants appeared devoid of wt plastome copies and the t^{Rst}LSA and t^{Rst}LSH$_6$A lines show an unexpected smaller fragment of ~3.4 kb. (Lower panel) DNA blot after a third round of regeneration on selective medium. The *Nco*I-*Bam*HI fragments hybridising with the *rbc*L probe are identical for the t^{Rst}LSA, t^{Rst}LSH$_6$A and t^{Rst}LA7 transformants indicating complete loss of all plastome copies containing the ^{cm}*rbc*S gene in the t^{Rst}LSA and t^{Rst}LSH$_6$A lines. (**b**) Confirmation of loss of ^{cm}*rbc*S by PCR with the product amplified from t^{Rst}LSA, t^{Rst}LSH$_6$A and t^{Rst}LA7 using primers SunrbcL1402 and LsM all being of identical size (in contrast to Fig. 1b) and sequence (not shown). m, 1 kb DNA ladder (sizes shown).

of *Nco*I / *Bam*HI digested DNA indicated that no plastome copies containing the ^{cm}*rbc*S and ^{cm}*rbc*SH$_6$ genes had been maintained (Fig. 2a).

B Why Were the ^{cm}rbcS^{S} Containing Plastome Copies Lost?

Loss of the ^{cm}*rbc*S^{S} (±H$_6$) genes in the t^{Rst}SLA13, t^{Rst}SLA17 and t^{Rst}SH$_6$LA3 lines occurred during the second round of regeneration on selective medium as a result of secondary recombination events between the replicated *psb*A 3′UTR sequences. These recombination events produced plastome copies identical to those in the t^{Rst}SL lines. Curiously only the plastome copies lacking the ^{cm}*rbc*S^{S} (±H$_6$) sequence were selected for in all three lines. Notably, similar recombination events between direct sequence repeats were reported by (Iamtham and Day, 2000), but at a much lower frequency. In that study 45 transplastomic lines transplanted with *uid*A (coding for GUS), *aad*A and the *bar* (coding for resistance to the herbicide glufosinate) genes were selected with each gene oriented in the same direction and controlled by the same 174-bp *rrn* promoter and 418-bp of the *psb*A 3′-sequence (including the 221-bp *psb*A 3′UTR used in pt^{Rst}SLA and pt^{Rst}SH$_6$LA).The transformants were selected on media containing glufosinate and therefore all maintained the *bar* gene. Curiously, at maturity two of the 45 T$_0$ lines were homoplasmic for the *bar* gene (derived through natural recombination between the 174-bp direct repeats) with the other 43 lines retaining heteroplasmic transplastome populations (two populations of plastomes that either contained all three transgenes or just the *bar* gene). This contrasts with the rapid selective elimination of all plastome copies containing the ^{cm}*rbc*S^{S}(±H$_6$) genes shown here. This selective loss suggests that either the repeated 221-bp *psb*A sequences caused adverse consequences to plastome stability or there was no benefit in retaining the ^{cm}*rbc*S(±H$_6$)^{S} genes. The latter implies that either there was no selective advantage to co-producing the cognate sunflower S- and L-subunits or, comparable to plastid-synthesized tobacco S-subunits, the plastid synthesized sunflower S-subunits were poorly expressed or unable to effectively compete with the cytosolic tobacco S-subunits for assembly. Strategies for more effectively co-engineering the sunflower Rubisco subunits in tobacco plastids are discussed in Section VIII.

IV Simple Removal of *aad*A in T$_0$ t^{Rst}SLA by Transient CRE Recombinase Expression

A Bacteriophage P1 CRE-lox Site-specific Recombination

A number of methods have been developed to facilitate removal of plastid marker genes (Lutz and Maliga, 2007). Various approaches using the bacteriophage P1 CRE-*lox* site-specific recombination system have been developed that rely on flanking the target sequence to be excised (typically the *aad*A gene) with two directly oriented 34-bp *loxP* sites. Typically a *cre* gene that codes for a plastid-targeted CRE site-specific recombinase is either stably introduced into the nucleus by *Agrobacterium* or pollination (Corneille et al., 2001; Hajdukiewicz et al., 2001) or transiently by introducing a CRE-expressing T-DNA by *Agrobacterium* infiltration (Lutz et al., 2006). Stable incorporation of *cre* into the nucleus by any of these methods necessitates its segregation in the seed progeny before being able to reuse the *lox* site-specific recombination system (Corneille et al., 2001; Lutz et al., 2006).

B Removing aadA by Bombarding with Plasmid pKO27

1 Selection and Screening for ΔaadA Lines

Marker-less (Δ*aad*A) transplastomic lines can be obtained using homoplasmic T_0 tissue during its regeneration on selective medium and biolistically bombarding it with the T-DNA plasmid pKO27 (Corneille et al., 2001) that codes plastid–targeted CRE. The transformation process is summarized in Fig. 3a and has been used to make Δ*aad*A transplastomic tobacco-*rubrum* lines called ^{cm}trL that produce the dimeric (L_2) form II *R. rubrum* Rubisco instead of tobacco L_8S_8 Rubisco (Whitney and Sharwood, 2008).

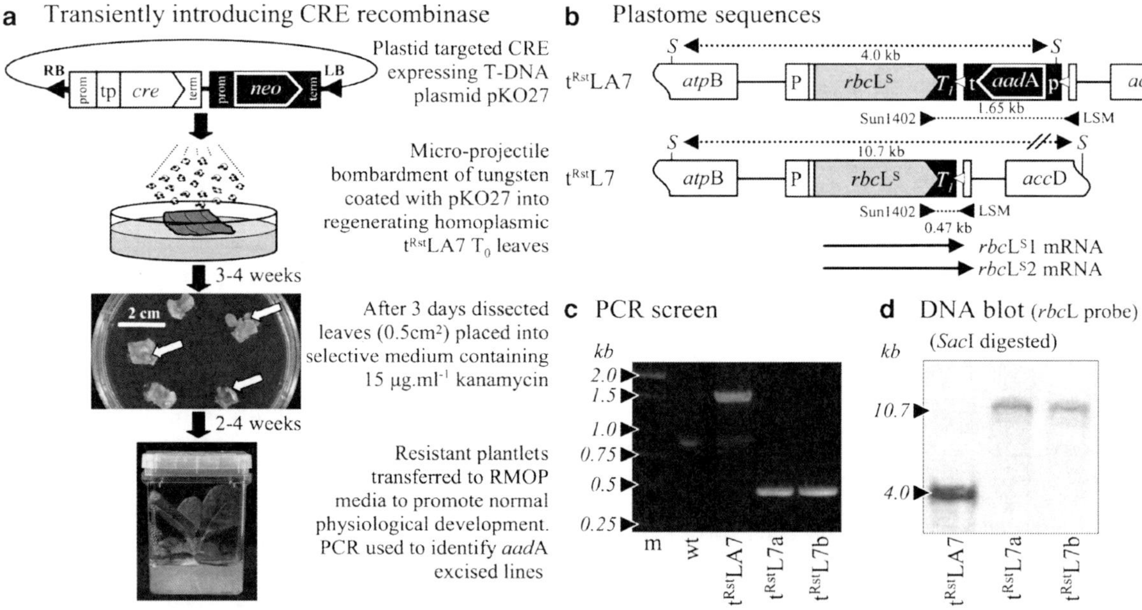

Fig. 3. Generation of the *aad*A-free tobacco-sunflower lines t^{Rst}L7a and t^{Rst}L7b. (**a**) The nucleus transforming T-DNA plasmid pKO27 (Corneille et al., 2001) harbours the CRE recombinase gene (*cre*) (equipped with the P2' *Agrobacterium* promoter, Pea *Rbc*S transit peptide (tp) and the nopaline synthase terminator) and the neomycin phosphotransferase II gene (*neo*) (equipped with the CAMV35S promoter and terminator sequences) between the left border (LB) and right border (RB) sequences. The pKO27 was coated onto tungsten and bombarded into sterile homoplasmic tsLA7 leaves (Svab and Maliga, 1993) and after 3 days the leaves were placed into selective medium containing kanamycin and grown at 25 °C and ~40 µmol photons $m^{-2} s^{-1}$. The first ten kan^R plantlets that developed were transferred to MS-medium in pots and leaf gDNA from the two fastest growing plants was screened by PCR and DNA blots. (**b**) Organisation of the t^{Rst}LA7 plastome sequence before and after (t^{Rst}L7) CRE-*lox* excision of *aad*A. The expected sizes of the amplified Sun1402 – LsM PCR products and *Sac*I (*S*) fragments that anneal with the rbcL DNA probe are shown. The *solid arrows* represent the two *rbc*L^S transcripts seen in the t^{Rst}L7 transformants. Refer to Fig. 1 for further details. (**c**) PCR confirmation of *aad*A excision from the two t^{Rst}L7 lines analysed which was (**d**) confirmed by DNA blots that showed the plants were homoplasmic. m, 1kb DNA ladder (sizes shown).

Here we applied the same process to remove *aad*A from the t^{Rst}LA7 line using leaves from a sterile 7 cm high homoplasmic T_0 t^{Rst}LA7 plant growing in MS medium (agar-solidified MS salts with 3% (w/v) sucrose) that had undergone two rounds of regeneration in selective medium. After transforming with pKO27 the bombarded leaves were regenerated on selective medium (containing 15 μg ml^{-1} kanamycin in place of spectinomycin) and after approximately 4 weeks green plantlets emerged from the bleached bombarded leaf sections (Fig. 3a). The first ten plantlets were transferred to growth in Magenta pots (Sigma) of MS medium (without kanamycin) in air where two plantlets (lines t^{Rst}L7a and t^{Rst}L7b) grew more rapidly and PCR analysis of total leaf DNA using primers SunrbcL1402 and LSM amplified a 470-bp product that was consistent with excision of the 1,180-bp *aad*A gene cassette (Fig. 3b and c). Loss of *aad*A was confirmed by DNA blot analysis of *Sac*I digested gDNA (Fig. 3d) where only a single 10.7 kb fragment hybridized with the *rbc*L DNA probe indicating both t^{Rst}L7 lines were homoplasmic. The remaining eight plantlets were discarded without analysis and both Δ*aad*A t^{Rst}L7 lines were grown to maturity in soil at high CO_2 pressures (see Section II.C) and their flowers backcrossed with wild-type tobacco pollen.

2 Screening the T_1 Progeny for aadA Loss and No Incorporation of the pKO27 T-DNA

Stable excision of *aad*A from the plastome of transgenic lines t^{Rst}L7a and t^{Rst}L7b was confirmed by the bleached phenotype of the T_1 progeny (~10^2 seed screened) on MS^{spec} medium which contrasted with the green phenotype of the *aad*A-containing T_1 t^{Rst}LA7 progeny (Fig. 4a). In comparison to the green phenotype of progeny from a transformed tobacco line that carried a nuclear neo transgene (*P. havana*^{KmR}), the T_1 progeny of t^{Rst}L7a, t^{Rst}L7b and t^{Rst}LA7 were sensitive to kanamycin indicating that the T-DNA from pKO27 had most likely not stably incorporated into the nucleus (Fig. 4b). Replicate PCR analyses using *cre* specific primers were unable to amplify a product from either t^{Rst}L7a or t^{Rst}L7b (data not shown) demonstrating that neither the *neo* nor the *cre* genes in the T-DNA from pKO27 had stably integrated into the nucleus.

V Growth Phenotypes of the tob^{Rst}, t^{Rst}LA and t^{Rst}L Lines

A Elevated CO_2 Partial Pressures Augment the Growth of the Juvenile Transformants

Consistent with earlier observations (Kanevski et al., 1999; Sharwood et al., 2008) the germination timing (~8 days after sowing) and cotyledon phenotype of the T_1 transgenic progeny grown in soil (see Section II.C for growth conditions) matched the wild-type controls. Further development of the tob^{Rst} and t^{Rst}LA lines could not be supported in ambient air (~400 ppm CO_2) due to limitations in L^{s}S^t production and source strength (Sharwood et al., 2008). In contrast the t^{Rst}L lines could grow in air, albeit extremely slowly such that after 8 weeks the plants were only ~3 cm tall with six leaves. To permit growth comparisons all lines were grown in air supplemented with 1% (v/v) CO_2.

B The Comparable Phenotype and Growth Rates of the Transgenic Lines

1 Differences in Leaf and Apical Meristem Development

When grown previously in air enriched with 0.5% (v/v) CO_2 the tob^{Rst} line produced clusters of 10–20 strappy oblanceolate leaves that lacked defined vascular structures (Sharwood et al., 2008). After ~35 days wild-type-like ovate leaves developed and a defined meristem (and sometimes more than one) emerged after which the oblanceolate leaves began to senesce. Here we found that under 1% (v/v) CO_2 the tob^{Rst} and t^{Rst}LA lines produced fewer oblanceolate leaves (typically 6–10, Fig. 4d) before developing wild-type like ovate leaves (see Fig. 4c) and only a single defined apical meristem. Under the same growth conditions the t^{Rst}L lines produced only a few oblancolate leaves (Fig. 4e).

The leaf phenotype in the transplastomic lines differed from tobacco controls (Fig. 4f). The older leaves of the tob^{Rst} and t^{Rst}LA lines in particular showed marginal curling and dimpling as observed previously (Sharwood et al., 2008, Fig. 4g). In contrast this phenotype was almost absent in t^{Rst}L leaves (Fig. 4h). Importantly, the comparable developmental phenotype of tob^{Rst} with t^{Rst}LA clearly confirms

Fig. 4. Comparative growth phenotype. The flowers of the T_0 transplastomic lines plants were pollinated with wild-type pollen and the T_2 $tob^{Rst}LA7$, T_1 $t^{Rst}L7$ and *neo*-carrying nucleus transformed tobacco (*P. havana*^{KmR}) progeny grown on sucrose-free MS medium containing (**a**) spectinomycin or (**b**) kanamycin. The $t^{Rst}L7a$ progeny (and $t^{Rst}L7b$ progeny, not shown) were sensitive to spectinomycin and kanamycin confirming they did not carry either a plastid *aad*A or a nuclear *neo*. The phenotype of (**c**) wild-type (*P. Havana*), (**d**) tob^{Rst} (T_2-generation) and (**e**) $t^{Rst}L7$ (T_1-generation) juvenile vegetative tissue grown in soil in a controlled environment cabinet with a 14 h light (25°C, 350 μmol quanta $m^{-2} s^{-1}$) and 10 h dark (20°C) photoperiod in air supplemented with 1% (v/v) CO_2 and 65% humidity. (**f–g**) Comparative phenotype of mature plants at a comparable physiological age (~65 cm in height), *Inset* photos highlight the reduced curling and dimpling in the $t^{Rst}L7$ leaves compared to tob^{Rst}.

that the phenotypic differences are a consequence of replacing *rbc*L with *rbc*L^S and not caused by other random genetic mutations.

2 Shoot Development

The onset of the exponential growth (shoot elongation) phase in the transplastomic lines was delayed relative to tobacco controls by approximately 15–20 days (Fig. 5a). This delay in shoot morphogenesis is ~10–15 days earlier than for tob^{Rst} plants grown at lower (0.5% (v/v)) CO_2 suggesting improvements in photosynthetic CO_2 assimilation under the higher CO_2 pressures. Whether this occurred due to improvements in L^sS^t content or its carbamylation level in the juvenile vegetative tissue remain to be addressed. However longer delays in shoot morphogenesis by anti-*Rbc*S lines correlate with larger reductions in Rubisco content (Rodermel, 1999). Notably, the rate of growth for tob^{Rst} (with respect to plant height) during the exponential growth phase in air containing 1% (v/v) CO_2 (6.8 ± 0.2 days, Fig. 5a) was equivalent to that with 0.5% (v/v) CO_2 (7.0 ± 0.4 days, Sharwood et al., 2008). Consistent with their improved shoot morphogenesis the exponential growth rate of the $t^{Rst}L$ lines was marginally faster than tob^{Rst} (6.4 ± 0.4 days, Fig. 5a) but still lagged approximately one-third behind the wild-type tobacco controls whose growth rate also matched that measured at 0.5% (v/v) CO_2 (4.1 ± 0.2 days).

C Leaf and Floral Development

Despite the delays in shoot morphogenesis and slower "fast-growth phase", the leaf number and height of the transplastomic lines at the onset of flowering were similar to wild-type (~approximately 65 cm tall, Figs. 4f–h and 5a). All the

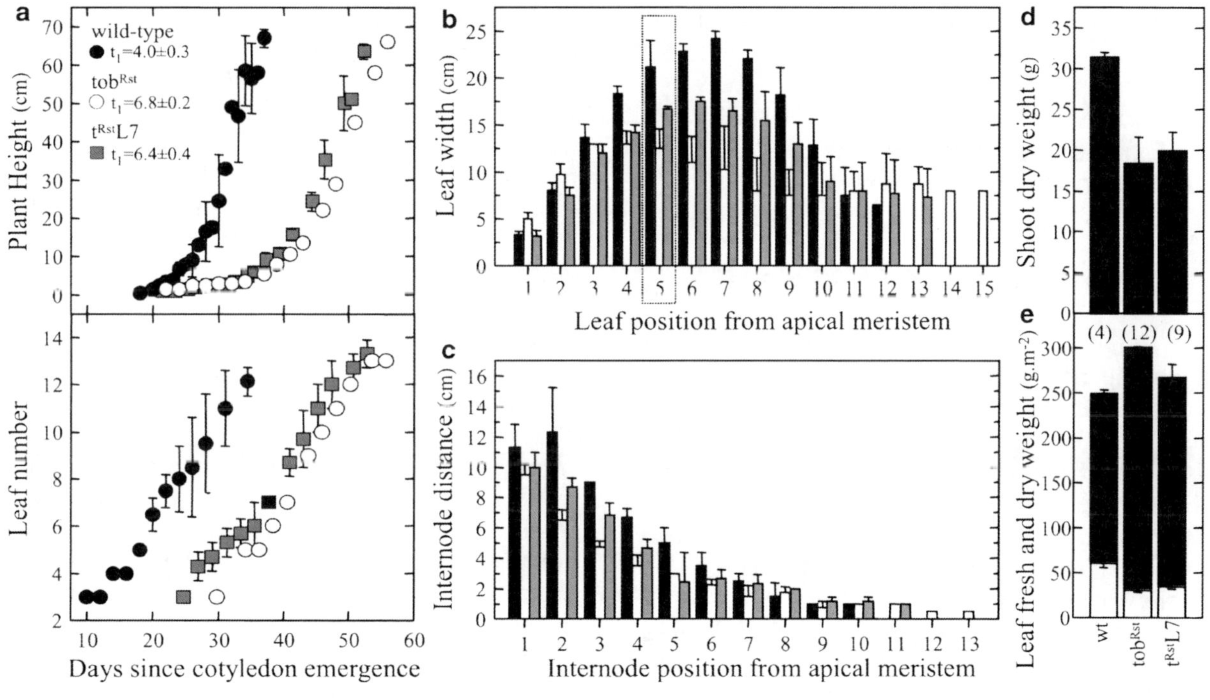

Fig. 5. Plant growth and leaf structural components. Wild-type tobacco (wt, n = 3, *black symbols*), T_1 generation $t^{Rst}L7$ lines (n = 3, *grey symbols*) and a T_3 generation tob^{Rst} line (*white symbols*) were grown as described in Fig. 4 and measured (±S.D.) were (**a**) the changes in plant height and the number of 'normal' ovate leaves and, at a similar physiological age (approximately 65 cm tall), their (**b**) leaf widths and (**c**) internodal distances. In a replicate experiment six wt and tob^{Rst} plants and three each of $t^{Rst}L7a$ and $t^{Rst}L7b$ ($t^{Rst}L7$) were grown and when ~65 cm tall (**d**) their above ground shoot dry weight determined after drying at 80°C and (**e**) the fresh weight (black + white + grey), dry weight (white + grey) and starch content (grey) of the expanding fifth leaf positioned equally in the canopy measured as previously described (Whitney and Andrews, 2001b). The fresh:dry weight ratio is shown in *brackets*. The rate constants (t_1) in panel (**a**) were determined from the change in plant height measurements during shoot elongation using the exponential rate equation $y = A1^{*}\exp^{(x|x_{t1})} + Y_0$ where A1 is the amplitude and Y_0 the offset.

transformants developed normal looking flowers at maturity that were fertile and produced viable seed. Differences in leaf width were found in the mature plants with the $t^{Rst}L$ leaves generally wider than tob^{Rst}, particularly the older leaves (numbers 5–9 from the apical meristem) where the marginal curling and dimpling phenotype was less pronounced (Fig. 5b). The leaves of wild-type controls were generally larger with the fully expanded mature leaves (number 6–8), having diameters almost twice that of tob^{Rst}. Analysis of the internodal distances in the plants showed that they were longer in the wild-type with the tob^{Rst} plants in particular having short internodal distances (Fig. 5c). Consistent with the larger leaves in wild-type plants, their overall above ground dry biomass at maturity exceeded that of both tob^{Rst} and $t^{Rst}L$ by about a third (Fig. 5d). In contrast, the fresh weight per unit area of the fifth leaf from tob^{Rst} and $t^{Rst}L$ was approximately 10–15% higher than wild-type, while their dry weight per unit area was reduced by ~50%, indicating that the transplastomic leaves were highly hydrated (Fig. 5e).

VI Expression of the Hybrid L^sS^t Rubisco in Mature Leaves

A Steady-State rbcL^S mRNA Levels

The abundance of the *rbc*L^S mRNAs in the expanding fifth leaf from tob^{Rst}, $t^{Rst}LA$ and $t^{Rst}L$ plants (boxed in Fig. 5b) of similar heights was found to vary by as much as fourfold (Fig. 6a). RNA blots were probed with the 575-bp 5′Δ*rbc*L DNA probe that encompasses the 5′UTR and first 26 nucleotides of *rbc*L that are common to both the *rbc*L and *rbc*L^S transcripts (Fig. 1a). As shown previously the tob^{Rst} line produces a monocistronic *rbc*L^S mRNA and a slightly less abundant bicistronic *rbc*L^S-*aad*A mRNA with the latter transcript produced due to inefficient transcription termination by the incomplete *rbc*L terminator sequence incorporated downstream of *rbc*L^S (Sharwood et al., 2008). In the $t^{Rst}LA$ lines, where *rbc*L^S is equipped with the full length *psb*A terminator sequence (and *aad*A inserted in the opposite orientation) only monocistronic *rbc*L^S mRNA (denoted *rbc*L^S1) was detected. In the Δ*aad*A $t^{Rst}L7$ lines both *rbc*L^S1 and a ~50% less abundant larger *rbc*L^S mRNA (denoted *rbc*L^S2) was detected (Fig. 6a).

Densitometry quantification of the *rbc*L^S mRNA's from replicate RNA blots showed that the steady-state message abundance in $t^{Rst}L7$ leaves exceeded that in the $t^{Rst}LA$ and tob^{Rst} leaves by ~fivefold and accumulated to levels equivalent to the *rbc*L mRNA pool in wild-type tobacco (Fig. 6b). On a leaf area basis little difference was found in the level of monocistronic *rbc*L^S transcript in the tob^{Rst} and $t^{Rst}LA$ leaves indicating that the 221-bp *psb*A terminator sequence did not improve *rbc*L^S transcript levels. However this changed upon excision of the *loxP-rrn* promoter-*aad*A-*rps*16 terminator cassette resulting in a three- to fourfold increase in the steady state pool of the comparable sized *rbc*L^S1 mRNA's in the $t^{Rst}L7$ leaves. This variation in *rbc*L^S1 mRNA abundance indicates that the mRNA's are not identical between $t^{Rst}LA$ and $t^{Rst}L7$. However the variation in their 3′ sequence has yet to be ascertained. Likewise, the 3′sequence in the larger *rbc*L^S2 transcript in $t^{Rst}L7$ remains undefined but presumably encompasses the 221-bp *psb*A terminator, the retained 34-bp *loxP* sequence and part of the *rbc*L terminator sequence that had previously resided upstream of *aad*A cassette prior to its excision (Fig. 3b). Typically the 3′UTR has a pervasive influence on plastid mRNA folding and maturation with inverted repeat sequences within the 3′sequence forming stem-loop (hairpin) structures that facilitate correct 3′end processing during maturation that ensure mRNA stability (Monde et al., 2000a; Bollenbach et al., 2004; Zicker et al., 2007). In particular, correct 3′-processing protects transcripts from exonucleases that rapidly degrade incorrectly folded mRNA molecules in the 3′–5′ direction (Rott et al., 1998; Monde et al., 2000b). Therefore it is possible that the *rbc*L^S1 and *rbc*L^S2 mRNA's either result from transcription terminating at different sequences or that *rbc*L^S1 is produced by exonuclease processing of the larger *rbc*L^S2 mRNA's during 3′-maturation.

B Rubisco and Protein Content

Rubisco content in soluble leaf protein extracts were measured using the [^{14}C]-carboxyarabinitol-P_2 binding assay as previously described (Butz and Sharkey, 1989; Ruuska et al., 1998). Like tobacco L_8S_8 Rubisco, [^{14}C]-carboxyarabinitol-P_2 binds

a RNA blot (5'Δ*rbc*L probe)

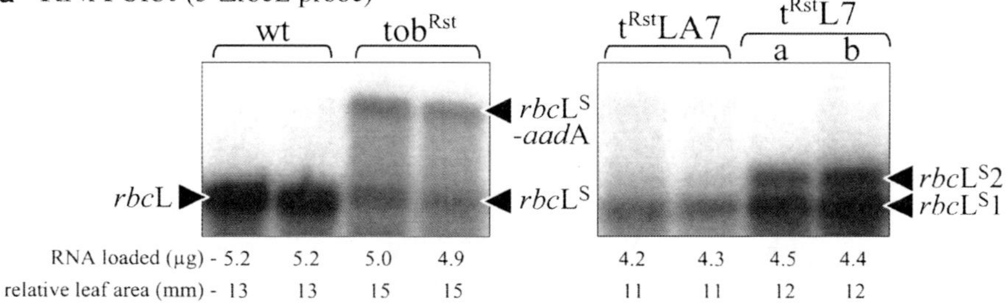

b Leaf protein and Rubisco content

Plant	Relative abundance (% of wt) protein	*rbc*L	L_8S_8 Rubisco	Rubisco translational efficiency (%wt)
wt (n=3)	100 ± 10 (7.2 ± 0.6 g.m^{-2})	100 ± 22	100 ± 2 (1.76 ± 0.03 g.m^{-2})	100%
tob^{Rst} (n=3)	65 ± 4 (4.7 ± 0.2 g.m^{-2})	15 ± 6 (*25 ± 9*)	11 ± 3 (0.2 ± 0.05 g.m^{-2})	76% (*45%*)
t^{Rst}LA7 (n=3)	59 ± 5 (4.3 ± 0.3 g.m^{-2})	17 ± 2	10 ± 3 (0.18 ± 0.05 g.m^{-2})	60%
t^{Rst}L7 (n=4)	83 ± 9 (6.0 ± 0.7 g.m^{-2})	60 ± 6 (*92 ± 9*)	17 ± 2 (0.29 ± 0.03 g.m^{-2})	27% (*18%*)

c Protein blot (spinach L-subunit antibody)

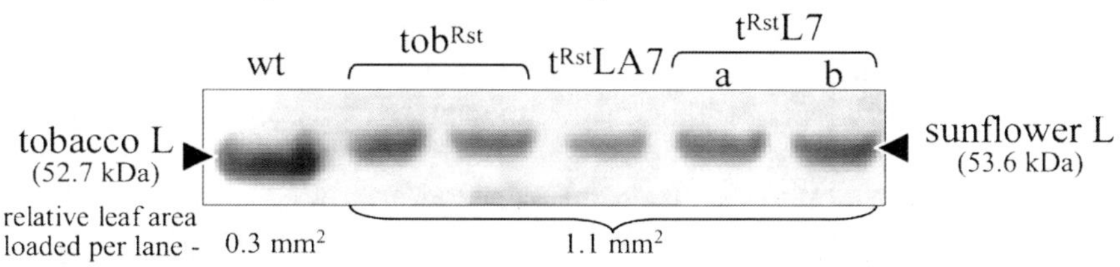

Fig. 6. Leaf Rubisco, *rbc*L mRNA and protein content. Measurements were made on the same leaves analysed in Fig. 5e. (**a**) Total leaf RNA was extracted, electrophoresed, blotted, hybridised with ^{32}P-labelled 5'Δ*rbc*L probe (see Fig. 1) and densitometry of the phosphor image performed as described previously (Whitney and Andrews, 2003). The RNA content loaded per lane and the corresponding leaf area sampled are shown. Soluble leaf protein was extracted as described (Sharwood et al., 2008) and (**b**) protein content measured using a dye-binding assay (Pierce Coomassie Plus kit) and the Rubisco content quantified by [^{14}C]-carboxyarabinitol-P$_2$ binding (Whitney and Andrews, 2001a). RNA from the same leaves was used to compare the *rbc*L and *rbc*L^S transcript abundance (see panel (**a**)) and used to calculate relative translational efficiency (RTE) by dividing relative Rubisco content by relative transcript abundance (expressed as a percentage of the wild-type (wt) average) after standardizing on a leaf area basis. For tob^{Rst} and t^{Rst}L7 (corresponding to measurements from two t^{Rst}L7a and two t^{Rst}L7b plants) the *rbc*L and RTE measurements in brackets are calculated from the sum of the abundances of the *rbc*L^S and *rbc*L^S-*aad*A transcripts and the *rbc*L^{S}1and *rbc*L^{S}2 transcripts respectively. (**c**) Immunoblot analysis (Whitney et al., 2001) of Rubisco expression in the soluble protein from the leaf areas shown. Some of these data were communicated in (Sharwood et al., 2008).

stoichiometrically to L^{s}S^t in essentially an irreversible manner enabling accurate quantification (Sharwood et al., 2008). Samples from the same leaves analyzed for *rbc*L^S abundance showed ~60–70% more L^{s}S^t in the t^{Rst}L7 leaves than the tob^{Rst} and t^{Rst}LA leaves but this was still more than fivefold less than the Rubisco content in the tobacco controls (Fig. 6b). These differences in Rubisco content were confirmed using immunoblot analyses (Fig. 6c). The increased L^{s}S^t levels in t^{Rst}L7 are consistent with its improved shoot morphogenesis and growth rate (Fig. 5a) and might explain its 'fitter' leaf phenotype (i.e. fewer oblanceolate leaves, reduced curling and dimpling of mature leaves, Fig. 4) if, as hypothesized (Sharwood et al., 2008), these phenotypic anomalies are a consequence of limitations in carbon assimilation and not due to changes in the efficiency of the photosystems which in tob^{Rst} are uncompromised at their growth illumination (Sharwood et al., 2008).

Soluble protein content in the fifth leaves was measured using a dye binding assay and standardized against bovine serum albumin (Fig. 6b). The soluble protein content on an area basis was reduced approximately 35%, 40% and 17% in the tob^{Rst}, $t^{Rst}LA$ and $t^{Rst}L7$ leaves, respectively, compared with wild-type controls. Comparable to that observed previously in transplastomic tobacco expressing 50% less tobacco Rubisco (Whitney and Andrews, 2003) or low amounts of *R. rubrum* Rubisco (~20% of wild-type (Whitney and Andrews, 2001b), the lower L^sS^t content in the tob^{Rst} and $t^{Rst}LA$ leaves did not fully account for the reduced protein abundance (respectively ~4.5 and ~4.1 g non-Rubisco protein m^{-2},) while in $t^{Rst}L7$ leaves the non-Rubisco protein content (~5.7 g m^{-2}) matched that in wild-type (~5.4 g m^{-2}) (Fig. 6b).

C Translational Efficiency and/or Folding and Assembly Limit L^sS^t Production

Despite the >threefold increase in *rbc*L^S mRNA content in $t^{Rst}L7$, only modest improvements in the level of L^sS^t were found, indicating that the enzyme's production is limited post-transcriptionally. The non-proportionate increase in L^sS^t levels in $t^{Rst}L7$ leaves is signified by its low *rbc*L^S relative translation efficiency (RTE, calculated as the amount of Rubisco relative to the mRNA abundance on a leaf area basis) which was 75% lower than the RTE of *rbc*L in wild-type tobacco and more than two-fold lower than the *rbc*L^S RTE in tob^{Rst} and $t^{Rst}LA$ (Fig. 6b). [^{35}S]-methionine labeling experiments confirmed that, as in tob^{Rst} leaves (Sharwood et al., 2008), the lower RTE in $t^{Rst}L7$ leaves was not due to increased turnover of L^sS^t as its stability matched that of tobacco Rubisco in wild-type leaves (data not shown). This suggests that problems with translational processing of the *rbc*L^S mRNA's could be limiting L^sS^t synthesis, particularly in $t^{Rst}L7$ leaves. As the variant *rbc*L^S transcripts in tob^{Rst}, $t^{Rst}LA$ and $t^{Rst}L7$share comparable 5′-regulatory and L-subunit coding sequences any difference in their translational processing would therefore result from the variations in their 3′UTR. Possibly the modifications to the 3′ sequence in the *rbc*L^S1 and *rbc*L^S2 mRNA's affect their interaction with the 5′UTR which is a necessary process for efficient recycling of mRNAs during translation (Kawaguchi and Bailey-Serres, 2002). The extent to which translation initiation or translation elongation is impeded in the different transplastomic lines remains to be examined using polysome-*rbc*L^S sedimentation analyses (Kim and Mullet, 1994).

Alternatively the paucity of L^sS^t in all the transplastomic lines might be attributable to chaperone incompatibility problems that either hinder the folding and assembly requirements of the sunflower L-subunit or/and perturb its assembly with tobacco S-subunits into hexadecamers. In contrast to the assembly of incompetent form I Rubisco subunits from non-green algae that accumulate in tobacco chloroplasts (Whitney et al., 2001), no insoluble unassembled sunflower L-subunits were detected. This may be due to their removal by chloroplast proteases that can rapidly degrade unassembled stromal proteins such as Rubisco S-subunits (Rodermel, 1999; Whitney and Andrews, 2001b, 2003).

VII Whole Leaf Gas Exchange Measurements of the L^sS^t Kinetics

Both photosynthetic gas exchange measurements with mature tob^{Rst} leaves and kinetic measurements made on extracted L^sS^t concurred that the catalytic properties of the hybrid enzyme were comparable to the catalytically similar wild-type tobacco and sunflower Rubiscos (Sharwood et al., 2008, Table 1). Analogous to that shown for tob^{Rst} (Sharwood et al., 2008), CO_2-assimilation in $t^{Rst}L7$ leaves remained carboxylase limited (rather than by ribulose-P_2 regeneration) at intercellular CO_2 partial pressures (C_i) below 2 mbar, even when the O_2 partial pressures (pO_2) were reduced from 21% to 5% (v/v). Like the Rubisco deficient anti-*Rbc*S tobacco plants (von Caemmerer et al., 1994) this made the $t^{Rst}L7$ lines suitable for measurement of the L^sS^t kinetic properties in vivo using the models outlined in von Caemmerer et al (1994) and Fig. 7.

A Measuring Gamma Star (Γ*)

Accurate measurements of Γ^* (the compensation point in the absence of mitochondrial respiration (R_d) that corresponds to the chloroplastic CO_2 partial pressure (C_c) where the rate of photorespiratory CO_2 release equals the rate of CO_2 assimilation) was determined using a whole leaf gas exchange system modified as described (Barbour et al., 2000). Matching Γ^* values of 37.2 and 36.6 μbar were measured for wild-type and $t^{Rst}L7$ leaves,

Table 1. Kinetic parameters for L^sS^t, tobacco and sunflower Rubisco measured in vivo (gas-exchange measurements) and in vitro (using leaf protein extract).

	L^sS^t		Tobacco rubisco		Sunflower rubisco
Parameter	In vivo	In vitro	In vivo	In vitro	In vitro
K_c^{app} (μM)[a]	13.2 ± 1.3 (394 ± 38 μbar)	n.d	8.6 ± 1.7 [d]	10.7 ± 0.6 [e]	n.d
K_o (μM) [a]	180 ± 43 (143 ± 34 mbar)	n.d	226 [d]	295 ± 71[e]	n.d
Γ* (μbar)	36.6	n.a	37.2 (38.6 [d])	n.a	n.a
$S_{c/o}$ [a,b]	103	84.9 ± 0.9 [c]	101 (98 [d])	82.2 ± 0.2 [c]	86.1 ± 1.5 [c]
V_c^{max} (s^{-1})	3.6 ± 0.2	3.3 ± 0.1[c]	3.53 [d]	3.2 ± 0.2 [c]	3.5 ± 0.1 [c]

[a] Values for partial pressures were converted to concentrations using the solubilities for CO_2 – 0.0334 mol (L bar)$^{-1}$ and O_2 – 0.00126 mol (L bar)$^{-1}$ (von Caemmerer et al., 1994)
[b] $S_{c/o}$ = 0.5 O/Γ* (von Caemmerer et al., 1994) where the ambient oxygen concentration is 200 mbar in Canberra
[c] From Sharwood et al., 2008
[d] From von Caemmerer et al., 1994
[e] From Whitney et al., 1999
n.d, not determined; n.a, not applicable

respectively (Fig. 7a) which were the same as that measured previously for tobacco (Γ* = 38.6 μbar (von Caemmerer et al., 1994)). Relative to the ratio of CO_2 and O_2 concentrations in solution these Γ* values equate to $S_{c/o}$ values of 101 and 103 for tobacco Rubisco and L^sS^t respectively (Table 1).

B Measuring the L^sS^t Michaelis Constants for CO_2 and O_2

Whole leaf gas exchange measurements of CO_2 assimilation rates (A) at varying CO_2 and O_2 partial pressures (pO_2)were used to measure the Michaelis constants (K_m) for O_2 (K_o), apparent K_m for CO_2 (K_c^{app}) and the substrate saturated carboxylase activity (V_c^{max}) for L^sS^t. CO_2-assimilation rates under varying CO_2 and O_2 pressures were plotted as gross assimilation ($A + R_d$) relative to C_c minus Γ* and estimates of K_c^{app} and V_c^{max} at the different pO_2 derived as described in Fig. 7 and von Caemmerer et al. (1994). As O_2 is a competitive inhibitor of carboxylation a negative linear correlation was found between the measured K_c^{app} at varying pO_2 (Fig. 7b). As well the modeled V_c^{max} rates (3.6 ± 0.2 s^{-1}, extrapolated saturated carboxylation rate divided by the Rubisco active site content measured by [^{14}C]-carboxyarabinitol-P_2 binding (Butz and Sharkey, 1989)) remained unchanged at the different pO_2 (Fig. 7c) and matched that measured previously using isolated L^sS^t enzyme (Sharwood et al., 2008) (Table 1). Likewise, the calculated K_c^{app} (Y-intercept in Fig. 7b: 392 μbar, 13.1 μM) and K_o (extrapolated from the slope of the fitted line in Fig. 7b; 143 mbar, 180 μM) were comparable to those measured in vitro using isolated tobacco Rubisco (Table 1) and that measured in vivo for tobacco using an anti-*Rbc*S line (von Caemmerer et al., 1994) (Table 1). As noted previously, variation in the in vivo measurements of K_c^{app} and K_o can reflect the dependence of these estimates on CO_2 conductance which was assumed to match that measured for wild-type tobacco (0.3 mol m^2 s^{-1} bar^{-1}) (Evans et al., 1986).

VIII Future Considerations for Transplanting Foreign Rubiscos into Tobacco Plastids

A Improving L^sS^t Synthesis

1 Limitations to Translational Processing of *rbc*L^S

Generally, expression of plastomic genes is regulated post-transcriptionally (Rochaix, 2001). Previously in tob^{Rst} plants grown in tissue culture there was a concomitant 70% decrease in both *rbc*L^S mRNA and L^sS^t content suggesting that expression of the hybrid enzyme was limited by the steady-state transcript abundance (Kanevski et al., 1999). This was not the case for tob^{Rst} (Sharwood et al., 2008) or the t^{Rst}LA and t^{Rst}L plants grown autotrophically in soil (Fig. 6b). Indeed variations in *rbc*L^S transcript abundance were not met by parallel changes in L^sS^t content. This in turn resulted in variably low translational efficiencies that indicated that the enzyme's production was limited post-transcriptionally. The reduced L^sS^t content

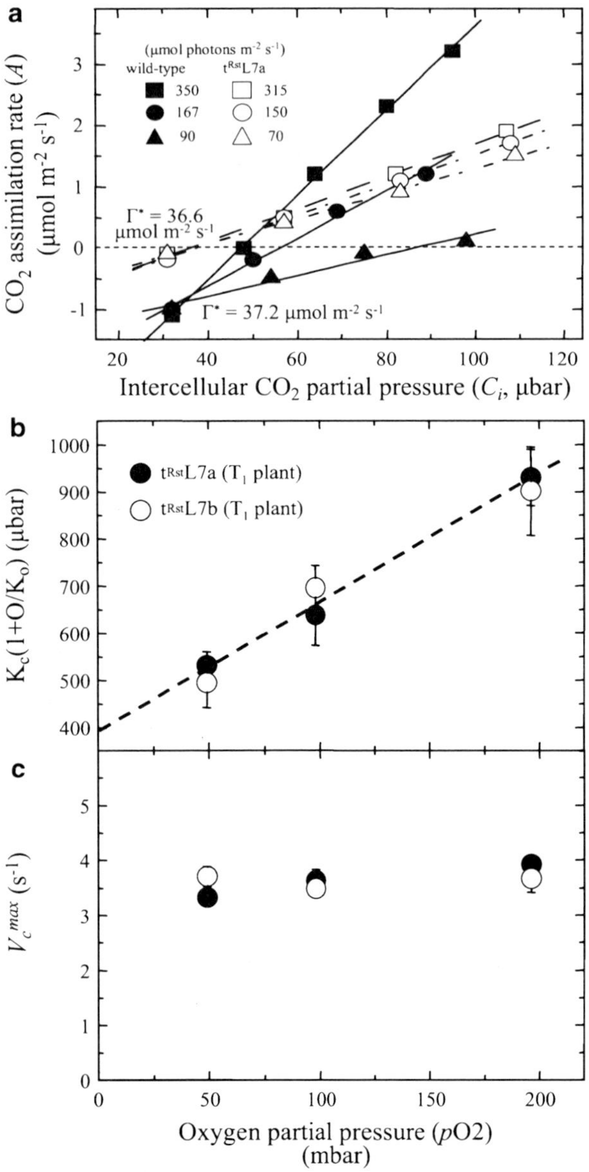

Fig. 7. Measurement Γ_*, K_c^{app}, K_o and V_c^{max} from CO_2 assimilation rates in t^Rst^L7 leaves under varying O_2 partial pressures. (**a**) Determination of Γ* (the CO_2 compensation point in the absence of nonphotorespiratory CO_2 release, R_d) from whole leaf gas exchange measurements of CO_2 assimilation rates (A) at 25°C under the illuminations shown according (Barbour et al., 2000). The response of gross CO_2 assimilation rate ($A+R_d$ where R_d = 1.29 μmol m^{-2} s^{-1}) to chloroplastic CO_2 partial pressure at the active site of Rubisco (C_c) minus Γ_* (36.6 μbar, panel (**a**)) measured for mature leaves (14 cm diameter) with L^sS^t contents of 5.1 μmol m^{-2} (*closed symbols*) and 4.5 μmol m^{-2} (*open symbols*) using a clamp on chamber LI-6400 (LI-COR), Lincon, NE) system at 950 μE m^2 s^{-1}, 25°C and the oxygen partial pressures shown. C_c was calculated using equation $C_c = C_i - A/g_i$ where C_i is the intercellular CO_2 partial pressure and g_i the CO_2 conductance between the intercellular airspaces and the chloroplasts

Fig. 7. (continued) (0.3 mol m^{-2} s^{-1} bar^{-1}; (Evans et al., 1986). The *solid lines* were modeled using the equation

$$A + Rd = \frac{(C_c - \Gamma^*)V_c^{max}}{(C_c - \Gamma^*) + (\Gamma^* + K_c(1 + O/K_o))}$$

(von Caemmerer et al., 1994). (**b**) The response of K_c^{app} (i.e. $K_c(1+O/K_o)$) and (**c**) V_c^{max} to O_2 partial pressures extrapolated from the data in panel (**b**). The slope of the linear fit (*dotted line*) in panel (**b**) (Y = 2.74X + 392, R^2 = 0.98) corresponds to K_c^{app}/K_o.

is unlikely the result of elevated degradation as pulse-chase analyses with ^{35}S-Methionine showed that once assembled the hybrid enzyme is equally stable as the tobacco L_8S_8 enzyme (Sharwood et al., 2008). Possibly translational processing of the foreign transcripts may limit L^sS^t production in the tob^Rst^, t^Rst^LA or t^Rst^L lines. This remains to be examined by comparing the sedimentation behavior of the polysomes associated with these mRNA's through sucrose gradients (Kim and Mullet, 1994). Notably in chloroplasts, translational initiation and processing requires correct folding of the mRNA to optimize interaction between the mRNA 5′UTR, the ribosomes and other *trans*-acting auxiliary factors (Monde et al., 2000b; Bollenbach et al., 2004; Zicker et al., 2007). As a result of this complexity, equipping foreign gene sequences with non-native regulatory sequences creates a substantial risk of perturbing mRNA folding thereby slowing translational processing in plastids (Koop et al., 2008). This problem continues to frustrate the plastid transformation community that typically attempts to overcome this difficulty by transplanting their gene(s) of interest under the control of different regulatory sequences (Maliga, 2002).

Also problems arise when using the tobacco *rbc*L 5′untranslated sequence because the translational control region extends into the *rbc*L coding sequence where nucleotide mutations within the first 14 codons can significantly perturb translation (Kuroda and Maliga, 2001b). Ideally this should pose no limitations to *rbc*L^S^ translation as it shares absolute sequence identity with *rbc*L for the first 56 nucleotides. However in tob^Rst^, t^Rst^LA and t^Rst^L the *rbc*L^S^ gene has two silent mutations incorporated at codons 9 and 10 that introduce a *Nhe*I site to facilitate cloning. Whether removing these two nucleotides mutations has any influence on sunflower-L translation and L^sS^t production has yet to be examined.

2 Subunit Assembly Limitations

It is possible that the post-translational limitations to L^sS^t production in tobacco leaves were compounded by folding and/or assembly incompatibilities between the sunflower L-subunits and one or more components of the plastid chaperone complexes. As chloroplasts are efficient at degrading unassembled Rubisco subunits (Rodermel, 1999; Whitney and Andrews, 2001b) it is likely that mis-folded sunflower L-subunits, or those inadequately associated with chaperones, would be rapidly degraded by stromal proteases (along with unassembled S-subunits) thus reducing L^sS^t production. One solution is to provide complementary chaperones to improve folding and assembly, as successfully demonstrated previously in tobacco chloroplasts (De Cosa et al., 2001). Such a feat may be more problematic with Rubisco given the apparent complexity of Rubisco subunit assembly in higher plant plastids and our rudimentary understanding of the process (Roy and Andrews, 2000). However a good starting point may be to target chaperones shown to have specificity for the Rubisco L-subunits such as the BSDII protein that shares homology with the DnaJ chaperone and is needed for Rubisco assembly in wheat (Brutnell et al., 1999) and tobacco (Wostrikoff and Stern, 2007). Notably the poor sequence homology between BSDII proteins from different plants (*Arabidopsis* and wheat BSDII show only ~50% sequence identity) is indicative that they may show L-subunit specificity. It may therefore be useful to examine whether L^sS^t production can be improved (a) by introducing a *bsd*II gene from sunflower into the $t^{Rst}L7$ plastome using *aad*A selection and (b) by examining if subsequently silencing production of the endogenous tobacco BSDII affects assembly of the Rubisco subunits.

The RbcX peptide whose gene, *rbc*X, is often situated juxtaposed to *rbc*L and *rbc*S in cyanobacteria genomes has been shown to chaperone the assembly of cyanobacterial Rubisco subunits in *E. coli* (Emlyn-Jones et al., 2006: Li and Tabita, 1997). RbcX forms dynamic intermediate complexes with the L subunits after their folding by the chaperonins and assists in the assembly of the L_8 cores with the small subunits (Saschenbrecker et al., 2007). Peptide binding studies showed that the *Synechococcus* PCC6301 RbcX dimer specifically interacted with the EIKFEFD motif located in the mobile C-terminal sequence of the L-subunit. Sequence comparisons showed that genes coding putative RbcX homologs are identifiable in higher plant genomes. However their function has yet to be defined. If plant and cyanobacterial RbcX share common functionality then possibly the ability of tobacco RbcX to interact with sunflower L-subunits may be compromised by the amino acid differences in the putative C-terminal binding motif (EIKFEFQ in the sunflower L compared with EIVFNFA in the tobacco L). Replacing the sunflower L-subunit motif with that from tobacco may provide a simple means to examine the potential importance of this sequence on L^sS^t production. However this may reduce catalytic efficiency due to the pervasive influence residues in this region have on the conformation and dynamics of loop 6 (Gutteridge et al., 1993).

B The Assembly and Kinetic Capacity of Other Hybrid Rubiscos

The finding that the catalytic properties of the L^sS^t enzyme closely mirror those of its source enzymes contrasts with that found for hybrid Rubiscos. These hybrid enzymes comprised cyanobacterial L-subunits and were invariably impaired kinetically with little, or no, change in their CO_2/O_2 selectivity (Spreitzer, 2003). Possibly there is greater structural flexibility between heterologous S- and L- subunits from higher plants which allow hybrid enzyme assembly with little or no perturbation in catalytic performance. This complementarily may stem from the high amino acid similarity between higher plant L-subunit sequences (generally > 90%) which may facilitate their interchangeability without significant compromise to the conformation of active site residues in the L-subunits. Clearly there are limitations to this interchangeability since tobacco L-subunits are unable to assemble with S-subunits from non-green algae (which show less than 30% sequence identity to tobacco S-subunits (Whitney et al., 2001). Conversely the assembly of cyanobacterial L- with tobacco S-subunits appear constrained in chloroplasts (Kanevski et al., 1999) although partly feasible in *E. coli* (Wang et al., 2001). Future investigations therefore appear warranted to define the range of form I Rubiscos whose L-subunits can effectively assemble with tobacco S-subunits and to examine the consequences to their kinetic phenotype. The realization that the catalytic activities of 'known' Rubiscos from other C_3 and C_4 species is unlikely to provide significant improvements to CO_2-assimilation rates in tobacco questions the merit of pursuing such an undertaking without first kinetically surveying variant higher plant Rubiscos for catalytically 'more efficient' candidates.

C Constraints on S-Subunit Engineering in Tobacco

In contrast to the unicellular green algae *Chlamydomonas reihardtii* where engineering of the S-subunit is possible due to the availability of a ΔRubisco mutant that lacks both *Rbc*S copies (Khrebtukova and Spreitzer, 1996; Spreitzer, 2003), genetic manipulation of *Rbc*S in tobacco is restricted by the multiple functional copies in the nucleus that preclude their synchronized deletion. Accordingly a viable method(s) for introducing complementary *rbc*L and *rbc*S genes into tobacco plastids has not been identified. One engineering approach might be to specifically knock down (possibly entirely) the endogenous *Rbc*S mRNA's using RNAi or anti-sense method and then introduce candidate foreign or mutated tobacco *Rbc*S sequences into the nucleus or plastome. While the feasibility of such an approach remains to be examined, the question remains as to whether the limitations of assembling plastid synthesized tobacco S-subunits into tobacco Rubisco (Whitney and Andrews, 2001a; Zhang et al., 2002; Dhingra et al., 2004) is equally applicable to the assembly of heterologous S-subunits when co-transplanted with their cognate L-subunits. We attempted to examine whether assembly incompatibilities between the sunflower L and tobacco S-subunits may afford sunflower S-subunits a selective advantage for assembly. Though unsuccessful due to the selective segregation of transformed plastomes that had excised ^{cm}*rbs*S^{S}, testing this hypothesis is now feasible with the generation of the marker-less (*aad*A-free) t^{Rst}L7 lines that allows for the ^{cm}*rbc*S($\pm H_6$)^{S} genes to be transformed either into the t^{Rst}L7 plastome using *aad*A selection or into the nucleus after equipping it with plastid targeting sequences. Analyzing the population of S-subunits that assemble with the sunflower L-subunits should establish the extent to which the plastid synthesised sunflower S-subunits out-compete the tobacco S-subunits for assembly and provides guidance as to which genome the introduced *rbc*S genes are best targeted to. As shown by (Dhingra et al., 2004) it may be of benefit to integrate duplicate copies of the ^{cm}*rbc*S^{S} ($\pm H_6$) genes into the tobacco plastome by targeted insertion into the 16SrDNA *trn*I-*trn*A region of the inverted repeat sequences where transcriptional processing is highly elevated.

D Rubisco Activase Compatibility

In higher plants Rubisco activase acts as an accessory (or helper) protein to Rubisco by removing bound sugar phosphates from its active site to maintain catalytic competency (Portis, 2003). Consequently an incompatibility between a transplanted Rubisco and the host Rubisco activase would hinder removal of inhibitory compounds from the active site perturbing Rubisco activation and / or catalysis. Early in vitro assays using purified enzymes demonstrated the inability of tobacco Rubisco activase to properly regulate some Rubiscos (e.g. those from non-Solanaceae species (Wang et al., 1992)). Since then studies have shown that residues between Arg-89 and Lys-94 strongly regulate, and specify, the interactivity of Rubisco with Rubisco activase (Larson et al., 1997; Ott et al., 2000). Curiously, within this region there is significant divergence between the tobacco L-subunit (RIERVVGEKDQY, R89 to K94 underlined) and sunflower L-subunit (GLEPVPGEDNQF). While the significance of this variation on the capacity of tobacco Rubisco activase to regulate L^sS^t has yet to be examined using in vitro assays, light transient gas exchange analyses showed that L^sS^t could be appropriately regulated by tobacco activase (Sharwood et al., 2008). It is possible however that the >fourfold higher stoichiometry of activase to L^sS^t in tob^{Rst} leaves may mask any regulatory incompatibilities.

IX Quicker Screening of the Assembly and Kinetics of Genetically Modified L_8S_8 Enzymes in Tobacco Chloroplasts

Addressing many of the Rubisco engineering challenges listed in Section VIII are complicated by the lengthy transformation and regeneration processes required for obtaining homoplasmic tobacco transformants and the propensity of unwanted recombination events between short regions of homologous sequence (Kanevski et al., 1999; Whitney et al., 1999). These limitations prompted the development of marker-free (Δ*aad*A) ^{cm}trL tobacco-*rubrum* lines where: (a) the *R. rubrum* Rubisco gene shows <25% sequence homology to form I Rubiscos eliminating unwanted recombination events (Andrews and Whitney, 2003); (b) their transformation efficiency is >threefold higher than wild-type tobacco; (c) the L_2 Rubisco produced by ^{cm}trL is antigenically distinct thereby enabling the production of transplanted form I L- and S- subunits to be assessed by immuno-detection (Whitney et al., 2001);and (d) the production and kinetics of L_8S_8 enzymes can be rapidly screened within 7–9 weeks post-transformation. The usefulness of ^{cm}trL for curtailing the process of screening the assembly and kinetics of mutated and foreign L_8S_8 Rubiscos in tobacco plastids is discussed in Whitney and Sharwood (2008).

Epilogue

Engineering Rubisco in higher plant chloroplasts still faces numerous challenges. In particular, the limited capacity of plastid synthesized S-subunits to assemble into hexadecamers. This currently frustrates Rubisco engineering endeavours, as does the assembly incompatibilities that preclude the simple transplantation of the kinetically more efficient Rubiscos from non-green algae into higher plant plastids. In the future there is a need to elucidate ways to circumvent these folding and assembly problems. Presented in Section VIII are some ideas for future directions and for identifying or circumventing factors that currently diminish, or preclude, efficient subunit translation and assembly in tobacco plastids. In the meantime, two questions remain to be resolved with regard to engineering sunflower Rubisco into tobacco chloroplasts. Firstly, is the paucity of L^sS^t due mostly to perturbations in translational processing of *rbc*L^S or primarily a consequence of chaperone-related folding and assembly problems between the sunflower L-and tobacco S-subunits? Secondly, to what extent can sunflower S-subunits whose *Rbc*S gene has been transplanted into the plastome (or nucleus) compete with tobacco S-subunits for assembly with their cognate sunflower L-subunit?

Acknowledgements

This research was supported by a Discovery grant (DP0450564) awarded to SW by the Australian Research Council.

References

Andersson I and Taylor TC (2003) Structural framework for catalysis and regulation in ribulose-1,5-bisphosphate carboxylase/oxygenase. Arch Biochem Biophys 414: 130–140

Andrews TJ and Whitney SM (2003) Manipulating ribulose bisphosphate carboxylase/oxygenase in the chloroplasts of higher plants. Arch Biochem Biophys 414: 159–169

Barbour MM, Schurr U, Henry BK, Wong SC and Farquhar GD (2000) Variation in the oxygen isotope ratio of phloem sap sucrose from castor bean. Evidence in support of the peclet effect. Plant Physiol 123: 671–680

Bollenbach TJ, Schuster G and Stern DB (2004) Cooperation of endo- and exoribonucleases in chloroplast mRNA turnover. In: Moldave K (ed) Progress in Nucleic Acid Research and Molecular Biology, Vol 78. Academic, New York, pp. 305–337

Brutnell TP, Sawers RJH, Mant A and Langdale JA (1999) Bundle sheath defective 2, a novel protein required for post-translational regulation of the *rbc*L gene of maize. Plant Cell 11: 849–864

Butz ND and Sharkey TD (1989) Activity ratios of ribulose-1,5-bisphosphate carboxylase accurately reflect carbamylation ratios. Plant Physiol 89: 735–739

Cleland WW, Andrews TJ, Gutteridge S, Hartman FC and Lorimer GH (1998) Mechanism of Rubisco – the carbamate as general base. Chem Rev 98: 549–561

Corneille S, Lutz K, Svab Z and Maliga P (2001) Efficient elimination of selectable marker genes from the plastid genome by the CRE-lox site-specific recombination system. Plant J 27: 171–178

De Cosa B, Moar W, Lee SB, Miller M and Daniell H (2001) Overexpression of the Bt cry2Aa2 operon in chloroplasts leads to formation of insecticidal crystals. Nat Biotechnol 19: 71–74

Dhingra A, Portis AR Jr and Daniell H (2004) Enhanced translation of a chloroplast-expressed RbcS gene restores small subunit levels and photosynthesis in nuclear *Rbc*S antisense plants. Proc Natl Acad Sci USA 101: 6315–6320

Emlyn-Jones D, Woodger FJ, Price GD and Whitney SM (2006) RbcX can function as a Rubisco-chaperonin, but is non essential in *Synechococcus* PCC7942. Plant Cell Physiol 47

Evans JR, Sharkey TD, Berry JA and Farquhar GD (1986) Carbon-isotope discrimination measured concurrently with gas exchange to investigate CO_2 diffusion in leaves of higher plants. Aust J Plant Physiol 13: 281–292

Ghannoum O, Evans JR, Chow WS, Andrews TJ, Conroy JP and von Caemmerer S (2005) Faster Rubisco is the key to superior nitrogen-use efficiency in NADP-malic enzyme relative to NAD-malic enzyme C_4 grasses. Plant Physiol 137: 638–650

Greene DN, Whitney SM and Matsumura I (2007) Artificially evolved *Synechococcus* PCC6301 Rubisco variants exhibit improvements in folding and catalytic efficiency. Biochem J 404: 517–524

Gutteridge S, Rhoades DF and Herrmann C (1993) Site-specific mutations in a loop region of the C-terminal domain of the large subunit of ribulose bisphosphate carboxylase/oxygenase that influence substrate partitioning. J Biol Chem 268: 7818–7824

Hajdukiewicz PTJ, Gilbertson L and Staub JM (2001) Multiple pathways for Cre/lox-mediated recombination in plastids. Plant J 27: 161–170

Iamtham S and Day A (2000) Removal of antibiotic resistance genes from transgenic tobacco plastids. Nat Biotechnol 18: 1172–1176

Kanevski I, Maliga P, Rhoades DF and Gutteridge S (1999) Plastome engineering of ribulose-1,5-bisphosphate carboxylase/oxygenase in tobacco to form a sunflower large subunit and tobacco small subunit hybrid. Plant Physiol 119: 133–141

Kawaguchi R and Bailey-Serres J (2002) Regulation of translational initiation in plants. Curr Opin Plant Biol 5: 460–465

Khrebtukova I and Spreitzer RJ (1996) Elimination of the Chlamydomonas gene family that encodes the small subunit of ribulose-1,5-bisphosphate carboxylase oxygenase. Proc Natl Acad Sci USA 93: 13689–13693

Kim J and Mullet JE (1994) Ribosome-binding sites on chloroplast *rbc*L and *psb*A mRNAs and light-induced initiation of D1 translation. Plant Mol Biol 25: 437–448

Koop H-U, Herz S and Golds TJ (2008) The genetic transformation of plastids. In: Bock R (ed) Cell and Molecular Biology of Plastids, Vol 19. Springer-Verlag, Berlin, Heidleberg, pp. 457–510

Kuroda H and Maliga P (2001a) Complementarity of the 16S rRNA penultimate stem with sequences downstream of the AUG destabilizes the plastid mRNAs. Nucleic Acids Res 29: 970–975

Kuroda H and Maliga P (2001b) Sequences downstream of the translation initiation codon are important determinants of translation efficiency in chloroplasts. Plant Physiol 125: 430–436

Larson EM, O'Brien CM, Zhu GH, Spreitzer RJ and Portis AR Jr (1997) Specificity for activase is changed by a Pro-89 to Arg substitution in the large subunit of ribulose-1,5-bisphosphate carboxylase/oxygenase. J Biol Chem 272: 17033–17037

Li LA and Tabita FR (1997) Maximum activity of recombinant Ribulose 1,5-bisphosphate carboxylase/oxygenase of Ananbaena Sp. strain Ca requires the product of the RbcX gene. J Bacteriol 179: 3793–3796

Lutz K and Maliga P (2007) Construction of marker-free transplastomic plants. Curr Opin Biotechnol 18: 107–114

Lutz K, Bosacchi MH and Maliga P (2006) Plastid marker-gene excision by transiently expressed CRE recombinase. Plant J 43: 447–456

Maliga P (2002) Engineering the plastid genome of higher plants. Curr Opin Plant Biol 5: 164–172

Maliga P (2003) Progress towards commercialization of plastid transformation technology. Trends Biotechnol 21: 20–28

Monde RA, Greene JC and Stern DB (2000a) The sequence and secondary structure of the 3'-UTR affect 3'-end maturation, RNA accumulation, and translation in tobacco chloroplasts. Plant Mol Biol 44: 529–542

Monde RA, Schuster G and Stern DB (2000b) Processing and degradation of chloroplast mRNA. Biochimie 82: 573–582

Murashige T and Skoog F (1962) A revised medium for the growth and bioassay with tobacco tissue culture. Physiol Plant 15: 473–497

Ott CM, Smith BD, Portis AR Jr and Spreitzer RJ (2000) Activase region on chloroplast ribulose-1,5-bisphosphate carboxylase/oxygenase – nonconservative substitution in the large subunit alters species specificity of protein interaction. J Biol Chem 275: 26241–26244

Parry MAJ, Madgwick PJ, Carvalho JFC and Andralojc PJ (2007) Prospects for increasing photosynthesis by overcoming the limitations of Rubisco. J Agric Sci 145: 31–43

Portis AR Jr (2003) Rubisco activase – Rubisco's catalytic chaperone. Photosynth Res 75: 11–27

Portis AR Jr and Parry MAJ (2007) Discoveries in Rubisco (Ribulose 1,5-bisphosphate carboxylase/oxygenase): a historical perspective. Photosynth Res 94: 121–143

Raines CA (2006) Transgenic approaches to manipulate the environmental responses of the C-3 carbon fixation cycle. Plant Cell Environ 29: 331–339

Read BA and Tabita FR (1994) High substrate specificity factor ribulose bisphosphate carboxylase/oxygenase from eukaryotic marine algae and properties of recombinant cyanobacterial rubisco containing "algal" residue modifications. Arch Biochem Biophys 312: 210–218

Rochaix JD (2001) Posttranscriptional control of chloroplast gene expression. From RNA to photosynthetic complex. Plant Physiol 125: 142–144

Rodermel S (1999) Subunit control of Rubisco biosynthesis – a relic of an endosymbiotic past? Photosynth Res 59: 105–123

Rott R, Liveanu V, Drager RG, Stern DB and Schuster G (1998) The sequence and structure of the 3'-untranslated regions of chloroplast transcripts are important determinants of mRNA accumulation and stability. Plant Mol Biol 36: 307–314

Roy H and Andrews TJ (2000) Rubisco: assembly and mechanism. In: Leegood RC, Sharkey TD and von Caemmerer S (eds) Photosynthesis: Physiology and Metabolism. Kluwer, Dordrecht, pp. 53–83

Ruuska S, Andrews TJ, Badger MR, Hudson GS, Laisk A, Price GD and von Caemmerer S (1998) The interplay between limiting processes in C-3 photosynthesis studied by rapid-response gas exchange using transgenic tobacco impaired in photosynthesis. Aust J Plant Physiol 25: 859–870

Saschenbrecker S, Bracher A, Rao KV, Rao BV, Hartl FU and Hayer-Hartl M (2007) Structure and function of RbcX, an assembly chaperone for hexadecameric Rubisco. Cell 129: 1189–1200

Sharwood RE, von Caemmerer S, Maliga P and Whitney SM (2008) The catalytic properties of hybrid Rubisco comprising tobacco small and sunflower large subunits mirror the kinetically equivalent source Rubiscos and can support tobacco growth. Plant Physiol 146: 83–96

Shimada H and Sugiura M (1991) Fine structural features of the chloroplast genome: comparison of the sequenced chloroplast genomes. Nucleic Acids Res 19: 983–995

Shinozaki K, Ohme M, Tanaka M, Wakasugi T, Hayashida N, Matsubayashi T, Zaita N, Chunwongse J, Obokata J, Yamaguchi-Shinozaki K, Ohto C, Torawaza.K., Meng BY, Sugita M, Deno H, Kamogashira T, Yamada K, Kusuda J, Takaiwa F, Kato A, Tohdoh N, Shimada H and Sugiura M (1986) The complete nucleotide sequence of the tobacco chloroplast genome: its gene organization and expression. EMBO J 5: 2043–2049

Spreitzer RJ (2003) Role of the small subunit in ribulose-1,5-bisphosphate carboxylase/oxygenase. Arch Biochem Biophys 414: 141–149

Spreitzer RJ, Peddi SR and Satagopan S (2005) Phylogenetic engineering at an interface between large and small subunits imparts land-plant kinetic properties to algal Rubisco. Proc Natl Acad Sci USA 102: 17225–17230

Staub JM, Garcia B, Graves J, Hajdukiewicz PTJ, Hunter P, Nehra N, Paradkar V, Schlittler M, Carroll JA, Spatola L, Ward D, Ye GN and Russell DA (2000) High-yield production of a human therapeutic protein in tobacco chloroplasts. Nat Biotechnol 18: 333–338

Svab Z and Maliga P (1993) High-frequency plastid transformation in tobacco by selection for a chimeric aadA gene. Proc Natl Acad Sci USA 90: 913–917

Tcherkez GGB, Farquhar GD and Andrews TJ (2006) Despite slow catalysis and confused substrate specificity, all ribulose bisphosphate carboxylases may be nearly perfectly optimized. Proc Natl Acad Sci USA 103: 7246–7251

Tregoning JS, Nixon P, Kuroda H, Svab Z, Clare S, Bowe F, Fairweather N, Ytterberg J, van Wijk KJ, Dougan G and Maliga P (2003) Expression of tetanus toxin Fragment C in tobacco chloroplasts. Nucleic Acids Res 31: 1174–1179

von Caemmerer S, Evans JR, Hudson GS and Andrews TJ (1994) The kinetics of ribulose-1,5-bisphosphate carboxylase/oxygenase in vivo inferred from measurements of photosynthesis in leaves of transgenic tobacco. Planta 195: 88–97

Waksman G, Lebrun M and Freyssinet G (1987) Nucleotide sequence of a gene encoding sunflower ribulose-1,5-bisphosphate carboxylase/oxygenase small subunit (*RbcS*). Nucleic Acids Res 15: 7181

Wang Z-Y, Snyder GW, Esau BD, Portis AR Jr and Ogren WL (1992) Species-dependent variation in the interaction of substrate- bound ribulose-1,5-bisphosphate carboxylase/oxygenase (Rubisco) and Rubisco activase. Plant Physiol 100: 1858–1862

Wang YL, Zhou JH, Wang YF, Bao JS and Chen HB (2001) Properties of hybrid enzymes between Synechococcus large subunits and higher plant small subunits of ribulose-1,5-bisphosphate carboxylase/oxygenase in *Escherichia coli*. Arch Biochem Biophys 396: 35–42

Whitney SM and Andrews TJ (2001a) The gene for the ribulose-1,5-bisphosphate carboxylase/oxygenase (Rubisco) small subunit relocated to the plastid genome of tobacco directs the synthesis of small subunits that assemble into Rubisco. Plant Cell 13: 193–205

Whitney SM and Andrews TJ (2001b) Plastome-encoded bacterial ribulose-1,5-bisphosphate carboxylase/oxygenase (RubisCO) supports photosynthesis and growth in tobacco. Proc Natl Acad Sci USA 98: 14738–14743

Whitney SM and Andrews TJ (2003) Photosynthesis and growth of tobacco with a substituted bacterial Rubisco mirror the properties of the introduced enzyme. Plant Physiol 133: 287–294

Whitney SM and Sharwood RE (2007) Linked rubisco subunits can assemble into functional oligomers without impeding catalytic performance. J Biol Chem 282: 3809–3818

Whitney SM and Sharwood RE (2008) Construction of a tobacco master line to improve Rubisco engineering in chloroplasts. J Exp Bot 59: 1909–1921

Whitney SM, von Caemmerer S, Hudson GS and Andrews TJ (1999) Directed mutation of the Rubisco large subunit of tobacco influences photorespiration and growth. Plant Physiol 121: 579–588

Whitney SM, Baldet P, Hudson GS and Andrews TJ (2001) Form I Rubiscos from non-green algae are expressed abundantly but not assembled in tobacco chloroplasts. Plant J 26: 535–547

Wingler A, Lea PJ, Quick WP and Leegood RC (2000) Photorespiration: metabolic pathways and their role in stress protection. Phil Trans R Soc Lond 355: 1517–1529

Wostrikoff K and Stern DB (2007) Rubisco large-subunit translation is autoregulated in response to its assembly state in tobacco chloroplasts. Proc Natl Acad Sci USA 104: 6466–6471

Ye GN, Colburn SM, Xu CW, Hajdukiewicz P and Staub JM (2003) Persistence of unselected transgenic DNA during a plastid transformation and segregation approach to herbicide resistance. Plant Physiol 133: 402–410

Zhang XH, Ewy RG, Widholm JM and Portis AR Jr (2002) Complementation of the nuclear antisense rbcS-induced photosynthesis deficiency by introducing an rbcS gene into the tobacco plastid genome. Plant Cell Physiol 43: 1302–1313

Zhu XG, Portis AR Jr and Long SP (2004) Would transformation of C-3 crop plants with foreign Rubisco increase productivity? A computational analysis extrapolating from kinetic properties to canopy photosynthesis. Plant Cell Environ 27: 155–165

Zicker A, Kadakia C and Herrin D (2007) Distinct roles for the 5' and 3' untranslated regions in the degradation and accumulation of chloroplast *tuf*A mRNA: identification of an early intermediate in the in vivo degradation pathway. Plant Mol. Biol 63: 689–702

Chapter 20

Engineering Photosynthetic Enzymes Involved in CO_2–Assimilation by Gene Shuffling

Genhai Zhu*
Pioneer Hi-Bred International Inc., 4010 Point Eden Way, Hayward, CA 94545-3721

Itzhak Kurek, and Lu Liu
Pioneer Hi-Bred International Inc., DuPoint Agriculture and Nutrition, 700A Bay Road, Redwood City, CA 94063, USA

Summary

Enhancing photosynthesis is a promising approach for increasing plant productivity. Advances in plant transformation technology make it possible to manipulate photosynthesis by overexpressing particular genes for alleviating bottleneck steps, diverting the flux of Calvin cycle intermediates and photoassimilates, or introducing new enzymes and pathways that can positively influence photosynthesis. Furthermore, directed molecular evolution makes it possible to target selected key enzymes in photosynthetic pathways for modifying their specific catalytic or protein properties and tailoring them to best function under specified growth conditions. In this chapter, advances in directed molecular evolution technology and the use of gene shuffling methodology to modify Rubisco and Rubisco activase to enhance plant photosynthesis and growth are described. By shuffling the *Chlamydomonas reinhardtii* Rubisco large subunit and utilizing competitive growth selection, several mutated Rubisco enzymes with increased carboxylase activity or CO_2/O_2 specificity were identified. The mutations identified in the modified *Chlamydomonas* Rubisco variants were then introduced into the tobacco enzyme by site-directed mutagenesis. Enzyme kinetic assays indicated that the modified tobacco Rubisco enzymes displayed increased CO_2/O_2 specificity, carboxylase activity and reduced K_m for CO_2. Similarly, gene shuffling technology was used to generate several *Arabidopsis thaliana* Rubisco activase variants exhibiting improved thermostability in order to alleviate the

* Author for Correspondence, e-mail: genhai.zhu@pioneer.com

Constantin A. Rebeiz et al. (eds.), The Chloroplast: Basics and Applications, pp. 307–322.

inhibition of plant photosynthesis by elevated temperatures. The thermostable activase variants were then expressed in an *Arabidopsis* Rubisco activase deletion line created by fast-neutron mutagenesis. The positive effects of the shuffled thermostable Rubisco activase variants on Rubisco activation state, rates of photosynthesis, and growth under moderate heat stress were demonstrated.

I Introduction

Increasing plant productivity is necessary to meet future worldwide food demands and may also be needed to alleviate the dependence on fossil fuels by providing more plant- based alternative energy sources. Photosynthesis, the process through which plants accumulate biomass by using light energy to convert inorganic carbon to carbohydrates, is a major target for improving plant productivity via conventional breeding practices and crop biotechnology (Richards, 2000; Sinclair et al., 2004). In the past century conventional breeding increased many crop yields by more than double. These achievements were accomplished mainly through: (1) genetic selection and agronomic management improvements, including increased photosynthesis per unit land area by maximizing leaf area index (LAI) via optimizing leaf orientation within a canopy; (2) extended duration of leaf photosynthesis by increasing disease resistance combined with utilizing inorganic fertilizer and improving agronomic practice; and (3) increased partitioning of crop biomass to the harvested product (harvest index – the ratio of yield biomass to the total cumulative biomass at harvest). Since the selection for increased yield by plant breeders has not resulted in a genetic increase in photosynthetic rate per leaf area (Richards, 2000), increasing genetic yield potential through such an approach is perhaps approaching its ceiling. The leaf area index is already high in many crop plants and the harvest index for many major crops, such as corn and rice has reached or exceeded 0.5 (Sinclair, 1998; Peng et al., 2000). Improving the net photosynthetic rate per leaf area to increase the inherent crop yield potential is a logical target for the next stage of agricultural research (Horton, 2000).

Realizing yield potential in an agricultural setting is often limited by environmental stress. Tollenaar and Lee (2002) believe that most of the improvement in corn yield has resulted from increased stress resistance. Yield loss from many stress conditions is directly or indirectly caused by effects on plant photosynthesis. In the field, drought is a very common stress which affects plant photosynthesis almost instantly by limiting CO_2 diffusion from the atmosphere into the chloroplasts by reducing stomatal opening. A tight positive relationship between the grain yield of wheat and maize and stomatal conductance has been observed (Evans and Fischer, 1999). Other factors such as light intensity, low or high temperatures, and high salinity can affect plant photosynthetic performance and hence, crop yield. To avoid a plateau in crop yield potential and to realize a higher percentage of yield potential under farming conditions, increasing plant net photosynthetic rate at the leaf level under normal conditions and improving the stability of photosynthesis under stress conditions are becoming two major challenges.

II Potential Targets for Improving Plant Photosynthesis

In the biochemical model of photosynthesis (Farquhar et al., 1980), Rubisco (ribulose-1,5-bisphosphote carboxylase/oxygenase) plays a central role in the determination of leaf photosynthetic rate and is often a rate-limiting enzyme under many physiological conditions. Much of the limitation can be attributed to the catalytic properties of the Rubisco enzyme. Rubisco is notorious for its low turnover number (k^c_{cat}) and catalyzes a wasteful oxygenation reaction which competes with its CO_2 fixing activity, the carboxylation reaction

Abbreviations: *aadA* – aminoglycoside 3" adenyltransferase gene; *ble* – bleomycin resistant gene; GFP – green fluorescent protein; HTP – high-throughput; *ictB* – A gene involved in HCO_3^- accumulation within the cyanobacterium *Synechococcus* sp. PCC 7942; k^c_{cat} – Rubisco k_{cat} for carboxylation; K_m – the substrate concentration at which an enzyme yields one half maximum velocity; k^o_{cat} – Rubisco k_{cat} for oxygenation; LAI – the ratio of total leaf surface area of a crop to the surface area of the land on which the crop grows; LSU – Rubisco large subunit; *otsA* – trehalose synthase gene; PGA – 3-phosphoglyceric acid; *rbc*L – Rubisco large subunit gene; *Rbc*S – Rubisco small subunit gene; RCA – Rubisco activase; RuBP – ribulose-1, 5-bisphosphate; SBPase – sedoheptulose-1,7-bisphosphatase; SSU – Rubisco small subunit; V_c – maximum velocity of Rubisco carboxylation reaction; Ω – Rubisco CO_2/O_2 specificity

(Laing et al., 1974). The oxygenation reaction product enters into the photorespiratory pathway through which 25% of the fixed carbon is released (Ogren, 1984). The ratio of Rubisco's carboxylation catalytic efficiency (k^c_{cat} of carboxylation over K_m for CO_2) to its oxygenation catalytic efficiency (k^o_{cat} of oxygenation over K_m for O_2) is defined as Rubisco CO_2/O_2 specificity (Ω). As predicted by the model, improvement in k^c_{cat} without altering its Ω value and K_m for CO_2 or vice versa will benefit photosynthetic CO_2 fixation under ambient growth condition. Significant work to genetically modify Rubisco proteins has been performed in the past decades with the aim of improving Rubisco k^c_{cat} and Ω (reviewed by Spreitzer, 1993, 1999; Hartman and Harpel, 1994; Tabita, 1999; Spretzer and Salvucci, 2002). Due to constraints in expressing a functional higher plant Rubisco in microbial hosts (Cloney et al., 1993; Gutteridge and Gatenby, 1995), Rubisco engineering has mainly focused on enzymes from only a few photosynthetic microorganisms, such as *Rhodospirillum rubrum*, cyanobacteria or the eukaryotic green algae, *Chlamydomonas reinhardtii*. The kinetic information generated from analyzing various genetically modified Rubisco mutants assisted in establishing the catalytic mechanisms and helped to identify some structural regions that may determine a specific catalytic parameter or structural stability, as well as Rubisco-activase specificity (Larson et al., 1997). It is hoped that the accumulation of the knowledge on Rubisco structure-function relationships will finally enable the engineering of a better plant Rubisco for improving crop photosynthesis.

Investigation of the natural variation in Rubisco catalytic properties from different species revealed that the Rubisco from red alga exhibited a surprisingly high Ω value, approximately two to three times that of crop plant Rubiscos (Read and Tabita, 1994; Uemura et al., 1997; Whitney et al., 2001). A general inverse relationship between Ω and K^c_{cat} among Rubiscos existing in nature has also been observed, although the data are considerably scattered along the trend line (Bainbridge et al., 1995; Zhu et al., 2004; Tcherkez et al., 2006). Recently, Tcherkez et al. (2006) hypothesized that a conflict may exist between the structural requirements for a higher catalytic turnover rate and increased discrimination between CO_2 and O_2. As such, a compromise has to be made between Ω and k^c_{cat}. Through long periods of natural selection, such a compromise between Ω and k^c_{cat} for a particular Rubisco may be nearly perfectly optimized for adaptation to its ecosystem, especially to the gaseous and thermal environments where the organism lives (Tcherkez et al., 2006). Tcherkez et al. (2006) further suggested that the potential for improving Rubisco catalytic efficiency may only be modest (within the range of the scatter). If this hypothesis is true, possible improvements from engineering Rubisco alone in order to enhance crop photosynthesis at leaf level might not be dramatic. A 15–20% increase in photosynthesis on a leaf area basis, however, could still have significant impacts on plant growth and yield.

Besides directly targeting Rubisco, there are some alternative approaches to increase plant photosynthesis. One possibility is to increase the Rubisco activation state under certain conditions. Rubisco must be "activated" in order to fix CO_2 (Lorimer and Miziorko, 1980). The Rubisco activation state is the ratio of catalytically competent sites to total Rubisco sites. The net photosynthetic rate is proportional to Rubisco activation state, but not necessarily to total Rubisco sites (Perchorowicz et al., 1981; Crafts-Brandner and Salvucci, 2000a). Rubisco activation in vivo is controlled by Rubisco activase (Portis, 1992), which is a thermolabile protein (Feller et al., 1998). Inhibition of plant photosynthesis by moderately elevated temperatures appears primarily due to temperature damage to Rubisco activase and perhaps also the specific activity of activase via influencing ATP/ADP ratio, which results in the loss of Rubisco activation state (Crafts-Brandner and Salvucci, 2000a). Rubisco from crop plants is considerably more thermostable and its catalytic activity increases with increases in temperature beyond 40–45°C. At these temperatures, Rubisco activase activity for most crop plants is significantly reduced or diminished (Feller et al., 1998; Crafts-Brandner and Salvucci, 2000a). Engineering a thermostable Rubisco activase that will stabilize or even increase plant photosynthesis at moderately elevated temperatures (Crafts-Brandner and Salvucci, 2000a, b) will be discussed separately in the section V.

Recent advances in plant transformation technology make it possible to manipulate photosynthesis by overexpressing particular genes or introducing new enzymes or pathways that can positively influence photosynthesis (reviewed by Parry et al., 2003; Raines, 2006). It has been reported that overexpression of the Calvin cycle enzymes, fructose-1,6-bisphosphtase or sedoheptulose-1,7-bisphosphatase (SBPase) in tobacco plants, not only increased RuBP concentration, but also Rubisco activation state (Miyagawa et al., 2001; Tamoi et al., 2006). A 1.2-fold higher activation state than that of the untransformed wild-type resulted in both photosynthetic rate per leaf area basis and growth of the transgenic plants being significantly increased (Miyagawa et al., 2001). Although it is not clear whether the increased SBPase activity in the chloroplast enhances Rubisco activation state due to elevated RuBP concentrations, the research demonstrates an alternative approach to manipulate plant photosynthesis by other key Calvin cycle enzymes besides Rubisco. Other research showing positive effects on plant photosynthesis by introducing a single enzyme include: (1) the overexpression of sucrose-phosphate synthase, which influences partitioning of photoassimilates and has resulted in the extended duration of older leaf photosynthesis and increased the biomass of transgenic tobacco and tomato plants (Baxter et al., 2003; Lunn et al., 2003); and (2) the overexpression of a *Escherichia coli* gene, *otsA*, for trehalose synthesis in tobacco that enhanced Rubisco activity and photosynthesis as well as biomass (Pellny et al., 2004). In attempts to increase the CO_2 concentration at the Rubisco site, overexpression of a C_4 cycle enzyme, phosphoenolpyruvare carboxylase, in rice, a C3 plant, as well as a cyanobacterial gene, *ictB*, involved in HCO_3^- accumulation in *Arabidopsis* and tobacco showed positive effects on photosynthesis in the transformed plants (Ku et al., 1999; Lieman-Hurwitz et al., 2003). A successful example for introducing an *E. coli* glycolate catabolic pathway into *Arabidopsis* chloroplasts aimed to alleviate photorespiratory losses has been recently published by Kebeish et al. (2007). In their approach, three *E. coli* enzymes, i.e. glucolate dehydrogenase, glyoxylate carboligase and tartronic semialdehyde reductase, expressed with chloroplast targeting peptides in transgenic *Arabidopsis*, convert glycolate directly into glycerate within the chloroplast. This short-circuited photorespiratory pathway releases CO_2 around Rubisco site which facilitates CO_2 refixation without extra energy input, and reduces NH_3 release that saves energy for NH_3 refixation, resulting in better plant growth and biomass accumulation.

III Directed Molecular Evolution Provides a Useful Tool to Engineer Selected Enzymes

An alternative approach to overexpression of naturally existing genes to manipulate photosynthetic pathways is by modifying selected endogenous or exogenous photosynthetic enzymes for best function under preferred growth conditions. Directed evolution, a powerful method to evolve proteins by generating libraries of mutants (variants) and selecting/screening for desirable properties not found in nature enables this approach. It mimics the natural evolution process in which protein variants are generated and tested for their improved properties in vitro or in vivo every cycle/generation (Fig. 1).

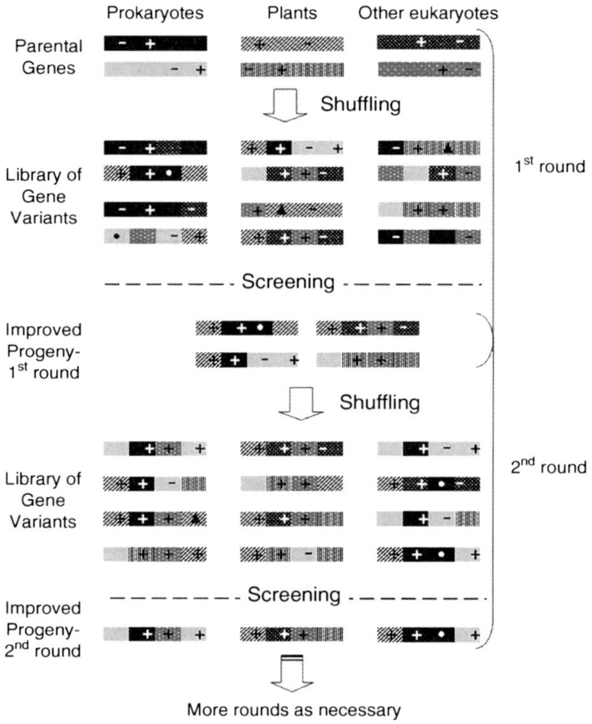

Fig. 1. Directed evolution through gene shuffling. Parental genes related by sequence from species of different kingdoms that possess beneficial mutations (+) and deleterious mutations (–) are fragmented and reassembled using the PCR reaction. During the gene shuffling process, random mutations with positive (*circle*) and negative (*triangle*) mutations are introduced into the library. The resultant library of shuffled variants is screened for desired properties and the selected variants with improved fitness are used as parents for the next round of directed evolution. Each round increases the amount of beneficial and positive random mutations and eliminates deleterious and negative random mutations. The process can be repeated until shuffled variants with the desired properties are identified.

Libraries of variants are generated through a variety of mutagenesis techniques or by gene shuffling in which multiple parental genes generate chimeric sequences. The libraries are then screened for desired properties such as improved catalytic activity, substrate specificity, pH or temperature stability etc. During the screening process, useful mutations and random beneficial mutations are accumulated and deleterious ones are discarded. The resultant progenies that exhibit higher fitness for the desired properties can be used as parental genes for additional rounds of directed evolution by gene shuffling. By consolidating the beneficial mutations in poolwise recombination and discarding the deleterious mutations in each round, directed evolution dramatically accelerates the rate of improvement compared to sexual evolution.

The concept of directed evolution was first demonstrated by evolving the *E. coli* enzyme, EbgA (evolved β-galactosidase activity), to hydrolyze o-nitrophenyl-β-D-galactoside (Campbell et al., 1973). An *E. coli* LacZ- deletion strain that expresses all the other genes involved in lactose uptake and metabolism was adapted to grow on lactose as the carbon source by evolving a novel lactose hydrolyzing enzyme. Campbell and colleagues demonstrated the major advantage of directed evolution, namely, the ability to generate a new protein with improved activity regardless of structure-function information. Moreover, directed evolution can overcome the lack of three-dimensional structural information and can enhance the activity of proteins that are linked to poorly folded polypeptides by improving their folding properties. This was recently demonstrated by generating an active green fluorescent protein (GFP) fused to a poorly folded polypeptide that interfered with the correct folding of the GFP (Pedelacq et al., 2006). Four rounds of shuffling and screening for bright fluorescent variants that can still fold in the presence of the misfolded fusion peptide resulted in an active super-folder GFP that is unaffected by the poorly folded polypeptide.

Over the last 3 decades direct protein evolution has become the key technology in protein engineering and is widely used in academic laboratories and industry (Minshull and Stemmer, 1999; Yuan et al., 2005; Matsuura and Yomo, 2006). Enhanced enzymatic performance such as protein specific activity, stability in extreme conditions such as temperature and pH, and new enzymatic functions by altering substrate specificity for nucleic acid modifying enzymes, reporter genes, biochemical catalysts and cellulolytic enzymes have been reported (Arnold and Moore, 1997; Minshull and Stemmer, 1999; Powell et al., 2001; Yuan et al., 2005; Kaur and Sharma, 2006). While changing substrate specificity by site-directed mutagenesis often negatively affects the specific activity on the natural substrate, enhancing catalytic activity and expanding the substrate selectivity of enzymes by directed evolution are frequently linked and provide a rapid and powerful method to optimize enzymes. For example, evolved β-fucosidase from the *E. coli lacZ* β-galactosidase exhibited over tenfold improvement of catalytic efficiency (k_{cat}/K_m) and dramatic improvement for fucose substrates compared to the parent activity (Zhang et al., 1977). While the native β-galactosidase acts only weakly on β-D-fucosyl moieties, the evolved β-fucosidase exhibits high specificity for *o*-nitrophenyl substrates and *p*-nitrophenyl substrates. Similarly, two rounds of shuffling of the glycosynthase β-glucosidase (Abg) increased the catalytic efficiency 27-fold and significantly expanded the repertoire of acceptable substrates (Kim et al., 2004).

The three major advantages of directed evolution in comparison to natural evolution are: larger diversity pool, a rapid screening process and increased selective pressure. While natural evolution is limited to two parental genomes per generation, directed evolution can incorporate a large number of genes from different species through gene shuffling methodology (Fig. 1). Random fragmentation of multiple genes and then reassembly into full-length chimeric sequences through PCR was first demonstrated by Stemmer (1994a, b) as an efficient gene shuffling method that generates direct recombination of beneficial mutations. Additional diversity can be introduced into the library during the reassembly process by controlling the fidelity of the DNA polymerase. Recognizing the significance of the size of the diversity pool, scientists have developed methods that allow the introduction of non-homologous and very small crossover fragments (reviewed by Yuan et al., 2005).

Each round of screening during the directed evolution process is equivalent to a single generation of an organism in natural evolution. Therefore, rapid HTP screens enhance the process and provides faster results compared with natural evolution in which the screen is determined by the life cycle of the organism. However, HTP assays for protein function are the major bottleneck in directed evolution. They are labor-intensive with limited screening capacity of about 10^4 variants per library (Boersma et al., 2007). Methods for recombinant protein production such as *E. coli* expression and purification systems, phage display and cell surface display are widely used for the screening and selection processes (Lin and Cornish, 2002). Therefore, increasing screening capacity depends on the optimization of automatic HTP liquid handling for processing and assay monitoring.

Positive genetic selection enables the organism to survive only in the presence of improved target protein variants under certain desired conditions. It is the most demanding approach that significantly increases the selection capacity to 10^{10}–10^{13} variants per cycle. This is in contrast to a screening approach in which one has to analyze each individual variant in the library. The power of a selection system for identifying improved prokaryotic Rubisco variants was successfully demonstrated by Smith and Tabita (2003) using a Rubisco deletion mutant host of the photosynthetic bacterium *Rhodobacter capsulatus* (SBI-II$^-$). *Rhodobacter capsulatus* SBI-II$^-$ was unable to grow photoautotrophically in the presence of either 1.5 or 5% CO_2. Complementation of this deletion host with the *Synechococuss* PCC6301 *rbc*LS allowed photoautotrophic growth in the presence of 5% CO_2 but not 1.5% CO_2. Mutant variants of *rbc*LS with improved kinetics properties that enable the host to grow photoautotrophically in the presence of 1.5% CO_2 were identified using the deletion host. This system was also used for negative selection of many *rbc*LS mutants that could not complement photoautotrophically the deletion host growth in the presence of 5% CO_2. Further biochemical analysis indicated that different kinetics properties were affected for positive and negative clones. Similarly, positive selection was used to screen for increased resistance against β-lactam antibiotics (Stemmer, 1994a, b) and moxalactam degradation by recursive shuffling of the cephalosporinase enzyme (Crameri et al., 1998).

More recently, a modified *E. coli* host in which only the active Rubisco enzyme can restore the growth of the mutant cell line was used as a selection host to identify improved prokaryotic Rubisco (Parikh et al., 2006). Expression of the phosphoribulokinase in *E. coli* converts irreversibly D-ribulose-5-phosphate into RuBP. Since *E. coli* can not use RuBP, the carbon flux diverts from the pentose phosphate shunt into a metabolic dead end that causes growth arrest. Co-expression of functional Rubisco in this genetically engineered *E. coli* strain will rescue the bacteria by converting RuBP into PGA that serves as a metabolic intermediate in glycolysis. Selection of three rounds of randomly mutagenized libraries of the *Synechococcus* PCC6301 LSU and co-expression in the host system with its wild type SSU resulted in identification of improved LSU variants. The mutant variants exhibited four- to fivefold improvement in specific activity and produced significant amounts of Rubisco proteins relative to the wild-type enzyme.

Recombinant DNA techniques were successfully utilized to improve herbicide and fungicide control in agricultural biotechnology during the twentieth century (Miflin, 2000).Thus, directed evolution technology offers great opportunity in the transgenic plant approach to study structure–function relationships and to produce commercially viable genetically modified (GM) products (Lassner and Bedbrook, 2001; Lassner and McElroy, 2002). In this respect, Castle et al. (2004) demonstrated a novel catalytic activity of glyphosate *N*-acetyltransferase (GAT) that provides herbicide tolerance by gene shuffling technology. Detoxification of the herbicide glyphosate (*N*-phosphonomethylglycine) can be achieved by *N*-acetylation. Screening of a microbial diversity collection consisting of predominantly *Bacillus licheniformis* identified three genes encoding glyphosate *N*-acetyltransferase (GAT) enzymes with poor glyphosate acetylation activity. Eleven iterations of gene shuffling improved the enzymatic efficiency by 9,000-fold. Transgenic maize lines expressing the improved GAT variants tolerate six times the concentration that causes severe symptoms to untransformed plants. This is the first agricultural product developed by gene shuffling technology that will be commercialized in the nearest future.

IV Improving Rubisco Catalytic Efficiency by Gene Shuffling

A Attempts to Express *Arabidopsis thaliana* Rubisco in *Chlamydomonas reinhardtii*

Our ultimate goal for Rubisco engineering is to improve crop plant productivity. It is therefore desired to directly evolve higher plant Rubisco through gene shuffling and the selection/screen process. Unfortunately this approach is greatly limited by the lack of a host system suitable for library screening. *Chlamydomonas* is the eukaryotic green alga that is often viewed as a plant cell model system (Weeks, 1992). The amino acid sequence of the *Chlamydomonas* Rubisco large subunit (LSU) shares nearly 90% identity to that of higher plant enzymes. To test if a higher plant Rubisco LSU can be expressed in *Chlamydomonas*, we generated a Rubisco LSU deficient mutant strain named MX3312 from wild type strain 2137 provided by Dr. Spreitzer, University of Nebraska. MX3312 has its entire *rbc*L coding sequence replaced by a bacterial *aadA* gene through homologous recombination and antibiotic selection on spectinomycin. This strain can heterotrophically grow on an acetate containing medium, but dies after withdrawal of acetate from the medium. The photoautotrophic growth of MX3312 can be easily restored by transforming the construct of *Chlamydomonas* wild type *rbc*L (*Cr-rbc*L) with both 5' and 3' flanking sequences and the transformation efficiency is high. Namely, one bombardment with approximate 1 μg plasmid carrying wild type *Cr-rbc*L and 6×10^7 MX3312 cells could typically generate 200–300 photosynthesis-competent colonies. If a plant Rubisco LSU expressed in *Chlamydomonas* can form a functional holoenzyme with the *Chlamydomonas* small subunit (SSU), the mutant phenotype of MX3312 can then be complemented. To test such possibility, the construct of *Arabidopsis rbc*L (*At-rbc*L) coding region linked to 5' (2.3 kb) and 3' (1 kb) flanking sequences of *Cr-rbc*L was delivered to MX3312 chloroplasts by particle bombardment (PDS 1000-He Biolistic Delivery System-BioRad). The transformed cells were plated on minimal medium to select for photosynthesis-competent colonies. After extensive transformation and selection, we were unable to recover any photoautotrophic colonies. There are at least three possible reasons to explain this outcome: first, *At-rbc*L was not expressed in *Chlamydomonas* at either the transcriptional or translational level; second, *At-rbc*L was expressed at the protein level, but could not fold correctly into a functional form; third, *Arabidopsis* LSU was not compatible with *Chlamydomonas* SSU to form a functional Rubisco. This approach, however, cannot explore any of these possibilities because of the life-or-death selection. To address this shortcoming, we used a cell wall-less strain cc349/CW15 (from *Chlamydomonas* genetic center, Duke University), which is suitable for both chloroplast (particle bombardment, Boynton et al., 1988) and nuclear (electroporation, Shimogawara et al., 1998) transformations subsequently, to further test the possibility of expressing *Arabidopsis* Rubisco in *Chlamydomonas*. A construct containing *At-rbc*L flanked with *Cr-rbc*L 5' and 3' sequences followed by an *aadA* cassette was transformed into the cc349/CW15 strain. Hundreds of spectinomycin resistant colonies were recovered and the replacement of *Cr-rbc*L by *At-rbc*L was confirmed by DNA analysis. The transformants could not grow photoautotrophically, which is to be expected based on the outcome of MX3312 transformed with *At-rbc*L. RT-PCR analysis with *At-rbc*L specific primers, indicated that the mRNA levels in the *At-rbc*L transformants were normal (Fig. 2). Western analysis, however, could not detect either LSU or SSU, indicating no Rubisco holoenzyme was formed in the transformants. By analyzing a Rubisco SSU deficient mutant strain, T60-3, Khrebtukova and Spreitzer (1996) have observed in a pulse labeling experiment that the Rubisco LSU could not be produced in T60-3 cells in the absence of SSU even though the *rbc*L mRNA level was normal, an indication that *rbc*L expression is suppressed at the translational level. To test if the lack of both LSU and SSU in cc349/CW15-*At-rbc*L transformants is due to incompatibility between *Arabidopsis* LSU and *Chlamydomonas* SSU, a construct containing a *At-RbcS* cDNA with a *Cr-RbcS* promoter and 5'-transit peptide followed after by a 3' *ble* cassette (from Dr. Saul Purton, University College London) conferring zeocin resistance, was constructed. Since *Cr-RbcS* genomic DNA contains 3 introns, a separate experiment was performed complementing T60-3 (From Dr. Spreitzer, University of Nebraska) with two *Cr-RbcS* constructs containing either all three introns or only intron 1. We found that the T60-3 strain could not be complemented by *Cr-RbcS* cDNA, but was complemented by both *Cr-RbcS* intron-containing constructs, indicating intron 1 is essential. But it was also observed that the construct containing 3-intron recovered at least five-fold more photosynthesis-competent transformants than the construct carrying 1-intron by using the

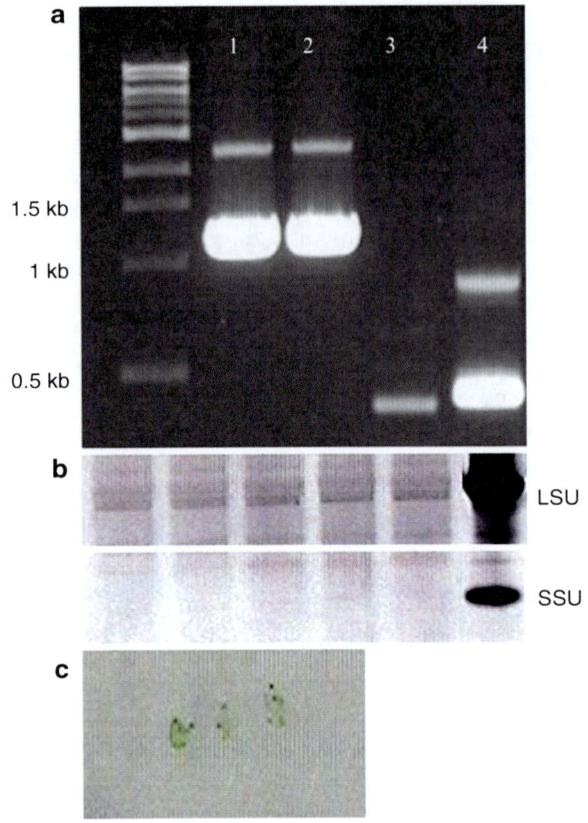

Fig. 2. Expression of *At-rbc*L and *At-Rbc*S in *Chlamydomonas* strain cc349/CW15. Panel A. RT-PCR results. Lane 1: RT-PCR product using *At-rbc*L specific primers; lane 2: RT-PCR product using *Cr-rbc*L specific primers; lane 3: RT-PCR product using *At-Rbc*S specific primers; lane 4: RT-PCR product using *Cr-Rbc*S specific primers. Panel B. Western analysis of five independent transformants containing *At-rbc*L and *At-Rbc*S. The cells were grown in acetate medium under continuous light (150 μE m^{2} s^{2}) at 23°C. Right lane: LSU, SSU standards. Panel C. *At-rbc*L and *At-Rbc*S transformants growing on minimal medium under light for 8 weeks after transferring from antibiotic selection plates.

same amount of DNA for transformation. Based on this information, we also inserted the *Cr-Rbc*S intron 1 into *At-Rbc*S cDNA in the corresponding position. After transforming the *At-Rbc*S construct into cc349/CW15-*At-Rbc*L cells, many zeocin resistant colonies were recovered. When the transformed cells containing both *At-rbc*L and *At-Rbc*S were transferred from acetate containing antibiotic selection media to the minimal media, the cells survived under light for months but hardly grew (Fig. 2). RT-PCR with *At-Rbc*S specific primers indicated the existence of mRNA, but at a highly reduced level compared to *Cr-Rbc*S mRNA (Fig. 2). Western analysis detected a very faint band at the expected LSU position after prolonged color development (Fig. 2), suggesting only a trace amount of Rubisco formation, but certainly not enough to support photoautotrophic growth even under elevated CO_2. In the transformants of *At-rbc*L and *At-Rbc*S, the endogenous *Cr-rbc*L was completely replaced by *At-rbc*L, but the native *Cr-Rbc*S genes in the nucleus were retained. Although it is not clear whether or not the extremely low expression of the *At-Rbc*S transgene in the transformants is due to the presence of the native *Cr-Rbc*S product or the lack of introns 2 and 3 in the *At-Rbc*S construct, it seems that further optimization of *At-Rbc*S expression in *Chlamydomonas* cells is needed in order to succeed. However, we also cannot rule out the possibility that the lack of a compatible chaperone for correct folding of *At*-Rubisco subunits in *Chlamydomonas* is responsible for the minimal holoenzyme accumulation. In this case, co-expression of a *At*-chaperone and/or *At*-RCA may be necessary.

B Shuffling the Chlamydomonas reinhardtii Rubisco Large Subunit

Because of the technical difficulties discussed above in developing a plant Rubisco expression and selection system, we have shuffled *Chlamydomonas* Rubisco in order to test: (1) if the catalytic properties of a eukaryotic Rubisco can be improved by gene shuffling; and (2) if changes in the catalytic properties resulting from the substitutions introduced into *Chlamydomonas* Rubisco variants can be achieved with the plant enzyme.

To shuffle *Chlamydomonas* LSU, the *Cr-rbc*L coding region with 2.3 kb 5' and 1 kb 3' flanking sequences was cloned into the pBluescript plasmid. The libraries were constructed according to Stemmer (1994a) and Crameri et al (1998). Single gene shuffling and semi-synthetic shuffling (Ness et al., 2002), in which oligos containing some of the natural-occurring diversity of the Rubisco gene family were spiked into the *Cr-rbc*L fragments during assembly, were performed in the 1st round shuffling. The parental genes for the 2nd and 3rd shuffling rounds were selected from the previous round's hits. The library variants were transformed into the *rbc*L deletion mutant strain, MX3312, by particle bombardment (PDS 1000-He Biolistic Delivery

System-BioRad). Approximately 150 bombardments for the 1st round and 120 bombardments each for 2nd and 3rd rounds were performed for subsequent selection and screening.

In order to identify Rubisco variants with improved catalytic properties from the shuffled libraries, we developed a three-tier selection/screen procedure (Fig. 3). The 1st tier is based on functional complementation. As discussed above, MX3312 can grow on acetate containing medium but not on minimal medium. The photoautotrophic growth of MX3312 can only be restored by introducing a functional Rubisco LSU. After transforming the shuffled *Cr-rbc*L variants into MX3312, only Rubisco LSU variants which are functional can be recovered as photosynthesis-competent colonies obtained from selection on minimal medium. With this single selection step, all non-functional variants in the library and those with inadequate catalytic activity to support photoautotrophic growth are eliminated. The 2nd tier screen relies on competitive growth. The photosynthesis-competent clones recovered from the 1st tier selection were pooled (usually 30 clones per group) with similar amounts of cells and grown together in a liquid culture for >30 generations as monitored by OD_{600} changes. As a single cell organism, the growth response of *Chlamydomonas* to photosynthesis is more sensitive than that of plants. It can be expected that the clones containing Rubisco variants with improved catalytic properties will grow better (faster) than those with less capable Rubisco variants. The consequence of the competitive growth is that the fast growing clones will become the dominant population in the resulting culture after a sufficient number of growth cycles (It is necessary to increase the number of growth cycles to enrich clones with only slightly improved growth rates). To increase the selection pressure, we also included 25–50 μM of a carbonic anhydrase inhibitor (6-ethoxy-2-benzothiazole-sulfonamide) in the competitive growth medium in the later rounds of shuffling to disrupt the CO_2 concentrating mechanism existent in *Chlamydomonas* cells. The resulting culture was plated on solid minimal agar medium to obtain single cell clones. The enriched variants after competitive growth were identified by *rbc*L sequence analysis and photosynthesis measurements for O_2 evolution using an O_2 electrode. The 3rd tier screen is an enzyme kinetic property assay. The cell crude extracts and ion exchange column purified Rubisco enzymes of the clones from competitive growth were used to measure Rubisco V_c and Ω to identify variants with improved catalytic properties.

After three rounds of shuffling, approximately 80,000 library variants were selected and screened. We were able to identify multiple clones showing increased in vitro carboxylase activity up to 56% greater and Ω values up to 18% greater as compared to the wild-type (Zhu et al., 2005). Sequence analysis of these clones displayed on average around three residue substitutions per variant. Most clones contained a substitution in the hydrophobic core of the N-terminal domain and another in the C-terminal tail region. A few substitutions at the surface of the C-terminal domain were also found. Examination of the crystallographic structure indicates that the substituted residue in C-terminal tail interacts with the loop between the αB and βC helices in the N-terminal domain.

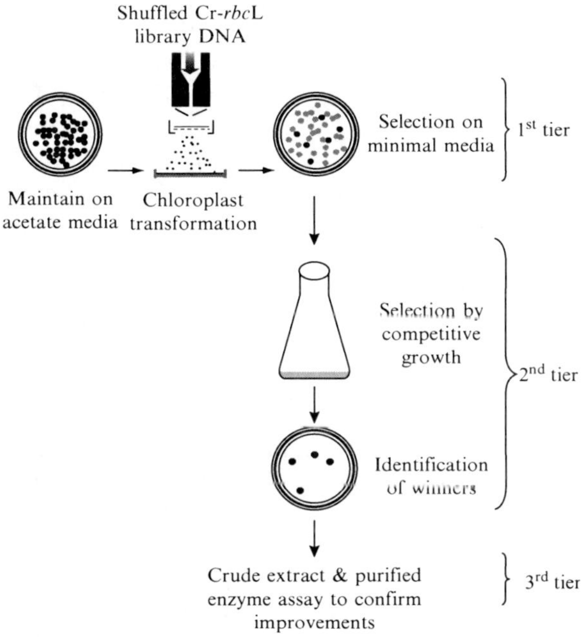

Fig. 3. Three-tier selection/screening procedure designed for identifying improved enzyme variants from shuffled *Chlamydomonas rbc*L library. For 3rd tier assay, Rubisco from *Chlamydomonas* crude extracts was purified by ammonium sulfate fractionation (35–55% of saturation) and polyethylene glycol precipitation (20%) followed by an anion exchange chromatographic separation (Poros HQ/20 column). Rubisco carboxylase activity (V_c) was determined by $^{14}CO_2$ incorporation. The Ω value was determined by quantifying 3-phosphoglyceric acid and 2-phosphoglycolic acid directly from Rubisco reaction mixtures with a LC/MS method.

Three active site residues (E60, T65 and W66) are located in this loop region. The substituted residues at the surface of the C-terminal domain are usually involved in interactions between the subunits, either within a L2 dimer or between L2 dimers, but not with SSU. Most of the substitutions resulting in improved catalytic properties are located at non-conserved regions.

To test if the mutations that positively impacted *Chlamydomonas* Rubisco catalytic properties can produce the same effect with a plant Rubisco enzyme, several mutation sets identified from the *Chlamydomonas* Rubisco variants were introduced into tobacco (Petite Havana) *rbc*L by site-directed mutagenesis and chloroplast transformation. One mutation set contains three mutations which are all novel to both *Chlamydomonas* and plant Rubiscos. Kinetic analysis indicated that the modified tobacco Rubisco with the triple mutations exhibited an increase in both V_c (1.5 ± 0.08 μmol·mg^{-1}·min^{-1}) and Ω (89 ± 2.98) by 15% and 14%, respectively by comparison to wild type (1.3 ± 0.03 μmol·mg^{-1}·min^{-1} and 78 ± 0.98 respectively). The triple mutant also displayed lower K_m for CO_2 (7.55 μM) than the wild type enzyme (12.63 μM). Another mutation set contains four mutations, but two already exist in tobacco. The V_c of the mutant tobacco Rubisco with this mutation set increased by 20%, but its Ω value remained unchanged or slightly reduced. This preliminary data suggested that it was possible to engineer higher plant Rubisco by using substitutions identified from shuffling *Chlamydomonas* Rubisco *rbc*L.

V Improving Rubisco Activase Thermostability by Gene Shuffling

Identifying ways to maintain high photosynthetic CO_2 fixation rates in plants exposed to moderately elevated temperatures remains a challenging task for academic labs, breeders and agriculture biotechnology companies. At elevated temperatures that are slightly higher than optimum and with sufficient water availability, plants maintain the stomatas open in order to cool their leaves by evapotranspiration. Under these conditions, the inhibition of photosynthesis is reversible for short periods (hours) of stress, while permanent inhibition occurs under more severe heat stress due to irreversible damage of the photosynthetic apparatus (Salvucci and Crafts-Brandner, 2004a). The inhibition of photosynthesis under moderate heat stress conditions in both C3 and C4 plants is hypothesized to be due to the extreme heat sensitivity of the Rubisco activase (RCA) enzyme, which constantly maintains Rubisco at a high activation state. The *Arabidopsis rca* mutant reported by Somerville et al. (1982) was the key to the discovery of RCA and its role in photosynthesis. This mutant, which requires high CO_2 concentrations to survive, lacks two polypeptides (Salvucci et al., 1985) that were later purified and shown to promote the activation of Rubisco at physiological concentrations of CO_2, Mg, and RuBP (Portis et al., 1987). RCA was subsequently characterized as a member of the AAA$^+$ family of ATPases associated with diverse cellular activities that interacts with inactive Rubisco and removes sugar-phosphate inhibitors from Rubisco's catalytic sites (Portis, 2003).

The hypothesis that photosynthesis is limited by inactivation of Rubisco due to heat sensitivity of RCA to moderately elevated temperature was first suggested by Feller et al. (1998). The extreme sensitivity of RCA to elevated temperatures is now well characterized for both C3 (Crafts-Brandner and Salvucci, 2000a; Salvucci and Crafts-Brandner, 2004a, b) and C4 plants (Crafts-Brandner and Salvucci, 2000b). Therefore, generating a thermostable RCA that can effectively activate Rubisco under moderately elevated temperatures is a potential target for agriculture biotechnology. Maintaining Rubisco at a high activation state under moderately elevated temperatures can be achieved by recombinant DNA technologies that stabilize RCA through: (1) overexpression of thermostable RCA from plants grown in warm regions such as creosote bush, or cotton; (2) increasing the concentrations of osmoprotectants such as glycine betaine in the chloroplasts; or (3) directed evolution of a thermostable RCA. Overexpression of a thermostable activase is a possibly limited solution due to the species dependence of Rubisco and RCA, namely, the incompatibility between Solanaceae RCA to fully activate non-solanaceae Rubisco and vice versa (Wang et al., 1992). In addition, a foreign gene typically possesses different GC content and thus requires codon optimization in order to ensure high expression levels.

Increasing glycine betaine concentration in chloroplasts is an indirect strategy to stabilize RCA and maintain Rubisco at a high activation state during heat stress (Yang et al., 2005). The osmoprotectant nature of glycine betaine was shown to prevent inactivation of RCA and positively affected Rubisco activation state and photosynthetic rate at elevated temperatures. However, glycine betaine levels are limited by the levels of the precursor, choline, in the chloroplast and high levels can affect other metabolic pathways. Therefore, directed evolution through gene shuffling of the endogenous *rca* gene was deemed a more favorable method to modulate the thermal properties of RCA. This approach will allow us to identify shuffled variants with enhanced thermostability and/or improved specific activity of the endogenous enzyme. The expression of the mutagenized genes coding for improved RCA variants could be controlled using the endogenous *rca* promoter and untranslated regulatory sequences without the needs for codon optimization and worries of incompatibility of the expressed RCA variants with the host Rubisco.

Arabidopsis contains two RCA polypeptides: the 43 kDa short (β) form and the redox-regulated 46 kDa long (α) form (Zhang et al., 2002). The two forms are generated by alternative splicing of a single pre-mRNA (Werneke et al., 1989). In order to enhance photosynthesis under moderately elevated temperatures, we have increased the thermostability of *Arabidopsis* RCA short form through directed evolution (Kurek et al., 2007). Variants of *Arabidopsis* RCAβ that effectively maintain Rubisco at high activation state under normal and elevated temperature were generated using diversity provided by relatively thermostable RCAs such as wheat, barley, maize and cotton. To increase the diversity pool, additional random mutations were incorporated. The libraries containing the shuffled variants were screened using a Rubisco activation assay that monitored the incorporation of ^{14}C into the Rubisco product, PGA, through the activation of inactive Rubisco (decarbamylated) in the presence of the shuffled RCAβ. This was achieved with an ATP-regenerating system, RuBP and [^{14}C] $NaHCO_3$ (Fig. 4a). Assaying the activation of Rubisco by the purified RCAβ or by the soluble fraction of *E. coli* cell lysates expressing the shuffled variants in a HTP format was achieved in three tiers (Fig. 4b). The first tier is a HTP screen using *E. coli* cell lysates expressing RCAs exposed to

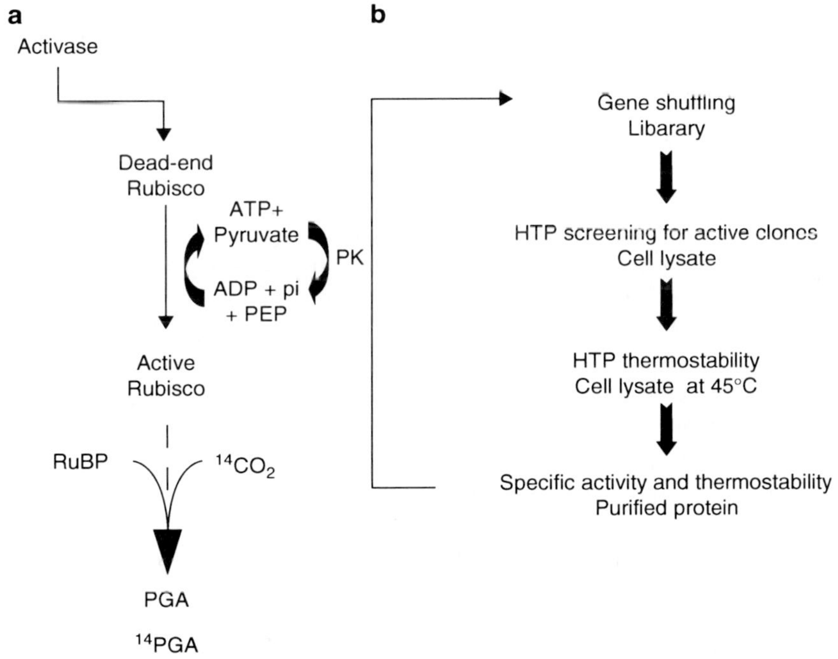

Fig. 4. Overview of the directed evolution of thermostable RCA. Schematic presentation of (**a**) Rubisco activation assay and (**b**) three tier shuffling cascade. PK – pyruvate kinase; PEP – phosphoenolpyruvate; PGA – 3-phosphoglyceric acid.

normal temperature prior to the Rubisco activation assay. The second tier screens for thermostability of the *E. coli* cell lysates of the selected variants from the first tier, exposed to higher temperatures prior to the Rubisco activation assay. The third tier is the characterization of selected RCAβ variants from the second tier for activation of Rubisco at normal and moderately elevated temperature by purified shuffled RCA. That in turn confirmed the superiority of the shuffled variants and determined the specific activity and thermostability. Two rounds of shuffling enhanced the thermostability of the shuffled RCA variants by 80% at 45°C compared to the activity the wild type RCA at the same temperature. The activity at 45°C was only 10% less than the activity of wild type RCA at 25°C. When monitoring the ability of wild type and shuffled variants to maintain Rubisco in an active state during heat treatment (Rubisco activation under catalytic conditions; Crafts-Brandner and Salvucci, 2000a), wild type RCA maintained a Rubisco activation state of 0.5 at 40°C, while the three selected shuffled leads were able to maintain activation states of 0.62–0.72 under the same condition. Relative to reactions at 25°C, the activation state of Rubisco maintained by the thermostable variants at 40°C was in the range of 78–98%, versus 70% for the wild type enzyme (Kurek et al., 2007).

Sequence analysis revealed that one amino acid substitution (T274R) was sufficient to improve activity and thermostability. Three amino acid substitutions (F168L, V257I, K310N and M131V, V257I, K310N) resulted in a 10°C increase in stability of *Arabidopsis* RCA1. The variant containing the mutations F168L, V257I, T274R, and K310N exhibited relatively low activity at 25°C (82% of the T274R activity), but maintained high activity at 40°C (102% of F168L, V257I, K310N activity). The substitutions V257I and K310N shared by two selected leads are also present as natural variation in plant species: the former is present in the cucumber enzyme and the later is conserved in wheat, rice, spinach and maize.

To test the effect of the improved RCA *in planta*, the shuffled RCA variants (sRCA) were expressed in an *Arabidopsis* mutant that lacks the endogenous *rca* gene (ΔRCA). This mutant line was selected in order to demonstrate that the phenotype, photosynthesis and growth rates under normal and moderately elevated temperatures were directly and solely affected by the properties of the shuffled variants and not by the presence of the wild-type endogenous RCAβ. In addition, the presence of endogenous RCA α and β (in wild-type plants) that potentially forms heterocomplexes with the shuffled variants (the active complex RCA-Rubisco consists of multiple RCA subunits) could affect the shuffled variants' properties. Finally the absence of endogenous RCAs mimics the screening for improved recombinant shuffled variants that was performed in the absence of the wild type RCA forms. Transgenic ΔRCA lines expressing wild-type RCAβ (wRCAβ) and sRCA exhibited normal photosynthetic rates and phenotypes under ambient growth conditions. Daily exposure of 2-week-old transgenic lines to moderately elevated temperature (30°C for 4 h day^{-1} in the middle of the light cycle), which mimics moderate heat stress during the day, demonstrated the positive effect of thermostable sRCA on growth and photosynthetic performance (Fig. 5). sRCA plants exhibited higher leaf area (about 15–20%) (Fig. 5a) than wRCAβ plants (Fig. 5b) and higher photosynthetic rates (about 10%) during the heat stress period (Fig. 5c–e). The leaves of wRCAβ grown under moderately elevated temperature were severely damaged and displayed discoloration. When wRCAβ and sRCA plants were grown continuously at 26°C under higher light intensity and humidity, the sRCA plants possessed 50–100 more siliques per plant than wRCAβ plants. In addition, the siliques of sRCA plants were larger and produced more seeds, higher seed weight and better seed viability (higher germination rate) than the wRCAβ plants (Kurek et al., 2007). The phenotype and photosynthetic performance of wRCAβ and sRCA strongly support the hypothesis that RCA limits Rubisco activity and therefore photosynthesis under elevated temperatures. The improved phenotype of the transgenic lines expressing the shuffled RCA under moderately elevated temperatures is most likely due to the improved thermostability of RCA, minimizing the negative effect on photosynthetic performance and the inhibition of biomass accumulation.

VI Future Prospects

Genetic adaptation of a crop plant to its growth environment is a key determining factor for yield potential. Current crop elites have experienced extensive genetic selection under their growth conditions to achieve higher yield potential. It is

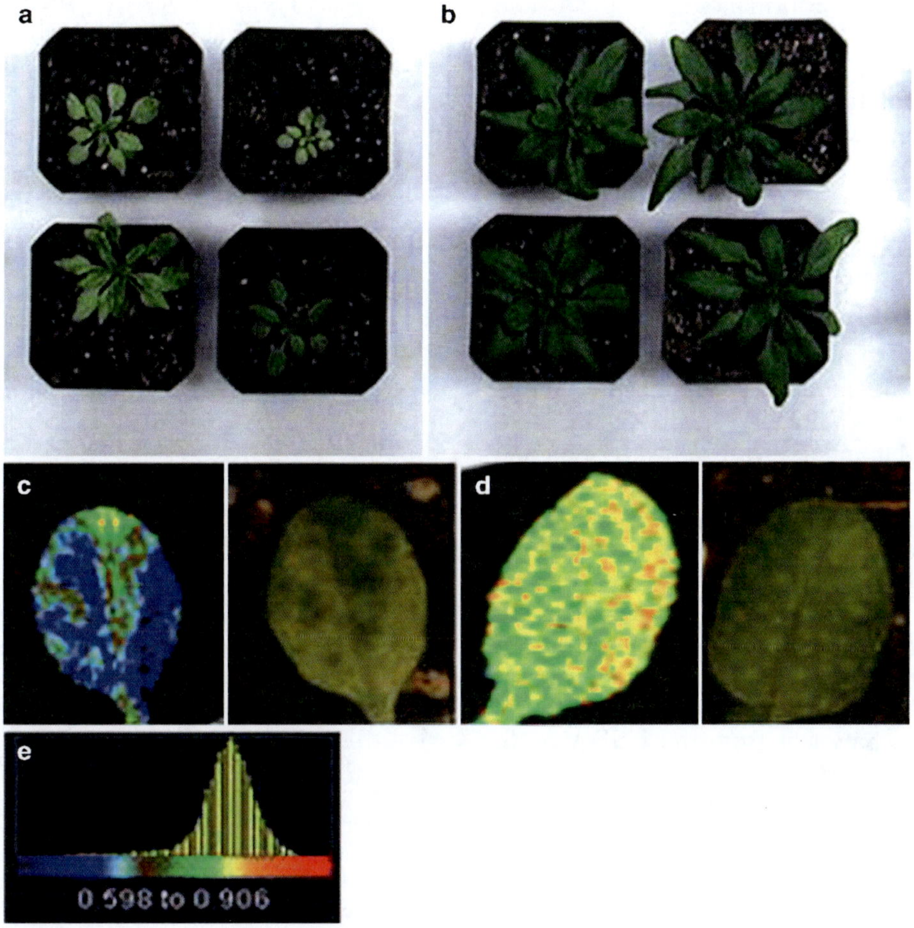

Fig. 5. Expression of thermostable RCA in ΔRCA mutant. Phenotype of 4-week-old transgenic plants overexpressing RCAβ. (**a**) and sRCA (**b**) grown under 16 h light (225 μmol photons m^{-2} s^{-1}) 8 h dark regime exposed daily to moderately elevated temperatures (6 h 22°C; 4 h 30°C; 6 h 22°C during the day cycle) for 2 weeks. Photosynthetic performance (Fq'/Fm') of a single leaf from RCAβ (**c**) and sRCA (**d**) was measured using fluorescence image analysis. The Photosynthetic performance scale from low (blue) to high (red) Fq'/Fm' values is indicated (**e**).

most likely that the potential enhancement of photosynthetic rate at the leaf level by modifying any single enzyme will not be dramatic under normal growth conditions because of other genetic constraints. The improvements from modifying a single enzyme or enhancing a single catalytic step for alleviating limitations under certain growth conditions, however, might generate significant positive impacts on photosynthesis. Environmental conditions change during the growth season. For instance, temperature and light, which are two key environmental factors for photosynthesis, fluctuate significantly within a growth season and even within a day. Enzymes have their own optimal temperature range within which they perform best. The optimal temperature ranges for some photosynthesis enzymes are quite narrow compared to the magnitude of temperature fluctuations during the growth season. This can often become limiting for the overall photosynthetic light use efficiency at suboptimal growth temperature. Adjusting the expression level of relevant enzymes is one of the common strategies used by plants to acclimate to a changed environment. Selection with genetic markers and gene overexpression could be an option to address some specific limitations, but both are limited to the use of naturally existing enzyme properties. With the aid of directed molecular evolution technology, it is possible to generate novel enzymes with catalytic properties that cannot be found in naturally existing enzymes. For example it is possible to maximize enzyme performance over the whole growth season, by broadening enzyme optimum temperature ranges. In this way, overall light use efficiency might be improved significantly and the sum of photosynthesis can be increased, as can biomass and yield.

Acknowledgements

This work was supported in part by grant from the National Institute of Standards and Technology-Advanced Technology Program. The authors would like to thank Thom True, Xin Zhang, Tiger Hu, Monica Majumdar, Eva Lin, Lik Hsueh and Alfred Madrigal for technical assistance; Shelly A. Straight for isolating the *Arabidopsis* activase deletionmutant; Dr. Cornelia Stettner of Icon Genetics AG for tobacco transformation; Dr. John A. Kiser for statistic analysis; Dr. Daniel Siehl for scientific discussions; Hoa M. Giang and Ingrid Udranszky for help in preparing the manuscript; Dr. Michael Lassner for scientific guidance; and Dr. Michael E. Salvucci from the USDA-ARS, Western Cotton Research Laboratory, Phoenix, AZ and Dr. Robert J. Spreitzer from the University of Nebraska for consultation to the project.

References

Arnold FH and Moore JC (1997) Optimizing industrial enzymes by directed evolution. Adv Biochem Eng Biotechnol 58: 1–14

Bainbridge G, Madgwick P, Parmer S, Mitchell R, Paul M, Pitts J, Keys AJ, and Parry AJ (1995) Engineering Rubisco to change its catalytic properties. J Exp Bot 46: 1269–1276

Baxter CJ, Foyer CH, Turner J, Rolfe SA, and Quick WP (2003) Elevated sucrose–phosphate synthase activity in transgenic tobacco sustains photosynthesis in older leaves and alters development. J Exp Bot 54: 1813–1820

Boersma YL, Dröge MJ, and Quax WJ (2007) Selection strategies for improved biocatalysts. FEBS J 274: 2181–2195

Boynton JE, Gillham NW, Harris EH, Hosler JP, Johnson AM, Jones AR, Randolph-Anderson BL, Robertson D, Klein TM, Shark KB, and Stanford JC (1988) Chloroplast transformation in *Chlamydomonas* with high velocity microprojectiles. Science 240: 1534–1538

Campbell JH, Lengyel JA, and Langridge J (1973) Evolution of a second gene for beta-galactosidase in Escherichia coli. Proc Natl Acad Sci USA 70: 1841–1845

Castle LA, Siehl DL, Gorton R, Patten PA, Chen YH, Bertain S, Cho HJ, Duck N, Wong J, Liu D, and Lassner MW (2004) Discovery and directed evolution of a glyphosate tolerance gene. Science 304: 1151–1154

Cloney LP, Bekkaoui DR, and Hemmingsen SM (1993) Co-expression of plastid chaperonin genes and a synthetic plant Rubisco operon in Escherichia coli. Plant Mol Biol 23: 1285–1290

Crafts-Brandner SJ and Salvucci ME (2000a) Rubisco activase constrains the photosynthetic potential of leaves at high temperature and CO_2. Proc Natl Acad Sci USA 97: 13430–13435

Crafts-Brandner SJ and Salvucci ME (2000b) Sensitivity of photosynthesis in a C4 plant, maize, to heat stress. Plant Physiol 129: 1773–1780

Crameri A, Raillard S, Bermudez E, and Stemmer WPC (1998) DNA shuffling of a family of genes from diverse species accelerates directed evolution. Nature 391: 288–291

Evans LT and Fischer RA (1999) Yield potential: definition, measurement, and significance. Crop Sci 39: 1544–1551

Farquhar GD, von Caemmerer S, and Berry JA (1980) A biochemical model of photosynthetic CO_2 assimilation in leaves of C_3 species. Planta 149: 78–90

Feller U, Crafts-Brandner SJ, and Salvucci ME (1998) Moderately high temperatures inhibit ribulose-1,5-bisphosphate carboxylase/oxygenase (Rubisco) activase-mediated activation of Rubisco. Plant Physiol 116: 539–546

Gutteridge S and Gatenby AA (1995) Rubisco synthesis, assembly, mechanism, and regulation. Plant Cell 7:809–819

Hartman FC and Harpel MR (1994) Structure, function, regulation, and assembly of D-ribulose-1,5-bisphosphate carboxylase/oxygenase. Annu Rev Biochem 63: 197–234

Horton P (2000) Prospects for crop improvement through the genetic manipulation of photosynthesis: morphological and biochemical aspects of light capture. J Exp Bot 51: 475–485

Kaur J and Sharma R (2006) Directed evolution: an approach to engineer enzymes. Crit Rev Biotechnol 26: 165–199

Kebeish R, Niessen M, Thiruveedhi K, Bari R, Hirsh HJ, Rosenkranz R, Stabler N, Schonfield B, Kreuzaler F, and Peterhansel C (2007) Chloroplastic photorespiratory bypass increases photosynthesis and biomass production in *Arabidopsis thaliana*. Nature Biotech 25: 593–599

Khrebtukova I and Spreitzer RJ (1996) Elimination of the Chlamydomonas gene family that encodes the small subunit of ribulose-1,5–bisphosphate carboxylase/oxygenase. Proc Natl Acad Sci USA 93: 13689–13693

Kim YW, Lee SS, Warren RA, and Withers SG (2004) Directed evolution of a glycosynthase from Agrobacterium sp. increases its catalytic activity dramatically and expands its substrate repertoire. J Biol Chem 279: 42787–42793

Ku MSB, Agarie S, Nomura M, Fukayama H, Tsuchida H, Ono K, Hirose S, Toki S, Miyao M, and Matsuoka M (1999) High-level expression of maize phosphoenolpyruvate carboxylase in transgenic rice plants. Nat Biotech 17: 76–80

Kurek I, Chang TK, Bertain SM, Madrigal A, Liu L, Lassner MW, and Zhu G (2007) Enhanced thermostability of

Arabidopsis Rubisco activase improves photosynthesis and growth rates under moderate heat stress. Plant Cell 19: 3230–3241

Laing WA, Ogren WL, and Hageman RH (1974) Regulation of soybean net photosynthetic CO_2 fixation by the interaction of CO_2, O_2 and ribulose 1,5-bisphosphate carboxylase. Plant Physiol 54: 678–685

Larson EM, O'Brien CM, Zhu G, Spreitzer RJ, and Portis AR Jr (1997) Specificity for activase is changed by a Pro-89 to Arg substitution in the large subunit of ribulose-1,5-bisphosphate carboxylase/oxygenase. J Biol Chem 272: 17033–17037

Lassner MW and McElroy D (2002) Directed molecular evolution: bridging the gap between genomics leads and commercial products. Omics 6: 153–162

Lassner M and Bedbrook J (2001) Directed molecular evolution in plant improvement. Curr Opin Plant Biol 4: 152–156

Lieman-Hurwitz J, Rachmilevitch S, Mittler R, Marcus Y, and Kaplan A (2003) Enhanced photosynthesis and growth of transgenic plants that express ictB, a gene involved in HCO_3 accumulation in cyanobacteria. Plant Biotech J 1: 43–50

Lin H and Cornish VW (2002) Screening and selection methods for large-scale analysis of protein function. Angew Chem 41: 4402–4425

Lorimer GH and Miziorko HM (1980) Carbamate formation on the ε-amino group of a lysyl residue as the basis for the activation of ribulosebisphosphate carboxylase by CO_2 and Mg^{2+}. Biochemistry 19: 5321–5328

Lunn JE, Gillespie VJ, and Furbank RT (2003) Expression of a cyanobacterial sucrose-phosphate synthase from Synechocystis sp. PCC 6803 in transgenic plants. J Exp Bot 54: 223–237

Matsuura T and Yomo T (2006) In vitro evolution of proteins. J Biosci Bioeng 101: 449–456

Miflin B (2000) Crop improvement in the 21st century. J Exp Bot 51: 1–8

Minshull J and Stemmer PC (1999) Protein evolution by molecular breeding. Curr Opin Chem Biol 3: 284–290

Miyagawa Y, Tamoi M, and Shigeoka S (2001) Overexpression of a cyanobacterial fructose-1,6-/sedoheptulose-1,7-bisphosphatase in tobacco enhances photosynthesis and growth. Nat Biotech 19: 965–969

Ness JE, Kim S, Gottman S, Pak R, Krebber A, Borchert TV, Govindarajan S, Mundorff EC, and Minshull J (2002) Synthetic shuffling expands functional protein diversity by allowing amino acids to recombine independently. Nat Biotech 20: 1251–1255

Ogren WL (1984) Photorespiration: pathways, regulation, and modification. Annu Rev Plant Physiol 35: 415–442

Parikh MR, Greene, DN, Woods, KK, and Matsumura I (2006) Directed evolution of RuBisCO hypermorphs through genetic selection in engineered E. coli. Prot Eng Des Sel 3: 113–119

Parry MAJ, Andralojc PJ, Mithell RA, Madgwick PJ, and Keys AJ (2003) Manipulation of Rubisco: the amount, activity, function and regulation. J Exp Bot 54: 1321–1333

Pedelacq JD, Cabantous S, Tran T, Terwilliger TC, and Waldo GS (2006) Engineering and characterization of a superfolder green fluorescent protein. Nat Biotech 24: 79–88

Pellny TK, Ghannoum O, Conroy JP, Schluepmann H, Smeekens S, Andralojc KP, Goddijn O, and Paul MJ (2004) Genetic modification of photosynthesis with E. coli genes for trehalose synthesis. Plant Biotech J 2: 71–82

Peng S, Laza RC, Visperas RM, Sanico AL, Cassman KG, and Khush GS (2000) Crop breeding, genetics and cytology. Crop Sci 40: 307–314

Perchorowicz JT, Raynes DA, and Jensen RG (1981) Light limitation of photosynthesis and activation of ribulose bisphosphate carboxylase in wheat seedlings. Proc Natl Acad Sci USA 78: 2985–2989

Portis AR Jr, Salvucci ME, and Ogren WL (1987) Activation of ribulose bisphosphate carboxylase/oxygenase at physiological CO_2 and ribulose bisphosphate concentrations by rubisco activase. Plant Physiol 82: 967–991

Portis AR Jr (1992) Regulation of ribulose 1,5-bisphosphate carboxylase/oxygenase activity. Annu Rev Plant Physiol 43: 415–437

Portis AR Jr (2003) Rubisco activase – Rubisco's catalytic chaperone. Photosynth Res 75: 11–27

Powell KA, Ramer SW, Del Cardayre SB, Stemmer WP, Tobin MB, Longchamp PF, and Huisman GW (2001) Directed evolution and biocatalysis. Angew Chem Int Ed Engl 40: 3948–3959

Raines CA (2006) Transgenic approaches to manipulate the environmental responses of the C3 carbon fixation cycle. Plant Cell Environ 29: 331–339

Read BA and Tabita FR (1994) High substrate-specificity factor ribulose-bisphosphate carboxylase oxygenase from eukaryotic marine-algae and properties of recombinant cyanobacterial Rubisco containing algae residue modifications. Arch Biochem Biophys 312: 210–218

Richards RA (2000) Selectable traits to increase crop photosynthesis and yield of grain crops. J Exp Bot 51: 447–458

Salvucci ME, Portis AR Jr, and Ogren WL (1985) A soluble chloroplast protein catalyzes ribulose bisphosphate carboxylase/oxygenase activation *in vivo.* Photosynth Res 7: 193–201

Salvucci ME and Crafts-Brandner SJ (2004a) Inhibition of photosynthesis by heat stress: the activation state of Rubisco as a limiting factor in photosynthesis. Physiol Plant 120: 179–186

Salvucci ME and Crafts-Brandner SJ (2004b) Relationship between the heat tolerance of photosynthesis and the thermal stability of Rubisco activase in plants from contrasting thermal environments. Plant Physiol 134: 1460–1470

Shimogawara K, Fujiwara S, Grossman A, and Usuda H (1998) High-efficiency transformation of *Chlamydomonas reinhardtii* by electroporation. Genetics 148: 1821–1828

Sinclair TR (1998) Historical changes in harvest index and crop nitrogen accumulation. Crop Sci 38: 638–643

Sinclair TR, Purcell LC, and Sneller CH (2004) Crop transformations and the challenge to increase yield potential. Trends Plant Sci 9: 70–75

Smith SA and Tabita FR (2003) Positive and negative selection of mutant forms of prokaryotic (cyanobacterial)Riboulose-1, 5-bisphosphate carboxylase/oxigenase. J Mol Biol 331: 557–569

Somerville CR, Portis AR Jr, and Ogren WL (1982) A mutant of *Arabidopsis thaliana* which lacks activation of RuBP carboxylase *in vivo*. Plant Physiol 70: 381–387

Spreitzer RJ (1993) Genetic dissection of Rubisco structure and function. Annu Rev Plant Physiol Plant Mol Biol 44: 411–434

Spreitzer RJ (1999) Questions about the complexity of chloroplast ribulose-1,5–bisphosphate carboxylase/oxygenase. Photosynth Res 60: 29–42

Spretzer RJ and Salvucci ME (2002) Rubisco: structure, regulatory interactions, and possibilities for a better enzyme. Annu Rev Plant Biol 53: 449–475

Stemmer WP (1994a) DNA shuffling by random fragmentation and reassembly: in vitro recombination for molecular evolution. Proc Natl Acad Sci USA 91: 10747–10751

Stemmer WP (1994b) Rapid evolution of a protein in vitro by DNA shuffling. Nature 370: 389–391

Tabita FR (1999) Microbial ribulose-1,5-bisphosphate carboxylase/oxygenase: a different perspective. Photosynth Res 60: 1–28

Tamoi M, Nagaoka M, Miyagawa Y, and Shigeoka S (2006) Contribution of fructose-1,6-bisphosphatase and sedoheptulose-1,7-bisphosphatase to the photosynthetic rate and carbon flow in the Calvin cycle in transgenic plants. Plant Cell Physiol 47: 380–390

Tcherkez GGB, Farquhar GD, and Andrews TJ (2006) Despite slow catalysis and confused substrate specificity, all ribulose bisphosphate carboxylases may be nearly perfectly optimized. Proc Natl Acad Sci USA 103: 7246–7251

Tollenaar M and Lee EA (2002) Yield potential, yield stability and stress tolerance in maize. Field Crops Res 75: 161–169

Uemura K, Anwaruzzaman, Miyachi S, and Yokota A (1997) Ribulose-1,5-bisphosphate carboxylase/oxygenase from thermophilic red algae with a strong specificity for CO_2 fixation. Biochem Biophys Res Commun 233: 568–571

Wang Z-Y, Snyder GW, Esau BD, Portis AR, and Ogren WL (1992) Species-dependent variation in the interaction of substrate-bound ribulose-1,5-biphosphate carboxylase/oxygenase (Rubisco) and Rubisco activase. Plant Physiol 100: 1858–1862

Weeks D (1992) *Chlamydomonas*: an increasingly powerful model plant system. Plant Cell 4: 871–878

Werneke JM, Chatfield JM, and Ogren WL (1989) Alternative mRNA splicing generates the two ribulosebisphosphate carboxylase/oxygenase activase polypeptides in spinach and Arabidopsis. Plant Cell 1: 815–825

Whitney SM, Baldet P, Hudson GS, and Andrews TJ (2001) Form I Rubiscos from non–green algae are expressed abundantly but not assembled in tobacco chloroplasts. Plant J 26: 535–547

Yang X, Liang Z, and Lu C (2005) Genetic engineering of the biosynthesis of glycinebetaine enhanced photosynthesis against high temperature stress in transgenic tobacco plants. Plant Physiol 138: 2299–2309

Yuan L, Kurek I, English J, and Keenan R (2005) Laboratory-directed protein evolution. Microb Mol Biol Rev 69: 373–392

Zhang YX, Dawes G, and Stemmer WP (1997) Directed evolution of a fucosidase from a galactosidase by DNA shuffling and screening. Proc Natl Acad Sci USA 94: 4504–4509

Zhang N, Kallis RP, Ewy RG, and Portis AR Jr (2002) Light modulation of Rubisco in Arabidopsis requires a capacity for redox regulation of the larger Rubisco activase isoform. Proc Natl Acad Sci USA 99: 3330–3334

Zhu G, Kurek I, True T, Zhang X, Majumdar M, Liu L, and Lassner M (2005) Enhancing photosynthesis by improving Rubisco carboxylase activity and specificity, and Rubisco activase thermostability through DNA shuffling. In: Photosynthesis: fundamental aspects to global perspectives. Proceedings of 13th international congress on photosynthesis, pp. 841–843. Lawrence, KS, USA

Zhu X-G, Portis AR Jr, and Long SP (2004) Would transformation of C3 crop plants with foreign Rubisco increase productivity? A computational analysis extrapolating from kinetic properties to canopy photosynthesis. Plant Cell Environ 27: 155–165

Chapter 21

Elevated CO_2 and Ozone: Their Effects on Photosynthesis

Ruth Grene*
Department of Plant Pathology, Virginia Tech, Blacksburg, VA 24061, USA

Pinghua Li† and Hans J. Bohnert
Departments of Plant Biology and of Crop Sciences, University of Illinois at Urbana-Champaign, Urbana, IL 61801, USA

Hans J. Bohnert
Center for Comparative and Functional Genomics, University of Illinois at Urbana-Champaign, Urbana, IL 61801, USA

Summary

Elevated carbon dioxide [CO_2] and ozone [O_3] are increasing worldwide in part due to human activities. Each gas affects plant cells, initially primarily by interfering with photosynthesis, in ways that are only partially understood. It appears that many plant species experience some stress in this altered atmosphere. This stress, defined as the deviation from normal, evolutionarily shaped homeostatic conditions. Particular consequences

*Author for Correspondence, e-mail: grene@vt.edu
† Present Address: Department of Molecular Biology and Genetics, Cornell University, 317 Biotechnology Building, Ithaca, NY 14853, USA

Constantin A. Rebeiz et al. (eds.), The Chloroplast: Basics and Applications, pp. 323–346.

ensue for the functioning of chloroplasts. Regulatory mechanisms that influence the operation of the photosynthetic machinery will be discussed, as will the current state of physiological and molecular genetic information concerning impacts, both direct and indirect, of the two gases on the chloroplast machinery. Recent reports are examined in light of that knowledge, and working hypotheses and interpretations and future experimental approaches are suggested that make use of the accumulated data on gene co-expression networks in model species, *Arabidopsis thaliana* in particular and its relative *Thellungiella halophila*, which is characterized by a very different lifestyle, and on the improved understanding of biochemical pathways and metabolism relating to interactions between plastids and the cytosol. Evidence is discussed arguing that the presence and increase of tropospheric ozone levels countermand or at least reduce the fertilizing effects of elevated [CO_2]. The necessity to increase food production worldwide will require enhanced efforts in breeding ozone-resistant crops.

I Introduction

Accelerated by or based on world-wide industrial and societal developments, atmospheric [CO_2] has steadily increased over the last century and a half. Although the potential for global change and sea-level change based on greenhouse gas effects has been recognized and countermeasures are gradually being implemented, global increases in [CO_2] will be difficult to contain and are likely to continue. Based on model studies the trajectory of the CO_2 concentration changes can be expected to lead to increased crop productivity. This would occur irrespective of constraints operating in the carbon fixation capacity of chloroplasts and a plant's ability to transport and utilize the excess fixed carbon and less in the generation of chemical energy in the light reactions. Stimulation of the CO_2-assimilation capacity on a biosphere-wide scale would counteract further increases of the greenhouse gas. This fact has provided a powerful justification for studying how the ability of plants could be enhanced to use this more readily available substrate effectively. According to global photosynthesis models (Peart et al., 1989; Parry et al., 2005) the expected increase of [CO_2] from the present level of ~375–550 ppm and possibly even higher by the middle of this century could be expected to increase plant productivity by more than 30%, with profound consequences for agriculture and planetary food security (Parry et al., 2004). However, such an increase has not been observed to date, while, in contrast, several meta-analyses of published data have indicated only minimal gains in yield or biomass, although it is becoming clear that differences exist between species and crops (Long et al., 2006; Ainsworth and Long, 2005). Effects on plant performance by increases in [CO_2] to levels predicted for later in this century have been intensively studied and analyzed (reviews by Nowak et al., 2004; Ainsworth and Rogers, 2007; Ziska and Bunce, 2007), and the somewhat controversial results have been extensively debated.

In addition to elevated [CO_2], plants face a second, equally man-made challenge that affects plant performance differently and could result in the reduction or at least cancellation of expected biomass and yield benefits resulting from increased [CO_2]. This challenge is presented by an increase in the concentration of ozone, O_3, in the troposphere. Combustion of fossil fuels, in manufacturing and significantly also in transportation, continues to increase nitric oxides which subsequently give rise to O_3 catalyzed by solar irradiation, and this

Abbreviations: [CO_2] – CO_2 concentration; ABA – abscisic acid; AGPase – ADP-glucose pyrophosphorylase; AOX1a – mitochondrial alternative oxidase gene/protein; APX1 – cytosolic ascorbate peroxidase gene/protein; AtSR – plants genes/proteins homologous of a mammalian redox-responsive pathway; C_2H_2 – proteins zinc-finger family gene/protein; COL-0 – *Arabidopsis thaliana* ecotype Columbia-0; CV –*Arabidopsis thaliana* ecotype Cape Verde Island; FACE – Free Air CO_2 Enrichment; GGM – graphical Gaussian model; GUN1 – regulatory gene encoding a step in chlorophyll biosynthesis; HSF21 – heat shock transcription factor; LE-ETR3 – ethylene receptor gene/protein in tomato; MAPK3 – mitogen activated protein kinase-3; MEX1 – maltose exporter gene/protein; PCD – programmed cell death; PP2C – protein phosphatase 2C family protein; PR – proteins several families of pathogenesis-related proteins; RCD1/CEO1 – protein with a "radical-induced cell death" phenotype, over-expression conferring tolerance to radical oxygen stress; ROS – radical oxygen species; SEX1 – starch excess gene/protein; SPS – sucrose phosphate synthase gene/protein; SRO1 – similar to RCD1 with NAD ± radical ADP-ribosylation activity; tAPX – thylakoid-located ascorbate peroxidase; WRKY – transcription factor family; WS – *Arabidopsis thaliana* ecotype Wassilewskija; ZAT10 – redox-responsive transcription factor

chemistry adds increasing amounts to the natural diurnal fluctuation of this gas (Zhang et al., 2004). Already at current levels, tropospheric ozone affects vegetation at concentrations, measured over the growing season, that commonly occur in or exceed agricultural regions in Western Europe and the United States. Also higher levels of ozone are expected in a number of rapidly industrializing developing countries (Heck et al., 1983; Pell et al., 1997; Vahala et al., 2003). Crops, particularly when grown downwind from industrialized areas, are subjected to varying ozone levels that range from natural levels of approximately 20 ppb to concentrations reaching or exceeding 150 ppb. Increases of just 20% over the natural concentration have been shown to result in yield reductions in a number of crops (Fiscus et al., 2005). The decline in yield is species-dependent. For example in soybeans, it may be substantial when O_3 exposure integrated over the season reaches levels that are by now quite common (Ashmore, 2005; Morgan et al., 2003). Potential problems that may arise from the increase in O_3 were foreseen many years ago but the severity of the increases has reached public debate and scientific attention (Giles, 2005) only during the last decade.

Scientific attention has been targeted to the predicted and now clearly emerging problem for approximately 20 years. This was achieved mostly through comparison of measurements with photosynthesis models and physiological and biochemical studies in growth chambers and greenhouses under controlled conditions in elevated [CO_2]. This led to field studies in open-top chambers, and more recently to experimentation with so-called FACE-facilities (Free Air CO_2 Enrichment) in an attempt to model the reality of plant growth in the field. In particular, FACE facilities – large ring structures that provide CO_2 depending on the direction of air flow such that a certain concentration is maintained within the rings, provide elevated CO_2 (and/or ozone) without confining the plants (Miglietta et al., 1996; Morgan et al., 2003; Bernacchi et al., 2006; Ziska and Bunce, 2007).

We will briefly review the substantial physiological and biochemical datasets and analyze approaches used to model the photosynthetic performance of plants in the changing atmosphere. As a clear message from the analyses, we suggest that molecular concepts must be progressively introduced as a tool for advancing understanding (e.g., Li et al., 2007). We argue for the use of genetic, molecular, including transgenic, and genomics-type approaches. To reveal the underlying mechanisms that give rise to the physiological parameters that are already well-documented, and focusing on the power provided by model species, ecotypes/breeding lines and mutants and transgenic crops, is a logical and necessary next step. *Arabidopsis thaliana,* with its many ecotypes and their evolutionary adaptations to different habitats, is a species that can advance understanding because of the tools available for this species. Bringing more of the genetic and genomics-type resources of *Arabidopsis* into play for a comprehensive view in the context of responses on the genome level, will enable us to understand the molecular events that underlie plant reactions in [CO_2]- and [O_3]-enriched environments.

We include a summary of the current understanding of chloroplast responses to environmental cues in general, then the discussion is narrowed to effects of elevated greenhouse gases. With an emphasis on chloroplast physiology and biochemistry, we will discuss the few studies that used molecular genetic tools for the analysis of plants in elevated [CO_2] and [O_3]. These recent studies focused on transcript profiling in poplar, soybean and *Arabidopsis* (Ainsworth et al., 2006; Liberloo et al., 2006; Li et al., 2006a, b) and, increasingly, on metabolite profiling as well (Ainsworth et al., 2006; Li et al., 2006a,b). We then present a recently developed bioinformatics tool which can be used to reveal networks of genes that co-respond to given stimuli, in this case, to greenhouse gases. Finally, we will consider how understanding of the consequences of elevated (greenhouse) gases on plant performance might be used in strategies that may improve plant performance and crop yields in an atmosphere that did not exist when many of these species emerged.

II Regulation of the Photosynthetic Apparatus: Metabolic and Environmental Signals

Chloroplasts continually report their shifting metabolic state to the nucleus (Lopez-Juez, 2007). Thus, studies querying responses of green cells to environmental changes constitute, of necessity, a search for the mechanisms that regulate a plastid's metabolic state, and the modes of reporting this state to the nuclear machinery that compares reality with genetically set values. The fundamental problem of the photosynthetically active plant cell, is to balance energy capture with downstream, consumption, distribution or storage processes, and to protect structures, and pathways that deal with carbon, and macromolecules from damage during periods of re-adjustment of that balance due to changes in the environment (Wilson et al., 2006). Not surprisingly, adjustments of the photosynthetic machinery to environmental cues takes the form of regulatory mechanisms that are manifest at the transcript, protein and metabolite levels. Temperature and light shifts, CO_2 availability (with attendant effects on photorespiration), sugar-responsive signaling, end-product inhibition, the tight coupling of carbon and nitrogen metabolism, and changes in the cellular redox environment, exert major and overlapping controls (Geigenberger et al., 2005; Wormuth et al., 2006; Schrader et al., 2007). Importantly also, it has now become accepted that mitochondrial functions exert an essential influence on chloroplast activities in vivo (Paul and Foyer, 2001; Noctor et al., 2007; Plaxton and Podesta, 2006; Nunes-Nesi et al., 2007).

Environmental changes induce protein kinase-mediated alterations in antenna pigment size or state transitions between the photosystems (Bellafiore et al., 2005), in the rapid protease-mediated turnover of D1 protein of PSII, in photoinhibition (Sun et al., 2007), and changes in the redox poise of the plastoquinone pool (Pfannschmidt, 2003), which influences photosynthetic gene expression. These responses serve as one layer of sensing mechanisms for the regulation of energy disposition within the thylakoid membrane, the ATP synthase and ATP/ADP exchange functions. Further downstream, evidence has been provided for redox-mediated adjustments *via* transcriptional and post-transcriptional changes of the entire photosynthetic machinery to light, temperature, end-product accumulation, or oxidative challenges (Buchanan and Balmer, 2005; Fey et al., 2005; Dietz et al., 2006). Redox signaling has been attributed by some to a "master switch", *via GUN1*, a regulatory gene encoding a step in the chlorophyll biosynthetic pathway (Nott et al., 2006; Koussevitzky et al., 2007). Simultaneously, redox signaling pathways trigger the expression of defense processes (Mittler et al., 2004), and the biosynthesis of antioxidants such as thioredoxins (Balmer et al., 2006; Lemaire et al., 2007) and peroxiredoxins (Dietz et al., 2006). The regeneration part of the Calvin cycle is also under redox control, being modulated at several enzymatic steps by the thioredoxin/ferredoxin couple, for example in redox-dependent assemblies of enzymes (Howard et al., 2008). This makes this pathway especially vulnerable to environmental influences that result in or aggravate oxidative stress. Transport of assimilate across the chloroplast envelope is also subject to metabolite and redox control (Weber et al., 2005).

In a whole cell context, Davletova et al. (2005) and Rizhsky et al. (2004) present convincing evidence for an ROS-responsive signaling pathway involving cytosolic ascorbate peroxidase (APX1), the transcription factor Zat12, and heat shock transcription factor 21 (HSF21). ROS originating in the chloroplast can act as a trigger for this signaling pathway. In Drosophila, HSFs have been demonstrated to undergo a redox-modulated conformational change which leads to altered DNA binding (Ahn and Thiele, 2003), thus suggesting one model for the translation of redox signal into gene activation.

Starch and sugar metabolism respond to many signals. Sugar signaling (e.g., *via* SNF1-related kinases and hexokinase), redox (e.g. *via* the redox modulation of ADPG-pyrophosphorylase activity), light, ABA, and ethylene exert overlapping controls on the two pathways (Tiessen et al., 2003; Koch, 2004; Geigenberger et al., 2005; Rolland et al., 2006; Rook et al., 2006). Trehalose-6-phosphate, which has been studied intensively in the past few years, is now known to exert profound control over environmental responses of carbohydrate metabolism (e.g. by conveying sucrose status from the cytosol to the chloroplast) and plant development, e.g. cell wall structure and cell wall development (Gomez et al., 2006; Paul, 2007). Furthermore, the regulation of expression of genes associated with carbon and nitrogen metabolism are tightly linked through induction of nitrogen metabolism genes in response to sugar supply (Blasing et al., 2005). Cross et al. (2006) showed specifically that it is, in fact, the metabolic intersection between carbon and nitrogen metabolism that responds to changes in sugar supply.

III Possible Scenarios Explaining Effects of Elevated [CO_2] and [O_3] on Plant Behavior in the Altered Earth Atmosphere

In response to increased [CO_2], which may be viewed as non-physiological, several scenarios can be envisaged. The use of the term "non-physiological" could be justified by pointing to the long term much lower concentration of [CO_2], which amounted to 250 ppm for approximately 20 million years (Houghton, 2001). Species would have evolved biochemical machineries that made effective use of the gas available at the time. First, one could imagine a control system that adjusts the photosynthetic and CO_2 fixation machineries to levels, either reduced or enhanced, that are synchronized to other metabolic parameters. This may be termed the "weakest link model". While initially more carbon will be fixed, an eventual decline could be actively promoted because of an imbalance imposed by the overall biosynthetic capacity of the species or line. One such imbalance could be, for example, the regeneration phase of the Calvin cycle. Increased investment in (some) Calvin cycle enzymes and ADP-glucose pyrophosphorylase appears to be incomplete or impossible for most C_3 plants in elevated [CO_2] (Zhu et al., 2007). Another weak link could be due to a perturbation of the complex interrelationships surrounding the carbon and nitrogen balance, i.e. the C/N ratio. In this case, legume species capable of N-fixation would experience advantages over other species in elevated [CO_2], if N-fixation is, or can be made responsive, to an increased amount of organic acids/carbohydrates.

However, other scenarios possibly preventing efficient utilization of the increased substrate for primary C-fixation cannot be excluded. Several possibilities can be contemplated. First, a response that correlates stomatal aperture and [CO_2] concentration would limit c_i and, thus, curtail CO_2 fixation. Plants indeed show restricted stoma aperture in elevated [CO_2] (Buckley, 2008), which in itself will then increase osmotic balance and prevent water loss. In other words, elevated [CO_2] would generate benefits only under conditions of water scarcity. Thus it could be argued that any increases in productivity in the field may be based on water conservation, rather than on [CO_2]. Based on evolutionary history, this behavior may be termed a "physiological trap". Second, several other factors may be imagined that could negate a beneficial effect by elevated [CO_2], which might summarily be labeled as "bottlenecks of distribution". These may be based on the capacity of chloroplasts to store transient starch, to export/exchange triose-phosphates, sugars, or nitrogen-containing compounds. Deficiencies may also be due to long-distance transport processes, e.g., from mesophyll cells to the phloem, to size and capacity of phloem tissues, and to unloading of carbohydrates or other end-products of primary metabolism into transient or permanent sinks (Lalonde et al., 2003, 2004). Also, insufficient capacity to sustain accelerated development could be invoked, possibly based on limited hormone or cofactor production, or on osmotic constraints based on ion imbalances that could impose stringent control over carbohydrate-relevant biochemical pathways.

Finally, two other factors, based in part on already existing data, are in our opinion likely candidates for explaining differences between the modeled, expected gain in productivity caused by elevated [CO_2] and reality. These factors are, first, the sensing of an imbalance in the C:N ratio that is perceived as a stress by the plants, and/or, second, the effect of tropospheric ozone that appears to counteract the positive effect of elevated [CO_2]. The documented and possible effects of ozone are discussed separately.

In a physiological context most easily rationalized can be the reduction in Rubisco content accompanied by higher activation of the remaining Rubisco enzyme (Parry et al., 2003). Irrespectively, elevated [CO_2] leads to increases in organic acids, sugars and sugar alcohols in *Arabidopsis* (Ainsworth et al., 2006; Li et al., 2006b). In C_3 plants incapable of nitrogen fixation, one consequence that is poorly understood seems to be insufficient nitrogen provision (Ainsworth and Rogers, 2007). Thus, "extra" carbon skeletons becoming available in the cytoplasm may, especially at night, support glycolysis and the mitochondrial TCA cycle (Ainsworth et al., 2006; Li et al., 2007). Increased respiration has indeed been observed in elevated [CO_2] in the light while photorespiration was reduced (Pärnik et al., 2007). Any subsequent conversion into amino acids may, however, be impeded by a lagging capacity of cells to adjust and increase organic nitrogen through nitrate uptake, reduction or incorporation. In favor of such an interpretation is the observation that N-fixing soybeans appear to gain from growing in elevated [CO_2] (Rogers et al., 2006).

A Plant Responses to Elevated [CO_2]

Plant growth and functioning in an altered environment leads to adjustments at the molecular genetic, biochemical, and, eventually, physiological and developmental levels. Clearly, short-time exposure to high [CO_2] in controlled environments showed substantial increases in CO_2-assimilation rates, but prolonged exposures often revealed "acclimation" (Wong, 1979; Jifon and Wolfe, 2002), a process leading to decreased activity of the photosynthetic machinery (Wong, 1979; von Caemmerer and Farquhar, 1984; Stitt, 1991; Bowes, 1993). The acclimation process is accompanied by a decrease of Rubisco transcripts, loss of Rubisco protein, and a general down-regulation of the photosynthetic light reaction machinery (Long et al., 2004; Li et al., 2007). In many studies, the importance of nitrogen supply and interdependence of photosynthetic activity and nitrogen nutrition have been pointed out (e.g., Stitt and Krapp, 1999; Smith et al., 2003). Developmental and structural effects of growth in elevated [CO_2] include a decrease in the number of mitochondria and chloroplasts (Wang et al., 2004) and changes in chloroplast ultrastructure (Griffin et al., 2001; Oksanen et al., 2005).

To paraphrase the results, the consequences for developmental, and physiological and metabolic adjustments to growth in increased [CO_2] have been rationalized in various contexts. Effects on chloroplast functioning are the central theme. Long distance source–sink relations, Rubisco as a nitrogen sink, an altered C/N ratio, stomatal behavior and the engagement of control circuits for the expression of photosynthetic genes and proteins influenced by sugar sensing represent the crucial mechanisms invoked to explain the physiological effects of elevated [CO_2] (Jang et al., 1997; Ainsworth et al., 2006; Rachmilevitch et al., 2004; Bernacchi et al., 2005; Bloom, 2006; Rolland et al., 2006; Young et al., 2006; Li et al., 2007).

However, these explanations in their majority have been prompted by measurements of physiological and phenotypic parameters, which reported medium- or long-term consequences of sensing and signaling events typically far removed in time from the effect that is recorded. More complete understanding should be possible after focusing on gene expression characteristics, protein composition, destination and activity, and influences exerted by metabolite pools and composition on the genetic machinery of plants in time series that span a season or seasons.

Long-term studies of the effects of elevated [CO_2] have been carried out in poplar and other tree species (LaDeau and Clark, 2001; Bernacchi et al., 2003; Karnovsky, 2003; Liberloo et al., 2006). Sustained increases in growth under elevated [CO_2] in poplar were correlated with sink capacity and nutrient supply (Liberloo et al., 2006).

Several reports have become available that focus on ecological aspects of plants in the FACE atmosphere and insect and pathogen predation have been studied (Hamilton et al., 2005; Knepp et al., 2005, 2007). Growth in elevated [CO_2] has been reported to result in increased defenses against herbivory (Bidart-Bouzat et al., 2005; Knepp et al., 2005). While the increased sugar content in elevated [CO_2] might serve as an attractant for insects and pathogens, increases in plant secondary products, e.g., in the phenylpropanoid and flavonoid pathways (Knepp et al., 2005, 2007), could lead to enhanced protection. A study of defense responses in three genotypes of *Arabidopsis* showed that herbivory resulted in increases in glucosinolate levels under elevated CO_2 but not under ambient CO_2 (Bidart-Bouzat et al., 2005). This response was genotype-specific, and it serves as an illustration of the genetic diversity underlying environmental responses in plants. In soybeans, on the other hand, growth in elevated CO_2 or ozone under FACE conditions resulted in increased susceptibility to insect herbivory (Hamilton et al., 2005; Zavala et al., 2008).

A significant way in which plants grown in FACE environments differ from plants raised in growth chambers or greenhouses is the presence of ambient ozone. Studies in controlled environments generally modified [CO_2] but overlooked ozone in general, or at concentrations that can be found in fields or natural ecosystems. We argue that this difference may be the important factor contributing to the observed differences in plant behavior under open-air versus controlled experimental conditions. Also, the combination of ozone and elevated [CO_2] in FACE rings appears to have a compounding effect on the biochemical reactions of plants that are exposed to higher ozone in ambient [CO_2] in the field. O_3 and CO_2 appear to have cumulative effects, which quite possibly constitute the major factor preventing most plants, at least those that lack an exceptionally large carbon sink, from reaching the benefits predicted by models of photosynthesis.

B Plant Responses to Tropospheric [O_3]

The information on plant-ozone interactions is difficult to place into a unified framework mostly due to differences in treatment conditions. Many studies have been conducted in controlled environments, often using acute treatments with high, unrealistic doses of O_3 that generated visual damage, with control plants in filtered air. Such treatments resulted in the activation of a programmed cell death (PCD) response which is very similar to the PCD that occurs during the hypersensitive response in an incompatible plant-pathogen interaction (Kangasajärvi et al., 1994, 2003; Pellinen et al., 1999; Wohlgemuth et al., 2002; Baier et al., 2005). The ozone-induced PCD is characterized by an oxidative burst, i.e. the production of more ROS, the induction of pathogenesis-related (PR) proteins (Rao and Davis, 2001; Langebartels et al., 2002), as well as the induction of defense genes (Tamaoki et al., 2003). Evidence for the participation of a signaling pathway involving MAPK3 and 6 in the response to acute ozone exposure was provided by Ahlfors et al. (2004).

Antioxidant responses to acute ozone exposure at the transcriptional and enzyme activity levels have been extensively documented for some time (Kliebenstein et al., 1998; Bernardi et al., 2004; Torres et al., 2007). For example, antioxidants play a key role in resistance to ozone (Conklin and Last, 1995; Pitcher and Zilinskas, 1996). Overall, such effects, however crucial they may be for resistance to ozone, are most likely downstream from the initial ozone sensing and signal transduction events located at the cell wall-plasma membrane continuum. Baier et al. (2005) present several possible alternative routes for ozone sensing from the apoplast across the cell membrane to the intracellular transmission of redox-responsive signals, involving lipid-based signaling, calcium responses, antioxidant signaling, and the downstream generation of ROS.

Hormonal participation downstream of ROS production in response to acute ozone exposure has been well documented (Kangasjarvi et al., 2005), including the identification of hormone response elements in the upstream regions of Arabidopsis genes that are induced under acute ozone (Mahalingam et al., 2006). Jasmonic acid appears to ameliorate ozone-induced PCD (Rao et al., 2000), while both salicylic acid and ethylene appear to enhance the damaging effect of high levels of ozone (300 ppb, 4h) (Rao et al., 2002; Baier et al., 2005). Nitric acid and ethylene also appear to interact to induce the mitochondrial alternative oxidase *AOX1a* under acute ozone exposure (Ederli et al., 2006). Castagna et al. (2007) reported, however, that the tomato ethylene receptor LE-ETR3 is not involved in ozone sensitivity responses.

Acute ozone exposure in the 300 ppb range or higher rarely occur outside the laboratory. As is stated explicitly by Mahalingam et al. (2006), brief exposures to high ozone levels serves the purpose of elucidating existing ROS response pathways in plant cells. In contrast, the question of how low level, chronic, ozone levels affect plant cell function has not received much attention (Baier et al., 2005). Another difficulty in generalizing from the results of acute ozone exposure studies in controlled environments, to conditions in the field is that ambient atmospheric ozone is never close to zero in the field, as it is in the controls used for chamber studies (~22 ppb). Indeed the effect of existing ambient ozone, is known to suppress soybean yield by 5–15% (Burkey et al., 2005; Fiscus et al., 2005). Furthermore, Miyazaki et al. (2004) demonstrated that exposure of *Arabidopsis* plants to ambient conditions in the field, have resulted in increased expression of a large number of genes that are commonly considered biotic and abiotic stress-related, in comparison to plants exposed to control conditions in growth chambers. A subset of these stress genes, heat-shock proteins in particular, was further up-regulated when the plants were grown in high O_3 or elevated [CO_2]. These two sets of data unmistakably illustrate the specificity of the problem, where the effects on plant cell function of small incremental changes, rather than acute shocks, in ozone levels in the field, must be measured to advance our understanding.

Ozone effects observed in the field are the result of marginal elevations over ambient levels.

Consequently attendant cellular changes may be relatively subtle, and not always reflect the same response pathways as in acute chamber exposures. The latter indeed record significant differences as compared to ambient ozone conditions. Much information is now available concerning the physiological and yield effects of long-term exposure to lower ozone levels. The ozone concentration balance to which plants are exposed to in a diurnal cycle that may peak at 60–100 ppb is very different in the field. The effects of ozone on soybean have been well studied at the physiological level. Morgan et al. (2003) used meta-analysis to quantitatively summarize the response of soybean to chronic ozone exposure from 53 reports. Overall, exposure to chronic ozone levels resulted in significant decreases in biomass, seed production and photosynthesis. At higher exposure levels, decreases in leaf area were also observed. The impact of ozone was greatest after seed filling. Yield and growth decreases and negative effects on photosynthesis in general, and on Photosystem II in particular, have been reported for other crop species (see the review by Fiscus et al., 2005, and references therein).

Mechanistic studies, as opposed to growth- or yield-orientated work, of the effects of chronic, season-long exposure to ozone at only marginally elevated concentrations closer to reality in natural settings have been rare (Vahala et al., 2003; Ainsworth et al., 2006; Li et al., 2006b). Ethylene was found by Vahala et al. (2003) to accelerate senescence under chronic ozone exposure.

Li et al. (2006b) compared responses of three *Arabidopsis* ecotypes with differing sensitivities to chronic ozone in FACE. Two ecotypes appeared as sensitive, one showed resistance. JA biosynthesis and JA responsive genes responded to ozone exposure only in the most sensitive ecotype. It would appear that the threshold for ozone-mediated induction of JA-related processes was only reached in this one ecotype. Further evidence for hormone-regulated, ozone-induced changes in gene expression is seen by the finding that both RCD1/CEO1, a WWE domain protein, and the closely related SRO1 (Ahlfors et al., 2004), were negatively affected in the most sensitive ecotype, while there was no effect of ozone exposure in RCD1 or SRO1 in the other, more resistant, ecotypes. Over-expression of the *RCD1* gene has been shown to confer resistance specifically to apoplastic ROS and to modulate methyl jasmonate and ABA signaling. The gene had been discovered underlying the phenotype of an ozone-sensitive mutant (Ahlfors et al., 2004; Overmyer et al., 2005).

Members of the AtSR gene family, which are homologs of the redox-responsive NF kappa Beta pathway that is well-studied in animals, were also repressed in the most ozone-sensitive Arabidopsis ecotype, with no response in the case of the other ecotypes. Since the sensitive ecotype showed foliar lesions as a result of ozone exposure, either or both of these redox-responsive genes and their pathways could have been involved in PCD. In contrast, in the case of those ecotypes that showed less damage the PCD-signaling pathway did not appear to have been stimulated.

Interestingly, the most resistant ecotype, showed an increase in the expression of another member of the AtSR gene family (At2g22300), and also showed higher "constitutive" expression of several stress resistance genes, such as members of the glutaredoxin family. From these ecotype-specific expression profiles a hypothesis emerged concerning the existence of a more sensitive pathway conferring oxidative stress resistance in Cvi-0 which is independent of the less sensitive and established hormone-related pathway leading to PCD. The pathway is documented for the typically used high levels of ozone exposure. In this context, a novel redox-responsive pathway in *Arabidopsis* was detected in double mutants of cytosolic and thylakoid-bound ascorbate peroxidases (Miller et al., 2007). The existence of higher constitutive levels of key stress-related genes in the acclimating plants, as compared with the two that exhibited injury, is in agreement with other data demonstrating the contribution of stress "readiness" to resistance mechanisms (Inan et al., 2004).

Urgently required are studies of plant behavior under natural conditions, because paradigms and models based on growth chamber studies are not sufficient. Even with the little information that is available about plant behavior at low chronic ozone levels, it is clear that further research will uncover novel pathways of redox signaling operating in field-grown plants that are masked by, and likely different from, the responses to acute levels of ozone exposure under controlled conditions.

C Combined Effects of [CO_2] and [O_3]

In our opinion, the aspect that is the most important determinant for the physiological behavior of plants in the field, and an important - though largely neglected - component in realistic models of crop performance in an altered atmosphere are the combined, synergistic or antagonistic effects, of increased [CO_2] and ozone levels that are higher than in pre-industrial times. To date, this essential information is still missing.

Long term exposure to chronic ozone levels in the field have generally been found to counteract stimulatory effects on growth by elevated [CO_2] (Ainsworth and Long, 2005). However, it was also reported that the presence of elevated [CO_2] did not prevent ozone-induced premature senescence in soybean (Fiscus et al., 2005). In contrast, ozone-mediated effects that lead to reductions in yield have been reported to be counteracted or prevented by growth in elevated [CO_2] (Morgan et al., 2003; Booker et al., 2007). The tendency of plants to close stomata more tightly in the presence of both gases could be of advantage because it would restrict evapotranspiration (Kim et al., 2006). An altered stomatal aperture would lead to increased leaf and canopy temperature which could induce heat shock responses. Indeed, increased expression of a number of HSP-related functions in the field under both elevated [CO_2] and ozone levels have been reported (Miyazaki et al., 2004; Li et al., 2006a, b).

Irrespectively, a comparison of *Arabidopsis* ecotypes grown in FACE rings also indicated variability of or limitations in the plants' ROS-scavenging capacity. Assuming that such variability will also be found in other species and in crop species it seems safe to assume that breeding efforts directed towards such an improvement will be successful. In any case, natural ecosystems are likely to experience changes that are difficult to predict, or to experience more chaotic changes in species distributions than at present, if the increases in [CO_2] and tropospheric ozone should persist or accelerate.

IV Benefits from Model Species: *Arabidopsis thaliana* and *Thellungiella halophila*

Listing resources and tools that are available for *Arabidopsis thaliana* need not be repeated here. It is clear however that they include the necessary tools for in depth studies of the plant's development, biochemistry and physiology. Most importantly, molecular tools are available from genome and gene sequences for every one of the roughly 30,000 genes that can be mutated. We have used *Arabidopsis* in FACE experiments (see above). In addition, we have used *Thellungiella halophila*, a close relative of *Arabidopsis*, that presently lacks a genome sequence. However molecular tools, such as ESTs for many treatment conditions, a set of full-length cDNAs (Shinozaki K., personal communication, 2007), and a set of tagged mutants are available (Bressan et al., 2001; Amtmann et al., 2005; Volkov and Amtmann, 2006; Wong et al., 2006; Griffith et al., 2007). *Thellungiella* is of interest because of its close relationship to Arabidopsis. It also shows extremely high tolerance to abiotic stresses, drought, salinity and – in particular – cold. In addition, the species includes a number of ecotypes collected from extreme ecosystems, such as the estuary of the Yellow River (Shandong, China), the Yukon (Canada), or the dry deserts of Xinjian (China). An additional attractive aspect of *Thellungiella's* lifestyle is that the plants are effective scavengers for ions and nutrients that are in short supply in their natural habitats (Griffith et al., 2007). *Thellungiella* (ecotype Shandong) in FACE experiments provided a strong counterpoint to the behavior of *Arabidopsis* and we observed significant differences in the reaction to elevated CO_2 characterizing *Arabidopsis* ecotypes as well (Li et al., 2006a, b, 2007).

A view of metabolite concentrations after the plants had been exposed to elevated [CO_2] for ~2 weeks exemplifies the differences between the two species (Table 1). Total metabolite concentration in *T. halophila* was unchanged in elevated [CO_2], but the composition changed, with a smaller proportion of the carbon present as organic acids. Sugars and sugar alcohols were higher in elevated [CO_2] than in ambient air in *Thellungiella*, and were higher than in *Arabidopsis*. Significantly, metabolites for all of carbon metabolism in the plastid increased, including starch, sugar alcohols and organic acids.

A major contributor to sugar content in *T. halophila*came from raffinose/galactinol and inositol. These sugars were present at higher levels than in *Arabidopsis*. Sorbose also contributed highly, in *T. halophila*, but was present only in trace amounts in the *Arabidopsis* ecotypes. Inorganic phosphate, inositol, glutamic-, aspartic-, malic- and citric acids, were also significantly higher under all conditions

Table 1. Effects of elevated carbon dioxide on metabolite levels in three ecotypes of *Arabidopsis thaliana (CV, WS, COL)* and *Thellungiella halophila (TH)*.

	CV-A	CV-CO_2	WS-A	WS-CO_2	Col-A	Col-CO_2	Th-A	Th-CO_2
Organic acids	3,964.1	2,453.1a	6,476.4	7,655.2b	8,791.3	7,364.9b	8,723.1	7,124.3a
Amino acids	721.1	451.5a	1,619.7	2,042.5	4,572	2,988a	8,189.9	7,634.5b
Sugars	2,959.1	2,375.7b	5,959.9	7,182.4a	8,616.4	7,526.1a	12,035.4	14,201.6b
Sugar alcohols	520.5	221a	993.7	961.4	1,578.3	1,442.3	2,369.4	3,224.4
Total	8,164.8	5,501.3a	15,049.7	17,841.5b	23,558	19,321.3a	31,317.8	32184.8

For each sample >10 individual plants (above-soil material) were combined. Metabolites were measured by GC/MS in ambient and elevated [CO_2]. Values for Arabidopsis ecotypes have been reported previously (Li et al., 2006a)
a – Significantly different from the corresponding control at p = 0.01; b – significantly different from the corresponding control at p = 0.05; no letter – not significant
A = ambient conditions, CO_2 = elevated carbon dioxide
Units are expressed as µg g^{-1} fresh weight

in *T. halophila* than the levels in the *Arabidopsis* ecotypes. In contrast, soybeans grown in elevated [CO_2] exhibited higher foliar amino acids and sugars than control plants by season's end, but lower levels of amino acids at the beginning of the season (Rogers et al., 2006).

Transcripts in elevated [CO_2] mirrored to a large degree the metabolite characteristics (Fig. 1; Table 2). Results for genes involved in sucrose and starch metabolism are shown, exemplified by genes involved in sucrose and starch metabolism. Sucrose phosphate synthase (SPS) genes were up-regulated in Col-0, WS, and *T. halophila*. Genes encoding ADP-glucose pyrophosphorylase (AGPase) were up-regulated in all lines with one AGPase down-regulated in Col-0 only. Both soluble and granule-bound starch synthases were down-regulated in Col-0 and *T. halophila*. At least one sugar transporter responded in each of the four lines. For example, up-regulation of the *Arabidopsis* MEX1 (Niittyla et al., 2004), a maltose exporter, in *T. halophila* is of note, and UDP-glucose dehydrogenase, a crucial enzyme in the allocation of carbon for cell wall biosynthesis, showed regulatory differences among lines. Gene expression patterns are consistent with an increase in the flow of carbon to the cell wall in *T. halophila* with increases in two UDP-Glc dehydrogenase genes, as well as AXS1 and NAD-dependent epimerase/dehydratase, which catalyze apiose and xylose synthesis. The UDG3 expression patterns observed in WS were similar to those in *T. halophila*, although increased transcripts for apiose and xylose-related genes were not observed. Col-0 showed generally weaker reactions for all transcripts in this category. A particular lucid example of species- and ecotype-specific differences is the regulation of sucrose synthase (SUS) genes. In each *Arabidopsis* ecotype, one of the SUS genes was significantly up-regulated, while in *Thellungiella* the SUS isoforms were regulated in a less obvious manner.

Transcripts increased for many proteins in several pathways that generate and transport carbon compounds. Also, transcripts increased for cytosolic and mitochondrial carbon channeling into energy production. By that criterion, transcript abundance in comparison to ambient [CO_2], for glycolysis and respiration were enhanced in elevated [CO_2] (Li et al., 2006a, b; and unpublished), as reported by others for soybean (Ainsworth et al., 2006). Ecotype Cvi-0, and especially *Thellungiella*, appeared to be less affected than other *Arabidopsis* ecotypes.

In contrast to sugars, increases for the amino acid biosynthesis pathways in the Arabidopsis ecotypes was less pronounced or missing, while *Thellungiella* showed up-regulation in genes defining N-metabolism including amino acid that reflect the changes in amounts (Table 1). Particularly, increases in the aromatic amino acids, F, W and Y, were pronounced (data not shown). Significantly, the overall behavior of transcripts and metabolites strongly suggested that glycolysis and mitochondrial functions in respiration were up-regulated, but this appears to exclude a corresponding diversion of carbon and ammonia into amino acid biosynthesis pathways at least in *Arabidopsis*.

Under ozone exposure, control levels of gene expression in *Thellungiella* showed the greatest number of differences compared to control values

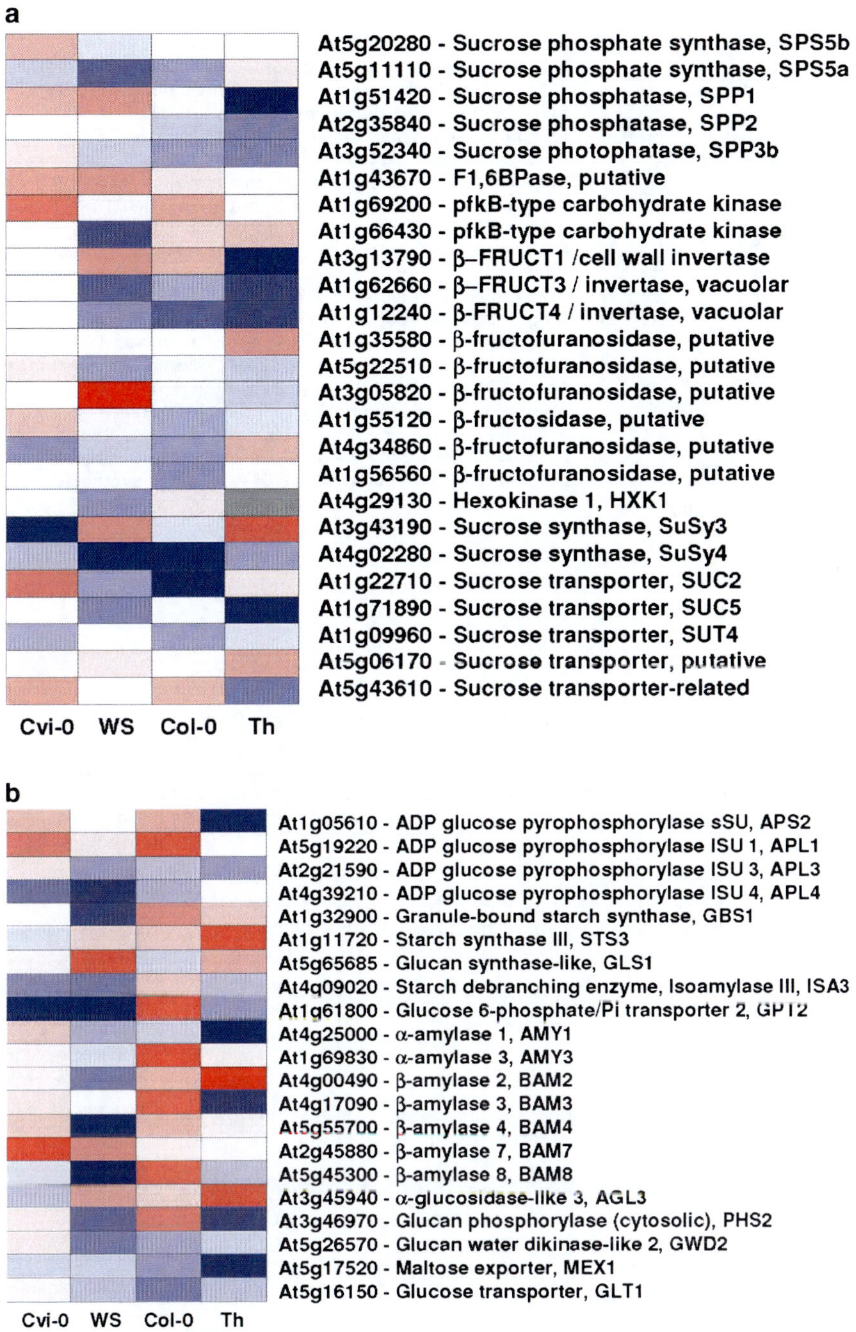

Fig. 1. Transcript analysis comparing genes involved in carbon and cell wall metabolism in three *Arabidopsis* ecotypes and *Thellungiella*. Transcripts are listed in (a) Sucrose, (b) starch

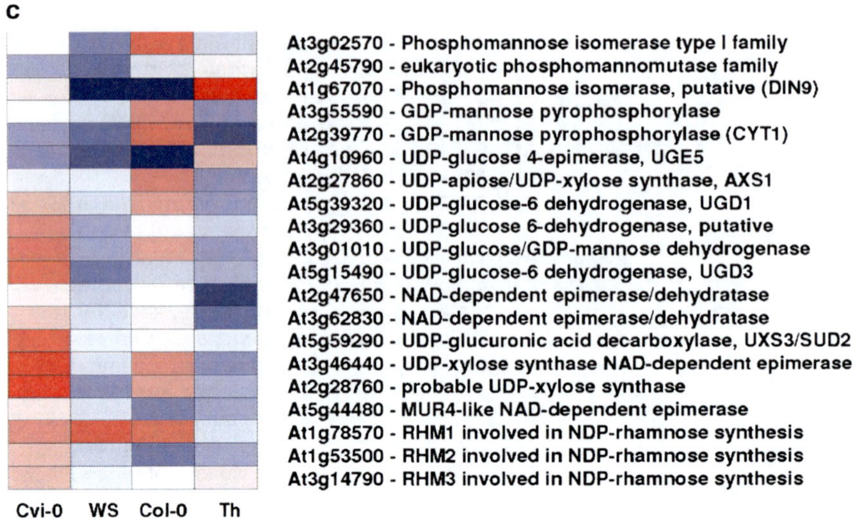

Fig. 1. (continued) and (c) cell wall metabolism that show significant changes in at least one *Arabidopsis* ecotype or in *Thellungiella* in elevated [CO_2] in FACE rings in the field. The range of up-regulation (blue) and down-regulation (red) is from +3 to -3. Values for *Arabidopsis* have been reported before (Li et al., 2006a).

Table 2. Genes transcriptionally correlated with the transcript of Starch Excess Protein 1 (SEX1).[a]

AAP1	Amino acid permease I (AAP1)
ACD32.1	31.2 kDa small heat shock family protein/hsp20 family protein
APL3	Glucose-1-phosphate adenylyltransferase large subunit 3 (APL3)
APRR3	Pseudo-response regulator 3 (APRR3)
AT1G03310	Protein with strong similarity to isoamylase (EC:3.2.1.68)
AT1G04620	Coenzyme F420 hydrogenase family / dehydrogenase, beta subunit family
AT1G69830	Plastid localized alpha-amylase. Expression is reduced in the SEX4 mutant
AT1G70820	Phosphoglucomutase, putative/glucose phosphomutase, putative
AT1G73470	Expressed protein
AT1G80130	Expressed protein
AT1G80480	PRLI-interacting factor L, putative, Cobalamin synthesis protein/P47K
AT2G15890	Expressed protein
AT2G25730	Expressed protein
AT2G28900	Mitochondrial inner membrane translocase subunit Tim17/Tim22/Tim23 family protein
AT2G42530	Cold-responsive protein / cold-regulated protein (cor15b)
AT2G47390	Expressed protein
AT3G01310	Expressed protein,
AT3G05660	Similar to disease resistance family protein [Arabidopsis thaliana]
AT3G29320	Glucan phosphorylase, putative
AT3G46970	A cytosolic alpha-glucan phosphorylase
AT3G47860	Apolipoprotein D-related
AT3G52180	Protein tyrosine phosphatase/kinase interaction sequence protein (PTPKIS1)
AT3G55760	Expressed protein
AT4G04330	Expressed protein
AT4G09020	Isoamylase, putative / starch debranching enzyme, putative
AT4G17110	Expressed protein
AT4G19120	Early-responsive to dehydration stress protein (ERD3)
AT4G24450	Similar to starch excess protein (SEX1)
AT4G30650	Hydrophobic protein, putative/low temperature and salt responsive protein, putative
AT4G30660	Hydrophobic protein, putative / low temperature and salt responsive protein, putative
AT4G32340	Expressed protein
AT4G33490	Similar to nucellin protein, putative

(continued)

Table 2. (continued)

AAP1	Amino acid permease I (AAP1)
AT5G23240	DNAJ heat shock N-terminal domain-containing protein
AT5G24300	Starch synthase, putative
AT5G26570	Similar to starch excess protein (SEX1)
AT5G27930	Protein phosphatase 2C, putative / PP2C, putative
AT5G52310	Low-temperature-responsive protein 78 (LTI78); desiccation-responsive 29A (RD29A)
AT5G64860	4-Alpha-glucanotransferase, putative/disproportionating enzyme, putative
ATHVA22A	ABA-responsive protein (HVA22a)
ATTPS5	Glycosyl transferase family 20 protein/trehalose-phosphatase family protein
COR15A	Cold-responsive protein / cold-regulated protein (cor15a)
COR314_TM2	Stress-responsive protein, putative
COR414_TM1	Similar to cold acclimation WCOR413-like protein gamma
DPE2	Glycoside hydrolase family 77 protein
KIN1	[AT5G15960, stress-responsive protein (KIN1) / stress-induced protein (KIN1)
NHL12	Harpin-induced family protein/HIN1 family protein
PGM	Phosphoglucomutase, chloroplast (PGM)
SBE2.1	1,4-Alpha-glucan branching enzyme / starch branching enzyme class II (SBE2-1)
SEX1	Starch excess protein (SEX1)

[a] Abbreviated annotations according to TAIR (http://www.TAIR.org); the corresponding network graph is shown in Fig. 2

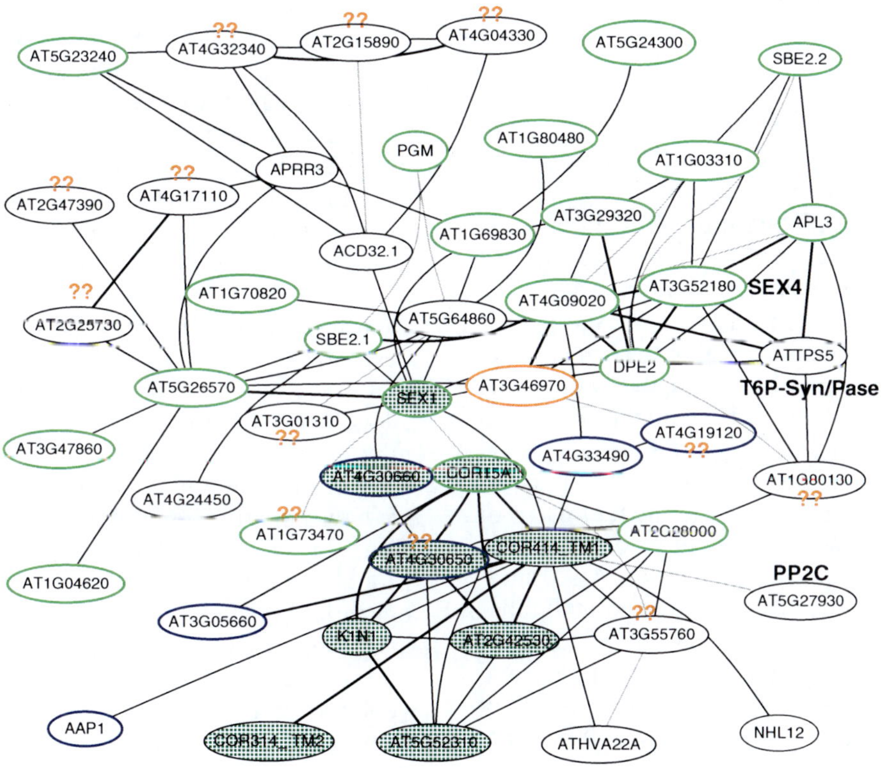

Fig. 2. Gene network centered on SEX1 (starch excess 1). The network graph, based on GGM (Ma et al., 2007), is centered on the SEX1 gene (AT1G10760), which showed up-regulation in all lines (Fig. 1). The genes shown are transcriptionally connected by one extension to the SEX1 transcript. The location of proteins encoded by these genes is indicated by colored lines: chloroplast (green), membrane systems (blue), known cytosolic (orange). Many non-colored transcripts are most likely also cytosolically located. Transcripts encoding unknown functions are indicated (??). A sub-group of the genes in this network is also regulated by low temperature (blue). T6P-synthase – trehalose-6-phosphate synthase; PP2C – putative protein phosphatase 2C; SEX4 – similar to SEX1. Genes are listed in Table 2.

for the three Arabidopsis ecotypes. This in turn suggested that differences in constitutive gene expression constituted a major contributor to its superior resistance in comparison to *Arabidopsis*. O_3-response diversity across the four lines was clearly related to a different response in components of signaling and transcriptional response networks. The most ozone-resistant Arabidopsis ecotype exhibited changes in a pathway involving AtSR, a homolog of the mammalian NFB family of redox-sensitive transcription factorsas well as changes in chaperones, WRKY and C_2H_2 proteins, and antioxidants, as discussed above. In contrast, the most ozone-sensitive Arabidopsis ecotype displayed ozone-mediated decreases in the expression of two AtSR/NFkappaB family members, in RCD1, a key regulator in the PCD pathway (Ahlfors et al., 2004, Kangasjarvi et al., 2005), in C2 domain proteins, and genes associated with cell wall processes, as well as changes in the expression of marker genes for PCD.

V Discussion

A The Importance of Model Species

The current understanding of chloroplast responses to environmental signals is mostly based on indirect physiological observations that measured to what extent photosynthesis or CO_2 fixation were capable of adjusting to such external deviations from homeostasis. At the same time, much less work has been directed to the mitochondrial compartment, and very little is understood about the molecular interactions between plastids and mitochondria, although the biochemistry of photorespiration that includes three organelles is well studied (Taiz and Zeiger, 2006). Also, studies are available that record biochemical functions of these organelles, such as fatty acid, lipid, hormone or secondary product biosynthesis and the availability of ATP (Wise and Hoober, 2006; Plaxton and Podesta, 2006). Known functional genomic consequences of exposure of plants to elevated [CO_2] and/or ozone can be placed in that biochemical context.

At present, the plant community can rely on a few genome sequences, while several more have appeared in draft form, and more will be forthcoming in rapid succession owing to the fast progress in DNA sequencing. Such sequences are an absolute necessity but they are not sufficient in casting enough light on the way sequences determine the lifestyle of a species. Needed are tissue and cell-specific gene and protein expression pattern, as well as mutants, and expression profiles over time and under a variety of external factors, namely chemical treatments, and biotic and abiotic challenges. At present, only work with *Arabidopsis* has provided the requisite complex data that determine its model character. However, comparative genomics/transcriptomics begins to provide information indicating that the majority of the genes in *Arabidopsis* and, for example, maize are transcribed in a very similar manner (Guillaumie et al., 2007).

B Gene Networks Explaining Transcript Behavior

We include here the discussion and demonstration of a tool that has become possible because for model plants such as *Arabidopsis*, large datasets are now available. The *Arabidopsis* community has generated several thousand transcript profiling experiments, the results of which are publicly available (Zimmermann et al., 2004). The experiments cover development, biotic and abiotic stress treatments, and treatments with chemicals, such as hormones, heavy metals, or ozone. The data may be scrutinized to detect patterns as has been done, for example, through the GeneInvestigator programs (https://www.genevestigator.ethz.ch; Zimmermann et al., 2004). Using these microarray data, based on the Affymetrix ATH1 platform that includes probes for ~22,000 genes, we have constructed an Arabidopsis gene expression network, based on a Graphical Gaussian Model (GGM) for which a detailed description is available (Ma et al., 2007; Li et al., 2008; Ma and Bohnert, 2008). It includes data from >2,400 ATH1 microarray chip experiments covering a variety of treatments extracted from NASCArrays (http://affymetrix.arabidopsis.info/narrays/help/usefulfiles.html). This adaptation and modification of classical GGM is based on calculating partial correlation coefficients between gene pairs, estimated *via* a shrinkage approach as originally proposed by Schäfer and Strimmer (2005). Calculations, conducted *via* the software package "GeneNet", version 1.0.1, implemented in Bioconductor (Gentleman et al., 2004; Opgen-Rhein et al., 2006), were limited to around ~2,000 genes at a time. To include the entire dataset for >22,000 genes (nodes), an iterative strategy coupled with random sampling was used until each gene had been compared multiple times against all other genes in the dataset. This then expanded the network to cover the entire transcriptome included on ATH1 (Ma et al., 2007; Ma and Bohnert, 2008). Graphs were extracted from the resulting network by specifying seed-nodes (genes) and the number of connections (edges) by which a network extended from the chosen node. Visualization used the neato-program, embedded in the software package Rgraphviz, a component of Bioconductor http://www.bioconductor.org. For detailed descriptions of the network features see Ma et al. (2007) and Ma and Bohnert (2008). The underlying concept of the GGM is similar to other programs that use correlation coefficients, such as Pearson relevance correlation networks. However, GGM derives additional significance from the fact that partial correlations were calculated, i.e., correlations between pairs of genes were conditional to the correlations of the genes in the pair to all other genes in the dataset.

As examples, we have extracted two sub-networks for genes that have been associated by our work and that of others with elevated carbon dioxide Starch Excess 1 (SEX1, At1g10760), encoding an alpha-glucan water dikinase (Fig. 2; Table 3), and a second sub-network associated with thylakoid ascorbate peroxidase (tAPX, At1g77490), an antioxidant defense protein that responded positively in ozone-susceptible Arabidopsis ecotypes (Fig. 3).

Forty eight genes are included in the network associated with SEX1, a gene that responded in the FACE elevated carbon dioxide experiment (Fig. 2). Of these genes, eighteen are chloroplast-associated, and sixteen are involved in carbohydrate metabolism, inside and outside the chloroplast. No genes associated with other chloroplast metabolic functions, such as genes associated with the photosystems or pigment biosynthesis, are present in the network, suggesting a tight regulatory mechanism, centered specifically on starch metabolism. Significantly over-represented are (17) abiotic stress responsive genes, associated with a range of stresses, including salt, water deprivation, ABA, cold, and heat. Only one gene, At3g05650, an ER gene product associated with disease resistance appears in the network. Transcriptional responses to abiotic stresses often overlap (or are a consequence of imprecise annotations) with the overlaps either reflecting common downstream targets of the stress response or cross talk among various signaling pathways.

At4g17770, encoding a trehalose-6-phosphate phosphatase, is present in the network. Trehalose is known to be a major regulator of plant cell function (Paul, 2007), including the regulation of starch metabolism *via* ADP-glucose pyrophosphorylase (Kolbe et al., 2005). The result points to the existence of interconnecting pathways in

Table 3. Genes transcriptionally associated with tAPXa.

A.P.R.2	5'-Adenylylsulfate reductase 2, chloroplast (APR2) (APSR)
AT1G13650	Expressed protein
AT1G14700	Purple acid phosphatase, putative
AT1G17650	Similar to 6-phosphogluconate dehydrogenase
AT1G26560	Glycosyl hydrolase family 1 protein
AT1G30520	Acyl-activating enzyme 14 (AAE14)
AT1G31920	Pentatricopeptide (PPR) repeat-containing protein,
AT1G32080	Membrane protein, putative, contains 12 transmembrane domains
AT1G51110.b	Plastid-lipid associated protein PAP; fibrillin family
AT1G54820	Protein kinase family protein
AT1G64150	Expressed protein
AT1G68190	Zinc finger (B-box type) family protein
AT1G68520	Zinc finger (B-box type) family protein
AT1G75460	ATP-dependent protease La (LON) domain-containing protein
AT1G77490	L-ascorbate peroxidase, thylakoid-bound (tAPX)
AT1G78450	SOUL heme-binding family protein
AT2G15290	Expressed protein

(continued)

Table 3. (continued)

A.P.R.2	5'-Adenylylsulfate reductase 2, chloroplast (APR2) (APSR)
AT2G21385	Expressed protein
AT2G29630	Thiamine biosynthesis family protein; thiC family
AT2G32500	Expressed protein , stress responsive alpha-beta barrel
AT2G34620	Mitochondrial transcription termination factor-related; mTERF-related
AT2G36145	Expressed protein
AT2G36885	Expressed protein
AT3G01510	5'-AMP-activated protein kinase beta-1 subunit-related, similar to SEX4
AT3G04340	FtsH protease family protein, similar to chloroplast FtsH protease
AT3G04550	Expressed protein
AT3G15520	Peptidyl-prolyl cis-trans isomerase TLP38, chloroplast; luminal PPIase
AT3G16670	Expressed protein
AT3G18890	NAD-dep. epimerase/dehydratase, responsive to UV-B & oxidative stress
AT3G19000	Oxidoreductase, 2OG-Fe(II) oxygenase family protein
AT3G23080	Expressed protein, weak similarity to Phosphatidylcholine transfer protein
AT3G46780	Expressed protein
AT3G47560	Esterase/lipase/thioesterase family protein
AT3G47860	Apolipoprotein D-related, weak similarity to Apolipoprotein D (ApoD)
AT3G53900	Uracil phosphoribosyltransferase, putative; UMP pyrophosphorylase
AT3G58140	Phenylalanyl-tRNA synthetase class IIc family protein
AT3G59300	Expressed protein
AT3G63490	Ribosomal protein L1 family protein
AT4G15560	1-Deoxy-D-xylulose 5-phosphate synthase, putative
AT4G17300	Asparaginyl-tRNA synthetase, chloroplast/mitochondrial
AT4G19830	Immunophilin; FKBP-type peptidyl-prolyl cis-trans isomerase family
AT4G24460	Expressed protein
AT4G28706	pfkB-type carbohydrate kinase family protein
AT4G33666	Expressed protein
AT4G39460	Mitochondrial substrate carrier family protein
AT5G04900	Short-chain dehydrogenase/reductase (SDR) family protein
AT5G06290	2-cys peroxiredoxin, chloroplast, putative
AT5G35790	Plastidic glucose-6-phosphate dehydrogenase
AT5G39830	DegP protease, putative
AT5G40950	50S ribosomal protein L27, chloroplast, putative (RPL27)
AT5G48490	Protease inhibitor/seed storage/lipid transfer protein (LTP) family
AT5G52520	RNA synthetase class II (G, H, P and S) family protein
AT5G55220	Trigger factor type chaperone family protein
AT5G62720	Integral membrane HPP family protein
AT5G64290	Oxoglutarate/malate translocator, putative
AT5G66530	Aldose 1-epimerase family protein
CH1	Chlorophyll a oxygenase (CAO); chlorophyll b synthase
CXIP2	CAX-interacting protein, putative, glutaredoxin-related (AT2G38270)
EMB506	Ankyrin repeat family protein
FSD2	Superoxide dismutase (Fe), putative; iron superoxide dismutase
FTSZ2_2	Chloroplast division protein, putative
GLB1	P II nitrogen sensing protein (GLB I)
HCF107	Tetratricopeptide repeat (TPR)-containing protein
ISPD	4-Diphosphocytidyl-2-C-methyl-D-erythritol synthase (CMS or IspD)
LUT2	Lycopene epsilon cyclase
OASB	Cysteine synthase, chloroplast;O-acetylserine (thiol)-lyase
PGM	Phosphoglucomutase, chloroplast (PGM)
RPL9	50S ribosomal protein L9, chloroplast (CL9)
RPS9	Ribosomal protein S9 (RPS9)
SIGA	RNA polymerase, subunit SigA (sigA)

[a] Abbreviated annotations according to TAIR (http://www.TAIR.org); the corresponding network graph is shown in Fig. 3

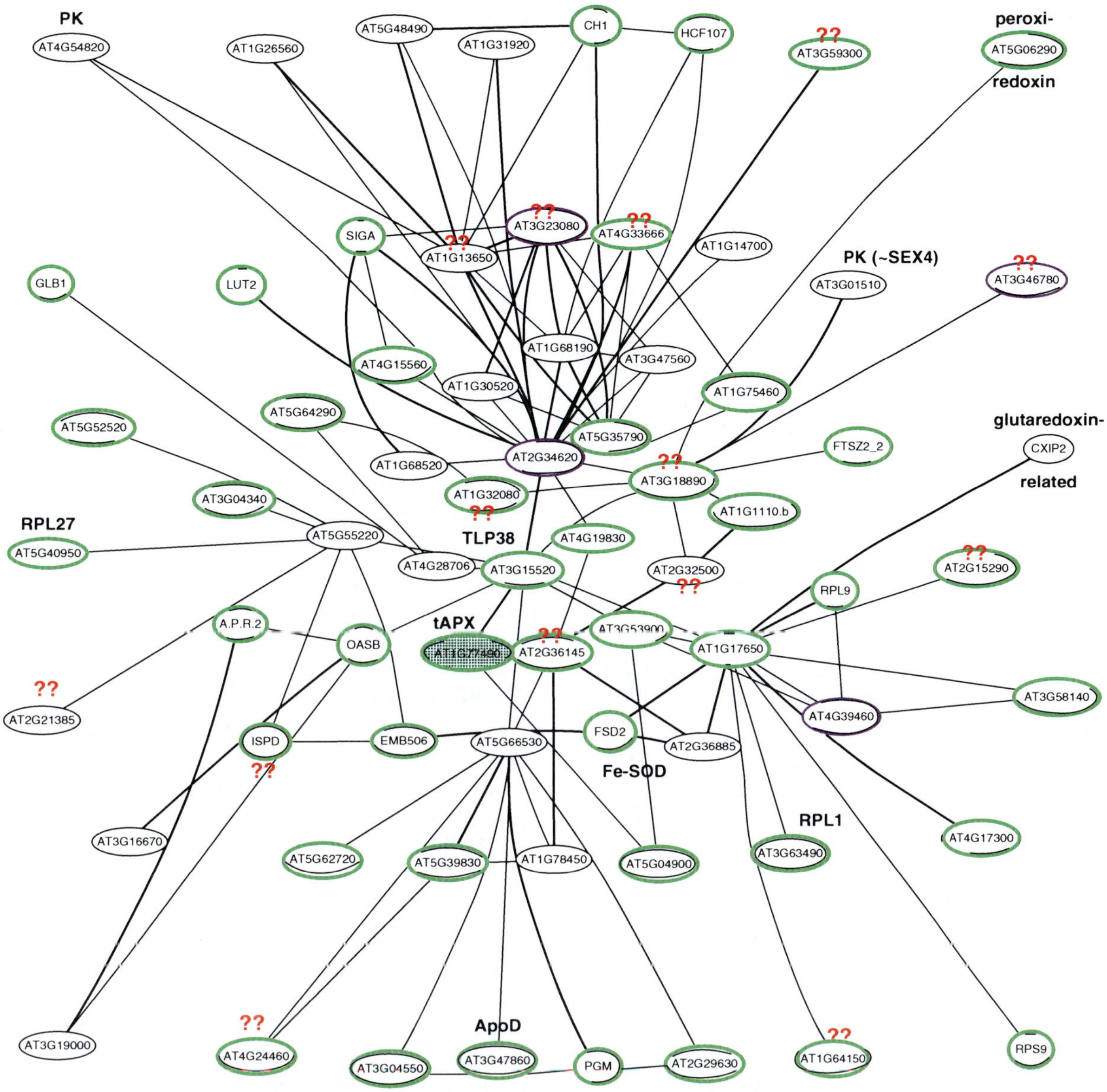

Fig. 3. Gene network centered on tAPX (thylakoid-located ascorbate peroxidase). The network graph, based on GGM (Ma et al., 2007), is centered on the tAPX gene (AT1G77490), which was down-regulated in WS, the most susceptible Arabidopsis line, under ozone exposure (not shown). The genes shown are transcriptionally connected by two extensions to the tAPX transcript. The known location of proteins encoded by these genes is indicated by colored lines: chloroplast (green), mitochondria (blue). Many non-colored transcripts are most likely cytosolically located. Transcripts encoding unknown functions are also indicated. PK – uncharacterized protein kinases, one with similarity to SEX4; RPL1/27 encode chloroplast ribosomal proteins; ApoD – lipase; TLP38 – peptidyl-prolyl *cis-trans* isomerase. Genes are listed in Table 3.

the regulation of starch metabolism and abiotic stress effects. In addition, the appearance in this network graph centered on SEX1 of only one biotic stress-responsive gene suggests that regulatory mechanisms may be specific to abiotic stress(es) and carbohydrate metabolism. Five genes encoding regulatory functions appear in the network, among them, At1g80480 (pTAC17), encoding a plastid-chromosome binding protein.

For Fig. 3 we have chosen the thylakoid-bound ROS scavenger ascorbate peroxide, tAPX.

The network associated with tAPX lists sixty-eight genes, forty-six of which encode chloroplast-localized products. Of the gene products located in the chloroplast, twelve encode plastid enzymes, involved in, for example, starch metabolism, chlorophyll, carotenoid and anthocyanin biosynthesis, and the non-mevalonate isoprenoid pathway, suggesting a stress-mediated coordination of chloroplast-based metabolic activity. The identification of functions in this network graph is fundamentally different from the functions associated with SEX1. It is quite striking to realize how a network of genes associated with metabolic functions in carbohydrate biochemistry is transcriptionally separated from a ROS-dependent network that affects several, and possibly all, chloroplast specific biochemical pathways.

Gene expression, associated with chloroplast function, is also well-represented, with nine genes included in the tAPX network. Six defense and repair genes, associated with redox and chaperone functions in the plastid, appear, including Fe-SOD2, which has been previously reported to respond specifically during recovery from ozone stress (Kliebenstein et al., 1998), and the stress-responsive 2-cys peroxiredoxins (At5g06290; Dietz et al., 2002). It is interesting to note that APX1, a central gene in the well-studied ROS signaling network described by Rizhsky et al. (2004) is not present, suggesting the existence of an ROS-associated regulatory network with tAPX as a "hub" that is independent of APX1. The network provides a hypothesis-generating tool for future studies, in particular with respect to stress biology and biochemical pathways. The predicted connections often suggest new studies using RNAi and knockout lines of *Arabidopsis*.

These results show that plants exposed to a combination of the two greenhouse gases in the field will exhibit a range of distinct molecular and metabolic responses that are not simply additive. Furthermore, the two examples of chosen networks illustrate the relevance of such an approach in identifying possible regulatory pathways, which in this case are centered on chloroplast functions.

VI Conclusions

Transcriptome and metabolite profiles in FACE experiments that used the model species *Arabidopsis* and *Thellungiella* have, we think, provided a picture of the plants that is fairly consistent and explains responses to elevated [CO_2] and higher O_3, under circumstances that seem to mimic stressful conditions. These responses may be summarized by a few points.

(1) FACE experiments that record metabolite and transcript behavior monitor the actual micro-climate, essentially local weather, at the time of sampling. It responds mainly to temperature and irradiation intensity. They appear to be of limited predictive quality with respect to long-term developmental processes that may be induced by elevated [CO_2]. In our experiments, *Arabidopsis* and *Thellungiella* showed no differences in the progression of development in elevated [CO_2] relative to ambient air. In supplemented O_3, however, the plants showed signs of increased activity of the phenylpropanoid/flavonoid pathways. Larger variability existed between ecotypes with respect to growth retardation and lesion formation, apparently due to chronic ROS increases or changes in ROS signaling. The observed acceleration of development in the field affected plants in ambient air the same way as plants in elevated [CO_2], bearing in mind that the plants in just over 2 weeks progressed from the rosette to post-flowering stages with ripening siliques, and senescing leaves..

(2) When analyzing metabolite and transcript profiles of several *Arabidopsis* ecotypes and its close relative *Thellungiella*, the problem the plants faced during chronic exposure to elevated greenhouse gases seemed to be centered on the discrepancy between an excess of carbon and a less than corresponding increase of N-compounds, amino acids in particular. Increased amounts of C-compounds of primary metabolism appeared to be a consequence of bottlenecks in transport and/or carbon storage. This tended to distort the C/N ratio sensing, or the responses aimed at correcting the discrepancy, at least in the short term.

(3) *Arabidopsis* ecotypes revealed considerable genetic variability and flexibility that allowed them to engage strategies leading to acclimation and gains in biomass or yield in elevated [CO_2]. Among the ecotypes in our experiments, Cvi-0 (Cape Verde Island), while not exhibiting an accelerated growth and senescence as much as ecotypes Col-0 and WS, showed the best acclimation strategies and the least disturbance. Cvi-0 is a moderately drought and high irradiance adapted ecotype. It seems that screening and selection

of crop species with resilience to the altered atmospheric conditions should not pose significant problems. Germplasms, ultimately enhanced by transgenic means, and already in breeder's hands will lead to changes in plant performance.

Large cultivar differences in soyabean yield in the response to elevated [CO_2] have been observed (Nelson, R., NSGC-UIUC, personal communication; see also Ziska and Bunce, 2007). As a point in favor of the above view, soybean lines exist that perform close to the theoretical value in elevated [CO_2]. This may be due to, or enhanced by, the N-fixation capability of soybeans. Indeed, soybean could more easily recruit metabolites that re-establish the C/N ratio under which performance can make maximal use of the fertilizing effects of [CO_2].

This is in contrast to studies using re-analyzed FACE experimental results arguing that crop responses to elevated [CO_2] may be lower than expected, thus requiring a re-evaluation of current models (Long et al., 2005; 2006). Other reports of reliably observed stimulations in yield in FACE experiments (Tubiello et al., 2007) exist. Simulations of growth and yield in response to elevated [CO_2] in several crop-simulation models, e.g., AFRC-Wheat, APSIM, CERES, CROPGRO, CropSyst, LINTULC and SIRIUS, which can project crop yield increases of 5–20% at 550 ppm [CO_2], have been validated by some experimental data (Tubiello et al., 2007).

Nonetheless, plant physiologists and modellers alike recognize that the effects of elevated CO_2 measured in experimental settings and implemented in models may overestimate actual field- and farm-level responses, due to many limiting factors such as pests, weeds, competition for resources, soil, and water and air quality, which are not well understood at large scales (Tubiello and Ewert, 2002; Karnosky, 2003; Ziska et al., 2004; Ainsworth and Long, 2005; Tubiello et al., 2007). In fact a recent assessment report by the Intergovernmental Panel on Climate Change (IPCC) points to deficits in a number of models, which failed to include the effects of biotic and abiotic stress factors such as elevated ozone, or competition for nutrients on plant performance under realistic field conditions (http.www.ipcc.ch/ipccreports/assessments-reports.htm). Clearly, such factors must be incorporated into future model calculations. In addition, we are confident in breeding programs and breeders to accomplish genetic adaptation of crops to changes in air composition and quality, in a changing climate.

(4) Elevated [CO_2] is, in principle, a bonus for the plants; but the increased load of ROS based on increases in ozone takes its toll either by an influence on stomatal behavior, by the damaging effects of the gas or by altering ROS signaling pathways, including as yet unstudied pathways that respond to low levels of ROS. This is also, as is suggested by our transcriptomics data, and by the tAPX network discussed above. The challenge posed to plant cells by chronic, as opposed to acute, exposure to ozone has scarcely been addressed in the context of underlying mechanisms.

Until such information has been collected and integrated into our current understanding of defense responses and signaling pathways, little progress will be made by way of engineering crop plants with superior ozone resistance characteristics. A focus on ROS scavenging capacity should also be attached to breeding programs of crop species.

(5) Understanding the growth of plants in the field requires different approaches compared to those that apply to plants grown in a growth chamber. The field situation includes multiple environmental (stress) factors that are typically separated in experiments in growth chambers (Miyazaki et al., 2004). Physiological or biochemical models that have been established for plants grown under controlled conditions may not apply to field conditions as expected. They might better be replaced by more realistic perspectives and paradigms.

Acknowledgments

We wish to thank Drs. Lisa Ainsworth, Cecilia Vasquez-Robinet, Shrinivasrao Mane, Qingqiu Gong and Shisong Ma for discussions. We are grateful to VT and UIUC for institutional support.

References

Ahlfors R, Lang S, Overmyer K, Jaspers P, Brosche M, Tauriainen A, Kollist H, Tuominen H, Belles-Boix E, Piippo M, Inze D, Palva ET and Kangasjarvi J (2004) Arabidopsis RADICAL-INDUCED CELL DEATH1 belongs to the WWE protein-protein interaction domain protein family and modulates abscisic acid, ethylene, and methyl jasmonate responses. Plant Cell 16: 1925–1937

Ahn S-G and Thiele DJ (2003) Redox regulation of mammalian heat shock factor 1 is essential for Hsp gene activation and protection from stress. Genes and Development 17: 516–528

Ainsworth EA and Long SP (2005) What have we learned from 15 years of free-air CO_2 enrichment (FACE)? A meta-analytic review of the responses of photosynthesis, canopy. New Phytologist 165: 351–371

Ainsworth EA and Rogers A (2007) The response of photosynthesis and stomatal conductance to rising [CO2]: mechanisms and environmental interactions. Plant Cell and Environment 30: 258–270

Ainsworth EA, Rogers A, Nelson R and Long SP (2004) Testing the "source-sink" hypothesis of down-regulation of photosynthesis in elevated [CO_2] in the field with single gene substitutions in *Glycine max*. Agricultural and Forest Meteorology 122: 85–94

Ainsworth EA, Rogers A, Vodkin LO, Walter A and Schurr U (2006) The effects of elevated CO_2 concentration on soybean gene expression. An analysis of growing and mature leaves. Plant Physiology 142: 135–147

Amtmann A, Bohnert HJ and Bressan RA (2005) Abiotic stress and plant genome evolution. Search for new models. Plant Physiology 138: 127–130

Ashmore MR (2005) Assessing the future global impacts of ozone on vegetation. Plant Cell and Environment 28: 949–964

Baier M, Kandlbinder A, Golldack D and Dietz KJ (2005) Oxidative stress and ozone: perception, signalling and response. Plant Cell and Environment 28: 1012–1020

Balmer Y, Vensel WH, Hurkman WJ and Buchanan BB (2006) Thioredoxin target proteins in chloroplast thylakoid membranes. Antioxidants & Redox Signaling 8: 1829–1834

Bellafiore S, Bameche F, Peltier G, and Rochaix JD (2005) State transitions and light adaptation require chloroplast thylakoid protein kinase STN7. Nature 433: 892–895

Bernacchi CJ, Calfapietra C, Davey PA, Wittig VE, Scarascia-Mugnozza GE, Raines CA and Long SP (2003) Photosynthesis and stomatal conductance responses of poplars to free-air CO_2 enrichment (PopFACE) during the first growth cycle and immediately following coppice. New Phytologist 159: 609–621

Bernacchi CJ, Leakey ADB, Heady LE, Morgan PB, Dohleman FG, McGrath JM, Gillespie KM, Wittig VE, Rogers A, Long SP and Ort DR (2006) Hourly and seasonal variation in photosynthesis and stomatal conductance of soybean grown at future CO_2 and ozone concentrations for 3 years under fully open-air field conditions. Plant Cell and Environment 29: 2077–2090

Bernacchi CJ, Morgan PB, Ort DR and Long SP (2005) The growth of soybean under free air [CO_2] enrichment (FACE) stimulates photosynthesis while decreasing in vivo Rubisco capacity. Planta 220: 434–446

Bernardi R, Nali C, Ginestri P, Pugliesi C, Lorenzini G and Durante M (2004) Antioxidant enzyme isoforms on gels in two poplar clones differing in sensitivity after exposure to ozone. Biologia Plantarum 48: 41–48

Bidart-Bouzat MG, Mithen R and Berenbaum MR (2005) Elevated CO_2 influences herbivory-induced defense responses of *Arabidopsis thaliana*. Oecologia 145: 415–424

Blasing OE, Gibon Y, Gunther M, Hohne M, Morcuende R, Osuna D, Thimm O, Usadel B, Scheible WR and Stitt M (2005) Sugars and circadian regulation make major contributions to the global regulation of diurnal gene expression in Arabidopsis. Plant Cell 17: 3257–3281

Bloom AJ (2006) Rising carbon dioxide concentrations and the future of crop production. Journal of the Science of Food and Agriculture 86: 1289–1291

Bohnert HJ, Ayoubi P, Borchert C, Bressan RA, Burnap RL, Cushman JC, Cushman MA, Deyholos M, Fischer R, Galbraith DW, Hasegawa PM, Jenks M, Kawasaki S, Koiwa H, Kore-eda S, Lee BH, Michalowski CB, Misawa E, Nomura M, Ozturk N, Postier B, Prade R, Song CP, Tanaka Y, Wang H and Zhu JK (2001) A genomics approach towards salt stress tolerance. Plant Physiology and Biochemistry 39: 295–311

Booker FL, Burkey KO, Pursley WA and Heagle AS (2007) Elevated carbon dioxide and ozone effects on peanut: I. Gas-exchange, biomass, and leaf chemistry. Crop Science 47: 1475–1487

Bowes G (1993) Facing the Inevitable - Plants and Increasing Atmospheric CO_2. Annual Review of Plant Physiology and Plant Molecular Biology 44: 309–332

Bressan RA, Zhang C, Zhang H, Hasegawa PM, Bohnert HJ and Zhu JK (2001) Learning from the Arabidopsis experience. The next gene search paradigm. Plant Physiol. 127:1354–60

Buchanan BB and Balmer Y (2005) Redox regulation: A broadening horizon. Annual Review of Plant Biology 56: 187–220

Buckley TN (2008) The role of stomatal acclimation in modelling tree adaptation to high CO_2. Journal of Experimental BotanyJournal of Experimental Botany, 59:1951–1961

Burkey KO, Miller JE and Fiscus EL (2005) Assessment of ambient ozone effects on vegetation using snap bean as a bioindicator species. Journal of Environmental Quality 34: 1081–1086

Castagna A, Ederli L, Pasqualini S, Mensuali-Sodi A, Baldan B, Donnini S and Ranieri A (2007) The tomato ethylene receptor LE-ETR3 (NR) is not involved in mediating ozone sensitivity: causal relationships among ethylene emission, oxidative burst and tissue damage. New Phytologist 174: 342–356

Conklin PL and Last RL (1995) Differential accumulation of antioxidant mRNAs in *Arabidopsis thaliana* exposed to ozone. Plant Physiol. 109: 203–212

Cross JM, von Korff M, Altmann T, Bartzetko L, Sulpice R, Gibon Y, Palacios N and Stitt M (2006) Variation of enzyme activities and metabolite levels in 24 Arabidopsis accessions growing in carbon-limited conditions. Plant Physiology 142: 1574–1588

Davletova S, Rizhsky L, Liang HJ, Zhong SQ, Oliver DJ, Coutu J, Shulaev V, Schlauch K and Mittler R (2005)

Cytosolic ascorbate peroxidase 1 is a central component of the reactive oxygen gene network of Arabidopsis. Plant Cell 17: 268–281

Dietz KJ, Horling F, Konig J, and Baier M (2002) The function of the chloroplast 2-cysteine peroxiredoxin in peroxide detoxification and its regulation. Journal of Experimental Botany 53: 1321–1329

Dietz KJ, Jacob S, Oelze ML, Laxa M, Tognetti V, de Miranda SMN, Baier M, and Finkemeier I (2006) The function of peroxiredoxins in plant organelle redox metabolism. Journal of Experimental Botany 57: 1697–1709

Ederli L, Morettini R, Borgogni A, Wasternack C, Miersch O, Reale L, Ferranti F, Tosti N, and Pasqualini S (2006) Interaction between nitric oxide and ethylene in the induction of alternative oxidase in ozone-treated tobacco plants. Plant Physiology 142: 595–608

Fey V, Wagner R, Brautigam K, Wirtz M, Hell R, Dietzmann A, Leister D, Oelmuller R, and Pfannschmidt T (2005) Retrograde plastid redox signals in the expression of nuclear genes for chloroplast proteins of *Arabidopsis thaliana*. Journal of Biological Chemistry 280: 17572–17572

Fiscus EL, Booker FL, and Burkey KO (2005) Crop responses to ozone: uptake, modes of action, carbon assimilation and partitioning. Plant Cell and Environment 28: 997–1011

Fuhrer J (1994) Effects of ozone on managed pasture: I. Effects of open-top chambers on microclimate, ozone flux, and plant growth. Environ Pollut. 86:297–305

Geigenberger P, Kolbe A, and Tiessen A (2005) Redox regulation of carbon storage and partitioning in response to light and sugars. Journal of Experimental Botany 56: 1469–1479

Gentleman RC, Carey VJ, Bates DM, Bolstad B, Dettling M, Dudoit S, Ellis B, Gautier L, Ge YC, Gentry J, Hornik K, Hothorn T, Huber W, Iacus S, Irizarry R, Leisch F, Li C, Maechler M, Rossini AJ, Sawitzki G, Smith C, Smyth G, Tierney L, Yang JYH, and Zhang JH (2004) Bioconductor: open software development for computational biology and bioinformatics. Genome Biology 5: R80

Giles J (2005) Hikes in surface ozone could suffocate crops. Nature 435: 7

Gomez LD, Baud S, Gilday A, Li Y, and Graham IA (2006) Delayed embryo development in the Arabidopsis trehalose-6-phosphate synthase 1 mutant is associated with altered cell wall structure, decreased cell division and starch accumulation. Plant Journal 46: 69–84

Griffin KL, Anderson OR, Gastrich MD, Lewis JD, Lin G, Schuster W, Seemann JR, Tissue DT, Turnbull MH, and Whitehead D (2001) Plant growth in elevated CO2 alters mitochondrial number and chloroplast fine structure. Proc Natl Acad Sci U S A. 98:2473–8

Griffith M, Timonin M, Wong ACE, Gray GR, Akhter SR, Saldanha M, Rogers MA, Weretilnyk EA, and Moffatt B (2007) Thellungiella: an Arabidopsis-related model plant adapted to cold temperatures. Plant Cell and Environment 30: 529–538

Guillaumie S, San-Clemente H, Deswarte C, Martinez Y, Lapierre C, Murigneux A, Barriere Y, Pichon M, and Goffner D (2007) MAIZEWALL. Database and developmental gene expression profiling of cell wall biosynthesis and assembly in maize. Plant Physiology 143: 339–363

Hamilton JG, Dermody O, Aldea M, Zangerl AR, Rogers A, Berenbaum MR, and DeLucia EH (2005) Anthropogenic changes in tropospheric composition increase susceptibility of soybean to insect herbivory. Environmental Entomology 34: 479–485

Heck WW, Adams RM, Cure WW, Heagle AS, Heggestad HE, Kohut RJ, Kress LW, Rawlings JO, and C. TO (1983) A reassessment of crop loss from ozone. Environmental Science and Technology 17: 573–581

Houghton RA (2001) Counting terrestrial sources and sinks of carbon. Climatic Change 48: 525–534

Howard TP, Metodiev M, Lloyd JC, and Raines CA (2008) Thioredoxin-mediated reversible dissociation of a stromal multiprotein complex in response to changes in light availability. Proc Natl Acad Sci USA. 105:4056–6

Inan G, Zhang Q, Li PH, Wang ZL, Cao ZY, Zhang H, Zhang CQ, Quist TM, Goodwin SM, Zhu JH, Shi HH, Damsz B, Charbaji T, Gong QQ, Ma SS, Fredricksen M, Galbraith DW, Jenks MA, Rhodes D, Hasegawa PM, Bohnert HJ, Joly RJ, Bressan RA, and Zhu JK (2004) Salt cress. A halophyte and cryophyte Arabidopsis relative model system and its applicability to molecular genetic analyses of growth and development of extremophiles. Plant Physiology 135: 1718–1737

Jang JC, Leon P, Zhou L, and Sheen J (1997) Hexokinase as a sugar sensor in higher plants. Plant Cell 9: 5–19

Jifon JLW and Wolfle D (2002) Photosynthetic acclimation to elevated CO_2 in *Phaseolus vulgaris* L. is altered by growth response to nitrogen supply. Global Change Biology 8: 1018–1027

Karnosky DF (2003) Impacts of elevated atmospheric CO(2) on forest trees and forest ecosystems: knowledge gaps. Environ Int. 29:161–169

Kangasjarvi J, Jaspers P, and Kollist H (2005) Signalling and cell death in ozone-exposed plants. Plant Cell and Environment 28: 1021–1036

Kangasjarvi J, Talvinen J, Utrianen M, and Karjalainen R (1994) Plant defence systems induced by ozone. Plant Cell Environ 17: 783–794

Kim SH, Sicher RC, Bae H, Gitz DC, Baker JT, Timlin DJ, and Reddy VR (2006) Canopy photosynthesis, evapotranspiration, leaf nitrogen, and transcription profiles of maize in response to CO_2 enrichment. Global Change Biology 12: 588-600

Kliebenstein DJ and Last RL (1998) Superoxide dismutase in Arabidopsis: an eclectic enzyme family with disparate regulation and protein localization. Plant Physiol 118: 637–650

Knepp RG, Hamilton JG, Mohan JE, Zangerl AR, Berenbaum MR, and DeLucia EH (2005) Elevated CO_2 reduces leaf damage by insect herbivores in a forest community. New Phytologist 167: 207–218

Knepp RG, Hamilton JG, Zangerl AR, Berenbaum MR, and DeLucia EH (2007) Foliage of oaks grown under elevated CO_2 reduces performance of *Antheraea polyphemus*

(Lepidoptera : Saturniidae). Environmental Entomology 36: 609–617

Koch K (2004) Sucrose metabolism: regulatory mechanisms and pivotal roles in sugar sensing and plant development. Current Opinion in Plant Biology 7: 235–246

Kolbe A, Tiessen A, Schluepmann H, Paul M, Ulrich S and Geigenberger P (2005) Trehalose 6-phosphate regulates starch synthesis via posttranslational redox activation of ADP-glucose pyrophosphorylase. Proceedings of the National Academy of Sciences of the United States of America 102: 11118–11123

Koussevitzky S, Nott A, Mockler TC, Hong F, Sachetto-Martins G, Surpin M, Lim IJ, Mittler R and Chory J (2007) Signals from chloroplasts converge to regulate nuclear gene expression. Science 316: 715–719

LaDeau SL and Clark JS (2001) Rising CO_2 levels and the fecundity of forest trees. Science 292: 95–98

Lalonde S, Tegeder M, Throne-Holst M, Frommer WB and Patrick JW (2003) Phloem loading and unloading of sugars and amino acids. Plant Cell and Environment 26: 37–56

Lalonde S, Wipf D and Frommer WB (2004) Transport mechanisms for organic forms of carbon and nitrogen between source and sink. Annual Review of Plant Biology 55: 341–372

Langebartels C, Wohlgemuth H, Kschieschan S, Grun S and Sandermann H (2002) Oxidative burst and cell death in ozone-exposed plants. Plant Physiology and Biochemistry 40: 567–575

Lemaire SD, Michelet L, Zaffagnini M, Massot V and Issakidis-Bourguet E (2007) Thioredoxins in chloroplasts. Current Genetics 51: 343–365

Li P, Ma S and Bohnert HJ (2008) Co-expression charcteristics of trehalose 6-phosphate phosphatase subfamily gene reveal different functions in a network context. Physiologia Plantarum in press

Li PH, Bohnert HJ and Grene R (2007) All about FACE - plants in a high-[CO_2] world. Trends in Plant Science 12: 87–89

Li PH, Sioson A, Mane SP, Ulanov A, Grothaus G, Heath LS, Murali TM, Bohnert HJ and Grene R (2006a) Response diversity of *Arabidopsis thaliana* ecotypes in elevated [CO_2] in the field. Plant Molecular Biology 62: 593–609

Li PH, Mane SP, Sioson AA, Robinet CV, Heath LS, Bohnert HJ and Grene R (2006b) Effects of chronic ozone exposure on gene expression in *Arabidopsis thaliana* ecotypes and in Thellungielia halophila. Plant Cell and Environment 29: 854–868

Liberloo M, Calfapietra C, Lukac M, Godbold D, Luos ZB, Polle A, Hoosbeek MR, Kull O, Marek M, Raines C, Rubino M, Taylor G, Scarascia-Mugnozza G and Ceulemans R (2006) Woody biomass production during the second rotation of a bio-energy Populus plantation increases in a future high CO_2 world. Global Change Biology 12: 1094–1106

Long SP, Ainsworth EA, Leakey AD, Nosberger J and Ort DR (2006) Food for thought: lower-than-expected crop yield stimulation with rising CO_2 concentrations. Science 312: 1918–1921

Long SP, Ainsworth EA, Leakey ADB and Morgan PB (2005) Global food insecurity. Treatment of major food crops with elevated carbon dioxide or ozone under large-scale fully open-air conditions suggests recent models may have overestimated future yields. Philosophical Transactions of the Royal Society B - Biological Sciences 360: 2011–2020

Long SP, Ainsworth EA, Rogers A and Ort DR (2004) Rising atmospheric carbon dioxide: Plants face the future. Annual Review of Plant Biology 55: 591–628

Lopez-Juez E (2007) Plastid biogenesis, between light and shadows. Journal of Experimental Botany 58: 11–26

Ma S and Bohnert HJ (2008) Gene networks in Arabidopsis thaliana for metabolic and environmental functions. Molecular Biosystems 4: 199–204

Ma S, Gong QQ and Bohnert HJ (2007) An Arabidopsis gene network based on the graphical Gaussian model. Genome Research 17: 1614–1625

Mahalingam R, Jambunathan N, Gunjan SK, Faustin E, Weng H and Ayoubi P (2006) Analysis of oxidative signalling induced by ozone in *Arabidopsis thaliana*. Plant Cell and Environment 29: 1357–1371

Miglietta F, Giuntoli A and Bindi M (1996) The effect of free air carbon dioxide enrichment (FACE) and soil nitrogen availability on the photosynthetic capacity of wheat. Photosynthesis Research 47: 281–290

Miller G, Suzuki N, Rizhsky L, Hegie A, Koussevitzky S and Mittler R (2007) Double mutants deficient in cytosolic and thylakoid ascorbate peroxidase reveal a complex mode of interaction between reactive oxygen species, plant development, and response to abiotic stresses. Plant Physiology 144: 1777–1785

Mittler R, Vanderauwera S, Gollery M and Van Breusegem F (2004) Reactive oxygen gene network of plants. Trends in Plant Sciences 9: 490–498

Miyazaki S, Fredricksen M, Hollis KC, Poroyko V, Shepley D, Galbraith DW, Long SP and Bohnert HJ (2004) Transcript expression profiles of *Arabidopsis thaliana* grown under controlled conditions and open-air elevated concentrations of CO_2 and of O_3. Field Crops Research 90: 47–59

Morgan PB, Ainsworth EA and Long SP (2003) How does elevated ozone impact soybean? A meta-analysis of photosynthesis, growth and yield. Plant Cell and Environment 26: 1317–1328

Niittyla T, Messerli G, Trevisan M, Chen J, Smith AM and Zeeman SC (2004) A previously unknown maltose transporter essential for starch degradation in leaves. Science 303: 87–89

Noctor G, De Paepe R and Foyer CH (2007) Mitochondrial redox biology and homeostasis in plants. Trends in Plant Science 12: 125–134

Nott A, Jung HS, Koussevitzky S and Chory J (2006) Plastid-to-nucleus retrograde signaling. Annual Review of Plant Biology 57: 739–759

Nowak RS, Ellsworth DS and Smith SD (2004) Functional responses of plants to elevated atmospheric CO_2 - do photosynthetic and productivity data from FACE experiments support early predictions? New Phytologist 162: 253–280

Nunes-Nesi A, Sweetlove LJ and Fernie AR (2007) Operation and function of the tricarboxylic acid cycle in the illuminated leaf. Physiologia Plantarum 129: 45–56

Oksanen E, Freiwald V, Prozherina N and Rousi M (2005) Photosynthesis of birch (*Betula pendula*) is sensitive to springtime frost and ozone. Canadian Journal of Forest Research 35: 703–712

Opgen-Rhein R, Schafer J, and Strimmer K (2006) GeneNet: Modeling and Inferring Gene Networks R Package version 101

Overmyer K, Brosche M, Pellinen R, Kuittinen T, Tuominen H, Ahlfors R, Keinanen M, Saarma M, Scheel D and Kangasjarvi J (2005) Ozone-induced programmed cell death in the Arabidopsis radical-induced cell death1 mutant. Plant Physiology 137: 1092–1104

Parnik T, Ivanova H and Keerberg O (2007) Photorespiratory and respiratory decarboxylations in leaves of C3 plants under different CO_2 concentrations and irradiances. Plant Cell and Environment 30: 1535–1544

Parry M, Rosenzweig C and Livermore M (2005) Climate change, and risk global food supply of hunger. Philosophical Transactions of the Royal Society B - Biological Sciences 360: 2125–2138

Parry MAJ, Andralojc PJ, Mitchell RAC, Madgwick PJ and Keys AJ (2003) Manipulation of Rubisco: the amount, activity, function and regulation. Journal of Experimental Botany 54: 1321–1333

Parry ML, Rosenzweig C, Iglesias A, Livermore M and Fischer G (2004) Effects of climate change on global food production under SRES emissions and socio-economic scenarios. Global Environmental Change Human and Policy Dimensions 14: 53–67

Paul M (2007) Trehalose 6-phosphate. Current Opinion in Plant Biology 10: 303–309

Paul MJ and Foyer CH (2001) Sink regulation of photosynthesis. J Exp Bot. 52:1383–400

Peart RM, Jones JW, Curry RB, Boote K, Allen LH and Jr 1989 (1989) Impact of climate change on crop yield in the south-eastern USA. *In* JB Smith, Tirpak, D.A., ed, The potential effects of global climate change on the United States. US Environmental Protection Agency, Washington, DC, pp. 118–140

Pell E, Schlagnhaufer C and Arteca R (1997) Ozone-induced oxidative stress: mechanisms of action and reaction. Plant Physiology 100: 264–273

Pellinen R, Palva T and Kangasjarvi J (1999) Subcellular localization of ozone-induced hydrogen peroxide production in birch (*Betula pendula*) leaf cells. Plant Journal 20: 349–356

Pfannschmidt T (2003) Chloroplast redox signals: how photosynthesis controls its own genes. Trends in Plant Science 8: 33–41

Pitcher LH and Zilinskas BA (1996) Overexpression of copper/zinc superoxide dismutase in the cytosol of transgenic tobacco confers partial resistance to ozone-induced foliar necrosis. Plant Physiology 110: 583–588

Plaxton WC and Podesta FE (2006) The functional organization and control of plant respiration. Critical Reviews in Plant Sciences 25: 159–198

Rachmilevitch S, Cousins AB and Bloom AJ (2004) Nitrate assimilation in plant shoots depends on photorespiration. Proc Natl Acad Sci USA 101: 11506–11510

Rao M and Davis K (2001) The physiology of ozone induced cell death. Planta 213

Rao MV, Lee HI, Creelman RA, Mullet JA and Davis KR (2000) Jasmonic acid signalling modulates ozone-induced hypersensitive cell death. Plant Cell 12: 1633–1646

Rao MV, Lee HI and Davis KR (2002) Ozone-induced ethylene production is dependent on salicylic acid, and both salicylic acid and ethylene act in concert to regulate ozone-induced cell death. The Plant Journal 32: 447–456

Rizhsky L, Davletova S, and Liang HRM (2004) The zinc finger protein Zat12 is required for cytosolic ascorbate peroxidase 1 expression during oxidative stress in Arabidopsis. Journal of Biological Chemistry 279: 11736–11743

Rogers A, Gibon Y, Stitt M, Morgan PB, Bernacchi CJ, Ort DR and Long SP (2006) Increased C availability at elevated carbon dioxide concentration improves N assimilation in a legume. Plant Cell and Environment 29: 1651–1658

Rolland F, Baena-Gonzalez E and Sheen J (2006) Sugar sensing and signaling in plants: Conserved and novel mechanisms. Annual Review of Plant Biology 57: 675–709

Rook F, Hadingham SA, Li Y and Bevan MW (2006) Sugar and ABA response pathways and the control of gene expression. Plant Cell and Environment 29: 426–434

Schäfer J and Strimmer K (2005) An empirical Bayes approach to inferring large-scale gene association networks. Bioinformatics 21: 754–764

Schrader SM, Kleinbeck KR, and Sharkey TD (2007) Rapid heating of intact leaves reveals initial effects of stromal oxidation on photosynthesis. Plant Cell and Environment 30: 671–678

Smith CS, Weljie AM and Moorhead GBG (2003) Molecular properties of the putative nitrogen sensor PII from *Arabidopsis thaliana*. Plant Journal 33: 353–360

Stitt M (1991) Rising CO_2 Levels and Their Potential Significance for Carbon Flow in Photosynthetic Cells. Plant Cell and Environment 14: 741–762

Stitt M and Krapp A (1999) The interaction between elevated carbon dioxide and nitrogen nutrition: the physiological and molecular background. Plant Cell and Environment 22: 583–621

Sun XW, Peng LW, Guo JK, Chi W, Ma JF, Lu CM and Zhang LX (2007) Formation of DEG5 and DEG8 complexes and their involvement in the degradation of photodamaged photosystem II reaction center D1 protein in Arabidopsis. Plant Cell 19: 1347–1361

Taiz L and Zeiger E (2006) Plant Physiology, Ed Fourth. Sinauer Associates, Sunderland, MA

Tamaoki M, Matsuyama T, Kanna M, Nakajima N, Kubo A, Aono M and Saji H (2003) Differential ozone sensitivity among Arabidopsis accessions and its relevance to ethylene synthesis. Planta 216: 552–560

Tiessen A, Prescha K, Branscheid A, Palacios N, McKibbin R, Halford NG and Geigenberger P (2003) Evidence that SNF1-related kinase and hexokinase are involved in separate sugar-signalling pathways modulating post-translational redox activation of ADP-glucose pyrophosphorylase in potato tubers. Plant Journal 35: 490–500

Torres NL, Cho K, Shibato J, Hirano M, Kubo A, Masuo Y, Iwahashi H, Jwa NS, Agarwal GK and Rakwal R (2007) Gel-based proteomics reveals potential novel protein markers of ozone stress in leaves of cultivated bean and maize species of Panama. Electrophoresis 28: 4369–4381

Tubiello FN, Soussana J-F and Howden SM (2007) Crop and pasture response to climate change. Proceedings of the National Academy of Sciences of the United States of America 104: 19686–19690

Vahala J, Keinanen M, Schutzendubel A, Polle A and Kangasjarvi J (2003) Differential effects of elevated ozone on two hybrid aspen genotypes predisposed to chronic ozone fumigation. Role of ethylene and salicylic acid. Plant Physiology 132: 196–205

Volkov V and Amtmann A (2006) *Thellungiella halophila*, a salt-tolerant relative of Arabidopsis thaliana, has specific root ion-channel features supporting K+/Na+ homeostasis under salinity stress. Plant Journal 48: 342–353

Von Caemmerer S and Farquhar GD (1984) Effects of partial defoliation, changes of Irradiance during growth, short-term water-stress and growth at enhanced P(Co2) on the photosynthetic capacity of leaves of *Phaseolus vulgaris* L. Planta 160: 320–329

Wang XZ, Anderson OR and Griffin KL (2004) Chloroplast numbers, mitochondrion numbers and carbon assimilation physiology of *Nicotiana sylvestris* as affected by CO_2 concentration. Environmental and Experimental Botany 51: 21–31

Weber APM, Schwacke R and Flugge UI (2005) Solute transporters of the plastid envelope membrane. Annual Review of Plant Biology 56: 133–164

Wilson KE, Ivanov AG, Oquist G, Grodzinski B, Sarhan F, and Huner NPA (2006) Energy balance, organellar redox status, and acclimation to environmental stress. Canadian Journal of Botany 84: 1355–1370

Wise RR and Hoober HJ (2006) The Structure and Function of Plastids, Advances in Photosynthesis and Respiration Series, Vol 23, Springer, Dordecht, pp.

Wohlgemuth H, Mittelstrass K, Kschieschan S, Bender J and Weigel H-J, Overmyer K, Kangasjarvi J, Sandermann H, and Langebartels C (2002) Activation of an oxidative burst is a general feature of sensitive plants exposed to the air pollutant ozone. Plant Cell and Environment 25: 717–726

Wong CE, Li Y, Labbe A, Guevara D, Nuin P, Whitty B, Diaz C, Golding GB, Gray GR, Weretilnyk EA, Griffith M, and Moffatt BA (2006) Transcriptional profiling implicates novel interactions between abiotic stress and hormonal responses in Thellungiella, a close relative of Arabidopsis. Plant Physiology 140: 1437–1450

Wong SC (1979) Elevated atmospheric partial pressure of CO_2 and plant plant growth I. Interactions of nitrogen nutrition and photosynthetic capacity in C3 and C4 plants. . Oecologia 44: 68–74

Wormuth D, Baier M, Kandlbinder A, Scheibe R, Hartung W, and Dietz KJ (2006) Regulation of gene expression by photosynthetic signals triggered through modified CO_2 availability. BMC Plant Biology 6: 15

Young JJ, Mehta S, Israelsson M, Godoski J, Grill E and Schroeder JI (2006) CO_2 signaling in guard cells: Calcium sensitivity response modulation, a Ca^{2+}-independent phase, and CO_2 insensitivity of the gca2 mutant. PNAS 103: 7506–7511

Zavala JA, Casteel C, DeLucia EH and Berenbaum MR (2008) Anthropogenic increase in carbon dioxide compromises plant defense against invasive insects. Proc. Nat Acad Sci USA 105: 5129–5133

Zhang RY, Lei WF, Tie XX and Hess P (2004) Industrial emissions cause extreme urban ozone diurnal variability. Proceedings of the National Academy of Sciences of the United States of America 101: 6346–6350

Zhu XG, de Sturler E and Long SP (2007) Optimizing the distribution of resources between enzymes of carbon metabolism can dramatically increase photosynthetic rate: A numerical simulation using an evolutionary algorithm. Plant Physiology 145: 513–526

Zimmermann P, Hirsch-Hoffmann M, Hennig L and Gruissem W (2004) Genevestigator. Arabidopsis Microarray Database and Analysis Toolbox. Plant Physiology 136: 2621–2632

Ziska LH and Bunce JA (2007) Predicting the impact of changing CO_2 on crop yields: some thoughts on food. New Phytologist 175: 607–618

Ziska LH, Bunce JA and Goins EW (2004) Characterization of an urban-rural CO2/temperature gradient and associated changes in initial plant productivity during secondary succession Oecologia 2004 139: 454–458

Chapter 22

Regulation of Photosynthetic Electron Transport

Toshiharu Shikanai
Department of Botany, Graduate School of Science, Kyoto University, Kitashirakawa, Sakyo-ku, Kyoto, 606-8502, Japan

Summary

The photosynthetic machinery of plants has two conflicting functions. It has to perform at maximum efficiency under light-limited conditions, but has to avoid photo-damage when the incoming light energy exceeds the capacity for utilization. To survive in the field and compete with other species, it is essential for plants to develop multiple strategies for responding to fluctuating light conditions. Thermal dissipation is triggered by acidification of the thylakoid lumen, which allows absorbed excessive light energy to be discarded as heat from photosystem II. The alternative electron transport pathways, photosystem I cyclic electron transport and the water-water cycle, are thought to regulate the induction of thermal dissipation. In higher plants, PSI cyclic electron transport consists of two partly redundant pathways: the PGR5- and NAD(P)H dehydrogenase-dependent pathways. On the other hand, state transition regulates the balance of excitation between the two photosystems by monitoring the reduction level of the plastoquinone pool. Recent genetic investigations have contributed to identifying mutants specifically defective in each pathway, making it possible to evaluate their physiological significance. However, the actual regulatory system is likely to consist of complex interactions between each component, and unfortunately our knowledge is still limited to discussing this network rather than the molecular details.

*Author for Correspondence, e-mail: shikanai@pmg.bot.kyoto-u.ac.jp

Constantin A. Rebeiz et al. (eds.), The Chloroplast: Basics and Applications, pp. 347–361.

I Introduction

Although light energy is essential for photosynthesis, the absorption of excessive light causes irreversible damage to the photosynthetic apparatus via production of reactive oxygen species (ROS). To avoid this photoinhibition, plants have evolved multiple mechanisms to control the efficiency of light energy utilization (reviewed in Niyogi, 1999). These strategies are classified into two categories based on the timescale of the response to changes in light conditions (Fig. 1). Long-term response (photo-acclimation) is a process of modifying the photosynthetic apparatus so that it is mostly suitable for ambient light conditions (reviewed in Mullineaux and Karpinski, 2002). This process is accompanied by changes in gene expression, controlled by long-term monitoring of light conditions.

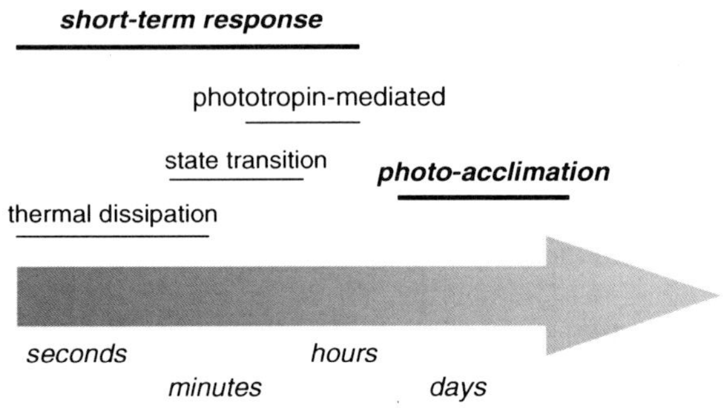

Fig. 1. Classification of plant responses to changes in light conditions based on their timescales of induction and relaxation.

Abbreviations:–AOX–alternative oxidase; CCD–charge-coupled device; *crr2–chlororespiratory reduction 2*; *cyt–cytochrome*; Fd–ferredoxin; FNR–ferredoxin-NADP$^+$ oxidoreductase; LHCII–light harvesting complex II; *hcf–high chlorophyll fluorescence*; MDA–monodehydroascorbate; MDAR–monodehydroascorbate reductase; NDH–NAD(P)H dehydrogenase; NPQ–non-photochemical quenching; PAM–pulse-amplitude modulation; *pgr5*–proton gradient regulation 5; *pgrl1–proton gradient regulation 5-like 1*; *PPR–pentatricopeptide repeat*; PQ–plastoquinone; PSI and PSII–photosystem I and II; PTOX–plastid terminal oxidase; qE, qT, qI–quenching depending on thylakoid energization, state transition and photoinhibition; qP–photochemical quenching; ROS–reactive oxygen species; Rubisco–ribulose-1,5-bisphosphate carboxylase/oxygenase; RuBP–ribulose-1, 5-bisphosphate; SOD–superoxide dismutase; stt7–*state transition deficient mutant 7*; UV–ultraviolet;

Photo-acclimation is a strategy used by plants to respond to long-term changes in the light environment, such as seasonal variation. However, light environments tend to fluctuate on a timescale of seconds to minutes. Thermal dissipation of excess absorbed light energy in photosystem II (PSII) can be induced and disengaged on this timescale (reviewed in Horton et al., 1996; Müller et al., 2001). State transition is the reversible movement of the light-harvesting complex II (LHCII) between two photosystems, and functions on a timescale of minutes (reviewed in Allen, 1992; Haldrup et al., 2001). Phototropins also maximize the photosynthetic activity on an intermediate timescale of minutes to hours (Takemiya et al., 2005), and this regulatory process includes chloroplast movement within a cell (Jarillo et al., 2001, Kagawa et al., 2001, Kasahara et al., 2002), opening of stomata (Kinoshita et al., 2001), and leaf expansion (Sakamoto and Briggs, 2002).

The induction of thermal dissipation of excessive absorbed light energy is regulated by monitoring the pH of the thylakoid lumen. If the generation of ΔpH depends solely on linear electron transport from water to NADP$^+$, the lumen pH would simply be determined by light intensity, provided that the ATP consumption that relaxes ΔpH is constant (although it is highly variable under fluctuating light conditions). However, there has been a long-running debate as to whether alternative electron transport, including photosystem I (PSI) cyclic electron transport (reviewed in Heber and Walker, 1992) and water–water cycle (reviewed in Asada, 1999), contributes to the adjustment of the lumen pH to optimize photosynthetic efficiency (Fig. 2).

In linear electron transport, in which electron transport is mediated by the two photosystems arranged in tandem, the yields of PSII and PSI are theoretically equal. Operation of PSI cyclic electron transport, however, requires a higher yield from PSI than PSII, since it is mediated solely by PSI. A state transition that regulates the balance of excitation between the two photosystems may be involved in this regulation by transferring more absorbed light energy to the PSI reaction center. As described below, the movements of LHCII are regulated by monitoring the redox state of the plastoquinone (PQ) pool.

In this review, I summarize current knowledge on the regulation of photosynthetic electron trans-

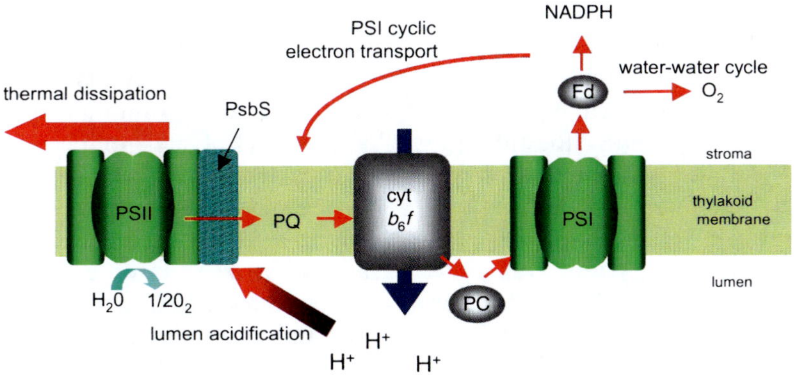

Fig. 2. Regulation of thermal dissipation via thylakoid lumen pH. Alternative electron transport pathways, PSI cyclic electron transport and water-water cycle lower the thylakoid lumen pH without accumulation of NADPH. PsbS is a pH sensor that regulates induction of thermal dissipation from PSII.

port, focusing on PSI cyclic electron transport and its interaction with other regulatory processes. In this decade, molecular genetics are significantly contributing to our knowledge of the regulatory machinery. This new information opens the way to revisiting the physiological significance of regulation.

II Chlorophyll Fluorescence: A Non-disruptive Tool for Electron Transport Analysis

Chlorophyll fluorescence reflects even subtle changes in photosynthetic electron transport and can be quantitatively analyzed using pulse-amplitude modulation (PAM) chlorophyll fluorometry in intact leaves (reviewed in Krause and Weis, 1991). Imaging techniques for this fluorescence under a CCD (charge-coupled device) camera make it possible to identify mutants defective in the regulation of photosynthesis. At room temperature, chlorophyll fluorescence is emitted from PSII and indicates the status of PSII photochemistry. The level of chlorophyll fluorescence is decreased by two separate processes. Photochemical quenching (qP) is due to a photochemical reaction in the PSII reaction center. Higher reduction of a primary quinone-type acceptor, Q_A, in PSII by electrons results in the emission of higher chlorophyll fluorescence. In mutants defective in electron transport activity, Q_A is highly reduced by electrons due to a restriction in downstream electron transport, leading to the *high chlorophyll fluorescence* (*hcf*) phenotype. In contrast, non-photochemical quenching (NPQ) is caused by several events that are not directly related to qP. NPQ is subdivided into three components related to the thermal dissipation of excessive absorbed light energy depending on thylakoid energization (qE), state transition (qT) and photoinhibition (qI). These three components are distinguishable by their different timescales of relaxation. When NPQ is induced, the efficiency of electron transport to the PSII reaction center is reduced by regulatory processes or photoinhibition, resulting in a lower fluorescence level.

Chlorophyll fluorescence analysis is a non-destructive method of monitoring the status of photosynthetic electron transport and can be used to screen for mutants defective in photosynthetic electron transport. Pioneer studies focused on *hcf* mutants, identified by exposure to ultraviolet (UV) light in the dark in maize (*Zea mays*) (Miles, 1980) and in Arabidopsis (*Arabidopsis thaliana*) (Meurer et al., 1996). Since the light intensity of UV light is very low, the exposure does not induce any regulatory processes. Therefore, the quenching of chlorophyll fluorescence depends exclusively on qP, resulting in identification of mutants severely defective in photosynthetic electron transport.

In contrast to the *hcf* mutant screening, mutants can be selected at relatively high light intensity using a video-imaging system (Niyogi et al., 1997; Shikanai et al., 1999; Peterson and Havir, 2000). Therefore, mutants defective in both qP and NPQ are the targets of this screening. Although PAM chlorophyll fluorometry can monitor the fluorescence level quantitatively under light conditions, it was originally difficult to record the fluorescence level as two-dimensional information. A CCD camera equipped with a filter transmitting chlorophyll fluorescence (>680 nm) but not actinic light (<590 nm) facilitated the visualization of fluorescence images. More recently, the PAM system has been developed to collect two-dimensional information on chlorophyll fluorescence and has facilitated the screening of mutants based on the precise monitoring of individual chlorophyll fluorescence parameters (Barbagallo et al., 2003; Walters et al., 2003).

III Thermal Dissipation of Absorbed Excessive Light Energy from PSII

Light energy, absorbed by chlorophylls and carotenoids which bind to LHCII, is funneled to the PSII reaction centers, where the energy is utilized for charge separation. The efficiency of this energy transfer among pigments is extremely high. Conversely, PSII has another function: that of dissipating absorbed excessive light energy safely as heat. Under excessive light conditions, over 75% of the absorbed light energy can be eliminated by this thermal dissipation (reviewed in Demmig-Adams et al., 1996). However, under limited light conditions, thermal dissipation reduces the efficiency of photosynthesis, and thus its induction needs to be regulated by monitoring the excess of light energy. To detect the degree of excess light, plants monitor the pH of the thylakoid lumen, which falls to below the threshold under high light conditions.

Chlorophyll fluorescence reflects the status of PSII photochemistry and can be utilized to monitor NPQ as a two-dimensional image under a CCD camera. Using a video imaging system, *npq1* (*non-photochemical quenching*) mutants defective in violaxanthin de-epoxidase activity were isolated in both Chlamydomonas (*Chlamydomonas reinhardtii*) (Niyogi et al., 1997) and Arabidopsis (Niyogi et al., 1998). Absorption of excessive light energy causes a buildup of ΔpH, resulting in acidification of the thylakoid lumen. This lowering of pH activates an enzyme, violaxanthin de-epoxidase, which converts violaxanthin to zeaxanthin via antheraxanthin (the xanthophyll cycle; reviewed in Demmig-Adams and Adams, 1996). Under low light conditions or in the dark, however, zeaxanthin and antheraxanthin are converted back to violaxanthin via the activity of zeaxanthin epoxidase. The size of qE is correlated with the accumulation of zeaxanthin and antheraxanthin (reviewed in Demmig-Adams and Adams, 1996), suggesting the involvement of specific carotenoids in qE induction. Genetic investigations have confirmed that the xanthophyll cycle is required for efficient qE induction (Niyogi et al., 1997, 1998).

Much more important progress was made with the discovery of the Arabidopsis mutant *npq4* (Li et al., 2000). Although its xanthophyll cycle activity is normal, *npq4* lacks qE induction. *NPQ4* encodes a PSII subunit, PsbS, which is a member of the LHC superfamily containing four rather than three trans-membrane domains. PsbS is not required for PSII photochemistry but it is essential for qE induction. Since thermal dissipation is triggered by lumen acidification, PsbS is believed to be involved in this pH sensing process (Li et al., 2004).

Plant response to fluctuating light conditions via thermal dissipation is essential, especially in the field (Külheim et al., 2002). In the current decade, our knowledge on the mechanism of NPQ induction has much improved, but it may have made the story even more complex. This topic has been discussed in recent reviews (Horton and Ruban, 2005; Niyogi et al., 2005). This review further focuses on another topic: how the thylakoid lumen pH is regulated to optimize the extent of thermal dissipation under fluctuating light conditions.

IV Balancing Excitation Energy Between Photosystems by State Transition

State transition is a process that redistributes light energy between the two photosystems to optimize photosynthetic performance. It was independently discovered by Murata (1969), and Bonaventura and Myers (1969). Light absorption is balanced through the reversible association of the major LHCII between PSII and PSI. In Chlamydomonas, 80% of the light energy absorbed by PSII is transferred to PSI under specific conditions, whereas the efficiency is much lower (15–20%) in higher plants (Allen, 1992). Imbalance of excitation between the two photosystems is monitored at the Q_o site of the cyt b_6f complex (Vener et al., 1997; Zito et al., 1999). Excess excitation of PSII relative to PSI leads to a reduction of the PQ pool, resulting in the activation of a kinase, the phosphorylation of LHCII, and the movement of LHCII from PSII to PSI (state 2). In contrast, excess excitation of PSI leads to oxidation of the PQ pool, resulting in the reverse process (state 1) (Allen et al., 1981).

A mutant specifically defective in state transition was identified in Chlamydomonas using chlorophyll fluorescence imaging (Fleischmann et al., 1999; Kruse et al., 1999). These mutants were defective in a gene encoding a thylakoid-associated serine-threonine protein kinase, Stt7 (Depege et al., 2003). The remaining question is whether Stt7 directly phosphorylates LHCII or acts upstream in the signaling cascade from the cyt b_6f complex. In the Arabidopsis genome, two genes, *STN7* and *STN8*, encode homologues to the Chlamydomonas *Stt7*. Characterization of knockout plants showed that only *STN7* is essential for state transition (Bellafiore et al., 2005). *STN8* is involved in phosphorylation of PSII core proteins (Bonardi et al., 2005; Vainonen et al., 2005).

A kinase involved in the signal transduction of state transition has finally been identified in Chlamydomonas and Arabidopsis. However, the entire signal pathway from the Q_o site of the cyt b_6f complex to LHCII remains unclear. Furthermore, the events that take place after the phosphorylation of LHCII are a matter of debate (Haldrup et al., 2001). Knockout of PSI-H severely impaired state transition in Arabidopsis (Lunde et al., 2000), suggesting that LHCII binds PSI in state 2 and that PSI-H is a binding site. In Chlamydomonas, two minor monomeric LHCII proteins, CP26 and CP29, and a type II major LHCII, LhcbM5, are associated with PSI in state 2, and probably act as a docking site for the major LHCII trimer (Takahashi et al., 2006).

Although the *stn7* mutant plants showed the same growth rate as the wild-type plants under constant light conditions, their CO_2 assimilation rate and growth were reduced under fluctuating light conditions (Bellafiore et al., 2005). A lack of state transition modifies the composition of the thylakoid proteins, implying that *stn7* mutant plants use an alternative strategy to adapt to fluctuating light conditions (Tikkanen et al., 2006). Interestingly, a shift from low- to high-light intensity transiently induces an oscillation in LHCII phosphorylation, resulting in the movement of LHCII between the two photosystems (Tikkanen et al., 2006). This oscillation may be a result of a buffering reaction against a sudden change in light intensity, during which plants monitor the dynamics of light intensity over a timescale of hours. If the dynamics exceed the plant's capacity for short-term response, it starts to remodel its photosynthetic apparatus via the process of photo-acclimation. This discovery may provide an important clue to understanding the physiological function of state transition in higher plants.

V Photorespiration and the Water–Water Cycle: Alternative Electron Sinks?

Rubisco (ribulose-1,5-bisphosphate carboxylase/oxygenase) catalyses both the carboxylation and the oxygenation of RuBP (ribulose-1,5-bisphosphate) as a result of competition between CO_2 and O_2 for the RuBP intermediate at the active site of the enzyme. The photorespiration pathway starts with the oxygenation of RuBP, and the resulting 2-phosphoglycolate is ultimately fed into the Calvin cycle as 3-phosphoglycerate. This metabolic process utilizes both ATP and NADPH and produces CO_2 and NH_3 in a series of reactions that take place in chloroplasts, peroxisomes and mitochondria. An early Arabidopsis genetics study contributed significantly to elucidating the photorespiration pathway (reviewed in Somerville, 2001). The mutants could not survive in air, but were able to grow at higher CO_2 concentrations (Somerville and Ogren, 1979).

It is generally accepted that photorespiration provides a major alternative sink for electrons during photosynthesis in C_3 plants (Wingler et al., 2000). This idea is consistent with the evidence generated by genetic manipulation of glutamine synthetase activity: a rate-limiting step in photorespiration can modify tolerance to photo-damage in tobacco (Kozaki and Takeba, 1996). Additional

ATP is required along the recycling pathway of 2-phosphoglycolate into 3-phosphoglycerate. Thus, under drought stress conditions, activation of photorespiration increases the requirement for ATP. To support the physiological function of photorespiration as an electron sink, this flexible requirement of the ATP/NADPH ratio needs to be fulfilled by light reactions. As discussed below, both the water-water cycle and PSI cyclic electron transport are candidates for this regulatory machinery.

In addition to photorespiration, O_2 can directly accept electrons from PSI in a process termed the Mehler reaction (Mehler, 1951). The resulting O_2^- is disproportionated to H_2O_2 and O_2 by superoxide dismutase (SOD), and subsequently H_2O_2 is reduced to water by ascorbate peroxidase in the chloroplast, a reaction accompanied by the generation of monodehydroascorbate (MDA). The water-water cycle consists of these scavenging reactions of ROS plus recycling of ascorbate. The reduction of MDA by ferredoxin (Fd) doubles the electron flux as a result of the water-water cycle (reviewed in Asada, 2006). Although the pathway has already been identified, controversy remains about the physiological function of the water-water cycle as an electron sink (Badger et al., 2000; Ort and Baker, 2002). It is believed that the Mehler reaction occurs particularly when the electron acceptors from PSI are depleted, allowing linear electron transport to continue. In this way, the water-water cycle may sustain high ΔpH, which is essential to maintain thermal dissipation of excessive absorbed light energy from PSII (Makino et al., 2002). It is also possible that the water-water cycle maintains the balance of ATP and NADPH, since it generates ΔpH without reduction of $NADP^+$. Despite numerous suggestions, almost nothing has been clarified concerning the physiological function of the water–water cycle except for the fact that scavenging of ROS is essential, even at low light intensities (Rizhsky et al., 2003).

Although the entire pathway for scavenging ROS has been clarified, the factor that mediates O_2 photoreduction remains unidentified. Although Fd and monodehydroascorbate reductase (MDAR) are hypothesized as mediators (Furbank and Badger, 1983; Miyake et al., 1998), it still remains a matter of debate. In cyanobacteria, A-type flavoprotein is essential for photoreduction of O_2 (Helman et al., 2003). In contrast to higher plants, the Mehler reaction operates via direct reduction of O_2 to water, and genes for flavoprotein are not conserved in higher plants. It is necessary to identify the factors involved in photoreduction of O_2 in higher plants to be able to apply reverse genetics to clarify the physiological function of the water-water cycle.

PTOX (plastid terminal oxidase) was discovered in the Arabidopsis variegated mutants *immutans* (Carol et al., 1999; Wu et al., 1999). PTOX is homologous to mitochondrial AOX (alternative oxidase) and is believed to mediate PQ-dependent O_2 reduction to H_2O independently of proton translocation across the thylakoid membrane. Characterization of the mutant phenotype indicates that PTOX is essential for carotenoid synthesis during chloroplast development (Josse et al., 2000). Phytoene desaturase, an essential enzyme in carotenoid biosynthesis, requires oxidized PQ that is produced via PTOX in the absence of PSI during early chloroplast development. PTOX may also act as an alternative electron sink during photosynthesis in mature chloroplasts, but its contribution is estimated to be minimal (Joët et al., 2002; Rosso et al., 2006). Because of the variegated nature of the *immutans* leaves, however, it is not simple to assess this issue conclusively.

VI The Discovery of PGR5-Dependent PSI Cyclic Electron Transport

In the light reactions of photosynthesis, electrons excised from water at PSII are ultimately transferred to $NADP^+$ via PSI. Coupled with electron transport through the cytochrome (cyt) b_6f complex, protons are translocated from the stroma to the thylakoid lumen (ΔpH generation). In contrast to this linear electron transport, PSI cyclic electron transport is solely driven by PSI and generates ΔpH without accumulating NADPH (reviewed in Bendall and Manasse, 1995; Munekage and Shikanai, 2005; Shikanai, 2007a) (Fig. 3). Although PSI cyclic electron transport was discovered more than 50 years ago, our knowledge on the machinery involved is still limited, and its physiological function has been overlooked.

The Arabidopsis mutant *pgr5* (*proton gradient regulation 5*), defective in PSI cyclic electron transport, was discovered on the basis of its defect in NPQ induction at high light intensity (Shikanai et al., 1999) by screening similar

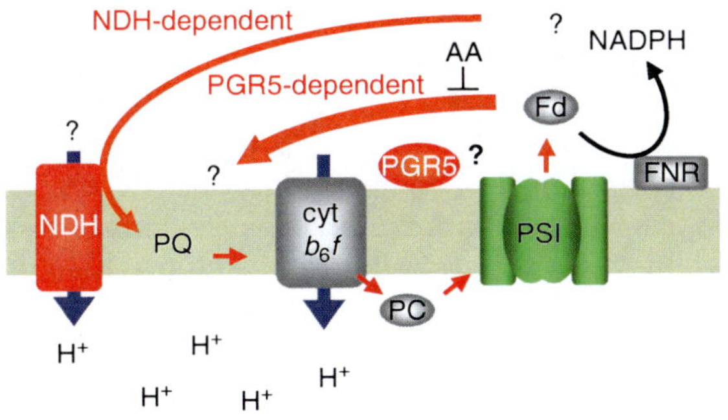

Fig. 3. Schematic representation of PSI cyclic electron transport. In higher plants, PSI cyclic electron transport consists of two partly redundant pathways: PGR5- and NDH-dependent pathways. The electron donor to NDH is unclear. The NDH complex may have proton-pumping activity. The PGR5-dependent pathway is sensitive to antimycin A (AA). The exact route taken by electrons is unclear in PGR5-dependent pathway. The exact localization and function of PGR5 are unclear. PC indicates plastocyanin.

to that used to identify *npq1* and *npq4* (Niyogi et al., 1998; Li et al., 2000). PGR5 is essential to induce thermal dissipation by lowering the lumen pH (Munekage et al., 2002). In ruptured chloroplasts, PGR5-dependent PSI cyclic activity is monitored as Fd-dependent PQ reduction, which mediates PSI cyclic electron transport in leaves (Munekage et al., 2002). High activity of PGR5-dependent PSI cyclic electron transport was detected even in the light, where PSI cyclic electron transport competes with linear electron transport (Okegawa et al., 2008). In *pgr5* this PQ reduction is markedly impaired. This phenotype, which is mimicked by addition of antimycin A that inhibits Fd-dependent PSI cyclic electron transport, was discovered by Arnon's group (Tagawa et al., 1963). Most probably, *pgr5* is defective in Arnon's pathway of PSI cyclic electron transport, so I do not rigorously distinguish them in this review.

In addition to a defect in NPQ induction, *pgr5* exhibits a remarkable phenotype with a low ratio of $P700^+/P700$ (oxidized/reduced PSI reaction centers) even at moderate light intensity (Munekage et al., 2002). In contrast, in the wild type, the ratio rises in response to increasing light intensity. PGR5-dependent PSI cyclic electron transport is required for balancing the ATP/NADPH production ratio during steady-state photosynthesis, and ATP synthesis is likely to limit photosynthesis at high light intensity in *pgr5*. This imbalance of light reactions causes accumulation of NADPH in the stroma (over-reduction of the stroma), leading to a shortage of electron acceptors from PSI. Because of this stromal over-reduction, PSI is sensitive to high light intensity in *pgr5* (Munekage et al., 2002). PSI photodamage caused by stromal over-reduction was characterized in more detail by chilling sensitive plants (Sonoike and Terashima, 1994). In summary, PGR5-dependent PSI cyclic electron transport is essential for photoprotection via two mechanisms: (1) Induction of NPQ, and (2) Protection of PSI by balancing the ATP/NADPH production ratio. The latter mechanism is directly related to photosynthesis rather than photoprotection.

Despite a long history of interest in the route taken by electrons in PGR5-dependent PSI cyclic electron transport, it is still not known. It is known, however, that electron transport is sensitive to antimycin A and that a small thylakoid protein, PGR5, is required, although its precise function is unclear (Tagawa et al., 1963; Munekage et al., 2002). A long-standing debate continues about whether or not the electron flow from Fd to PQ passes via the Q-cycle in the cyt b_6f complex (Bendall and Manasse, 1995; Shikanai 2007a), where a new heme *x* (also referred to as heme c_i) was discovered (Kurisu et al., 2003; Stroebel et al., 2003). This heme is not conserved in the cyt bc_1 complex in mitochondria, implying that it is involved in PGR5-dependent PSI cyclic electron transport. However, we could not obtain any results suggesting this possibility by characterization of *pgr1* (Okegawa et al., 2005), in which the activity of cyt b_6f activity is hypersensitive to lumen acidification (Munekage et al., 2001; Jahns et al., 2002). Recently, a protein with two transmembrane domains, PGRL1, was shown to be essential for stabilizing PGR5 in the thylakoid membranes (DalCorso et al., 2008). Over-accumulation of PGR5 in the thylakoid membraned increases the maximum rate of PSI cyclic electron transport under fluctuating light conditions, suggesting that the protein complex including PGR5 and PGRL1 determines the rate of PSI cyclic electron transport (Okegawa et al., 2007; Long et al., 2008). In future research, priority should be given to identifying the route taken by electrons in PGR5-dependent PSI cyclic electron transport. It takes place probably through the complex including PGR5.

VII PSI Cyclic Electron Transport Mediated by Chloroplast NAD(P)H Dehydrogenase

Prior to the rediscovery of PGR5-dependent PSI cyclic electron transport, the chloroplast NAD(P)H dehydrogenase (NDH) complex was shown to be also involved in PSI cyclic electron transport (Burrows et al., 1998; Kofer et al., 1998; Shikanai et al., 1998). As described in detail in other reviews (Shikanai and Endo 2000; Shikanai, 2007b), the following three breakthroughs have significantly advanced our knowledge of chloroplast NDH in the last 20 years: (1) Discovery of 11 *ndh* genes encoding proteins homologous to the subunits of the mitochondrial NADH dehydrogenase (complex I) in the plastid genome (Matsubayashi et al., 1987). It was previously a mystery as to why the plastid genome encodes homologs of the respiratory machinery. (2) Identification of the cyanobacterial M55 mutant that requires high CO_2 levels for growth. M55 is defective in *ndhB* encoding a subunit of cyanobacterial NDH (Ogawa, 1991). Physiological characterization of M55 demonstrated that the cyanobacterial NDH complex mediates PSI cyclic electron transport (Mi et al., 1992, 1994, 1995). For cyanobacterial NDH, refer to the most recent review of Ogawa (Ogawa and Mi, 2007). (3) Establishment of the plastid transformation technique in tobacco (*Nicotiana tabacum*) (Svab and Maliga, 1993), which enabled us to directly knock out the plastid-encoded *ndh* genes (Burrows et al., 1998; Kofer et al., 1998; Shikanai et al., 1998).

Chlorophyll fluorescence analysis of the knockout lines of NDH in tobacco revealed that the chloroplast NDH mediates PQ reduction via the stromal electron pool, which functions in PSI cyclic electron transport in the light (Burrows et al., 1998; Kofer et al., 1998; Shikanai et al., 1998). However, the knockout did not cause any strong phenotypes, at least under the greenhouse conditions. Chloroplast NDH may act as machinery that provides stress resistance. To test this possibility, the stress sensitivity of the NDH knockout lines was extensively studied under various conditions. These tobacco lines are sensitive to high light intensities (Endo et al., 1999), low humidity stress (Horváth et al., 2000), drought stress (Munné-Bosch et al., 2005) and high and low temperatures (Wang et al., 2006). The chloroplast NDH complex alleviates oxidative stress in chloroplasts, although its exact molecular mechanism is unclear.

Even chloroplast NDH seems to be dispensable, at least under greenhouse conditions. This may due to the fact that PGR5-dependent PSI cyclic electron transport complements the function of NDH. On the basis of mutant phenotypes, the PGR5-dependent pathway appears to be a main route of PSI cyclic electron transport in higher plants (Munekage et al., 2004). To characterize the mutant phenotype of the double mutants defective in both pathways of PSI cyclic electron transport, it was essential to identify the NDH mutants in Arabidopsis. However, plastid transformation is not routinely feasible in Arabidopsis. Thus, our strategy was to screen Arabidopsis nuclear mutants specifically defective in NDH activity using a chlorophyll fluorescence imaging technique (Hashimoto et al., 2003). We focussed on the subtle changes in chlorophyll fluorescence after actinic light illumination, which is impaired in tobacco *ndh* knockout lines (Burrows et al., 1998; Kofer et al., 1998; Shikanai et al., 1998). Arabidopsis *crr2* and *crr4* (*chlororespiratory reduction*) mutants are defective in the expression of plastid-encoded *ndh* genes and are equivalent to the direct knockout lines of the plastid-encoded *ndh* genes. Both genes encode members of the pentatricopeptide repeat (PPR) family, which is extraordinarily large in higher plants (Lurin et al., 2004). Members of the PPR family are involved in RNA maturation processes in plastids and mitochondria (Shikanai, 2006). CRR2 is essential for intergenic RNA cleavage between *rps7* and *ndhB*, which is probably essential for *ndhB* translation (Hashimoto et al., 2003). On the other hand, CRR4 is required for the RNA editing process that creates the translational initiation codon of *ndhD* (Kotera et al., 2005; Okuda et al., 2006).

In the double mutants *crr2 pgr5* and *crr4 pgr5*, photosynthesis is severely affected and the seedlings are sensitive even to the very low light intensity of 50 μmol photons m^{-2} s^{-1} (Munekage et al., 2004). This result indicates that PSI cyclic electron transport is essential for photosynthesis, even at very low light intensities. In other words, the chloroplast NDH complex is required for survival against the mutant background of *pgr5* by preventing stromal over-reduction. Since plants in the field often encounter the risk of stromal over-reduction (Endo et al., 2005), the chloroplast NDH complex is likely to function as a safety valve in the wild type, in which the PGR5-dependent PSI cyclic electron transport is operating.

Compared to the PGR5-dependent PSI cyclic electron transport, much more information has accumulated on the machinery involved in NDH-dependent electron transport. However, we are still not sure about the entire subunit composition of the NDH complex, especially of the electron donor-binding module. Furthermore the electron donor is also unclear (Shikanai, 2007b). The same story is true in the cyanobacterial NDH complex (Ogawa and Mi, 2007). Our genetic approach, which focuses on *crr* phenotypes, has contributed to addressing this problem by identifying proteins essential for the accumulation of NDH (Munshi et al., 2005; 2006; Muraoka et al., 2006; Shimizu and Shikanai, 2007) and also the novel subunit, CRR23/NdhL (Shimizu et al., 2008). The NDH complex has a fragile nature, especially in its unknown electron donor-binding module. Genetic investigations may provide biochemical insights to this problem. Proteomics strategies in chloroplasts (Rumeau et al., 2005) and cyanobacteria (Battchikova et al., 2004; Prommeenate et al., 2004; Zhang et al., 2004) have also identified several novel subunits, NdhL-O, which are not homologous to the bacterial NDH-1 subunits (Shikanai, 2007b). Recently, we showed that the NDH complex stably interacts with PSI to form a super-complex (Peng et al., 2008). It is possible that electrons move in this super-complex directly from PSI to NDH.

VIII PSI Cyclic Electron Transport and Thermal Dissipation

pgr5 was isolated based on its phenotype, which is defective in qE at high light intensities, which indicates that PSI cyclic electron transport is essential to acidify the thylakoid lumen and trigger qE (Munekage et al., 2002). qE is induced not only at high light intensities but also at low light intensities when Calvin cycle activity is reduced under stressed conditions, such as drought and low temperatures. *pgr5* is defective in qE induction in CO_2-free air, which mimics the closure of stomata under drought stress (Munekage et al., 2002). Under stressed conditions, reduction of Calvin cycle activity results in a decreased rate of linear electron transport. The question that remains a subject of hot debate is how ΔpH is maintained to induce qE under these excessive light conditions (Kramer et al., 2003; Johnson, 2004).

Classically, the conflicting requirements of slowing linear electron transport and generating a greater ΔpH to induce qE were explained by the contribution of PSI cyclic electron transport (Heber and Walker, 1992). The *pgr5* phenotype in CO_2-free air strongly supports this idea (Munekage et al., 2002). The simplest version of this idea is that PSI cyclic electron transport is directly essential to acidify the thylakoid lumen to below pH 6.0 in CO_2-free air by generating additional ΔpH. We estimate that 20% of total ΔpH generation depends on PGR5-dependent PSI cyclic electron transport during steady-state photosynthesis (Okegawa et al., 2005). If the ratio of linear and PSI cyclic electron transport is constant even in CO_2-free air, where the rate of linear electron transport decreases, can the lumen pH be simply explained by the activity of PSI cyclic electron transport? In this calculation, the decrease in the relaxation of ΔpH, which is directly involved in the reduction of ATP consumption, should be included. The rate of PSI cyclic electron transport may also increase in CO_2-free air (Golding and Johnson, 2003; Golding et al., 2004). Recent publications support the classical idea that PSI cyclic electron transport is essential for qE induction (Miyake et al., 2005), as well as the *pgr5* mutant phenotype (Munekage et al., 2002). However, the validity of in vivo measurements needs to be confirmed.

An alternative idea is that qE induction in CO_2-free air is mainly attributable to a decrease in the proton conductivity in ATPase (Kanazawa and Kramer, 2002; Kramer et al., 2003). Without a doubt, ATPase significantly contributes to qE induction in CO_2-free air by decreasing the relaxation activity of ΔpH due to reduced ATP consumption. The question is whether a specific regulatory process is applied during this process.

Avenson et al. (2005) estimated a 13% reduction in the steady-state proton influx to the thylakoid lumen in *pgr5* and suggested that it is too small to explain the drastic defect in qE. In contrast, proton conductivity is still high at high light intensities in *pgr5*, although it is significantly reduced in the wild type. These authors suggest that a decreased ratio of ATP/NADPH generation may indirectly affect the regulatory process that tunes the sensitivity of qE to the rate of electron transport. If this regulation really exists, the *pgr5* phenotypes would involve two independent effects: the direct effect of decreased ATP/NADPH production ratio leading to stromal over-reduction, and a secondary effect of qE induction via aberrant modification of proton conductivity in ATPase. A mutation in the regulatory machinery of H^+-ATPase may specifically suppress the qE phenotype in *pgr5*, while the P700 phenotype remains.

IX PSI Cyclic Electron Transport and State Transition

In Chlamydomonas, a state transition is involved in the switch between linear and PSI cyclic electron flow (Finazzi et al., 2002). During a state 1 to state 2 transition, redistribution of LHCII from PSII to PSI may modify the dynamics of PQ diffusion in the thylakoid membranes, allowing the PSI cyclic electron transport to operate. Since the contribution of state transition to the regulation of light harvesting is smaller in higher plants than in Chlamydomonas, it may not be possible to assume that the same process takes place in higher plants. Because the phenotype of *stn7* is much less severe than that of *pgr5*, it is evident that PGR5-dependent PSI cyclic electron transport is operating in *stn7*.

Even though the effect of state transition on the induction of PSI cyclic electron transport is subtle, it may be physiologically significant. The *stn7* phenotype suggests that state transition is essential to adapting to fluctuating light conditions (Bellafiore et al., 2005; Tikkanen et al., 2006). Via activation of PSI cyclic electron transport, the effect of the state transition may be amplified in the dynamics of the thylakoid membranes. If a reliable in vivo measuring technique is established for PSI cyclic electron transport (reviewed in Johnson, 2005), it could be applied to assess whether the rate of PSI cyclic electron transport is enhanced in state 2 in higher plants.

X The Water–Water Cycle and PSI Cyclic Electron Transport

In linear electron transport, reduced Fd is oxidized by $NADP^+$ via FNR (Fd-$NADP^+$ oxidoreductase). Reduced Fd is alternatively oxidized by PGR5-dependent PSI cyclic electron transport and via the Mehler reaction (water-water cycle). Although this topic has been of interest to many plant physiologists, we are not sure how and when two pathways contribute as alternative electron sinks. At least on the basis of the mutant phenotype, PGR5-dependent PSI cyclic electron transport is required for efficient photosynthesis (Munekage et al., 2002, 2004), and this result is consistent with physiological estimates of the rate of electron transport (Avenson et al., 2005; Okegawa et al., 2005). It is necessary to isolate mutants specifically defective in the Mehler reaction rather than the scavenging process of ROS to evaluate the physiological significance of the water-water cycle as an alternative electron sink. To identify mutants in minor alternative electron sinks, we performed screening of *hcf* mutants in Arabidopsis at low light intensity in air containing 10% O_2 but without CO_2. We identified a weak allele of *pgr5*, several alleles of *crr* mutants, and a mutant defective in an unknown minor electron sink (Munekage, Horiguchi, Narumiya and Shikanai, unpublished results). No potential candidates that were defective in the water–water cycle were identified.

XI Concluding Remarks

When plants are exposed to high-intensity light, flexible operation of alternative electron transport acidifies the thylakoid lumen, resulting in the induction of thermal dissipation of excessive absorbed light energy. Even though this process is certain to take place, what actually happens in the chloroplasts remains mysterious. What is the contribution of each alternative pathway to ΔpH generation and/or as an electron sink? What are the routes taken by the electrons? Is there any mechanism for balancing the contribution of each separate pathway? What really happens during state transition in the thylakoid membranes? At least our knowledge of the physiological function of each regulatory process has progressively improved in the past decade. The regulation of photosynthesis clearly takes place and is essential for plant survival. Despite their physiological significance, the mechanisms of regulation are still poorly understood. Photosynthetic research continuously delivers new results that throw previously held conclusions into question.

Acknowledgements

This work was supported by a grant-in-aid for Scientific Research on Priority Areas (16085296) and for Creative Scientific Research (17GS0316) from the Ministry of Education, Culture, Sports, Science and Technology, Japan.

References

Allen JF (1992) Protein phosphorylation in regulation of photosynthesis. Biochim Biophys Acta 1098: 273–335

Allen JF, Bennett J, Steinback KE and Arntzen CJ (1981) Chloroplast protein phosphorylation couples plastoquinone redox state to distribution of excitation energy between photosystems. Nature 291: 25–29

Asada K (1999) The water–water cycle in chloroplasts: scavenging of active oxygens and dissipation of excess photons. Annu Rev Plant Physiol Plant Mol Biol 50: 601–639

Asada K (2006) Production and scavenging of reactive oxygen species in chloroplasts and their functions. Plant Physiol 141: 391–396

Avenson TJ, Cruz JA, Kanazawa A and Kramer DM (2005) Regulating the proton budget of higher plant photosynthesis. Proc Natl Acad Sci USA 101: 5530–5555

Badger MR, von Caemmerer S, Ruuska S and Nakano H (2000) Electron flow to oxygen in higher plants and algae: rates and control of direct photoreduction (Mehler reaction) and rubisco oxygenase. Philos Trans R Soc Lond B 355: 1433–1446

Barbagallo RP, Oxborough K, Pallett KE and Baker NR (2003) Rapid, noninvasive screening for perturbations of metabolism and plant growth using chlorophyll fluorescence imaging. Plant Physiol 132: 485–493

Battchikova N, Zhang P, Rudd S, Ogawa T and Aro E-M (2004) Identification of NdhL and Ssl1690 (NdhO) in NDH-1L and NDH-1M complexes of *Synechocystis* sp. PCC 6803. J Biol Chem 280: 2587–2595

Bellafiore S, Barneche F, Peltier G and Rochaix J-D (2005) State transitions and light adaptation require chloroplast thylakoid protein kinase STN7. Nature 433: 892–895

Bendall D and Manasse RS (1995) Cyclic photophosphorylation and electron transport. Biochim Biophys Acta 1229: 23–38

Bonardi V, Pesaresi P, Becker T, Schleiff E, Wagner R, Pfannschmidt T, Jahns P and Leister D (2005) Photosystem II core phosphorylation and photosynthetic acclimation require two different protein kinases. Nature 437: 1179–1182

Bonaventura C and Myers J (1969) Fluorescence and oxygen evolution from *Chlorella pyrenoidosa*. Biochim Biophys Acta 3: 366–383

Burrows PA, Sazanov LA, Svab Z, Maliga P and Nixon PJ (1998) Identification of a functional respiratory complex in chloroplasts through analysis of tobacco mutants containing disrupted plastid *ndh* genes. EMBO J 17: 868–876

Carol P, Stevenson D, Bisanz C, Breitenbach J, Sandmann G, Mache R, Coupland G and Kuntz M (1999) Mutations in the Arabidopsis gene *IMMUTANS* cause a variegated phenotype by inactivating a chloroplast terminal oxidase associated with phytoene desaturation. Plant Cell 11: 57–68

DalCorso G, Pesaresi P, Masiero S, Aseeva E, Schunemann D, Finazzi G, Joliot P, Barbato R and Leister D (2008) A complex containing PGRL1 and PGR5 is involved in the switch between linear and cyclic electron flow in Arabidopsis. Cell 132: 273–285

Demmig-Adams B and Adams WW III (1996) The role of xanthophylls cycle carotenoids in the protection of photosynthesis. Trends Plant Sci 1: 21–26

Demmig-Adams B, Adams WW III, Baker DH, Logan BA, Bowling DR and Verhoeven AS (1996) Using chlorophyll fluorescence to assess the fraction of absorbed light allocated to thermal dissipation of excessive excitation. Physiol Plant 98: 253–264

Depege N, Bellafiore S and Rochaix J-D (2003) Role of chloroplast protein kinase Stt7 in LHCII phosphorylation and state transition in *Chlamydomonas*. Science 299: 1572–1575

Endo T, Shikanai T, Takabayashi A, Asada K and Sato F (1999) The role of chloroplastic NAD(P)H dehydrogenase in photoprotection. FEBS Lett 457: 5–8

Endo T, Kawase D and Sato F (2005) Stromal over-reduction by high-light stress as measured by decreases in P700 oxidation by far-red light and its physiological relevance. Plant Cell Physiol 46: 775–781

Finazzi G, Rappaport F, Furia A, Fleischmann M, Rochaix J-D, Zito F and Forti G (2002) Involvement of state transitions in the switch between linear and cyclic electron flow in *Chlamydomonas reinhardtii*. EMBO Rep 3: 280–285

Fleischmann MM, Ravanel S, Delosme R, Olive J, Zito F, Zito F and Forti G (1999) Isolation and characterization of photoautotrophic mutants of *Chlamydomonas reinhardtii* deficient in state transition. J Biol Chem 274: 30987–30994

Furbank RT and Badger MR (1983) Oxygen exchange associated with electron transport and photophosphorylation in spinach thylakoids. Biochim Biophys Acta 723: 400–409

Golding AJ and Johnson GN (2003) Down-regulation of linear and activation of cyclic electron transport during drought. Planta 218: 107–114

Golding AJ, Finazzi G and Johnson GN (2004) Reduction of the thylakoid electron transport chain by stromal reductants – evidence for activation of cyclic electron transport upon dark adaptation or under drought. Planta 220: 356–363

Haldrup A, Jensen PE, Lunde C and Scheller HV (2001) Balance of power: a view of the mechanism of photosynthetic state transition. Trends Plant Sci 6: 301–305

Hashimoto M, Endo T, Peltier G, Tasaka M and Shikanai T (2003) A nucleus-encoded factor, CRR2, is essential for the expression of chloroplast *ndhB* in Arabidopsis. Plant J 36: 541–549

Heber U and Walker D (1992) Concerning a dual function of coupled cyclic electron transport in leaves. Plant Physiol 100: 1621–1626

Helman Y, Tchernov D, Reinhold L, Shibata M, Ogawa T, Schwarz R, Ohad I and Kaplan A (2003) Genes encoding A-type flavoproteins are essential for photoreduction of O_2 in cyanobacteria. Curr Biol 13: 230–235

Horton P and Ruban A (2005) Molecular design of the photosystem II light-harvesting antenna: photosynthesis and photoprotection. J Exp Bot 56: 365–373

Horton P, Ruban AV and Walters RG (1996) Regulation of light harvesting in green plants. Anne Rev Plant Physiol Plant Mol Biol 47: 655–684

Horváth EM, Peter SO, Joët T, Rumeau D, Cournac L, Horváth GV, Kavanagh TA, Schäfer C, Peltier G and Medgyesy P (2000) Targeted inactivation of the plastid *ndhB* gene in tobacco results in an enhanced sensitivity of photosynthesis to moderate stomatal closure. Plant Physiol 123: 1337–1350

Jahns P, Graf M, Munekage Y and Shikanai T (2002) Single point mutation in the Rieske iron-sulfur subunit of cytochrome b_6/f leads to an altered pH dependence of plastoquinol oxidation in *Arabidopsis*. FEBS Lett 519: 99–102

Jarillo JA, Gabrys H, Capel J, Alonso JM, Ecker JR and Cashmore AR (2001) Phototropin-related NPL1 controls chloroplast relocation induced by blue light. Nature 410: 952–954

Joët T, Genty B, Josse EM, Kuntz M, Cournac L and Peltier G (2002) Involvement of a plastid terminal oxidase in plastoquinone oxidation as evidenced by expression of the *Arabidopsis thaliana* enzyme in tobacco. J Biol Chem 277: 31623–31630

Johnson GN (2004) Controversy remains: regulation of pH gradient across the thylakoid membrane. Trends Plant Sci 9: 570–571

Johnson GN (2005) Cyclic electron transport in C_3 plants: fact or artefact? J Exp Bot 56: 407–416

Josse EM, Simkin AJ, Gaffe J, Laboure AM, Kuntz M and Carol P (2000) A plastid terminal oxidase associated with carotenoid desaturation during chromoplast differentiation. Plant Physiol 123: 1427–1436

Kanazawa A and Kramer DM (2002) *In vivo* modulation of nonphotochemical exciton quenching (NPQ) by regulation of the chloroplast ATP synthase. Proc Natl Acad Sci USA 99: 12789–12794

Kagawa T, Sakai T, Suetsugu N, Oikawa K, Ishiguro S, Kato T, Tabata S, Okada K and Wada M (2001) *Arabidopsis* NPL1: A phototropin homolog controlling the chloroplast high-light avoidance response. Science 291: 2138–2141

Kasahara M, Kagawa T, Oikawa K, Suetsugu N, Miyao M and Wada M (2002) Chloroplast avoidance movement reduces photodamage in plants. Nature 420: 829–832

Kinoshita T, Doi M, Suetsugu N, Kagawa T, Wada M and Shimazaki K (2001) *phot1* and *phot2* mediate blue light regulation of stomatal opening. Nature 414: 656–660

Kramer DM, Cruz JA and Kanazawa A (2003) Balancing the central roles of the thylakoid proton gradient. Trends Plant Sci 8: 27–32

Krause GH and Weis E (1991) Chlorophyll fluorescence and photosynthesis: the basics. Annu Rev Plant Physiol Plant Mol Biol 42: 313–349

Kofer W, Koop H-U, Wanner G and Steinmüller K (1998) Mutagenesis of the genes encoding subunits A, C, H, I, J and K of the plastid NAD(P)H-plastoquinone-oxidoreductase in tobacco by polyethylene glycol-mediated plastome transformation. Mol Gen Genet 258: 166–173

Kotera E, Tasaka M and Shikanai T (2005) A pentatricopetide repeat protein is essential for RNA editing in chloroplasts. Nature 433: 326–330

Kozaki A and Takeba G (1996) Photorespiration protects C_3 plants from photooxidation. Nature 384: 557–560

Kruse O, Nixon P, Schmid GH and Mullineaux CW (1999) Isolation of state transition mutants of *Chlamydomonas reinhardtii* by fluorescence video imaging. Photosynth Res 61: 43–51

Külheim C, Ågren J and Jansson S (2002) Rapid regulation of light harvesting and plant fitness in the field. Science 297: 91–93

Kurisu G, Zhang H, Smith JL and Cramer WA (2003) Structure of the cytochrome b_6f complex of oxygenic photosynthesis: tuning the cavity. Science 302: 1009– 014

Li X-P, Björkman O, Shih C, Grossman AR, Rosenquist M, Jansson S and Niyogi KK (2000) A pigment-binding protein essential for regulation of photosynthetic light harvesting. Nature 403: 391–395

Li X-P, Gilmore AM, Caffarri S, Bassi R, Golan T, Kramer D and Niyogi KK (2004) Regulation of photosynthetic light harvesting involves intrathylakoid lumen pH sensing by the PsbS protein. J Biol Chem 279: 22866–22874

Long TA, Okegawa Y, Shikanai T, Schmidt GW and Covert SF (2008) Conserved role of PROTON GRADIENT REGULATOR 5 (PGR5) in the regulation of PSI cyclic electron transport. Planta 228: 907–918

Lunde C, Jensen PE, Haldrup A, Knoetzel J and Scheller HV (2000) The PSI-H subunit of photosystem I is essential for state transitions in plant photosynthesis. Nature 408: 613–615

Lurin C, Andrés C, Aubourg S, Bellaoui M, Bitton F, Bruyere C, Caboche M, Debast C, Gualberto J, Hoffmann B, Lecharny A, Le Ret M, Martin-Magniette ML, Mireau H, Peeters N, Renou JP, Szurek B, Taconnat L and Small I (2004) Genome-wide analysis of Arabidopsis pentatricopeptide repeat proteins reveals their essential role in organelle biogenesis. Plant Cell 16: 2089–2103

Makino A, Miyake C and Yokota A (2002) Physiological functions of the water-water cycle (Mehler reaction) and the cyclic electron flow around PSI in rice leaves. Plant Cell Physiol 43: 1017–1026

Matsubayashi T, Wakasugi T, Shinozaki K, Yamaguchi-Shinozaki K, Zaita N, Hidaka T, Meng BY, Ohto C, Tanaka M, Kato A, Maruyama T and Sugiura M (1987) Six chloroplast genes (*ndhA-F*) homologous to human mitochondrial genes encoding components of the respiratory chain NADH dehydrogenase are actively expressed: determination of the splice sites in *ndhA* and *ndhB* pre-mRNAs. Mol Gen Genet 210: 385–393

Mehler AH (1951) Studies on reactivities of illuminated chloroplasts. I. Mechanism of the reduction of oxygen and other Hill reagents. Arch Biochem Biophys 33: 65–77

Meurer J, Meierhoff K and Westhoff P (1996) Isolation of high-chlorophyll-fluorescence mutants of *Arabidopsis thaliana* and their characterisation by spectroscopy, immunoblotting and northern hybridisation. Planta 198: 385–396

Mi H, Endo T, Schreiber U, Ogawa T and Asada K (1992) Electron donation from cyclic and respiratory flows to the photosynthetic intersystem chain is mediated by pyridine nucleotide dehydrogenase in the cyanobacterium *Synochocystis* PCC6803. Plant Cell Physiol 33: 1233–1237

Mi HL, Endo T, Schreiber U, Ogawa T and Asada K (1994) NAD(P)H-dehydrogenase-dependent cyclic electron flow around photosystem-I in the cyanobacterium *Synochocystis* – a study of dark-starved cells and spheroplasts. Plant Cell Physiol 35: 163–173

Mi H, Endo T, Ogawa T and Asada K (1995) Thylakoid membrane-bound pyridine nucleotide dehydrogenase complex mediates cyclic electron transport in the cyanobacteria *Synechocystis* PCC 6803. Plant Cell Physiol 36: 661–668

Miles D (1980) Mutants of higher plants: maize. Meth Enzymol 69: 3–23

Miyake C, Schreiber U, Hormann H, Sano S and Asada K (1998) The FAD-enzyme monodehydroascorbate radical reductase mediates photoproduction of superoxide radicals in spinach thylakoid membranes. Plant Cell Physiol 39: 821–829

Miyake C, Miyata M, Shinzaki Y and Tomizawa K (2005) CO_2 response of cyclic electron flow around PSI (CEF-PSI) in tobacco leaves – relative electron fluxes through PSI and PSII determine the magnitude of non-photochemical quenching (NPQ) of Chl fluorescence. Plant Cell Physiol 46: 629–637

Müller P, Li X-P and Niyogi KK (2001) Non-photochemical quenching. A response to excess light energy. Plant Physiol 125: 1558–1566

Mullineaux P and Karpinski S (2002) Signal transduction in response to excess light: getting out of the chloroplast. Curr Opin Plant Biol 5: 43–48

Munekage Y and Shikanai T (2005) Cyclic electron transport though photosystem I. Plant Biotech 22: 361–69

Munekage Y, Takeda S, Endo T, Jahns P, Hashimoto T and Shikanai T (2001) Cytochrome b_6f mutation specifically affects thermal dissipation of absorbed light energy in *Arabidopsis*. Plant J 28: 351–359

Munekage Y, Hojo M, Meurer J, Endo T, Tasaka M and Shikanai T (2002) *PGR5* is involved in cyclic electron flow around photosystem I and is essential for photoprotection in *Arabidopsis*. Cell 110: 361–371

Munekage Y, Hashimoto M, Miyake C, Tomizawa K, Endo T, Tasaka M and Shikanai T (2004) Cyclic electron flow around photosystem I is essential for photosynthesis. Nature 429: 579–582

Munné-Bosch S, Shikanai T and Asada K (2005) Enhanced ferredoxin-dependent cyclic electron flow around photosystem I and α-tocopherol quinone accumulation in water-stressed *ndhB*-inactivated tobacco mutants. Planta 222: 502–511

Murata N (1969) Control of excitation transfer in photosynthesis. I. Light-induced change of chlorophyll a fluorescence in *Porphyridium cruentum*. Biochim Biophys Acta 172: 242–251

Munshi MK, Kobayashi Y and Shikanai T (2005) Identification of a novel protein CRR7 required for the stabilization of the chloroplast NAD(P)H dehydrogenase complex in *Arabidopsis*. Plant J 44: 1036–1044

Munshi MK, Kobayashi Y and Shikanai T (2006) CRR6 is a novel factor required for accumulation of the chloroplast NAD(P)H dehydrogenase complex in Arabidopsis. Plant Physiol 141: 737–744

Muraoka R, Okuda K, Kobayashi Y and Shikanai T (2006) A eukaryotic factor required for accumulation of the chloroplast NAD(P)H dehydrogenase complex in Arabidopsis. Plant Physiol 142: 1683–1689

Niyogi KK (1999) Photoprotection revisited: Genetic and molecular approaches. Annu Rev Plant Physiol Plant Mol Biol 50: 333–359

Niyogi KK, Björkman O and Grossman AR (1997) The roles of specific xanthophylls in photoprotection. Proc Natl Acad Sci USA 84: 14162–14167

Niyogi KK, Grossman AR and Björkman O (1998) Arabidopsis mutants define a central role for the xanthophyll cycle in the regulation of photosynthetic energy conversion. Plant Cell 10: 1121–1134

Niyogi KK, Li X-P, Rosenberg V and Jung HS (2005) Is PsbS the site of non-photochemical quenching in photosynthesis? J Exp Bot 56: 375–382

Ogawa T (1991) A gene homologous to the subunit-2 gene of NADH dehydrogenase is essential to inorganic carbon transport of *Synechocystis* PCC6803. Proc Natl Acad Sci USA 88: 4275–4279

Ogawa T and Mi HL (2007) Cyanobacterial NADPH dehydrogenase complexes. Photosynth Res 93: 69–77

Okegawa Y, Tsuyama M, Kobayashi Y and Shikanai T (2005) The *pgr1* mutation in the Rieske subunit of the cytochrome b_6f complex does not affect PGR5-dependent cyclic electron transport around photosystem I. J Biol Chem 280: 28332–28336

Okegawa Y, Long TA, Iwano M, Takayama S, Kobayashi Y, Covert SF and Shikanai T (2007) Balanced PGR5 level is required for chloroplast development and optimum operation of cyclic electron transport around photosystem I. Plant Cell Physiol 48: 1462–1471

Okegawa Y, Kagawa Y, Kobayashi Y and Shikanai T (2008) Characterization of factors affecting the activity of photosystem I cyclic electron transport in chloroplasts. Plant Cell Physiol 49: 825–834

Okuda K, Nakamura T, Sugita M, Shimizu T and Shikanai T (2006) A pentatricopeptide repeat protein is a site-recognition factor in chloroplast RNA editing. J Biol Chem 281: 37661–37667

Ort DR and Baker NR (2002) A photoprotective role for O_2 as an alternative electron sink in photosynthesis? Curr Opin Plant Biol 5: 193–198

Peng L, Shimizu H and Shikanai T (2008) The chloroplast NAD(P)H dehydrogenase complex interacts with photosystem I in Arabidopsis. J Biol Chem 283: 34873–34879

Peterson RB and Havir EA (2000) A nonphotochemical-quenching-deficient mutant of *Arabidopsis thaliana* possessing normal pigment composition and xanthophyll-cycle activity. Planta 210: 205–214

Prommeenate P, Lennon AM, Markert C, Hippler M and Nixon PJ (2004) Subunit composition of NDH-1 complexes of *Synechocystis* sp. PCC 6803: identification of two new *ndh* gene products with nuclear-encoded homologues in the chloroplast Ndh complex. J Biol Chem 279: 28165–28173

Rizhsky L, Liang H and Mittler R (2003) The water-water cycle is essential for chloroplast protection in the absence of stress. J Biol Chem 278: 38921–38995

Rosso D, Ivanov AG, Fu A, Geisler-Lee J, Hendrickson L, Geisler M, Stewart G, Krol M, Hurry V, Rodermel SR, Maxwell DP and Hüner NPA (2006) IMMUTANS does not act as a stress-induced safety valve in the protection of the photosynthetic apparatus of Arabidopsis during steady-state photosynthesis. Plant Physiol 142: 574–585

Rumeau D, Bécuwe-Linka N, Beyly A, Louwagie M, Garin J and Peltier G (2005) New Subunits NDH-M, -N, and -O, encoded by nuclear genes, are essential for plastid Ndh complex functioning in higher plants. Plant Cell 17: 219–232

Sakamoto K and Briggs WR (2002) Cellular and subcellular localization of phototropin 1. Plant Cell 14: 1723–1735

Shikanai T (2006) RNA editing in plant organelles: machinery, physiological function and evolution. Cell Mol Life Sci 63: 698–708

Shikanai T (2007a) Cyclic electron transport around photosystem I: genetic approaches. Annu Rev Plant Biol 58: 199–217

Shikanai T (2007b) The NAD(P)H dehydrogenase complex in photosynthetic organisms: Subunit composition and physiological function. Func Plant Biol Biotech 1: 129–137

Shikanai T and Endo T (2000) Physiological function of a respiratory complex, NAD(P)H dehydrogenase in chloroplasts: dissection by chloroplast reverse genetics. Plant Biotech 17: 79–86

Shikanai T, Endo T, Hashimoto T, Yamada Y, Asada K and Yokota A (1998) Directed disruption of the tobacco *ndhB* gene impairs cyclic electron flow around photosystem I. Proc Natl Acad Sci USA 95: 9705–9709

Shikanai T, Munekage Y, Shimizu K, Endo T and Hashimoto T (1999) Identification and characterization of *Arabidopsis* mutants with reduced quenching of chlorophyll fluorescence. Plant Cell Physiol 40: 1134–1142

Shimizu H and Shikanai T (2007) Dihydrodipicolinate reductase-like protein, CRR1, is essential for chloroplast NAD(P)H dehydrogenase in Arabidopsis. Plant J 52: 539–547

Shimizu H, Peng L, Myouga F, Motohashi R, Shinozaki K and Shikanai T (2008) CRR23/NdhL is a subunit of the chloroplast NAD(P)H dehydrogenase complex in Arabidopsis. Plant Cell Physiol 49: 835–842

Somerville CR (2001) An early Arabidopsis demonstration. Resolving a few issues concerning photorespiration. Plant Physiol 125: 20–24

Somerville CR and Ogren WL (1979) Phosphoglycolate phosphatase-deficient mutant of Arabidopsis. Nature 280: 833–836

Sonoike K and Terashima I (1994) Mechanism of photosystem I photoinhibition in leaves of *Cucumis sativus* L. Planta 194: 287–293

Stroebel D, Choquet Y, Popot J-L and Picot D (2003) An atypical haem in the cytochrome b_6f complex. Nature 426: 413–418

Svab Z and Maliga P (1993) High-frequency plastid transformation in tobacco by selection for a chimeric *aadA* gene. Proc Natl Acad Sci USA 90: 913–917

Tagawa K, Tsujimoto HY and Arnon DI (1963) Role of chloroplast ferredoxin in the energy conversion process of photosynthesis. Proc Natl Acad Sci USA 49: 567–572

Takahashi H, Iwai M, Takahashi Y and Minagawa J (2006) Identification of the mobile light-harvesting complex II polypeptides for state transitions in *Chlamydomonas reinhardtii*. Proc Natl Acad Sci USA 103: 447–482

Takemiya A, Inoue S, Doi M, Kinoshita T and Shimazaki K (2005) Phototropins promote plant growth in response to blue light in low light environments. Plant Cell 17: 1120–1127

Tikkanen M, Piippo M, Suorsa M, Sirpio S, Mulo P, Vainonen J, Vener AV, Allahverdiyeva Y and Aro E-M (2006) State transitions revisited – a buffering system for dynamic low light acclimation of *Arabidopsis*. Plant Mol Biol 62: 779–793

Vainonen JP, Hansson M and Vener AV (2005) STN8 protein kinase in *Arabidopsis thaliana* is specific in phosphorylation of photosystem II core proteins. J Biol Chem 280: 33679–33686

Vener AV, van Kan PJ, Rich PR, Ohad I and Andersson B (1997) Plastoquinol at the quinol oxidation site of reduced cytochrome *bf* mediates signal transduction between light and protein phosphorylation: Thylakoid protein kinase deactivation by a single-turnover flash. Proc Natl Acad Sci USA 94: 1585–1590

Walters RG, Shephard F, Rogers JJ, Rolfe SA and Horton P (2003) Identification of mutants of *Arabidopsis* defective in acclimation of photosynthesis to the light environment. Plant Physiol 131: 472–481

Wang P, Duan W, Takabayashi A, Endo T, Shikanai T, Ye JY and Mi H (2006) Chloroplastic NAD(P)H dehydrogenase in tobacco leaves functions in alleviation of oxidative damage caused by temperature stress. Plant Physiol 141: 465–474

Wingler A, Lea PJ, Quick WP and Leegood, RC (2000) Photorespiration: metabolic pathways and their role in stress protection. Phylos Trans R Soc Lond B 355: 1517–1529

Wu D, Wright DA, Wetzel C, Voytas DF and Rodermel S (1999) The *IMMUTANS* variegation locus of *Arabidopsis* defines a mitochondrial alternative oxidase homolog that functions during early chloroplast biogenesis. Plant Cell 11: 43–55

Zhang P, Battchikova N, Jansen T, Appel J, Ogawa T and Aro E-M (2004) Expression and functional roles of the two distinct NDH-1 complexes and the carbon acquisition complex NdhD3/NdhF3/CupA/Sll1735 in *Synechocystis* sp PCC 6803. Plant Cell 16: 3326–3340

Zito F, Finazzi G, Delosme R, Nitschke W, Picot D and Wollman FA (1999) The Qo site of cytochrome b_6f complexes controls the activation of the LHCII kinase. EMBO J 18: 2961–2969

Chapter 23

Mechanisms of Drought and High Light Stress Tolerance Studied in a Xerophyte, *Citrullus lanatus* (Wild Watermelon)

Yoshihiko Nanasato, Chikahiro Miyake, Kentaro Takahara, Kaori Kohzuma, Yuri Nakajima Munekage, Akiho Yokota, and Kinya Akashi*

Graduate School of Biological Sciences, Nara Institute of Science and Technology (NAIST), Ikoma, Nara 630-0192, Japan

Yoshihiko Nanasato

Present address: Plant Genetic Engineering Research Unit, National Institute of Agrobiological Sciences, 2-1-2 Kannondai, Tsukuba, Ibaraki, 305-8602, Japan

Chikahiro Miyake

Department of Biological and Environmental Science, Faculty of Agriculture, Kobe University, 1-1 Rokkodai-cho, Nada-ku, Kobe, 657-8501, Japan

Kentaro Takahara

Institute of Molecular and Cellular Biosciences, University of Tokyo, 1-1-1, Yayoi, Bunkyo-ku, Tokyo, 113-0032, Japan

Kaori Kohzuma

Present address: Institute of Biological Chemistry, 289 Clark Hall, Washington State University, Pullman, WA 99164-6340, USA

*Author for Correspondence, e-mail: akashi@bs.naist.jp

Constantin A. Rebeiz et al. (eds.), The Chloroplast Basics: and Applications, pp. 363–377.

Summary

The majority of higher plants are unable to survive extreme drought in the presence of strong solar radiation. However, a small group of vascular plants termed ''xerophytes'' have evolved drought and high light stress tolerance, and successfully thrives in the arid areas. This chapter will focus on the physiological, biochemical and molecular responses of wild watermelon (*Citrullus lanatus*), a xerophyte which is indigenous to the Kalahari Desert despite carrying out C_3-type photosynthesis. The electrochromic shift of carotenoids in the thylakoid membranes was analyzed in vivo, which revealed that the proton efflux through chloroplast ATP synthase was strongly suppressed under drought and high light stresses. In addition, cyclic electron flow around photosystem I was significantly activated under the stress, suggesting the functional relevance of these processes to the build-up of large ΔpH across thylakoid membranes, for sustaining high qE quenching under excess light conditions. Biochemical analyses showed that key components for ROS metabolism, such as chloroplastic ascorbate peroxidase and monodehydroascorbate reductase, were markedly fortified in this plant. Moreover, unique responses of wild watermelon under the stress were described like metabolism and function of citrulline, a novel compatible solute with potent activity for scavenging hydroxyl radicals. Furthermore, characteristic gene expression patterns were observed in this plant under drought, which are exemplified by the induction of cytochrome b_{561}, a trans-plasma membrane protein for transferring reducing equivalents from cytosol to the apoplasts. Interestingly, unprecedentedly high activity of ascorbate oxidase was observed in the leaf apoplasts, suggesting the electron flux from cytosol to this terminal oxidase may be activated under drought. Taken together, these findings offer intriguing implications on how terrestrial plants can achieve effective adaptation to the harsh environmental conditions.

I Introduction

In C_3 plants, assimilation of atmospheric CO_2 through photosynthesis is accompanied with huge water expenses. The amount of water lost by leaf transpiration through stomata is estimated to be 500–1,000 times larger than that of CO_2 fixed in the leaves on a molar basis (Hopkins and Huner, 2004). Therefore, plants carrying out C_3-type photosynthesis are prone to suffer from water deficits under various circumstances. Water deficit occurring under strong solar radiation disturbs energy- and water-balance of a plant, and inhibit most C_3 mesophytes, including a majority of crop plants, from inhabiting arid regions (Ehleringer and Monson, 1993). More than one-third of the Earth's land area is dryland, which is characterized by limited precipitation and the persistent natural menace of recurrent drought (Midleton and Thomas, 1992).

To maintain the water status of the leaf tissue, stomatal closure is induced under water deficits, which also restricts CO_2 entry into the leaves. As a consequence, photosynthetic CO_2 fixation is severely inhibited under water deficits (Cornic and Massacci, 1996). Under these conditions, excessively absorbed light energy that cannot be consumed through photosynthetic metabolism is liable to bring an enhanced production of reactive oxygen species (ROS) (Asada, 1999; Price et al., 1989). This in turn leads to irreversible damages in the plant cells (Foyer et al., 1994). Moreover, leaf water transpiration is important for lowering leaf temperature (Nobel, 1999). Stomatal closure in the presence of high light inevitably leads to the elevation of leaf temperature, and a rise in temperature up to 60°C has been reported in wild plants inhabiting natural desert conditions (Larcher, 1995). Therefore, plants are challenged by the multiple physicochemical constraints

Abbreviations: AGK–*N*-acetylglutamate kinase; AGS–*N*-acetylglutamate synthase; Asc–ascorbate; AO–ascorbate oxidase; AOD–*N*-acetylornithine deacetylase; CEF1–cyclic electron flow around photosystem I; CLb561A–*Citrullus lanatus* cytochrome b_{561} A; Cyt b_{561}–Cytochrome b_{561}; Cyt b_6f–Cytochrome b_6f; DHA–dehydroascorbate; DHAR–dehydroascorbate reductase; DIRK–dark-induced relaxation kinetics; DRIP-1–drought-induced polypeptide 1; EST–expression sequence tag; GAT–glutamate *N*-acetyltransferase; GFP–green fluorescent protein; GR–glutathione reductase; GSH–glutathione; GSSH–glutathione disulphide; MDA–monodehydroascorbate; MDAR–monodehydroascorbate reductase; MDH–malate dehydrogenase; MT–metallothionein; OAA–oxaloacetate; P_{II}protein–Peak two protein; PQ–plastoquinone; PS-I–photosystem I; PS-II–photosystem II; ROS–reactive oxygen species

under drought conditions in the presence of high light that lead to dehydration and high temperature.

To adapt to drought and strong light environments, terrestrial plants have evolved various mechanisms. These mechanisms include those for stress-avoidance as well as stress-tolerant strategies, and are particularly manifest in wild plants inhabiting arid and semi-arid regions (Akashi et al., 2008; Smith et al., 1997). Wild watermelon (*Citrullus lanatus*) is a xerophyte indigenous to the Kalahari Desert, Africa (van Wyk and Gericke, 2000). Ethnobotanically, the fruits of wild watermelon have been the most important source of water for indigenous inhabitants of the Kalahari Desert during the 9 dry months of the year when no surface water is available. Despite carrying out C_3-type photosynthesis, wild watermelon is tolerant to drought in the presence of strong light (Kawasaki et al., 2000; Yokota et al., 2002), and is thus an attractive model for studying how C_3 plants can cope with excess light stress (Fig. 1). In this chapter, we would like to describe physiological, biochemical and molecular studies on this plant, in which unique responses to drought and high light stresses have been documented.

II Experimental Procedures

Seeds of wild watermelon (*Citrullus lanatus* sp. #101117-1) and domesticated watermelon (*Citrullus lanatus* L. cv. Sanki) were sown in a 500-ml paper pots filled with commercial peat-based compost. Plants were grown with a daily cycle consisting of 35°C and 40% relative humidity. In the growth chamber, light intensity was adjusted to 1,000 μmol m^{-2} s^{-1} for 16 h, and at 25°C and 60% relative humidity in the dark for 8 h. Plants were watered daily at 3 h after the onset of the light regime, and were fertilized (Hyponex 8-12-6; Hyponex Japan, Osaka, Japan) twice a week. Drought stress was imposed by withholding irrigation. All measurements were taken from the

Fig. 1. Wild watermelon and domesticated cucumber plants under drought in the presence of high light. Wild watermelon (**a**, **b**) and domesticated cucumber cv. Suuyoo (**c**, **d**) plants were grown in a chamber with 16/8 h light/dark regime at temperatures of 35/25°C, 50/60% humidity and at a light intensity of 700 μmol photons m^{-2} s^{-1}. The plants were grown for 2 weeks (**a**, **c**), then subjected to stress for 5 days by withholding irrigation (**b**, **d**). Although leaves of cucumber plant lost their water and wilted severely (**d**), leaves of wild watermelon remained turgid and healthy under the harsh conditions (**b**).

fourth fully expanded leaves 3 weeks after sowing. Harvest of the leaves was performed 7 h after the onset of the light period.

For the assays of ROS scavenging enzymes, fully expanded fourth leaves were harvested 7 h after the light period had began, ground with a chilled mortar and pestle in a buffer containing 50 mM potassium phosphate, pH 7.0, 1 mM EDTA, 1 mM sodium ascorbate, 1%(w/v) 3-[(3-chloramidopropyl)dimethyl-ammonio]-2-hydroxyl-1-propanesulfonate (CHAPSO) and 2% polyvinylpolypyrrolidone (PVPP). The homogenates were filtered through two layers of cheese cloth, and the cleared lysates after centrifugation at 12,000 g for 20 min were used for enzyme assay. APX activity was determined as described by (Nakano and Asada, 1981). Activity of chloroplast APX including thylakoid-bound and stromal APX, and that of cytosolic APX were calculated from the difference in the sensitivity to hydrogen peroxide by the method of (Amako et al., 1994). MDAR activity was examined as described by (Hossain and Asada, 1985). In the present assay conditions, the rate of MDA generation was 0.14 mM min^{-1} and its steady state concentration was 2.1 μM, as determined from the absorbance at 360 nm using the absorption coefficient of MDA of 2.64 mM^{-1} cm^{-1}. When necessary, correction was made for the oxidation of NADH by diaphorase, which is likely to have been present in the crude extract, before the addition of ascorbate oxidase. DHAR was assayed as described by (Shimaoka et al., 2000). GR was assayed as described by Sgherri et al (1994).

For measurements of the Asc and GSH contents, leaves were frozen in liquid N_2 and ground in 1 M $HClO_4$. The homogenate was centrifuged for 5 min at 10,000*g*. The supernatant was placed on ice, the pH was adjusted with 5 M K_2CO_3 to either 6.0 for Asc assay or 7.0 for GSH assay, and the solution was centrifuged at 10,000*g* for 1 min at 4°C to remove insoluble potassium perchlorate. The clear supernatant was used for measuring Asc and GSH. Enzymatic methods as described by (Anderson, 1985) were used to determine both total glutathione [GSH + oxidized form of GSH (GSSG)] as GSH equivalents and GSSG, respectively. Asc and dehydroascorbate were measured as described by (Foyer et al., 1983).

III Physiological Response of Wild Watermelon

One of the noticeable traits in wild watermelon is its very rapid vegetative growth. Within 10 days after germination of the seed, wild watermelon plant develops up to four matured true leaves with 50–100 cm^2 leaf area for each leaf, in a growth chamber at a light intensity of 750–1,000 μmol photons m^{-2} s^{-1} and temperature of 35°C (Kawasaki et al., 2000; Takahara et al., 2005). Development of root tissues is also very rapid at the juvenile stage (Yoshimura et al., 2008). Vigorous growth at the juvenile stage should offer many advantages to this plant inhabiting dry regions, where precipitation is notoriously variable in both time and space, and competition for water is very keen among plants in the field (Larcher, 1995). Interestingly, root growth was promoted in wild watermelon at the early phase of drought stress, in comparison to that under the well-irrigated condition (Yoshimura et al., 2008). This observation suggests that this plant possesses excellent drought-avoidance mechanisms for absorbing water from deep soil layers upon drought imposition.

Under severe drought conditions, net CO_2 assimilation rate and leaf water conductance were strongly suppressed in wild watermelon due to stomatal closure (Kawasaki et al., 2000; Kohzuma et al., 2009). Under these conditions, wild watermelon has an ability to withstand severe drought conditions by maintaining its water content, especially in the upper leaf tissues (Kawasaki et al., 2000; Yokota et al., 2002). In the previous study using pot cultures, the relative water content of fully-expanded uppermost leaf was found to be virtually unaffected after 8 days of severe drought treatment, whereas other plants used as references in this experiment, e.g., domesticated watermelon, cucumber and maize, lost their tissue water signifi cantly under the same condition (Kawasaki et al., 2000). Since soil water content dropped to the minimal value already during the third day of the stress employed in this experiment, the excellent maintenance of leaf water status in wild watermelon may be attributable to the mechanism(s) for minimizing water evaporation from the leaves. This includes, the efficiency of stomatal closure (Larcher, 1995) and/or the suppression of cuticle transpiration by fortification of the cuticle (Cameron et al., 2006; Nawrath, 2006).

Another distinguishing feature of wild watermelon, in comparison to other C_3 plants, is that this plant shows persistent tolerance to strong light conditions after stomatal closure (Kawasaki et al., 2000; Yokota et al., 2002). Chlorophyll fluorescence analysis revealed that the F_v/F_m ratio (an indicator of PS II maximal activity) was largely unaffected during prolonged drought in the presence of high light (Kohzuma et al., 2008). By resuming irrigation of wild watermelon plants after severe drought and high light stresses, photosynthesis recovered almost to the original level within a few days. These observations suggested that this plant may possess effective mechanisms for protecting/repairing the photosynthetic apparatus and cellular components from photodamages.

During drought and high light stresses, plants are exposed to excessive solar radiation that cannot be used for photosynthetic CO_2 assimilation. Under these conditions, photochemical reactions in the thylakoid membranes have to be balanced with downstream metabolic reactions (Takizawa et al., 2007). The energy-dependent quenching mechanism (qE) is a process which is induced by the acidification of the thylakoid lumen, and is important for dissipating excessively absorbed light energy as heat (Müller et al., 2001). Although short-term regulations of the light reactions of photosynthesis under excess light conditions have been extensively investigated in many plants (Baker et al., 2007), mechanisms for the long-term acclimation of photosynthesis to prolonged drought have been poorly understood. Wild watermelon is an interesting experimental model system for addressing this issue. Indeed, tolerance of this plant to severe drought conditions enable us to investigate the true acclimation responses of photosynthesis rather than the degenerative disorder caused by the stress.

The in vivo spectroscopy probing of the electrochromic shift of carotenoids is a useful tool for examining the regulation of the proton motive force (*pmf*) and proton flux across thylakoid membranes (Cruz et al., 2005; Kramer et al., 2003; Sacksteder et al., 2000). Analysis of dark-induced relaxation kinetics (DIRK) (Sacksteder and Kramer, 2000) was employed for wild watermelon, demonstrating that proton conductivity across the thylakoid membranes was approximately fivefold lower in plants under drought than in the irrigated controls (Kohzuma et al., 2009) (Fig. 2). Since most of the proton efflux from thylakoid membranes is via chloroplast ATP synthase (Kramer and Crofts, 1989), this result suggested that suppression of proton translocation through ATP synthase is one of the critical processes for sustaining large *pmf* across thylakoid membranes for elevated qE under drought. Western blotting of the representative photosynthetic components showed that the content of the ATP synthase complex decreased

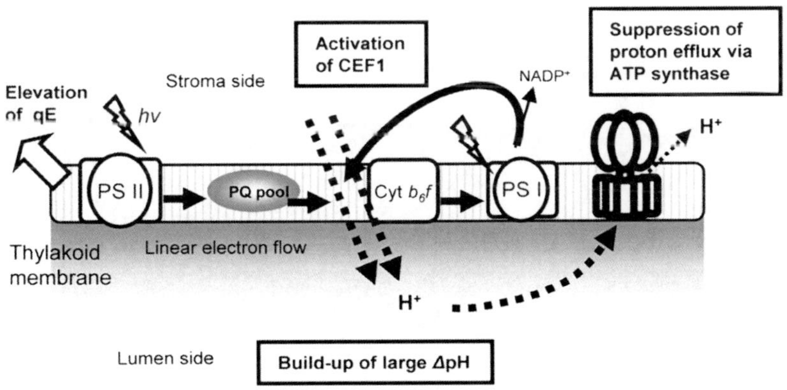

Fig. 2. Schematic diagram for the dynamics of electron and proton fluxes in the thylakoid membranes in wild watermelon under drought conditions in the presence of high light. Electron and proton flows are shown by solid and broken arrows, respectively. Under stress, proton efflux via the ATP synthase complex is strongly suppressed. Moreover, cyclic electron flow around PS I (CEF1) and the concomitant proton influx are enhanced under the stress. These processes are suggested to contribute to the buildup of large ΔpH across thylakoid membranes, thereby sustaining high qE for excess energy dissipation under stress conditions.

under drought, whereas those of photosystems I and II remained unchanged, suggesting that the quantitative control of the ATP synthase complex may play a role in the down-regulation of proton efflux from thylakoid membranes. The DIRK experiments also revealed that proton influx via cyclic electron flow around photosystem I (CEF1) (Munekage and Shikanai, 2005; Shikanai, 2007) was substantially enhanced under drought conditions (Kohzuma et al., 2009), suggesting that the defensive role of this electron flux for sustaining high qE. These observations demonstrated that acclimation to long-term drought stress involves dynamic changes in the regulations of proton and electron fluxes across thylakoid membranes in this xerophyte (Fig. 2).

IV Enzymes for Scavenging Reactive Oxygen Species

Under drought conditions in the presence of high light, excessive photon energy absorbed in the leaves is prone to enhance the generation of ROS (Asada, 1999; Price et al., 1989). One of the major sources for ROS generation in chloroplasts is in PS I, where the excessive photon energy tends to promote the photoreduction of O_2 to produce O_2^- (Biehler and Fock, 1996; Miyake and Yokota, 2000; Quartacci and Navari-Isso, 1992). Damaging effects of ROS to the photosynthetic machineries and other cellular components have been extensively described in plants, such as the damage to the D1 protein in the PS II complex (Miyao, 1994), peroxidation of membrane lipids (Krause et al., 1985; Meyer et al., 1992; Mishra and Singhal, 1992; Roberts et al., 1991), inactivation of enzymes for carbon reduction cycle (Shikanai et al., 1998; Tanaka et al., 1982), and protein translation machinery in chloroplasts (Nishiyama et al., 2006). Therefore, the prevention of oxidative injuries, and/or the efficient repair from oxidative damages, should be of critical importance for plants to survive under drought in the presence of high light conditions.

Plants are equipped with an array of antioxidative defense systems, which are composed of ROS scavenging enzymes and small antioxidative compounds (Apel and Hirt, 2004). To investigate the characteristics of antioxidative system in wild watermelon, changes in the activity of representative antioxidative enzymes were examined in the course of drought/high light stress (Fig. 3, also see section II). In this experiment, drought treatment caused the gradual decrease in the rates of leaf water transpiration and net CO_2 assimilation, and resulted in the nearly complete suppression of these parameters during the third day of stress (manuscript in preparation). Activities of glutathione reductase (GR), dehydroascorbate

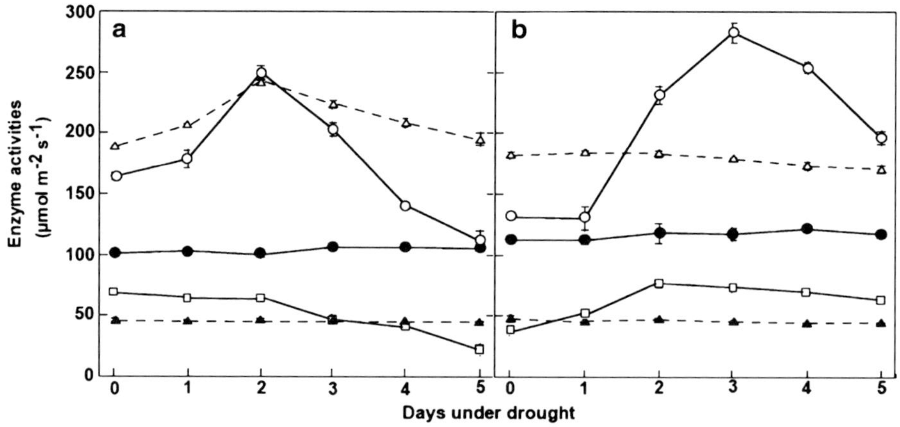

Fig. 3. Changes in the activities of chloroplast and cytosolic ascorbate peroxidase (APX) and of the enzymes involved in the regeneration of ascorbate (Asc) in the wild watermelon (**a**) and domesticated watermelon cv. Sanki (**b**) under drought in the presence of high light. Symbols are as follows: chloroplast APX, open circles; cytosolic APX, closed circles; monodehydroascorbate reductase (MDAR), open triangles; dehydroascorbate reductase (DHAR), open squares; glutathione reductase (GR), closed triangles. The data are the means of three independent experiments and SD is shown by vertical bars.

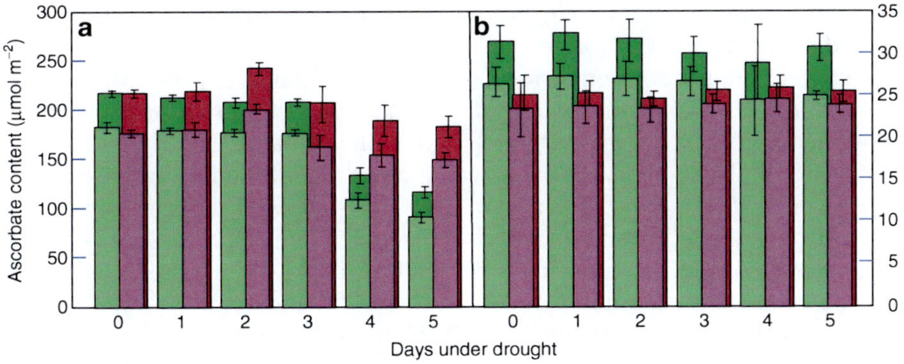

Fig. 4. Amounts of ascorbate (Asc) (**a**) and glutathione (GSH) (**b**) in the leaves of two species of watermelon plants under drought in the presence of high light. Pale-green and pink bars represent the amount of reduced form of the compounds for wild watermelon and domesticated watermelon cv. Sanki, respectively. Bars colored by dark-green and purple represent the amount of total (reduced + oxidized) compounds for wild watermelon and domesticated watermelon, respectively. The values are the means from three independent experiments and vertical thin bars are SD.

reductase (DHAR) and cytosolic ascorbate peroxidase (APX) either remained constant or slightly decreased under the stress. Interestingly, the activity of chloroplast APX transiently increased by a factor of 1.6 from approximately 160–250 μmol ascorbate m^{-2} s^{-1} in the first 2 days of drought treatment, then decreased gradually to 100 μmol ascorbate m^{-2} s^{-1} (Fig. 3a). Up-regulation of chloroplast APX activity was observed also in domesticated watermelon, suggesting that this behavior may be a common response in the watermelon family (Fig. 3b). Although up-regulation of cytosolic APX under various environmental stress is well documented in plants (Karpinski et al., 1997; Shigeoka et al., 2002), activation of chloroplast APX under drought/high light stresses is unprecedented, and is in marked contrast to a report in spinach, where high light stress led to the decrease in the activity of chloroplast APX (Yoshimura et al., 2000). Chloroplast APX is very susceptible to H_2O_2 in the absence of ascorbate (Kitajima et al., 2006; Miyake and Asada, 1996), and is known to be one of the initial targets for oxidative damages in plant leaves under the stress conditions (Mano et al., 2001; Shikanai et al., 1998). Fortification of chloroplast APX activity observed in the watermelon plants may therefore suggest that it is beneficial for resisting excess light conditions.

The activity for monodehydroascorbate reductase (MDAR) also increased during the first 2 days of the stress, then decreased marginally in the following 3 days (Fig. 3). More notably, the level of MDAR activity in wild watermelon was three- to tenfold larger than those reported in many other plants (Dat et al., 1998; Miyake and Asada, 1992; Moran et al., 1994). These observations suggested that watermelon plants are equipped with robust systems for regenerating ascorbate and glutathione, which are the two major antioxidants in plant cells. Consistent with this view, ratios of the contents of reduced forms to the total contents of both ascorbate and glutathione remained at higher values between 80% and 95% during stress (Fig. 4).

V Cytochrome b_{561} and Ascorbate Oxidase

In response to drought stress, expression of a large number of genes is known to be up-regulated in plants (Bray, 2004; Seki et al., 2002). Therefore it is interesting to examine to what extent the transcriptional regulation is diversified between mesophyte models such as Arabidopsis and xerophytes such as wild watermelon. Drought-responsive genes in the leaves of wild watermelon were analyzed by the technique of mRNA differential display (Akashi et al., 2004). Among ~2,700 cDNA bands detected in this study, approximately 14% of the transcripts were up-regulated, while ~3% of the cDNA were found to be down-regulated under drought stress. One of the up-regulated genes, CLb561A, has deduced amino acid sequence homology with cytochrome

b_{561} (cyt b_{561}) (Njus et al., 1987). Plant cyt b_{561} is known to be localized in the plasma and tonoplast membranes, and catalyzes transmembrane electron transport from ascorbate (Asc) on one side of the membranes to monodehydroascorbate (MDA) on the other side (Asard et al., 2001; Bashtovyy et al., 2003; Griesen et al., 2004) (Fig. 5). Notably, Asc in the leaf apoplasts is of crucial importance as the major antioxidant since no significant amount of glutathione exists in this compartment (Pignocchi and Foyer, 2003). The level of Asc in the leaf apoplasts is known to reach millimolar concentrations (Horemans et al., 2000).

Immunoblot analysis has revealed that CLb561A was induced at the protein level in the leaves of wild watermelon at the later stage of drought and strong-light stresses, whereas the same stress did not induce the protein in domesticated watermelon (Nanasato et al., 2005). GENEVESTIGATOR database reported that CLb561A orthologs in Arabidopsis were not induced by drought (Zimmermann et al., 2004), thus suggesting that the induction of cyt b_{561} under drought is a unique response for wild watermelon. Under low light conditions, the CLb561A protein did not accumulate after 5 days of drought treatment (Nanasato et al., 2005). Therefore, the induction of CLb561A protein appears to be associated with the occurrence of excess light rather than water deficit *per se*.

A transient expressing assay using GFP-fusion and cell fractionation experiments demonstrated that CLb561A was localized in the plasma membranes. Interestingly, biochemical analysis showed that the activity of the apoplastic ascorbate oxidase (AO) (Pignocchi et al., 2003), a possible terminal electron acceptor for CLb561A-mediated redox flow, and was unexpectedly high in the leaves of wild watermelon (Nanasato et al., 2005), as compared to published reports in other plants (Lin and Varner, 1991; Pignocchi et al., 2003). Moreover, AO activity that was detected in the leaves correlated positively with the growth light intensity (Nanasato et al., 2005). These observations raised the possibility that the putative cyt b_{561}-AO redox chain, which utilizes intracellular reducing energy for the reduction of molecular oxygen to yield water in the apoplasts, may contribute to excess energy dissipation under drought in this plant (Nanasato et al., 2005) (Fig. 5).

Emerging evidence has pointed out that the networks of redox systems between various intracellular organelles, as well as the interplay of ROS-scavenging systems among them, are of fundamental importance in responding

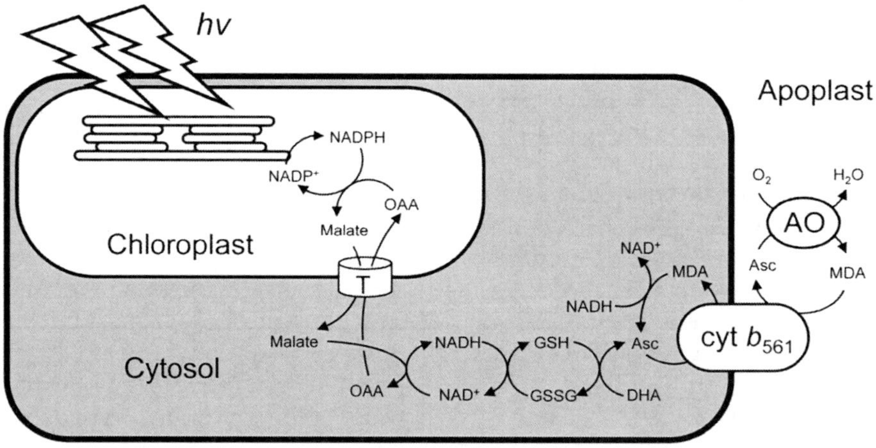

Fig. 5. Schematic representation of the putative link between the cyt b_{561}-AO chain and cytosolic/chloroplastic redox networks in wild watermelon. In this model, reducing equivalent transported by trans-plasma membrane cyt b_{561} is consumed by ascorbate oxidase with the concomitant reduction of molecular oxygen, yielding water in the apoplasts. This redox chain may be linked with cytosolic/chloroplastic redox networks, which potentially consume excess light energy absorbed in the chloroplasts. See text for discussion. AO, ascorbate oxidase; Asc, ascorbate; cyt b_{561}, cytochrome b_{561}; DHA, dehydroascorbate; GSH, glutathione; GSSG, glutathione disulphide; MDA, monodehydroascorbate; OAA, oxaloacetate ; T, dicarboxylase translocator (Scheibe, 2004).

to the biotic and abiotic stresses in plants (Van Breusegem et al., 2008). Redox processes are interlinked between the chloroplast stroma and the cytosol via malate-OAA shuttle and the triose-phosphate-3-phosphoglycerate shuttle (Heineke et al., 1991). It is reported that gene expression of cytosolic APXs were up-regulated under high light stress (Karpinski et al., 1997; Shigeoka et al., 2002). Moreover, the entire chloroplastic H_2O_2-scavenging system collapsed in an Arabidosis mutant defective in cytosolic ascorbate peroxidase 1, suggesting the importance of cytosolic ROS-scavenging system in protecting the chloroplasts during light stress (Davletova et al., 2005). Furthermore, the delicate metabolic interactions among chloroplasts, mitochondria, peroxisomes and cytosol have been documented, suggesting that it played pivotal roles for dissipating excessive photon energy (Noguchi and Yoshida, 2008; Raghavendra and Padmasree, 2003). Potential fortification of cyt b_{561}-AO redox chain in wild watermelon suggests that regulations of the redox networks under stress is not confined to the intracellular compartments, but may be extended to the extracellular compartments in this xerophyte.

VI Global Changes in the Proteomes

To address molecular mechanisms underlying the drought and high light stress tolerance of wild watermelon, large-scale proteome analyses, using two-dimensional electrophoresis and mass spectrometry, were undertaken for both leaf and root tissues in this plant. In the roots, up-regulated proteins were grouped into two classes according to their timing of induction: the one which was induced at the early stage of drought stress where vigorous root growth was observed, and the other induced at the later stage of the stress when soil water content decreased below 20% and root growth ceased (Yoshimura et al., 2008). Proteins induced at the earlier stage included the ones involved in root morphogenesis and carbon/nitrogen metabolism, which may be related to enhanced root growth at this stage. In contrast, at the later stage, proteins related to lignin synthesis and molecular chaperons were induced, suggesting that the proteins synthesized at this stage were mostly devoted to cellular defense and/or fortification of the mechanical strength in the roots. These observations suggested that wild watermelon regulates its root proteome according to the extent of water stress, which may be related to the switching of survival strategies from drought-avoidance such as enhanced root growth, to drought-tolerance such as the fortification of defensive mechanisms.

Unique responses in the proteome of wild watermelon were also observed in the leaf tissue. Among ~500 protein spots detected from leaf soluble proteins, the intensity of 44 spots increased under drought, whereas the intensity of 13 spots decreased (manuscript in preparation). One of the unique responses was for the DRIP-1 protein, which accumulated massively in the leaves of wild watermelon under drought (Kawasaki et al., 2000). Analysis of the corresponding cDNA revealed that DRIP-1 is homologous to *N*-acetylornithine deacetylase (AOD), which catalyzes the fifth step of arginine and citrulline biosynthesis in *Enterobacteria*. As will be described in the next section, citrulline metabolism appears to be particularly involved in the adaptation to the strong light stress conditions in this plant.

VII Citrulline Metabolism and Function

Analysis of free amino acid composition in the leaves of wild watermelon revealed that citrulline, a nitrogen-rich amino acid, accumulated massively under drought conditions (Kawasaki et al., 2000). In the previous report it was pointed out that, citrulline accumulated up to approximately 24 μmol $(gFW)^{-1}$ at the eighth day of drought stress, which amounts to ~50% of the total free amino acid in the leaves. Citrulline did not exhibit any inhibitory effects upon metabolic enzymes in vitro even at 600 mM, suggesting that this amino acid functions as a compatible solute in wild watermelon (Akashi et al., 2001). Moreover, in vitro analyses revealed that citrulline is an efficient scavenger for hydroxyl radicals, which is known to be one of the most toxic ROS (Akashi et al., 2001). The rate constant of the reaction between citrulline and hydroxyl radicals was 3.9×10^9 M^{-1} s^{-1}, which exceeded the value for mannitol (2.1×10^9 M^{-1} s^{-1}), a well-known scavenger of hydroxyl radicals. These results suggested that

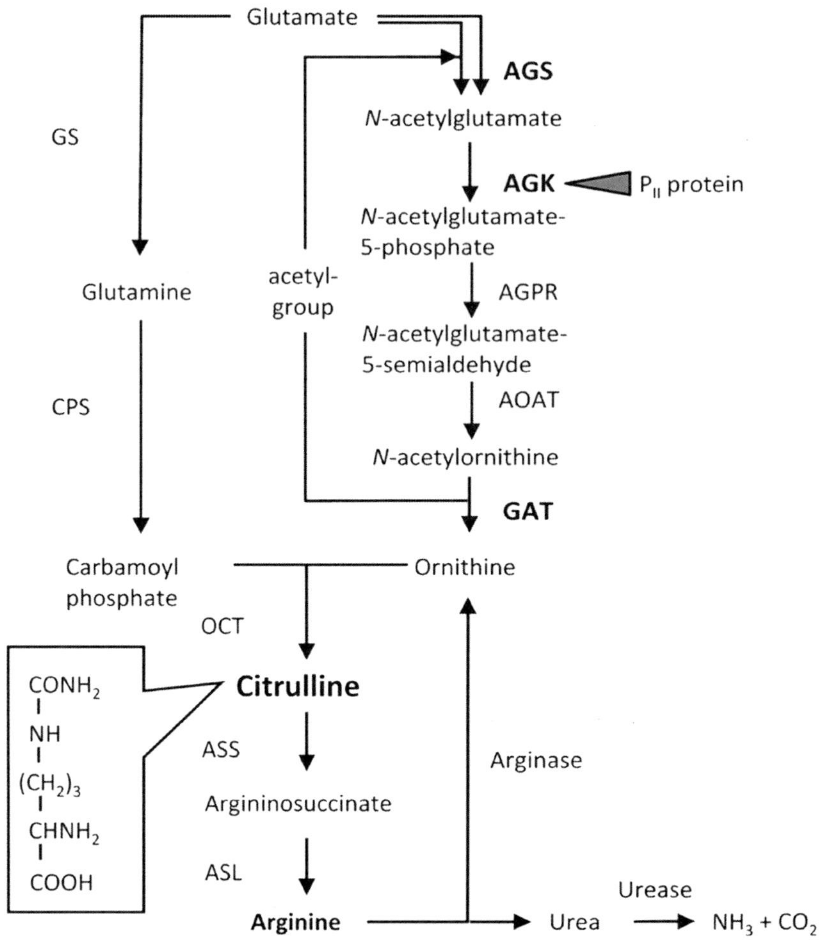

Fig. 6. Proposed citrulline metabolic pathway. GS, glutamine synthetase; CPS, carbamoyl phosphate synthetase; AGS, *N*-acetylglutamate synthase; AGK, *N*-acetylglutamate kinase; AGPR, *N*-acetylglutamate 5-phosphate reductase; AOAT, *N*-acetylornithine aminotransferase; GAT, glutamate *N*-acetyltransferase; OCT, ornithine carbamoyltransferase; ASS, argininosuccinate synthase; ASL, argininosuccinate lyase. The possible interaction between AGK and P_{II} protein is shown by a gray triangle.

citrulline accumulation may provide survival advantage to this xerophyte under oxidative stress conditions.

Citrulline is indeed an intermediate in the arginine biosynthetic pathway (Fig. 6). Although regulation of this pathway has been extensively investigated in prokaryotes and yeast (Cunin et al., 1986; Davis, 1986), the regulatory mechanism in plants has remained elusive (Slocum, 2005). As described above, proteome analysis revealed the up-regulation of DRIP-1 protein, with a deduced amino acid sequence homologous to AOD, the enzyme catalyzing the fifth step of the arginine biosynthetic pathway in *Enterobacteriaceae* such as *Escherichia coli*. AOD catalyses deacetylation of *N*-acetylornithine to yield ornithine and acetate (Vogel and Bonner, 1956). However, biochemical analyses suggested that extracts of wild watermelon leaves did not show any AOD activity, but instead possessed glutamate *N*-acetyltransferase (GAT) activity (Takahara et al., 2005). GAT occurs in various microorganisms such as *Bacillus subtilis* and *Saccharomyces cerevisiae*, and catalyzes transfer reaction of acetyl group from *N*-acetylornithine to glutamate to yield ornithine and *N*-acetylglutamate. GAT therefore catalyzes the first and fifth steps of arginine and citrulline biosynthesis simultaneously (Liu et al., 1995; Sakanyan et al., 1992). The GAT enzyme purified 7,000-fold from wild watermelon leaves is composed of two distinct subunits, which are suggested to be derived from the cleavage of a single precursor after transport into chloroplasts (Takahara et al., 2005). The enzyme exhibits thermostability up to 70°C, and is insensitive to feedback inhibition by downstream products such as citrulline and arginine, which suggests that this enzyme possesses favorable enzymatic properties for citrulline accumulation under high temperature and drought conditions. On the other hand, the function of DRIP-1 has remained unclear. It is interesting to note that DRIP-1 was induced not only in the leaves but also in the root tissue at the later stage of drought stress (Yoshimura et al., 2008), suggesting the importance of this protein in both leaves and roots under drought conditions.

The citrulline-arginine metabolic pathway consists of 12 enzymatic steps (Fig. 6). Based on sequence information, putative genes for the enzymes in the arginine biosynthetic pathway have been assigned in the Arabidopsis genome (Slocum, 2005). In plants, computer predictions from their deduced sequences (Slocum, 2005), GFP-fusion transient assay (Sugiyama et al., 2004), subcellular fractionation studies (de Ruiter and Kolloffel, 1985; Jain et al., 1987), and proteome analysis (Kleffmann et al., 2004) suggested that most of the enzymes of the arginine biosynthetic pathway are localized in the plastid compartment. Nevertheless, comprehensive analyses of the metabolic pathway and its regulation have been hampered partly due to the lack of versatile enzyme assay systems for this pathway. In particular, knowledge of the enzymatic properties of *N*-acetylglutamate synthase (AGS) and *N*-acetylglutamate kinase (AGK), which catalyze the first and second steps in arginine synthesis and are known for their feedback regulation by arginine in microorganisms (Cunin et al., 1986; Pauwels et al., 2003), will be of pivotal importance to understand the citrulline metabolism. Moreover, AGK is known to be regulated by P_{II} protein, which functions as a sensor for carbon and nitrogen status of the cells (Chen et al., 2006; Forchhammer, 2008). This raises the interesting possibility that citrulline biosynthesis is coordinately regulated by C/N metabolic pathways via the action of the P_{II} protein. To resolve these issues, continuous spectrophotometric assays for three regulatory enzymes of citrulline and arginine biosynthetic pathway were established (Takahara et al., 2007). In these systems, activities for the regulatory enzymes of citrulline and arginine synthesis, i.e., AGS, AGK and GAT were coupled with the downstream coupling enzymes, which enabled to monitor the target activities by the redox-induced change in the absorbance of NADPH at 340 nm. Characterization of these enzymes which are involved in citrulline biosynthesis in wild watermelon is eagerly awaited for.

VIII Concluding Remarks

Earth's land accommodates an estimated ~250,000 terrestrial plant species, which are extremely diversified in their life form, morphology and physiology (Pearson, 1995). The distribution of diversified plants in various ecological niches must have been accompanied by effective acclimation mechanisms to the habitats, which should have been acquired in the evolution of respective plant species. Watermelon plants are thought to have originated in the Kalahari Desert (Whitaker and Bemis, 1976). Historical records suggest that watermelon was cultivated in the Nile Valley in the Northern Africa at least since the start of the second millennium BC (Zohary and Hopf, 2000), and from there spread to the Mediterranean and Asian regions. While cultivation processes resulted in lowered tolerance to drought and to high light stresses wild, watermelon inhabiting the Kalahari Desert still possesses the strong ability to maintain the water status of the leaves as well as to avoid photoinhibition, and to effectively adapt to severe environmental conditions.

Taking advantages of this excellent traits, physiological, biochemical and molecular responses of wild watermelon have been investigated. The in vivo spectroscopy probing of the electrochromic shift of carotenoids revealed the dynamic regulation of photochemical reactions under long-term drought and high light stresses. In these processes, the down-regulation of chloroplast ATP synthase and the activation of cyclic electron flow around photosystem I are suggested to be involved in the build-up of high ΔpH across thylakoid membranes which sustains high qE for excess energy dissipation. The key ROS scavenging enzymes such as chloroplast APX and MDAR are significantly fortified in wild watermelon. Proteome and transcriptome analyses showed unique molecular responses of this plant under drought conditions, which was exemplified by the induction of cyt b_{561}, that transfers reducing equivalents from cytosol to the apoplasts, a process potentially involved in energy dissipation. In wild watermelon, citrulline is a unique compatible solute that appears to be involved in metabolic adaptation to the harsh environmental conditions. Citrulline possesses the efficient activity for scavenging hydroxyl radicals, which may provide survival advantages under oxidative conditions. Taken together, studies on wild watermelon have offered intriguing insights about the potential of C_3 plants, which are normally thought to be susceptible to drought and excess light stresses. Further attempts to develop the experimental systems using wild watermelon are underway. Stable genetic transformation of wild watermelon was established by *Agrobacterium*-mediated technique (Akashi et al., 2005) and a large-scale EST project was conducted (manuscript in preparation). These initiatives will expand the versatility of this plant as one of the xerophyte experimental model systems, and will be helpful for elucidating the secret of this plant for successful adaptation to desert conditions.

Acknowledgements

This work was supported by a Grant-in-Aid for Scientific Research, and Asia-Africa Science and Technology Strategic Cooperation Promotion Program from Japan Science and Technology Agency, Foundation for Nara Institute of Science and Technology (NAIST), NAIST Global COE program, and Nissan Science Foundation.

References

Akashi, K., Miyake, C. and Yokota, A. (2001) Citrulline, a novel compatible solute in drought-tolerant wild watermelon leaves, is an efficient hydroxyl radical scavenger. FEBS Lett., 508: 438–442.

Akashi, K., Morikawa, K. and Yokota, A. (2005) *Agrobacterium*-mediated transformation system for the drought and excess light stress-tolerant wild watermelon (*Citrullus lanatus*). Plant Biotech., 22: 13–18.

Akashi, K., Nishimura, N., Ishida, Y. and Yokota, A. (2004) Potent hydroxyl radical-scavenging activity of drought-induced type-2 metallothionein in wild watermelon. Biochem Biophys Res Commun., 323: 72–78.

Akashi, K., Yoshimura, K., Nanasato, Y., Takahara, K., Munekage, Y. and Yokota, A. (2008) Wild plant resources for studying mkolecular mechanisms of drought/strong light stress tolerance. Plant Biotechnol., 25: 257–263.

Amako, K., Chen, G.-X. and Asada, K. (1994) Separate assays specific for ascorbate peroxidase and guaiacol peroxidase and for the chloroplastic and cytosolic isozymes of ascorbate peroxidase in plants. Plant Cell Physiol., 35: 497–504.

Anderson, M.E. (1985) Determination of glutathione and glutathione disulfide in biological samples. Methods Enzymol., 113: 548–555.

Apel, K. and Hirt, H. (2004) Reactive oxygen species: Metabolism, oxidative stress, and signal transduction. Annu Rev Plant Biol., 55: 373–399.

Asada, K. (1999) The water-water cycle in chloroplasts: scavenging of active oxygens and dissipation of excess photons. Annu Rev Plant Physiol Plant Mol Biol., 50: 601–639.

Asard, H., Kapila, J., Verelst, W. and Bérczi, A. (2001) Higher-plant plasma membrane cytochrome b_{561}: a protein in search of a function. Protoplasma, 217: 77–93.

Baker, N.R., Harbinson, J. and Kramer, D.M. (2007) Determining the limitations and regulation of photosynthetic energy transduction in leaves. Plant Cell Environ., 30: 1107–1125.

Bashtovyy, D., Bérczi, A., Asard, H. and Páli, T. (2003) Structure prediction for the di-heme cytochrome b_{561} protein family. Protoplasma, 221: 31–40.

Biehler, K. and Fock, H. (1996) Evidence for the contribution of the Mehler peroxidase reaction in dissipating excess electrons in drought-stressed wheat. Plant Physiol., 112: 265–272.

Bray, E.A. (2004) Genes commonly regulated by water-deficit stress in Arabidopsis thaliana. J Exp Bot., 55: 2331–2341.

Cameron, K.D., Teece, M.A. and Smart, L.B. (2006) Increased accumulation of cuticular wax and expression of lipid transfer protein in response to periodic drying events in leaves of tree tobacco. Plant Physiol., 140: 176–183.

Chen, Y.M., Ferrar, T.S., Lohmeier-Vogel, E.M., Morrice, N., Mizuno, Y., Berenger, B., Ng, K.K., Muench, D.G. and Moorhead, G.B. (2006) The P_{II} signal transduction protein of Arabidopsis thaliana forms an arginine-regulated complex with plastid *N*-acetyl glutamate kinase. J Biol Chem., 281: 5726–5733.

Cornic, G. and Massacci, A. (1996) Leaf photosynthesis under drought stress. In: Baker, N.R. (ed.), Photosynthesis and the Environment, Kluwer, Dordrecht, pp. 347–366.

Cruz, J.A., Kanazawa, A., Treff, N. and Kramer, D.M. (2005) Storage of light-driven transthylakoid proton motive force as an electric field ($\Delta\Psi$) under steady-state conditions in intact cells of *Chlamydomonas reinhardtii*. Photosynth Res., 85: 221–233.

Cunin, R., Glansdorff, N., Pierard, A. and Stalon, V. (1986) Biosynthesis and metabolism of arginine in bacteria. Microbiol Rev., 50: 314–352.

Dat, J.F., Foyer, C.H. and Scott, I.M. (1998) Changes in salicylic acid and antioxidants during induced thermotolerance in mustard seedlings. Plant Physiol., 118: 1455–1461.

Davis, R.H. (1986) Compartmental and regulatory mechanisms in the arginine pathways of *Neurospora crassa* and *Saccharomyces cerevisiae*. Microbiol Rev., 50: 280–313.

Davletova, S., Rizhsky, L., Liang, H., Shengqiang, Z., Oliver, D.J., Coutu, J., Shulaev, V., Schlauch, K. and Mittler, R. (2005) Cytosolic ascorbate peroxidase 1 is a central component of the reactive oxygen gene network of Arabidopsis. Plant Cell, 17: 268–281.

de Ruiter, H. and Kolloffel, C. (1985) Properties of Ornithine Carbamoyltransferase from *Pisum sativum* L. Plant Physiol., 77: 695–699.

Ehleringer, J.R. and Monson, R.K. (1993) Evolutionary and ecological aspects of photosynthetic pathway variation. Annu Rev Ecol Syst., 24: 411–439.

Forchhammer, K. (2008) P(II) signal transducers: novel functional and structural insights. Trends Microbiol., 16: 65–72.

Foyer, C.H., Descourvieres, P. and Kunert, K.J. (1994) Protection against oxygen radicals: an important defence mechanism studied in transgenic plants. Plant Cell Environ., 17: 507–523.

Foyer, C.H., Rowell, H. and Walker, D. (1983) Measurement of the ascorbate content of spinach leaf protoplasts and chloroplasts during illumination. Planta, 157: 239–244.

Griesen, D., Su, D., Bérczi, A. and Asard, H. (2004) Localization of an ascorbate-reducible cytochrome b_{561} in the plant tonoplast. Plant Physiol., 134: 726–734.

Heineke, D., Riens, B., Grosse, H., Hoferichter, P., Peter, U., Flugge, U.I. and Heldt, H.W. (1991) Redox transfer across the inner chloroplast envelope membrane. Plant Physiol., 95: 1131–1137.

Hopkins, W.G. and Huner, N.P.A. (2004) Energy Conservation in Photosynthesis: CO_2 Assimilation. *In:* Introduction to Plant Physiology. Wiley, New Jersey, pp. 89–122.

Horemans, N., Foyer, C.H. and Asard, H. (2000) Transport and action of ascorbate at the plant plasma membrane. Trends Plant Sci., 5: 263–267.

Hossain, M.A. and Asada, K. (1985) Monodehydroascorbate reductase from cucumber is a flavin adenine dinucleotide enzyme. J Biol Chem., 250: 12920–12926.

Jain, J.C., Shargool, P.D. and Chung, S. (1987) Compartmentation studies on enzymes of ornithine biosynthesis in plant cells. Plant Sci., 51: 17–20.

Karpinski, S., Escobar, C., Karpinska, B., Creissen, G. and Mullineaux, P.M. (1997) Photosynthetic electron transport regulates the expression of cytosolic ascorbate peroxidase genes in Arabidopsis during excess light stress. Plant Cell, 9: 627–640.

Kawasaki, S., Miyake, C., Kohchi, T., Fujii, S., Uchida, M. and Yokota, A. (2000) Responses of wild watermelon to drought stress: accumulation of an ArgE homologue and citrulline in leaves during water deficits. Plant Cell Physiol., 41: 864–873.

Kitajima, S., Tomizawa, K., Shigeoka, S. and Yokota, A. (2006) An inserted loop region of stromal ascorbate peroxidase is involved in its hydrogen peroxide-mediated inactivation. FEBS J., 273: 2704–2710.

Kleffmann, T., Russenberger, D., von Zychlinski, A., Christopher, W., Sjolander, K., Gruissem, W. and Baginsky, S. (2004) The Arabidopsis thaliana chloroplast proteome reveals pathway abundance and novel protein functions. Curr Biol., 14: 354–362.

Kohzuma, K., Akashi, K., Munekage, Y., Sanda, S., Hisabori, T. and Yokota, A. (2008) Preferential decay of the CF1-ε subunit induces thylakoid uncoupling in wild watermelon under drought stress. In: Allen, J.F., Gantt, E., Golbeck, J.H. and Osmond, B. (eds.), Photosynthesis. Energy from the Sun: 14th International Congress on Photosynthesis. Springer, Dordrecht, The Netherlands, pp. 617–621.

Kohzuma, K., Cruz, J.A., Akashi, K., Hoshiyasu, S., Munekage, Y., Yokota, A. and Kramer, D.M. (2009) The long-term responses of the photosynthetic proton circuit to drought. Plant Cell Environ., 32: 209–219.

Kramer, D.M. and Crofts, A.R. (1989) Activation of the chloroplast ATPase measured by the electrochromic change in leaves of intact plants. Biochim Biophys Acta, 976, 28–41.

Kramer, D.M., Cruz, J.A. and Kanazawa, A. (2003) Balancing the central roles of the thylakoid proton gradient. Trends Plant Sci., 8: 27–32.

Krause, G.H., Koster, S. and Wong, S.C. (1985) Photoinhibition of photosynthesis under anaerobic conditions studied with leaves and chloroplasts of *Spinacia oleracea* L. Planta, 165: 430–438.

Larcher, W. (1995) Plants under stress. *In:* Physiological Plant Ecology. Springer-Verlag, Berlin, pp. 321–448.

Lin, L.S. and Varner, J.E. (1991) Expression of ascorbic acid oxidase in Zucchini squash (*Cucurbita pepo* L.). Plant Physiol., 96: 159–165.

Liu, Y., Van Heeswijck, R., Hoj, P. and Hoogenraad, N. (1995) Purification and characterization of ornithine acetyltransferase from *Saccharomyces cerevisiae*. Eur J Biochem., 228: 291–296.

Mano, J., Ohno, C., Domae, Y. and Asada, K. (2001) Chloroplastic ascorbate peroxidase is the primary target of methylviologen-induced photooxidative stress in spinach leaves: its relevance to monodehydroascorbate radical detected with *in vivo* ESR. Biochim Biophys Acta, 1504: 275–287.

Meyer, S., Phung, N.H.S., Tremolieres, A. and De Kouchkovsky, Y. (1992) Energy coupling, membrane lipids and structure of thylakoids of Lupin plants subjected to water stress. Photosynth Res., 32: 95–107.

Midleton, N.J. and Thomas, D.S.G. (1992) World Atlas of Desertification. UNEP (United Nations Environment Programme), Hodder & Stoughton Ltd., Kent, UK, pp. 1–5.

Mishra, R.K. and Singhal, G.S. (1992) Function of photosynthetic apparatus of intact wheat leaves under high light and heat stress and its relationship with peroxidation of thylakoid lipids. Plant Physiol., 98: 1–6.

Miyake, C. and Asada, K. (1992) Thylakoid-bound ascorbate peroxidase in spinach chloroplasts and photoreduction of its primary oxidation product monodehydroascorbate radicals in thylakoids. Plant Cell Physiol., 33: 541–553.

Miyake, C. and Asada, K. (1996) Inactivation mechanism of ascorbate peroxidase at low concentrations of ascorbate; hydrogen peroxide decomposes Compound I of ascorbate peroxidase. Plant Cell Physiol., 37: 423–430.

Miyake, C. and Yokota, A. (2000) Determination of the rate of photoreduction of O_2 in the water-water cycle in watermelon leaves and enhancement of the rate by limitation of photosynthesis. Plant Cell Physiol., 41: 335–343.

Miyao, M. (1994) Involvement of active oxygen species in degradation of the D1 protein under strong illumination in isolated subcomplexes of photosystem II. Biochemistry, 33: 9722–9730.

Moran, J.F., Becana, M., Iturbe-Ormaetxe, I., Frechilla, S., Klucas, R.V. and Aparicio-Tejo, P. (1994) Drought induces oxidative stress in pea plants. Planta, 194: 346–352.

Müller, P., Li, X.P. and Niyogi, K.K. (2001) Non-photochemical quenching. A response to excess light energy. Plant Physiol., 125: 1558–1566.

Munekage, Y. and Shikanai, T. (2005) Cyclic electron transport through photosystem I. Plant Biotechnol., 22: 361–369.

Nakano, Y. and Asada, K. (1981) Hydrogen peroxide is scavenged by ascorbate-specific peroxidase in spinach chloroplasts. Plant Cell Physiol., 22: 867–880.

Nanasato, Y., Akashi, K. and Yokota, A. (2005) Co-expression of cytochrome b_{561} and ascorbate oxidase in leaves of wild watermelon under drought and high light conditions. Plant Cell Physiol., 46: 1515–1524.

Nawrath, C. (2006) Unraveling the complex network of cuticular structure and function. Curr Opin Plant Biol., 9: 281–287.

Nishiyama, Y., Allakhverdiev, S.I. and Murata, N. (2006) A new paradigm for the action of reactive oxygen species in the photoinhibition of photosystem II. Biochim Biophys Acta, 1757: 742–749.

Njus, D., Kelley, P.M., Harnadek, G.J. and Pacquing, Y.V. (1987) Mechanism of ascorbic acid regeneration mediated by cytochrome b_{561}. Ann N Y Acad Sci., 493: 108–119.

Nobel, P.S. (1999) Temperature-Energy Budgets. *In:* Physicochemical & Environmental Plant Physiology. Academic Press, San Diego, CA, pp. 257–292.

Noguchi, K. and Yoshida, K. (2008) Interaction between photosynthesis and respiration in illuminated leaves. Mitochondrion, 8: 87–99.

Pauwels, K., Abadjieva, A., Hilven, P., Stankiewicz, A. and Crabeel, M. (2003) The *N*-acetylglutamate synthase/*N*-acetylglutamate kinase metabolon of *Saccharomyces cerevisiae* allows co-ordinated feedback regulation of the first two steps in arginine biosynthesis. Eur J Biochem., 270: 1014–1024.

Pearson, L.C. (1995) Plant Diversity and Classification. *In:* The Diversity and Evolution of Plants. CRC Press, Boca Raton, FL, pp. 3–38.

Pignocchi, C., Fletcher, J.M., Wilkinson, J.E., Barnes, J.D. and Foyer, C.H. (2003) The function of ascorbate oxidase in tobacco. Plant Physiol., 132: 1631–1641.

Pignocchi, C. and Foyer, C.H. (2003) Apoplastic ascorbate metabolism and its role in the regulation of cell signalling. Curr Opin Plant Biol., 6: 379–389.

Price, A.M., Atherton, N.M. and Hendry, G.A.F. (1989) Plants under drought-stress generate activated oxygen. Free Rad Res Commun., 8: 61–66.

Quartacci, M.F. and Navari-Isso, F. (1992) Water stress and free radical-mediated changes in sunflower seedlings. J Plant Physiol., 139: 621–625.

Raghavendra, A.S. and Padmasree, K. (2003) Beneficial interactions of mitochondrial metabolism with photosynthetic carbon assimilation. Trends Plant Sci., 8: 546–553.

Roberts, D.R., Kristie, D.N., Thompson, J.E., Dumbroff, E.B. and Gepstein, S. (1991) *In vitro* evidence for the involvement of activated oxygen in light-induced aggregation of thylakoid proteins. Physiol Plant., 82: 389–396.

Sacksteder, C.A., Kanazawa, A., Jacoby, M.E. and Kramer, D.M. (2000) The proton to electron stoichiometry of steady-state photosynthesis in living plants: A proton-pumping Q cycle is continuously engaged. Proc Natl Acad Sci USA, 97: 14283–14288.

Sacksteder, C.A. and Kramer, D.M. (2000) Dark-interval relaxation kinetics (DIRK) of absorbance changes as a quantitative probe of steady-state electron transfer. Photosynth Res., 66: 145–158.

Sakanyan, V., Kochikyan, A., Mett, I., Legrain, C., Charlier, D., Pierard, A. and Glansdorff, N. (1992) A re-examination of the pathway for ornithine biosynthesis in a thermophilic and two mesophilic *Bacillus* species. J Gen Microbiol., 138: 125–130.

Scheibe, R. (2004) Malate valves to balance cellular energy supply. Physiol Plant 120: 21–26.

Seki, M., Narusaka, M., Ishida, J., Nanjo, T., Fujita, M., Oono, Y., Kamiya, A., Nakajima, M., Enju, A., Sakurai, T., Satou, M., Akiyama, K., Taji, T., Yamaguchi-Shinozaki, K., Carninci, P., Kawai, J., Hayashizaki, Y. and Shinozaki, K. (2002) Monitoring the expression profiles of 7000 Arabidopsis genes under drought, cold and high-salinity stresses using a full-length cDNA microarray. Plant J., 31: 279–292.

Sgherri, C.L.M., Loggini, B., Puliga, S. and Navari-Izzo, F. (1994) Antioxidant system in Sporobolus stapfianus: Changes in response to desiccation and rehydration. Phytochemistry, 35: 561–565.

Shigeoka, S., Ishikawa, T., Tamoi, M., Miyagawa, Y., Takeda, T., Yabuta, Y. and Yoshimura, K. (2002) Regulation and function of ascorbate peroxidase isoenzymes. J Exp Bot., 53: 1305–1319.

Shikanai, T. (2007) Cyclic electron transport around photosystem I: genetic approaches. Annu Rev Plant Biol., 58: 199–217.

Shikanai, T., Takeda, T., Yamauchi, H., Sano, S., Tomizawa, K.I., Yokota, A. and Shigeoka, S. (1998) Inhibition of ascorbate peroxidase under oxidative stress in tobacco having bacterial catalase in chloroplasts. FEBS Lett., 428: 47–51.

Shimaoka, T., Yokota, A. and Miyake, C. (2000) Purification and characterization of chloroplast dehydroascorbate reductase from spinach leaves. Plant Cell Physiol., 41: 1110–1118.

Slocum, R.D. (2005) Genes, enzymes and regulation of arginine biosynthesis in plants. Plant Physiol Biochem., 43: 729–745.

Smith, S.D., Monson, R.K. and Anderson, J.E. (1997) Plant processes and responses to stress. In: Smith, S.D., Monson, R.K. and Anderson, J.E. (eds.), Physiological Ecology of North American Desert Plants. Springer-Verlag, Heidelberg, pp. 45–72.

Sugiyama, K., Hayakawa, T., Kudo, T., Ito, T. and Yamaya, T. (2004) Interaction of *N*-acetylglutamate kinase with a P_{II}-like protein in rice. Plant Cell Physiol., 45: 1768–1778.

Takahara, K., Akashi, K. and Yokota, A. (2005) Purification and characterization of glutamate *N*-acetyltransferase involved in citrulline accumulation in wild watermelon. FEBS J., 272: 5353–5364.

Takahara, K., Akashi, K. and Yokota, A. (2007) Continuous spectrophotometric assays for three regulatory enzymes of the arginine biosynthetic pathway. Anal Biochem., 368: 138–147.

Takizawa, K., Cruz, J.A., Kanazawa, A. and Kramer, D.M. (2007) The thylakoid proton motive force *in vivo*. Quantitative, non-invasive probes, energetics, and regulatory consequences of light-induced *pmf*. Biochim Biophys Acta, 1767: 1233–1244.

Tanaka, K., Otsubo, T. and Kondo, N. (1982) Participation of hydrogen peroxide in the inactivation of Calvin-cycle sulfhydride enzymes in sulfur dioxide fumigated spinach (*Spinacia oleracea* cultivar New Asia). Plant Cell Physiol., 23: 1009–1018.

Van Breusegem, F., Bailey-Serres, J. and Mittler, R. (2008) Unraveling the tapestry of networks involving reactive oxygen species in plants. Plant Physiol., 147: 978–984.

van Wyk, B.-E. and Gericke, N. (2000) Fruits & Berries. *In:* People's plants. A Guide to Useful Plants of Southern Africa. Briza Publications, Pretoria, pp. 33–62.

Vogel, H.J. and Bonner, D.M. (1956) Acetylornithinase of *Escherichia coli*: partial purification and some properties. J Biol Chem., 218: 97–106.

Whitaker, T.W. and Bemis, W.P. (1976) Cucurbits. In: Simmonds, N.W. (ed.), Evolution of Crop Plants. Longman, New York, pp. 64–69.

Yokota, A., Kawasaki, S., Iwano, M., Nakamura, C., Miyake, C. and Akashi, K. (2002) Citrulline and DRIP-1 protein (ArgE homologue) in drought tolerance of wild watermelon. Ann Bot (Lond), 89: 825–832.

Yoshimura, K., Masuda, A., Kuwano, M., Yokota, A. and Akashi, K. (2008) Programmed proteome response for drought avoidance/tolerance in the root of a C_3 xerophyte (wild watermelon) under water deficits. Plant Cell Physiol., 49: 226–241.

Yoshimura, K., Yabuta, Y., Ishikawa, T. and Shigeoka, S. (2000) Expression of spinach ascorbate peroxidase isozymes in response to oxidative stresses. Plant Physiol., 123: 223–233.

Zimmermann, P., Hirsch-Hoffmann, M., Hennig, L. and Gruissem, W. (2004) GENEVESTIGATOR. Arabidopsis microarray database and analysis toolbox. Plant Physiol., 136: 2621–2632.

Zohary, D. and Hopf, M. (2000) Vegetables and Tubers. *In:* Domestication of Plants in the Old World. Oxford University, New York, pp. 192–204.

Chapter 24

Antioxidants and Photo-oxidative Stress Responses in Plants and Algae

Anchalee Sirikhachornkit and Krishna K. Niyogi*
Department of Plant and Microbial Biology, University of California, Berkeley, CA 94720-3102, USA

Summary

Ever since the origin of oxygenic photosynthesis and the accumulation of molecular oxygen in the earth's atmosphere over 2.2 billion years ago, living organisms have needed to adapt to life in an aerobic atmosphere and cope with reactive oxygen species (ROS). The term ROS refers to various forms of harmful oxygen excited states, radicals, and peroxides. Specifically, these molecules include singlet oxygen,

*Author for Correspondence, e-mail: niyogi@berkeley.edu

Constantin A. Rebeiz et al. (eds.), The Chloroplast Basics: and Applications, pp. 379–396.

hydroxyl radical, hydrogen peroxide, and superoxide anion, which can readily react with and damage cell membrane lipids, nucleic acids, and proteins. Living organisms have evolved several ways to protect themselves from oxidative damage, including synthesis of various antioxidants and enzymes that neutralize ROS. When the ability to prevent and/or cope with oxidative damage is insufficient to protect cells, oxidative stress occurs. A number of antioxidants exist in photosynthetic organisms, and their roles in protecting cells against oxidative stress have been extensively studied. Some antioxidants seem to have overlapping functions, which might act as a backup mechanism under conditions when the capacity of one antioxidant is overwhelmed. Different antioxidants have roles in protecting cells in specific compartments and in particular conditions. Other than protecting cells from oxidative damage, these antioxidants also have been shown to have other roles such as cell signaling or maintaining cellular redox state. This chapter summarizes knowledge about oxidative damage in oxygenic photosynthetic organisms with a focus on ROS scavenging mechanisms and the roles of antioxidants and related enzymes.

I Types of Reactive Oxygen Species

The reduction of molecular oxygen to water requires four electrons. However, oxygen cannot accept all four electrons at the same time, making it inevitable to generate reactive oxygen intermediates (Eq. 1).

$$O_2 \xrightarrow{e^-} O_2^- \xrightarrow[2H^+]{e^-} H_2O_2 \xrightarrow[H^+]{e^-} H_2O$$
$$+OH^\bullet \xrightarrow[H^+]{e^-} 2H_2O \qquad (1)$$

Several types of ROS exist as by-products of different biological processes. Most molecules exist in singlet ground state with paired electron spins. However, ground state molecular oxygen is unusual because it exists in a triplet state. Its two outermost electrons are in parallel spin, which makes it difficult to react with most molecules. Singlet oxygen is formed when molecular oxygen has acquired enough energy to reverse its spin to antiparallel (Paterson et al., 2006). One major mechanism that generates single oxygen is energy transfer from a photosensitizer. After a photosensitizer absorbs a photon, it becomes excited but still reserves its outermost electron spin (Eq. 2). This state is called the singlet-excited state. It may decay back to its ground state through fluorescence emission. It may also decay to a lower energy form of triplet excited state through intersystem crossing (Eq. 3), with spontaneous spin inversion of excited electrons. The triplet-excited state can transfer its energy to ground state triplet molecular oxygen, generating singlet oxygen (Eq. 4). To return to the triplet ground state, singlet oxygen needs to have its electron spin reversed. However, according to the rules of physical chemistry, the relaxation of singlet oxygen is "spin forbidden". Therefore, this species needs to transfer its energy to another molecule to return to the ground state.

$$S \xrightarrow[\text{fluorescence}]{h\upsilon} {}^1S^* \qquad (2)$$

$${}^1S^* \xrightarrow{isc} {}^3S^* \qquad (3)$$

$${}^3S^* + O_2 \xrightarrow{energytransfer} S + {}^1O_2 \qquad (4)$$

Superoxide anion is formed when molecular oxygen accepts one electron. It spontaneously reacts with itself to form hydrogen peroxide and oxygen (Eq. 5). This process is called dismutation. This reaction can also be catalyzed by the enzyme superoxide dismutase (SOD). Superoxide can also act as a reducing agent, donating an electron to metal ions such as ferric ion, producing ferrous ion and molecular oxygen (Eq. 6). The ferrous ion produced can further catalyze the Fenton reaction, breaking hydrogen peroxide into

Abbreviations: γ-GCS – γ-glutamylcysteine synthetase; APX – ascorbate peroxidase; DHA – dehydroascorbate; DMPBQ – 2,3-dimethyl-6-phytyl-1,4-benzoquinone; ER – endoplasmic recticulum; GPX – glutathione peroxidase; GPXH – glutathione peroxidase homolog; GSH – reduced glutathione; GSSG – oxidized glutathione; LHC – light-harvesting complex; MDA – monodehydroascorbate MPBQ – 2-methyl-6-phytyl-1,4-benzoquinone; MV – methyl viologen; NADPH – nicotinamide adenine dinucleotide phosphate; NPQ – non-photochemical quenching; PRX – peroxiredoxin; PSI – photosystem I; PSII – photosystem II; PUFA – polyunsaturated fatty acid; ROS – reactive oxygen species; SOD – superoxide dismutase; TRX – thioredoxin

a hydroxide anion and a hydroxyl radical (Eq. 7). The hydroxyl radical can react with superoxide to make singlet oxygen and a hydroxide anion (Eq. 8). Superoxide can also react with hydrogen peroxide, generating molecular oxygen, a hydroxide anion, and a hydroxyl radical. This process is called the Haber-Weiss reaction (Eq. 9).

$$2O_2^- + 2H^+ \xrightarrow{dismutation} H_2O_2 + {}^3O_2 \quad (5)$$

$$O_2^- + Fe^{3+} \rightarrow {}^3O_2 + Fe^{2+} \quad (6)$$

$$Fe^{2+} + H_2O_2 \rightarrow Fe^{3+} + HO^\bullet + HO^- \quad (7)$$

$$O_2^- + HO^\bullet \rightarrow {}^1O_2 + HO^- \quad (8)$$

$$O_2^- + H_2O_2 \rightarrow O_2 + OH^- + OH^\bullet \quad (9)$$

Because biological membranes can contain a large amount of polyunsaturated fatty acids (PUFAs), which are very susceptible to oxidation (Blokhina et al., 2003), another very important ROS in biological systems is lipid hydroperoxide. A radical such as hydroxyl radical can abstract a proton from a PUFA, creating a PUFA radical (Eq. 10). The PUFA radical is highly reactive and it will react with molecular oxygen to form a peroxyl radical (Eq. 11). The peroxyl radical can then propagate a lipid peroxidation chain reaction by abstracting a proton from another PUFA (Eq. 12). The peroxyl radical can also generate singlet oxygen by disproportionation (Eq. 13).

$$PUFA - H + OH^\bullet \rightarrow PUFA^\bullet + H_2O \quad (10)$$

$$PUFA^\bullet + O_2 \rightarrow PUFA - OO^\bullet \quad (11)$$

$$PUFA - OO^\bullet + PUFA - H \rightarrow PUFA - OOH + PUFA^\bullet \quad (12)$$

$$2PUFA - OO^\bullet \rightarrow {}^1O_2 + PUFA - O + PUFA - OH \quad (13)$$

II Sources of Reactive Oxygen Species in Algae and Plants

In photosynthetic eukaryotes, a major site for ROS generation is chloroplasts. Both biotic and abiotic stresses such as wounding, senescence, pathogen attack, excess light, drought, heat, and cold can trigger generation of ROS. The photosystem II (PSII) reaction center contains D1 and D2 proteins as a heterodimer. Together, they bind several cofactors, including the primary electron donor chlorophyll P680, a primary electron acceptor pheophytin, and the quinone electron acceptors Q_A and Q_B. In the presence of excess light energy, the plastoquinone pool can be overreduced. As a result, charge recombination between oxidized chlorophyll $P680^+$ and reduced Q_A^- can generate triplet chlorophyll P680. Triplet chlorophyll then can transfer its excitation energy to molecular oxygen, forming singlet oxygen (Krieger-Liszkay, 2005). In contrast to PSII, PSI is not thought to be a major source of singlet oxygen production. It is, however, a site for superoxide generation through its primary electron acceptor and ferredoxin. Superoxide anions then disproportionate to produce hydrogen peroxide, which can be further decomposed into hydroxyl radical, a powerful oxidant (Redmond and Kochevar, 2006). Organelles other than chloroplasts also contribute to the ROS pool of plant cells (Mittler, 2002; Vranová et al., 2002). Peroxisomes are a site of H_2O_2 during photorespiration. Xanthine oxidase, which catalyzes the oxidation of xanthine to uric acid, also produces H_2O_2. Cytochromes in the cytoplasm and ER produce superoxide anion during detoxification. NADPH-dependent oxidase in the plasma membrane also generates ROS during the oxidative burst (reviewed in Apel and Hirt, 2004). Other oxidases such as amine oxidases that catalyze the oxidation of amines to aldehydes, oxalate oxidase that catalyzes the conversion of oxalate to H_2O_2 and CO_2, and pH-dependent cell wall peroxidase have been proposed as sources of ROS in the apoplast (Bolwell and Wojtaszek, 1997).

III Functions of Reactive Oxygen Species

Even though ROS can cause oxidative damage, ROS are also involved in a number of cellular regulatory functions. ROS have been shown to act as signaling molecules in defense responses against both biotic and abiotic stress, programmed cell death, redox signaling and gene regulation, and hormonal responses. For example, the *flu* mutant of *Arabidopsis*, which accumulates excess protochlorophyllide in the dark, generates singlet oxygen in the chloroplasts when exposed to light (op den Camp et al., 2003). After the release of singlet oxygen, the mutant exhibits growth arrest and necrotic lesions. The phenotype of this mutant, however, is not due to damage caused by singlet oxygen but by activation of a genetically controlled response as revealed by studies of the *flu executer1* double mutant (Wagner et al., 2004). In catalase-deficient tobacco plants, H_2O_2 accumulation causes an induction of defense proteins such as glutathione peroxidase and PR-1 (pathogenesis-related 1) protein (Chamnongpol et al., 1998). Maize seedlings that have been pretreated with H_2O_2 also showed increased chilling tolerance (Prasad et al., 1994). A discovery that an oxidative burst kills cells via activation of a signal transduction pathway rather than toxic levels of ROS came from an experiment in soybean culture (Levine et al., 1994). H_2O_2 is produced to kill pathogens, but it also acts as a signaling molecule to activate programmed cell death in the damaged cells, preventing the pathogens from spreading to neighboring cells (reviewed in Apel and Hirt, 2004). A similar situation is observed in animals where phagocytes produce ROS to defend against pathogen attack (Stafford et al., 2002).

Activity of several proteins is controlled by the redox state of the cells either directly by oxidation of the proteins or indirectly via redox-sensitive molecules such as glutathione or thioredoxin (Arrigo, 1999). Proteins can be modified at their thiol groups (-SH) to yield sulfenic (-SOH), sulfinic (-SO_2H) or sulfonic (-SO_3H) acid derivatives. Proteins can also be oxidized at their iron-sulfur (Fe-S) clusters. These modifications lead to changes in protein conformation and/or protein-protein interaction. The activity of many enzymes in the chloroplasts such as key enzymes in the Calvin-Benson cycle are activated via their disulfide bridges (Buchanan and Balmer, 2005). Many other metabolic enzymes contain Fe-S clusters, suggesting that they might be under redox regulation (Imsande, 1999).

IV Oxidative Damage in Chloroplasts

Several sites in chloroplasts are targets of oxidative damage. Singlet oxygen and hydroxyl radicals damage lipids, proteins, pigments, and DNA in close proximity (Redmond and Kochevar, 2006). Hydrogen peroxide inhibits enzymes that contain thiol groups such as fructose 1,6-biphosphatase and other enzymes in the Calvin-Benson cycle (Kaiser, 1976; Kaiser, 1979; Charles and Halliwell, 1980; Tanaka et al., 1982). Superoxide leads to production of peroxyl radical, which can start lipid peroxidation chain reactions. PSI has been shown to undergo oxidative damage at the Fe-S clusters at its reducing side (Sonoike et al., 1995). Specifically, hydroxyl radical has been shown to trigger a conformational change of the PSI complex, allowing access of a protease to degrade the PsaB subunit of PSI (Sonoike et al., 1997).

A major site of photo-oxidative damage in the chloroplast is PSII, especially the D1 protein. Under oxidative stress condition such as high light, high or low temperature, drought, or CO_2 deficiency, degradation of D1 protein is observed (Aro et al., 1993). PSII constantly undergoes damage and repair even under normal light condition, which does not normally lead to oxidative stress (Keren et al., 1995). When the rate of repair cannot keep up with the rate of damage, photosynthesis and plant growth are reduced. This condition, known as photoinhibition, is exacerbated by excess light. Several years of studies on the photoinhibition repair cycle have allowed us to understand the sequence of PSII repair (reviewed in Aro et al., 1993; Melis, 1999; Takahashi and Murata, 2008). It begins with partial disassembly of the PSII holocomplex, followed by proteolytic degradation of D1 protein and its removal from the membrane. New D1 protein is then synthesized as 33.5 kDa pre-D1 and inserted in the thylakoid membrane with the assembly of other PSII components. Carboxy-terminal processing of pre-D1 into the mature 32-kDa D1 protein and the assembly of the oxygen-evolving machinery complete the repair cycle.

By examining photodamage and repair processes separately in *Synechocystis* sp. PCC 6803, Nishiyama et al. (2006) concluded from several studies that oxidative stress caused by H_2O_2 or singlet oxygen inhibited the repair of PSII, presumably at the translation elongation step, but did not accelerate photodamage of PSII directly. Furthermore, the rate of electron transport had no effect on the photodamage of PSII. A two-step photodamage model was then proposed by Ohnishi et al. (2006). By using monochromatic light and thylakoid membranes from *Thermosynechococcus elongatus*, they found that the primary damage occurs at the oxygen-evolving complex by UV and strong blue light. Secondary damage occurs at the reaction center of PSII by light absorbed by photosynthetic pigments. In addition to the role of ROS in inhibiting PSII repair, ROS can also oxidize PSII directly. After the oxygen-evolving complex is damaged, it allows free access of oxygen to P680, thereby producing singlet oxygen and other ROS that damage the reaction center (Anderson and Chow, 2002; Nishiyama et al., 2006).

V Avoidance of Reactive Oxygen Species Production

Because ROS generated during photosynthesis can be harmful, photosynthetic organisms have evolved several ways to minimize their production (Fig. 1). For example, chloroplasts can realign themselves perpendicular to the direction of light to minimize the amount of light absorbed (Kasahara et al., 2002). Alternative oxidases divert electron flow through electron transport chains to reduce O_2 to water, thereby decreasing the amount of excess O_2 that could turn into ROS (Mittler, 2002). Photorespiration, which is initiated when the Rubisco enzyme uses O_2 instead of CO_2 as a substrate, also functions as an alternative sink for excess photosynthetic energy (Foyer, 1997). Cyclic electron flow around PSI helps in dissipating extra energy and preventing PSI overreduction (Asada, 2006). Non-photochemical quenching (NPQ) mechanisms by which excess energy in PSII is dissipated as heat are also thought to help reduce ROS generation (reviewed in Niyogi, 2000; Müller et al., 2001). Even when all these mechanisms are working together to minimize ROS generation, excess ROS can still be generated and some of them can escape from the thylakoids to cause damage at other sites or initiate signaling pathways. Therefore, a number of ROS scavenging mechanisms are needed to prevent oxidative stress and regulate signaling (Fig. 1).

VI Non-enzymatic Mechanisms for Scavenging Reactive Oxygen Species

Non-enzymatic ROS scavenging mechanisms involve hydrophilic antioxidants such as ascorbate and glutathione and hydrophobic antioxidants such as tocopherols and carotenoids, as well as flavonoids and alkaloids (Fig. 1). This section focuses on the major antioxidants and their roles in ROS detoxification.

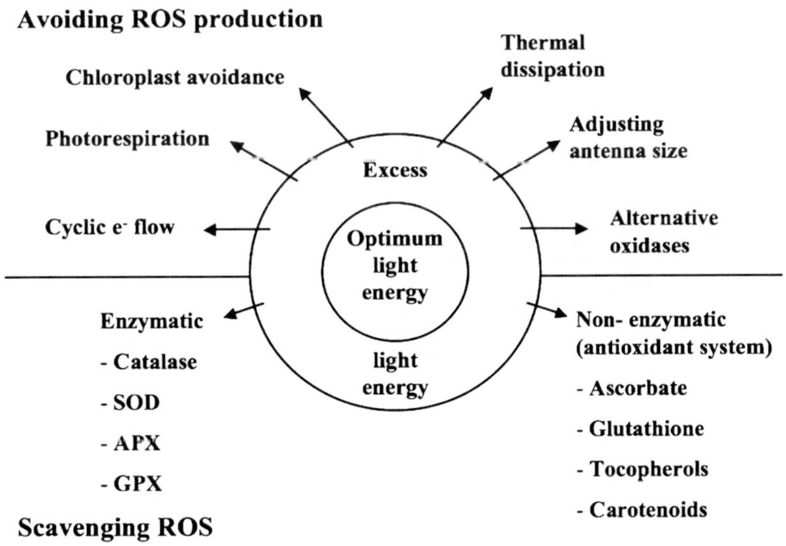

Fig. 1. Overview of mechanisms that decrease ROS levels.

A Hydrophilic Antioxidants

1 Ascorbate

Hydrophilic antioxidants such as ascorbate and glutathione are found in cytosolic, mitochondrial, chloroplastic, and nuclear aqueous compartments at concentrations up to the millimolar range in plants (Asada, 1999). Ascorbate is synthesized outside the chloroplasts and is imported through an ascorbate translocator in the envelope membrane (Beck et al., 1983). The import of ascorbate to the thylakoid lumen, however, is thought to occur by diffusion only (Foyer and Lelandais, 1996; Horemans et al., 2000). Ascorbate is a major line of defense against H_2O_2 with the help of ascorbate peroxidase (APX) (Smirnoff and Wheeler, 2000). APX uses two molecules of ascorbate to reduce H_2O_2 to water. In addition, ascorbate can react with other forms of ROS such as peroxyl and hydroxyl radicals and singlet oxygen (Smirnoff and Wheeler, 2000). The reaction of ascorbate with ROS produces monodehydroascorbate (MDA), an oxidized form. MDA can be reduced back to ascorbate by the enzyme monodehydroascorbate reductase using electrons from NADPH or ferredoxin (Fig. 2). If not reduced immediately, MDA will disproportionate to form dehydroascorbate (DHA) and ascorbate. DHA is not stable at physiological pH, so DHA must be reduced back to ascorbate by DHA reductase using glutathione as the reducing substrate.

Physiological evidence of the role of ascorbate in oxidative stress protection has come from analysis of *Arabidopsis* mutants (Conklin et al., 1996). For example, the *vtc1* mutant contains only 30% of wild-type ascorbate. *VTC1* encodes GDP-mannose pyrophosphorylase, an enzyme that converts mannose-1-phosphate to GDP-mannose (Conklin et al., 1999). The *vtc1* mutant is also sensitive to oxidative stress caused by UV and pollutants (Conklin et al., 1997). Sensitivity to high light combined with high-salt stress has also been reported (Smirnoff and Wheeler, 2000).

In addition to its antioxidant functions, other functions of ascorbate also contribute to ROS scavenging mechanisms. Ascorbate is believed to regenerate tocopherol, a lipophilic antioxidant, from its oxidized form (Smirnoff and Wheeler, 2000) (Fig. 2). It is also acts as a cofactor for violaxanthin de-epoxidase (Müller-Moulé et al., 2002), an enzyme that converts violaxanthin to zeaxanthin (Yamamoto et al., 1999), which is involved in NPQ (Demmig-Adams, 1990; Niyogi et al., 1998). Ascorbate also participates in maintaining cyclic electron flow around PSI (Ivanov et al., 2005). Other than its role in ROS scavenging, ascorbate also participates in a number of cellular processes such as cell wall synthesis and synthesis of plant hormones such as ethylene and gibberellins (Smirnoff and Wheeler, 2000).

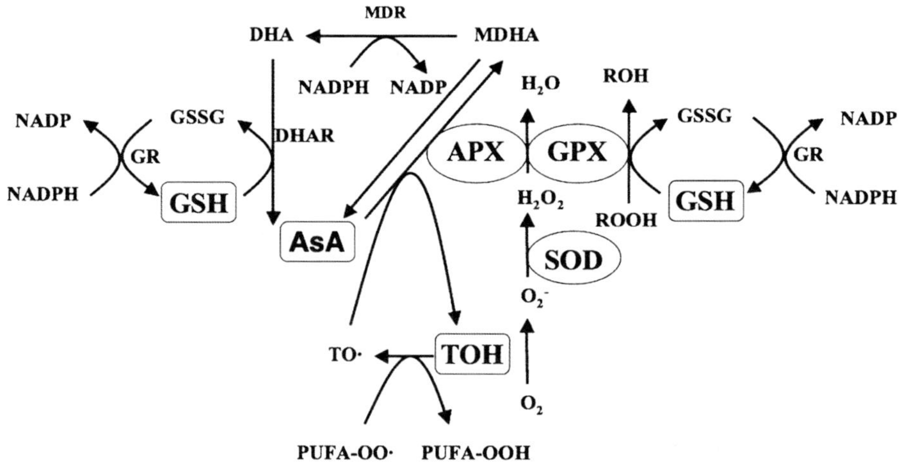

Fig. 2. Cooperative interactions among antioxidants and related enzymes.

2 Glutathione

The tripeptide glutathione is a water-soluble antioxidant composed of cysteine, glutamic acid, and glycine. Its synthesis occurs in two ATP-dependent steps (Hell and Bergmann, 1990). The first step is the formation of γ-glutamylcysteine from glutamic acid and cysteine, catalyzed by the enzyme γ-glutamylcysteine synthetase (γ-GCS). The second step is catalyzed by the enzyme glutathione synthetase, which forms a peptide bond between glycine and the cysteine of the γ-glutamylcysteine intermediate. Under physiological conditions, glutathione predominantly exists in the reduced form (GSH). Under oxidizing conditions, however, GSH is oxidized by several reactions, forming glutathione disulfide, GSSG. GSSG is reduced back to GSH by the enzyme glutathione reductase, utilizing NADPH as an electron donor (Fig. 2). The level of GSSG increases under oxidative stress conditions (Noctor and Foyer, 1998). Nevertheless, the ratio of GSH:GSSG is maintained relatively constant by various mechanisms (Filomeni et al., 2002).

Glutathione is involved in a number of cellular mechanisms related to oxidative stress. As an antioxidant, glutathione functions by reducing various types of ROS and is used as an electron donor for regeneration of ascorbate. Glutathione is also important in protein posttranslational modification via glutathionylation. Glutathionylation helps in preventing proteolysis under oxidative stress, modifying protein turnover, and regulating cellular redox state and gene transcription (Foyer and Noctor, 2000). Glutathionylation occurs in the presence of GSSG or GSH and oxidants. Several studies have suggested that glutathionylation is driven by higher production of ROS (Rouhier et al., 2008). Several targets for glutathionylation have been reported, including aldolase and triose-phosphate isomerase in sugar metabolism (Ito et al., 2003). Other functions of glutathione are detoxification of xenobiotics via glutathione-S transferase, heavy metal detoxification via phytochelatin, which uses glutathione as a precursor, and being an important pool of reduced sulfur (Noctor et al., 1998).

Because GSH is a major thiol in both prokaryotes and eukaryotes, studies on the role of GSH in protection against oxidative stress have been conducted in several model organisms such as *E. coli*, *S. cerevisiae*, and *Arabidopsis*. In *E. coli*, GSH is dispensable, and mutants that are devoid of GSH show wild-type growth under oxidative stress (reviewed in Carmel-Harel and Storz, 2000). In contrast, a *S. cerevisiae* mutant lacking GSH requires GSH for growth under minimal medium, and it is sensitive to H_2O_2 (Grant et al., 1996). Nevertheless, growth can be restored with reducing agents containing a sulfhydryl group such as dithiothreitol, β-mercaptoethanol, and cysteine, indicating that GSH is essential only as a reductant during normal cellular processes. The mutant is also sensitive to H_2O_2 and tert-butyl hydroperoxide. A mutant of *Arabidopsis*, *cad2*, affecting γ-GCS, was isolated from a screen for cadmium sensitivity. This mutant has a reduced accumulation of glutathione. It grows normally in the absence of cadmium (Cobbett et al., 1998), but it is sensitive to high light (Larkindale et al., 2005). Another allele of this gene, *pad2-1*, shows hypersensitivity to pathogen attack (Parisy et al., 2007). Overexpression studies of the enzymes in the glutathione biosynthesis pathway in higher plants such as poplar and cotton have also been carried out to test the level of stress tolerance (Noctor et al., 1998). Poplars overexpressing glutathione reductase in chloroplasts show higher resistance to oxidative stress than the ones overexpressing γ-GCS. Therefore, it is suggested that maintaining the GSH pool has a more important role in protection against oxidative stress than simply elevating glutathione content (Noctor et al., 1998). In response to oxidative stress caused by pathogen attack or herbicide, plants increase the activity of GSH biosynthetic enzymes (Vanacker et al., 2000; Batish et al., 2006) and GSH level (Noctor et al., 2002). In addition, transgenic *Arabidopsis* GSH-deficient plants are sensitive to heavy metal stress, photooxidative stress and ozone stress (Xiang et al., 2001).

B Lipophilic Antioxidants

1 Tocopherol

Tocopherols are abundant lipid-soluble antioxidants synthesized only in oxygenic photosynthetic organisms (Hofius and Sonnewald, 2003). They are amphipathic molecules that localize mainly in the plastid envelope, thylakoids, and plastoglobules. Tocopherol biosynthesis begins with condensation of a polar head group, homogentisic acid, derived from aromatic amino acid metabolism, with a phytyl diphosphate tail. This reaction is catalyzed by the enzyme homogentisate phytyltransferase, resulting in an intermediate, 2-methyl-6-phytyl-1,4-benzoquinone (MPBQ). MPBQ is a precursor of both δ- and β-tocopherols. Methylation of MPBQ results in 2,3-dimethyl-6-phytyl-1,4-benzoquinone (DMPBQ), which is a precursor of both γ- and α-tocopherols.

Four forms of tocopherols, α, β, γ, and δ, differ structurally only in the number and position of methyl groups on the chromanol head groups. α-tocopherol, which is the predominant form found in every part of plants except seeds, is thought to have the highest antioxidant activity in vivo (Burton and Traber, 1990; Kamal-Eldin and Appelqvist, 1996). The main evidence for this, however, comes from studies in liposomes (Fukuzawa et al., 1982). It has been suggested that the relative activities of tocopherols in plant cells may be different from liposomes (Munné-Bosch and Alegre, 2002). Transgenic tobacco plants that accumulate γ-tocopherol are more tolerant to sorbitol treatment compared to wild-type plants, showing that γ-tocopherol has a more important role in desiccation stress compared to α-tocopherol (Abbasi et al., 2007).

Tocopherols have two main antioxidant functions. First, they physically quench singlet oxygen (Trebst et al., 2002), and second, they can react with lipid peroxyl radicals, thereby terminating lipid peroxidation chain reactions (Eq. 14).

$$PUFA - OO^{\cdot} + TOH \rightarrow PUFA - OOH + TO^{\cdot} \qquad (14)$$

In this process, the chromanol head group of tocopherols donates a hydrogen atom to the lipid peroxyl radical, forming a hydroperoxide and a tocopheroxyl radical. The tocopheroxyl radical is thought to be recycled back to tocopherol by ascorbate and glutathione (Munné-Bosch and Alegre, 2002).

Under oxidative stress conditions, the amount of tocopherol increases in plants and algae such as *Arabidopsis*, tobacco, *Chlamydomonas*, and *Euglena gracilis* (Ruggeri et al., 1985; Porfirova et al., 2002; Ledford et al., 2004). However, the lack of tocopherol in the *Arabidopsis vte1* mutant, which still accumulates the intermediate DMPBQ, resulted in only a minor phenotype even under stress condition, suggesting that the intermediates have antioxidant activity that could protect the plants (Porfirova et al., 2002). Direct evidence of the role of tocopherol in lipid protection comes from mutants of *Arabidopsis* and *Synechocystis*. The *Arabidopsis vte2* mutant, which cannot synthesize any tocopherols and intermediates, shows severe seedling growth defects and high level of lipid hydroperoxide during germination (Sattler et al., 2004). In the case of *Synechocystis* tocopherol-deficient mutants, high light treatment alone does not reveal any growth phenotype (Maeda et al., 2005). However, the mutants are sensitive to linoleic or linolenic acid treatments combined with high light, which supports the idea that tocopherol plays a role in protecting the cells from lipid peroxidation. Under high light growth, blocking tocopherol synthesis by herbicides in *Chlamydomonas* leads to a quick disappearance of tocopherol, as well as of D1 protein and PSII activity, implying that tocopherol has a role as a singlet oxygen quencher in PSII (Trebst et al., 2002). This conclusion is supported by the fact that *Arabidopsis vte1 npq4*, which lacks both tocopherol and NPQ, exhibits a marked increase in PSII photoinhibition, suggesting that tocopherol is protecting PSII (Havaux et al., 2005).

2 Carotenoids

Carotenoids are C40 tetraterpene pigments derived from geranylgeranyl diphosphate. There are two classes of carotenoids, carotenes and xanthophylls. Carotenes are hydrocarbons that can be either linear or containing one or more cyclic β-ionone or ε-ionone rings. Xanthophylls are oxygenated derivatives of carotenes. Carotenoids are synthesized by plants and algae as well as in some bacteria and fungi. Carotenoids are essential for photosynthesis and photoprotection. In plants and algae, the majority of carotenoids are located in the chlorophyll-binding proteins embedded in the thylakoid membrane. They function in harvesting light energy, stabilizing the membrane, quenching triplet chlorophyll and singlet oxygen, participating in the NPQ process, and inhibiting lipid peroxidation (Baroli and Niyogi, 2000).

Carotenoids can quench triplet chlorophyll, thereby preventing the formation of singlet oxygen (reviewed in Baroli and Niyogi, 2000). This possibility only occurs in the antenna but not in P680. In the PSII reaction center core, there are 11 molecules of β-carotene near antenna chlorophylls and two near P680. However, the distance between β-carotene and P680 in PSII does not allow the quenching of ^{3}P680* by the β-carotenes (reviewed in Telfer, 2002). Carotenoids can also quench singlet oxygen directly (Cogdell and Frank, 1987). Therefore, the primary function of β-carotenes near P680 is probably quenching singlet oxygen generated from triplet chlorophyll in PSII (Telfer, 2002).

Xanthophyll cycle carotenoids have been studied extensively for roles in NPQ and as antioxidants, using a number of *Arabidopsis* and *Chlamydomonas* mutants. The xanthophyll cycle involves reversible interconversion of zeaxanthin and violaxanthin, with an intermediate antheraxanthin. In low light, violaxanthin is accumulated, but in high light, violaxanthin is converted to zeaxanthin, accompanied by the development of NPQ. This conversion is triggered by low thylakoid lumen pH, which activates the enzyme violaxanthin de-epoxidase (Yamamoto et al., 1999), using ascorbic acid as a cofactor (Bratt et al., 1995). The *Arabidopsis npq1* mutant carrying a mutation in the gene encoding violaxanthin de-epoxidase is unable to convert violaxanthin to zeaxanthin, and it exhibits a reduction in NPQ, confirming that zeaxanthin is necessary for normal NPQ in vivo (Niyogi et al., 1998). Another *Arabidopsis* mutant, *lut2*, which lacks lutein, also showed reduced NPQ (Pogson et al., 1998). The double mutant *npq1 lut2* completely lacks rapidly reversible NPQ, suggesting a role of lutein in NPQ (Niyogi et al. 2001). Evidence of the role of xanthophylls in protecting membranes has come from *Arabidopsis* strains overexpressing β-carotene hydroxylase. The overexpressors accumulate about two times the normal xanthophyll cycle pool (Davison et al., 2002). They are more resistant to oxidative stress conditions induced by high light and high temperature, shown by reduced amount of anthocyanin and lipid peroxidation, which are indicators of stress (Davison et al., 2002).

Chlamydomonas npq1, which lacks zeaxanthin, and *lor1* mutants, which lack lutein, exhibited partial NPQ defect but are able to grow in high light (Niyogi et al., 1997). The *npq1 lor1* double mutant, however, lacks NPQ completely and bleaches in high light (Niyogi et al., 1997). Growth of the single mutants in high light implied a redundancy of xanthophylls in photoprotection, because having either zeaxanthin or lutein is sufficient for survival (Niyogi et al., 1997). Bleaching of the double mutant is not solely due to the lack of NPQ because another NPQ-deficient mutant, *npq5*, with a normal xanthophyll composition is not as sensitive to high light stress (Elrad et al., 2002). Further characterization of the *npq1 lor1* double mutant revealed that its high light sensitivity is caused by higher production of reactive oxygen species, because loss of cell viability was partially rescued by lower oxygen tension (Baroli et al., 2004). Supplying a carbon source in the medium did not prevent cell death under high light, suggesting that the light sensitivity is probably due to lack of antioxidant capacity that causes general chloroplast damage rather than specifically affecting photosynthesis (Baroli et al., 2004).

C Antioxidant Interactions

Because plants and algae experience constant stress from ROS, synthesizing more than one type of antioxidant might be necessary for coping with various oxidative stresses. It is, therefore, very likely that a number of these antioxidants might have functional overlap or might be interacting. In fact, evidence that supports this notion comes from a number of studies in vitro and in vivo. In the case of tocopherols and carotenoids, α-tocopherol and zeaxanthin showed a synergistic effect on protection against lipid peroxidation in liposomes as shown by measuring the product of the reaction between cholesterol and singlet oxygen and free radicals (Wrona et al., 2004). The *Arabidopsis npq1* mutant, which is able to grow in high light, accumulates higher amounts of α-tocopherol during photoacclimation suggesting that tocopherol can compensate for the lack of zeaxanthin and antheraxanthin (Havaux et al., 2000). Conversely, the *vte1* mutant, which cannot synthesize tocopherol, accumulated higher amounts of zeaxanthin (Havaux et al., 2005). The mutant seedlings accumulate a high amount of lipid peroxides, but the mature plants showed only a slightly abnormal phenotype (Kanwischer et al., 2005).

Similarly, *Synechocystis* tocopherol-deficient mutants did not show any growth phenotype in high light or in the presence of ROS generators compared to the wild type (Maeda et al., 2005). Nevertheless, the phenotype was revealed in the presence of sublethal levels of norflurazon, an inhibitor of carotenoid synthesis, suggesting that carotenoids and tocopherols have overlapping or interacting function to protect *Synechocystis* from lipid peroxidation and high light stress (Maeda et al., 2005).

The interaction of tocopherols, ascorbate, and glutathione is thought to play a major role in protection against ROS in photosynthetic organisms. The ascorbate-glutathione cycle, in which glutathione maintains the reduced pool of ascorbate, is important for the regeneration of α-tocopherol from its α-chromanoxyl radical (Fig. 2). An ascorbate-deficient mutant of *Arabidopsis*, *vtc2*, showed an increase in glutathione level in high light, suggesting an overlapping function of these two antioxidants (Müller-Moulé et al., 2004). The *vtc1* and *cad2* mutants both have an elevated level of α-tocopherol (Kanwischer et al., 2005). Conversely, a mutation in *vte1* leads to tocopherol deficiency and concomitant increases of both ascorbate and glutathione, whereas overexpression of *VTE1* resulted in decreased levels of ascorbate and glutathione (Kanwischer et al., 2005).

Although antioxidants seem to be interacting and having overlapping function, the mechanisms by which their biosynthesis is regulated are still poorly understood. Having a large amount of one antioxidant is not always beneficial. Overexpressing the genes encoding enzymes for glutathione synthesis raised cellular glutathione level by threefold (Creissen et al., 1999). Surprisingly, these transformed plants exhibit continuous oxidative stress as seen by light dependent chlorosis or necrosis. Even though the total glutathione pool is increased in the transformants, most of the glutathione pool is in the oxidized form, giving a much lower redox state compared to the wild type (Creissen et al., 1999). Oxidized glutathione or oxidized ascorbate must be readily removed because they may inhibit enzyme reactions and protein synthesis (Wolosiuk and Buchanan, 1977; Morell et al., 1997). Therefore, the overall balance of antioxidants and between ROS synthesis and scavenging must be controlled because it could affect signaling pathways that minimize damage (Creissen et al., 1999).

VII Enzymatic Mechanisms for Scavenging Reactive Oxygen Species

A Superoxide Dismutase

SOD acts as a first line of defense in the enzymatic ROS scavenging system by dismutating superoxide to H_2O_2. Superoxide is generated in PSI in the chloroplast. There are no enzymes that catalyze the reduction of superoxide to H_2O_2 or the oxidation of superoxide to oxygen in any organisms (Asada, 1999). Moreover, superoxide does not spontaneously react with antioxidants such as ascorbate or glutathione due to its low reactivity with stromal components compared with that of SOD (Asada, 1999). Three main types of SODs have been found: CuZnSOD, MnSOD, and FeSOD. CuZnSOD is the major form in plant chloroplasts, but FeSOD is also found in the chloroplasts of some plants (Kurepa et al., 1997). MnSOD is thought to be only in the mitochondria but a recent study has found its activity also in the chloroplast (Allen et al., 2007). In spinach, over half of the CuZnSOD is attached to the stroma thylakoids where PSI is localized (Ogawa et al., 1995).

The importance of SOD in ROS scavenging became evident in a number of studies of algae and plants including mutants and transformants. The activity of SOD increases in *Chlamydomonas* cultures treated with paraquat (Vartak and Bhargava, 1999). Similarly, low temperature stress increases SOD activity in *Synechocystis* (Rady et al., 1994). Heavy metal stress in algae such as the marine microalga *Tetraselmis gracilis* or *Gonyaulax polyedra* also results in an increase of SOD activity (Okamoto et al., 1996; Okamoto and Colepicolo, 1998). Overexpression of CuZn SOD in tobacco results in higher resistance to high light and cold stress (Gupta et al., 1993). In contrast, antisense suppression of CuZnSOD results in photobleaching of tobacco under high light stress and inactivated PSI complex, even when MnSOD is overexpressed in the chloroplasts, showing that intrachloroplastic targeting of SOD is important for stress tolerance (Asada, 1999). Thus, it seems necessary for the membrane-bound

SOD to scavenge superoxide at its generation site before it can diffuse to stroma or cytosol (Asada, 1999).

B Catalase

Catalase is a heme-containing, tetrameric enzyme found in all aerobic organisms (Mallick and Mohn, 2000). It catalyzes the dismutation of two H_2O_2 molecules to water and molecular oxygen. Catalase is localized only in peroxisomes where it is found in the millimolar concentration range. It has a high V_{max} but low affinity for its substrate (Foyer and Noctor, 2000). This enzyme is unique among H_2O_2-scavenging enzymes because it does not consume cellular reducing equivalents (Mallick and Mohn, 2000). Catalase activity of a symbiotic photosynthetic prokaryote *Prochloron* sp. is directly proportional to irradiance (Lesser and Stochaj, 1990). Cu and UV-B stresses stimulate catalase activity in the N_2-fixing cyanobacterium *Anabaena doliolum* and in a green alga *Chlorella vulgaris* (Malanga et al., 1999; Mallick and Rai, 1999). Transgenic tobacco plants with suppressed catalase show enhanced ROS levels in response to both abiotic and biotic stresses (Willekens et al., 1997).

C Ascorbate Peroxidase

APX is a heme peroxidase in the same superfamily as cytochrome *c* peroxidase (Asada, 1999). It is the main enzyme in the chloroplast that scavenges H_2O_2, because chloroplasts do not contain catalase. APX is found only in the micromolar range in chloroplasts, cytosol, mitochondria, peroxisomes, and apoplast. Unlike catalase, APX has high affinity for its substrate, and it uses ascorbate as an electron donor (Asada, 1992). The reduction of H_2O_2 by APX produces water and monodehydroascorbate (MDA). There are two types of chloroplast APX: thylakoid-bound (tAPX) and stroma-localized forms (sAPX) (Asada, 2006). tAPX is bound in the vicinity of PSI; together with PSI-associated SOD, it functions as the first defense against ROS (Asada, 2006). Overexpression of tAPX in tobacco plants resulted in higher tolerance to oxidative stress (Yabuta et al., 2002; Kangasjärvi et al., 2008), whereas antisense *Arabidopsis* plants are sensitive to stress (Tarantino et al., 2005). *Arabidopsis* T-DNA insertion lines revealed functional redundancy of sAPX and tAPX in mature leaves (Giacomelli et al., 2007), but sAPX seems to be more important during the greening process (Kangasjärvi et al., 2008). After an acclimation period, however, neither single nor double mutants show stress phenotype due to increased level of 2-cysteine peroxiredoxin (PRX) (Kangasjärvi et al., 2008).

D Glutathione Peroxidase

Glutathione peroxidase (GPX) is an antioxidant enzyme capable of reducing a variety of organic and inorganic hydroperoxides to their hydroxyl counterpart, using glutathione and/or other equivalents (Kühn and Borchert, 2002). In animals, all types of GPX are selenium-dependent enzymes and have been shown to have antioxidant activity (Ursini et al., 1995). The reaction catalyzed by GPX is shown in Eq. 15.

$$2RSH + ROOH \rightarrow RSSR + H_2O + ROH \quad (15)$$

A number of GPX-like proteins have been found in plants, but they contain cysteine instead of selenocysteine, causing their activity to be lower than that of animal homologs (Yoshimura et al., 2004). Eight GPX isoforms have been found in the *Arabidopsis* genome and five isoforms in the *Chlamydomonas* genome (Rouhier and Jacquot, 2005; Dayer et al., 2008). The physiological characterization of plant GPX has been very limited. There have been several reports on increased GPX expression in plants due to various types of stress (Gaber et al., 2006). More specifically, the expression of GPXH in *Chlamydomonas* has been shown to be upregulated in response to singlet oxygen and high light stress (Fischer et al., 2006), and overexpression of this gene increased oxidative stress tolerance in the same organism (Ledford et al., 2007). Transgenic tobacco plants overexpressing a *Chlamydomonas* GPX-like protein results in higher tolerance to stress such as methyl viologen, moderate light, chilling stress under high light, and salt stress (Yoshimura et al., 2004). The transcript levels of *gpx1* and *gpx2* in *Synechocystis* increase under oxidative stress (Gaber et al., 2004). Knockout mutants of each of these genes also exhibit higher lipid peroxidation level (Gaber et al., 2004). The two genes seem to be necessary for survival even under normal conditions, because isolation of a double mutant was unsuccessful (Gaber et al., 2004). Overexpressing *Synechocystis GPX2* in *Arabidopsis* resulted in increased tolerance toward oxidative stress caused by H_2O_2, Fe ions, MV, and chilling with high light, high salt, and drought (Gaber et al., 2006).

E Thioredoxin

In addition to GSH and GPX, other thiol-containing molecules also play a significant role in the ROS scavenging system. In particular, members of the thioredoxin (TRX) fold superfamily carry out a number of diverse functions, which involve disulfide oxido-reduction and/or isomerization (Lemaire and Miginiac-Maslow, 2004). TRXs are small ubiquitous redox proteins that reduce disulfide bridges of numerous target proteins by thiol-disulfide exchange reactions. They all have a conserved WC[G/P]PC motif exposed at the protein surface that is used to carry out the reaction (Buchanan and Balmer, 2005). TRXs serve many functions in cells including regulation of target enzymes and transcription factors. In addition, they can act as electron donors for peroxiredoxin, another thiol protein that scavenges cellular hydroperoxides (Jacquot et al., 2002) (Fig. 3). TRXs react with disulfide bonds in target proteins, resulting in changes in enzymatic activity or protein stability that can affect gene expression level in the case of proteins that are transcription factors. Reacting with TRX can also protect thiol-containing proteins, because they are prone to ROS-induced modification (Dos Santos and Rey, 2006). In plants, there is a large multigene family of TRXs compared to algae. *Arabidopsis* is estimated to have more than 20 genes, whereas the numbers are only eight in *Chlamydomonas* and four in *Synechocystis* (Lemaire and Miginiac-Maslow, 2004). Different isoforms are expressed specifically in certain cellular compartments such as chloroplasts, mitochondria, and cytosol. The mitochondrial and cytosolic isoforms use NADPH as the electron donor, whereas the chloroplastic ones use photoreduced ferredoxin through a specific iron-sulfur containing enzyme called ferredoxin-thioredoxin reductase (Dai et al., 2000; Jacquot et al., 2002).

TRXs play a significant role in defense against oxidative stress (reviewed in Dos Santos and Rey, 2006). They can directly reduce H_2O_2, dehydroascorbate, and certain radicals (Copley et al., 2004). Potato transformants lacking TRX show increased sensitivity and higher lipid peroxidation levels upon photo-oxidative treatments (Broin et al., 2002). Expression of thioredoxin genes is elevated in different plant species upon oxidative stress (Dos Santos and Rey, 2006). Proteomic studies have identified nearly 300 putative TRX targets, 92 of which are localized in chloroplasts (reviewed in Michelet et al., 2006; Lemaire et al., 2007). Catalase has been

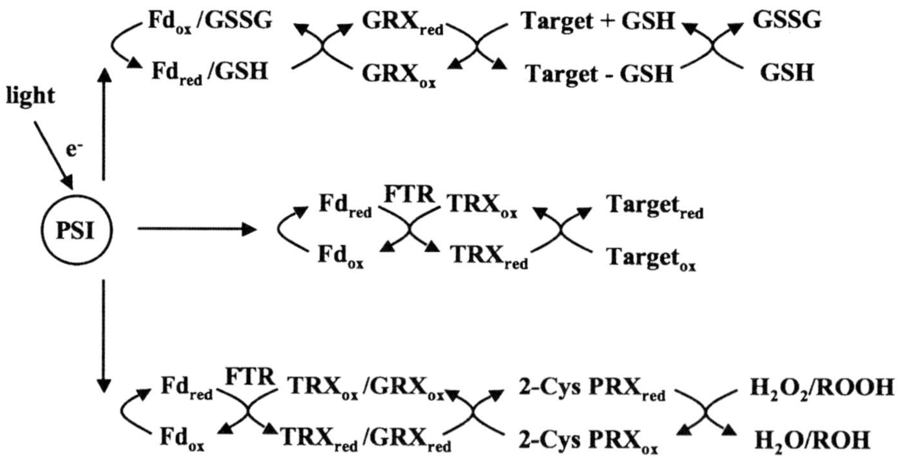

Fig. 3. Involvement of thioredoxin, glutaredoxin, and peroxiredoxin in ROS scavenging, thiol-disulfide redox reactions, and glutathionylation.

identified as a TRX target in *Chlamydomonas* (Lemaire et al., 2004) and *Arabidopsis* (Balmer et al., 2004). Other targets involved in antioxidant mechanisms are PRX, GPX, and SOD (Wong et al., 2003; Yamazaki et al., 2004; Dos Santos and Rey, 2006).

F Glutaredoxin

Glutaredoxin (GRX) is a small redox proteins related to TRX (reviewed in Lemaire, 2004). Approximately thirty isoforms of GRX are found in the *Arabidopsis* genome compared to eight in *Chlamydomonas* and three in *Synechocystis* sp. (Lemaire, 2004). They are classified into three groups based on their active site motif: CPYC, CGFS, and CC, the last of which is specific to higher plants (Rouhier et al., 2008). When using GSH as a reductant, GRXs catalyze thiol-disulfide redox reactions by either a monothiol mechanism or a dithiol mechanism (Rouhier et al., 2008). In the monothiol mechanism, the N-terminal active site cysteine of GRX reduces the mixed disulfide between glutathione and a target protein. Regeneration of oxidized GRX requires another glutathione molecule. In the dithiol mechanism, both cysteines of GRX are needed to reduce protein disulfide. The third mechanism of GRX does not require glutathione, but can be reduced in the light using photoreduced ferredoxin and ferredoxin-thioredoxin reductase (Zaffagnini et al., 2008) (Fig. 3). GRXs can also reduce dehydroascorbate, peroxides, and PRX (Rouhier et al., 2008). In addition, deglutathionylation is likely catalyzed by GRX (Rouhier et al., 2008). Very little is known about physiological role of GRXs in photosynthetic organisms. Nevertheless, it is clear that GRXs are involved in oxidative stress response. An *Arabidopsis* chloroplastic GRX4 mutant shows a defect in early seedling growth under oxidative stress (Cheng et al., 2006). Expressing the gene in a yeast *grx5* mutant suppressed H_2O_2 sensitivity phenotype (Cheng, 2008).

G Peroxiredoxin

PRXs are the most recently identified H_2O_2-scavenging antioxidant enzymes (reviewed in Dayer et al., 2008). Unlike APX, PRX has activity toward alkyl hydroperoxides and peroxynitrite in addition to H_2O_2 (Asada, 1999; Rouhier and Jacquot, 2002) (Fig. 3). Other chloroplastic enzymes that can reduce alkyl hydroperoxides are lipid hydroperoxide reductase bound to the chloroplast envelope and the stromal phospholipid hydroperoxide glutathione peroxidase bound in the thylakoids (Rouhier and Jacquot, 2002). Several isoforms of PRX exist in chloroplasts, mitochondria, and cytosol. The localization of thylakoid-bound PRX suggests that it could help prevent lipid peroxidation chain reactions in the thylakoids (Dietz, 2003). PRXs are categorized into four types, which are 1-Cys PRX, 2-Cys PRX, PRXQ, and type-I PRX, based on sequence similarity and catalytic mechanisms (Rouhier and Jacquot, 2002; Dietz, 2003). In all cases, the peroxide substrate reacts with an -SH group on the PRXs to a sulfenic acid, Cys-SOH, releasing H_2O in the case of H_2O_2 or the corresponding alcohol in the case of alkyl hydroperoxides. The sulfenic acid derivatives of cysteine efficiently react with other thiols to form disulfide bridges. The disulfide is regenerated by interaction with dithiols such a TRXs, GRXs, and cyclophilin (Dietz, 2003). Physiological characterization of PRX has been performed in a few model organisms. Under oxidative stress condition, upregulation of PRX genes has been observed in *Chlamydomonas* and *Arabidopsis* (Horling et al., 2003; Dietz et al., 2006). Knockout mutants of 2 Cys-PRX of *Synechocystis* are sensitive to high light stress, and *Arabidopsis* antisense mutants of the same gene show lower photosynthesis parameters and are sensitive to photoinhibition (Klughammer et al., 1998; Baier and Dietz, 1999).

In addition to general detoxification of H_2O_2 and alkyl hydroperoxide, Dietz et al. proposed that PRX, together with TRX, establish an alternative pathway for the water-water cycle (Dietz et al., 2002) (Fig. 3). Several lines of evidence led to this model. First, APX seems to be sensitive to ROS inactivation in severe stress conditions, so it is not sufficient for protecting photosynthesis (Shikanai et al., 1998; Yabuta et al., 2002). Second, PRXs are robust enzymes, capable of retaining activity even at high temperature (Dietz et al., 2006). Third, *Arabidopsis* antisense of 2-Cys PRX accumulated a more oxidized ascorbate pool (Baier et al., 2000), suggesting that PRX and APX participate in the same H_2O_2 detoxification pathway.

From these results, it appears that PRX significantly contributes to the detoxification of Mehler reaction-derived H_2O_2 in chloroplasts (Dietz et al., 2006). It is still not known under what conditions this systems functions as an alternative to the APX-dependent water-water cycle (Asada, 2006).

Acknowledgments

This work was supported by the National Institutes of Health (GM058799 and GM071908).

References

Abbasi AR, Hajirezaei M, Hofius D, Sonnewald U and Voll LM (2007) Specific roles of α-and γ-tocopherol in abiotic stress responses of transgenic tobacco. Plant Physiol 143: 1720–1738

Allen MD, Kropat J, Tottey S, Del Campo JA and Merchant SS (2007) Manganese deficiency in*Chlamydomonas* results in loss of photosystem II and MnSOD function, sensitivity toperoxides, and secondary phosphorus and iron deficiency. Plant Physiol 143: 263–277

Anderson JM and Chow WS (2002) Structural and functional dynamics of plant photosystem II.Phil Trans R Soc Lond B 357: 1421–1430

Apel K and Hirt H (2004) Reactive oxygen species: metabolism, oxidative stress, and signaltransduction. Ann Rev Plant Biol 55: 373–399

Aro EM, Virgin I and Andersson B (1993) Photoinhibition of Photosystem II. Inactivation, protein damage and turnover. Biochim Biophys Acta 1143: 113–134

Arrigo AP (1999) Gene expression and the thiol redox state. Free Radic Biol Med 27: 936–944

Asada K (1992) Ascorbate peroxidase: a hydrogen peroxide--scavenging enzyme in plants.Physiol Plant 85: 235–241

Asada K (1999) The water--water cycle in chloroplasts: Scavenging of active oxygens anddissipation of excess photons. Ann Rev Plant Physiol Plant Mol Biol 50: 601–639

Asada K (2006) Production and scavenging of reactive oxygen species in chloroplasts and theirfunctions. Plant Physiol 141: 391–396

Baier M and Dietz KJ (1999) Protective function of chloroplast 2-cysteine peroxiredoxin inphotosynthesis. Evidence from transgenic Arabidopsis. Plant Physiol 119: 1407–1414

Baier M, Noctor G, Foyer CH and Dietz KJ (2000) Antisense suppression of 2-cysteineperoxiredoxin in Arabidopsis specifically enhances the activities and expression ofenzymes associated with ascorbate metabolism but not glutathione metabolism. PlantPhysiol 124: 823–832

Balmer Y, Vensel WH, Tanaka CK, Hurkman WJ, Gelhaye E, Rouhier N, Jacquot JP, ManieriW, Schürmann P, Droux M and Buchanan BB (2004) Thioredoxin links redox to theregulation of fundamental processes of plant mitochondria. Proc Natl Acad Sci USA 101: 2642–2647

Baroli I, Gutman BL, Ledford HK, Shin JW, Chin BL, Havaux M and Niyogi KK (2004) Photo-oxidative stress in a xanthophyll--deficient mutant of *Chlamydomonas*. J Biol Chem 279:6337–6344

Baroli I and Niyogi KK (2000) Molecular genetics of xanthophyll--dependent photoprotection ingreen algae and plants. Phil Trans R Soc Lond B 355: 1385–1393

Batish DR, Singh HP, Setia N, Kaur S and Kohli RK (2006) 2-Benzoxazolinone (BOA) inducedoxidative stress, lipid peroxidation and changes in some antioxidant enzyme activities inmung bean (*Phaseolus aureus*). Plant Physiol Biochem 44: 819–827

Beck E, Burkert A and Hofmann M (1983) Uptake of L-ascorbate by intact spinachchloroplasts. Plant Physiol 73: 41–45

Blokhina O, Virolainen E and Fagerstedt KV (2003) Antioxidants, oxidative damage and oxygendeprivation stress: a review. Ann Bot 91: 179–194

Bolwell GP and Wojtaszek P (1997) Mechanisms for the generation of reactive oxygen speciesin plant defence – a broad perspective. Physiol Mol Plant Pathol 51: 347–366

Bratt CE, Arvidsson P-O, Carlsson M and Åkerlund H-E (1995) Regulation of violaxanthin de-epoxidase activity by pH and ascorbate concentration. Photosynth Res 45: 169–175

Broin M, Cuine S, Eymery F and Rey P (2002) The plastidic 2-cysteine peroxiredoxin is a targetfor a thioredoxin involved in the protection of the photosynthetic apparatus againstoxidative damage. Plant Cell 14: 1417–1432

Buchanan BB and Balmer Y (2005) Redox regulation: A broadening horizon. Ann Rev PlantBiol 56: 187–220

Burton GW and Traber MG (1990) Vitamin E – Antioxidant activity, biokinetics, andbioavailability. Ann Rev Nutr 10: 357–382

Carmel-Harel O and Storz G (2000) Roles of the glutathione- and thioredoxin-dependentreduction systems in the *Escherichia coli* and *Saccharomyces cerevisiae* responses tooxidative stress. Ann Rev Microbiol 54: 439–461

Chamnongpol S, Willekens H, Moeder W, Langebartels C, Sandermann H, Van Montagu A,Inzé D and Van Camp W (1998) Defense activation and enhanced pathogen toleranceinduced by H_2O_2 in transgenic tobacco. Proc Natl Acad Sci USA 95: 5818–5823

Charles SA and Halliwell B (1980) Effect of hydrogen-peroxide on spinach (*Spinacia oleracea*)chloroplast fructose biphosphatase. Biochem J 189: 373–376

Cheng NH (2008) AtGRX4, an Arabidopsis chloroplastic monothiol glutaredoxin, is able tosuppress yeast *grx5* mutant phenotypes and respond to oxidative stress. FEBS Lett 582: 848–854

Cheng NH, Liu JZ, Brock A, Nelsono RS and Hirschi KD (2006) AtGRXcp, an *Arabidopsis* chloroplastic glutaredoxin, is critical for protection against protein oxidative damage. J Biol Chem 281: 26280–26288

Cobbett CS, May MJ, Howden R and Rolls B (1998) The glutathione-deficient, cadmium-sensitive mutant, *cad2-1*, of *Arabidopsis thaliana* is deficient in γ-glutamylcysteine synthetase. Plant J 16: 73–78

Cogdell RJ and Frank HA (1987) How carotenoids function in photosynthetic bacteria. Biochim Biophys Acta 895: 63–79

Conklin PL, Norris SR, Wheeler GL, Williams EH, Smirnoff N and Last RL (1999) Genetic evidence for the role of GDP-mannose in plant ascorbic acid (vitamin C) biosynthesis. Proc Natl Acad Sci USA 96: 4198–4203

Conklin PL, Pallanca JE, Last RL and Smirnoff N (1997) L-ascorbic acid metabolism in the ascorbate-deficient Arabidopsis mutant *vtc1*. Plant Physiology 115: 1277–1285

Conklin PL, Williams EH and Last RL (1996) Environmental stress sensitivity of an ascorbic acid-deficient *Arabidopsis* mutant. Proc Natl Acad Sci USA 93: 9970–9974

Copley SD, Novak WRP and Babbitt PC (2004) Divergence of function in the thioredoxin fold superfamily: Evidence for evolution of peroxiredoxins from a thioredoxin-like ancestor. Biochemistry 43: 13981–13995

Creissen G, Firmin J, Fryer M, Kular B, Leyland N, Reynolds H, Pastori G, Wellburn F, Baker N, Wellburn A and Mullineaux P (1999) Elevated glutathione biosynthetic capacity in the chloroplasts of transgenic tobacco plants paradoxically causes increased oxidative stress. Plant Cell 11: 1277–1291

Dai S, Schwendtmayer C, Schürmann P, Ramaswamy S and Eklund H (2000) Redox signaling in chloroplasts: Cleavage of disulfides by an iron-sulfur cluster. Science 287: 655–658

Davison PA, Hunter CN and Horton P (2002) Overexpression of β-carotene hydroxylase enhances stress tolerance in *Arabidopsis*. Nature 418: 203–206

Dayer R, Fischer BB, Eggen RIL and Lemaire SD (2008) The peroxiredoxin and glutathione peroxidase families in *Chlamydomonas reinhardtii*. Genetics 179: 41–57

Demmig-Adams B (1990) Carotenoids and photoprotection in plants: A role for the xanthophyll zeaxanthin. Biochim Biophys Acta 1020: 1–24

Dietz K-J (2003) Plant peroxiredoxins. Ann Rev Plant Biol 54: 93–107

Dietz K-J, Horling F, König J and Baier M (2002) The function of the chloroplast 2-cysteine peroxiredoxin in peroxide detoxification and its regulation. J Exp Bot 53: 1321–1329

Dietz K-J, Jacob S, Oelze ML, Laxa M, Tognetti V, de Miranda SMN, Baier M and Finkemeier I (2006) The function of peroxiredoxins in plant organelle redox metabolism. J Exp Bot 57: 1697–1709

Dos Santos CV and Rey P (2006) Plant thioredoxins are key actors in the oxidative stress response. Trends Plant Sci 11: 329–334

Elrad D, Niyogi KK and Grossman AR (2002) A major light-harvesting polypeptide of photosystem II functions in thermal dissipation. Plant Cell 14: 1801–1816

Filomeni G, Rotilio G and Ciriolo MR (2002) Cell signalling and the glutathione redox system. Biochem Pharmacol 64: 1057–1064

Fischer BB, Eggen RIL, Trebst A and Krieger-Liszkay A (2006) The glutathione peroxidase homologous gene *Gpxh* in *Chlamydomonas reinhardtii* is upregulated by singlet oxygen produced in photosystem II. Planta 223: 583–590

Foyer CH (1997) Photosynthetic oxygen metabolism. In: Scandalios J (ed) Oxidative stress and the molecular biology of antioxidant defenses, pp. 587–621. Cold Spring Harbor Laboratory Press, Plainview, NY

Foyer CH and Lelandais M (1996) A comparison of the relative rates of transport of ascorbate and glucose across the thylakoid, chloroplast and plasmalemma membranes of pea leaf mesophyll cells. J Plant Physiol 148: 391–398

Foyer CH and Noctor G (2000) Oxygen processing in photosynthesis: regulation and signalling. New Phytol 146: 359–388

Fukuzawa K, Tokumura A, Ouchi S and Tsukatani H (1982) Antioxidant activities of tocopherols on Fe^{2+}-ascorbate-induced lipid-peroxidation in lecithin liposomes. Lipids 17: 511–513

Gaber A, Yoshimura K, Tamoi M, Takeda T, Nakano Y and Shigeoka S (2004) Induction and functional analysis of two reduced nicotinamide adenine dinucleotide phosphate-dependent glutathione peroxidase-like proteins in *Synechocystis* PCC 6803 during the progression of oxidative stress. Plant Physiol 136: 2855–2861

Gaber A, Yoshimura K, Yamamoto T, Yabuta Y, Takeda T, Miyasaka H, Nakano Y and Shigeoka S (2006) Glutathione peroxidase-like protein of *Synechocystis* PCC 6803 confers tolerance to oxidative and environmental stresses in transgenic *Arabidopsis*. Physiol Plant 128: 251–262

Giacomelli L, Masi A, Ripoli DR, Lee MJ and van Wijk KJ (2007) *Arabidopsis thaliana* deficient in two chloroplast ascorbate peroxidases shows accelerated light-induced necrosis when levels of cellular ascorbate are low. Plant Mol Biol 65: 627–644

Grant CM, MacIver FH and Dawes IW (1996) Glutathione is an essential metabolite required for resistance to oxidative stress in the yeast *Saccharomyces cerevisiae*. Curr Genet 29: 511–515

Gupta AS, Heinen JL, Holaday AS, Burke JJ and Allen RD (1993) Increased resistance to oxidative stress in transgenic plants that overexpress chloroplastic Cu/Zn superoxide dismutase. Proc Natl Acad Sci USA 90: 1629–1633

Havaux M, Bonfils J-P, Lütz C and Niyogi KK (2000) Photodamage of the photosynthetic apparatus and its dependence on the leaf developmental stage in the *npq1* *Arabidopsis* mutant deficient in the xanthophyll cycle enzyme violaxanthin de-epoxidase. Plant Physiol 124: 273–284

Havaux M, Eymery F, Porfirova S, Rey P and Dörmann P (2005) Vitamin E protects against photoinhibition and photooxidative stress in *Arabidopsis thaliana*. Plant Cell 17: 3451–3469

Hell R and Bergmann L (1990) γ-Glutamylcysteine synthetase in higher plants: catalytic properties and subcellular localization. Planta 180: 603–612

Hofius D and Sonnewald U (2003) Vitamin E biosynthesis: biochemistry meets cell biology. Trends Plant Sci 8: 6–8

Horemans N, Foyer CH and Asard H (2000) Transport and action of ascorbate at the plant plasma membrane. Trends Plant Sci 5: 263–267

Horling F, Lamkemeyer P, König J, Finkemeier I, Kandlbinder A, Baier M and Dietz K-J (2003) Divergent light-, ascorbate-, and oxidative stress-dependent regulation of expression of the peroxiredoxin gene family in Arabidopsis. Plant Physiol 131: 317–325

Imsande J (1999) Iron-sulfur clusters: Formation, perturbation, and physiological functions. Plant Physiol Biochem 37: 87–97

Ito H, Iwabuchi M and Ogawa K (2003) The sugar-metabolic enzymes aldolase and triose-phosphate isomerase are targets of glutathionylation in *Arabidopsis thaliana*: detection using biotinylated glutathione. Plant Cell Physiol 44: 655–660

Ivanov B, Asada K, Kramer DM and Edwards G (2005) Characterization of photosynthetic electron transport in bundle sheath cells of maize. I. Ascorbate effectively stimulates cyclic electron flow around PSI. Planta 220: 572–581

Jacquot JP, Gelhaye E, Rouhier N, Corbier C, Didierjean C and Aubry A (2002) Thioredoxins and related proteins in photosynthetic organisms: molecular basis for thiol dependent regulation. Biochem Pharmacol 64: 1065–1069

Kaiser W (1976) Effect of hydrogen peroxide on CO_2 fixation of isolated intact chloroplasts. BiochimBiophys Acta 440: 476–482

Kaiser WM (1979) Reversible inhibition of the calvin cycle and activation of oxidative pentose-phosphate cycle in isolated intact chloroplasts by hydrogen peroxide. Planta 145: 377–382

Kamal-Eldin A and Appelqvist LA (1996) The chemistry and antioxidant properties of tocopherols and tocotrienols. Lipids 31: 671–701

Kangasjärvi S, Lepistö A, Hännikäinen K, Piippo M, Luomala E-M, Aro E-M and Rintamäki E (2008) Diverse roles for chloroplast stromal and thylakoid-bound ascorbate peroxidases in plant stress responses. Biochem J 412: 275–285

Kanwischer M, Porfirova S, Bergmüller E and Dörmann P (2005) Alterations in tocopherol cyclase activity in transgenic and mutant plants of *Arabidopsis* affect tocopherol content, tocopherol composition, and oxidative stress. Plant Physiology 137: 713–723

Kasahara M, Kagawa T, Oikawa K, Suetsugu N, Miyao M and Wada M (2002) Chloroplast avoidance movement reduces photodamage in plants. Nature 420: 829–832

Keren N, Gong H and Ohad I (1995) Oscillations of reaction center II: D1 protein degradation in vivo induced by repetitive light flashes. J Biol Chem 270: 806–814

Klughammer B, Baier M and Dietz K-J (1998) Inactivation by gene disruption of 2-cysteine-peroxiredoxin in *Synechocystis* sp. PCC 6803 leads to increased stress sensitivity. Physiol Plant 104: 699–706

Krieger-Liszkay A (2005) Singlet oxygen production in photosynthesis. J Exp Bot 56: 337–346

Kühn H and Borchert A (2002) Regulation of enzymatic lipid peroxidation: The interplay of peroxidizing and peroxide reducing enzymes. Free Radic Biol Med 33: 154–172

Kurepa J, Hérouart D, Van Montagu M and Inzé D (1997) Differential expression of CuZn- and Fe-superoxide dismutase genes of tobacco during development, oxidative stress, and hormonal treatments. Plant Cell Physiol 38: 463–470

Larkindale J, Hall JD, Knight MR and Vierling E (2005) Heat stress phenotypes of Arabidopsis mutants implicate multiple signaling pathways in the acquisition of thermotolerance. Plant Physiol 138: 882–897

Ledford HK, Baroli I, Shin JW, Fischer BB, Eggen RIL and Niyogi KK (2004) Comparative profiling of lipid-soluble antioxidants and transcripts reveals two phases of photooxidative stress in a xanthophyll-deficient mutant of *Chlamydomonas reinhardtii*. Mol Genet Genomics 272: 470–479

Ledford HK, Chin BL and Niyogi KK (2007) Acclimation to singlet oxygen stress in *Chlamydomonas reinhardtii*. Eukaryot Cell 6: 919–930

Lemaire SD (2004) The glutaredoxin family in oxygenic photosynthetic organisms. Photosynth Res 79: 305–318

Lemaire SD, Guillon B, Le Marechal P, Keryer E, Miginiac-Maslow M and Decottignies P (2004) New thioredoxin targets in the unicellular photosynthetic eukaryote *Chlamydomonas reinhardtii*. Proc Natl Acad Sci USA 101: 7475–7480

Lemaire S, Michelet L, Zaffagnini M, Massot V and Issakidis-Bourguet E (2007) Thioredoxins in chloroplasts. Curr Genet 51: 343–365

Lemaire SD and Miginiac-Maslow M (2004) The thioredoxin superfamily in *Chlamydomonas reinhardtii*. Photosynth Res 82: 203–220

Lesser MP and Stochaj WR (1990) Photoadaptation and protection against active forms of oxygen in the symbiotic procaryote *Prochloron* sp. and its ascidian host. Appl Environ Microbiol 56: 1530–1535

Levine A, Tenhaken R, Dixon R and Lamb C (1994) H_2O_2 from the oxidative burst orchestrates the plant hypersensitive disease resistance response. Cell 79: 583–593

Maeda H, Sakuragi Y, Bryant DA and DellaPenna D (2005) Tocopherols protect *Synechocystis* sp strain PCC 6803 from lipid peroxidation. Plant Physiol 138: 1422–1435

Malanga G, Kozak RG and Puntarulo S (1999) *N*-acetylcysteine-dependent protection against UV-B damage in two photosynthetic organisms. Plant Sci 141: 129–137

Mallick N and Mohn FH (2000) Reactive oxygen species: response of algal cells. J Plant Physiol 157: 183–193

Mallick N and Rai LC (1999) Response of the antioxidant systems of the nitrogen fixing cyanobacterium *Anabaena doliolum* to copper. J Plant Physiol 155: 146–149

Melis A (1999) Photosystem-II damage and repair cycle in chloroplasts: what modulates the rate of photodamage *in vivo*? Trends Plant Sci 4: 130–135

Michelet L, Zaffagnini M, Massot V, Keryer E, Vanacker H, Miginiac-Maslow M, Issakidis-Bourguet E and Lemaire S (2006) Thioredoxins, glutaredoxins, and glutathionylation: new crosstalks to explore. Photosynth Res 89: 225–245

Mittler R (2002) Oxidative stress, antioxidants and stress tolerance. Trends Plant Sci 7: 405–410

Morell S, Follmann H, DeTullio M and Häberlein I (1997) Dehydroascorbate and dehydroascorbate reductase are phantom indicators of oxidative stress in plants. FEBS Lett 414: 567–570

Müller P, Li X-P and Niyogi KK (2001) Non-photochemical quenching. A response to excess light energy. Plant Physiol 125: 1558–1566

Müller-Moulé P, Conklin PL and Niyogi KK (2002) Ascorbate deficiency can limit violaxanthin de-epoxidase activity *in vivo*. Plant Physiol 128: 970–977

Müller-Moulé P, Golan T and Niyogi KK (2004) Ascorbate-deficient mutants of Arabidopsis grow in high light despite chronic photooxidative stress. Plant Physiol 134: 1163–1172

Munné-Bosch S and Alegre L (2002) The function of tocopherols and tocotrienols in plants. Crit Rev Plant Sci 21: 31–57

Nishiyama Y, Allakhverdiev SI and Murata N (2006) A new paradigm for the action of reactive oxygen species in the photoinhibition of photosystem II. Biochim Biophys Acta 1757: 742–749

Niyogi KK (2000) Safety valves for photosynthesis. Curr Op Plant Biol 3: 455–460

Niyogi KK, Björkman O and Grossman AR (1997) The roles of specific xanthophylls in photoprotection. Proc Natl Acad Sci USA 94: 14162–14167

Niyogi KK, Grossman AR and Björkman O (1998) Arabidopsis mutants define a central role for the xanthophyll cycle in the regulation of photosynthetic energy conversion. Plant Cell 10: 1121–1134

Niyogi KK, Shih C, Chow, WS, Pogson BJ, DellaPenna D and Björkman O (2001) Photoprotection in a zeaxanthin- and lutein-deficient double mutant of *Arabidopsis*. Photosynth Res 67: 139-145

Noctor G, Arisi ACM, Jouanin L, Kunert KJ, Rennenberg H and Foyer CH (1998) Glutathione: biosynthesis, metabolism and relationship to stress tolerance explored in transformed plants. J Exp Bot 49: 623–647

Noctor G and Foyer CH (1998) Ascorbate and glutathione: Keeping active oxygen under control. Ann Rev Plant Physiol Plant Mol Biol 49: 249–279

Noctor G, Gomez L, Vanacker H and Foyer CH (2002) Interactions between biosynthesis, compartmentation and transport in the control of glutathione homeostasis and signalling. J Exp Bot 53: 1283–1304

Ogawa K, Kanematsu S, Takabe K and Asada K (1995) Attachment of CuZn-superoxide dismutase to thylakoid membranes at the site of superoxide generation (PSI) in spinach chloroplasts: detection by immunogold labeling after rapid freezing and substitution method. Plant Cell Physiol 36: 565–573

Okamoto OK, Asano CS, Aidar E and Colepicolo P (1996) Effects of cadmium on growth and superoxide dismutase activity of the marine microalga *Tetraselmis gracilis* (Prasinophyceae). J Phycol 32: 74–79

Okamoto OK and Colepicolo P (1998) Response of superoxide dismutase to pollutant metal stress in the marine dinoflagellate *Gonyaulax polyedra*. Comp Biochem Physiol C Pharmacol Toxicol Endocrinol 119: 67–73

op den Camp RGL, Przybyla D, Ochsenbein C, Laloi C, Kim CH, Danon A, Wagner D, Hideg E, Göbel C, Feussner I, Nater M and Apel K (2003) Rapid induction of distinct stress responses after the release of singlet oxygen in Arabidopsis. Plant Cell 15: 2320–2332

Parisy V, Poinssot B, Owsianowski L, Buchala A, Glazebrook J and Mauch F (2007) Identification of PAD2 as a γ-glutamylcysteine synthetase highlights the importance of glutathione in disease resistance of Arabidopsis. Plant J 49: 159–172

Paterson MJ, Christiansen O, Jensen F and Ogilby PR (2006) Overview of theoretical and computational methods applied to the oxygen-organic molecule photosystem. Photochem Photobiol 82: 1136–1160

Pogson BJ, Niyogi KK, Björkman O and DellaPenna D (1998) Altered xanthophyll compositions adversely affect chlorophyll accumulation and nonphotochemical quenching in *Arabidopsis* mutants. Proc Natl Acad Sci USA 95: 13324–13329

Porfirova S, Bergmüller E, Tropf S, Lemke R and Dörmann P (2002) Isolation of an *Arabidopsis* mutant lacking vitamin E and identification of a cyclase essential for all tocopherol biosynthesis. Proc Natl Acad Sci USA 99: 12495–12500

Prasad TK, Anderson MD and Stewart CR (1994) Acclimation, hydrogen peroxide, and abscisic acid protect mitochondria against irreversible chilling injury in maize seedlings. Plant Physiol 105: 619–627

Rady AA, Elsheekh MM and Matkovics B (1994) Temperature shift-induced changes in the antioxidant enzyme system of cyanobacterium *Synechocystis* PCC 6803. Int J Biochem 26: 433–435

Redmond RW and Kochevar IE (2006) Spatially resolved cellular responses to singlet oxygen. Photochem Photobiol 82: 1178–1186

Rouhier N and Jacquot J-P (2005) The plant multigenic family of thiol peroxidases. Free Radic Biol Med 38: 1413–1421

Rouhier N and Jacquot JP (2002) Plant peroxiredoxins: alternative hydroperoxide scavenging enzymes. Photosynth Res 74: 259–268

Rouhier N, Lemaire SD and Jacquot JP (2008) The role of glutathione in photosynthetic organisms: emerging functions for glutaredoxins and glutathionylation. Ann Rev Plant Biol 59: 143–166

Ruggeri BA, Gray RJH, Watkins TR and Tomlins RI (1985) Effects of low-temperature acclimation and oxygen stress on tocopherol production in *Euglena gracilis*. Appl Environ Microbiol 50: 1404–1408

Sattler SE, Gilliland LU, Magallanes-Lundback M, Pollard M and DellaPenna D (2004) Vitamin E is essential for seed longevity, and for preventing lipid peroxidation during germination. Plant Cell 16: 1419–1432

Shikanai T, Takeda T, Yamauchi H, Sano S, Tomizawa K, Yokota A and Shigeoka S (1998) Inhibition of ascorbate peroxidase under oxidative stress in tobacco having bacterial catalase in chloroplasts. FEBS Lett 428: 47–51

Smirnoff N and Wheeler GL (2000) Ascorbic acid in plants: Biosynthesis and function. Crit Rev Plant Sci 19: 267–290

Sonoike K, Kamo M, Hihara Y, Hiyama T and Enami I (1997) The mechanism of the degradation of *psaB* gene product, one of the photosynthetic reaction center subunits of photosystem I, upon photoinhibition. Photosynth Res 53: 55–63

Sonoike K, Terashima I, Iwaki M and Itoh S (1995) Destruction of photosystem I iron-sulfur centers in leaves of *Cucumis sativus* by weak illumination at chilling temperatures. FEBS Lett 362: 235–238

Stafford JL, Neumann NF and Belosevic M (2002) Macrophage-mediated innate host defense against protozoan parasites. Crit Rev Microbiol 28: 187–248

Takahashi S and Murata N (2008) How do environmental stresses accelerate photoinhibition? Trends Plant Sci 13: 178–182

Tanaka K, Otsubo T and Kondo N (1982) Participation of hydrogen-peroxide in the inactivation of calvin cycle SH enzymes in SO_2-fumigated spinach leaves. Plant Cell Physiol 23: 1009–1018

Tarantino D, Vannini C, Bracale M, Campa M, Soave C and Murgia I (2005) Antisense reduction of thylakoidal ascorbate peroxidase in Arabidopsis enhances paraquat-induced photooxidative stress and nitric oxide-induced cell death. Planta 221: 757–765

Telfer A (2002) What is beta-carotene doing in the photosystem II reaction centre? Phil Trans R Soc Lond B 357: 1431–1439

Trebst A, Depka B and Hollander-Czytko H (2002) A specific role for tocopherol and of chemical singlet oxygen quenchers in the maintenance of photosystem II structure and function in *Chlamydomonas reinhardtii*. FEBS Lett 516: 156–160

Ursini F, Maiorino M, Brigeliusflohe R, Aumann KD, Roveri A, Schomburg D and Flohe L (1995) Diversity of glutathione peroxidases. *In* Biothiols, Pt B, Vol 252, pp. 38–53

Vanacker H, Carver TLW and Foyer CH (2000) Early H_2O_2 accumulation in mesophyll cells leads to induction of glutathione during the hyper-sensitive response in the barley-powdery mildew interaction. Plant Physiol 123: 1289–1300

Vartak V and Bhargava S (1999) Photosynthetic performance and antioxidant metabolism in a paraquat-resistant mutant of *Chlamydomonas reinhardtii*. Pestic Biochem Physiol 64: 9–15

Vranová E, Inzé D and Van Breusegem F (2002) Signal transduction during oxidative stress. J Exp Bot 53: 1227–1236

Wagner D, Przybyla D, op den Camp R, Kim C, Landgraf F, Lee KP, Würsch M, Laloi C, Nater M, Hideg E and Apel K (2004) The genetic basis of singlet oxygen-induced stress responses of *Arabidopsis thaliana*. Science 306: 1183–1185

Willekens H, Chamnongpol S, Davey M, Schraudner M, Langebartels C, Van Montagu M, Inzé D and Van Camp W (1997) Catalase is a sink for H_2O_2 and is indispensable for stress defence in C-3 plants. EMBO J 16: 4806–4816

Wolosiuk RA and Buchanan BB (1977) Thioredoxin and glutathione regulate photosynthesis in chloroplasts. Nature 266: 565–567

Wong JH, Balmer Y, Cai N, Tanaka CK, Vensel WH, Hurkman WJ and Buchanan BB (2003) Unraveling thioredoxin-linked metabolic processes of cereal starchy endosperm using proteomics. FEBS Lett 547: 151–156

Wrona M, Manowska M and Sarna T (2004) Zeaxanthin in combination with ascorbic acid or alpha-tocopherol protects ARPE-19 cells against photosensitized peroxidation of lipids. Free Radic Biol Med 36: 1094–1101

Xiang C, Werner BL, Christensen ELM and Oliver DJ (2001) The biological functions of glutathione revisited in Arabidopsis transgenic plants with altered glutathione levels. Plant Physiol 126: 564–574

Yabuta Y, Motoki T, Yoshimura K, Takeda T, Ishikawa T and Shigeoka S (2002) Thylakoid membrane-bound ascorbate peroxidase is a limiting factor of antioxidative systems under photo-oxidative stress. Plant J 32: 915–925

Yamamoto HY, Bugos RC and Hieber AD (1999) Biochemistry and molecular biology of the xanthophyll cycle. In: Frank HA, Young AJ, Britton G and Cogdel RJ (eds) The Photochemistry of Carotenoids, pp. 293–303. Kluwer Academic Publishers, Dordrecht

Yamazaki D, Motohashi K, Kasama T, Hara Y and Hisabori T (2004) Target proteins of the cytosolic thioredoxins in *Arabidopsis thaliana*. Plant Cell Physiol 45: 18–27

Yoshimura K, Miyao K, Gaber A, Takeda T, Kanaboshi H, Miyasaka H and Shigeoka S (2004) Enhancement of stress tolerance in transgenic tobacco plants overexpressing Chlamydomonas glutathione peroxidase in chloroplasts or cytosol. Plant J 37: 21–33

Zaffagnini M, Michelet L, Massot V, Trost P and Lemaire SD (2008) Biochemical characterization of glutaredoxins from *Chlamydomonas reinhardtii* reveals the unique properties of a chloroplastic CGFS-type glutaredoxin. J Biol Chem 283: 8868–8876

Chapter 25

Singlet Oxygen-Induced Oxidative Stress in Plants

Baishnab C. Tripathy* and Gopal K. Pattanayak†
School of Life Sciences, Jawaharlal Nehru University, New Delhi-110067, India

Summary

In the present oxidizing environment, metabolic processes such as photosynthesis and respiration are mostly responsible for the generation of reactive oxygen species (ROS) including singlet oxygen ($^1O_2$) in chloroplasts, mitochondria, peroxisomes and other sites of the plant cell. Imbalance between ROS production and their detoxification by enzymatic and non-enzymatic means results in higher net ROS formation and consequent oxidative damage which ultimately causes cell death. $^1O_2$ often acts as a signaling molecule involved in pathogen defense responses such as hypersensitive reactions and systemic acquired resistance, stress hormone production, acclimation and programmed cell death. In plants, $^1O_2$ is mainly produced by the chlorophyll (Chl) and its tetrapyrrole metabolic intermediates in the presence of light via type II photosensitization reactions. Diphenyl ethers cause the accumulation of protoporphyrin IX that generates $^1O_2$ via photosensitization reactions and causes necrotic spots and cell death by destroying the plasma membrane. Similarly, exogenous application of 5-aminolevulinc acid (ALA) results in the

*Author for Correspondence, e-mail: bctripathy@mail.jnu.ac.in
† Present Address: Department of Molecular Genetics and Cell Biology, University of Chicago, Chicago, IL 60637, USA

Constantin A. Rebeiz et al. (eds.), The Chloroplast: Basics and Applications, pp. 397–412.

accumulation of other Chl biosynthesis intermediates such as protochlorophyllide which generates $^1O_2$ in the light causing cell death. Genetic mutants that are deficient in Chl biosynthetic enzymes or regulatory proteins accumulate excess tetrapyrroles leading to excess $^1O_2$ generation and cell death. Similarly, plants that are genetically deficient in Chl degradation enzymes accumulate excess Chl catabolic products; this generates $^1O_2$, via photosensitization reactions, and causes cell death. As there is no enzymatic means available to detoxify $^1O_2$, it is essential to minimize its production rather than detoxifying it after it is generated. In this article the mechanisms of generation of $^1O_2$, its detoxification, its mode of cellular damage and ways to minimize its destructive potential and programmed cell death are discussed.

I Introduction

The activation or reduction of oxygen gives rise to reactive oxygen species (ROS) that includes the singlet oxygen ($^1O_2$), superoxide (O_2^-), hydrogen peroxide (H_2O_2) and hydroxyl radical ($HO^·$). Plants and other living organisms in the oxidizing environment constantly produce ROS in chloroplasts, mitochondria, peroxisomes and other sites of the cell because of their metabolic processes such as photosynthesis and respiration. The generation of ROS in plants is triggered by different kinds of environmental stresses, such as high light, high or low temperature, salinity, drought, nutrient deficiency and pathogen attack. (See Pareek et al. (eds), 2010 for a book on abiotic stress in plants.) Plants and other living organisms have evolved a host of anti-oxidants and anti-oxidative enzymes and other small molecules to harmlessly dissipate ROS. Imbalance between ROS production and their detoxification by enzymatic and non-enzymatic reactions causes oxidative stress. As a result of higher net ROS formation, there is photooxidative damage to DNA, Proteins and lipids and ultimately cell death. Recent studies also indicate that ROS can act as signaling molecules involved in growth and developmental processes, pathogen defense responses such as hypersensitive reaction and systemic acquired resistance, stress hormone production, acclimation and programmed cell death (Apel and Hirt, 2004).

Abbreviations: ACD – accelerated cell death; AF-Na – acifluorfen-sodium; ALA – 5-aminolevulinic acid; Chl – chlorophyll; DCMU – (3-(3,4-dichlorophenyl)-1,1-dimethylurea); DPE – diphenyl ether; exe – executer; F0 – minimum chlorophyll a fluorescence; flu – fluorescence; Fm – maximum chlorophyll a fluorescence; Fv – variable chlorophyll a fluorescence; GluTR – glutamyl-tRNA reductase; LHC – light harvesting complex; LHCP – light harvesting chlorophyll protein; lin2 – lesion initiation 2; LLS1 – lethal leaf spot 1; MDA – malondialdehyde; MPE – Mg-protoporphyrin IX monomethylester; $^1O_2$ – singlet oxygen; $^3O_2$ – triplet state oxygen; PAO – pheophorbide a oxygenase; Pchlide – protochlorophyllide; POR – protochlorophyllide oxido-reductase; PR – Pathogen related; PRL – pleiotropic response locus 1; Proto – IX protoporphyrin IX; PSI – photosystem I; PSII – photosystem II; RC1 – reaction center I; RCCR – red chlorophyll catabolite reductase; RNO – p-nitroso-N,N′-dimethylaniline; ROS – reactive oxygen species; Sens – sensitizers; TL – thermoluminescence

II Formation of Singlet Oxygen in Plants

Oxygen in its ground state is not very reactive and does not have any deleterious effect. The ground state molecular oxygen is a triplet state ($^3O_2$) and in fact a biradical, as it has two unpaired electrons. Its two unpaired electrons have parallel spins (↑↑) that do not allow them to react with most molecules. However, if the triplet oxygen absorbs sufficient energy, the spin restriction is removed and the spin of one of its unpaired electrons is reversed. As a result there is generation of singlet oxygen ($^1O_2$), whose outermost pair of electrons has antiparallel spins (↑↓). Singlet oxygen molecules are also formed when superoxide radicals interact with hydroxyl radicals. The excitation energy required to produce $^1O_2$ from the triplet oxygen is 94 kJ mol^{-1}. It has a short half-life of about 200 ns in cells with a possible diffusion distance of about 270 nm and could even diffuse out of the chloroplast into the cytosol (Skovsen et al., 2005).

In plants, $^1O_2$ is mainly produced by the chlorophyll (Chl) and its tetrapyrrole metabolic intermediates in the presence of light (Fig. 1). Chl, the most abundant pigment in land plants, is

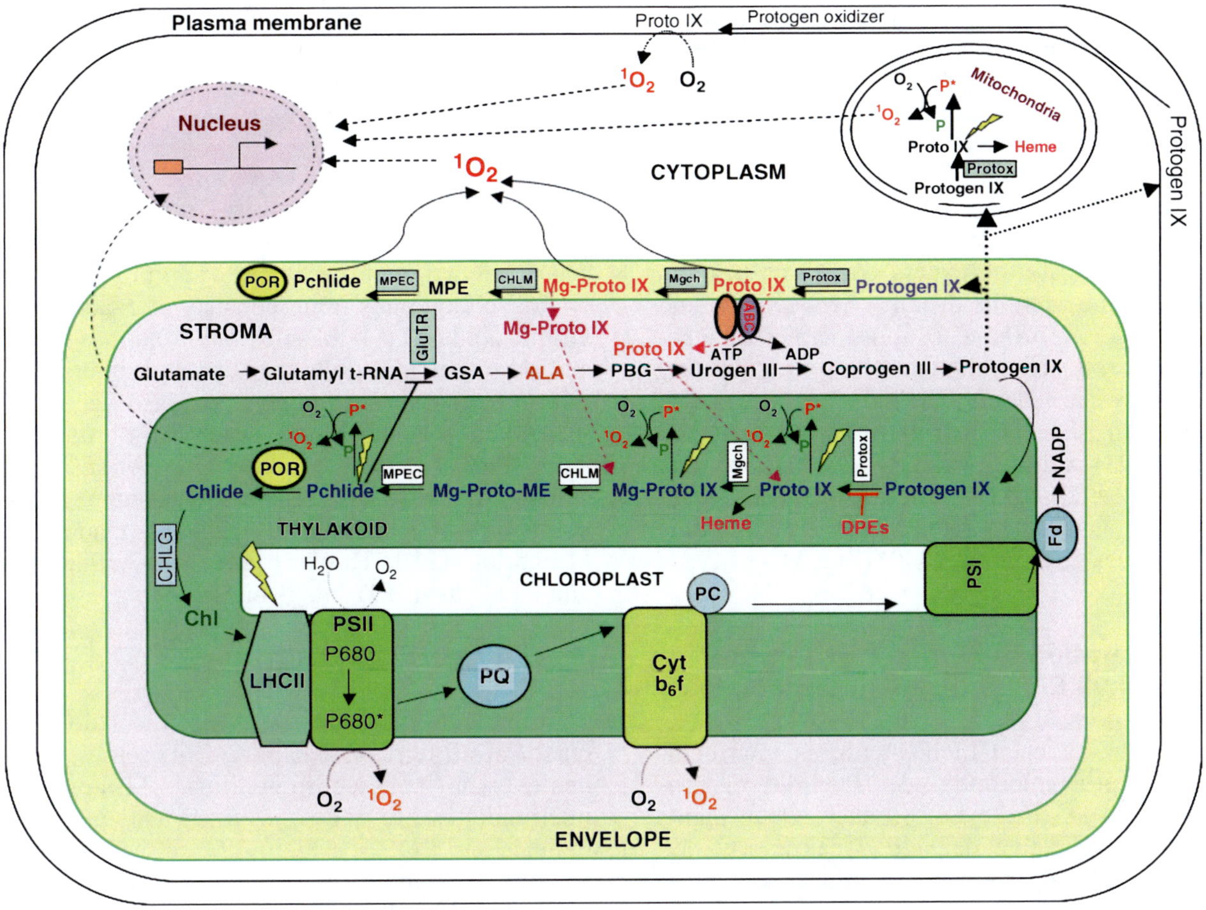

Fig. 1. Intracellular generation of $^1O_2$ from chlorophyll and heme biosynthesis intermediates and photosystems.

the main light absorbing pigment and is present both in the light harvesting complex (LHC) and the photosynthetic reaction centers. The excited state of these are long lived and allow conversion of the excitation energy to an electrochemical potential via charge separation. Inefficient transfer of energy results in the generation of triplet state Chl that reacts with triplet oxygen to produce the highly reactive $^1O_2$ (Fig. 1). In the light harvesting complex, the $^1O_2$ is quenched by the carotenoids. (For further information on the photochemistry of carotenoids, see Frank et al. (eds), 1999.)

Singlet oxygen ($^1O_2$) is produced near the reaction centers of the photosystems (Fig. 1). With increase in light intensity, i.e., from the early morning to noon, light absorption by leaves increases almost linearly. However, the rate of photosynthesis reaches its maximum value much before the linear increase in light absorption ceases. Therefore plants end up absorbing more light than they could utilize in photosynthesis. This results in the over excitation of the photosynthetic apparatus. In the presence of excess light energy, the Q_A and Q_B, the first and second plastoquinone electron acceptors of Photosystem II (PS II) in the electron transport chain, are over reduced (Barber and Andersson, 1992) and because of that, charge separation cannot be completed between P680 and pheophytin. As a result the triplet state of the reaction center Chl P680 (^{3}P680) is favored (Aro et al., 1993; Ohad et al., 1994) leading to the formation of $^1O_2$ (Foote et al., 1984). Normally when excess light is absorbed, an alternative dissipating pathway is activated that safely returns ^{1}Chl* to its ground state before it is converted to ^{3}Chl*. The excitation energy of excess ^{1}Chl* is dissipated by zeaxanthin or other carotenoids as heat in Chl and/or carotenoid binding protein complexes (Baroli and Niyogi, 2000; Baroli et al., 2003, 2004; Davison et al., 2002; Pogson and Rissler, 2000). The carotenoids, which quench the excited state of Chl, must be in close proximity with triplet Chl i.e., within a maximum distance of 3.6 Å. In this spin exchange reaction, the triplet state of carotenoids is formed that can dissipate the excess energy as heat. In the reaction center, the distance between Chl and carotenoid is too large to allow triplet quenching. $^1O_2$, produced in the reaction center, directly reacts with carotenoids. The release of $^1O_2$ is also detected in isolated PS II particles (Macpherson et al., 1993) and in thylakoids (Chakraborty and Tripathy, 1992a, b; Fryer et al., 2002, Hideg et al., 1998). $^1O_2$ is also generated from the cytochrome b6f complex (Suh et al., 2000) (Fig. 1).

III Generation of Singlet Oxygen from Chlorophyll Biosynthesis Intermediates

Upon illumination, Chl biosynthesis intermediates i.e., protochlorophyllide (Pchlide) or protoporphyrin IX (Proto IX) produce $^1O_2$ in plants and cause oxidative damage (Chakraborty and Tripathy, 1992a, b; Op den Camp et al., 2003; Tripathy et al., 2007). Formation of active oxygen species, from Chl biosynthesis intermediates, was proposed by several others (Jung et al., 2008; Lermontova and Grimm, 2006; Mock and Grimm, 1997; Rebeiz et al, 1984, 1988, 1990; Shalygo et al., 1998; Tripathy and Chakraborty 1991). The site of generation of $^1O_2$ is mostly in the thylakoids. This is because Chl biosynthesis intermediates are partially hydrophobic, and consequently are loosely attached to the thylakoid membranes (Mohapatra and Tripathy, 2002, 2007; Tripathy et al., 2004). Although they are associated with the thylakoid membranes, these tetrapyrroles do not form pigment protein complexes and hence are not connected to the reaction center. Although some of the carotenoids are present in the lipid bilayer, a lot more are located in the pigment-protein complexes and they are spatially too far from Chl biosynthesis intermediates to quench their triplet states (Havaux et al., 2007; Mozzo et al., 2008). Synthesis of Chl biosynthetic intermediates are highly regulated and are not overproduced in plants. However, Chl biosynthesis intermediates that are normally present in plants are capable of producing $^1O_2$ that cause oxidative damage in high light and several other stress conditions (Chakraborty and Tripathy, 1992a).

IV Porphyrin-Generating Compounds

There are two important types of porphyrin- generating compounds. One consists of 5-aminolevulinic acid (ALA), the substrate of tetrapyrroles and the other is a group of diphenyl ethers that inhibit protoporphyrinogen oxidase activity, thereby deregulating the tetrapyrrole metabolism (Fig. 1). Besides other compounds like cercosporin, rose bengal and several other compounds could generate $^1O_2$ and O_2^- (Daub and Hangarter, 1983; Haworth and Hess, 1988, Ito and Kobayashi, 1977; Wilson, 1966).

A 5-Aminolevulinic Acid

5-Aminolevulinic acid (ALA) is the sole precursor of all tetrapyrroles, i.e., Chl, hemes, sirohemes, and phytochromobilins. Tetrapyrrole intermediates are photosensitizers and generate radicals and ROS, especially $^1O_2$ in the presence of light. So plants regulate their own tetrapyrrole biosynthesis and degradation pathway to avoid the consequence of the excess generation of ROS. The major regulatory point is at the production of the initial precursor ALA. So ALA synthesis is the rate-limiting step of the tetrapyrrole biosynthetic pathway. ALA is formed from Glutamyl-tRNA by the enzyme Glutamyl-tRNA reductase (GluTR). In Arabidopsis this enzyme has three isoforms (HEMA1, HEMA2 and HEMA3) and their expression levels are different in different plant tissues. Pchlide, which accumulates in the dark, repress the ALA synthesis by downregulating the ALA synthesis by feed-back inhibition (Stobart and Ameen-Bukhari, 1984, 1986). However, when ALA is applied externally, green plants bypass the regulatory feedback inhibition of Pchlide pool and induce excess accumulation of Mg-tetrapyrroles in dark (Granik, 1959, Rebeiz et al., 1984, 1988). When only ALA is applied externally, Pchlide is the major porphyrin that accumulates. But in ALA + Modulator (structurally related to tetrapyrrole molecule) treatments, several other types of porphyrins accumulate, depending upon the target site of the modulator (Rebeiz et al., 1991). ALA is added along with the modulator for providing the carbon skeleton to accumulate porphyrins. Eight molecules of ALA are required to form one molecule of tetrapyrrole. The mode of action of some of the modulators has been attributed to their metal chelating properties. Enzymes in the porphyrin synthesizing pathway essentially require certain metal ions for their activity, which is made unavailable. Another way by which some of these modulating chemicals may be acting is by stimulating enzyme activity, i.e. by behaving as cofactor analogs.

B Diphenyl Ethers

The diphenyl ethers (DPEs) are inhibitors of the enzyme protoporphyrinogen oxidase (protox) (Fig. 1) (Matringe et al., 1989), an enzyme that converts protoporphyrinogen IX to Proto IX. The latter is an intermediate in Chl and heme biosynthetic pathway. DPEs cause plants to overaccumulate a large quantity of Proto IX (Duke et al., 1991). Pchlide, Mg-Proto IX and Mg-Proto IX monomethylester (MPE) are also found to be elevated, but to a significantly less extent as compared to Proto IX, in tissues with inhibited protox activity. Though DPE entry into cells is light-independent, light and pigments are mandatory for their herbicidal action. Initial symptom of DPE damage is seen as water soaked spots on leaf tissue, followed by loss of leaf turgidity, bleaching and necrosis. More lipophilic DPEs like oxyfluorfen and acifluorfen-methyl have greater potency as herbicides than the more polar acifluorfen. This correlation may also be partly due to the greater ease of penetration through the cuticle by the more lipophilic molecules. DPEs can inhibit carotenoid synthesis, ATP formation, photosynthetic electron transport, and induce membrane peroxidation by causing massive accumulation of Proto IX (Kunert et al., 1987; Lydon and Duke, 1988). Superoxide radical is not of primary importance in the development of DPE toxicity (Ensminger and Hess, 1985; Kunert and Boger, 1981; Yoshimoto and Matsunaka, 1982). Oxyfluorfen added to isolated thylakoid membranes, in vitro, generates $^1O_2$ during illumination (Haworth and Hess, 1988). Treatment with the protox inhibitor acifluorfen-sodium (AF-Na) in the light induced the overaccumulation of protoporphyrinogen IX that migrates out of the chloroplast to the cytoplasm where it is oxidized to Proto IX by plasma membrane bound AF-Na-insensitive protoporphyrinogen oxidase (Duke et al., 1991, Gupta and Tripathy, 1999). A part of Proto IX so generated in the plasma membrane migrates back to the chloroplast and partitions between the cytoplasm and the chloroplast (Gupta and Tripathy, 1999, 2000). The $^1O_2$ generated by the photosensitization reaction of Proto IX creates necrotic spots and cell death by destroying the plasma membrane (Tripathy et al., 2007). The DPE herbicide lactofen induces cell death and expression of PR1, PR5 and PR10 protein in soyabean plants. The anthocyanin biosynthesis pathway genes i.e., CHS (Chalcone synthase) and CHR (Chalcone reductase) are also induced in lactofen treated samples (Graham, 2005).

V Type I and Type II Photosensitization Reactions of Tetrapyrroles

We describe type II photosensitization first (type I sensitization is described below). In type II photosensitization reaction, the sensitizer can transfer its excitation energy to a ground state oxygen molecule, resulting in $^1O_2$. This highly reactive form of oxygen can oxidize substrates of biological importance such as lipids and proteins (Foote et al., 1984) and affect metabolic pathways. ALA or DPE compounds induce over-accumulation of non-phototransformable Pchlide or Proto IX respectively. In the presence of light, these tetrapyrroles generate $^1O_2$ through type II photosensitization reaction, which ultimately destroys the plant (Chakraborty and Tripathy, 1992a, b, Tripathy et al., 2007) (Fig. 1).$^1O_2$ is quite selective and fails to react with molecules that are not enough electron rich and simply returns to the ground state.

The type I photosensitization involves hydrogen atom or electron transfer from the sensitizer to the substrate (Spikes & Bommer, 1991). The resulting free radicals can subsequently react with O_2 to produce oxidized products or other reactive species. The products are often peroxides, which can in turn breakdown to induce free radical chain auto-oxidation, leading to further oxidation in a non-photochemical step.

LOO + LH \to LOOH + L

where L and LOO^{D} are lipid free radicals; LH is an unsaturated lipid and LOOH is a lipid hydroperoxide.

Sensitizers (Sens) can produce superoxide radical ($O_2^{-\cdot}$) by undergoing electron transfer processes with the substrate or O_2, as shown below.

$$^3Sens + Subs \rightarrow Sens^- + Subs_{ox}$$

$$Sens^- + O_2 \rightarrow Sens + O_2^-$$

or

$$^3Sens + O_2 \rightarrow Sens^+ + O_2^-$$

These reactions produce $O_2^{-\cdot}$, which can subsequently give rise to the highly reactive hydroxyl radical ($OH^{\cdot}$). The hydroxyl radicals thus produced can react with organic molecules in a variety of ways or can initiate radical chain auto-oxidation.

VI Intracellular Destruction of Singlet Oxygen

The most efficient mechanism of detoxification of $^1O_2$ in plants involves carotenoids. The carotenoids reach the triplet excited state by absorbing the excess energy of $^1O_2$ which returns it to its triplet ground state ($^3O_2$). They finally dissipates the excess acquired energy as heat (Edge et al., 1997). The physical quencher has to be lipid soluble and needs to be very close to the photosensitizer. Carotenoids, because of their conjugated double bonds, are the most abundant quenchers of $^1O_2$ in the pigment bed of the photosynthetic apparatus. Photosynthetic antenna systems have several xanthophylls i.e., lutein, violaxanthin, neoxanthin, and zeaxanthin. Out of these, lutein is the most abundant as it is needed for efficient quenching of $^3Chl^*$. Zeaxanthin is synthesized from violaxanthin under high-light stress by the violaxanthin deepoxidase enzyme and is involved in energy dependent quenching of Chl a fluorescence (Mozzo et al., 2008). Tocopherol is lipid soluble and is a minor but significant component of $^1O_2$ quenchers in the thylakoid membranes. The suppression of both zeaxanthin and tocopherol in the *npq1/ vte1* double mutant results in $^1O_2$-mediated lipid peroxidation in high light (Havaux et al., 2005; Triantaphylidès et al., 2008). In the PS II reaction center, especially under high light regime, $^1O_2$ is quenched by β-carotene and α-tocopherol (Trebst, 2003).

Scavengers of $^1O_2$ are usually water soluble and are themselves oxidized or destroyed while detoxifying the ROS. The oxidized scavenger is re-reduced by a set of biochemical reduction reactions at a great cost to the cell. The cell has only limited capability to resynthesize the destroyed scavengers. Therefore, the cells become extremely prone to $^1O_2$ attack. Ascorbate is an example of $^1O_2$ scavenger that is oxidized after detoxification. It is predominantly present in the plastids. $^1O_2$ reacts with ascorbate to produce dehydroascorbate (Kramarenko et al., 2006). The latter is converted back to ascorbate by the dehydroascorbate reductase, and the glutathione reductase enzymes involved in the *Halliwell-Asada pathway*. Vitamin B6 (pyridoxine, pyridoxal, pyridoxamine) can efficiently scavenge $^1O_2$ (Bilski et al., 2000; Ehrenshaft et al., 1999). The fungus Cercospora secretes cercosporin, a $^1O_2$-generating photosensitizer into the extracellular matrix during plant infection (Ehrenshaft et al., 1999). The cercosporin secreted to the host cell by the fungus absorbs solar energy and transfers its energy to oxygen to generate $^1O_2$ that kills the host cell. However, the fungus itself is protected against $^1O_2$-mediated damage by the $^1O_2$- scavenger vitamin B6. In the same vein, the pyridoxine synthase is involved in tolerance to oxidative stress (Chen and Xiong, 2005). Similarly, exogenous vitamin-B6 protects protoplasts of the *flu* (fluorescence) mutants of *Arabidopsis thaliana* that generated $^1O_2$ (Danon et al., 2005). Flavonoids that are present in plants in high concentrations in the cytoplasm and isoprene that is mostly synthesized in the chloroplasts could also function as $^1O_2$ quenchers (Affek and Yakir, 2002; Agati et al., 2007; Nagai et al., 2005; Velikova et al., 2004). The water soluble chlorophyll binding protein (WSCP) binds to free Chl molecules as well as to its biosynthetic intermediates and does not allow them to get photoactivated to produce $^1O_2$. It acts as a physical barrier between free Chl molecules and molecular oxygen (Schmidt et al., 2003).

VII Singlet Oxygen-Mediated Oxidative Damage to the Photosynthetic Apparatus

Although, plants may have the capacity to detoxify $^1O_2$, they have limited ability to do so. Moreover, unlike O_2^-, where extensive networks of enzymes are available for its detoxification, there are no known enzymatic means for the detoxification of $^1O_2$. Therefore, if excess $^1O_2$ is generated, plants fail to detoxify the system.

A Generation of Tetrapyrrole-Induced Singlet Oxygen in Chloroplasts

DPE compounds accumulate Proto IX, instead of Pchlide as photosensitizer (Ensminger and Hess, 1985; Gupta and Tripathy, 1999, 2000; Haworth and Hess, 1988; Kunert and Boger, 1981; Yoshimoto and Matsunakas, 1982; Witkowski and Halling, 1988). Pchlide (Chakraborty and Tripathy, 1992a) or Proto IX (Tripathy et al., 2007) produced in response to treatments by ALA or DPEs, respectively, generate excess $^1O_2$ via type II photosensitization reactions (Fig. 1). ALA-induced oxidative stress is photosensitized by the Chl biosynthetic intermediate Pchlide (Tripathy and Chakraborty, 1991) whereas that induced by AF-Na is mediated by the tetrapyrrole Proto IX (Duke et al., 1991; Gupta and Tripathy, 1999). $^1O_2$ is generated by both photo-sensitizers i.e., Pchlide (Fig. 2a) and Proto IX (Fig. 2b). As 98% of Pchlide and 65% of Proto IX are localized in the thylakoid membranes, they are the primary sites of generation of $^1O_2$ inside the plastid. As these photosensitizers are partially partitioned to the chloroplast envelope membranes (Mohapatra and Tripathy, 2002, 2007) they too could be a potential site for generation of $^1O_2$.

B Singlet Oxygen-Induced Impairment of the Electron Transport Chain

Photosystem I (PS I) and Photosystem II (PS II) are two major functional units of thylakoid membrane (see Golbeck, 2006; Wydrzynski and Satoh, 2005, for PS I and PS II, respectively). Therefore, any damage to the thylakoid membrane is likely to affect the activities of the above photosystems. $^1O_2$ impairs the PS I- and PS II-dependent electron transport reactions within a few hours of light exposure to ALA or DPE-treated plants. However, the ALA-induced damage to the thylakoid membranes is more severe than that induced by DPE. The differential extent of injury to the thylakoid membrane caused by the photosensitizer Pchlide or Proto IX is due to increased generation of $^1O_2$ induced by Pchlide as compared to Proto IX (compare Fig. 2a and b) and their different localization in the cell. Pchlide is mostly bound to thylakoid membrane, whereas Proto IX is distributed among soluble and

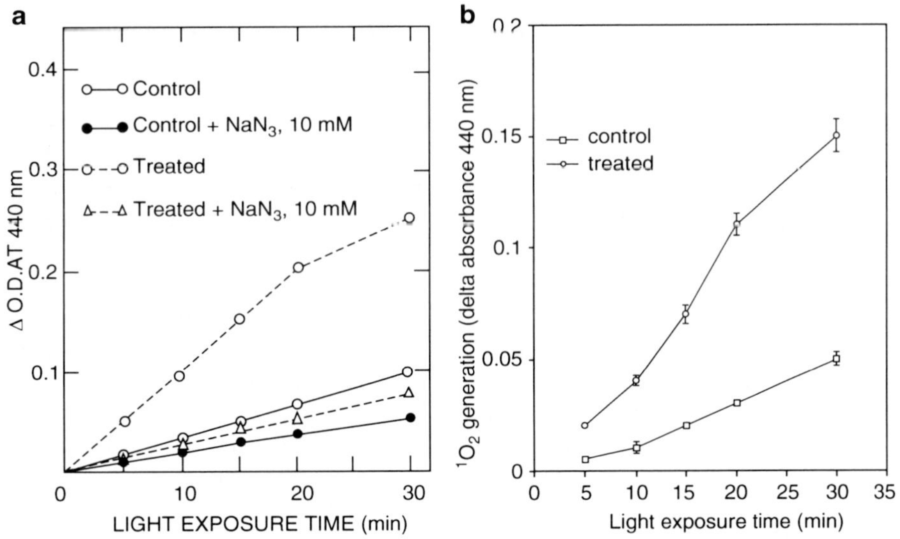

Fig. 2. RNO (p-nitroso-N,N′-dimethylaniline) bleaching reactions of thylakoids isolated from (**a**) control and ALA-treated seedlings and (**b**) control and AF-Na-treated seedlings exposed to incandescent light (1,000 μmol photons m^{-2} s^{-1}). Error bars represent standard deviation of three replicates and missing error bars indicate that they are smaller than the symbols.

membrane fractions (Tripathy et al., 2004; Mohapatra and Tripathy, 2007). $^1O_2$ generated by the thylakoid membrane-localized photosensitizer Pchlide could immediately penetrate the membrane and cause quick damage to the photosynthetic apparatus. On the other hand, $^1O_2$ produced in the stroma is significantly quenched in the aqueous phase before it reaches the membranes. Therefore, AF-Na-induced oxidative stress is probably slower than that of ALA.

C Role of Singlet Oxygen Scavengers

The PSII activity of thylakoid membranes, isolated from ALA-treated plants, when exposed to moderate light (250 μmoles photons m^{-2} s^{-1}) for 30 min is impaired by 60%. The exogenous scavengers of $^1O_2$, histidine and NaN_3, protect the PSII activity demonstrating that the damage to the thylakoid membrane is caused by this active oxygen species (Chakraborty and Tripathy, 1991). Scavengers of the hydroxyl radical formate and O_2^- detoxifier superoxide dismutase fail to protect the thylakoid membranes suggesting that these active oxygen species may not be involved in the tetrapyrrole-sensitized oxidative damage (Fig. 3).

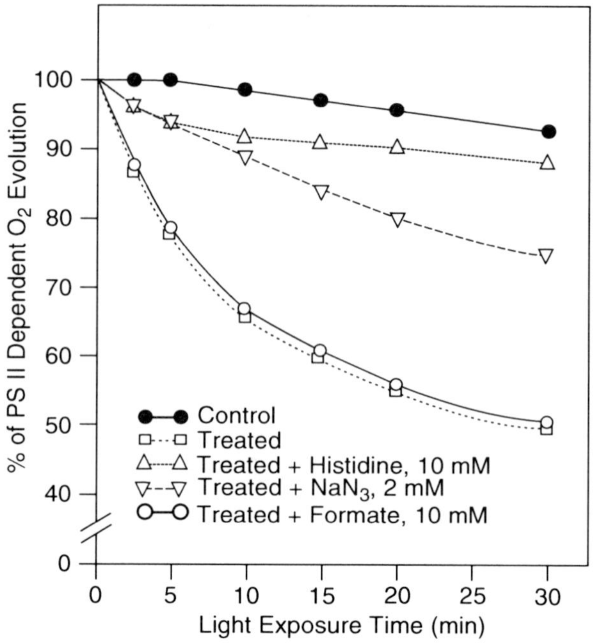

Fig. 3. Effect of scavengers of reactive O_2 species on light (250 μmol photons m^{-2} s^{-1}) induced damage to the thylakoid membrane-linked photosystem II photochemical reactions in intact chloroplasts isolated from control and ALA-treated cucumber plants. The scavengers did not have any inhibitory or protective effect on phenylenediamine-mediated photosystem II activity of control chloroplasts.

D Impact of $^1O_2$ on Chlorophyll a Fluorescence

The $^1O_2$ generated in the plastids, due to the photosensitization reaction of Chl biosynthesis intermediates, induces damage to the thylakoid membranes. In ALA-treated plants exposed to sunlight for 30 min and 1 h, the apparent minimum Chl *a* fluorescence level (Fo) remains almost the same as that of the control. However, the apparent variable Chl *a* fluorescence {Fv= F_{max} (or F_p) minus Fo} is reduced by 60% after 30 min and 78% after 1 h of exposure to sunlight (Tripathy and Chakraborty, 1991) (Fig. 4a). Plants exposed to 15 min of sunlight do not exhibit any physical symptoms of oxidative damage. However, the same plants when kept in dark for 12 h show physical injuries which include prominent necrotic patches. This results in decline in the Fo fluorescence (Fig. 4b). (For a description of Chl a fluorescence, see Papageorgiou and Govindjee, 2004.) The exogenous electron donors to PS II reaction center, $MnCl_2$, diphenylcarbazide and hydroxylamine fail to restore the loss of variable fluorescence (Fv) suggesting that the damage to PS II is very close to the reaction center (Tripathy and Chakraborty, 1991; Tripathy, 1993).

The PS II and PS I photochemical reactions are impaired due to oxidative stress induced by $^1O_2$ (Krieger-Liszkay et al., 2008) Due to gross perturbation of the thylakoid membrane structure, a spectral shift in the Chl a fluorescence spectra of thylakoid membranes is observed at low temperature (77 K). The low temperature fluorescence spectra have peaks at 685 nm (F685) and at 695 nm (F695), which mostly originate from the PS II CP 43 and CP 47 respectively (Govindjee, 1995, 2004) and a large F735 peak that originates mostly from PSI (Mullet et al., 1980) (Fig. 5). If LHCPI is removed from PS I by detergent treatment, the inner antenna of reaction center I (RCI) fluoresces at 722 nm (Kuang et al., 1984). Isolated LHCI fluoresces around 735 nm (Haworth et al., 1983). Thus it is apparent that the inner antenna of RCI emits F722 and LHCI emits F735 (Briantais et al., 1986). The relative decline in F735 by $^1O_2$ -induced damage may be

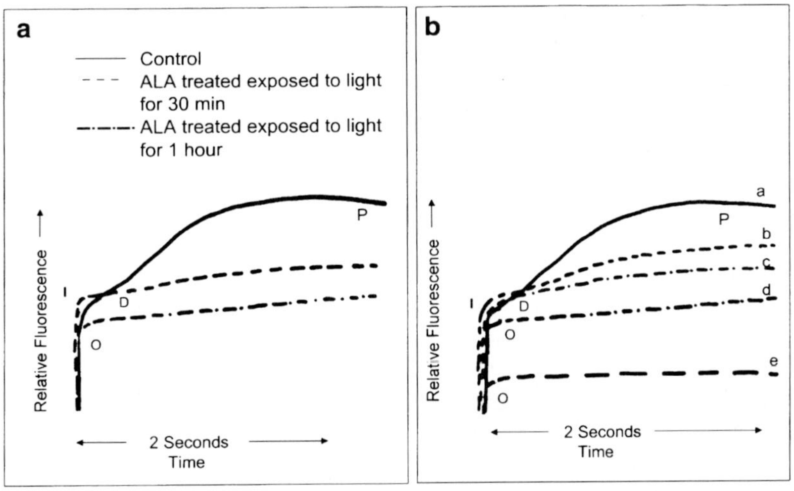

Fig. 4. (**a**) Chlorophyll a fluorescence transients of cucumber leaves in control and ALA-treated leaves exposed to sunlight for 30 min and 1 h; (**b**) chlorophyll a fluorescence transients of control and ALA-treated leaves exposed to 15 min of light and incubated in dark for various lengths of time; control (curve a), 15 min of light treatment (curve b), further incubation in dark for 2 h (curve c), 4 h (curve d), and 12 h (curve e).

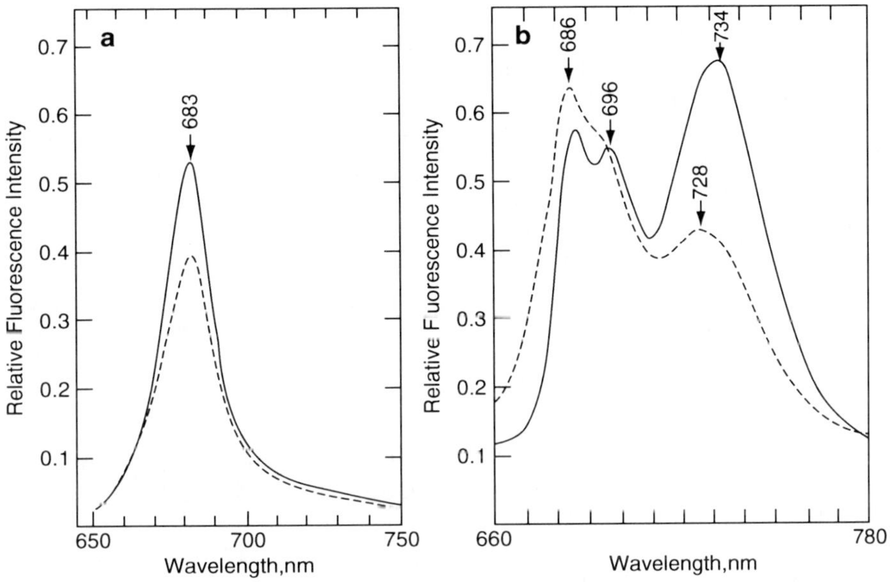

Fig. 5. Room temperature (**a**) and 77K (**b**) fluorescence spectra of chloroplasts isolated from control (-) and ALA-treated plants exposed to sunlight for 1 h (--). The spectra were not corrected for the photomultiplier tube response.

due to damage to LHCI. $^1O_2$ produced by photosensitization reaction of ALA-induced Pchlide or DPE-treated Proto IX cause almost similar structural alteration of the thylakoid membrane i.e., disintegration of LHCI (Tripathy and Chakraborty, 1991, Tripathy et al., 2007). This is evident from the peak shift from 735 nm (attributed to LHCI) to 728 nm (Fig. 5). The fluidity of thylakoid membranes isolated from ALA-treated leaves decreases. The decrease in delayed Chl fluorescence intensity coincides with an increase in $P700^+$ formation, which indicates disturbance in electron transfer between the two photosystems (Hartel et al., 1993a, b). Formation of $^1O_2$ from isolated LHCII was also observed (Rinalducci et al., 2004).

Due to membrane lipid peroxidation by $^1O_2$, the MDA (malondialdehyde) content in ALA-treated as well as DPE-treated light-exposed plants increases. Scavengers of $^1O_2$, i.e., L-histidine and sodium azide are able to protect the oxidative damage to the thylakoid membrane and consequently reduce the MDA production (Chakraborty and Tripathy, 1992a). Among superoxide detoxifying enzymes, present in the stroma, superoxide dismutase and ascorbate specific peroxidase are not affected by $^1O_2$, whereas glutathione reductase activity is impaired suggesting that the superoxide-detoxifying capability of chloroplasts is downregulated due to generation of $^1O_2$ (Tripathy and Singhal, 1999).

E Effect of Singlet Oxygen on Thermoluminiscence

Thermoluminescence (TL) from plants, discovered by Arnold and Sherwood (1957) is a useful probe of the back reaction of PS II (Demeter and Govindjee, 1989; Horváth, 1986; Sane and Rutherford, 1986; Vass, 2005; Vass and Govindjee, 1996). The TL bands at 12°C, 25°C and 48°C are due to Q, B and C bands respectively (Fig. 6). The Q (also called D) band is due to charge recombination of the reduced primary quinone acceptor Q_A^- and S2/3 states whereas the B band arises due to charge recombination of the reduced secondary quinone acceptor Q_B^- with S2/3 states and the C band emanates from charge recombination of Q_A^- with the S state, S1 (see, e.g., DeVault et al., 1983; Rutherford et al., 1982). The intensification of Q band by the addition of DCMU (3-(3,4-dichlorophenyl)-1,1-dimethylurea) is accompanied by the loss of the B band and increase of the C band. A residual B band, observed in the presence of DCMU, is due to its incomplete penetration into the leaves and/or due to the presence of Q_B^- in dark-adapted leaves (Rutherford et al., 1984). In cotyledons, harvested from light exposed AF-Na-treated seedlings, the B band intensity sharply decreases by around 50% and the Q and C bands also decline by 50% demonstrating uniform 50% damage to PS II reaction centers by Proto IX-sensitized $^1O_2$-induced oxidative damage (Tripathy et al., 2007). The characteristic oscillation of TL band is affected when water oxidation is damaged, for example, when 33 kDa protein involved in water oxidation is removed (Ono and Inoue, 1985). In both the control and the treated samples, the TL yield observes a periodicity of 4 (Inoue and Shibata, 1978) and the maximum yield is obtained on the 2nd and 6th flashes. This shows that non-damaged PS II centers behave normally. No significant changes are seen in any of the temperature maxima of the various bands (Tripathy et al., 2007) suggesting no change in the activation energies of the various back reactions of PS II (De Vault and Govindjee, 1990).

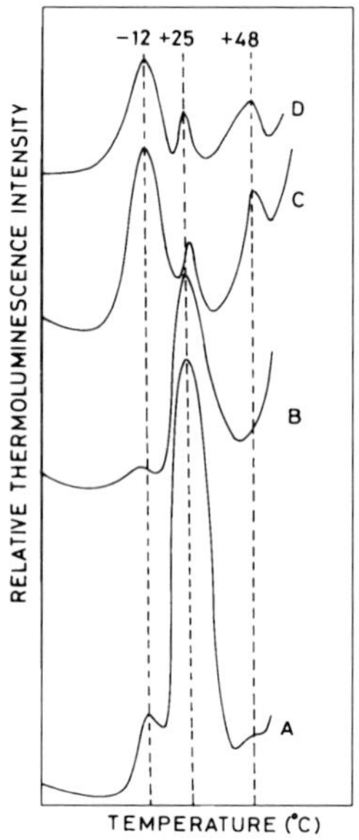

Fig. 6. Thermoluminescence (TL) of leaf discs taken from cotyledons of control (curve A, without DCMU; curve C, with DCMU) and AF-Na-treated (curve B, without DCMU; curve D, with DCMU) seedlings exposed to cool-white fluorescent light (75 µmol photons m^{-2} s^{-1}) for 24 h.

VIII Singlet Oxygen-induced Oxidative Damage in Mutants

A Chlorophyll Anabolic Mutants

Tigrina mutant of barley accumulates two to ten times more Pchlide in darkness than the wild type (Nielsen, 1974). Homozygous *tigrina-d* mutants are fully green and viable if grown in continuous weak light, but show a green-white banded phenotype, when grown under light/dark cycles. They have normal level of POR (Protochlorophyllide oxidoreductase) and when illuminated, the excess Pchlide causes photodynamic damage resulting in the formation of necrotic patches (Von Wettstein et al., 1995). Runge et al. (1995) isolated *xantha* mutants of Arabidopsis and classified them in two groups, mutants that are blocked in various steps of the Chl biosynthetic pathway prior to POR, and mutants that accumulate Pchlide in the dark. The etiolated PORA and PORB mutant seedlings accumulate significant amounts of non-phototransformable Pchlide in darkness and upon light exposure they show bleaching effect and germination defect (Armstrong et al., 1995). However, overexpression of PORA and PORB results in the efficient transformation of non-phototransformable Pchlide to chlorophyllide (Chlide). Transgenic seedlings grown under far red light when transferred to white light are more resistant to photobleaching because of high POR proteins.

The isolation and studies on Arabidopsis *flu* mutant by Klaus Apel's group confirm the role of Pchlide in $^1O_2$ generation; it leads to oxidative damage. In *flu* mutant there is a massive accumulation of Pchilde if those plants are grown under constant dark/light cycle and there is growth arrest and cell death because of generation of $^1O_2$. However if the plants are grown under continuous light, there is no accumulation of Pchlide, no generation of $^1O_2$ and plants behave like wild type plants. The FLU protein interacts with GluTR (HEMA1) and regulates the accumulation of Pchlide in darkness (Meskauskiene and Apel, 2002). In the *flu* mutant there is no regulatory point at the GluTR level and there is a massive accumulation of Pchlide. Lee et al. (2003) have revealed that the *TIGRINA d* gene of barley is an ortholog of the *FLU* gene of *Arabidopsis thaliana*. Pchlide-mediated $^1O_2$ formation leads to the induction of the early stress-responsive gene (Op den Camp et al., 2003). There is no change in amounts of other photosensitizers i.e, Proto IX, Mg -proto IX and MPE in the *flu* mutant. Two major stress reactions were observed when dark-grown *flu* plants were returned to light: a cell death response and a rapid inhibition of growth. Oxygenation derivatives of linolenic acid, by far the most prominent polyunsaturated fatty acid of chloroplast membrane lipids, start to accumulate rapidly in the *flu* mutant after the dark/light shift. The oxidation of linolenic acid is not caused by direct interaction with $^1O_2$ but instead occurs enzymatically. Thus, the development of stress symptoms in the *flu* mutant seems not to be attributable to cell damage caused by $^1O_2$ but rather appears to result from the more indirect role of this ROS. Vitamin B6 that quenches $^1O_2$ in fungi was able to protect *flu* protoplasts from cell death (Danon et al., 2005); further, protoplasts of *flu* mutant depleted of both ethylene and salicylic acid had reduced cell death. However, when the gene *Executer1* (exe1) was mutated in the *flu* background, the *exe1/flu* double mutant accumulated free Pchlide in the dark like the *flu* mutant, but unlike the wild type plants. After transfer to light, *exe1/flu* generated $^1O_2$ in amounts similar to those of *flu* but grew like wild type when kept under non-permissive light-dark cycles (Wagner et al., 2004). In *flu* plants, the growth rate was reduced immediately after the beginning of re-illumination. The *exe1/flu* plants, however, grew like wild-type plants. Growth inhibition of *flu* plants was particularly striking when plants were transferred to repeated light-dark cycles, whereas the *exe1/flu* mutant continued to grow like wild-type plants (Wagner et al., 2004). Both assays demonstrate that the rapid bleaching of *flu* seedlings and the inhibition of growth after the release of $^1O_2$ are not by the toxicity of this ROS; further, these do not reflect photooxidative damage and injury, but instead result from the activation of genetically controlled responses that require the activity of the *Executer1* gene. The isolation of the Executer 2 protein also shows a similar kind of response in *flu* background (Lee et al., 2007). Inactivation of executer proteins blocks the $^1O_2$-mediated signaling from the chloroplast to the nucleus that affects the normal plastid development in germinating seeds (Kim et al., 2009). Coll et al. (2009) have isolated another $^1O_2$-linked death activator (soldat8) that encodes the SIGMA6 factor of the plastid RNA polymerase, specifically abrogate $^1O_2$-mediated stress responses in young *flu* seedlings without grossly affecting $^1O_2$-mediated stress responses of mature *flu* plants. The other protein named PRL1 (Pleiotropic response locus 1) also affect the expression of $^1O_2$- responsive genes in Arabidopsis (Baruah et al., 2009).

Apart from Pchlide, early intermediates i.e., coproporphyrin also acts as a photosensitizer (Ishikawa et al., 2001; Kruse et al., 1995a, b). The antisense coproporphyrinogen oxidase (that converts coproporphyrinogen III to protoporphyrinogen IX) in tobacco plants, have an excessive amount of coproporphyrin. This oxidized porphyrin gives rise to photodynamic reactions, which affect cellular processes resulting in retarded growth and necrotic leaves (Kruse et al., 1995a, b). The Arabidopsis coproporphyrinogen oxidase mutants (*lin2*; lesion initiation 2) had pale leaves and developed lesions on the young leaf (Ishikawa et al., 2001). 3,3-Diamino benzidine and trypan blue staining of the mutant leaves shows H_2O_2 accumulation and cell death. Seedlings homozygous for a null mutation in the cpx1 gene of maize completely lack chlorophyll and develop necrotic lesions in the light (Williams et al., 2006). The accumulation of uroporphyrin I in the uroporphyrinogen III cosynthase antisense barley plants results in necrotic leaves and ultimately cell death because of accumulation of ROS (Ayliffe et al., 2009). Like uroporphyrin I, uroporphyrin III, an oxidized derivative of uroporphyrinogen III, an intermediate of the chlorophyll biosynthesis pathway, also acts as a photosensitizer. Accumulation of Uroporphyrin III leads to light-dependent necrosis in tobacco (Mock and Grimm, 1997; Mock et al., 1999) and in maize (Hu et al., 1998). Antisense tobacco plants of Uroporphyrinogen decarboxylase have stunted growth with necrotic leaves and high PR1 gene expression. The maize lesion mimic a mutant, coding for uroporphyrinogen decarboxylase, that has necrotic spots in the leaves. Inhibition of protox in Arabidopsis leads to production of lesion-mimic phenotype, high endogenous level of salicylic acid and PR1 gene expression (Molina et al., 1999). Overexpression of plastidic protox leads to resistance to the DPE herbicide acifluorfen. The overexpressed plants did not show any necrotic leaves (Lermontova and Grimm, 2000). Tobacco plants having reduced ferrochelatse activity also show necrotic leaves in a light intensity dependent manner (Papenbrock et al., 2001).

B Chlorophyll Catabolic Mutants

Intermediates involved in the Chl degradation pathway also produce ROS. Squash plants expressing the mature (lacking the N-terminal 21 amino acids) citrus chlorophyllase protein, display a lesion-mimic phenotype when grown under natural light. The phenotype is caused by the accumulation of chlorophyllide, which is a photodynamic porphyrin molecule (Harpaz-Saad et al., 2007). The Arabidopsis Pheophorbide a oxygenase (PAO, also called *acd1*, accelerated cell death 1) mutant shows a cell death phenotype because of the accumulation of the Chl degradation intermediate pheophorbide a. The latter gets photoactivated in the presence of light and generates ROS that forms lesions in the mutant plants. The lesions that form in *acd1* mutant leaves start mostly at the tip of the leaf and subsequently run down the leaf blade (Pruzinska et al., 2003). Hirashima et al, 2009, also observed that the accumulation of Pheophorbide a in dark grown *acd1* antisense plants caused cell death. LLS1 (lethal leaf spot 1), the homologue of ACD1 (Accelerated cell death 1/ Pheophorbide a oxygenase) in maize is responsible for Chl catabolism. The maize *lls1* mutant formed lesions when grown in the light (Gray et al., 1997). Similarly the Arabidopsis Red chlorophyll catabolite reductase (RCCR, also called *acd2*, Accelearated cell deah 2) mutant showed lesion formation in leaves and spontaneous cell death phenotype (Mach et al., 2001). It is observed that the accumulation of H_2O_2 in the *acd2* mitochondria is causal for its cell death phenotype (Yao and Greenberg, 2006). The lesion formation in *acd2* is caused by the accumulation of red chlorophyll catabolite (RCC) in darkness that generates $^1O_2$ in the presence of light (Pruzinska et al., 2007) Further work should be done to check whether the generation of H_2O_2 and $^1O_2$ are independent events or whether one leads to the other.

IX Future Prospects

Acclimation to $^1O_2$ has been shown in the green alga *Chlamydomonas reinhardtii* (Ledford et al., 2007). This approach could be further exploited to generate plants that could tolerate higher doses of $^1O_2$. As there is no enzymatic means available to detoxify $^1O_2$, it is essential to minimize its production rather than detoxifying it after it is generated. $^1O_2$-induced signaling events leading to 180 base pair nucleotide ladder formation in apoptosis will provide valuable clues for early responses of plants to $^1O_2$. There is bound to be cross-talk between $^1O_2$-mediated and other ROS-mediated signaling events leading to apoptosis.

Acknowledgements

This was supported by the Depatment of biotechology, Government of India, for research grant BT/PR6074/AGR/02/311/2005.

Thanks are due to Govindjee for critically reading the manuscript; his suggestions have been incorporated in this manuscript.

References

Affek HP and Yakir D (2002) Protection by isoprene against singlet oxygen in leaves. Plant Physiol 129: 269–277

Agati G, Matteini P, Goti A and Tattini M (2007) Chloroplast-located flavonoids can scavenge singlet oxygen. New Phytol 174: 77–89

Apel K and Hirt H (2004) Reactive oxygen species: metabolism, oxidative stress, and signal transduction. Annu Rev Plant Biol 55: 373–399

Armstrong GA, Runge S, Frick G, Sperling U and Apel K (1995) Identification of NADPH:protochlorophyllide oxidoreductases A and B: a branched pathway for light-dependent chlorophyll biosynthesis in Arabidopsis thaliana. Plant Physiol 108: 1505–1517

Aro EM, Virgin I and Andersson B (1993) Photoinhibition of photosystem II: inactivation, protein damage and turnover. Biochimica Biophysica Acta 1143: 113–134

Ayliffe MA, Agostino A, Clarke BC, Furbank R, von Caemmerer S and Pryor AJ (2009) Suppression of the Barley uroporphyrinogen III synthase gene by a Ds activation tagging element generates developmental photosensitivity. Plant Cell 21: 814–831

Barber J and Andersson B (1992) Too much of a good thing: Light can be bad for photosynthesis. Trends Biochem Sci 17: 61–66

Baroli I and Niyogi KK (2000) Molecular genetics of xanthophyll-dependent photoprotection in green algae and plants. Philos Trans R Soc Lond B Biol Sci 355: 1385–1394

Baroli I, Do AD, Yamane T and Niyogi KK (2003) Zeaxanthin accumulation in the absence of a functional xanthophyll cycle protects Chlamydomonas reinhardtii from photooxidative stress. Plant Cell 15: 992–1008

Baroli I, Gutman BL, Ledford HK, Shin JW, Chin BL, Havaux M and Niyogi KK (2004) Photo oxidative stress in a xanthophyll-deficient mutant of Chlamydomonas. J Biol Chem 279: 6337–6344

Baruah A, Simková K, Hincha DK, Apel K and Laloi C (2009) Modulation of O-mediated retrograde signaling by the PLEIOTROPIC RESPONSE LOCUS 1 (PRL1) protein, a central integrator of stress and energy signaling. Plant J [Epub ahead of print]

Bilski P, Li MY, Ehrenshaft M, Daub ME and Chignell CF (2000) Vitamin B6 (pyridoxine) and its derivatives are efficient singlet oxygen quenchers and potential fungal antioxidants. Photochem Photobiol 71: 129–134

Briantais JM, Vernotte C, Krause GH and Weis E (1986) Chlorophyll a fluorescence of higher plants, chloroplasts and leaves. In: Govindjee, Amesz J, Fork DC (eds) Light Emission by Plants and Bacteria, pp. 539–583. Academic, New York

Chakraborty N and Tripathy BC (1991) 5-Aminolevulinic acid induced photodynamic damage of the photosynthetic electron transport chain of cucumber (*Cucumis sativus* L.) Cotyledons. Plant Physiol 96: 761–767

Chakraborty N and Tripathy BC (1992a) Involvement of singlet oxygen in photodynamic damage of isolated chloroplasts of cucumber (Cucumis sativus L.) cotyledons. Plant Physiol 98: 7–11

Chakraborty N and Tripathy BC (1992b) 5-aminolevulinic acid-induced photodynamic reactions in thylakoid membranes of cucumber (Cucumis sativus L.) cotyledons. J Plant Biochem Biotech 1: 65–68

Chen H and Xiong L (2005) Pyridoxine is required for postembryonic root development and tolerance to osmotic and oxidative stresses. Plant J 44: 396–408

Coll NS, Danon A, Meurer J, Cho WK and Apel K (2009) Characterization of soldat8, a suppressor of singlet oxygen-induced cell death in Arabidopsis seedlings. Plant Cell Physiol 50: 707–718

Danon A, Miersch O, Felix G, Camp RG and Apel K (2005) Concurrent activation of cell death-regulating signaling pathways by singlet oxygen in Arabidopsis thaliana. Plant J 41: 68–80

Daub M and Hangarter RP (1983) Light-induced production of singlet oxygen and superoxide by the fungal toxin cercosporin. Plant Physiol 73: 855–857

Davison PA, Hunter CN and Horton P (2002) Overexpression of b-carotene hydroxylase enhances stress tolerance in Arabidopsis. Nature 418: 203–206

Demeter S and Govindjee (1989) Thermoluminescence in plants. Physiol Plant 75: 121–130

De Vault D and Govindjee (1990) Photosynthetic glow peaks and their relationship with their free energy changes. Photosynth Res 24: 175–181

DeVault D, Govindjee and Arnold W (1983) Energetics of photosynthetic glow peaks, Proc Natl Acad Sci USA 80: 983–987

Duke SO, Becerril JM, Sherman TD and Matsumoto H (1991) Photosensitizing porphyrins as herbicides. In: Hedin PA (eds) Naturally Occurring Pest Bioregulators. ACS Symposium series No. 449, pp. 371–386, ACS, Washington

Edge R, McGarvey DJ and Truscott TG (1997) The carotenoids as anti-oxidant-a review. J Photochem Photobiol B 41: 189–200

Ensminger MP and Hess FD (1985) Action spectrum of the activity of aciflurofen-methyl, a diphenyl ether herbicide, in Chlamydomonas eugametos. Plant physiol 77: 503–505

Ehrenshaft M, Bilski P, LI MY, Chignell CF and Daub ME (1999) A highly conserved sequence is a novel gene involved in de novo vitamin B6 biosynthesis. Proc Nat Acad Sci USA 96: 9374–9378

Foote CS, Shook FC and Abakerli RB (1984) Characterization of singlet oxygen. Methods Enzymol 105: 36–47

Frank HA, Young AJ, Britton G and Cogdell RJ (eds) (1999) The Photochemistry of Carotenoids. In: Advances in Photosynthesis [and Respiration], Volume 8, Kluwer Academic Publishers (now Springer), Dordrecht

Fryer MJ, Oxborough K, Mullineaux PM and Baker NR (2002) Imaging of photooxidative stress responses in leaves. J Exp Bot 53: 1249–1254

Golbeck JH (ed) (2006) Photosystem I: The Light-Driven Plastocyanin: Ferredoxin Oxidoreductase. Advances in Photosynthesis and Respiration, Volume 24, Springer, Dordrecht

Govindjee (1995) Sixty-three years since Kautsky: chlorophyll a fluorescence. Aust J Plant Physiol 22: 131–160

Govindjee (2004) Chlorophyll a fluorescence: a bit of basics and history. In: Papageorgiou GC, Govindjee (eds) Chlorophyll a Fluorescence a Signature of Photosynthesis, Springer, Dordrecht, pp. 1–41

Graham MY (2005) The diphenylether herbicide lactofen induces cell death and expression of defense-related genes in soybean. Plant Physiol 139: 1784–1794

Granik S (1959) Magnesium porphyrins formed by barley seedlings treated with δ-aminolevulinic acid. Plant Physiol 34: XVIII

Gray J, Close PS, Briggs SP and Johal GS (1997) A novel suppressor of cell death in plants encoded by the Lls1 gene of maize. Cell 89: 25–31

Gupta I and Tripathy BC (1999) Oxidative stress in cucumber (Cucumis sativus L.) Seedlings treated with Acifluorfen. Z Naturforsch 54C: 771–781

Gupta I and Tripathy BC (2000) Oxidative stress in cucumber (Cucumis sativus L) Seedlings treated with acifluorfen. Indian J Biochem Biophys 37: 498–505

Harpaz-Saad S, Azoulay T, Arazi T, Ben-Yaakov E, Mett A, Shiboleth Y M, Hörtensteiner S, Gidoni D, Gal-On A, Goldschmidt EE and Eyal Y (2007) Chlorophyllase is a rate-limiting enzyme in chlorophyll catabolism and is posttranslationally regulated. Plant Cell 19: 1007–1022

Hartel H, Walter G and Hanke T (1993a) Photoinduced damage in leaf segments of wheat (Triticum aestivum L.) and lettuce (Lactuca sativa L.) treated with 5-aminolevulinic acid. 1. effect on structural components of photosynthetic apparatus. J Plant Physiol 142: 230–236

Hartel H, Haseloff RF, Walter G and Hanke T (1993b) Photoinduced damage in leaf segments of wheat (Triticum aestivum L.) and lettuce (Lactuca sativa L.) treated with 5-aminolevulinic acid. 2. Characterization of photodynamic damage by means of delayed chlorophyll fluorescence and P700 photooxidation. J Plant Physiol 142: 237–243

Havaux M, Eymery F, Porfirova S, Rey P and Dormann P (2005) Vitamin E protects against photoinhibition and photooxidative stress in Arabidopsis thaliana. Plant Cell 17: 3451–3469

Havaux M, Dall'osto L and Bassi R (2007) Zeaxanthin has enhanced antioxidant capacity with respect to all other xanthophylls in Arabidopsis leaves and functions independent of binding to PSII antennae. Plant Physiol 145: 1506–1520

Haworth P, Watson JL and Arntzen CJ (1983) Orientation of pigments in the thylakoid membrane and in the isolated chlorophyll-protein complexes of higher plants. I. Determination of optimal conditions for linear dichroism measurement Biochim Biophy Acta 724: 151–158

Haworth P and Hess FD (1988) The generation of singlet oxygen (1O2) by the Nitrodiphenyl ether herbicide Oxyfluorfen is independent of photosynthesis. Plant Physiol 86: 672–676

Hideg E, Kalai T, Hideg K and Vass I (1998) Photoinhibition of photosynthesis in vivo results in singlet oxygen production. Detection via nitroxide-induced fluorescence quenching in broad bean leaves. Biochemistry 37: 11405–11411

Hirashima M, Tanaka R and Tanaka A (2009) Light-independent cell death induced by accumulation of pheophorbide a in Arabidopsis thaliana. Plant Cell Physiol 50: 719–729

Horváth G (1986) Usefulness of thermoluminescence in herbicide research. Crit Rev Plant Sci 4: 293–310

Hu G, Yalpani N, Briggs SP and Johal GS (1998) A porphyrin pathway impairment is responsible for the phenotype of a dominant disease lesion mimic mutant of maize. Plant Cell 10: 1095–1105

Inoue Y and Shibata K (1978) Oscillation of thermoluminescence at medium low temperature. FEBS Lett 85: 193–197

Ishikawa A, Okamoto H, Iwasaki Y and Asahi T (2001) A deficiency of coproporphyrinogen III oxidase causes lesion formation in Arabidopsis. Plant J 27: 89–99

Ito T and Kobayashi K (1977) A survey of *in vivo* photodynamic activity of xanthenes, thiazines, and acridines in yeast cells. Photochem Photobiol 26: 581–587

Jung S, Lee HJ, Lee Y, Kang K, Kim YS, Grimm B and Back K (2008) Toxic tetrapyrrole accumulation in protoporphyrinogen IX oxidase-overexpressing transgenic rice plants. Plant Mol Biol 67: 535–546

Kim C, Lee KP, Baruah A, Nater M, Göbel C, Feussner I and Apel K (2009) 1O2-mediated retrograde signaling during late embryogenesis predetermines plastid differentiation in seedlings by recruiting abscisic acid. Proc Natl Acad Sci USA May 29 [Epub ahead of print]

Kramarenko GG, Hummel SG, Martin SM and Buettner GR (2006) Ascorbate reacts with singlet oxygen to produce hydrogen peroxide. Photochem Photobiol 82: 1634–1637

Krieger-Liszkay A, Fufezan C and Trebst A (2008) Singlet oxygen production in photosystem II and related protection mechanisms. Photosynth Res 98: 551–564

Kruse E, Mock HP and Grimm B (1995a) Reduction of coproporphyrinogen oxidase level by antisense RNA synthesis leads to deregulated gene expression of plastid proteins and affects the oxidative defense system. EMBO J 14: 3712–3720

Kruse E, Mock HP and Grimm B (1995b) Coproporphyrinogen III oxidase from barley and tobacco-sequence analysis and initial expression studies. Planta 196: 796–803

Kuang TY, Argyroudi-Akoyunoglou JH, Nakatani HY, Watson J and Arntzen CJ (1984) The origin of the long-wavelength fluorescence emission band (77 degrees K) from photosystem I. Arch Biochem Biophys 235: 618–627

Kunert KJ and Boger P (1981) The bleaching effect of diphenyl ether oxyfluorefen. Weed Sci 29: 169–172

Kunert KJ, Sandmann G and Boger P (1987) Mode of action of diphenyl ethers. Weed Sci 3: 35–55

Ledford HK, Chin BL and Niyogi KK (2007) Acclimation to singlet oxygen stress in Chlamydomonas reinhardtii. Eukaryotic Cell 6: 919–930

Lee KP, Kim C, Lee DW and Apel K (2003) TIGRINA d, required for regulating the biosynthesis of tetrapyrroles in barley, is an ortholog of the FLU gene of Arabidopsis thaliana. FEBS Lett 55: 119–124

Lee KP, Kim C, Landgraf F and Apel K (2007) EXECUTER1- and EXECUTER2-dependent transfer of stress-related signals from the plastid to the nucleus of Arabidopsis thaliana. Proc Natl Acad Sci USA 104: 10270–10275

Lermontova I and Grimm B (2000) Overexpression of plastidic protoporphyrinogen IX oxidase leads to resistance to the diphenyl-ether herbicide acifluorfen. Plant Physiol 122: 75–84

Lermontova I and Grimm B (2006) Reduced activity of plastid protoporphyrinogen oxidase causes attenuated photodynamic damage during high-light compared to low-light exposure. Plant J 48: 499–510

Lydon J and Duke SO (1988) Porphyrin synthesis is required for photobleaching activity of the p-nitrosubstituted diphenyl ether herbicides. Pest Biochem Physiol 31: 74–83

Mach JM, Castillo AR, Hoogstraten R and Greenberg JT (2001) The Arabidopsis-accelerated cell death gene ACD2 encodes red chlorophyll catabolite reductase and suppresses the spread of disease symptoms. Proc Natl Acad Sci USA 98: 771–776

Macpherson AN, Telfer A, Barber J and Truscott TG (1993) Direct detection of singlet oxygen from isolated photosystem II reaction centres. Biochim Biophys Acta 1143: 301–309

Matringe M, Camadro JM, Labbe P and Scalla R (1989) Protoporphyrinogen oxidase as a molecular target for diphenyl ether herbicides. Biochem J 260: 231–235

Meskauskiene R and Apel K (2002) Interaction of FLU, a negative regulator of tetrapyrrole biosynthesis, with the glutamyl–tRNA reductase requires the tetratricopeptide repeat domain of FLU. FEBS Lett 532: 27–30

Mock HP and Grimm B (1997) Reduction of uroporphyrinogen decarboxylase by antisense RNA expression affects activities of other enzymes involved in tetrapyrrole biosynthesis and leads to light dependent necrosis. Plant Physiol 113: 1101–1112

Mock HP, Heller W, Molina A, Neubohn B, Sandermann H and Grimm B (1999) Expression of uroporphyrinogen decarboxylase or coproporphyrinogen oxidase antisense RNA in tobacco induces pathogen defence responses conferring increased resistance to tobacco mosaic virus. J Biol Chem 274: 4231–4238

Mohapatra A and Tripathy BC (2002) Detection of protoporphyrin IX in envelope membranes of pea chloroplasts Biochem Biophys Res Commun 299: 751–754

Mohapatra A and Tripathy BC (2007) Differential distribution of chlorophyll biosynthetic intermediates in stroma, envelope and thylakoid membranes in Beta vulgaris. Photosynth Res 94: 401–410

Molina A, Volrath S, Guyer D, Maleck K, Ryals J and Ward E (1999) Inhibition of protoporphyrinogen oxidase expression in Arabidopsis causes a lesion-mimic phenotype that induces systemic acquired resistance. Plant J 17: 667–678

Mozzo M, Dall'Osto L, Hienerwadel R, Bassi R and Croce R (2008) Photoprotection in the antenna complexes of photosystem II. Role of individual xanthophylls in chlorophyll triplet quenching. J Biol Chem 283: 6184–6192

Mullet JE, Burke JJ and Arntzen CJ (1980) Chlorophyll proteins of Photosystem I. Plant Physiol 65: 814–822

Nagai S, Ohara K and Mukai K (2005) Kinetic study of the quenching reaction of singlet oxygen by flavonoids in ethanol solution. J Phys Chem B 109: 4234–4240

Nielsen OF (1974) Macromolecular physiology of plastids XII Tigrina mutants of barley. Hereditas 76: 269–304

Ohad I, Keren N, Zer H, Gong H and Mor TS (1994) Light-induced degradation of the photosystem II reaction center D1 protein in vivo: an integrative approach. Baker NR, Bowyer JR, eds.. Photoinhibition of Photosynthesis: From Molecular Mechanisms to the Field. BIOS Scientific Publishers, Oxford, pp. 161–177

Ono T and Inoue Y (1985) S-state turnover in the O2 evolving system of CaCl2-washed photosystem II particles depleted of three peripheral proteins as measured by thermoluminescence. Removal of 33 kDa proteins inhibits S3 to S4 transitions. Biochim Biophys Acta 973: 443–449

Op den Camp RG, Przybyla D, Ochsenbein C, Laloi C, Kim C, Danon A, Wagner D, Hideg E, Göbel C, Feussner I, Nater M and Apel K (2003) Rapid induction of distinct stress responses after the release of singlet oxygen in Arabidopsis. Plant Cell 15: 2320–2332

Papageorgiou GC and Govindjee (eds) (2004) Chlorophyll a Fluorescence: A Signature of Photosynthesis. Advances in Photosynthesis and Respiration, Volume 19, Springer, Dordrecht

Papenbrock J, Mishra S, Mock HP, Kruse E, Schmidt EK, Petersmann A, Braun HP and Grimm B (2001) Impaired expression of the plastidic ferrochelatase by antisense RNA synthesis leads to a necrotic phenotype of transformed tobacco plants. Plant J 28: 41–50

Pareek A, Sopory S, Bohnert H and Govindjee (eds) (2010) Abiotic Stress Adaptation in Plants: Physiological, Molecular and Genomic Foundation, Springer, Dordrecht

Pogson BJ and Rissler HM (2000) Genetic manipulation of carotenoid biosynthesis and photoprotection. Philos Trans R Soc Lond B Biol Sci 355: 1395–1403

Pruzinska A, Tanner G, Anders I, Roca M and Hortensteiner S (2003) Chlorophyll breakdown: pheophorbide a oxygenase is a Rieske-type iron-sulfur protein, encoded by the accelerated cell death 1 gene. Proc Natl Acad Sci USA 100: 15259–15264

Pruzinská A, Anders I, Aubry S, Schenk N, Tapernoux-Lüthi E, Müller T, Kräutler B and Hörtensteiner S (2007) In vivo participation of red chlorophyll catabolite reductase in chlorophyll breakdown. Plant Cell 19: 369–387

Rebeiz CA, Montazer-ZouhoorA, Hopen HJ and Wu SM (1984) Photodynamic herbicides: 1. Concept and phenomenology. Enzymol Microb Technol 6: 390–401

Rebeiz CA, Montazer-Zouhoor A, Mayasich JM, Tripathy BC, Wu SM and Rebeiz CC (1988) Photodynamic herbicides. Recent development and molecular basis of selectivity. CRC Crit Rev Plant Sci 6: 385–436

Rebeiz CA, Reddy KN, Nandihalli UB and Velu J (1990) Tetrapyrrole-dependent photodynamic herbicides. Photochem Photobiol 52: 1099–1117

Rebeiz CA, Nandihalli UB and Reddy KN (1991) In: Topics in Photosynthesis -Volume 10, Herbicides. (Baker NR and Percival MP eds.), Elsevier, Amsterdam, pp. 173–208

Rinalducci S, Pedersen JZ and Zolla L (2004) Formation of radicals from singlet oxygen produced during photoinhibition of isolated light-harvesting proteins of photosystem II. Biochim Biophys Acta1608: 63–73

Runge S, van Cleve B, Lebedev N, Armstrong G and Apel K (1995) Isolation and classification of chlorophyll-deficient xantha mutants of Arabidopsis thaliana. Planta 197: 490–500

Rutherford AW, Crofts AR and Inoue Y (1982) Thermoluminescence as a probe of photosystem photochemistry. The origin of the flash-induced glow peaks. Biochim Biophys Acta 682: 457–467

Rutherford AW, Govindjee and Inoue Y (1984) Charge accumulation and photochemistry in leaves studied by thermoluminescence and delayed light emission. Proc Natl Acad Sci USA 81: 1107–1111

Sane PV and Rutherford AW (1986) Thermoluminescence from photosynthetic membranes. In: Govindjee et al. (eds) Light Emission by Plants and Bacteria. Academic, Orlando, pp. 329–360

Schmidt K, Fufezan C, Krieger-Liszkay A, Satoh H and Paulsen H (2003) Recombinant water-soluble chlorophyll protein from Brassica oleracea var. Botrys binds various chlorophyll derivatives. Biochemistry 42: 7427–7433

Shalygo NV, Mock HP, Averina NG and Grimm B (1998) Photodynamic action of uroporphyrin and protochlorophyllide in greening barley leaves treated with cesium chloride. J Photochem Photobiol B 42: 151–158

Skovsen E, Snyder JW, Lambert JD and Ogilby PR (2005) Lifetime and diffusion of singlet oxygen in a cell. J Phys Chem B 109: 8570–8573

Spikes DK and Bommer JC (1991) Chlorophyll and related pigments as photosensitizers in biology and medicine. In: Scheer H (ed.) Chlorophylls, pp. 1181–1204, CRC Press, Boca Raton, FL

Stobart AK and Ameen-Bukhari I (1984) Regulation of delta-aminolaevulinic acid synthesis and protochlorophyllide regeneration in the leaves of dark-grown barley (Hordeum vulgare) seedlings. Biochem J 222: 419–426

Stobart AK and Ameen-Bukhari I (1986) Photoreduction of protochlorophyllide and its relationship to delta-aminolaevulinic acid synthesis in the leaves of dark-grown barley (Hordeum vulgare) seedlings. Biochem J 236: 741–748

Suh HJ, Kim CS and Jung J (2000) Cytochrome b6/f complex as an indigenous photodynamic generator of singlet oxygen in thylakoid membranes. Photochem Photobiol 71: 103–109

Trebst A (2003) Function of β-carotene and tocopherol in photosystem II. Z Naturforsch C58: 609–620

Triantaphylidès C, Krischke M, Hoeberichts FA, Ksas B, Gresser G, Havaux M, Van Breusegem F and Mueller MJ (2008) Singlet oxygen is the major reactive oxygen species involved in photo-oxidative damage to plants. Plant Physiol 148: 960–968

Tripathy BC (1993) 5-Aminolevulinic acid-induced photodynamic damage to cucumber (*Cucumis sativus* L.) plants mediated by singlet oxygen In Porphyric Pesticides: Chemistry, Toxicology and Pharmaceutical Applications (S. O. Duke and C. A. Rebeiz, eds.) American Chemical Society Washington DC pp. 65–80

Tripathy BC and Singhal GS (1999) Oxidative stress in photodynamic herbicidal action of 5-aminolevulinic acid. *In* Concepts in Photobiology Photosynthesis and Photomorphogenesis. (G. S. Singhal, G. Renger, S. K. Sopory, K-D Irrgang and Govindjee eds.) Narosa Publishing house New Delhi, pp. 668–688

Tripathy BC, Mohapatra A and Pattanayak GK (2004) Subplastidic distribution of chlorophyll biosynthetic intermediates and characterization of protochlorophyllide oxidoreductase C. in Agricultural applications in green chemistry (W M. Nelson ed, ACS ss 887.) American Chemical Society, Washington DC pp. 117–128

Tripathy BC, Mohapatra A and Gupta I (2007) Impairment of the photosynthetic apparatus by oxidative stress induced by photosensitization reaction of protoporphyrin IX. Biochim Biophys Acta 1767: 860–868

Vass I (2005) The history of photosynthetic thermoluminescence. In: Govindjee, Beatty JT, Gest H, Allen JF (eds) Discoveries in Photosynthesis, pp. 665–680, Advances in Photosynthesis and Respiration, Volume 20, Springer, Dordrecht

Vass I and Govindjee (1996) Thermoluminescence from the photosynthetic apparatus. Photosynth Res 48: 117–126

Velikova V, Edreva A and Loreto F (2004) Endogenous isoprene protects Phragmites australis leaves against singlet oxygen. Physiol Plant 122: 219–225

Von Wettstein D, Gough S and Kannangara CG (1995) Chlorophyll biosynthesis. Plant Cell 7: 1039–1057

Wagner D, Przybyla D, op den Camp RGL, Kim C, Landgraf F, Lee KP, Wursch M, Laloi C, Nater M and Apel K (2004) The genetic basis of singlet oxygen–induced stress responses of Arabidopsis thaliana. Science 306: 1183–1185

Williams P, Hardeman K, Fowler J and Rivin C (2006) Divergence of duplicated genes in maize: evolution of contrasting targeting information for enzymes in the porphyrin pathway. Plant J 45: 727–739

Wilson T (1966) Excited singlet molecular oxygen in photooxidation. J Am Chem Soc 88: 2898–2902

Witkowski DA and Halling BP (1988) Accumulation of photodynamic tetrapyrroles induced by acifluorfen-methyl. Plant Physiol 87: 632–637

Wydrzynski T and Satoh K (eds) (2005) Photosystem II: The Light-Driven Water: Plastoquinone Oxidoreductase. Advances in Photosynthesis and Respiration, Volume 22, Springer, Dordrecht

Yao N and Greenberg JT (2006) Arabidopsis ACCELERATED CELL DEATH2 modulates programmed cell death. Plant Cell 18: 397–411

Yoshimoto T and Matsunakas (1982) The mode of action of diphenyl ether herbicides: The scavenging of active oxygen is not inhibited by diphenyl ethers. In: "Abstracts of the 5th International Congress of Pesticide chemistry", Vol. 5, p. 62

Subject Index

C

D

E

F

G

H

I

K

L

M

N

O

P

Q

R

S

W

X

Z